BIRKHÄUSER

Contemporary Mathematicians

Gian-Carlo Rota†
Joseph P.S. Kung

Editors

For other titles published in this series, go to
http://www.springer.com/series/4817

Misha E. Kilmer • Dianne P. O'Leary

G.W. Stewart

Selected Works with Commentaries

Linocut by Henk van der Vorst

Birkhäuser
Boston • Basel • Berlin

Misha E. Kilmer
Tufts University
Department of Mathematics
Medford, MA 02155
USA
misha.kilmer@tufts.edu

Dianne P. O'Leary
University of Maryland
Computer Science Department and
Institute for Advanced Computer Studies
College Park, MD 20742
USA
oleary@cs.umd.edu

ISBN 978-0-8176-4967-8 e-ISBN 978-0-8176-4968-5
DOI 10.1007/978-0-8176-4968-5
Springer New York Dordrecht Heidelberg London

Library of Congress Control Number: 2010929252

Mathematics Subject Classification (2010): 01A75, 01A70, 01A60, 15-03, 15A06, 15A12, 15A18, 15A22, A23, 15A24, 65-03, 65F05, 65F10, 65F15, 65F25, 65F30

Printed on acid-free paper

Birkhäuser is part of Springer Science+Business Media (www.birkhauser.com)

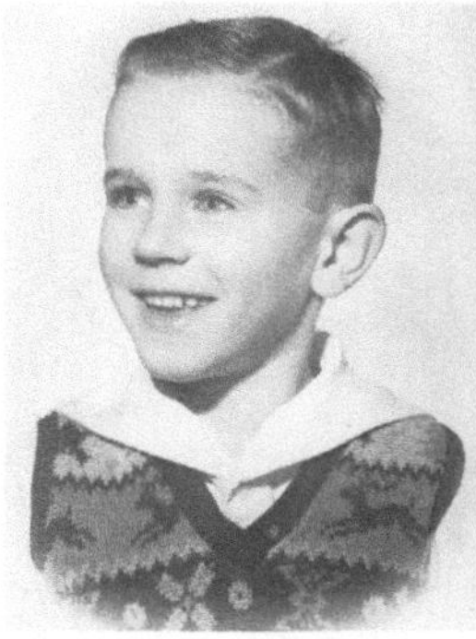

Pete circa 1944

Jack Dongarra, Cleve Moler, Pete Stewart, and Jim Bunch, with Cleve's car and license plate, late 1970s

Pete with Oak Ridge colleagues Michael Heath, Alston Householder (his Ph.D. advisor), and Robert Funderlic, circa 1970

Pete lecturing in the 1980s

Pete and his wife, Astrid Schmidt-Nielsen, circa 1987

Pete in his office at the University of Maryland, circa 1998

Contents

Foreword xi

List of Contributors xiii

Part I G. W. Stewart

1. Biography of G. W. Stewart
Iain S. Duff 3

2. Publications, Honors, and Students
. . . . 11

Part II Commentaries

3. Introduction to the Commentaries
Misha E. Kilmer and Dianne P. O'Leary 25

4. Matrix Decompositions: LINPACK and Beyond
Charles F. Van Loan, Misha E. Kilmer, and Dianne P. O'Leary 27

5. Updating and Downdating Matrix Decompositions
Lars Eldén, Misha E. Kilmer, and Dianne P. O'Leary 45

6. Least Squares, Projections, and Pseudoinverses
Misha E. Kilmer and Dianne P. O'Leary 59

7. The Eigenproblem and Invariant Subspaces: Perturbation Theory
Ilse C. F. Ipsen 71

8. The SVD, Eigenproblem, and Invariant Subspaces: Algorithms
James W. Demmel 95

9. The Generalized Eigenproblem
Zhaojun Bai 105

10. Krylov Subspace Methods for the Eigenproblem
Howard C. Elman and Dianne P. O'Leary 111

11. Other Contributions
Misha E. Kilmer and Dianne P. O'Leary 121

References 125

Index 133

Part III Reprints

12. Papers on Matrix Decompositions 137

12.1 GWS-B2 (with J. J. Dongarra, J. R. Bunch, and C. B. Moler) Introduction from LINPACK *Users Guide* 139

12.2 GWS-J17 (with R. H. Bartels), "Algorithm 432: Solution of the Matrix Equation $AX + XB = C$" 153

12.3 GWS-J32 "The Economical Storage of Plane Rotations" 161

12.4 GWS-J34 "Perturbation Bounds for the QR Factorization of a Matrix" 164

12.5 GWS-J42 (with A. K. Cline, C. B. Moler, and J. H. Wilkinson), "An Estimate for the Condition Number of a Matrix" 175

12.6 GWS-J49 "Rank Degeneracy" 184

12.7 GWS-J78 "On the Perturbation of LU, Cholesky, and QR Factorizations" 196

12.8 GWS-J89 "On Graded QR Decompositions of Products of Matrices" 202

12.9 GWS-J92 "On the Perturbation of LU and Cholesky Factors" 214

12.10 GWS-J94 "The Triangular Matrices of Gaussian Elimination and Related Decompositions" 221

12.11 GWS-J103 "Four Algorithms for the the (sic) Efficient Computation of Truncated Pivoted QR Approximations to a Sparse Matrix" 232

12.12 GWS-J118 (with M. W. Berry and S. A. Pulatova) "Algorithm 844: Computing Sparse Reduced-Rank Approximations to Sparse Matrices" 244

13. Papers on Updating and Downdating Matrix Decompositions 263
13.1 GWS-J29 (with W. B. Gragg), "A Stable Variant of the Secant Method for Solving Nonlinear Equations" 264
13.2 GWS-J31 (with J. W. Daniel, W. B. Gragg, L. Kaufman), "Reorthogonalization and Stable Algorithms for Updating the Gram-Schmidt QR Factorization" 280
13.3 GWS-J40 "The Effects of Rounding Error on an Algorithm for Downdating a Cholesky Factorization" 305
13.4 GWS-J73 "An Updating Algorithm for Subspace Tracking" 317
13.5 GWS-J77 "Updating a Rank-Revealing ULV Decomposition" 325
13.6 GWS-J87 "On the Stability of Sequential Updates and Downdates" 332

14. Papers on Least Squares, Projections, and Generalized Inverses 340
14.1 GWS-J4 "On the Continuity of the Generalized Inverse" 341
14.2 GWS-J35 "On the Perturbation of Pseudo-inverses, Projections and Linear Least Squares Problems" 355
14.3 GWS-J65 "On Scaled Projections and Pseudoinverses" 385

15. Papers on the Eigenproblem and Invariant Subspaces: Perturbation Theory 391
15.1 GWS-J15 "Error Bounds for Approximate Invariant Subspaces of Closed Linear Operators" 392
15.2 GWS-J19 "Error and Perturbation Bounds for Subspaces Associated with Certain Eigenvalue Problems" 406
15.3 GWS-J48 "Computable Error Bounds for Aggregated Markov Chains" 445
15.4 GWS-J70 "Two Simple Residual Bounds for the Eigenvalues of a Hermitian Matrix" 461
15.5 GWS-J71 (with G. Zhang) "Eigenvalues of Graded Matrices and the Condition Numbers of a Multiple Eigenvalue" 466
15.6 GWS-J108 "A Generalization of Saad's Theorem on Rayleigh-Ritz Approximations" 477
15.7 GWS-J109 "On the Eigensystems of Graded Matrices" 483
15.8 GWS-J114 "On the Powers of a Matrix with Perturbations" 506

16. Papers on the SVD, Eigenproblem and Invariant Subspaces: Algorithms . 521
16.1 GWS-J5 "Accelerating the Orthogonal Iteration for the Eigenvectors of a Hermitian Matrix" 522
16.2 GWS-J30 "Simultaneous Iteration for Computing Invariant Subspaces of Non-Hermitian Matrices" 538

16.3 GWS-J33 "Algorithm 506: HQR3 and EXCHNG: FORTRAN Subroutines for Calculating and Ordering the Eigenvalues of a Real Upper Hessenberg Matrix" 553
16.4 GWS-J37 (with C. A. Bavely) "An Algorithm for Computing Reducing Subspaces by Block Diagonalization" 560
16.5 GWS-J75 (with R. Mathias) "A Block QR Algorithm and the Singular Value Decomposition" 595
16.6 GWS-J102 "The QLP Approximation to the Singular Value Decomposition" 606
16.7 GWS-J107 (with Z. Jia) "An Analysis of the Rayleigh-Ritz Method for Approximating Eigenspaces" 620

17. Papers on the Generalized Eigenproblem 632
17.1 GWS-J16 "On the Sensitivity of the Eigenvalue Problem $Ax = \lambda Bx$" 633
17.2 GWS-J18 (with C. B. Moler), "An Algorithm for Generalized Matrix Eigenvalue Problems" 652
17.3 GWS-J27 "Gershgorin Theory for the Generalized Eigenvalue Problem $Ax = \lambda Bx$" 669
17.4 GWS-J38 "Perturbation Bounds for the Definite Generalized Eigenvalue Problem" 677

18. Papers on Krylov Subspace Methods for the Eigenproblem 695
18.1 GWS-J111 "A Krylov-Schur Algorithm for Large Eigenproblems" 696
18.2 GWS-J113 "Addendum to 'A Krylov-Schur Algorithm for Large Eigenproblems' " 711
18.3 GWS-J110 "Backward Error Bounds for Approximate Krylov Subspaces" 715
18.4 GWS-J112 "Adjusting the Rayleigh Quotient in Semiorthogonal Lanczos Methods" 722

Foreword

G.W. (Pete) Stewart is a world-renowned expert in computational linear algebra. It is widely accepted that he is the successor to James Wilkinson, the first giant in the field, taking up the perturbation theory research that Wilkinson so ably began and complementing it with algorithmic innovation. Stewart's results on rounding error in numerical computations provide basic understanding of floating-point computation. His results on perturbation of eigensystems, generalized inverses, least squares problems, and matrix factorizations are fundamental to numerical practice today. His algorithms for the singular value decomposition, updating and downdating matrix factorizations, and the eigenproblem broke new ground and are still widely used in an increasing number of applications. His papers, widely cited, are characterized by elegance in theorems and algorithms and clear, concise, and beautiful exposition. His six popular textbooks are excellent sources of knowledge and history.

Stewart's 60th birthday was celebrated with a meeting at College Park, MD. His 70th birthday will be observed by a meeting in Austin, TX, and by a special issue of the journal *Linear Algebra and its Applications* dedicated to him. It is fitting that there be a collection of his selected works published on this occasion, and we were happy to undertake the task of editing this volume.

Pete chose the papers to include here, and we are grateful for the permission to reprint these papers. The publishers are ACM, AMS, Elsevier, ETNA, IEEE, Oxford University Press, SIAM, and Springer. We are especially grateful to SIAM for the permission to reprint 19 papers.

We thank Iain S. Duff for providing a lively biography of Pete, based on interviews. We are very grateful to our collaborators in writing the commentaries: Zhaojun Bai, James W. Demmel, Lars Eldén, Howard C. Elman, Ilse C.F. Ipsen, and Charles F. Van Loan. These leading experts in their fields produced commentary with depth and breadth.

Each chapter of the commentary was reviewed for accuracy and completeness, and we are grateful to Jesse Barlow, Åke Björck, James Demmel, Nick Higham,

Chris Paige, Yousef Saad, Michael Saunders, Nick Trefethen, and David Watkins for doing these reviews. We are also grateful to a number of individuals who provided us with remarks regarding the impact of Pete's scholarship on their own research: M.W. Berry, R.A. Brualdi, J.R. Bunch, Z. Jia, C.-R. Lee, K.J.R. Liu, C.C. Paige, B.N. Parlett, Y. Saad, and M.A. Saunders. Comments and quotations from these individuals are interspersed within the commentaries.

Finally, we thank Henk van der Vorst for providing the wonderful linocut image used on the title page.

We present this volume as a gift to Pete, gathering some of his most important contributions. He is a good friend and an inspiring colleague, and we are honored to be associated with him.

Misha E. Kilmer
Medford, MA

Dianne P. O'Leary
College Park, MD

List of Contributors

Zhaojun Bai
Department of Computer Science
University of California
Davis, CA, USA
bai@cs.ucdavis.edu

James W. Demmel
Department of Mathematics
Computer Science Division
University of California
Berkeley CA, USA
demmel@cs.berkeley.edu

Iain S. Duff
Rutherford Appleton Laboratory
Chilton Didcot Oxfordshire, UK
iain.duff@stfc.ac.uk

Lars Eldén
Department of Mathematics
Linköping University
Linköping, Sweden
laeld@math.liu.se

Howard C. Elman
Computer Science Department
and Institute for Advanced Computer
Studies, University of Maryland
College Park, MD, USA
elman@cs.umd.edu

Ilse C. F. Ipsen
Department of Mathematics
North Carolina State University
Raleigh, NC, USA
ipsen@ncsu.edu

Misha E. Kilmer
Department of Mathematics
Tufts University
Medford, MA, USA
misha.kilmer@tufts.edu

Dianne P. O'Leary
Computer Science Department
and Institute for Advanced
Computer Studies
University of Maryland
College Park, MD, USA
oleary@cs.umd.edu

Charles F. Van Loan
Department of Computer Science
Cornell University
Ithaca, NY, USA
cv@cs.cornell.edu

Part I

G. W. Stewart

1

Biography of G. W. Stewart

Iain S. Duff

If one is asked to name the most influential people in numerical linear algebra, then Pete (G.W.) Stewart would come very high on the list. Pete has had a major influence on the field and, in several ways, on my own career. It is with great pleasure that I pen these words as a biography and tribute to him. I am grateful to Pete not only for spending the time to discuss his life with me but also for going carefully over a draft and finding many instances where accuracy had been sacrificed for the sake of the narrative. I should, however, stress that any rounding errors that remain are purely my responsibility but hopefully do not now contaminate the result.

Pete was born in Washington, DC in the fall of 1940 and spent his first 5 years in Arlington, Virginia, just across the Potomac river from Washington. His father was a journalist who spent the year when Pete turned six at Harvard as a Nieman Fellow. Thus Pete started first grade in Cambridge, coincidentally at the same time that David Young arrived at Harvard as a new graduate student. Pete's father then accepted a position in the State Department attached to the United Nations, and Pete attended elementary school for 3 years in Jamaica, NY, and 3 more years in Hempstead, Long Island.

Pete's long association with the sour mash state of Tennessee began in the fall of 1953, when his father accepted a position in Knoxville as assistant director of the Information Department of the Tennessee Valley Authority. Pete completed his high school education at Bearden High School. Although at this point it was not clear that a future star for our field was in the making, Pete discovered his talent for mathematics in high school. With the encouragement of his math teacher, Ms. Ival Aslinger, he completed the high school mathematics curriculum in 3 years and taught himself calculus. Indeed this self-teaching is characteristic of Pete; and it is fair to say that at most periods of his life, he has engaged in this form of learning, whether it be in mathematics, languages, or history. Perhaps one of his main interests outside of mathematics at this time (and to some extent even today)

M.E. Kilmer and D.P. O'Leary (eds.), *G.W. Stewart: Selected Works with Commentaries*, Contemporary Mathematicians, DOI 10.1007/978-0-8176-4968-5_1,

was in history. He still recalls reading Gibbon's *Decline and Fall* for the first time. He graduated from high school with good grades in 1957.

Pete matriculated at the University of Tennessee, where at one time or other, he declared several majors: engineering physics, psychology, pre-medicine, in addition to mathematics and physics. Perhaps, the first inkling of what was to happen came when at the end of his sophomore year in 1959 he became a summer student employee at the Gaseous Diffusion Plant in Oak Ridge. This brought him in contact with computational mathematics and with issues in numerical analysis, many of which he was to develop, refine, and define in his later works. He worked with the supercomputer of the day, an IBM 704 with 16K of 36-bit word memory. His work at the plant led to a fairly well-paid consulting contract that enabled him to enjoy life (another characteristic that endures to this day!) while completing his undergraduate studies. He graduated *summa cum laude* in 1962 with a major in mathematics and a minor in physics.

In his undergraduate work, Pete also took courses in pure mathematics, including topology. He was awarded a Woodrow Wilson/NSF fellowship to Princeton, where he intended to study algebraic topology. But he soon switched to logic and the foundations of mathematics, taking a class from Alonzo Church. Much to the later benefit of our field, however, he only stayed a year at Princeton, before returning to Oak Ridge on a full-time basis in 1963. At that time he married one of his co-workers, Lynn Tharp, and they moved to Phoenix, Arizona, where he worked in the General Electric Computer Division (1964–1965). The job was not particularly demanding, and Pete had plenty of time to hone his personal skills, especially when the company sent him to Fairbanks, Alaska for a summer of working at the Nimbus weather satellite tracking station in Fairbanks. During this time he devoted himself to mastering the classics by Feller on probability theory and Sheffé on the analysis of variance. Perhaps more importantly, Bob Funderlic, a friend and colleague from Oak Ridge, passed on some bootleg notes by Jim Wilkinson from the Michigan Engineering Summer Conference in Ann Arbor. This conference on Numerical Analysis, organized by Bob Bartels, consisted of 2 weeks of lectures by distinguished speakers to a class of about 40 students. After the lectures, the speakers adjourned for rounds of liquid conviviality at Ann Arbor's Old German Restaurant. In his spare time in Phoenix, Pete coded algorithms from Wilkinson's notes, including the beautiful, double implicit shift QR algorithm of Francis. Following a lead from W. Ross Burrus, who had been using plane rotations to solve least squares problems at Oak Ridge National Laboratory, Pete devised a least squares algorithm that used Householder transformations and wrote it up in what he acknowledges to be a thoroughly amateurish paper. Fortunately, Gene Golub saved Pete from embarrassing himself by publishing essentially the same algorithm (and much more) in his classic *Numerische Mathematik* paper.

Pete returned to Oak Ridge in 1965 and enrolled as a graduate student in mathematics at the University of Tennessee, where among other required courses he took one from Alston Householder on the theory of matrices in numerical analysis.

It was at this time that his first journal paper, on a derivative-free version of Davidon's method, appeared in the *Journal of the ACM*. As we might expect, this included a rigorous error analysis of the effect of rounding errors on the difference approximations of derivatives.

The defining moment in Pete's professional life was when he and another Oak Ridge friend, Bert Rust, attended the Michigan Summer Conference. The lecturers in this summer school included Alston Householder, Jim Wilkinson, Dick Varga, John Todd, and John Rice. Pete was entranced by the talks, and at that moment realized his destiny and embraced the topic of numerical linear algebra that was to define and guide his life.

In the beginning, Pete did not have too much interaction with Alston Householder, who was at that time the Director of the Mathematics and Computer Division of ORNL. He was also Ford Professor at the University of Tennessee, where he taught a class that met on Wednesday afternoons and Saturday mornings. Pete took two of these classes – one on numerical linear algebra, as mentioned above, and one on the solution of nonlinear equations. Although, in Pete's words, Householder's style of lecturing left something to be desired, his subject matter more than made up for it. Among the influential texts that Pete can recall reading as a graduate student were Wilkinson's *Rounding Errors in Algebraic Processes*, and his *Algebraic Eigenvalue Problem.* In the spring of 1967, Pete passed his comprehensives (qualifying exam) that comprised both written and oral exams in algebra, analysis, topology, and the foundations of mathematics, and Householder agreed to take him on as his student. His interaction with Householder was only at the level of once a month; but Pete was already quite advanced in his research, largely because of his good background in computation and computational analysis, and he finished his thesis in record time by the end of 1967. Pete was supported during his thesis year by an Oak Ridge Associates Fellowship that also entailed his moving from the Gaseous Diffusion Plant to Oak Ridge National Laboratory – from K25 to X10 in the local coordinate system.

Part of Pete's thesis concerned a rigorous analysis of Lehmer's method for solving polynomial equations, including a scaling algorithm to avoid overflow. He also translated the work of Bauer on *Treppeniteration* (a.k.a. subspace iteration), and was inspired to analyze a variant that used Rayleigh–Ritz approximations to speed up the calculations. Unknown to him at the time, Rutishauser had come up with the same algorithm, and the perturbation theory that Pete devised for the analysis was about to be published in a more general form by Chandler Davis and W. (Velvel) Kahan. It was around that time that Pete first met Gene Golub, who visited the department in the fall of 1967.

With his PhD assured, Pete went job hunting and was invited for interviews at Florida State University, the University of Virginia, and the University of Texas, accepting the offer of the latter to become an Assistant Professor in the fall of 1968. This was no mean achievement given the state of the academic market, which had just gone into a nosedive. A year later he accepted a half-time research appointment in David Young's Institute for Numerical Analysis. Tenure and promotion to Associate Professor followed rapidly in the following year, with Pete's abilities at that time being recognized by the fact that this took less than half the expected period for translating tenure track to tenured position. A glance at Pete's publication record at that time shows that he was becoming something of a paper-generating machine, with four or more journal papers a year being published in the late sixties and early seventies. It should be noted that these were substantial contributions in high quality journals. In addition to his research, Pete was for the first time heavily into teaching within the Math and Computer Science Departments, teaching not only mathematics but the theory of programming languages. Pete got to know Cleve Moler when the latter visited Texas, and in an afternoon session during his visit they sketched the outlines of the QZ algorithm for the generalized eigenvalue problem. Cleve extended his stay so that they could flesh out their ideas. While in Texas, Pete's two children were born: his son Michael (who has to some extent followed in his father's footsteps) in 1969 and his daughter Laura in 1970.

Pete then spent 2 years at Carnegie–Mellon University (1972–1974) with a joint appointment in the Department of Mathematics and the Department of Computer Science, the latter headed by Joe Traub. At the end of his stay in Pittsburgh, his *Introduction to Matrix Computations* was published, arguably the first of the great modern texts and the book of choice for supporting courses in numerical linear algebra until the publication of Golub and Van Loan's book in 1983. Another significant event during the Carnegie–Mellon period was Pete's divorce from Lynn in 1974.

Werner Rheinboldt, who with Jim Ortega had written an influential book on nonlinear equations and optimization, was the technical go-between for Pete and his publisher. In 1974, he arranged an appointment for Pete in the Computer Science Department at the University of Maryland with a half-time research appointment in the Institute for Fluid Dynamics and Applied Mathematics (later to become the Institute for Physical Sciences and Technology). Pete was promoted to Full Professor in 1976. At this point, it is impressive to think that Pete was still only 35 – a meteoric career indeed. It is fair to say that although Pete enjoyed the teaching aspects of being a professor, his dedication to research meant that his love of teaching was somewhat tempered and, in his words, he loved teaching in moderation.

Although Pete's parents were still living in Knoxville, he had good family connections in the Washington area through his uncle Bill and aunt Carleen. (Bill was a ground crew soldier in the Air Force during WWII who came to Washington on rotation and met and fell in love with Carleen, Pete's babysitter.) Shortly after Pete arrived at Maryland, he met Astrid Schmidt-Nielsen at a dinner party, and 3 years later they were married (with Alston Householder in attendance). Astrid has been a constant, influential, and inspiring partner to Pete from that day until today.

In the summer of 1975, Pete visited the Division of Mathematical Sciences headed by Jim Poole at Argonne National Laboratory. The Division had sponsored the development of EISPACK, a collection of routines to solve eigenvalue problems. Jim asked Pete to chair a public meeting on the possibility of producing software for linear systems – what eventually became LINPACK. It was decided to confine the effort to dense matrices with Pete, Cleve Moler, and Jim Bunch designing and coding the algorithms (Jack Dongarra came on board later). At that time granting agencies were not funding program development, arguing that it was not research. Pete returned to Argonne in the fall, and wrote a proposal that billed the effort as *research* in how to develop software. As we know, the proposal was successful. The efforts of the group were assisted by the visits of Jim Wilkinson to Argonne during the summer. Work on LINPACK was perhaps Pete's most visible contribution for the next 3 years, with annual visits to Argonne culminating in the release of LINPACK in the summer of 1978. Pete played a leading role in the coding of the routines, which was greatly helped by the decision to use the BLAS. These were of course just the Level 1 BLAS, and thus the decision was more important for clarity and modularity than for efficiency. An important aspect of the package was its *Users' Guide,* which was intended to explain the workings of the algorithms, not just how to use them. The introduction, which was largely written by Pete, resonated, I am sure, with the NSF and certainly contributed greatly into making the manual a SIAM best seller.

In 1977–1978, Pete and Astrid spent time in Minnesota, Astrid as a Post Doc and Pete holding joint appointments in the Computer Science Department, headed by Ben Rosen, and the Applied Statistics Department, headed by Steve Fienberg. This was the beginning of Pete's interest in statistical computing.

Around this time Pete first met Dianne O'Leary, a student of Gene Golub's. Sometime after her arrival as an Assistant Professor at Maryland in 1978, they worked and published together on parallel computing – a goodly feat given that it would be several more years before practical, commercial parallel computers were readily available. They became involved in many aspects of parallel computing, including developing numerical algorithms and designing an operating system for a home grown parallel computer at Maryland. Pete says that his friendship and collegial relationship with Dianne has been one of the most satisfying aspects of his professional career.

At that time, much of his research was performed at the National Bureau of Standards (later to become NIST) where he consulted once a week and where he spent a half-year sabbatical in 1988. Some of Pete's more exciting and significant work at that time was his collaboration with Sun on matrix perturbation problems, culminating in their book *Matrix Perturbation Theory* published in 1990. The two-body problem (Astrid had a job with the Naval Research Laboratory in Washington) has kept Pete more at home than many researchers, although they both spent 6 months back in Minneapolis in 1992.

Pete has had a long and distinguished association with the Gatlinburg conferences, later to be renamed the Householder Symposia in honor of Pete's mentor and friend, Alston. It is amazing to think that Alston's 65th birthday was celebrated in 1969 at the last of the conferences to be held in Gatlinburg (and the first attended by Pete). Pete was elected to the Householder Committee at the Asilomar conference in 1977 and remained an active and involved member until he stood down at the first meeting after his 60th birthday, almost 10 years ago. Pete well knew the meaning of the word symposium (Greek for drinking party), and I have many happy memories of the Gatlinburg/Householder meetings where high kudos was given to active attendance at the first talk of the morning after earnest early morning discussions fuelled by the waters of Scotland or Tennessee. This tradition was a staple of the community in all its meetings. In a sense, it is an embodiment of a singular aspect of the field of numerical linear algebra, where camaraderie, collaboration, and encouragement are the norm, rather than the cut-throat competitiveness of some other scientific and mathematical disciplines. This atmosphere was much appreciated by Pete and indeed is a major reason that he has spent his life working primarily in this area.

Pete was of course of an era when there were many fewer prizes and awards than are available today – no Fox, Householder, and Wilkinson prizes for example. Nevertheless, he has had recognition of his pioneering work in the award of the F.L. Bauer Prize by the Technical University of Munich in 1988 and his election to the National Academy of Engineering in 2004. He also holds the title of Distinguished University Professor, a signal honor at a university as prestigious as Maryland. He was an inaugural fellow of SIAM in 2009 with the citation "For contributions to numerical linear algebra."

So what can we say about the man who so influenced our field and our lives. He was certainly his own man, almost to the point of being a lone researcher (he has only 43 joint journal publications out of the 137 listed in his current curriculum vitae). He did not "father" many graduate students, with only six PhD students in his long career. Like Ramanujan, he ascribes some of his most creative work to an almost mystical crystallization of an incoherent muddle of ideas – the ideas bubbling around for some time before the eureka moment, perhaps as emerging from a dream. Unlike many of us, Pete actually enjoyed the duties of refereeing and found this at times to be another source of stimulation and inspiration. One thing that Pete shares with Don Knuth is that his magnum opus (*Matrix Algorithms* published by SIAM) was originally planned to be in four volumes but has stopped at two with no plans to continue the series. Pete is always happy to share his thoughts and was one of the first researchers to make his works available to all by anonymous ftp in the early 1990s. I recollect accessing his "Afternotes" on the web as early as 1993.

I am notoriously bad at remembering my first meeting with anybody, even someone as influential as Pete. However, for sure I met him at the Gatlinburg meeting at Asilomar in 1977, and we co-edited the SIAM Sparse Matrix Proceedings in 1978. In the latter exercise, I overcame one of the latent prejudices of the UK toward the Americans with respect to language, when it was quite apparent that Pete's knowledge of the mother tongue was equal to that of anybody from my side of the Atlantic. Indeed he introduced me to Strunk and White, which is very much a worthy competitor to Fowler's *Modern English Usage*, with which we in the UK were more familiar. Another early memory of Pete and Astrid was a visit to their house in January 1980 accompanied by our new baby, Catriona. We found that, as Pete apparently keeps UK time by his early start to the day, it was not the place to recover from any effects of jet lag. Pete and Astrid later attended more than one early birthday of Catriona, bringing with them on one trip a Fisher–Price boat named the SS Schmidt-Nielsen. That was in the days when Fisher–Price toys were little known in Europe, and so it was a rather special present that, having been later used by other village children, is now in our attic awaiting future grandchildren.

Pete has now added Emeritus to his title of Distinguished University Professor, but, like most of you, I do not believe that such an active and able mind will cease his research any time in the near future. I anticipate and hope that he will emulate Alston, who I recollect was still attending his eponymous meetings over the age of 90. So I thank the orchestrators of this volume, Misha Kilmer and Dianne O'Leary, for giving me the opportunity of wishing Pete (and Astrid) all the best for their "retirement" shared between their houses in Washington and Maine. You never know, he might even be persuaded to continue with the Matrix Algorithms series, and, given that I learned not only mathematics but some US Civil War history from the first volume, we could all gain much if this were to happen.

As a small postscript to my discussions with Pete while preparing this short biography, I can reveal a possible solution to the often discussed puzzle of why "G.W." became "Pete." Like his father and grandfather before him, he was christened Gilbert Wright, and like them he was also called Pete. This sobriquet was acquired by his preacher grandfather from cowboys in South Dakota just after he graduated from seminary. The story goes that when he reached town, he gave the saloon keeper an offer he couldn't refuse: if the keeper would close the bar for an hour on Sunday so his customers could attend church, he (Pete's grandfather) would not preach against liquor. Whether this inspired the cowboys to call him Pete and why they chose "Pete" in the first place is not clear. (Some family stories say that it was a college nickname). But apocryphal or not, his grandfather's compromise is one of which I am sure Pete (G.W.) the Third would approve.

Happy Birthday Pete (G.W.)

Iain Duff
Oxfordshire
11/11/2009

2

Publications, Honors, and Students

2.1. Publications of G. W. Stewart

2.1.1. Thesis

[T1] Dissertation: G. W. Stewart III, "Some Topics in Numerical Analysis," University of Tennessee. Published as Technical Report ORNL-4303, Oak Ridge National Laboratory, September 1968.
`http://www.ornl.gov/info/reports/1968/3445605155079.pdf`

2.1.2. Books

[B1] *Introduction to Matrix Computations*, Academic Press, New York (1973).
[B2] (with J. J. Dongarra, J. R. Bunch, and C. B. Moler), LINPACK *Users' Guide*, SIAM, Philadelphia (1979).
[B3] (with J.-G. Sun) *Matrix Perturbation Theory*, Academic Press, New York (1990).
[B4] Translation of Karl Friedrich Gauss, *Theoria Combinationis Observationum Erroribus Minimis Obnoxiae*, (Theory of the Combination of Observations Least Subject to Errors, Part One, Part Two, Supplement) SIAM, 1995.
[B5] *Afternotes on Numerical Analysis*, SIAM, 1996.
[B6] *Afternotes Goes to Graduate School*, SIAM, 1998.
[B7] *Matrix Algorithms Volume I: Basic Decompositions*, SIAM, 1998.
[B8] *Matrix Algorithms Volume II: Eigensystems*, SIAM, 2001.

2.1.3. Journal Publications

[J1] "A Modification of Davidon's Minimization Method to Accept Difference Approximations of Derivatives," *Journal of the ACM* **14** (1967) 72–83.

M.E. Kilmer and D.P. O'Leary (eds.), *G.W. Stewart: Selected Works with Commentaries*, Contemporary Mathematicians, DOI 10.1007/978-0-8176-4968-5_2,

[J2] "A Generalization of a Theorem of Fan on Gershgorin Disks," *Numerische Mathematik* **10** (1967) 162.

[J3] (with D. W. Lick) "Numerical Solution of a Thin Plate Heat Transfer Problem," *Communications of the ACM* **11** (1968) 639–640.

[J4] "On the Continuity of the Generalized Inverse," *SIAM Journal on Applied Mathematics* **17** (1969) 33–45.

[J5] "Accelerating the Orthogonal Iteration for the Eigenvectors of a Hermitian Matrix," *Numerische Mathematik* **13** (1969) 362–376.

[J6] "Some Iterations for Factoring a Polynomial," *Numerische Mathematik* **13** (1969) 458–470.

[J7] "On Lehmer's Method for Finding the Zeros of a Polynomial," *Mathematics of Computation* **23** (1969) 829–836, s24–s30. Corrigendum: **25** (1971) 203.

[J8] "On Samelson's Iteration for Factoring Polynomials," *Numerische Mathematik* **15** (1970) 306–314.

[J9] "Incorporating Origin Shifts into the QR Algorithm for Symmetric Tridiagonal Matrices," *Communications of the ACM* **13** (1970) 365–367.

[J10] "Algorithm 384: Eigenvalues and Eigenvectors of a Real Symmetric Matrix," *Communications of the ACM* **13** (1970) 369–371. Remark: *Communications of the ACM* **13** (1970) 750.

[J11] "On the Convergence of Sebastião E Silva's Method for Finding a Zero of a Polynomial," *SIAM Review* **12** (1970) 458–460.

[J12] (with A. S. Householder) "The Numerical Factorization of a Polynomial," *SIAM Review* **13** (1971) 38–46.

[J13] "Error Analysis of the Algorithm for Shifting the Zeros of a Polynomial by Synthetic Division," *Mathematics of Computation* **25** (1971) 135–139.

[J14] "On a Companion Operator for Analytic Functions," *Numerische Mathematik* **18** (1971) 26–43.

[J15] "Error Bounds for Approximate Invariant Subspaces of Closed Linear Operators," *SIAM Journal on Numerical Analysis* **8** (1971) 796–808.

[J16] "On the Sensitivity of the Eigenvalue Problem $Ax = \lambda Bx$," *SIAM Journal on Numerical Analysis* **9** (1972) 669–686.

[J17] (with R. H. Bartels) "Algorithm 432: Solution of the Matrix Equation $AX + XB = C$," *Communications of the ACM* **15** (1972) 820–826.

[J18] (with C. B. Moler) "An Algorithm for Generalized Matrix Eigenvalue Problems," *SIAM Journal on Numerical Analysis* **10** (1973) 241–256.

[J19] "Error and Perturbation Bounds for Subspaces Associated with Certain Eigenvalue Problems," *SIAM Review* **15** (1973) 727–764.

[J20] "Conjugate Direction Methods for Solving Systems of Linear Equations," *Numerische Mathematik* **21** (1973) 285–297.

[J21] "Some Iterations for Factoring a Polynomial. II A Generalization of the Secant Method," *Numerische Mathematik* **22** (1974) 33–36.

[J22] "The Convergence of Multipoint Iterations to Multiple Zeros," *SIAM Journal on Numerical Analysis* **11** (1974) 1105–1120.

[J23] (with T. Söderström) "On the Numerical Properties of an Iterative Method for Computing the Moore-Penrose Generalized Inverse," *SIAM Journal on Numerical Analysis* **11** (1974) 61–74.

[J24] (with M. M. Blevins) "Calculating the Eigenvectors of Diagonally Dominant Matrices," *Journal of the ACM* **21** (1974) 261–271.

[J25] "Modifying Pivot Elements in Gaussian Elimination," *Mathematics of Computation* **28** (1974) 537–542.

[J26] "The Convergence of the Method of Conjugate Gradients at Isolated Extreme Points of the Spectrum," *Numerische Mathematik* **24** (1975) 85–93.

[J27] "Gershgorin Theory for the Generalized Eigenvalue Problem $Ax = \lambda Bx$," *Mathematics of Computation* **29** (1975) 600–606.

[J28] "An Inverse Perturbation Theorem for the Linear Least Squares Problem," *(ACM) SIGNUM Newsletter* **10**(2–3) (1975) 39–40.

[J29] (with W. B. Gragg) "A Stable Variant of the Secant Method for Solving Nonlinear Equations," *SIAM Journal on Numerical Analysis* **13** (1976) 889–903.

[J30] "Simultaneous Iteration for Computing Invariant Subspaces of Non-Hermitian Matrices," *Numerische Mathematik* **25** (1976) 123–136.

[J31] (with J. W. Daniel, W. B. Gragg, and L. Kaufman) "Reorthogonalization and Stable Algorithms for Updating the Gram-Schmidt QR Factorization," *Mathematics of Computation* **30** (1976) 772–795.

[J32] "The Economical Storage of Plane Rotations," *Numerische Mathematik* **25** (1976) 137–138.

[J33] "Algorithm 506: HQR3 and EXCHNG: Fortran Subroutines for Calculating and Ordering the Eigenvalues of a Real Upper Hessenberg Matrix," *ACM Transactions on Mathematical Software* **2** (1976) 275–280.

[J34] "Perturbation Bounds for the QR Factorization of a Matrix," *SIAM Journal on Numerical Analysis* **14** (1977) 509–518.

[J35] "On the Perturbation of Pseudo-Inverses, Projections and Linear Least Squares Problems," *SIAM Review* **19** (1977) 634–662.

[J36] (with C. B. Moler) "On the Householder-Fox Algorithm for Decomposing a Projection," *Journal of Computational Physics* **28** (1978) 82–91.

[J37] (with C. A. Bavely) "An Algorithm for Computing Reducing Subspaces by Block Diagonalization," *SIAM Journal on Numerical Analysis* **16** (1979) 359–367.

[J38] "Perturbation Bounds for the Definite Generalized Eigenvalue Problem," *Linear Algebra and its Applications* **23** (1979) 69–85.

[J39] "A Note on the Perturbation of Singular Values," *Linear Algebra and its Applications* **28** (1979) 213–216.

[J40] "The Effects of Rounding Error on an Algorithm for Downdating a Cholesky Factorization," *Journal of the Institute of Mathematics and its Applications* **23** (1979) 203–213.

[J41] (with D. P. O'Leary and J. S. Vandergraft) "Estimating the Largest Eigenvalue of a Positive Definite Matrix," *Mathematics of Computation* **33** (1979) 1289–1292.

[J42] (with A. K. Cline, C. B. Moler, and J. H. Wilkinson) "An Estimate for the Condition Number of a Matrix," *SIAM Journal on Numerical Analysis* **16** (1979) 368–375.

[J43] "The Efficient Generation of Random Orthogonal Matrices with an Application to Condition Estimators," *SIAM Journal on Numerical Analysis* **17** (1980) 403–409.

[J44] "The Behavior of a Multiplicity Independent Root-finding Scheme in the Presence of Error," *BIT* **20** (1980) 526–528.

[J45] "On the Implicit Deflation of Nearly Singular Systems of Linear Equations," *SIAM Journal on Scientific and Statistical Computing* **2** (1981) 136–140.

[J46] "Constrained Definite Hessians Tend to be Well Conditioned," *Mathematical Programming* **21** (1981) 235–238.

[J47] "Computing the CS Decomposition of a Partitioned Orthonormal Matrix," *Numerische Mathematik* **40** (1982) 297–306.

[J48] "Computable Error Bounds for Aggregated Markov Chains," *Journal of the ACM* **30** (1983) 271–285.

[J49] "Rank Degeneracy," *SIAM Journal on Scientific and Statistical Computing* **5** (1984) 403–413.

[J50] "A Second Order Perturbation Expansion for Small Singular Values," *Linear Algebra and its Applications* **56** (1984) 231–235.

[J51] (with D. F. McAllister and W. J. Stewart) "On a Rayleigh-Ritz Refinement Technique for Nearly Uncoupled Stochastic Matrices," *Linear Algebra and its Applications* **60** (1984) 1–25.

[J52] "On the Invariance of Perturbed Null Vectors under Column Scaling," *Numerische Mathematik* **44** (1984) 61–65.

[J53] "On the Asymptotic Behavior of Scaled Singular Value and QR Decompositions," *Mathematics of Computation* **43** (1984) 483–489.

[J54] "A Note on Complex Division," *ACM Transactions on Mathematical Software* **11** (1985) 238–241. Corrigendum: **12** (1986) 285.

[J55] "A Jacobi-like Algorithm for Computing the Schur Decomposition of a Nonhermitian Matrix," *SIAM Journal on Scientific and Statistical Computing* **6** (1985) 853–864.

[J56] (with D. P. O'Leary) "Data-flow Algorithms for Parallel Matrix Computation," *Communications of the ACM* **28** (1985) 840–853.

[J57] (with P. E. Gill, W. Murray, M. A. Saunders, and M. H. Wright) "Properties of a Representation of a Basis for the Null Space," *Mathematical Programming* **33** (1985) 172–186.

[J58] (with D. P. O'Leary) "Assignment and Scheduling in Parallel Matrix Factorization," *Linear Algebra and its Applications* **77** (1986) 275–299.

[J59] (with G. H. Golub and A. Hoffman) "A Generalization of the Eckart-Young-Mirsky Matrix Approximation Theorem," *Linear Algebra and its Applications* **88–89** (1987) 317–327.

[J60] "Collinearity and Least Squares Regression," *Statistical Science* **2** (1987) 68–100 (including commentary).

[J61] (with D. P. O'Leary) "From Determinacy to Systaltic Arrays," *IEEE Transactions on Computers* **C-36** (1987) 1355–1359.

[J62] "A Parallel Implementation of the QR Algorithm," *Parallel Computing* **5** (1987) 187–196.

[J63] "A Curiosity Concerning the Representation of Integers in Noninteger Bases," *Mathematics of Computation* **51** (1988) 755–756.

[J64] (with C. D. Meyer) "Derivatives and Perturbations of Eigenvectors," *SIAM Journal on Numerical Analysis* **25** (1988) 679–691.

[J65] "On Scaled Projections and Pseudoinverses," *Linear Algebra and its Applications* **112** (1989) 189–193.

[J66] (with G. N. Stenbakken and T. M. Souders) "Ambiguity Groups and Testability," *IEEE Transactions on Instrumentation and Measurement* **38** (1989) 941–947.

[J67] (with D. P. O'Leary) "Computing the Eigenvalues and Eigenvectors of Symmetric Arrowhead Matrices," *Journal of Computational Physics* **90** (1990) 497–505.

[J68] "Communication and Matrix Computations on Large Message Passing Systems," *Parallel Computing* **16** (1990) 27–40.

[J69] "Stochastic Perturbation Theory," *SIAM Review* **32** (1990) 579–610.

[J70] "Two Simple Residual Bounds for the Eigenvalues of a Hermitian Matrix," *SIAM Journal on Matrix Analysis and Applications* **12** (1991) 205–208.

[J71] (with G. Zhang) "Eigenvalues of Graded Matrices and the Condition Numbers of a Multiple Eigenvalue," *Numerische Mathematik* **58** (1991) 703–712.

[J72] (with G. Zhang) "On a Direct Method for the Solution of Nearly Uncoupled Markov Chains," *Numerische Mathematik* **59** (1991) 1–11.

[J73] "An Updating Algorithm for Subspace Tracking," *IEEE Transactions on Signal Processing* **40** (1992) 1535–1541.

[J74] "Error Analysis of QR Updating with Exponential Windowing," *Mathematics of Computation* **59** (1992) 135–140.

[J75] (with R. Mathias) "A Block QR Algorithm and the Singular Value Decomposition," *Linear Algebra and its Applications* **182** (1993) 91–100.

[J76] (with P. Schweitzer) "The Laurent Expansion of Pencils that are Singular at the Origin," *Linear Algebra and its Applications* **183** (1993) 237–254.

[J77] "Updating a Rank-Revealing ULV Decomposition," *SIAM Journal on Matrix Analysis and Applications* **14** (1993) 494–499.

[J78] "On the Perturbation of LU, Cholesky, and QR Factorizations," *SIAM Journal on Matrix Analysis and Applications* **14** (1993) 1141–1145.

[J79] "On the Early History of the Singular Value Decomposition," *SIAM Review* **35** (1993) 551–566.

[J80] "On the Perturbation of Markov Chains with Nearly Transient States," *Numerische Mathematik* **65** (1993) 135–141.

[J81] (with A. Edelman) "Scaling for Orthogonality," *IEEE Transactions on Signal Processing* **41** (1993) 1676–1677.

[J82] "Updating URV Decompositions in Parallel," *Parallel Computing* **20** (1994) 151–172.

[J83] "On the Convergence of Multipoint Iterations," *Numerische Mathematik* **68** (1994) 143–147.

[J84] "Perturbation Theory for Rectangular Matrix Pencils," *Linear Algebra and its Applications* **208-209** (1994) 297–301.

[J85] (with K. J. R. Liu, D. P. O'Leary, and Y.-J. J. Wu) "URV ESPRIT for Tracking Time-Varying Signals," *IEEE Transactions on Signal Processing* **42** (1994) 3441–3448.

[J86] "On the Solution of Block Hessenberg Systems," *Numerical Linear Algebra with Applications* **2** (1995) 287–296.

[J87] "On the Stability of Sequential Updates and Downdates," *IEEE Transactions on Signal Processing* **43** (1995) 2642–2648.

[J88] "Gauss, Statistics, and Gaussian Elimination," *Journal of Computational and Graphical Statistics* **4** (1995) 1–11.

[J89] "On Graded QR Decompositions of Products of Matrices," *Electronic Transactions on Numerical Analysis* **3** (1995) 39–49.

[J90] (with U. von Matt) "Rounding Errors in Solving Block Hessenberg Systems," *Mathematics of Computation* **65** (1996) 115–135.

[J91] (with X.-W. Chang and C. C. Paige) "New perturbation analyses for the Cholesky factorization," *IMA Journal of Numerical Analysis* **16** (1996) 457–484.

[J92] "On the Perturbation of LU and Cholesky Factors," *IMA Journal of Numerical Analysis* **17** (1997) 1–6.

[J93] (with X.-W. Chang and C. C. Paige) "Perturbation Analyses for the QR Factorization," *SIAM Journal on Matrix Analysis and Applications* **18** (1997) 775–791.

[J94] "The Triangular Matrices of Gaussian Elimination and Related Decompositions," *IMA Journal of Numerical Analysis* **17** (1997) 7–16.

[J95] "On Markov Chains with Sluggish Transients," *Stochastic Models* **13** (1997) 85–94.

[J96] (with Z. Bai) "Algorithm 776: SRRIT: A Fortran Subroutine to Calculate the Dominant Invariant Subspace of a Nonsymmetric Matrix," *ACM Transactions on Mathematical Software* **23** (1997) 494–513.

[J97] "On the Weighting Method for Least Squares Problems with Linear Equality Constraints," *BIT* **37** (1997) 961–967.

[J98] (with M. Stewart) "On Hyperbolic Triangularization: Stability and Pivoting," *SIAM Journal on Matrix Analysis and Applications* **19** (1998) 847–860.

[J99] (with D. P. O'Leary) "On the Convergence of a New Rayleigh Quotient Method with Applications to Large Eigenproblems," *Electronic Transactions on Numerical Analysis* **7** (1998) 182–189.

[J100] "On the Adjugate Matrix," *Linear Algebra and its Applications* **283** (1998) 151–164.

[J101] (with R. F. Boisvert, J. J. Dongarra, R. Pozo, and K. A. Remington) "Developing numerical libraries in Java," *Concurrency: Practice and Experience* **10** (1998) 1117–1129.

[J102] "The QLP Approximation to the Singular Value Decomposition," *SIAM Journal on Scientific Computing* **20** (1999) 1336–1348.

[J103] "Four Algorithms for the the (sic) Efficient Computation of Truncated Pivoted QR Approximations to a Sparse Matrix," *Numerische Mathematik* **83** (1999) 313–323.

[J104] (with M. Kilmer) "Iterative Regularization and MINRES," *SIAM Journal on Matrix Analysis and Applications* **21** (2000) 613–628.

[J105] "The Decompositional Approach to Matrix Computation," *Computing in Science and Engineering* **2** (2000) 50–59.

[J106] (with R.-C. Li) "A New Relative Perturbation Theorem for Singular Subspaces," *Linear Algebra and its Applications* **313** (2000) 41–51.

[J107] (with Z. Jia) "An Analysis of the Rayleigh–Ritz Method for Approximating Eigenspaces," *Mathematics of Computation* **70** (2001) 637–647.

[J108] "A Generalization of Saad's Theorem on Rayleigh-Ritz Approximations," *Linear Algebra and its Applications* **327** (2001) 115–119.

[J109] "On the Eigensystems of Graded Matrices," *Numerische Mathematik* **90** (2001) 349–370.

[J110] "Backward Error Bounds for Approximate Krylov Subspaces," *Linear Algebra and its Applications* **340** (2002) 81–86.

[J111] "A Krylov–Schur Algorithm for Large Eigenproblems," *SIAM Journal on Matrix Analysis and Applications* **23** (2002) 601–614.

[J112] "Adjusting the Rayleigh Quotient in Semiorthogonal Lanczos Methods," *SIAM Journal on Scientific Computing* **24** (2002) 201–207.

[J113] "Addendum to 'A Krylov–Schur Algorithm for Large Eigenproblems'," *SIAM Journal on Matrix Analysis and Applications* **24** (2002) 599–601.

[J114] "On the Powers of a Matrix with Perturbations," *Numerische Mathematik* **96** (2003) 363–376.

[J115] "Memory Leaks in Derived Types Revisited," *(ACM) SIGPLAN Fortran Forum* **22** (2003) 25–27.

[J116] "An Elsner-Like Perturbation Theorem for Generalized Eigenvalues," *Linear Algebra and its Applications* **390** (2004) 1–5.

[J117] "Error Analysis of the Quasi-Gram–Schmidt Algorithm," *SIAM Journal on Matrix Analysis and Applications* **27** (2005) 493–506.

[J118] (with M. W. Berry and S. A. Pulatova) "Algorithm 844: Computing Sparse Reduced-Rank Approximations to Sparse Matrices," *ACM Transactions on Mathematical Software* **31** (2005) 252–269.

[J119] "A Note on Generalized and Hypergeneralized Projectors," *Linear Algebra and its Applications* **412** (2006) 408–411.

[J120] (with C.-R. Lee) "Algorithm 879: EIGENTEST: A Test Matrix Generator for Large-Scale Eigenproblems," *ACM Transactions on Mathematical Software* **35** (2008) article 7, 1–11.

[J121] "Block Gram–Schmidt Orthogonalization," *SIAM Journal on Scientific Computing* **31** (2008) 761–775.

2.1.4. Other Notable Publications

[N1] (with A. S. Householder), "Bigradients, Hankel Determinants, and the Padé Table," in *Constructive Aspects of the Fundamental Theorem of Algebra,* B. Dejon and P. Henrici, eds., John Wiley & Sons, New York (1969) 131–150.

[N2] "A Set Theoretic Formulation of Backward Rounding Error Analysis," University of Texas at Austin Computation Center Report TNN-92, 1969.

[N3] (with L. L. Hoberock), "Input Requirements and Parametric Errors for Systems Identification under Periodic Excitement," *Transactions of the AMSE* **94** (1972) 296–302.

[N4] "The Numerical Treatment of Large Eigenvalue Problems," in *Information Processing 74: Proceedings of the IFIP Congress, Stockholm,* J. L. Rosenfeld, ed., North Holland, Dordrecht (1974) 666–672.

[N5] "Methods of Simultaneous Iteration for Calculating Eigenvectors of Matrices," in *Topics in Numerical Analysis II. Proceedings of the Royal Irish Academy Conference on Numerical Analysis, 1974,* John J. H. Miller, ed., Academic Press, New York (1974) 185–196.

[N6] (with G. H. Golub and Virginia Klema) "Rank Degeneracy and Least Squares Problems," University of Maryland Computer Science TR-751, 1976.

[N7] "A Bibliographical Tour of the Large, Sparse Generalized Eigenvalue Problem," in *Sparse Matrix Computations,* J. R. Bunch and D. J. Rose, eds., Academic Press, New York (1976) 113–130.

[N8] "Compilers, Virtual Memory, and Matrix Computations," in *Computer Science and Statistics: Proceedings of the 9th Symposium on the Interface,* D. C. Hoaglin and R. E. Welsch, eds., Boston (1976) 85–88.

[N9] "Research, Development, and LINPACK," in *Mathematical Software III,* John Rice, ed., Academic Press, New York (1977) 1–14.

[N10] "Perturbation Theory for the Generalized Eigenvalue Problem," in *Recent Advances in Numerical Analysis,* C. de Boor and G. H. Golub, eds., Academic Press, New York (1978) 193–206.

[N11] (with Iain Duff, ed.) *Sparse Matrix Proceedings,* SIAM, Philadelphia, 1978.

[N12] "A Method for Computing the Generalized Singular Value Decomposition," in *Matrix Pencils,* B. Kagstrom and A. Ruhe, eds., Springer Verlag, New York (1983) 207–220.

[N13] (with J. Dongarra) "LINPACK – a Package for Solving Linear Systems," in *Sources and Development of Mathematical Software,* W. R. Cowell, ed., Prentice Hall, Englewood Cliffs, NJ (1984) 20–48.

[N14] "On the Structure of Nearly Uncoupled Markov Chains," in *Mathematical Computer Performance and Reliability,* G. Iazeolla, P. J. Courtois, A. Hordijk, eds., North-Holland, Dordrecht (1984) 287–302.

[N15] "Collinearity, Scaling, and Rounding Error," in *Computer Science and Statistics: Proceedings of the Seventeenth Symposium on the Interface,* D. M. Allen, ed., North Holland, New York (1985) 195–198.

[N16] (with D. P. O'Leary and R. van de Geijn) "DOMINO: A Message Passing Environment for Parallel Computation," University of Maryland Computer Science TR-1648, 1986 (documentation of a system distributed over NETLIB).

[N17] (with Dianne P. O'Leary, Roger Pierson, and Mark Weiser), "The Maryland Crab: A module for building parallel computers," University of Maryland Computer Science Report CS-1660, Institute for Advanced Computer Studies Report UMIACS-86-9, 1986.

[N18] "Communication in Parallel Algorithms: An Example," in *Computer Science and Statistics: Proceedings of the 18th Symposium on the Interface,* T. J. Boardman, ed., ASA, Washington, D.C. (1986) 11–14.

[N19] "Numerical Linear Algebra in Statistical Computing," in *The State of the Art in Numerical Analysis,* A. Iserles and M. J. D. Powell, eds., Oxford University Press, Oxford (1987) 41–58.

[N20] (with D. A. Buell et al.) "Parallel Algorithms and Architectures. Report of a Workshop," *Journal of Supercomputing* **1** (1988) 301–325.

[N21] "Parallel Linear Algebra in Statistical Computing," in *COMPSTAT Proceedings in Computational Statistics, 8th Symposium, 1988,* D. Edwards and N. E. Raun, eds., Physica-Verlag, Heidelberg (1988) 3–14.

[N22] (with D. P. O'Leary and R. van de Geijn) "DOMINO: A Transportable Operating System for Parallel Computation," in *Parallel Processing and Medium-Scale Multiprocessors (Proceedings of a 1986 Conference),* Arthur Wouk, ed., SIAM Press, Philadelphia (1989) 25–34.

[N23] "An Iterative Method for Solving Linear Inequalities," in *Reliable Numerical Computation,* M. G. Cox and S. Hammarling, eds., Oxford University Press, Oxford (1990) 241–247.

[N24] "Perturbation Theory and Least Squares with Errors in the Variables," in *Contemporary Mathematics 112: Statistical Analysis of Measurement Error Models and Applications,* P. J. Brown and W. A. Fuller, eds., American Mathematical Society, Providence RI (1990) 171–181.

[N25] "On the Sensitivity of Nearly Uncoupled Markov Chains," in *Numerical Solutions of Markov Chains,* W. J. Stewart, ed., Dekker, New York (1990) 105–119.

[N26] "Perturbation Theory for the Singular Value Decomposition," in *SVD and Signal Processing, II*, R. J. Vacarro, ed., Elsevier, Amsterdam (1991) 99–109.

[N27] (with G. Adams and M. F. Griffin) "Direction-of-Arrival Estimation Using the Rank-Revealing URV Decomposition," *1991 IEEE Conference on Acoustics, Speech, and Signal Processing ICASSP-91* 2 (1991) 1385–1388.

[N28] "Jeep: A General Purpose Style File," *TeX and TUG News,* **0**(0) (1991) 3–4. http://tug.ctan.org/tex-archive/digests/ttn/ttn0n0.tex

[N29] (with M. F. Griffin and E. C. Boman) "Minimum-Norm Updating with the Rank-Revealing URV Decomposition." *1992 IEEE Conference on Acoustics, Speech, and Signal Processing ICASSP-92* 5 (1992) 293–296.

[N30] "Lanczos and Linear Systems," in *Proceedings of the Cornelius Lanczos International Centenary Conference,* J. D. Brown, M. T. Chu, D. C. Ellison, and R. J. Plemmons, eds., SIAM Philadelphia (1993) 134–139.

[N31] "Gaussian Elimination, Perturbation Theory, and Markov Chains." in *Linear Algebra, Markov Chains, and Queuing Models* (Proceedings of an IMA workshop), C. D. Meyer and R. J. Plemmons, eds., Springer, New York (1993) 59–69.

[N32] "Determining Rank in the Presence of Error," in *Linear Algebra for Large Scale and Real-Time Applications,* M. S. Moonen, G. H. Golub, and B. L. R. De Moor, eds., Kluwer Academic Publishers, Dordrecht (1993) 275–291.

[N33] (with W. J. Stewart and D. F. McAllister) "A Two-Stage Iteration for Solving Nearly Completely Decomposable Markov Chains," in *Recent Advances in Iterative Methods,* G. Golub, A. Greenbaum, and M. Luskin, eds., Springer, New York (1994) 201–216.

[N34] "UTV Decompositions," *Proceedings of the 15th Biennial Conference on Numerical Analysis, Dundee,* D. F. Griffiths and G. A. Watson, eds., Longman Scientific, Harlow Essex (1994) 225–236.

[N35] (with G. Latouche) "Numerical Methods for M/G/1 Type Queues," in *Computations with Markov Chains,* W. J. Stewart, ed., Kluwer, Boston, 1995, 571–581

[N36] "Errors in Variables for Numerical Analysts," in *Recent Advances in Total Least Squares Techniques and Errors-in-Variables Modeling,* S. Van Huffel, ed., SIAM, Philadelphia (1997) 3–10.

[N37] "Building an Old-Fashioned Sparse Solver," University of Maryland Computer Science TR-4527, UMIACS TR-2003-95, 2003.

2.2. Major Honors of G. W. Stewart

- The F. L. Bauer prize, awarded by the Technical University of Munich, 1998.
- Elected to the National Academy of Engineering, 2004.
- Distinguished University Professor, University of Maryland, 2006.
- Distinguished Editor of *Linear Algebra and its Applications.*
- SIAM Fellow (inaugural group), 2009.

2.3. Ph.D. Students of G. W. Stewart

- Eric Hill, "Computer Solution of Large Dense Linear Problems," 1977.
- Nancy David, "A First Order Theory of Hypothesis Testing," 1982.
- Robert van de Geijn, "Implementing the QR-Algorithm on an Array of Processors," 1987.
- Xiaobai Sun, "An Unified Analysis of Numerical Methods for Nearly Uncoupled Markov Chains," 1991.
- Misha Kilmer, "Regularization of Ill-Posed Problems," 1997 (jointly directed by Dianne P. O'Leary).
- Che-Rung Lee, "Residual Arnoldi Methods: Theory, Package, and Experiments," 2007.

Part II

Commentaries

3

Introduction to the Commentaries

Misha E. Kilmer and Dianne P. O'Leary

In research spanning over 40 years, G.W. (Pete) Stewart has made foundational contributions to numerical linear algebra.

A major theme in this research is understanding the effects of small perturbations in the matrix on key quantities derived from it: its eigenvalues and eigenvectors, its invariant subspaces, and solutions to linear systems or least squares problems involving the matrix.

A second major theme is the development of efficient matrix algorithms. His insights range from the clever (e.g., economical storage of rotation matrices) to the elegant (the QZ algorithm for solving the generalized eigenproblem and the rank-revealing QR decomposition), and they are grounded in stable matrix decompositions and hands-on computational experience.

The following seven chapters of this commentary outline some of Stewart's important contributions to these two areas.

Notation

Throughout the commentaries, matrices and vectors will be indicated with boldface type, and all vectors are column vectors. Transpose is indicated by a superscript "T," and complex conjugate transpose by a superscript "H."

A specific norm will be identified by a subscript (e.g., 1, 2, F), while a norm without a subscript indicates any of a class of norms.

References to [GWS...] refer to the list of Stewart's publications in Sect. 2.1.

M.E. Kilmer and D.P. O'Leary (eds.), *G.W. Stewart: Selected Works with Commentaries*, Contemporary Mathematicians, DOI 10.1007/978-0-8176-4968-5_3,

4

Matrix Decompositions: LINPACK and Beyond

Charles F. Van Loan, Misha E. Kilmer, and Dianne P. O'Leary

---

1. [GWS-B2] (with J. J. Dongarra, J. R. Bunch, and C. B. Moler), Introduction from LINPACK *Users Guide,* SIAM, Philadelphia (1979).
2. [GWS-J17] (with R. H. Bartels), "Algorithm 432: Solution of the Matrix Equation $AX + XB = C$," *Communications of the ACM* **15** (1972) 820–826.
3. [GWS-J32] "The Economical Storage of Plane Rotations," *Numerische Mathematik* **25** (1976) 137–138.
4. [GWS-J34] "Perturbation Bounds for the QR Factorization of a Matrix," *SIAM Journal on Numerical Analysis* **14** (1977) 509–518.
5. [GWS-J42] (with A. K. Cline, C. B. Moler, and J. H. Wilkinson), "An Estimate for the Condition Number of a Matrix," *SIAM Journal on Numerical Analysis* **16** (1979) 368–375.
6. [GWS-J49] "Rank Degeneracy," *SIAM Journal on Scientific and Statistical Computing* **5** (1984) 403–413.
7. [GWS-J78] "On the Perturbation of LU, Cholesky, and QR Factorizations," *SIAM Journal on Matrix Analysis and Applications,* **14** (1993) 1141–1145.
8. [GWS-J89] "On Graded QR Decompositions of Products of Matrices," *Electronic Transactions in Numerical Analysis* **3** (1995) 39–49.
9. [GWS-J92] "On the Perturbation of LU and Cholesky Factors," *IMA Journal of Numerical Analysis,* **17** (1997) 1–6.
10. [GWS-J94] "The Triangular Matrices of Gaussian Elimination and Related Decompositions," *IMA Journal of Numerical Analysis,* **17** (1997) 7–16.
11. [GWS-J103] "Four Algorithms for the the (sic) Efficient Computation of Truncated Pivoted QR Approximations to a Sparse Matrix," *Numerische Mathematik,* **83** (1999) 313–323.

M.E. Kilmer and D.P. O'Leary (eds.), *G.W. Stewart: Selected Works with Commentaries,* Contemporary Mathematicians, DOI 10.1007/978-0-8176-4968-5_4,

12. [GWS-J118] (with M. W. Berry and S. A. Pulatova) "Algorithm 844: Computing Sparse Reduced-Rank Approximations to Sparse Matrices," *ACM Transactions on Mathematical Software (TOMS)* **31** (2005) 252–269.

Stewart's thesis advisor Alston Householder was a strong and effective advocate for using factorizations in explaining and in solving matrix problems [73]. Stewart adopted a similar viewpoint in his expository work; see [GWS-B1,GWS-B3], etc. Through his research, Stewart brought new theoretical insights into matrix factorizations and worked on definitive software implementations of factorization algorithms. In this chapter we focus on some of his key contributions to LU, QR, Cholesky, and singular value decompositions (SVD). Additional contributions to eigendecompositions, SVD, and updating of factorizations will be discussed in later chapters.

4.1. The LINPACK Project

The LINPACK project's aim was to create a portable and efficient set of FORTRAN-callable codes to solve computational linear algebra problems using matrix decompositions. It was meant to complement the EISPACK codes, which focused on the eigendecomposition and the singular value decomposition.

The LINPACK project [GWS-B2] was a major milestone in the history of mathematical software. Every aspect of software development was taken to new heights by the participants: modularity, reliability, documentation, testing, distribution, etc. Although it was driven by just four researchers, Jack Dongarra, Cleve Moler, Jim Bunch, and Pete Stewart, LINPACK became a focal point of research effort for the entire matrix computation community. Indeed, it helped shape that community by setting forth an agenda that captured the imagination of numerical analysts in academia, the national laboratories, and the great corporate research centers. Jim Bunch [21] recalls the novelty of the project:

> Pete was the "chair" of the group (Stewart, Moler, Bunch, and Dongarra) involved in the LINPACK Project. The object of the project was research into the mechanics of mathematical software production and to produce a computer package for solving linear algebra problems, in particular, linear equations and least squares. We used structured FORTRAN and indentation conventions, and also common nomenclature and commenting conventions ... [We] also decided to use the BLAS (Basic Linear Algebra Subroutines), which had been developed by Lawson, Hanson, Kincaid, and Krogh. Furthermore, the codes were to be completely machine independent. No package had been done in this way before.

Dealing with linear equation solving and least squares data fitting, the LINPACK implementations of the LU, Cholesky, QR, and singular value decompositions solidified the matrix factorization paradigm as a superior alternative to computing matrix inverses or pseudoinverses. Later factorization software for structured problems was benchmarked against general LINPACK counterparts.

Although the production of every code in the package was a team effort, Stewart was principally in charge of QR, SVD, and factorization updating subroutines. Software development in the linear algebra area forces one to "be up close and personal" with the underlying mathematics. Without question, Stewart's involvement with the LINPACK project helps explain why his perturbation theory research is so numerically informed. Many of his greatest algorithmic contributions involve connections between QR, updating, and the SVD – precisely his scope of primary responsibility in LINPACK.

LINPACK was built upon the level-1 Basic Linear Algebra Subroutines (BLAS),[1] a major innovation that allowed low-level vector operations such as dot-product and norm to be hidden in modules that could be executed by generic FORTRAN code or optimized by various computer manufacturers. The ability to encapsulate such operations in a single line of code foreshadowed the development of languages such as MATLAB. LINPACK (together with EISPACK [53]) was superseded in the early 1990s by LAPACK [1], which addressed memory-traffic issues through the implementation of block algorithms that in turn relied upon level-2 and level-3 BLAS implementations of matrix operations, building upon LINPACK technology in a very natural way. It is through LAPACK (and successor packages) that the high standards established by the LINPACK team live on.

The careful attention to detail in LINPACK influenced other software as well. Michael Saunders [124] notes that the LINPACK paper [GWS-N9] "presented a computable estimate of the backward error for an approximate solution to min $\|\mathbf{b}-\mathbf{Ax}\|$. This led to a practical method for terminating the iterative solver LSQR [107] .... This rule has served us well for nearly 30 years. We thank Pete for making it possible."

The authors of LINPACK hoped that it would transform matrix computation, and indeed it did. But it had another unforeseen consequence: LINPACK played a major role in dictating floating-point hardware design. Through the efforts of Jack Dongarra, performance of new machines on the LINPACK benchmark [39], a test suite for solving dense linear systems built upon LINPACK's Gaussian elimination modules, has become a required datapoint for the evaluation of new machines and architectures.

4.2. Some Algorithmic Insights

The success of a project with a scope as wide as LINPACK depended upon getting major and minor design principles correct. In this section we discuss two examples

[1] Stewart insisted upon the "s."

of Stewart's attention to such principles: a later proposal for a storage scheme for Givens rotations and the use of condition number estimators.

4.2.1. The Economical Storage of Plane Rotations

When a Householder transformation is used to zero all but the top component of a vector, it is possible to overwrite the zeroed locations with a representation of the transformation. In a very brief paper [GWS-J32], Stewart shows how to do basically the same thing with plane rotations, thus making Givens rotations competitive in storage with Householder transformations.[2]

When

$$\mathbf{Q} = \begin{bmatrix} c & s \\ -s & c \end{bmatrix}$$

is used to zero a component of a 2-vector, then in that location one can store a single-number representation of $\mathbf{Q}$ and (later on) reliably reconstruct $\mathbf{Q}$ from that number:

Represent $\mathbf{Q}$ with ρ	Reconstruct $\pm\mathbf{Q}$ from ρ
if $c = 0$	**if** $\rho = 1$
$\quad\rho = 1$	$\quad c = 0;\ s = 1$
else	**else**
$\quad$**if** $\lvert s\rvert < \lvert c\rvert$	$\quad$**if** $\rho < 1$
$\quad\quad\rho = \mathrm{sgn}(c)s/2$	$\quad\quad s = 2\rho;\ c = \sqrt{1-s^2}$
$\quad$**else**	$\quad$**else**
$\quad\quad\rho = 2 \cdot \mathrm{sgn}(s)/c$	$\quad\quad c = 2/\rho;\ s = \sqrt{1-c^2}$
$\quad$**end**	$\quad$**end**
end	**end**

A potential pitfall in storing only the sine s or the cosine c is that precision is lost in the calculation $\sqrt{1-x^2}$ if x is close to unity. This never happens with the above scheme because the formula $c = \sqrt{1-s^2}$ is never invoked unless $s^2 \leq 1/2$ and $s = \sqrt{1-c^2}$ is never invoked unless $c^2 \leq 1/2$.

The 2ρ and $2/\rho$ computations make possible the correct determination of whether ρ encodes s or c. The "$\pm$" is not a concern; if $\mathbf{Q}$ is used to solve a zeroing problem, then $-\mathbf{Q}$ can also be used.

This contribution has important practical value, for example, in reducing the storage required for the QR factorization of a sparse matrix. Additionally, this paper illustrates how to wiggle out of a representation problem while maintaining respect for the constraints of floating-point arithmetic. It is an improvement over an earlier proposal by Wilkinson [158, p. 347] which sacrificed floating-point bits to store the sign information.

---

[2] This idea was conceived during a morning shower, and the paper was completed by the end of the day.

4.2.2. An Estimate for the Condition Number of a Matrix

A fringe benefit of using the SVD to solve $\mathbf{Ax} = \mathbf{b}$, where $\mathbf{A}$ is an $n \times n$ nonsingular matrix, is that a reliable estimate $\hat{\kappa}_2$ of the condition number

$$\kappa_2(\mathbf{A}) = \|\mathbf{A}\|_2 \|\mathbf{A}^{-1}\|_2$$

is available as the ratio of the largest singular value to the smallest, with error arising only from roundoff. This yields an inexpensive 2-norm bound on the relative error in the computed solution $\widehat{\mathbf{x}}$ as an approximation to the true solution $\mathbf{x}$:

$$\frac{\|\widehat{\mathbf{x}} - \mathbf{x}\|_2}{\|\mathbf{x}\|_2} \leq \kappa_2(\mathbf{A}) \frac{\|\mathbf{r}\|_2}{\|\mathbf{b}\|_2} \approx \hat{\kappa}_2 \frac{\|\mathbf{r}\|_2}{\|\mathbf{b}\|_2},$$

where $\mathbf{r} = \mathbf{b} - \mathbf{A}\widehat{\mathbf{x}}$. If a cheaper $O(n^3)$ solution process is invoked such as LU, Cholesky, or QR, then an estimate of the condition number, and thus a bound on the relative error, is not so obvious. Can we reliably estimate the condition number in some norm as a cheap by-product of the underlying factorization? In this context, "cheap" means $O(n^2)$ and "reliable" means (for example) that the estimate $\hat{\kappa}$ is no smaller than 90% of the true value κ. The 1-norm is a good choice because computing $\|\mathbf{A}\|_1$ costs $O(n^2)$, thereby reducing the condition estimation problem to the problem of finding a cheap, reliable estimate of $\|\mathbf{A}^{-1}\|_1$. The stakes are high because an underestimate of the true condition can inspire a dangerous false confidence in the quality of the computed solution $\widehat{\mathbf{x}}$.

Cline, Moler, Stewart, and Wilkinson [GWS-J42] developed a reliable condition estimation framework, and their paper is a landmark contribution in three regards. First, it highlighted the importance of computable error bounds to the broad scientific community. Second, it connected perturbation theory and analysis to software development in the best possible way. Third, it helped make heuristically based procedures acceptable within numerical linear algebra, a field that is driven by rigorous testing and analysis.

We should note, however, that although [GWS-J42] gives the most famous condition number estimator, it was not the first. Buried in a paper that Stewart wrote with Gragg on the secant algorithm [GWS-J29][3] is a proposal for estimating an approximate right null vector of a matrix, using the inverse power method, in order to see how close the matrix is to being rank deficient. This also leads to a condition number estimate.

The main idea behind condition estimation is to produce (cheaply) a vector $\mathbf{c}$ so that the solution to $\mathbf{Ay} = \mathbf{c}$ is relatively large in a suitable norm. From the inequality

$$\frac{\|\mathbf{y}\|}{\|\mathbf{c}\|} \leq \|\mathbf{A}^{-1}\|,$$

[3] See Sects. 5.1 and 5.7 in this commentary.

it follows that the quality of

$$\hat{\kappa} := \frac{\|\mathbf{y}\|}{\|\mathbf{c}\|}\|\mathbf{A}\|$$

as a condition estimator improves the more we succeed in making the ratio $\|\mathbf{y}\|/\|\mathbf{c}\|$ as large as possible for the given matrix **A**. SVD theory tells us that in order to accomplish this, **c** should "look like" the right singular vector of **A** corresponding to the smallest singular value.

To fix ideas, assume that we have a pivoted factorization $\mathbf{PAQ} = \mathbf{LU}$ where **P** and **Q** are permutation matrices. Since the 1-norm, 2-norm, and ∞-norm condition numbers (our favorite choices) are not affected by permutation, we assume hereafter that $\mathbf{P} = \mathbf{Q} = \mathbf{I}$. The estimation of $\|\mathbf{A}^{-1}\|$ proceeds as follows:

Step 1. Choose **d** so that the solution to $\mathbf{U}^T\mathbf{z} = \mathbf{d}$ is relatively large in norm.
Step 2. Solve $\mathbf{L}^T\mathbf{c} = \mathbf{z}$ and note that $\mathbf{A}^T\mathbf{c} = \mathbf{d}$.
Step 3. Solve $\mathbf{Ay} = \mathbf{c}$ via $\mathbf{Lw} = \mathbf{c}$ and $\mathbf{Uy} = \mathbf{w}$, and approximate $\|\mathbf{A}^{-1}\|$ with $\|\mathbf{y}\|/\|\mathbf{c}\|$.

The rationale behind this sequence of steps has to do with the expected conditioning of the **L** and **U** matrices. In a pivoted LU factorization, **U** typically inherits **A**'s condition and **L** is typically modestly conditioned. (More on this later.) It follows that if we succeed in Step 1, then **c** will be a large norm solution to $\mathbf{A}^T\mathbf{c} = \mathbf{d}$. This means that **d** will be rich in the direction of the left singular vector of $\mathbf{A}^T$ corresponding to the smallest singular value. It follows that **c** will have a strong component in the direction of the corresponding right singular vector, as required.

The authors' implementation of Step 1 involves an interesting "greedy" algorithm in which **d** is a vector of 1's and -1's determined one component at a time. Consider the following partitioning of the system $\mathbf{U}^T\mathbf{z} = \mathbf{d}$:

$$\begin{bmatrix} \mathbf{U}_{11}^T & \mathbf{0} & \mathbf{0} \\ \mathbf{f}^T & u_{kk} & \mathbf{0} \\ \mathbf{U}_{13}^T & \mathbf{g} & \mathbf{U}_{33}^T \end{bmatrix} \begin{bmatrix} \mathbf{z}_- \\ \gamma \\ \mathbf{z}_+ \end{bmatrix} = \begin{bmatrix} \mathbf{d}_- \\ \delta \\ \mathbf{d}_+ \end{bmatrix}.$$

Assume that the $(k-1)$-vector $\mathbf{d}_-$ has been determined so that $\mathbf{z}_-$ is large in norm. It follows that for a given δ we have

$$\gamma = \frac{\delta - \mathbf{f}^T\mathbf{z}_-}{u_{kk}}$$

and

$$\mathbf{U}_{33}^{\mathrm{T}}\mathbf{z}_{+} = \mathbf{d}_{+} - \mathbf{U}_{13}^{\mathrm{T}}\mathbf{z}_{-} - \gamma\mathbf{g}.$$

The scalar $\delta \in \{+1, -1\}$ is chosen to maximize

$$\left|\frac{\delta - \mathbf{f}^{\mathrm{T}}\mathbf{z}_{-}}{u_{kk}}\right| + \|\mathbf{U}_{13}^{\mathrm{T}}\mathbf{z}_{-} + \gamma\mathbf{g}\|_{1}.$$

The first term encourages growth in the kth component of $\mathbf{z}$, while the second term is a heuristic way of encouraging the emergence of "big numerators" during the subsequent computation of $\delta_{k+1}, \ldots, \delta_{n}$.

Experimental evidence affirms that the method is reliable. Condition estimators based on the ideas set forth in this paper were incorporated in LINPACK [GWS-B2] and subsequently LAPACK [1]. Over the years, a host of structured-matrix versions have been developed (e.g., [160]). There have also been extensions to least squares problems (e.g., [68]), eigenvalue problems (e.g., [146]), and matrix functions (e.g., [83]). Condition estimation is now an integral part of numerical linear algebra computations, and it all started with this seminal contribution and the inclusion of the method in LINPACK.

4.3. The Triangular Matrices of Gaussian Elimination and Related Decompositions

The triangular matrices that emerge from the pivoted LU or QR factorizations are special. In particular, it had been observed for decades that if $\mathbf{T}$ is such a triangular factor, then the computed solution to $\mathbf{Tx} = \mathbf{b}$ is considerably more accurate than what $\mathbf{T}$'s condition number would ordinarily predict. In [GWS-J94], Stewart demystifies this remarkable property through a slick recursion that connects the ratios $\sigma_{\min}(\mathbf{T}_k)/t_{kk}$, $k = 1, \ldots, n$. Here, the subscript k on a matrix denotes its leading $k \times k$ principal submatrix, and $\sigma_{\min}(\cdot)$ denotes smallest singular value.

To fix ideas, let $\mathbf{U}$ be an $n \times n$ upper triangular matrix obtained via a pivoted LU factorization, and let $\mathbf{D}$ contain the main diagonal of $\mathbf{U}$. For $k = 1, \ldots, n$, define

$$\beta_k = \sigma_{\min}(\mathbf{U}_k)/|u_{kk}|.$$

The β_k are always less than 1 and can be regarded as measuring the "rank-revealing quality" of u_{kk}: if β_k is close to unity, then $|u_{kk}|$ does a good job of approximating the smallest singular value of $\mathbf{U}_k$.

This paper provides a model of how to go about explaining observed numerical behavior:

1. *Start with a simple why-is-the-sky-blue type of question.* What does it mean to say that the pivoted LU factorization produces rank-revealing $\mathbf{U}$ matrices?

2. *Set the stage for a quantitative answer by identifying the key quantities.* The β_k are key and apparently they do *not* converge to zero. If they did, then small singular values would sneak under the u_{kk} "radar."
3. *Prove something illuminating about these quantities.* What do "nicely behaved" β_k imply? Stewart derives the recursion

$$\mu_k \geq \mu_{k-1} \frac{\beta_k}{\sqrt{\beta_k^2 + \mu_{k-1}^2}}, \qquad \mu_k \equiv \sigma_{\min}(\mathbf{D}_k^{-1}\mathbf{U}_k),$$

from which it can be concluded that the row-equilibrated matrices $\mathbf{D}_k^{-1}\mathbf{U}_k$ are well conditioned.
4. *Connect to the existing literature.* Higham [69] has shown that the accuracy of upper triangular system solving is independent of row scalings. Thus, if a row scaling of $\mathbf{U}$ makes it well conditioned, then $\mathbf{Uz} = \mathbf{b}$ can be solved accurately.

Given the central role that LU and Cholesky play in scientific computing, it is vitally important that we understand the forces at work when we compute these fundamental factorizations. Stewart's paper [GWS-J94] succeeds in providing just that sort of insight.

4.4. Solving Sylvester Equations

Other sections of this commentary discuss some of Stewart's very clever algorithmic insights into problems such as storing plane rotations (Sect. 4.2.1), updating matrix factorizations (Chap. 5), and computing eigensystems (Chap. 8). The algorithm of Richard Bartels and G.W. Stewart for solving Sylvester equations [GWS-J17] is yet another illustration of the key role played by the ability to recognize exploitable structure.

The problem that Bartels and Stewart consider is the solution of the matrix equation

$$\mathbf{AX} + \mathbf{XB} = \mathbf{C}, \tag{4.1}$$

for $\mathbf{X}$, where the matrices are real valued, $\mathbf{A}$ is $m \times m$, $\mathbf{B}$ is $n \times n$, and $\mathbf{C}$ is $m \times n$. This is called a *Sylvester equation*, or, if $\mathbf{B} = \mathbf{A}^{\mathrm{H}}$, a *Lyapunov equation.* They note that equations of the form (4.1) arise, for example, in the discrete solution to Poisson's equation

$$-u_{xx} - u_{yy} = f(x, y)$$

over the unit square with Dirichlet boundary conditions. This can be observed by letting $\mathbf{X}$ contain estimates of u at equally spaced grid points in the square, setting $\mathbf{C}$ to the values of f at these grid points, and letting $\mathbf{A}$ and $\mathbf{B}$ be finite difference matrices for the second derivatives in the x and y directions, respectively.

Bartels and Stewart use the standard principle of reducing the problem (4.1) to one for which the solution is straightforward, as well as the unfortunately not-entirely-standard principle of making sure the reduction is stable. At the time they were working on their article, a previously published algorithm had relied on the eigendecompositions of $\mathbf{A}$ and $\mathbf{B}$ to compute $\mathbf{X}$. This approach has two disadvantages: first, without an assumption that the matrices $\mathbf{A}$ and $\mathbf{B}$ are symmetric, stability is not assured, and second, complex arithmetic may be necessary. Instead, the algorithm proposed by Bartels and Stewart uses the real Schur decompositions of $\mathbf{A}$ and $\mathbf{B}$, respectively:

$$\mathbf{L} = \mathbf{U}^{\mathrm{T}}\mathbf{A}\mathbf{U}, \quad \mathbf{R} = \mathbf{V}^{\mathrm{T}}\mathbf{B}\mathbf{V},$$

where $\mathbf{U}$ and $\mathbf{V}$ are orthogonal, $\mathbf{L}$ is real block lower triangular, $\mathbf{R}$ is real block upper triangular, and the size of the diagonal blocks is at most 2×2. This factorization proves to be exactly the right "hammer" to complete the job of computing $\mathbf{X}$. To see this, we multiply our equation $\mathbf{AX} + \mathbf{XB} = \mathbf{C}$ by $\mathbf{U}^{\mathrm{T}}$ on the left and $\mathbf{V}$ on the right to obtain

$$\mathbf{U}^{\mathrm{T}}\mathbf{A}\mathbf{U}\mathbf{U}^{\mathrm{T}}\mathbf{X}\mathbf{V} + \mathbf{U}^{\mathrm{T}}\mathbf{X}\mathbf{V}\mathbf{V}^{\mathrm{T}}\mathbf{B}\mathbf{V} = \mathbf{U}^{\mathrm{T}}\mathbf{C}\mathbf{V}.$$

If we then let

$$\mathbf{F} = \mathbf{U}^{\mathrm{T}}\mathbf{C}\mathbf{V}, \quad \mathbf{Y} = \mathbf{U}^{\mathrm{T}}\mathbf{X}\mathbf{V},$$

the original equation (4.1) transforms to

$$\mathbf{LY} + \mathbf{YR} = \mathbf{F}.$$

This is still a Sylvester equation, with unknown matrix $\mathbf{Y}$, but now the matrices $\mathbf{L}$ and $\mathbf{R}$ are triangular. To see that the triangular matrices lead to an efficient algorithm, let us examine the equation in the case of three blocks:

$$\begin{bmatrix} \mathbf{L}_{11} & \mathbf{0} & \mathbf{0} \\ \mathbf{L}_{21} & \mathbf{L}_{22} & \mathbf{0} \\ \mathbf{L}_{31} & \mathbf{L}_{32} & \mathbf{L}_{33} \end{bmatrix} \begin{bmatrix} \mathbf{Y}_{11} & \mathbf{Y}_{12} & \mathbf{Y}_{13} \\ \mathbf{Y}_{21} & \mathbf{Y}_{22} & \mathbf{Y}_{23} \\ \mathbf{Y}_{31} & \mathbf{Y}_{32} & \mathbf{Y}_{33} \end{bmatrix}$$

$$+ \begin{bmatrix} \mathbf{Y}_{11} & \mathbf{Y}_{12} & \mathbf{Y}_{13} \\ \mathbf{Y}_{21} & \mathbf{Y}_{22} & \mathbf{Y}_{23} \\ \mathbf{Y}_{31} & \mathbf{Y}_{32} & \mathbf{Y}_{33} \end{bmatrix} \begin{bmatrix} \mathbf{R}_{11} & \mathbf{R}_{12} & \mathbf{R}_{13} \\ \mathbf{0} & \mathbf{R}_{22} & \mathbf{R}_{23} \\ \mathbf{0} & \mathbf{0} & \mathbf{R}_{33} \end{bmatrix} = \begin{bmatrix} \mathbf{F}_{11} & \mathbf{F}_{12} & \mathbf{F}_{13} \\ \mathbf{F}_{21} & \mathbf{F}_{22} & \mathbf{F}_{23} \\ \mathbf{F}_{31} & \mathbf{F}_{32} & \mathbf{F}_{33} \end{bmatrix}.$$

Now, equating each block on the right with the corresponding expression on the left, we see that we can solve for $\mathbf{Y}_{11}$, then $\mathbf{Y}_{21}$, $\mathbf{Y}_{31}$, $\mathbf{Y}_{12}$, $\mathbf{Y}_{22}$, $\mathbf{Y}_{32}$, $\mathbf{Y}_{31}$, $\mathbf{Y}_{32}$, and finally $\mathbf{Y}_{33}$. For each block $\mathbf{Y}_{k\ell}$ we obtain an equation of the form

$$\mathbf{L}_{kk}\mathbf{Y}_{k\ell} + \mathbf{Y}_{k\ell}\mathbf{R}_{\ell\ell} = \widehat{\mathbf{F}}_{k\ell}, \tag{4.2}$$

where $\widehat{\mathbf{F}}_{k\ell}$ is known. If the blocks $\mathbf{L}_{kk}$ and $\mathbf{R}_{\ell\ell}$ are 1×1, we solve by dividing through by $\mathbf{L}_{kk} + \mathbf{R}_{\ell\ell}$. If the blocks are 2×2, we can write (4.2) as a system of four linear equations for the four entries of $\mathbf{Y}_{k\ell}$. In either case, a unique solution exists if and only if $\mathbf{L}$ and $\mathbf{R}$ have no eigenvalues of equal magnitudes but opposite signs, and this was well known to be the correct existence and uniqueness condition for (4.1). After $\mathbf{Y}$ is computed, $\mathbf{X}$ is recovered by orthogonal transformation.

The authors note that the reduction to triangular form can be accomplished by applying the QR algorithm, and they provide FORTRAN software implementing the algorithm. They discuss the use of iterative refinement to improve the accuracy, the reduction in work if additional data matrices $\mathbf{C}$ are provided for the same matrices $\mathbf{A}$ and $\mathbf{B}$, simplification if $\mathbf{B} = \mathbf{A}^{\mathrm{T}}$, and the use of an upper triangular $\mathbf{L}$ instead of a lower triangular one.

The paper concludes with a brief discussion of operations counts and a discussion of the software. While the authors note that they are unable to establish a forward error bound on the difference between the computed $\mathbf{X}$ and the true one, they do touch on the backward error properties of their proposed algorithm.

In addition to its use in developing fast solvers for elliptic partial differential equations, the Sylvester equation arises in a variety of applications. Its symmetric version, the Lyapunov equation, occurs in many branches of control theory, including stability analysis and optimal control. The ubiquity of these equations is partly due to their relation to matrix problems with Kronecker product structure [147]: If we define vec($\mathbf{X}$) to be the vector formed by stacking the columns of the matrix $\mathbf{X}$, then the Sylvester equation $\mathbf{AX} + \mathbf{XB} = \mathbf{C}$ is equivalent to

$$(\mathbf{I}_n \otimes \mathbf{A} + \mathbf{B}^{\mathrm{T}} \otimes \mathbf{I}_m)\mathrm{vec}(\mathbf{X}) = \mathrm{vec}(\mathbf{C}),$$

where $\mathbf{I}_k$ is the $k \times k$ identity matrix. Problems having Kronecker product structure also arise in restoration of two-dimensional images after convolution with separable point-spread functions [64].

This paper of Bartels and Stewart has nearly 400 citations and the algorithm has been widely used. Golub et al. [57] proposed a faster alternative method, based on the reduction of $\mathbf{A}$ to Hessenberg, rather than triangular, form, and provided a forward-error bound that also applies to the Bartels–Stewart algorithm. Hammarling [62] proposed a variant of the Bartels–Stewart algorithm to compute a Cholesky factor of the solution to a Lyapunov equation.

Stewart later used the triangular solution portion of the algorithm above as an example of how to develop dataflow algorithms for parallel matrix computations [GWS-J56]. Subsequent papers that focused on algorithms for solving large-scale [120] and constrained Sylvester equations [7,139] have used the Bartels and Stewart algorithm as a component of their solution approaches [85].

4.5. Perturbation Bounds for Matrix Factorizations

More than 20 of Stewart's published works have the word *perturbation* in their title. This theme is closely coupled with his other major research direction, algorithmics. As an example of this coupling, the successful use of the matrix factorization algorithms in LINPACK depended on guarantees that the computed factors are the exact factors of a nearby matrix. This invokes the *backward error analysis* that was a central theme of Wilkinson's work [156, 158], asking how close the problem we solved is to the problem we wanted to solve. Beginning in the 1970s, and continuing into the 1990s, Stewart worked out rigorous perturbation bounds for LU, QR, and Cholesky factorizations.

4.5.1. Perturbation of QR Factors

Suppose $\mathbf{A}$ is an $m \times n$ matrix with full column rank. Its QR factorization $\mathbf{A} = \mathbf{QR}$ is unique up to column scalings of $\mathbf{Q}$ (or row scalings of $\mathbf{R}$). How do the $\mathbf{Q}$ and $\mathbf{R}$ factors change if we perturb $\mathbf{A}$? In particular, if

$$\mathbf{A}+\mathbf{E} \;=\; (\mathbf{Q}+\mathbf{W})(\mathbf{R}+\mathbf{F})$$

is the QR factorization of $\mathbf{A}+\mathbf{E}$ with $\mathbf{Q}+\mathbf{W}$ unitary and $\mathbf{R}+\mathbf{F}$ upper triangular, then what can we say about $\|\mathbf{W}\|_F$ and $\|\mathbf{F}\|_F$? Rigorous bounds are provided in [GWS-J34].

The analysis requires a pair of assumptions. First, we need $\mathbf{A}+\mathbf{E}$ to have full rank, which is guaranteed if $\|\mathbf{A}^\dagger\|_F\|\mathbf{E}\|_F$ is small enough, where $\mathbf{A}^\dagger$ is the pseudoinverse of $\mathbf{A}$. (This ensures that $\sigma_{\min}(\mathbf{A}+\mathbf{E}) > 0$.) The analysis also requires that the linear transformation[4]

$$\mathcal{T}: \mathbf{X} \;\mapsto\; \mathbf{R}^{\mathrm{T}}\mathbf{X}+\mathbf{X}^{\mathrm{T}}\mathbf{R} \qquad (\mathbf{X} \text{ upper triangular})$$

be sufficiently nonsingular. The $\mathcal{T}$ transformation arises because the perturbation matrix $\mathbf{F}$ satisfies

$$(\mathbf{A}+\mathbf{E})^{\mathrm{T}}(\mathbf{A}+\mathbf{E}) \;=\; (\mathbf{R}+\mathbf{F})^{\mathrm{T}}(\mathbf{R}+\mathbf{F});$$

i.e.,

$$\mathbf{R}^{\mathrm{T}}\mathbf{F}+\mathbf{F}^{\mathrm{T}}\mathbf{R} \;=\; \left(\mathbf{E}^{\mathrm{T}}\mathbf{A}+\mathbf{A}^{\mathrm{T}}\mathbf{E}\right) \;+\; \left(\mathbf{E}^{\mathrm{T}}\mathbf{E}+\mathbf{F}^{\mathrm{T}}\mathbf{F}\right).$$

This is nonlinear in $\mathbf{F}$, but it becomes tractable if the second-order terms on the right-hand side are properly handled. It then follows that

$$\|\mathbf{F}\|_F \;=\; O\left(\|\mathcal{T}^{-1}\|_F\|\mathbf{A}\|_F\|\mathbf{E}\|_F\right).$$

After carefully analyzing $\mathcal{T}$, Stewart proceeds to give rigorous bounds for both $\mathbf{W}$ and $\mathbf{F}$. If

$$\kappa_2(\mathbf{A})\frac{\|\mathbf{E}\|_F}{\|\mathbf{A}\|_F} \;=\; \|\mathbf{A}^\dagger\|_F\|\mathbf{E}\|_F \;\le\; \frac{1}{4},$$

and $\mathcal{T}$ is not too badly conditioned, then

$$\|\mathbf{W}\|_F \;\le 6\kappa_2(\mathbf{A})\frac{\|\mathbf{E}\|_F}{\|\mathbf{A}\|_F} \qquad \text{and} \qquad \|\mathbf{F}\|_F \;\le\; 8\kappa_2(\mathbf{A})\frac{\|\mathbf{E}\|_F}{\|\mathbf{A}\|_F}.$$

Aside from providing useful and informative bounds, the thoughtful organization of this paper sends a powerful message about exposition: always share your insights and intuition with the reader *before* you get precise and rigorous.

[4] A similar operator is studied in [GWS-J19]; see Sect. 7.1.

4.5.2. Perturbation of LU Factors

If the norm of a square matrix $\mathbf{F}$ is small enough, then $\mathbf{I}+\mathbf{F}$ is guaranteed to have an LU factorization, and our intuition tells us that both $\mathbf{L}$ and $\mathbf{U}$ should be small departures from the identity. In [GWS-J78], Stewart makes this precise with bounds valid for norms satisfying[5] $\| |\mathbf{A}| \| = \|\mathbf{A}\|$.

Let $\mathcal{L}_p(\mathbf{M})$ denote the strictly lower triangular portion of a matrix $\mathbf{M}$ plus p times its main diagonal, and define $\mathcal{U}_p(\mathbf{M})$ to be the strictly upper triangular portion plus the same diagonal matrix. If $\|\mathbf{F}\| \leq 1/4$, then there is a matrix $\mathbf{G}$ with $\|\mathbf{G}\| \leq \|\mathbf{F}\|^2$ that satisfies

$$\mathbf{I}+\mathbf{F} = (\mathbf{I}+\mathcal{L}_0(\mathbf{F})+\mathcal{L}_0(\mathbf{G}))\,(\mathbf{I}+\mathcal{U}_1(\mathbf{F})+\mathcal{U}_1(\mathbf{G}))\,.$$

To first order we therefore have $\mathbf{I}+\mathbf{F} \approx (\mathbf{I}+\mathcal{L}_0(\mathbf{F}))(\mathbf{I}+\mathcal{U}_1(\mathbf{F}))$. We will use this formula to understand perturbation of the LU factors.

If $\mathbf{A} = \mathbf{LU}$, then

$$\mathbf{A}+\mathbf{E} = \mathbf{L}(\mathbf{I}+\mathbf{L}^{-1}\mathbf{E}\mathbf{U}^{-1})\mathbf{U}.$$

It follows that if $\mathbf{F} = \mathbf{L}^{-1}\mathbf{E}\mathbf{U}^{-1}$ is small enough, $\mathbf{A}+\mathbf{E} \approx \widetilde{\mathbf{L}}\widetilde{\mathbf{U}}$ where

$$\widetilde{\mathbf{L}} = \mathbf{L}(\mathbf{I}+\mathcal{L}_0(\mathbf{F})), \qquad \widetilde{\mathbf{U}} = (\mathbf{I}+\mathcal{U}_1(\mathbf{F}))\mathbf{U}.$$

From this, first-order bounds can be derived for the relative changes in $\mathbf{L}$ and $\mathbf{U}$; for example,

$$\frac{\|\widetilde{\mathbf{L}}-\mathbf{L}\|}{\|\mathbf{L}\|} \lesssim \|\mathbf{L}^{-1}\|\|\mathbf{U}^{-1}\|\|\mathbf{A}\|\frac{\|\mathbf{E}\|}{\|\mathbf{A}\|}.$$

Using similar manipulations, Stewart derives bounds for the QR and Cholesky factorizations.

From a pedagogic point of view, [GWS-J78] is valuable because it shows how to conduct a first-order perturbation analysis *without* sweeping the second-order terms under the rug!

If $\mathbf{A} = \mathbf{LU}$ and $\mathbf{E}$ is small enough, then $\mathbf{A}+\mathbf{E}$ has an LU factorization

$$\mathbf{A}+\mathbf{E} = (\mathbf{L}+\Delta\mathbf{L})(\mathbf{U}+\Delta\mathbf{U}).$$

In [GWS-J92], Stewart turns his attention to deriving informative first-order approximations to $\Delta\mathbf{L}$ and $\Delta\mathbf{U}$. Let $\mathbf{D}_L$ and $\mathbf{D}_U$ be arbitrary nonsingular diagonal matrices. Stewart shows that if

$$\breve{\mathbf{F}}_L = (\mathbf{L}\mathbf{D}_L)\,\mathcal{L}_p\left((\mathbf{L}\mathbf{D}_L)^{-1}\mathbf{E}\mathbf{U}^{-1}\right)$$

[5] The 1-norm, ∞-norm, and Frobenius norm satisfy this equality, but the 2-norm does not.

and

$$\breve{\mathbf{F}}_U = \mathcal{U}_{1-p}\left(\mathbf{L}^{-1}\mathbf{E}(\mathbf{D}_U\mathbf{U})^{-1}\right)(\mathbf{D}_U\mathbf{U}),$$

then

$$\frac{\|\breve{\mathbf{F}}_L\|}{\|\mathbf{L}\|} \leq \kappa(\mathbf{L}\mathbf{D}_L)\kappa(\mathbf{U})\frac{\|\mathbf{E}\|}{\|\mathbf{A}\|}$$

and

$$\frac{\|\breve{\mathbf{F}}_U\|}{\|\mathbf{U}\|} \leq \kappa(\mathbf{L})\kappa(\mathbf{D}_U\mathbf{U})\frac{\|\mathbf{E}\|}{\|\mathbf{A}\|}.$$

(Again, any norm for which $\||\mathbf{A}|\| = \|\mathbf{A}\|$ may be used here.) The parameter p determines how much of the perturbation is attached to the diagonals of $\mathbf{L}$ and $\mathbf{U}$, respectively; for example, if $p = 0$ then the diagonal of $\mathbf{L}$ does not change. Notice that the relative error bounds are independent of the scaling matrices $\mathbf{D}_L$ and $\mathbf{D}_U$ in the following sense: for situations where $\mathbf{L}$ (or $\mathbf{U}$) is ill-conditioned because of poorly scaled columns (or rows), the above bounds can be minimized by choosing $\mathbf{D}_L$ and $\mathbf{D}_U$ accordingly.

When perturbation results consistently overestimate what is observed in practice, it is appropriate to revisit the analysis and identify the factor that has been overlooked. In this case, Stewart refined his earlier analysis [GWS-J78] to include the scaling matrices $\mathbf{D}_L$ and $\mathbf{D}_U$. He was prompted by Chang and Paige [24], who had similar concerns.

4.6. Rank Degeneracy

Perhaps more than any other single subject, the treatment of rank is what distinguishes numerical from theoretical linear algebra. In theoretical linear algebra, rank is a crisp, easy concept. A given $n \times p$ matrix $\mathbf{X}$ with $n \geq p$ either has full column rank or it does not. And if it does not, then just count the number of independent columns and get on with it! In the numerical linear algebra context, $\mathbf{X}$ may not even be directly available. Typically, we are in possession of an error-contaminated version $\widetilde{\mathbf{X}} = \mathbf{X} + \mathbf{E}$ and are challenged to make inferences about $\mathbf{X}$'s rank, a task that is made even more difficult because *every* matrix is arbitrarily close to a full rank matrix. [GWS-J49] is about using $\tilde{\mathbf{X}}$ to estimate $k = \text{rank}(\mathbf{X})$ and constructing a matrix $\widehat{\mathbf{X}}$ that approximates $\mathbf{X}$ with the property that $\text{rank}(\widehat{\mathbf{X}}) = k$. It mainly targets the statistics community and contrasts methods based on the SVD, QR with column pivoting, and the inverse of the cross-product matrix

$$\mathbf{A}^{-1} = (\mathbf{X}^{\mathrm{T}}\mathbf{X})^{-1}.$$

The volume of work and the effect of rounding errors are considered in each case. Intelligent comments are offered that pertain to "how small is small," since each method ultimately requires a judgment call about when to declare a tiny value as zero.

The paper has something to offer both the statistician and the numerical analyst and is a tribute to Stewart's ability to bridge the gap between the two communities. For example, he points out connections between the diagonal element r_{kk} in the R-factor of the QR decomposition and the diagonal entries in $\mathbf{A}^{-1}$. Researchers in matrix computations are well aware of what r_{kk} "says," while regression experts have a statistical appreciation for $(\mathbf{A}^{-1})_{kk}$. Another interesting discussion point in the paper pertains to the structure of the error matrix $\mathbf{E}$ and how its statistical properties have a bearing on intelligent rank determination. In particular, if the elements of $\mathbf{E}$ are independent, then it is critical to scale $\widetilde{\mathbf{X}}$ so that the resulting scaled errors are of roughly the same size. Otherwise, small computed singular values might be associated with a kind of artificial ill-conditioning.

By sensitizing statisticians to numerical issues and numerical analysts to the statistical undercurrents of data analysis, Stewart did much to raise the quality of statistical computation during the 1970s, 1980s, and beyond.

4.7. Pivoted QR as an Alternative to SVD

The QR decomposition with column pivoting is a successful alternative to the singular value decomposition for computing a low rank approximation to a given $m \times n$ matrix $\mathbf{A}$:

$$\mathbf{A}\mathbf{P} = \begin{bmatrix} \mathbf{Q}_k & \bar{\mathbf{Q}}_k \end{bmatrix} \begin{bmatrix} \mathbf{R}_k & \mathbf{S}_k \\ \mathbf{0} & \mathbf{G}_k \end{bmatrix},$$

where $\mathbf{P}$ is a permutation matrix. In particular, if $\mathbf{R}_k$ is $k \times k$ and nonsingular, then the *truncated pivoted QR approximation*

$$\widehat{\mathbf{A}}_k = \mathbf{Q}_k \begin{bmatrix} \mathbf{R}_k & \mathbf{S}_k \end{bmatrix} \mathbf{P}^{\mathrm{T}}$$

is a rank-k approximation that reproduces exactly k (strongly) independent columns of $\mathbf{A}$, and we will collect these columns in a matrix called $\mathbf{B}_k$. The quality of the approximation $\widehat{\mathbf{A}}_k$ depends upon $\|\mathbf{G}_k\|$, but in many important applications, this is small.

Stewart has devoted considerable attention to exploring the use of the pivoted-QR decomposition as an inexpensive alternative to the SVD. In this section, we consider three of his papers on this subject.

4.7.1. Truncated Pivoted QR for Sparse Matrices

If $\mathbf{B}_k = \mathbf{Q}_k\mathbf{R}_k$ where $\mathbf{R}_k$ is square and nonsingular, then

$$\mathbf{Q}_k = \mathbf{B}_k\mathbf{R}_k{}^{-1}.$$

This is one of the most obvious facts in linear algebra, but its clever application is fundamental to Stewart's work on sparse matrix approximation.

The ideas presented in Stewart's 2005 joint paper with Berry and Pulatova [GWS-J118] date back to a 1999 paper [GWS-J103], and we discuss both papers together.

In [GWS-J103], Stewart notes that if $\mathbf{B}_k$ is a sparse $m \times k$ matrix with m much bigger than k, then the factors in the QR factorization of $\mathbf{B}_k$ may be much more dense than $\mathbf{B}_k$. This is not a problem for the upper triangular factor $\mathbf{R}_k$ because it is only $k \times k$, but the matrix $\mathbf{Q}_k$, whose columns are orthonormal, may have almost mk nonzeros, and this may be too much to store. Instead, however, we can store $\mathbf{B}_k$ and $\mathbf{R}_k$, and, whenever we need to apply $\mathbf{Q}_k$ to a vector $\mathbf{z}$, just form $\mathbf{y}$ from $\mathbf{R}_k\mathbf{y} = \mathbf{z}$ and then $\mathbf{Q}_k\mathbf{z} = \mathbf{B}_k\mathbf{y}$.

Stewart then uses this fact in his development of two algorithms for producing sparse, rank-revealing approximations of the larger matrix $\mathbf{A}$. Specifically, he continues the QR factorization to the largest value of k so that $\mathbf{B}_k$ is well conditioned. The pivoted QR factorization chooses the permutation matrix $\mathbf{P}$ so that $\mathbf{B}_k$ is well conditioned and so that

$$E_k = \|\mathbf{AP} - \mathbf{Q}_k \begin{bmatrix} \mathbf{R}_k & \mathbf{S}_k \end{bmatrix} \|_F$$

is as small as possible. In particular, the factorization reproduces k columns of $\mathbf{A}$ exactly, and approximates the others optimally. Thus, Stewart has constructed a sparse rank-k approximation to $\mathbf{A}$. Stewart computes this factorization using the pivoted Gram–Schmidt algorithm, where

$$\begin{bmatrix} \mathbf{R}_k & \mathbf{S}_k \end{bmatrix} = \mathbf{Q}_k^{\mathrm{T}}\mathbf{AP}.$$

The next column to include can be determined from the entries in $\mathbf{S}_k$, as described in Sect. 2 of [GWS-J118], and updates are performed on only that column, using the trick described in the preceding paragraph, to save storage. Berry et al.[6] call this the *sparse pivoted QR approximation (SPQR)*. Stewart [GWS-J103] called the algorithm *quasi-Gram–Schmidt* for computing a *truncated pivoted QR approximation*. It belongs to a class of approximation methods called *CU* approximations, where $\mathbf{C}$ represents a matrix of columns chosen from $\mathbf{A}$ and $\mathbf{U}$ is determined to make the approximation good. Michael Berry [13] says, "Pete's ingenuity in noticing that a efficient sparse QR approximation could be an effective alternative to the Lanczos- and Arnoldi-based approaches for the sparse SVD problem was very ground breaking. His SPQR algorithm has drawn a lot of attention and use."

In [GWS-J103], Stewart proposes an alternative algorithm, using Householder transformations to compute the truncated pivoted QR factorization. Stewart modifies this classical approach in two ways to accommodate sparsity. First, the Householder transformations are applied only as needed. In particular, to compute the

[6] Berry [13] notes that "Pete insisted that he not be first on our joint paper with one of my MS students. He was definitely the catalyst for the paper yet very gracious in putting my student and me first .... I was so impressed with his generosity and his genuine interest in others as well as good science."

"next" Householder transformation, all the previous Householder transformations are applied to the relevant column of **A**. The "next" row of **R** can be computed by an efficient recursion, eliminating the need for a complete, filled-in update of **A**. Second, the product of the Householder transformations can be represented in the compact block form proposed by Schreiber and Van Loan [125]. This leads to a particularly efficient implementation.

In [GWS-J103], Stewart also shows how to construct what came to be known as a *CUR* approximation (see further discussion below), where **R** (not to be confused with the right triangular matrix $\mathbf{R}_k$ above) represents a matrix of rows chosen from **A**. He chooses the rows by applying quasi-Gram–Schmidt to $\mathbf{A}^{\mathrm{T}}$ and derives a formula for the optimal choice of **U**. In the Berry et al. paper, the authors call this the *sparse column–row approximation (SCR).* [GSW-J103] clearly demonstrates that intelligent pivoted-QR is a worthy alternative to the SVD in many applications.

The Berry et al. paper [GWS-J118] reviews the Gram–Schmidt SPQR and SCR algorithms of [GWS-J103], describing them in pseudo-MATLAB. The main disadvantage of these algorithms is that the columns in the computed matrix $\mathbf{Q}_k$ may fail to be orthogonal to working precision, and a reorthogonalization process is advocated. After the computation is well defined, however, the paper goes one step further. The entirety of Sect. 5 is devoted to discussion of the nonmathematical part of the software design. In particular, they consider what options a user might want to control and what output a user might desire. This paper, therefore, is an excellent example of the need for careful thought about the user interface.

After some numerical experiments for SPQR, the paper concludes with a lengthy discussion of careful implementation of the algorithm for matrices in sparse row- or column-oriented packed storage. An error analysis for the algorithm is provided in [GWS-J117].

An important aspect of the work in the Berry, Pulatnova, and Stewart paper is its impact on applications. As mentioned in the paper, there is a wide range of applications of this algorithm in various areas of information retrieval. For example, suppose we construct a *term-document matrix* where the rows correspond to terms, the columns to documents, and the entries measure the importance of a term in a document. Such matrices can be very large (since there are billions of documents on the Web), but quite sparse. Low-storage approximations can be used to identify documents relevant to a query. As the amount of data becomes larger and larger over time, algorithms such as these and those born of ideas in this paper will become increasingly important.

In the literature, the distinction between Stewart's CU and CUR algorithms and others [40–43] is often described as deterministic vs. nondeterministic. However, the more critical difference is that Stewart chooses columns *without replacement*, while the randomized algorithms often choose the columns *with replacement*, and this partially accounts for the need in nondeterministic algorithms for larger values of k in order to achieve error comparable to E_k. Stewart's algorithm is generally faster than those based on sampling with probability distributions determined by the singular vectors, and somewhat slower than those based on distributions determined from the elements of **A**.

4.7.2. Pivoted QR for Products of Matrices

One thing you should *not* do when asked to compute the singular values of a matrix product $\mathbf{M}_m = \mathbf{A}_1 \cdots \mathbf{A}_m$ is form the explicit product. Information is lost, making it very difficult to compute accurately the small singular values. The *product* SVD of Bojanczyk et al. [17] circumvents this problem but is demanding of storage and does not gracefully handle the update situation, i.e., the situation when you have the singular values of $\mathbf{M}_m$ and now require the singular values of $\mathbf{M}_{m+1} = \mathbf{M}_m\mathbf{A}_{m+1}$.

[GWS-J89] further demonstrates Stewart's knack for solving SVD problems through the intelligent manipulation of triangular matrices and the pivoted QR decomposition. We first need a definition. An upper triangular matrix **R** is *graded* if $\mathbf{R} = \mathbf{D}\widehat{\mathbf{R}}$ where

$$\mathbf{D} = \operatorname{diag}(d_1, \ldots, d_n), \qquad d_1 \geq \cdots \geq d_n > 0,$$

and the upper triangular entries in $\widehat{\mathbf{R}}$ are (at most) $O(1)$. Note that if $\mathbf{R}_1 = \mathbf{D}_1\widehat{\mathbf{R}}_1$ and $\mathbf{R}_2 = \mathbf{D}_2\widehat{\mathbf{R}}_2$ are graded, then

$$\mathbf{R}_1\mathbf{R}_2 = \mathbf{D}_1\widehat{\mathbf{R}}_1\mathbf{D}_2\widehat{\mathbf{R}}_2 = (\mathbf{D}_1\mathbf{D}_2)(\mathbf{D}_2^{-1}\widehat{\mathbf{R}}_1\mathbf{D}_2)\widehat{\mathbf{R}}_2.$$

It follows that $\mathbf{R}_1\mathbf{R}_2$ is graded because

$$[\mathbf{D}_2^{-1}\widehat{\mathbf{R}}_1\mathbf{D}_2]_{ij} \leq [\widehat{\mathbf{R}}_1]_{ij} = O(1), \qquad i \leq j,$$

and so (heuristically) the entries in $(\mathbf{D}_2^{-1}\widehat{\mathbf{R}}_1\mathbf{D}_2)\widehat{\mathbf{R}}_2$ are $O(1)$.

We say that $\mathbf{A} = \mathbf{QRP}^{\mathrm{T}}$ is a graded QR factorization if **P** is a permutation, **Q** is orthogonal, and **R** is a graded upper triangular matrix. Note that the pivoted-QR algorithm produces a graded QR factorization and that (generally excellent) small singular value estimates can be deduced from the trailing principal submatrices of **R**.

The main idea behind the paper can be gleaned from the problem of computing a graded QR decomposition of **AB** given that

$$\mathbf{A} = \mathbf{Q}_{\mathrm{A}}\mathbf{R}_{\mathrm{A}}\mathbf{P}_{\mathrm{A}}^{\mathrm{T}}$$

is a graded QR decomposition of **A**. It is basically a two-step process. First, compute the graded decomposition

$$(\mathbf{P}_{\mathrm{A}}^{\mathrm{T}}\mathbf{B}) = \mathbf{Q}_{\mathrm{B}}\mathbf{R}_{\mathrm{B}}\mathbf{P}_{\mathrm{C}}^{\mathrm{T}}$$

of $\mathbf{P}_{\rm A}^{\rm T}\mathbf{B}$ and observe that

$$\mathbf{AB} \;=\; (\mathbf{Q}_{\rm A}\mathbf{R}_{\rm A}\mathbf{P}_{\rm A}^{\rm T})\mathbf{B} \;=\; \mathbf{Q}_{\rm A}\mathbf{R}_{\rm A}(\mathbf{P}_{\rm A}^{\rm T}\mathbf{B}) \;=\; \mathbf{Q}_{\rm A}\mathbf{R}_{\rm A}(\mathbf{Q}_{\rm B}\mathbf{R}_{\rm B}\mathbf{P}_{\rm C}^{\rm T}).$$

If $\mathbf{U}\widehat{\mathbf{R}}_{\rm A} = \mathbf{R}_{\rm A}\mathbf{Q}_{\rm B}$ is the QR factorization of $\mathbf{R}_{\rm A}\mathbf{Q}_{\rm B}$, then

$$\mathbf{AB} \;=\; (\mathbf{Q}_{\rm A}\mathbf{U})(\widehat{\mathbf{R}}_{\rm A}\mathbf{R}_{\rm B})\mathbf{P}_{\rm C}^{\rm T}$$

is the required decomposition, if we can show that the upper triangular product $\widehat{\mathbf{R}}_{\rm A}\mathbf{R}_{\rm B}$ is graded. Since $\mathbf{R}_{\rm B}$ is graded, it suffices to show that $\widehat{\mathbf{R}}_{\rm A}$ is graded from our remarks above about products of graded matrices. Stewart proceeds to do just that with a heuristic argument based on Givens rotations. His conjectures are confirmed through careful testing.

The single-product ideas outlined above extend naturally to the case of multiple products $\mathbf{A}_1 \cdots \mathbf{A}_m$. Once again, Stewart points the way to an effective QR-based update procedure that can be used to provide information about the small singular values without computing the update-hostile SVD itself.

4.8. Summary

Stewart's algorithmic insights are both sweeping and detailed, ranging from making QR decomposition practical for large, sparse matrices, to giving attention to whether a rotation matrix can be well characterized by storing a single number. His algorithmic decisions were determined by his detailed error analyses, and his error analyses were informed by his algorithmic experience. This double-barreled approach has been key to the success of LINPACK and other software projects throughout his career.

5

Updating and Downdating Matrix Decompositions

Lars Eldén, Misha E. Kilmer, and Dianne P. O'Leary

---

1. [GWS-J29] (with W. B. Gragg), "A Stable Variant of the Secant Method for Solving Nonlinear Equations," *SIAM Journal on Numerical Analysis* **13** (1976) 889–903.
2. [GWS-J31] (with J. W. Daniel, W. B. Gragg, and L. Kaufman), "Reorthogonalization and Stable Algorithms for Updating the Gram-Schmidt QR Factorization," *Mathematics of Computation* **30** (1976) 772–795.
3. [GWS-J40] "The Effects of Rounding Error on an Algorithm for Downdating the Cholesky Factorization," *Journal of the Institute for Mathematics and its Applications* (IMA), *Applied Mathematics* **23** (1979) 203–213.
4. [GWS-J73] "An Updating Algorithm for Subspace Tracking," *IEEE Transactions on Signal Processing* **40** (1992) 1535–1541.
5. [GWS-J77] "Updating a Rank-Revealing ULV Decomposition," *SIAM Journal on Matrix Analysis and Applications* **14** (1993) 494–499.
6. [GWS-J87] "On the Stability of Sequential Updates and Downdates," *IEEE Transactions on Signal Processing* **43** (1995) 2642–2648.

The Sherman–Morrison–Woodbury formula ([GWS-B7], p. 328) is a recipe for constructing the inverse of a matrix after it has been modified by a low-rank correction. For a matrix of size $n \times n$ that has been so modified, it enables the inverse of this matrix to be updated in time proportional to kn^2, where k is the rank of the correction, rather than the n^3 time usually necessary to compute the inverse directly. This important fact has enabled a variety of algorithms, from early implementations of the simplex method for linear optimization [29] to algorithms for solving least squares problems when new data arrive.

M.E. Kilmer and D.P. O'Leary (eds.), *G.W. Stewart: Selected Works with Commentaries*, Contemporary Mathematicians, DOI 10.1007/978-0-8176-4968-5_5,

This formula, however, can suffer from instability because it is akin to applying Gaussian elimination without pivoting. Therefore, starting in the 1970s, considerable effort was devoted to finding efficient and stable update formulas for matrix factorizations. We summarize in this chapter some of Stewart's work in developing update algorithms and his use of such algorithms in applications.

In addition to his technical contributions, it was Stewart who coined the name *downdating* to denote the process of replacing a symmetric matrix $\mathbf{A}$ by $\mathbf{A} - \mathbf{x}\mathbf{x}^{\mathrm{T}}$, in analogy to the *updating* process of replacing it by $\mathbf{A} + \mathbf{x}\mathbf{x}^{\mathrm{T}}$. The distinction is important, since only in the downdating case can a symmetric positive definite matrix become singular.

5.1. Solving Nonlinear Systems of Equations

Much of Stewart's early work, including collaboration with his thesis advisor Alston Householder, was related to finding roots of polynomials and solutions of more general nonlinear equations, and he returned to this familiar subject throughout his career [GWS-J1,GWS-J6,GWS-J7,GWS-J8,GWS-J11,GWS-J12,GWS-J13,GWS-J14, GWS-J21,GWS-J22,GWS-J29,GWS-J44,GWS-J83]. His paper [GWS-J29] with Bill Gragg, discussed in this section, is an example of this work. First, we provide some historical context for this paper.

At the time that Gragg and Stewart were working on their paper, a rebellion of sorts was occurring in the optimization community. Since 1957 [95], simplex implementations had advanced to updating basis factorizations in product form:

$$\begin{aligned}\mathbf{B}_0 &= \mathbf{LU}\\ \mathbf{B}_k &= \mathbf{B}_{k-1}\mathbf{T}_k\\ &= \mathbf{LUT}_1\mathbf{T}_2\cdots\mathbf{T}_k, \quad k = 1, 2, \ldots,\end{aligned}$$

where each basis matrix $\mathbf{B}_k$ differs from $\mathbf{B}_{k-1}$ in only one column, and each $\mathbf{T}_k$ is a permuted triangular matrix that differs from the identity in only one column.[7] In the early 1970s, a troublesome fact was being broadcast by the numerical analysis community: the product-form update was not guaranteed to be numerically stable. This had been recognized by the optimizers, who used "reinversion" (factorizing $\mathbf{B}_k$ from scratch) to condense the factors $\mathbf{T}_1$, $\mathbf{T}_2$, ... and to try to avoid breakdown, but there was strong resistance to any remedy that would raise the cost of a simplex iteration. Questions of cost were further complicated by the need to take advantage of sparsity in the data.

The rebellion was begun by Gene Golub and his student Richard Bartels [10],[8] who proposed implementing the simplex method using *stable* updates of LU factors.

[7] Most authors spoke in terms of the basis *inverse*, but their matrices $\mathbf{T}_k^{-1}$ involved the same numbers as $\mathbf{T}_k$, and we know that the triangular factors $\mathbf{L}$, $\mathbf{U}$, $\mathbf{T}_1$, $\mathbf{T}_2$, ... allow easy solution of equations involving $\mathbf{B}_k$ or $\mathbf{B}_k^{\mathrm{T}}$.

[8] Bartels's work is also discussed in Sect. 5.4.

Walter Murray, his student Philip Gill, Golub's student Michael Saunders, and Golub and Murray's student Margaret Wright [54, 55] continued the movement, focusing on dense matrix problems. It took several decades and some compromises, but eventually the matrix factorization approach became preeminent in the simplex-type (active set) algorithms as well as in the newer interior-point methods.

The paper by Gragg and Stewart is in this same matrix factorization spirit. They consider the problem of solving the nonlinear system of equations

$$\mathbf{f}(\mathbf{x}) = \mathbf{0},$$

where $\mathbf{f}$ is a mapping from a domain in $\mathcal{R}^n$ to $\mathcal{R}^n$ (for example, the gradient of a function to be optimized) and where the Jacobian matrix $\mathbf{J}(\mathbf{x})$ of partial derivatives of $\mathbf{f}$ is not available. Suppose $\mathbf{J}$ were available, and let $\mathbf{x}_1$ denote the current approximate solution.[9] Then we might use the Newton algorithm, updating the approximate solution $\mathbf{x}_1$ by

$$\mathbf{x}_* = \mathbf{x}_1 - \alpha \mathbf{J}(\mathbf{x}_1)^{-1}\mathbf{f}(\mathbf{x}_1), \tag{5.1}$$

where α is a steplength parameter, until $\mathbf{f}(\mathbf{x})$ is small enough. Gragg and Stewart assume that $\alpha = 1$, although this is not necessary.

Now assume that $n+1$ previous estimates $\mathbf{x}_j$ are available.[10] Since the Jacobian is not available, it is natural, once we have a history of values $\mathbf{f}_j = \mathbf{f}(\mathbf{x}_j)$, $j = 1, \ldots, n+1$, to try to approximate the Jacobian using finite difference approximations. Gragg and Stewart present this as taking the Jacobian matrix of the affine function that interpolates $\mathbf{f}$ at $\mathbf{x}_j$. Defining

$$\mathbf{X} = [\mathbf{x}_1, \mathbf{x}_2, \cdots, \mathbf{x}_{n+1}],$$
$$\mathbf{F} = [\mathbf{f}_1, \mathbf{f}_2, \cdots, \mathbf{f}_{n+1}],$$

they seek an approximation of $\mathbf{J}(\mathbf{x}_1)$ as follows. Defining

$$\mathbf{\Delta X} = [\mathbf{x}_2 - \mathbf{x}_1, \mathbf{x}_3 - \mathbf{x}_1, \cdots, \mathbf{x}_{n+1} - \mathbf{x}_1],$$
$$\mathbf{\Delta F} = [\mathbf{f}_2 - \mathbf{f}_1, \mathbf{f}_3 - \mathbf{f}_1, \cdots, \mathbf{f}_{n+1} - \mathbf{f}_1],$$

they obtain

$$\mathbf{J}(\mathbf{x}_1)^{-1} \approx \mathbf{\Delta X}(\mathbf{\Delta F})^{-1}.$$

We can use this approximation in the Newton iteration (5.1), and this variant of the *secant algorithm* is studied in this paper.

Let us consider the operations involved in implementing this algorithm. At each iteration, there is a linear system in (5.1) that must be solved to obtain a new

---

[9] Denoting $\mathbf{x}_1$ as the most recent iterate is nonstandard.

[10] Initially, the indexing is in reverse chronological order, with $\mathbf{x}_1$ referring to the most recent estimate and $\mathbf{x}_{n+1}$ referring to the first estimate. However, as newer estimates replace older ones according to the mechanism in the paper, the subscripts are not necessarily consistent with the order in which they are computed via the secant update.

$\mathbf{x}$-iterate, and the matrix $\mathbf{X}$ (and $\mathbf{F}$) must be updated by inserting the new $\mathbf{x}_*$ (and $\mathbf{f}(\mathbf{x}_*)$) and removing some old column. A method for determining which column to replace at a given iteration is described in detail in the paper. Computing the secant approximation

$$\mathbf{x}_* = \mathbf{x}_1 - \mathbf{\Delta X}(\mathbf{\Delta F})^{-1}\mathbf{f}(\mathbf{x}_1) \tag{5.2}$$

will in general cost $O(n^3)$ operations, unless we exploit the fact that the matrix $\mathbf{F}$, and hence $\mathbf{\Delta F}$, changes by only one column from iteration to iteration.

Gragg and Stewart propose storing and updating two matrix factorizations. Let $\mathbf{X}_k$ denote the $n \times (n+1)$ matrix with the current n estimates of $\mathbf{x}$ and $\mathbf{F}_k$ denote the $n \times (n+1)$ matrix whose columns are given by evaluating $\mathbf{f}$ at each of the respective columns of $\mathbf{X}_k$. The two factorizations they propose are

$$\begin{aligned} \mathbf{X}_k &= \mathbf{P}_k^{\mathrm{T}}\mathbf{Y}_k, \\ \mathbf{G}_k &= \mathbf{Q}_k^{\mathrm{T}}\mathbf{F}_k, \end{aligned}$$

where $\mathbf{P}_k$ and $\mathbf{Q}_k$ are nonsingular and $\mathbf{Y}_k$ and $\mathbf{G}_k$ are upper trapezoidal (i.e., zero below the main diagonal). Then they make the observation that these factorizations induce a factorization of $\mathbf{\Delta X}$ and $\mathbf{\Delta F}$:

$$\begin{aligned} \mathbf{\Delta X}_k &= \mathbf{P}_k^{\mathrm{T}}\mathbf{\Delta Y}_k, \\ \mathbf{\Delta G}_k &= \mathbf{Q}_k^{\mathrm{T}}\mathbf{\Delta F}_k, \end{aligned}$$

where $\mathbf{\Delta Y}_k$ and $\mathbf{\Delta G}_k$ are upper Hessenberg (i.e., zero below the first subdiagonal). Upper Hessenberg systems can be solved in $O(n^2)$ computation, so we can compute the new iterate $\mathbf{x}_*$ efficiently. Finally, to prepare for the next iteration, we need to replace one column in $\mathbf{X}_k$ by $\mathbf{x}_*$ and make the corresponding change to $\mathbf{F}_k$. This introduces a *stalactite* of nonzeros into $\mathbf{Y}$, and these nonzeros can be eliminated by orthogonal rotation matrices in zero-chasing algorithms in $O(n^2)$ operations, completing the algorithm.

The authors present a concise argument that the method is backward stable, in the sense that the step we have taken is exactly the secant step for a slightly perturbed pair $\tilde{\mathbf{X}}_k$, $\tilde{\mathbf{F}}_k$. They also offer practical suggestions, including forcing each column of $\mathbf{X}_k$ to be replaced periodically to avoid error buildup; storing the factors instead of the matrices $\mathbf{X}_k$ and $\mathbf{F}_k$; ordering points so that the first column of $\mathbf{G}_k$ has the smallest norm; bootstrapping the factorization algorithm to obtain the initial factors; and proposing a delicate procedure for using a good initial approximation to $\mathbf{J}$ while avoiding scaling problems.

The main impediment to the algorithm is diagnosis of linear dependence among the columns of $\mathbf{\Delta Y}$. To this end, the authors suggest an algorithm for finding an approximate right null vector of the matrix. Interestingly, this one piece of their paper has become quite well known (see Sect. 5.7). They then suggest using this vector to replace a column of $\mathbf{Y}$, deciding which column to replace based on the maximal magnitude entry in the left null vector.

Also included in their paper is a discussion regarding methods for reducing the dimension of the problem, which would be necessary if some of the functions $\mathbf{f}(\mathbf{x})$ were linear, by using a QR factorization of the linear constraint matrix.

The idea of using the matrix $\widehat{\mathbf{J}}$ that satisfies

$$\widehat{\mathbf{J}}^{-1}\Delta\mathbf{F}_k = \Delta\mathbf{X}_k$$

to approximate the Jacobian matrix, as Gragg and Stewart propose, is one approach to constructing a *quasi-Newton* method for solving nonlinear equations, and each column of this relation forms a *secant condition* imposed on $\widehat{\mathbf{J}}$. Gragg and Stewart draw their terminology from the book of Ortega and Rheinbolt [101, Sect. 7.2], which discusses several variants of the algorithm and credits the original version to Gauss. Broyden [18, 19] used a single secant condition, while authors [9, 159] besides Gragg and Stewart investigated the imposition of multiple secant conditions. Fang and Saad [50] call these *generalized Broyden's methods*, and propose using a variable number of secant conditions to update the Jacobian approximation at each step.

5.2. More General Update Formulas for QR

The Gragg–Stewart paper showed how to replace a column in a QR factorization of a square matrix. This problem was also considered in the paper by Gill et al. [54]. In contrast, the Daniel–Gragg–Kaufman–Stewart paper [GWS-J31] considers updates of the QR factorization of a matrix $\mathbf{A}$ of size $m \times n$, where $m \geq n$, and focuses on the compact factorization, where $\mathbf{Q}$ is $m \times n$ with orthonormal columns, and $\mathbf{R}$ is $n \times n$ and upper triangular. The Gragg–Stewart paper left the algorithm for determination of the original factorization indeterminate and used (Givens) plane rotations or (Householder) reflection matrices for the update. The Daniel–Gragg–Kaufman–Stewart paper works with an initial Gram–Schmidt process with reorthogonalization, updating using Gram–Schmidt along with Givens or Householder matrices.

The body of the paper begins with a discussion of the Gram–Schmidt process and the need for reorthogonalization: if one column of $\mathbf{A}$ is almost linearly dependent upon the others, then the corresponding column of $\mathbf{Q}$ can be quite inaccurate and fail to be close to orthogonal to the preceding columns. The remedy is to repeatedly reorthogonalize this candidate column against the preceding columns until the candidate has a significant component orthogonal to them.

The updating algorithms are presented in Sect. 3 of the paper. The authors consider a general rank-1 update (as in Sherman–Morrison), but also the important special cases of deleting and inserting a row or a column. The paper concludes with six pages of Algol code and some numerical experiments. This paper has been quite influential over the years: at time of this writing, the ISI index logged nearly 150 citations.

5.3. Effects of Rounding Error on Downdating Cholesky Factorizations

This paper [GWS-J40] gives an error analysis of an algorithm taken from the 1972 Ph.D. dissertation of Michael Saunders [123], also presented in the paper by Gill et al. [54], and ultimately used in the LINPACK software [GWS-B2]. Saunders [124] recalls, "Long ago when he was inventing the name 'downdating' ... Pete wrote to ask for my thoughts about an algorithm he was proposing. I replied by letter from New Zealand that the method I had developed in my 1972 thesis seemed to be rather simpler. Pete very graciously wrote back 'Thank you for saving me from myself!' Ultimately, Pete implemented my simpler downdate and extended it to downdating the right-hand side and $\|\mathbf{r}\|$ of the associated least-squares problem. This has been called the LINPACK downdate ever since."

The paper opens with a review of the relation between the QR factorization of an $m \times n$ matrix[11] $\mathbf{A} = \mathbf{QR}$ ($\mathbf{Q}$ of dimension $m \times n$ and $m \geq n$) and the Cholesky factorization of

$$\mathbf{A}^{\mathrm{T}}\mathbf{A} = \mathbf{R}^{\mathrm{T}}\mathbf{Q}^{\mathrm{T}}\mathbf{Q}\mathbf{R} = \mathbf{R}^{\mathrm{T}}\mathbf{R}.$$

Since $\mathbf{R}^{\mathrm{T}}$ is lower triangular, Stewart observes that this "shows that the triangular part of the QR factorization [of $\mathbf{A}$] is the Cholesky factor [of $\mathbf{A}^{\mathrm{T}}\mathbf{A}$]."[12]

As Stewart notes, a primary use of this relation is in solving least squares problems

$$\min_{\mathbf{x}} \|\mathbf{b} - \mathbf{A}\mathbf{x}\|_2.$$

Defining $\mathbf{z} = \mathbf{Q}^{\mathrm{T}}\mathbf{b}$, then the solution $\mathbf{x}$ is the solution to the linear system $\mathbf{R}\mathbf{x} = \mathbf{z}$ and the residual norm ρ is computed as

$$\rho^2 = \|\mathbf{b} - \mathbf{A}\mathbf{x}\|_2^2 = \|\mathbf{b}\|_2^2 - \|\mathbf{z}\|_2^2.$$

Assuming $\mathbf{b}$ is available at the start of the QR factorization stage, $\mathbf{z}$ can be computed by updating without storing $\mathbf{Q}$ explicitly.

Alternatively, we can compute $\mathbf{x}$ by solving the system of *normal equations* $\mathbf{A}^{\mathrm{T}}\mathbf{A}\mathbf{x} = \mathbf{A}^{\mathrm{T}}\mathbf{b}$ by forward and backward substitution on $\mathbf{R}^{\mathrm{T}}\mathbf{R}\mathbf{x} = \mathbf{A}^{\mathrm{T}}\mathbf{b}$. In this case we do not need the $\mathbf{Q}$ factor at all, and this can be an advantage if $m \gg n$. The disadvantage is that the normal equations algorithm is somewhat less stable than the QR algorithm [15, Chap. 2].

Since the columns of $\mathbf{Q}$ are orthonormal, $\mathbf{Q}$ is quite stable under perturbation. Thus, to assess the stability of the solution to the least squares problem via the triangular version of the normal equations, Stewart chooses to perform a rounding error analysis of the Cholesky factor $\mathbf{R}$ only, and only in the case where data are

[11] In the paper, $\mathbf{A}$ actually refers to a normal equations matrix $\mathbf{X}^{\mathrm{T}}\mathbf{X}$. We use $\mathbf{A}$ in place of $\mathbf{X}$ in our discussion for consistency.

[12] The Cholesky factor is usually defined to have non-negative entries on its main diagonal, and the matrix $\mathbf{R}$ may violate this convention. If desired, nonnegativity of the kth diagonal entry can be restored by multiplying the kth column of $\mathbf{Q}$ and the kth row of $\mathbf{R}$ by -1.

deleted, since adding data cannot decrease the smallest singular value. Thus, he reformulates the problem as computing the Cholesky factor $\tilde{\mathbf{R}}$ so that

$$\tilde{\mathbf{R}}^{\mathrm{T}}\tilde{\mathbf{R}} = \mathbf{R}^{\mathrm{T}}\mathbf{R} - \mathbf{a}\mathbf{a}^{\mathrm{T}},$$

where $\mathbf{a}$ might be a row of $\mathbf{A}$.

Stewart first considers the conditioning of this problem, demonstrating the possible pitfalls in downdating. If the Cholesky factor is perturbed, so that we compute $\bar{\mathbf{R}}^{\mathrm{T}}\bar{\mathbf{R}} = (\mathbf{R}+\mathbf{E})(\mathbf{R}+\mathbf{E})^{\mathrm{T}} - \mathbf{a}\mathbf{a}^{\mathrm{T}}$, Stewart notes the lower bound

$$\|\tilde{\mathbf{R}} - \bar{\mathbf{R}}\|_2 \geq \max_i |\sigma_i(\tilde{\mathbf{R}}) - \sigma_i(\bar{\mathbf{R}})|,$$

where $\sigma_i(\mathbf{B})$ denotes the ith singular value of the matrix. Combining this with other results (on p. 205 of [GWS-J40]) Stewart notes, "one cannot tolerate spread of singular values of half the computational precision without losing all precision in the smallest singular value." He then shows that

$$\frac{\|\tilde{\mathbf{R}}^{\mathrm{T}}\tilde{\mathbf{R}} - \bar{\mathbf{R}}^{\mathrm{T}}\bar{\mathbf{R}}\|_2}{\|\tilde{\mathbf{R}}^{\mathrm{T}}\tilde{\mathbf{R}}\|_2} \leq \epsilon_{\mathrm{M}}(1-\epsilon_{\mathrm{M}})\frac{\|\mathbf{R}^{\mathrm{T}}\mathbf{R}\|_2}{\|\tilde{\mathbf{R}}^{\mathrm{T}}\tilde{\mathbf{R}}\|_2},$$

where ϵ_{M} is the relative perturbation in the nonzero elements of $\mathbf{R}$, assumed to be all of about the same magnitude. Therefore, the relative perturbation is well controlled unless the last factor in the equation above is large. Although the normal equation matrix is well behaved, this does not necessarily imply that there is a matrix $\tilde{\mathbf{A}}$ that is close to $\bar{\mathbf{A}}$:

$$\|\tilde{\mathbf{A}} - \bar{\mathbf{A}}\| \geq \max_i |\sigma_i(\tilde{\mathbf{R}}) - \sigma_i(\bar{\mathbf{R}})|,$$

where the term on the right[13] can be as large as $\sqrt{\epsilon_{\mathrm{M}}}\sigma_1$, where σ_1 is the largest singular value of $\mathbf{R}$. Stewart presents similar results for perturbations in $\mathbf{a}$.

In the third section of the paper, Stewart presents a downdating algorithm due to Saunders for computing $\tilde{\mathbf{R}}$. The algorithm requires the triangular solve $\mathbf{R}^{\mathrm{T}}\mathbf{r} = \mathbf{a}$ and then uses plane rotations to transform $[\mathbf{r}^{\mathrm{T}}, \alpha]^{\mathrm{T}}$ to $[\mathbf{0}, 1]^{\mathrm{T}}$. Applying these same plane rotations to $[\mathbf{R}^{\mathrm{T}}, \mathbf{0}]^{\mathrm{T}}$ gives the desired $\tilde{\mathbf{R}}$ in the top block. He then derives the resulting formulas for updating $\mathbf{x}$ and ρ. In the fourth section, Stewart performs a rounding error analysis of this algorithm, using the standard Wilkinson conventions of ignoring higher-order terms in ϵ_{M}, by approximating, for example, $(1+\epsilon_{\mathrm{M}})^2 \approx 1 + 2\epsilon_{\mathrm{M}}$. We would expect an algorithm built around orthogonal rotations to be quite stable, and the error analysis confirms this expectation. Indeed, Stewart concludes that "Any inaccuracies observed in the results cannot be attributed to the algorithm; they must instead be due to the ill-conditioning of the problem (p. 212)."

[13] A typo has been corrected in this equation.

Finally, Stewart addresses the diagnosis of ill-conditioning. If the vector $\mathbf{r}$ that solves $\mathbf{R}^{\mathrm{T}}\mathbf{r} = \mathbf{a}$ has norm greater than 1, then the updated matrix is indefinite and the Cholesky factor cannot be updated; in fact, if the norm is 1, then the updated matrix is singular. Stewart demonstrates that a value of $\|\mathbf{r}\|_2$ near 1 will signal ill-conditioning and "cannot cry wolf."

This was the first analysis of an updating algorithm. Saunders [124] notes that "Pan [110] later gave a new error analysis and showed that the same downdate can be implemented with about half the operations." Saunders also notes the important special case when a downdate is guaranteed to make the matrix rank deficient, as happens, for example, in the simplex method for linear programming [29], and had hoped that Stewart would study this case. "In the end it was Pete's son, Michael Stewart, who resolved the question during a 1997 visit to Stanford. He showed that if row $\mathbf{x}^{\mathrm{T}}$ has just been *added*, a downdate to delete it should be reliable, but if there are intervening downdates in which $\mathbf{R}$ is ill-conditioned, it probably will not work well."

5.4. Stability of a Sequence of Updates and Downdates

In this 1995 paper [GWS-J87], Stewart picks up the theme from his 1979 paper [GWS-J40], discussed in Sect. 5.3, of the stability of modifying a Cholesky factor. Rather than analyzing the stability of a single downdate, in this paper he considers a sequence of updates and downdates. Stewart packs many results into this paper, published in a rather unlikely venue.

First, Stewart considers the stability of the Saunders algorithm [54, 123], which uses orthogonal rotations. Stewart analyzed a single application of this algorithm in 1979, and here he contrasts it with that of an algorithm due to Golub and a variant of it due to Chambers, which rely on hyperbolic transformations. Although none of the algorithms is backward stable, the Saunders algorithm and the Chambers algorithm have *relational stability* (a term which Stewart coined), which means that the same mathematical relations hold (approximately) for the true and the perturbed quantities. In particular, for these two algorithms, after an update or downdate, there exists a matrix $\mathbf{Q}$ such that

$$\mathbf{Q}^{\mathrm{T}} \begin{bmatrix} \tilde{\mathbf{R}} \\ \mathbf{0} \end{bmatrix} = \begin{bmatrix} \mathbf{R} \\ \mathbf{r}^{\mathrm{T}} \end{bmatrix} + \mathbf{E},$$

where

$$\|\mathbf{E}\|_F \leq c\|\tilde{\mathbf{R}}\|_F \, \epsilon_{\mathrm{M}},$$

and c is a constant that depends on the size of the matrix and the details of the computer arithmetic.

This provides a backward error bound in terms of the Cholesky factors, but not in terms of the original data matrix $\mathbf{A}$ or $\mathbf{A}^{\mathrm{T}}\mathbf{A}$. So Stewart makes a digression, using a result from [GWS-J78] to relate the errors back to the matrix $\mathbf{A}^{\mathrm{T}}\mathbf{A}$.

Returning to relational stability, Stewart shows that the property is preserved in a *sequence* of updates and downdates, obtaining a bound

$$\frac{\|\mathbf{S}_n - \mathbf{R}_n\|_F}{\|\mathbf{R}_n\|_F} \preceq \sqrt{4.5}nc\hat{\rho}^2\|\mathbf{R}_n{}^{-1}\|_F^2\epsilon_{\mathrm{M}}, \tag{5.3}$$

where $\mathbf{R}_n$ is the numerical result after the sequence of updates and downdates and $\mathbf{S}_n$ is the true Cholesky factor. This holds under the assumption that $nc\epsilon_{\mathrm{M}} < 1$, and $\hat{\rho}$ is defined to be the maximum of the Frobenius norm of all the intermediate Cholesky factors and that of the original Cholesky factor augmented by all of the downdates.

Stewart contrives a simple numerical example in which a well-conditioned factor is updated to form an ill-conditioned one and then updated again to restore well conditioning. As expected, the Saunders algorithm and the Chambers algorithm produce good final results, even though the intermediate factor is inaccurate, while Golub's algorithm does not recover as well.

Stewart then extends the analysis to updating a rank-revealing URV decomposition, assuming that one of the relationally stable algorithms is used in the updating. He produces a relation analogous to (5.3), shows that any $\mathbf{V}$ that satisfies such a relation must produce accurate approximate null spaces, and discusses the accuracy of the upper-triangular factor.

In the conclusion, Stewart makes some interesting observations. First, he notes that if the norm of the update is too small, the update does not change the factor at all. Second, both updating and downdating of an ill-conditioned factor can produce inaccuracies, but well-conditioned factors are computed accurately. Third, exponential windowing, which downweights old data without downdating (see Sect. 5.5), is "simpler and has better numerical properties" than rectangular windowing, which downdates to drop old data, but since they produce different results, it is important to understand the stability of downdates.

5.5. An Updating Algorithm for Subspace Tracking

In the early 1990s much research in numerical linear algebra focused on applications in signal processing. A standard task is to decompose a signal into k components, separating it from noise. If p detectors sample the signal discretely at n times, the result is a matrix $\mathbf{A}$ of dimension $n \times p$. The task can be accomplished by observing that:

- The k left singular vectors corresponding to the largest singular values are an approximate basis for the *signal subspace*.
- The $p - k$ left singular vectors corresponding to the smallest singular values constitute an approximate basis for the *noise subspace*.

Often, in dynamic situations, we would like to base the subspace estimates on only the n most recent observations ($n \geq p$). This can be done by *rectangular windowing*, where the older observations are simply deleted, or by *exponential windowing*, where all observations are preserved but weighted by a factor β^j where $0 < \beta < 1$ and j denotes the age of the observation. Stewart considers exponential windowing in [GWS-J73].

The SVD is the most stable way of finding the signal and noise subspaces. Unfortunately, the cost of computing a singular value decomposition from scratch is $O(np^2)$ operations, which is quite high. One would prefer an algorithm that updates the bases of the signal and noise subspaces in an order of magnitude fewer operations, i.e., $O(np)$. Unfortunately, despite many attempts at developing efficient algorithms for updating the SVD, there still does not exist any algorithm that does it *quite* this efficiently [61].

This problem was important in the early 1990s, because it had become possible to use embedded processors in different "mobile" devices. In real-time applications it is essential that algorithms be stable and robust, since it is difficult to recover from numerical failures. At a conference in the early 1990s, a representative of a major manufacturer of consumer electronics stated that for robustness they would prefer to use SVD and other methods based on orthogonal transformations in some of their products, but they could not do it because the batteries could not sustain such expensive computations.

Stewart's paper [GWS-J73] on the rank-revealing URV decomposition was a major breakthrough because it presented a decomposition algorithm that is almost as reliable as the SVD for computing the subspaces and can accommodate updates efficiently. The development of the algorithm is based on the observation that the SVD really gives more information than is needed for subspace estimates: it is not necessary to diagonalize the matrix. If the matrix $\mathbf{A}$ has rank k *exactly*, then it has a *complete orthogonal decomposition* [49, 65].

$$\mathbf{A} = \mathbf{U} \begin{bmatrix} \mathbf{R} & \mathbf{0} \\ \mathbf{0} & \mathbf{0} \end{bmatrix} \mathbf{V}^{\mathrm{T}},$$

where $\mathbf{U}$ and $\mathbf{V}$ are orthogonal, and $\mathbf{R}$ is upper triangular and of order k. In signal-noise subspace applications, the matrix has *numerical rank* equal to k, which means that it has k singular values that are significantly larger than the rest (see [GWS-N6]). The basic idea of Stewart's paper is that in this case, the matrix can be transformed to

$$\mathbf{A} = \mathbf{U} \begin{bmatrix} \mathbf{R} & \mathbf{F} \\ \mathbf{0} & \mathbf{G} \end{bmatrix} \mathbf{V}^{\mathrm{T}}, \tag{5.4}$$

where, in terms of the singular values σ_i of $\mathbf{A}$, $\sqrt{\|\mathbf{F}\|_F^2 + \|\mathbf{G}\|_F^2} \approx \sqrt{\sigma_{k+1}^2 + \cdots + \sigma_p^2}$, and maintained in this form when updated. For obvious reasons, this is called the *rank-revealing* URV decomposition. The SVD is the special case in which $\mathbf{F}$ and $\mathbf{G}$ are zero and $\mathbf{R}$ is diagonal.

An absolute prerequisite for the URV algorithm is a good *condition estimator*. Such methods had been developed in the 1970s and 1980s, starting from

[GWS-J29] (discussed in Sect. 5.1) and the seminal paper by Cline, Moler, Stewart and Wilkinson [GWS-J42] (see Sect. 4.2.2), mainly as diagnostic tools for the solution of linear systems of equations. In 1987, Nick Higham [68] had written a survey that demonstrated that they are very reliable. As a by-product of the condition estimation, an approximation of the right singular vector corresponding to the smallest singular value is produced, and this is an approximate null-vector if **A** is ill-conditioned.

In order to maintain the partitioning in (5.4) when the matrix is updated, we need to be able to reduce the dimension of **R** when it has less-than-full numerical rank. Assume that an initial QR decomposition of **A** has been computed with upper triangular **R**, and that a condition estimator has given an approximate right singular vector **w** (of length 1) corresponding to the smallest singular value, i.e.,

$$\mathbf{b} = \mathbf{R}\mathbf{w}, \qquad \eta = \|\mathbf{b}\| \approx \sigma_{\min}(\mathbf{R}).$$

If η is small, then the matrix **R** can be deflated as a first step towards (5.4). The algorithm uses plane rotations from the left and right that reduce **w** to $\mathbf{e}_p$ while preserving the triangular structure. The result is

$$\begin{bmatrix} \widehat{\mathbf{R}} & \mathbf{e} \\ \mathbf{0} & \epsilon \end{bmatrix},$$

where the vector **e** and the scalar ϵ are small. By another procedure based on plane rotations, the norm of the column can be reduced, thereby refining the decomposition. The refinement procedure is important because reducing the norm of **F** gives improved estimates of the subspaces (see [47, 51]).

If $\widehat{\mathbf{R}}$ has a small singular value, then the deflation is repeated. After $p - k$ deflation steps, we obtain the rank-revealing decomposition (5.4).

The computation of the URV decomposition requires $O(k^2)$ operations, as does the updating procedure. When a new observation $\mathbf{z}^{\mathrm{T}}$ is appended, the problem is to reduce

$$\begin{bmatrix} \mathbf{R} & \mathbf{F} \\ \mathbf{0} & \mathbf{G} \\ \mathbf{x}^{\mathrm{T}} & \mathbf{y}^{\mathrm{T}} \end{bmatrix},$$

where $\mathbf{z}^{\mathrm{T}}\mathbf{V} = [\mathbf{x}^{\mathrm{T}}, \ \mathbf{y}^{\mathrm{T}}]$, to upper triangular form without destroying the structure, but taking into account the possibility of a rank increase or decrease. This again is done by a carefully chosen sequence of plane rotations from the left and right.

The paper includes an analysis of the potential parallelism in the algorithm. Stewart uses *precedence diagrams* [GWS-J56] to demonstrate that each step in the process can be performed in $O(p)$ time given p processors, in contrast to the p^2 processors needed for SVD.

The paper concludes with a few observations about the self-starting nature of the algorithm and its backward error analysis. Stewart notes that the reliability of the method is entirely dependent on the reliability of the condition number estimate chosen. He also provides a pointer to a FORTRAN implementation.

5.6. From the URV to the ULV

Stewart observed that the URV is an excellent decomposition for approximating the subspace spanned by left singular vectors corresponding to large singular values. In some applications, however, we want an approximate null space for **A**, and the URV is not so well suited for this computation. From (5.4), we see that the presence of the nonzero matrix **F** introduces errors into the use of the trailing columns, $\mathbf{V}_2$, of **V** as a basis for the approximate null space of **A**. In fact,

$$\|\mathbf{A}\mathbf{V}_2\|_2 = \left\| \begin{bmatrix} \mathbf{F} \\ \mathbf{G} \end{bmatrix} \right\|_2 \equiv \delta_1 \,.$$

Thus, **A** is within δ_1 of a matrix for which the columns of $\mathbf{V}_2$ span the null space. Now $\|\mathbf{G}\|_2$ is bounded below by σ_{k+1}, and this is the only term we would like to have in the product of **A** with the matrix of basis vectors, but instead, we need to be concerned about the size of **F**.

In the URV paper [GWS-J73], Stewart mitigates this fault by refinement steps that reduce the size of **F**. In his ULV paper [GWS-J77], he takes an alternate approach, computing a decomposition of the form

$$\mathbf{A} = \mathbf{U} \begin{bmatrix} \mathbf{L} & \mathbf{0} \\ \mathbf{H} & \mathbf{E} \end{bmatrix} \mathbf{V}^{\mathrm{T}}, \tag{5.5}$$

where **U** and **V** are orthogonal, and **L** is lower triangular, well conditioned, and of order k. With this decomposition,

$$\|\mathbf{A}\mathbf{V}_2\|_2 = \left\| \begin{bmatrix} \mathbf{0} \\ \mathbf{E} \end{bmatrix} \right\|_2 \equiv \delta_2 \,,$$

and this is a much cleaner bound.

Stewart considers two types of changes to **A**:

- Adding a row to **A** (Sect. 3 of the paper).
- Downweighting **A** by a factor β (Sect. 2 of the paper).

Addition of a row is handled by a series of rotations, and the zero-chasing is illustrated in a sequence of diagrams that carefully distinguish small entries (h and e) from those of average size (x and y). The update is followed by a deflation procedure (when necessary) to restore the rank-revealing nature of the decomposition, and this deflation is also needed to handle downweighting.

Deflation is handled by estimating the condition number of **L**, and, if it is not well conditioned, using the approximate left null vector from the estimate to reduce the size of **L**, similar to the procedure in the preceding paper.

The paper concludes with a few observations comparing URV and ULV. Stewart notes that the URV updating algorithm is simpler, but the URV deflation algorithm is more expensive (if refinement is used), and both algorithms can be efficiently implemented on parallel architectures.

5.7. Impact

The paper by Gragg and Stewart [GWS-J29] focuses on the numerical solution of nonlinear equations using a secant algorithm. Two of the main contributions of this paper are a numerically stable update algorithm and a procedure for determining which data to drop. However, their algorithm for finding an approximate right null vector has become quite famous in its own right, and is closely related to algorithms for estimating the condition number of a matrix; see the discussion of [GWS-J42] in Sect. 4.2.2. Indeed, the paper is as often cited for the condition number estimate as for the secant algorithm. Thus, this paper is an excellent example of how attention to algorithmic detail can lead to useful insights into other important areas.

The follow-on Daniel–Gragg–Kaufman–Stewart paper [GWS-J31] has been heavily cited as well, indicating the impact these results have had in the numerical linear algebra community.

Stewart's rounding analysis of the Cholesky downdating algorithm [GWS-J40], along with other 1970s papers on the sensitivity of the eigenvalue problem [GWS-J16, GWS-J19, GWS-J38], singular values [GWS-J39], QR factors [GWS-J34], and pseudoinverses, projections, and least squares problems [GWS-J35], established Stewart as Wilkinson's successor for error analysis and perturbation theory of matrix problems.

When the URV paper [GWS-J73] was published in 1992, rank-revealing QR decompositions had been studied for some time. As they allowed only for permutations from the right, they were not able to give subspace estimates as good as those from the URV. Stewart was the first to use orthogonal transformations from the right to compute a "*numerical complete orthogonal decomposition*" in a finite number of steps, i.e., one that gives good estimates of both the signal and noise subspace. This was the first in a series of papers on rank-revealing decompositions, including the ULV paper [GWS-J77] also discussed above. Just as the URV and ULV are related to the singular value decomposition, the ULLV by Luk and Qiao [94] gives subspace approximations for the generalized SVD. The downdating problem for the URV decomposition was treated in [8, 111].

Stewart used the URV decomposition for tracking time-varying signals. He used it to speed up the ESPRIT algorithm [GWS-J85] and the root-MUSIC algorithm [GWS-N29]. Stewart also proposed a parallel variant of the URV update [GWS-J82].

The URV paper [GWS-J73] has had great influence in the signal processing community, providing just the right hammer for a very important nail. K.J.R. Liu [93] says:

> This paper changes the development and thinking of the signal processing community in that the signal subspace can now be updated and tracked. It is a major breakthrough in array processing and spectral estimation. A Google search finds 252 citations to this paper. That itself speaks for the influence of this paper.

Perhaps even more significant is that only about half of the citations are in signal processing journals, indicating an even broader impact. The ULV paper currently has about 1/3 as many citations as the URV paper, which is still quite a respectable number. This difference is probably due to the difference in publication venues as well as the computational problems each factorization was meant to address. As more applications surface that require the computation of an approximate null space, the ULV paper will undoubtedly see a surge in references.

Surprisingly, at the time of this writing, the sequential update paper [GWS-J87] does not seem to have the citations in the signal processing literature one might expect to have seen, given the venue and date of publication. On the other hand, the citation indices indicate that the numerical community has made good use of it, with reference being made to Stewart's paper as recently as 2008.

6

Least Squares, Projections, and Pseudoinverses

Misha E. Kilmer and Dianne P. O'Leary

1. [GWS-J4] "On the Continuity of the Generalized Inverse," *SIAM Journal on Applied Mathematics* **17** (1969) 33–45.
2. [GWS-J35] "On the Perturbation of Pseudo-inverses, Projections and Linear Least Squares Problems," *SIAM Review* **19** (1977) 634–662.
3. [GWS-J65] "On Scaled Projections and Pseudoinverses," *Linear Algebra and its Applications* **112** (1989) 189–193.

These papers form a small sample of Stewart's work on least squares, projections, and generalized inverses; see also [GWS-J23, GWS-J28, GWS-J60, GWS-J97, GWS-J119, GWS-N6, GWS-N24], for example. Stewart's focus in this area was to put the understanding of generalized inverses on as firm a footing as that of the usual matrix inverse. He proceeded by establishing continuity properties, effects of scaling, and perturbation theory, using orthogonal projectors as a unifying framework.

Least squares, projections, and pseudoinverses (a.k.a. generalized inverses) are closely intertwined. Suppose we have observed some noisy data $\mathbf{b}$ and we believe that it can be explained by a set of parameters $\mathbf{x}$ in a model of the form

$$\mathbf{b} = \mathbf{A}\mathbf{x} + \boldsymbol{\eta},$$

where the model matrix $\mathbf{A}$ is $m \times n$ with $m \geq n$, the observed data $\mathbf{b}$ is $m \times 1$, and $\boldsymbol{\eta}$ accounts for the noise. One way to fit the model to the data is to solve the so-called least squares problem

$$\min_{\mathbf{x}} \|\mathbf{b} - \mathbf{A}\mathbf{x}\|_2 . \qquad (6.1)$$

M.E. Kilmer and D.P. O'Leary (eds.), *G.W. Stewart: Selected Works with Commentaries*, Contemporary Mathematicians, DOI 10.1007/978-0-8176-4968-5_6,

Such a formulation goes back to Gauss and Legendre[14] and is quite appropriate when noise is Gaussian and the model has been renormalized if necessary so that $\eta_i \sim N(0, \alpha)$ for $i = 1, \ldots, m$, where α is a constant. The unique solution $\mathbf{x}^*$ to (6.1) of minimum norm is defined to be

$$\mathbf{x}^* = \mathbf{A}^\dagger \mathbf{b},$$

where $\mathbf{A}^\dagger$ is the *pseudoinverse* or *generalized inverse* of $\mathbf{A}$. The resulting residual

$$\mathbf{r} = \mathbf{b} - \mathbf{A}\mathbf{x}^*$$

is the *orthogonal projection* of $\mathbf{b}$ onto the complement of the range of $\mathbf{A}$ and can be expressed in terms of a projector matrix $\mathbf{I} - \mathbf{P}_{\mathrm{A}}$:

$$\mathbf{r} = (\mathbf{I} - \mathbf{P}_{\mathrm{A}})\mathbf{b},$$

where $\mathbf{P}_{\mathrm{A}}$ denotes the orthogonal projector onto the range of $\mathbf{A}$. By setting the derivative of (6.1) to zero, we find that if $\mathbf{A}$ has rank n, then $\mathbf{x}^*$ satisfies

$$\mathbf{x}^* = (\mathbf{A}^{\mathrm{T}}\mathbf{A})^{-1}\mathbf{A}^{\mathrm{T}}\mathbf{b},$$

so

$$\mathbf{A}^\dagger \equiv (\mathbf{A}^{\mathrm{T}}\mathbf{A})^{-1}\mathbf{A}^{\mathrm{T}}.$$

Direct computation shows that the orthogonal projector onto the range of $\mathbf{A}$ is given by $\mathbf{P}_{\mathrm{A}} = \mathbf{A}\mathbf{A}^\dagger$, and hence the orthogonal projector onto the complement of the range of $\mathbf{A}$ is

$$\mathbf{I} - \mathbf{P}_{\mathrm{A}} = \mathbf{I} - \mathbf{A}\mathbf{A}^\dagger = \mathbf{I} - \mathbf{A}(\mathbf{A}^{\mathrm{T}}\mathbf{A})^{-1}\mathbf{A}^{\mathrm{T}}.$$

Thus, the behavior of the pseudoinverse under perturbation determines the behavior of the solution to the least squares problem, and the behavior of the projector $\mathbf{I} - \mathbf{P}_{\mathrm{A}}$ dictates the behavior of the residual.

A main message of Stewart's work in this area is that the fundamental properties of the pseudoinverse and of solutions to least squares problems are best understood by studying the projectors $\mathbf{P}_{\mathrm{A}}$ and $\mathbf{I} - \mathbf{P}_{\mathrm{A}}$.

6.1. Continuity of the Pseudoinverse

At the time Stewart was writing [GWS-J4], it was well known that the inverse of a matrix $\mathbf{A}$ was a continuous function of the elements of the matrix. In [GWS-J4],

[14] See Gauss's papers "Theoria Combinationis," translated from Latin to English by Stewart [GWS-B4].

Stewart presents this as a corollary on the following perturbation bound, well known at the time [158]:

$$\frac{\|\mathbf{B}^{-1} - \mathbf{A}^{-1}\|}{\|\mathbf{A}^{-1}\|} \leq \frac{\kappa(\mathbf{A})\|\mathbf{E}\|/\|\mathbf{A}\|}{1 - \kappa(\mathbf{A})\|\mathbf{E}\|/\|\mathbf{A}\|}, \tag{6.2}$$

where

$$\mathbf{B} = \mathbf{A} + \mathbf{E}$$

and $\kappa(\mathbf{A}) = \|\mathbf{A}\|\,\|\mathbf{A}^{-1}\|$. The bound holds (for any norm for which $\|\mathbf{I}\| = 1$) whenever $\|\mathbf{A}^{-1}\|\,\|\mathbf{E}\| < 1$. As $\mathbf{E} \to \mathbf{0}$, the bound goes to zero and continuity is established.

On the other hand, Stewart noted that when $\mathbf{A}$ is not invertible, the pseudoinverse $\mathbf{A}^\dagger$ need not be a continuous function of the matrix elements. As evidence, Stewart exhibits the example

$$\mathbf{A} = \begin{bmatrix} 1 & 0 \\ 0 & 0 \end{bmatrix}, \quad \mathbf{E} = \begin{bmatrix} 0 & 0 \\ 0 & 1 \end{bmatrix}.$$

Then

$$(\mathbf{A} + \epsilon\mathbf{E})^\dagger = \begin{bmatrix} 1 & 0 \\ 0 & \epsilon^{-1} \end{bmatrix},$$

which has no limit as $\epsilon \to 0$.

A large part of the remainder of the paper, therefore, is devoted to examining what conditions are necessary for the continuity of the pseudoinverse. Prior to the publication of [GWS-J4], Ben-Israel [12] had showed that if

$$\lim_{n\to\infty} \mathbf{E}_n = \mathbf{0}, \tag{6.3}$$

and if

$$\|\mathbf{A}^\dagger\|\,\|\mathbf{E}_n\| < 1 \tag{6.4}$$

for sufficiently large n, then a sufficient condition for convergence

$$\lim_{n\to\infty} (\mathbf{A} + \mathbf{E}_n)^\dagger = \mathbf{A}^\dagger$$

is that the rows and columns of $\mathbf{E}_n$ lie within the row and column spaces of $\mathbf{A}$ for sufficiently large n. His resulting bound is similar to (6.2).

In [GWS-J4], Stewart generalizes this result in important ways. He shows that under conditions (6.3) and (6.4), convergence holds if and only if, for sufficiently large n,

$$\operatorname{rank}(\mathbf{A} + \mathbf{E}_n) = \operatorname{rank}(\mathbf{A}).$$

The price of this generality is a somewhat more complicated bound than that which appeared in the Ben-Israel paper, which we now describe.

The main result of [GWS-J4], Theorem 5.2, gives the perturbation result analogous to (6.2), but for pseudoinverses. Let

$$\mathbf{R}_{\mathrm{A}} = \mathbf{A}^{\dagger}\mathbf{A}$$

be the orthogonal projector onto the rowspace of $\mathbf{A}$ (i.e., the range of $\mathbf{A}^{\mathrm{T}}$), and let

$$\begin{aligned} \mathbf{E}_{11} &= \mathbf{P}_{\mathrm{A}}\mathbf{E}\mathbf{R}_{\mathrm{A}}, \\ \mathbf{E}_{12} &= \mathbf{P}_{\mathrm{A}}\mathbf{E}(\mathbf{I}-\mathbf{R}_{\mathrm{A}}), \\ \mathbf{E}_{21} &= (\mathbf{I}-\mathbf{P}_{\mathrm{A}})\mathbf{E}\mathbf{R}_{\mathrm{A}}, \\ \mathbf{E}_{22} &= (\mathbf{I}-\mathbf{P}_{\mathrm{A}})\mathbf{E}(\mathbf{I}-\mathbf{R}_{\mathrm{A}}). \end{aligned}$$

Assume that

$$\|\mathbf{A}^{\dagger}\|\|\mathbf{E}_{11}\| < 1. \tag{6.5}$$

If $\mathrm{rank}(\mathbf{B}) = \mathrm{rank}(\mathbf{A})$, then

$$\frac{\|\mathbf{B}^{\dagger} - \mathbf{A}^{\dagger}\|}{\|\mathbf{A}^{\dagger}\|} \leq \beta_{11} + \gamma \sum_{(i,j)\neq(1,1)} \left(\frac{\beta_{ij}^2}{1+\beta_{ij}^2}\right)^{1/2},$$

where

$$\begin{aligned} \gamma &= \left(1 - \kappa(\mathbf{A})\frac{\|\mathbf{E}_{11}\|}{\|\mathbf{A}\|}\right)^{-1}, \\ \beta_{ij} &= \frac{\gamma\kappa(\mathbf{A})\|\mathbf{E}_{ij}\|}{\|\mathbf{A}\|}, \quad i,j = 1,2. \end{aligned}$$

Thus, the pseudoinverse of the perturbed matrix under these conditions satisfies a continuity condition. On the other hand, if $\mathrm{rank}(\mathbf{B}) > \mathrm{rank}(\mathbf{A})$ (note that if (6.5) holds, the rank of $\mathbf{B}$ cannot decrease), then the pseudoinverse of $\mathbf{B}$ is large:

$$\|\mathbf{B}^{\dagger}\| \geq \|\mathbf{E}\|^{-1}.$$

In the final section of the paper, Stewart derives an error bound for solutions to least squares problems. He notes that a "new" method, Householder QR, has recently been developed for solving these problems, and that Golub and Wilkinson [59] showed that, in the presence of rounding error, it produces a solution[15]

$$(\mathbf{A}+\mathbf{E})(\mathbf{x}^* + \mathbf{h}) = \mathbf{b} + \mathbf{k}, \tag{6.6}$$

[15] In the following equation, we correct a misprint in the first equation on p. 44 of [GWS-J4].

where $\|\mathbf{E}\|$ and $\|\mathbf{k}\|$ are small. They derived a first-order bound for $\|\mathbf{h}\|$. Stewart takes $\mathbf{k} = \mathbf{0}$ and derives a different bound

$$\frac{\|\mathbf{h}\|}{\|\mathbf{x}^*\|} \leq \beta_1 + \kappa(\mathbf{A})\gamma \left(\frac{\beta_2^2}{1+\beta_2^2} + \left(\frac{\beta_2^2}{1+\beta_2^2} \right)^{1/2} \frac{\|(\mathbf{I} - \mathbf{P}_{\mathrm{A}})\mathbf{b}\|}{\|\mathbf{P}_{\mathrm{A}}\mathbf{b}\|} \right), \tag{6.7}$$

where

$$\begin{aligned}
\gamma &= \left(1 - \frac{\kappa(\mathbf{A})\|\mathbf{E}_1\|}{\|\mathbf{A}\|} \right)^{-1}, \\
\beta_i &= \frac{\gamma\kappa(\mathbf{A})\|\mathbf{E}_i\|}{\|\mathbf{A}\|}, \quad i = 1, 2, \\
\mathbf{E}_1 &= \mathbf{P}_{\mathrm{A}}\mathbf{E}, \\
\mathbf{E}_2 &= (\mathbf{I} - \mathbf{P}_{\mathrm{A}})\mathbf{E}.
\end{aligned}$$

As Stewart notes, an important feature of this bound is that the third term indicates that when the residual norm is large, the perturbation can be proportional to $\kappa(\mathbf{A})^2$, while if the model is good, the first two terms indicate that the perturbation bound is proportional to $\kappa(\mathbf{A})$. The most widely used bound for understanding perturbations to least squares problems [15, Theorem 1.4.6] is closely related to this bound but does not split the effects of $\mathbf{E}$:

$$\frac{\|\mathbf{h}\|}{\|\mathbf{x}^*\|} \leq \frac{\kappa(\mathbf{A})}{1 - \epsilon_{\mathrm{A}}\kappa(\mathbf{A})} \left(\epsilon_{\mathrm{A}} + \frac{\|\mathbf{b}\|_2}{\|\mathbf{A}\|_2\|\mathbf{x}\|_2}\epsilon_b + \frac{\|\mathbf{r}\|_2}{\|\mathbf{A}\|_2\|\mathbf{x}\|_2}\epsilon_{\mathrm{A}}\kappa(\mathbf{A}) \right) + \epsilon_{\mathrm{A}}\kappa(\mathbf{A}),$$

where $\epsilon_{\mathrm{A}} = \|\mathbf{E}\|_2/\|\mathbf{A}\|_2$, $\epsilon_b = \|\mathbf{r}\|_2/\|\mathbf{b}\|_2$, and $\epsilon_{\mathrm{A}}\kappa(\mathbf{A})$ is assumed to be less than 1.

This important paper concludes with an acknowledgment to his Oak Ridge colleague Bert Rust and his colleague and thesis adviser Alston Householder for their helpful comments.

6.2. Perturbation Theory

Stewart's *SIAM Review* paper [GWS-J35] on the subject of perturbation theory for pseudoinverses, projections, and linear least squares problems systematically presents known results, beautifully summarizing them and filling in some gaps. The real surprise in the paper, though, is that hidden in the appendix is an astounding innovation, now known as the *CS decomposition.*

Stewart begins by reviewing some material on unitarily invariant norms, perturbations of matrix inverses, and projectors, providing references and notes, before proceeding to the central results. In particular, he notes that to obtain useful perturbation results when $m > n$, it is important not only that the norm be unitarily invariant, but also that it be *uniformly generated*, meaning that the norm of a matrix depends only on the singular values, and that the matrix norm, applied to a vector, reduces to the Euclidean norm. These properties ensure, for example, that $\|\mathbf{A}\| = \|\mathbf{A}^H\|$, which is not true of all unitarily invariant norms.

In all of Stewart's least squares work, the unifying concept is projectors, and he takes time in this introductory material to establish several key properties. He shows that if rank($\mathbf{B}$) = rank($\mathbf{A}$), then the singular values of $\mathbf{P}_A(\mathbf{I} - \mathbf{P}_B)$ and of $\mathbf{P}_B(\mathbf{I} - \mathbf{P}_A)$ are the absolute values of the eigenvalues of $\mathbf{P}_B - \mathbf{P}_A$. This allows him to compute the norms of these various matrices, which will contribute to the later bounds.

Stewart states (p. 643) that these results "are well known to people who work closely with orthogonal projectors." Yet, Stewart presents a new proof in the appendix, based on a factorization of a unitary matrix which has since come to be known (see [GWS-J47]) as a CS decomposition. Stewart states in his paper: Every $n \times n$ unitary matrix $\mathbf{W}$ can be factored as

$$\mathbf{U}^H\mathbf{W}\mathbf{V} = \begin{bmatrix} \mathbf{\Gamma} & -\mathbf{\Sigma} & \mathbf{0} \\ \mathbf{\Sigma} & \mathbf{\Gamma} & \mathbf{0} \\ \mathbf{0} & \mathbf{0} & \mathbf{I} \end{bmatrix}$$

where the $r \times r$ matrices $\mathbf{\Gamma}$ and $\mathbf{\Sigma}$ are square, non-negative, and diagonal and $r \leq n/2$. Stewart establishes the existence of this decomposition, noting (p. 661) that it has "apparently not been explicitly stated before; however, it is implicit in the works of Davis and Kahan (1970) and Björck and Golub (1973)." (See [16, 31] and also [30], of which Stewart was unaware.) He explains that the diagonal elements of $\mathbf{\Gamma}$ are the cosines of the *canonical angles* between the range of $\mathbf{A}$ and the range of $\mathbf{B}$. Because of this fact, it has become more common to use the notation $\mathbf{C}$ in place of $\mathbf{\Gamma}$ and $\mathbf{S}$ in place of $\mathbf{\Sigma}$, to represent *cosine* and *sine*, motivating Stewart's naming of the decomposition [GWS-J47]. Stewart uses the CS decomposition to construct canonical bases, as well as to prove the bounds he needs on the projectors.

The importance of the aforementioned result cannot be overemphasized. Indeed, Paige and Wei [108] say, "This contribution was extremely important, not so much because it appears to be the first complete proof given for a CSD, but because it simply stated the CSD as a general decomposition of unitary matrices, rather than just a tool for analyzing angles between subspaces – this was something [Davis and Kahan in [30]] had not quite done. This elegant and unequivocal statement brought the CSD to the notice of a wider audience, as well as emphasizing its broader usefulness."

In the last part of the introduction, Stewart carefully defines the concept of two matrices $\mathbf{A}$ and $\mathbf{B}$ being *acute*, meaning that $\|\mathbf{P}_A - \mathbf{P}_B\|_2 < 1$, which implies that no vector in the range of $\mathbf{A}$ can be orthogonal to a vector in the range of $\mathbf{B}$. He shows that $\mathbf{A}$ and $\mathbf{B}$ are acute if and only if

$$\text{rank}(\mathbf{A}) = \text{rank}(\mathbf{B}) = \text{rank}(\mathbf{P}_A\mathbf{B}\mathbf{R}_A),$$

where $\mathbf{R}_\mathrm{A}$ denotes the orthogonal projector onto the row space of $\mathbf{A}$ (i.e., $\mathbf{R}_\mathrm{A} = \mathbf{A}^\dagger\mathbf{A}$). He then proceeds (Sect. 3) to present a unified development of bounds for $\mathbf{B}^\dagger - \mathbf{A}^\dagger$ (see [155] for the following three bounds):

- If $\mathbf{A}$ and $\mathbf{B}$ are not acute, then

$$\|\mathbf{B}^\dagger - \mathbf{A}^\dagger\| \geq \frac{1}{\|\mathbf{E}\|_2}.$$

-

$$\|\mathbf{B}^\dagger - \mathbf{A}^\dagger\| \leq \mu \max\{\|\mathbf{A}\|_2^2, \|\mathbf{B}\|_2^2\} \,\|\mathbf{E}\|,$$

where μ is a modest constant depending on the choice of norm.
- If $\operatorname{rank}(\mathbf{A}) = \operatorname{rank}(\mathbf{B})$, then

$$\|\mathbf{B}^\dagger - \mathbf{A}^\dagger\| \leq \mu\|\mathbf{A}\|_2\|\mathbf{B}\|_2\|\mathbf{E}\|.$$

Stewart notes that all of these bounds are too loose if $\mathbf{A}$ and $\mathbf{B}$ are acute, so he devotes a subsection to deriving better bounds for this case, based on the canonical form

$$\mathbf{U}^H\mathbf{A}\mathbf{V} = \begin{bmatrix} \mathbf{A}_{11} & \mathbf{0} \\ \mathbf{0} & \mathbf{0} \end{bmatrix},$$

where $\mathbf{U}$ and $\mathbf{V}$ are unitary and $\mathbf{A}_{11}$ is $r \times r$ and nonsingular. By applying these unitary transformations to $\mathbf{E}$, he can isolate the pieces of the resulting matrix that are relevant, thus tightening the bounds (Theorem 3.8).

He then computes derivatives under the assumptions that $\mathbf{E}$ is a function of a single variable τ and that $\mathbf{B}(\tau)^\dagger \to \mathbf{A}^\dagger$ as $\tau \to 0$. This implies, of course, that $\operatorname{rank}(\mathbf{B}(\tau)) = \operatorname{rank}(\mathbf{A})$ for τ sufficiently small. Since $\mathbf{E}(\tau)$ becomes arbitrarily small, in some neighborhood of $\mathbf{A}$, the matrices $\mathbf{A}$ and $\mathbf{B}$ must be acute. The resulting formula is

$$\frac{\mathrm{d}\mathbf{A}^\dagger}{\mathrm{d}\tau} = -\mathbf{A}^\dagger\mathbf{P}_\mathrm{A}\frac{\mathrm{d}\mathbf{A}}{\mathrm{d}\tau}\mathbf{R}_\mathrm{A}\mathbf{A}^\dagger + (\mathbf{A}^H\mathbf{A})^\dagger\mathbf{R}_\mathrm{A}\frac{\mathrm{d}\mathbf{A}^H}{\mathrm{d}\tau}(\mathbf{I} - \mathbf{P}_\mathrm{A}) - (\mathbf{I} - \mathbf{R}_\mathrm{A})\frac{\mathrm{d}\mathbf{A}^H}{\mathrm{d}\tau}\mathbf{P}_\mathrm{A}(\mathbf{A}\mathbf{A}^H)^\dagger.$$

He notes that this is a special case of a formula by Golub and Pereyra [58], who allow $\mathbf{E}$ to be a function of several variables.

Following the establishment of these results, Stewart then proceeds to: (a) derive new bounds for $\|\mathbf{P}_\mathrm{B} - \mathbf{P}_\mathrm{A}\|$ (Theorem 4.1); (b) present the new insight that a sufficient condition for convergence of $\mathbf{P}_\mathrm{B}$ to $\mathbf{P}_\mathrm{A}$ is that $\mathbf{A}$ and $\mathbf{B}$ are acute (Cor. 4.2); (c) establish a formula for $\mathrm{d}\mathbf{P}_\mathrm{A}/\mathrm{d}\tau$; and (d) present an asymptotic bound of Golub and Pereyra on $\mathbf{P}_\mathrm{B} - \mathbf{P}_\mathrm{A}$ (his eqn. (4.4)).

In the final section of the paper, Stewart discusses the implications of these results for the solution of linear least squares problems. Under the assumption that

$\mathbf{B}$ and $\mathbf{A}$ are acute, he rederives a bound for the error in the solution to (6.6) similar to (6.7) as well as the simpler formulation

$$\frac{\|\mathbf{h}\|_2}{\|\mathbf{x}^*\|_2} \leq \kappa(\mathbf{A})\eta \frac{\|\mathbf{P}_{\mathrm{A}}\mathbf{k}\|_2}{\|\mathbf{P}_{\mathrm{A}}\mathbf{b}\|_2},$$

where

$$\eta = \frac{\|\mathbf{P}_{\mathrm{A}}\mathbf{b}\|_2}{\|\mathbf{A}\|_2\|\mathbf{x}^*\|_2}.$$

He presents bounds on the perturbation to the residual, as well as a backward error bound.

At the end of Sect. 5, he considers what he calls an "inverse perturbation theorem." Suppose that $\tilde{\mathbf{x}}$ is an *approximate* solution to the least squares problem (6.1). Stewart poses the problem of finding a matrix $\mathbf{E}$ so that $\tilde{\mathbf{x}}$ is the exact solution to the least squares problem

$$\min_{\mathbf{x}} \|\mathbf{b} - (\mathbf{A} + \mathbf{E})\mathbf{x}\|_2 .$$

It took less than a page to establish the formula

$$\mathbf{E} = \frac{\mathbf{A}(\mathbf{x}^* - \tilde{\mathbf{x}})\tilde{\mathbf{x}}^{\mathrm{T}}}{\|\tilde{\mathbf{x}}\|_2^2}. \tag{6.8}$$

This result is in the spirit of the *total least squares problem*, in which perturbations are allowed in both the matrix and the data vector. Stewart also discusses this result in [GWS-J28] and notes that if $\tilde{\mathbf{x}}$ is $\mathbf{x}^*$ rounded to t digits, then the numerator is of order $10^{-t}\|\mathbf{A}\|_2\|\mathbf{x}\|_2$, so $\mathbf{E}$ itself can be interpreted as a t-digit rounding of $\mathbf{A}$. He remarks, therefore, that since $\mathbf{A}(\tilde{\mathbf{x}} - \mathbf{x}^*)$ is the difference between the residual corresponding to $\mathbf{x}^*$ and that corresponding to $\tilde{\mathbf{x}}$, that if the residual is nearly optimal and the elements of $\mathbf{A}$ are not known exactly, then the computed solution $\tilde{\mathbf{x}}$ is as good as we can expect to compute.

In summary, this paper is an excellent compilation of the perturbation theory of pseudoinverses, projectors, and least squares problems. It is written in a clear style, with good taste in its choice of results, and it introduces several important new bounds.

Undoubtedly, however, the most influential development in this paper is the CS decomposition. It would be difficult to list all of the work that has been built on this decomposition. It is indeed amazing what one can find in a seemingly minor appendix!

6.3. Weighted Pseudoinverses

In 1988, Pete Stewart was trying to establish a uniform bound on a weighted pseudoinverse, the solution operator for the problem

$$\min_{\mathbf{x}} \|\mathbf{D}^{1/2}(\mathbf{b} - \mathbf{A}\mathbf{x})\|_2, \tag{6.9}$$

where $\mathbf{A}$ is $m \times n$ with rank n and $\mathbf{D} \in \mathcal{D}_+$, the set of diagonal matrices with positive eigenvalues. The solution operator for this problem is

$$\mathbf{A}_{\mathrm{D}}^{\dagger} = (\mathbf{A}^{\mathrm{T}}\mathbf{D}\mathbf{A})^{-1}\mathbf{A}^{\mathrm{T}}\mathbf{D},$$

and the projector

$$\mathbf{P}_{\mathrm{D}} = \mathbf{A}\mathbf{A}_{\mathrm{D}}^{\dagger}$$

applied to $\mathbf{b}$ determines the data that $\mathbf{A}\mathbf{x}$ models. Since $\mathbf{A}_{\mathrm{D}}^{\dagger}$ can be arbitrarily large as $\mathbf{D}$ is varied, the fact that a uniform bound on $\mathbf{P}_{\mathrm{D}}$ exists for all choices of $\mathbf{D}$ is not at all obvious.

In the acknowledgment section of [GWS-J65], Stewart thanks Michael Powell for "an elegant proof based on considerations from mathematical programming" which caused him to rethink an original "messy" induction argument. The argument in the paper is geometrically inspired, showing that the orthogonal complement of range ($\mathbf{DA}$) cannot "lean too far" toward range($\mathbf{A}$); thus, the norm of $\mathbf{P}_{\mathrm{D}}$ is controlled, even though $\mathbf{A}_{\mathrm{D}}^{\dagger}$ can be arbitrarily large. Using this insight, Stewart in this paper proves two results. The first is that the spectral norms of these matrices are bounded independently of $\mathbf{D}$ as

$$\sup_{\mathbf{D}\in\mathcal{D}_+} \|\mathbf{P}_{\mathrm{D}}\|_2 \leq \rho^{-1},$$

and

$$\sup_{\mathbf{D}\in\mathcal{D}_+} \|\mathbf{A}_{\mathrm{D}}^{\dagger}\|_2 \leq \rho^{-1}\|\mathbf{A}^{\dagger}\|_2,$$

where

$$\rho \equiv \inf_{\substack{\mathbf{y}\in\mathcal{Y}\\ \mathbf{x}\in\mathcal{X}}} \|\mathbf{y}-\mathbf{x}\|_2 > 0,$$

$$\mathcal{X} = \{\mathbf{x} \in \mathcal{R}(\mathbf{A}) : \|\mathbf{x}\|_2 = 1\},$$

and

$$\mathcal{Y} = \{\mathbf{y} : \exists \mathbf{D} \in \mathcal{D}_+ \text{ with } \mathbf{A}^{\mathrm{T}}\mathbf{D}\mathbf{y} = \mathbf{0}\}.$$

His second result is that if the columns of $\mathbf{U}$ form an orthonormal basis for $\mathcal{R}(\mathbf{A})$, then

$$\rho \leq \min \inf{}_+(\mathbf{U}_I),$$

where $\mathbf{U}_I$ denotes any submatrix formed from a nonempty set of rows of $\mathbf{U}$ and $\inf_+(\mathbf{Z})$ denotes the smallest nonzero singular value of $\mathbf{Z}$. Stewart was a bit unsatisfied that he had not proved that the bound is tight, and the paper concludes with a remark that the proof of this fact remained to be established.

As it turned out, an opportunity for tightening this bound did present itself. In January of 1989, Dianne O'Leary told Pete Stewart that she was working on this uniform bound problem but was stuck on a difficult part. Stewart mentioned his recently completed paper, and said that he, too, had left one part unfinished. The prospect of being scooped by your colleague down the hall was indeed discouraging to O'Leary, but in further conversation it became clear that what seemed difficult to O'Leary was easy to Stewart. Taking the results in the two papers [GWS-J65] and [100] together, they had in fact proved the complete result that

$$\rho = \min \inf_{+}(U_I).$$

As Stewart noted in the introduction section of his paper, the results he proves have several applications, two of which we mention here.

- In robust regression, $\mathbf{D}$ is chosen iteratively in order to reduce the influence of outliers in the data. It is important to know a priori that ill-conditioning due to poor intermediate choices of $\mathbf{D}$ will not be an issue, as long as a stable algorithm is used.
- In interior-point methods for solving linear programming problems, each iteration can be performed by solving (6.9) with a different matrix $\mathbf{D}$, and these matrices become arbitrarily ill-conditioned. Again it is crucial to the success of these methods that the solution to these problems remains well defined.

The prevalence of these two applications alone suggests the work has the potential for large impact. Stewart's paper is becoming more and more widely known in the optimization community, in particular, its relation to earlier results by Dikin [37], although the results are periodically rediscovered.

6.4. Impact

This collection of papers contains everything one would want to know about normwise perturbation bounds for least squares problems, projectors, and pseudoinverses. Stewart's contributions, in some cases building on the work of Wedin [154,155], are fundamental. Stewart considered these results again in his book with Sun (Sec III.3–5 of [GWS-B3]), basing the discussion primarily on the *SIAM Review* paper [GWS-J35]. More recently, attention has been given to *componentwise* perturbation bounds for these problems; see, for example, [15, Sect. 1.4.5].

The *SIAM Review* paper [GWS-J35] is widely cited in a variety of contexts; its insight into convergence of pseudoinverses has been useful to researchers in fields such as super-resolution image reconstruction [152] and methods for solving transport equations [141].

And, of course, the CS decomposition presented in this work has had impact far beyond its usefulness in numerical linear algebra. For example, it has broad application in signal processing [142] and it is fundamental to quantum circuit design [151]. It is a special case of the *generalized singular value decomposition* [144], in case the columns of the two matrices of interest are orthonormal. Stewart named it and developed it for nonsquare matrices in [GWS-J47], and computational algorithms were proposed in that paper as well as in [145]. It was generalized to any 2×2 block partitioning in [109]. Its early history is summarized in [15,108]. To conclude, we quote from Paige and Wei [108]: "[The CS decomposition] is slowly being recognized as one of the major tools of matrix analysis."

7

The Eigenproblem and Invariant Subspaces: Perturbation Theory

Ilse C. F. Ipsen

1. [GWS-J15] "Error Bounds for Approximate Invariant Subspaces of Closed Linear Operators," *SIAM Journal on Numerical Analysis* **8** (1971) 796–808.
2. [GWS-J19] "Error and Perturbation Bounds for Subspaces Associated with Certain Eigenvalue Problems," *SIAM Review* **15** (1973) 727–764.
3. [GWS-J48] "Computable Error Bounds for Aggregated Markov Chains," *Journal of the ACM* **30** (1983) 271–285.
4. [GWS-J70] "Two Simple Residual Bounds for the Eigenvalues of a Hermitian Matrix," *SIAM Journal on Matrix Analysis and Applications* **12** (1991) 205–208.
5. [GWS-J71] (with G. Zhang) "Eigenvalues of Graded Matrices and the Condition Numbers of a Multiple Eigenvalue," *Numerische Mathematik* **58** (1991) 703–712.
6. [GWS-J108] "A Generalization of Saad's Theorem on Rayleigh-Ritz Approximations," *Linear Algebra and its Applications* **327** (2001) 115–120.
7. [GWS-J109] "On the Eigensystems of Graded Matrices," *Numerische Mathematik* **90** (2001) 349–370.
8. [GWS-J114] "On the Powers of a Matrix with Perturbations," *Numerische Mathematik* **96** (2003) 363–376.

In this collection of papers, Pete Stewart established the foundations for the perturbation theory of invariant subspaces. He introduced two crucial concepts that allow a systematic approach toward such a perturbation theory: subspace rotation and operator separation. These two concepts form the guiding principle in most of these papers.

M.E. Kilmer and D.P. O'Leary (eds.), *G.W. Stewart: Selected Works with Commentaries*, Contemporary Mathematicians, DOI 10.1007/978-0-8176-4968-5_7,

7.1. Perturbation of Eigenvalues of General Matrices

Pete Stewart introduced perturbation bounds for invariant subspaces first for linear operators in [GWS-J15], and then for matrices in [GWS-J19]. Since there is substantial overlap between these papers, we will discuss both of them together. For simplicity, we concentrate on matrices of finite dimension, and omit the issues that can arise for linear operators in infinite dimensional spaces.

We say that a subspace $\mathcal{S}$ is an invariant subspace of a matrix $\mathbf{A}$ if $\mathbf{A}\mathbf{y} \in \mathcal{S}$ for every $\mathbf{y} \in \mathcal{S}$. The simplest example is the subspace spanned by a single eigenvector of $\mathbf{A}$. Stewart starts out his paper [GWS-J19] with a small example to illustrate the high sensitivity of eigenvectors associated with nearby eigenvalues. A small perturbation in the matrix can lead to drastic changes in such eigenvectors. However, Stewart emphasizes that although individual eigenvectors can change dramatically, the space spanned by the eigenvectors corresponding to a cluster of eigenvalues can be quite stable. This is Stewart's motivation for studying the behavior of spaces rather than individual vectors.

At the time of Stewart's paper [GWS-J15] little work existed on subspace perturbations. Notable prior work consisted of Kato's power series of spectral projectors for matrices that are analytic functions of a scalar parameter [80], Davis and Kahan's extensive treatment of Hermitian linear operators [30, 31], and preliminary work by Ruhe [118] and Varah [149] for non-normal matrices.

Previous Results for Hermitian Matrices

To appreciate Pete Stewart's contributions, it is helpful to be aware of what was known at that time about subspace perturbations for Hermitian matrices. It is likely that the work of Davis and Kahan had the most influence on Stewart. Davis and Kahan [30, 31] studied the following problem: If a Hermitian linear operator is slightly perturbed then how much can its invariant subspaces change?

Let $\mathbf{A}$ be a Hermitian matrix whose invariant subspace of interest is spanned by the columns of $\mathbf{X}_1$, so that[16] $\mathbf{A}\mathbf{X}_1 = \mathbf{X}_1\mathbf{A}_{11}$. Let us choose $\mathbf{X}_2$ so that $\mathbf{X} \equiv \begin{bmatrix}\mathbf{X}_1 & \mathbf{X}_2\end{bmatrix}$ is unitary. Then $\mathbf{X}$ block diagonalizes $\mathbf{A}$,

$$\mathbf{X}^{\mathrm{H}}\mathbf{A}\mathbf{X} = \begin{bmatrix}\mathbf{A}_{11} & \mathbf{0} \\ \mathbf{0} & \mathbf{A}_{22}\end{bmatrix}.$$

Suppose that the perturbed matrix $\mathbf{A}+\mathbf{E}$ is also Hermitian with a corresponding subspace[17] $\widehat{\mathbf{X}}_1$, so that $(\mathbf{A}+\mathbf{E})\widehat{\mathbf{X}}_1 = \widehat{\mathbf{X}}_1\widehat{\mathbf{A}}_{11}$. Here $\widehat{\mathbf{X}}_1$ has the same number of columns as $\mathbf{X}_1$. Let $\widehat{\mathbf{X}} = \begin{bmatrix}\widehat{\mathbf{X}}_1 & \widehat{\mathbf{X}}_2\end{bmatrix}$ be unitary, so that

[16] In Sects. 7.1 and 7.2, $\mathbf{A}_{ij}$ denotes the (i,j) subblock of $\mathbf{X}^{\mathrm{H}}\mathbf{A}\mathbf{X}$ rather than the (i,j) subblock of $\mathbf{A}$.

[17] For brevity we will say "the subspace $\mathbf{Y}$" when we mean "the subspace spanned by the columns of $\mathbf{Y}$."

$$\widehat{\mathbf{X}}^{\mathrm{H}}(\mathbf{A}+\mathbf{E})\widehat{\mathbf{X}} = \begin{bmatrix} \widehat{\mathbf{A}}_{11} & \mathbf{0} \\ \mathbf{0} & \widehat{\mathbf{A}}_{22} \end{bmatrix}.$$

The goal now is to express the deviation between the subspaces $\mathbf{X}_1$ and $\widehat{\mathbf{X}}_1$ in terms of the perturbation $\mathbf{E}$.

Since $\mathbf{X}$ and $\widehat{\mathbf{X}}$ are unitary, we can find a unitary transformation that rotates $\mathbf{X}$ into $\widehat{\mathbf{X}}$. That is, $\widehat{\mathbf{X}} = \mathbf{X}\mathbf{U}$, where

$$\mathbf{U} \equiv \begin{bmatrix} \mathbf{C} & -\widehat{\mathbf{S}}^{\mathrm{H}} \\ \mathbf{S} & \widehat{\mathbf{C}} \end{bmatrix} = \mathbf{X}^{\mathrm{H}}\widehat{\mathbf{X}}. \tag{7.1}$$

From $\mathbf{C} = \mathbf{X}_1^{\mathrm{H}}\widehat{\mathbf{X}}_1$ we see that the singular values of $\mathbf{C}$ are the cosines of the canonical angles between the two spaces $\widehat{\mathbf{X}}_1$ and $\mathbf{X}_1$; see also the discussion of the CS decomposition in Sect. 6.2 and in [GWS-J35]. Similarly, $\mathbf{S} = \mathbf{X}_2^{\mathrm{H}}\widehat{\mathbf{X}}_1$ shows that the singular values of $\mathbf{S}$ represent the sines of the canonical angles. The deviation between $\mathbf{X}_1$ and $\widehat{\mathbf{X}}_1$ can now be described by the angles required to rotate one subspace into the other. Bounding the sine of the largest angle $\|\mathbf{S}\|_2$ gives a quantitative measure of this deviation.

For Hermitian matrices, one can bound $\|\mathbf{S}\|_2$ as follows [30, 31]. Multiplying $(\mathbf{A}+\mathbf{E})\widehat{\mathbf{X}}_1 = \widehat{\mathbf{X}}_1\widehat{\mathbf{A}}_{11}$ on the left by $\mathbf{X}^{\mathrm{H}}$ and rearranging gives, for the second block component,

$$\mathbf{E}_{21} \equiv \mathbf{X}_2^{\mathrm{H}}\mathbf{E}\widehat{\mathbf{X}}_1 = \mathbf{S}\widehat{\mathbf{A}}_{11} - \mathbf{A}_{22}\mathbf{S}.$$

If $\widehat{\mathbf{A}}_{11}$ is nonsingular with all eigenvalues in an interval $[\alpha, \beta]$, while all eigenvalues of $\mathbf{A}_{22}$ lie outside $(\alpha - \delta, \beta + \delta)$, or vice versa, then one can show that

$$\|\mathbf{E}_{21}\|_2 \geq \frac{\|\mathbf{S}\|_2}{\delta}.$$

This is the $\sin\Theta$ theorem [31]:

$$\|\mathbf{S}\|_2 \leq \frac{\|\mathbf{E}_{21}\|_2}{\delta}.$$

It says that a small Hermitian perturbation in a Hermitian matrix causes a small rotation of the invariant subspace $\mathbf{X}_1$ – but only if the eigenvalues of $\widehat{\mathbf{A}}_{11}$ are well separated from those of $\mathbf{A}_{22}$. Conversely, if the eigenvalues associated with an invariant subspace are poorly separated from the remaining eigenvalues, then a small perturbation in the matrix can cause a large rotation of the subspace.

Non-Hermitian Matrices

Stewart extends Davis and Kahan's results to non-Hermitian matrices $\mathbf{A}$. Instead of comparing given subspaces for $\mathbf{A}$ and $\mathbf{A}+\mathbf{E}$, he changes the problem slightly and asks: Given some subspace, can we construct from it an invariant subspace of $\mathbf{A}$, and if so, how far is this invariant subspace from our original subspace?

More specifically, suppose we are given a complex square matrix $\mathbf{A}$ and a matrix $\mathbf{X}_1$ with orthonormal columns. The columns of $\mathbf{X}_1$ span an invariant subspace of $\mathbf{A}$ if and only if $\mathbf{A}\mathbf{X}_1 = \mathbf{X}_1\mathbf{B}$ for some matrix $\mathbf{B}$. Let us choose $\mathbf{X}_2$ so that $\mathbf{X} \equiv \begin{bmatrix}\mathbf{X}_1 & \mathbf{X}_2\end{bmatrix}$ is unitary. Then the columns of $\mathbf{X}_1$ span an invariant subspace of $\mathbf{A}$ if and only if $\mathbf{X}_2^{\mathrm{H}}\mathbf{A}\mathbf{X}_1 = \mathbf{0}$.

However, if $\mathbf{X}_2^{\mathrm{H}}\mathbf{A}\mathbf{X}_1$ is merely "small," then how close is $\mathbf{X}_1$ to an invariant subspace of $\mathbf{A}$? Stewart answers this question and in the process establishes the central importance of two concepts: the subspace rotation $\mathbf{U}$, and the separation of operators.

Subspace Rotation

Like Davis and Kahan, Stewart rotates subspaces, but he expresses the rotation in terms of a tangent. If the cosine $c \neq 0$, then we can express a 2×2 rotation in terms of the tangent $p = s/c$,

$$\begin{bmatrix} c & -s \\ s & c \end{bmatrix} = \begin{bmatrix} 1 & -p \\ p & 1 \end{bmatrix} \begin{bmatrix} (1+p^2)^{-1/2} & 0 \\ 0 & (1+p^2)^{-1/2} \end{bmatrix}.$$

Extending this to higher dimensions gives an expression for the rotation $\mathbf{U}$ in (7.1)

$$\mathbf{U} = \begin{bmatrix} \mathbf{I} & -\mathbf{P}^{\mathrm{H}} \\ \mathbf{P} & \mathbf{I} \end{bmatrix} \begin{bmatrix} \left(\mathbf{I} + \mathbf{P}^{\mathrm{H}}\mathbf{P}\right)^{-1/2} & \mathbf{0} \\ \mathbf{0} & \left(\mathbf{I} + \mathbf{P}\mathbf{P}^{\mathrm{H}}\right)^{-1/2} \end{bmatrix},$$

where $\mathbf{P}$ has the same number of columns as $\mathbf{X}_1$. Here $\left(\mathbf{I} + \mathbf{P}^{\mathrm{H}}\mathbf{P}\right)^{-1/2}$ is the Hermitian positive definite square root of the Hermitian positive definite matrix $\left(\mathbf{I} + \mathbf{P}^{\mathrm{H}}\mathbf{P}\right)^{-1}$, and similarly for $\left(\mathbf{I} + \mathbf{P}\mathbf{P}^{\mathrm{H}}\right)^{-1/2}$. Stewart creates a new submatrix $\widehat{\mathbf{X}}_1$ by applying the rotation $\mathbf{U}$ to $\mathbf{X}$, so that $\widehat{\mathbf{X}} = \mathbf{X}\mathbf{U}$. This allows him to derive an explicit expression of the new subspace $\widehat{\mathbf{X}}_1$ in terms of the old,

$$\widehat{\mathbf{X}}_1 = (\mathbf{X}_1 + \mathbf{X}_2\mathbf{P})(\mathbf{I} + \mathbf{P}^{\mathrm{H}}\mathbf{P})^{-1/2}. \tag{7.2}$$

It also allows him to make a connection to the earlier work by Davis and Kahan. From $\mathbf{X}_2^{\mathrm{H}}\widehat{\mathbf{X}}_1 = \mathbf{P}(\mathbf{I} + \mathbf{P}^{\mathrm{H}}\mathbf{P})^{-1/2}$ it follows that $\|\mathbf{X}_2^{\mathrm{H}}\widehat{\mathbf{X}}_1\|_2 \leq \|\mathbf{P}\|_2$. This means that the sine of the largest canonical angle between $\widehat{\mathbf{X}}_1$ and $\mathbf{X}_1$ is bounded by its tangent $\|\mathbf{P}\|_2$. While Davis and Kahan bound the sine of the largest canonical angle, Stewart finds a bound on the tangent.

Invariant Subspaces

Now let $\mathbf{X}_1$ be only an approximation[18] to an invariant subspace, so that

$$\mathbf{X}^{\mathrm{H}}\mathbf{A}\mathbf{X} = \begin{bmatrix} \mathbf{A}_{11} & \mathbf{A}_{12} \\ \mathbf{A}_{21} & \mathbf{A}_{22} \end{bmatrix},$$

where $\mathbf{A}_{21} = \mathbf{X}_2^{\mathrm{H}}\mathbf{A}\mathbf{X}_1$ is "small." Stewart's idea is to construct an invariant subspace of $\mathbf{A}$ from the columns of the matrix (7.2) and to estimate the distance to the invariant subspace by bounding[19] $\|\mathbf{P}\|$.

Suppose, as in the Hermitian case, that there is a matrix $\mathbf{U}$ that can indeed rotate the space $\mathbf{X}_1$ into an invariant subspace $\widehat{\mathbf{X}}_1$, so that $\widehat{\mathbf{X}} = \mathbf{X}\mathbf{U}$. We know that $\widehat{\mathbf{X}}_1$ is an invariant subspace of $\mathbf{A}$ if and only if $\widehat{\mathbf{X}}_2^{\mathrm{H}}\mathbf{A}\widehat{\mathbf{X}}_1 = \mathbf{0}$. Stewart expresses $\widehat{\mathbf{X}}_1$ in terms of $\mathbf{X}_1$, $\mathbf{X}_2$, and $\mathbf{U}$ to conclude that $\widehat{\mathbf{X}}_1$ is an invariant subspace of $\mathbf{A}$ if and only if $\mathbf{P}$ is a solution of the Riccati equation

$$\mathbf{P}\mathbf{A}_{11} - \mathbf{A}_{22}\mathbf{P} = \mathbf{A}_{21} - \mathbf{P}\mathbf{A}_{12}\mathbf{P}. \tag{7.3}$$

To establish conditions for the existence of $\mathbf{P}$, Stewart defines a linear operator $\mathcal{T}$ by

$$\mathcal{T}\mathbf{P} = \mathbf{P}\mathbf{A}_{11} - \mathbf{A}_{22}\mathbf{P}.$$

A similar operator is used in Sect. 4.5.1 and in [GWS-J34]. The identification of this operator paved the way for a bound that has become the basis for many other subspace perturbation results. Here is a weaker but simpler version of the bound:

> If $\mathcal{T}$ is invertible and $(4\|\mathbf{A}_{21}\|\,\|\mathbf{A}_{12}\|\,\|\mathcal{T}^{-1}\|^2) < 1$, then there exists a matrix $\mathbf{P}$ with
>
> $$\|\mathbf{P}\| < 2\|\mathbf{A}_{21}\|\,\|\mathcal{T}^{-1}\| \tag{7.4}$$
>
> so that the columns of the matrix
>
> $$\widehat{\mathbf{X}}_1 = (\mathbf{X}_1 + \mathbf{X}_2\mathbf{P})(\mathbf{I} + \mathbf{P}^{\mathrm{H}}\mathbf{P})^{-1/2}$$
>
> span an invariant subspace of $\mathbf{A}$.

Stewart derives (7.4) by writing the Riccati equation (7.3) as $\mathcal{T}\mathbf{P} = \mathbf{A}_{21} - \phi(\mathbf{P})$ where $\phi(\mathbf{P}) = \mathbf{P}\mathbf{A}_{12}\mathbf{P}$. In a formal proof he shows that the iterates from successive substitution converge to a unique solution of this nonlinear equation. However, he also shares his intuition with the reader: If the quadratic $\phi(\mathbf{P})$ is small, then $\mathbf{P} \approx \mathcal{T}^{-1}\mathbf{A}_{21}$, so that $\|\mathbf{P}\|$ is approximately bounded by $\|\mathbf{A}_{21}\|\,\|\mathcal{T}^{-1}\|$.

---

[18] From now on, except where obvious from the discussion, $\widehat{\mathbf{X}}_1$ refers to an exact invariant subspace while $\mathbf{X}_1$ refers to its approximation.

[19] In this section, $\|\cdot\|$ denotes a subordinate matrix norm.

When viewed in the two-norm or Frobenius norm, the bound (7.4) implies that the maximum angle between $\mathbf{X}_1$ and $\widehat{\mathbf{X}}_1$ becomes smaller as $\|\mathbf{A}_{21}\|$ becomes smaller. In other words, the two spaces get closer as $\|\mathbf{A}_{21}\| \to 0$.

Stewart gives a detailed interpretation for the factors $\|\mathbf{A}_{21}\|$ and $\|\mathcal{T}^{-1}\|$ in the bound (7.4). Let us start with $\|\mathbf{A}_{21}\|$. For any matrix $\mathbf{B}$, the residual $\mathbf{A}\mathbf{X}_1 - \mathbf{X}_1\mathbf{B}$ gives a bound for how much $\mathbf{X}_1$ deviates from an invariant subspace. Writing

$$\|\mathbf{A}\mathbf{X}_1 - \mathbf{X}_1\mathbf{B}\|_{2,F} = \left\| \begin{bmatrix} \mathbf{A}_{11} - \mathbf{B} \\ \mathbf{A}_{21} \end{bmatrix} \right\|_{2,F}$$

shows that the residual is minimized when $\mathbf{B} = \mathbf{A}_{11}$, so that

$$\min_{\mathbf{B}} \|\mathbf{A}\mathbf{X}_1 - \mathbf{X}_1\mathbf{B}\|_{2,F} = \|\mathbf{A}_{21}\|_{2,F}\,.$$

This means that $\|\mathbf{A}_{21}\|_{2,F}$ is the norm of the smallest possible residual for $\mathbf{X}_1$, which, Stewart argues, makes the quantity $\|\mathbf{A}_{21}\|_{2,F}$ in itself a measure for how far $\mathbf{X}_1$ deviates from an invariant subspace.

The second factor in the bound (7.4) is $\|\mathcal{T}^{-1}\|$; the assumptions require that it should not be too large. Since $\|\mathcal{T}^{-1}\|$ is a keystone in Stewart's approach, he discusses it in detail.

Operator Separation

Let us simplify the notation for a moment and define $\mathcal{T}$ by $\mathcal{T}\mathbf{P} \equiv \mathbf{P}\mathbf{B} - \mathbf{C}\mathbf{P}$ for given square matrices $\mathbf{B}$ and $\mathbf{C}$.

The eigenvalues of $\mathcal{T}$ are $\lambda_i(\mathbf{B}) - \lambda_j(\mathbf{C})$. One can see this by expressing the Sylvester equation $\mathbf{P}\mathbf{B} - \mathbf{C}\mathbf{P} = \mathbf{R}$ as a Kronecker product [150],

$$\begin{bmatrix} \mathbf{B} - c_{11}\mathbf{I} & -c_{21}\mathbf{I} & \cdots & -c_{m1}\mathbf{I} \\ -c_{12}\mathbf{I} & \mathbf{B} - c_{22}\mathbf{I} & & \vdots \\ \vdots & & \ddots & \vdots \\ -c_{1m}\mathbf{I} & \cdots & \cdots & \mathbf{B} - c_{mm}\mathbf{I} \end{bmatrix} \begin{bmatrix} \mathbf{p}_1 \\ \vdots \\ \vdots \\ \mathbf{p}_m \end{bmatrix} = \begin{bmatrix} \mathbf{r}_1 \\ \vdots \\ \vdots \\ \mathbf{r}_m \end{bmatrix},$$

where $\mathbf{p}_i$ is the ith column of $\mathbf{P}$ and $\mathbf{r}_i$ is the ith column of $\mathbf{R}$.

Thus, if $\mathbf{B}$ and $\mathbf{C}$ have no eigenvalues in common, then $\mathcal{T}$ is invertible, and $1/\|\mathcal{T}^{-1}\|$ is a lower bound for the eigenvalue separation

$$\frac{1}{\|\mathcal{T}^{-1}\|} \le \min_{i,j} |\lambda_i(\mathbf{B}) - \lambda_j(\mathbf{C})|\,.$$

Stewart interprets $1/\|\mathcal{T}^{-1}\|$ as a measure of *separation between the operators* $\mathbf{B}$ and $\mathbf{C}$. He formally defines

$$\operatorname{sep}(\mathbf{B},\mathbf{C}) = \inf_{\|\mathbf{P}\|=1} \|\mathcal{T}\mathbf{P}\| = \inf_{\|\mathbf{P}\|=1} \|\mathbf{P}\mathbf{B} - \mathbf{C}\mathbf{P}\|, \tag{7.5}$$

which is equal to $1/\|\mathcal{T}^{-1}\|$ if $\mathbf{B}$ and $\mathbf{C}$ have no common eigenvalues and 0 otherwise. For instance, if the second matrix is equal to a scalar c, and c is not an eigenvalue of $\mathbf{B}$, then the operator separation has the explicit expression

$$\mathrm{sep}(\mathbf{B}, c) = \left\|(\mathbf{B} - c\mathbf{I})^{-1}\right\|^{-1}.$$

Stewart derives a number of properties for the operator separation [GWS-J19, Sect. 4.3]. For instance, in the special case of block diagonal matrices $\mathbf{B}$ and $\mathbf{C}$, where

$$\mathbf{B} = \begin{bmatrix} \mathbf{B}_1 & & \\ & \ddots & \\ & & \mathbf{B}_l \end{bmatrix}, \qquad \mathbf{C} = \begin{bmatrix} \mathbf{C}_1 & & \\ & \ddots & \\ & & \mathbf{C}_m \end{bmatrix},$$

Stewart demonstrates that the Frobenius norm operator separation equals the minimal separation between any pair of diagonal blocks,

$$\mathrm{sep}_F(\mathbf{B}, \mathbf{C}) = \min_{i,j}\{\mathrm{sep}(\mathbf{B}_i, \mathbf{C}_j)\}.$$

Consequently, for normal, and in particular Hermitian $\mathbf{B}$ and $\mathbf{C}$, the Frobenius norm operator separation is equal to the eigenvalue separation,

$$\mathrm{sep}_F(\mathbf{B}, \mathbf{C}) = \min_{i,j} \left|\lambda_i(\mathbf{B}) - \lambda_j(\mathbf{C})\right|.$$

Unfortunately, this is not true for non-normal matrices; in this case the operator separation $\mathrm{sep}(\mathbf{B}, \mathbf{C})$ can be much smaller than the eigenvalue separation. To illustrate this, Stewart first shows how the operator separation can change under a similarity transformation

$$\mathrm{sep}(\mathbf{B}, \mathbf{C}) \geq \frac{\mathrm{sep}(\mathbf{V}^{-1}\mathbf{B}\mathbf{V},\ \mathbf{W}^{-1}\mathbf{C}\mathbf{W})}{\kappa(\mathbf{V})\,\kappa(\mathbf{W})},$$

where $\kappa(\mathbf{Y}) = \|\mathbf{Y}\|\,\|\mathbf{Y}^{-1}\|$ is the condition number of $\mathbf{Y}$ with respect to inversion. Now assume that $\mathbf{B}$ and $\mathbf{C}$ are diagonalizable and that the similarity transformation reveals the eigenvalues so that $\mathbf{V}^{-1}\mathbf{B}\mathbf{V}$ and $\mathbf{W}^{-1}\mathbf{C}\mathbf{W}$ are diagonal. By means of the above inequality, Stewart can then relate the Frobenius norm operator separation to the eigenvalue separation,

$$\mathrm{sep}_F(\mathbf{B}, \mathbf{C}) \geq \frac{\min_{i,j} |\lambda_i(\mathbf{B}) - \lambda_j(\mathbf{C})|}{\kappa_2(\mathbf{V})\,\kappa_2(\mathbf{W})}.$$

Thus the operator separation can be much smaller than the eigenvalue separation if the eigenvectors of $\mathbf{B}$ or $\mathbf{C}$ are ill-conditioned with respect to inversion.

Furthermore, in contrast to the eigenvalue separation, the operator separation is essentially perfectly conditioned with respect to perturbations in the matrices,

$$\mathrm{sep}(\mathbf{B}, \mathbf{C}) - \|\mathbf{E}\| - \|\mathbf{F}\| \leq \mathrm{sep}(\mathbf{B}+\mathbf{E}, \mathbf{C}+\mathbf{F}) \leq \mathrm{sep}(\mathbf{B}, \mathbf{C}) + \|\mathbf{E}\| + \|\mathbf{F}\|. \qquad (7.6)$$

Back to Invariant Subspaces

With the help of the operator separation, Stewart [GWS-J19, Sect. 4.1] can express the bound (7.4) entirely in terms of $\mathbf{A}$:

If $\delta = \text{sep}(\mathbf{A}_{11}, \mathbf{A}_{22}) > 0$ and $(4\,\|\mathbf{A}_{21}\|\,\|\mathbf{A}_{12}\|\,) < \delta^2$, then there exists a matrix $\mathbf{P}$ with

$$\|\mathbf{P}\| < 2\,\frac{\|\mathbf{A}_{21}\|}{\delta} \tag{7.7}$$

so that the columns of the matrix

$$\widehat{\mathbf{X}}_1 = (\mathbf{X}_1 + \mathbf{X}_2\mathbf{P})(\mathbf{I} + \mathbf{P}^{\mathrm{H}}\mathbf{P})^{-1/2}$$

span an invariant subspace of $\mathbf{A}$.

Thus $\mathbf{X}_1$ can be rotated into an invariant subspace of $\mathbf{A}$ – provided the diagonal blocks $\mathbf{A}_{11}$ and $\mathbf{A}_{22}$ are sufficiently well separated compared to the size of the off-diagonal blocks.

Moreover, armed with the robustness of the operator separation (7.6), Stewart [GWS-J19, Sect. 4.4] is able to quantify how an invariant subspace $\mathbf{X}_1$ can change under a perturbation $\mathbf{E}$. Partition

$$\mathbf{X}^{\mathrm{H}}\mathbf{A}\mathbf{X} = \begin{bmatrix} \mathbf{A}_{11} & \mathbf{A}_{12} \\ \mathbf{0} & \mathbf{A}_{22} \end{bmatrix}, \qquad \mathbf{X}^{\mathrm{H}}\mathbf{E}\mathbf{X} = \begin{bmatrix} \mathbf{E}_{11} & \mathbf{E}_{12} \\ \mathbf{E}_{21} & \mathbf{E}_{22} \end{bmatrix}.$$

Let $\delta \equiv \text{sep}(\mathbf{A}_{11}, \mathbf{A}_{22}) - \|\mathbf{E}_{11}\| - \|\mathbf{E}_{22}\| > 0$. If $4\|\mathbf{E}_{21}\|\,(\|\mathbf{A}_{12}\| + \|\mathbf{E}_{12}\|) < \delta^2$, then there exists a matrix $\mathbf{P}$ with

$$\|\mathbf{P}\| \leq 2\|\mathbf{E}_{21}\|/\delta \tag{7.8}$$

so that the columns of

$$(\mathbf{X}_1 + \mathbf{X}_2\mathbf{P})(\mathbf{I} + \mathbf{P}^{\mathrm{H}}\mathbf{P})^{-1/2}$$

span an invariant subspace of $\mathbf{A} + \mathbf{E}$.

This means that if the operator separation between $\mathbf{A}_{11}$ and $\mathbf{A}_{22}$ is large enough compared to the perturbation $\mathbf{E}$, then $\mathbf{X}_1$ can be rotated into an invariant subspace for $\mathbf{A} + \mathbf{E}$. For the angle of rotation to be small, $\|\mathbf{E}_{21}\|$ needs to be small, and $\mathbf{A}_{11}$ and $\mathbf{A}_{22}$ must be well separated.

Comparing the last two bounds to Davis and Kahan's $\sin\Theta$ theorem shows that, for non-normal matrices, the operator separation has taken the place of the eigenvalue separation.

Eigenvalues

Stewart also employed his subspace perturbation results to obtain eigenvalue bounds [GWS-J19, Sect. 4.5]. Consider the similarities induced by the approximate and exact invariant subspaces,

$$\mathbf{X}^{\mathrm{H}}\mathbf{A}\mathbf{X} = \begin{bmatrix} \mathbf{A}_{11} & \mathbf{A}_{12} \\ \mathbf{A}_{21} & \mathbf{A}_{22} \end{bmatrix}, \qquad \widehat{\mathbf{X}}^{\mathrm{H}}\mathbf{A}\widehat{\mathbf{X}} = \begin{bmatrix} \widehat{\mathbf{A}}_{11} & \widehat{\mathbf{A}}_{12} \\ \mathbf{0} & \widehat{\mathbf{A}}_{22} \end{bmatrix}.$$

The eigenvalues of $\mathbf{A}$ consist of the eigenvalues of $\widehat{\mathbf{A}}_{11}$ and $\widehat{\mathbf{A}}_{22}$. Stewart shows that if the assumptions for the bound (7.7) hold, then the eigenvalues of these two matrices are disjoint. His argument goes as follows. There is a matrix $\mathbf{U}$ so that $\widehat{\mathbf{X}} = \mathbf{X}\mathbf{U}$ and

$$\begin{bmatrix} \widehat{\mathbf{A}}_{11} & \widehat{\mathbf{A}}_{12} \\ \mathbf{0} & \widehat{\mathbf{A}}_{22} \end{bmatrix} = \mathbf{U}^{\mathrm{H}} \begin{bmatrix} \mathbf{A}_{11} & \mathbf{A}_{12} \\ \mathbf{A}_{21} & \mathbf{A}_{22} \end{bmatrix} \mathbf{U}.$$

The Riccati equation (7.3) gives for the diagonal blocks

$$\widehat{\mathbf{A}}_{11} = \left(\mathbf{I} + \mathbf{P}^{\mathrm{H}}\mathbf{P}\right)^{1/2} (\mathbf{A}_{11} + \mathbf{A}_{12}\mathbf{P}) \left(\mathbf{I} + \mathbf{P}^{\mathrm{H}}\mathbf{P}\right)^{-1/2}$$

and

$$\widehat{\mathbf{A}}_{22} = \left(\mathbf{I} + \mathbf{P}\mathbf{P}^{\mathrm{H}}\right)^{-1/2} (\mathbf{A}_{22} - \mathbf{P}\mathbf{A}_{12}) \left(\mathbf{I} + \mathbf{P}\mathbf{P}^{\mathrm{H}}\right)^{1/2}.$$

Since $\widehat{\mathbf{A}}_{11}$ is similar to $\mathbf{A}_{11} + \mathbf{A}_{12}\mathbf{P}$, the two matrices have the same eigenvalues. Analogously $\widehat{\mathbf{A}}_{22}$ and $\mathbf{A}_{22} - \mathbf{P}\mathbf{A}_{12}$ have the same eigenvalues. From (7.7) it follows that $\|\mathbf{A}_{12}\mathbf{P}\|$ and $\|\mathbf{P}\mathbf{A}_{12}\|$ are bounded above by $2\|\mathbf{A}_{12}\|\,\|\mathbf{A}_{21}\|/\delta$ where $\delta = \mathrm{sep}\,(\mathbf{A}_{11}, \mathbf{A}_{22})$. Combined with the robustness of the operator separation (7.6) we obtain

$$\mathrm{sep}\,(\mathbf{A}_{11} + \mathbf{A}_{12}\mathbf{P},\ \mathbf{A}_{22} - \mathbf{P}\mathbf{A}_{12}) > 0,$$

which means that the eigenvalues of $\widehat{\mathbf{A}}_{11}$ and $\widehat{\mathbf{A}}_{22}$ must be disjoint. Stewart presents a similar argument for the eigenvalues associated with the perturbation bound (7.8). Now one can use standard perturbation bounds, such as the Bauer–Fike Theorem, to bound the distance between eigenvalues of $\mathbf{A}_{11}$ and $\mathbf{A}_{11} + \mathbf{A}_{12}\mathbf{P}$, and between eigenvalues of $\mathbf{A}_{22}$ and $\mathbf{A}_{22} - \mathbf{P}\mathbf{A}_{12}$.

In the special case when the approximate subspace is spanned by a unit vector $\mathbf{x}_1$, Stewart is able to obtain a simple bound. In this case. $\alpha \equiv \mathbf{A}_{11} = \mathbf{x}_1^{\mathrm{H}}\mathbf{A}\mathbf{x}_1$ is a Rayleigh quotient and $\lambda \equiv \widehat{\mathbf{A}}_{11}$ is an eigenvalue of $\mathbf{A}$. The distance between the approximate eigenvalue and the perturbed one is

$$|\alpha - \lambda| \le \|\mathbf{A}_{12}\mathbf{P}\| \le 2\frac{\|\mathbf{A}_{12}\|\,\|\mathbf{A}_{21}\|}{\delta}, \qquad \delta = \mathrm{sep}(\alpha, \mathbf{A}_{22}). \tag{7.9}$$

Stewart emphasizes that this type of bound is too pessimistic for eigenvalues that are poorly separated, because the separation is a part of the bound, and he considers the dependence of eigenvalue bounds on $\text{sep}(\mathbf{A}_{11}, \mathbf{A}_{22})$ unsatisfactory. Instead, he suggests the use of left invariant subspaces to reduce the influence of the operator separation on the eigenvalue bounds. He derives a bound for $\|\mathbf{A}_{11} - \widehat{\mathbf{A}}_{11}\|$ in terms of the secant of the angle between range($\mathbf{X}_1$) and its left invariant counterpart. This bound generalizes the condition number of a simple eigenvalue, since this condition number is the secant of the angle between the associated left and right eigenvectors.

Other Topics

Stewart extends his approach to perturbation bounds for deflating pairs of subspaces in the generalized eigenvalue problem [GWS-J19, Sect. 5] and to perturbation bounds for singular vector subspaces in the singular value decomposition [GWS-J19, Sect. 6].

7.2. Further Results for Hermitian Matrices

In [GWS-J70], Stewart returns to eigenvalue perturbation bounds from [GWS-J19], but now he concentrates on Hermitian matrices.

Let $\mathbf{A}$ be an $n \times n$ Hermitian matrix with eigenvalues $\lambda_1 \geq \cdots \geq \lambda_n$. Let $\mathbf{X}_1$ be a matrix with k orthonormal columns whose span approximates an invariant subspace of $\mathbf{A}$, and define $\mathbf{A}_{11} = \mathbf{X}_1^{\mathrm{H}} \mathbf{A} \mathbf{X}_1$. The eigenvalues $\alpha_1 \geq \cdots \geq \alpha_k$ of $\mathbf{A}_{11}$ are called *Ritz values*. The accuracy of the Ritz values as approximations to the eigenvalues of $\mathbf{A}$ can be expressed in terms of the residual $\mathbf{R} = \mathbf{A}\mathbf{X}_1 - \mathbf{X}_1\mathbf{A}_{11}$. At that time it was known [114, Sect. 11-5] that there are k eigenvalues $\lambda_{j_1}, \ldots, \lambda_{j_k}$ of $\mathbf{A}$ that are close to those of $\mathbf{A}_{11}$,

$$|\alpha_i - \lambda_{j_i}| \leq \|\mathbf{R}\|_2, \qquad 1 \leq i \leq k.$$

Moreover, a few bounds that are quadratic in $\|\mathbf{R}\|_2$ were also in existence; see [114]. In fact, Stewart's previous bound (7.9) is one of them. To see this, let the approximate subspace be spanned by a unit-norm vector $\mathbf{x}_1$ with Rayleigh quotient $\alpha \equiv \mathbf{x}_1^{\mathrm{H}} A \mathbf{x}_1$ and residual $\mathbf{r} \equiv \mathbf{A}\mathbf{x}_1 - \alpha\mathbf{x}_1$. If α approximates an eigenvalue λ_j, then the eigenvalue separation is $\delta = \min_{i \neq j} |\alpha - \lambda_i|$. If λ_j is well separated from all other eigenvalues of A so that the assumptions for (7.7) hold, then we can apply (7.9) to a Hermitian matrix and get

$$|\alpha - \lambda_j| \leq 2\|\mathbf{A}_{21}\|_2^2/\delta.$$

With unitary $\mathbf{X} = \begin{bmatrix} \mathbf{x}_1 & \mathbf{X}_2 \end{bmatrix}$, we have $\|\mathbf{A}_{12}\|_2 = \|\mathbf{X}_2^{\mathrm{H}} \mathbf{A}_{12}^{\mathrm{H}}\|_2 = \|\mathbf{r}\|_2$, so that the above bound is equal to the quadratic residual bound

$$|\alpha - \lambda_j| \leq 2\|\mathbf{r}\|_2^2/\delta.$$

A slightly better bound, without the factor of 2, was also known at the time [114, Sect. 11-7].

The New Bound

Stewart wanted to find an equally simple quadratic residual bound for eigenvalues associated with higher dimensional subspaces. The start of this endeavor is a bound by Kahan [79] for Hermitian matrices $\mathbf{H}$ and non-Hermitian perturbations $\mathbf{F}$. When applied to matrices $\mathbf{H}+\mathbf{F}$ with real eigenvalues Kahan's bound implies

$$|\lambda_j(\mathbf{H}+\mathbf{F}) - \lambda_j(\mathbf{H})| \leq \|\mathbf{F}\|_2.$$

Stewart applied this bound to quantities created from the similarity transformation with the rotation $\mathbf{U}$ in Sect. 7.1,

$$\mathbf{H} = \widehat{\mathbf{A}}_{11}, \qquad \mathbf{F} = -(\mathbf{I}+\mathbf{P}^{\mathrm{H}}\mathbf{P})^{1/2}\,\mathbf{A}_{12}\mathbf{P}\,(\mathbf{I}+\mathbf{P}^{\mathrm{H}}\mathbf{P})^{-1/2}$$

so that

$$\mathbf{H}+\mathbf{F} = (\mathbf{I}+\mathbf{P}^{\mathrm{H}}\mathbf{P})^{1/2}\,\mathbf{A}_{11}\,(\mathbf{I}+\mathbf{P}^{\mathrm{H}}\mathbf{P})^{-1/2}.$$

The k eigenvalues of $\mathbf{H}$ are eigenvalues of $\mathbf{A}$, while $\mathbf{H}+\mathbf{F}$ has the same eigenvalues as $\mathbf{A}_{11}$, which are the Rayleigh quotients $\alpha_1 \geq \cdots \geq \alpha_k$. Application of the above bound by Kahan bound gives

$$|\alpha_j - \lambda_j(\widehat{\mathbf{A}}_{11})| \leq \left\|\left(\mathbf{I}+\mathbf{P}^{\mathrm{H}}\mathbf{P}\right)^{1/2}\right\|_2 \left\|\left(\mathbf{I}+\mathbf{P}^{\mathrm{H}}\mathbf{P}\right)^{-1/2}\right\|_2 \|\mathbf{P}\|_2\,\|\mathbf{A}_{12}\|_2.$$

Since $\|\mathbf{A}_{12}\|_2 = \|\mathbf{R}\|_2$, this bound already contains one power of $\|\mathbf{R}\|_2$. Stewart produced a second power of $\|\mathbf{R}\|_2$ by realizing that the tangent $\|\mathbf{P}\|_2$ is close to the sine, and if the eigenvalues are well separated, then Davis and Kahan's $\sin\Theta$ theorem applies.

To see this, assume that the eigenvalues corresponding to the subspace $\mathbf{X}_1$ are well separated from all others, i.e., for some $\delta > 0$ exactly k eigenvalues $\lambda_j, \ldots, \lambda_{j+k-1}$ of $\mathbf{A}$ lie inside the interval $(\alpha_k - \delta, \alpha_1 + \delta)$, and also assume that the residual is sufficiently small compared to this eigenvalue gap, $\|\mathbf{R}\|_2 < \delta$. These assumptions allow Stewart to apply the $\sin\Theta$ theorem to $\mathbf{A}$ and $\mathbf{X}_1$, and then to transition from sine to tangent,

$$\rho \equiv \frac{\|\mathbf{R}\|_2}{\delta} \geq \left\|\mathbf{P}\left(\mathbf{I}+\mathbf{P}^{\mathrm{H}}\mathbf{P}\right)^{-1/2}\right\|_2 \geq \frac{\|\mathbf{P}\|_2}{\sqrt{1+\|\mathbf{P}\|_2^2}}.$$

Hence $\|\mathbf{P}\|_2 \leq \rho/\sqrt{1-\rho^2}$. With

$$\left\|\left(\mathbf{I}+\mathbf{P}^{\mathrm{H}}\mathbf{P}\right)^{1/2}\right\|_2 \leq \frac{1}{\sqrt{1-\rho^2}} \qquad \text{and} \qquad \left\|\left(\mathbf{I}+\mathbf{P}^{\mathrm{H}}\mathbf{P}\right)^{-1/2}\right\|_2 \leq 1$$

we obtain the desired quadratic bound for the Rayleigh quotients,

$$|\alpha_i - \lambda_{j+i-1}| \leq \frac{\|\mathbf{R}\|_2^2}{(1-\rho^2)\,\delta}, \qquad 1 \leq i \leq k.$$

For approximate subspaces $\mathbf{X}_1$ associated with well-separated eigenvalues, the eigenvalue error is therefore bounded by the square of the residual norm. Stewart also presents an analogous bound in the Frobenius norm.

7.3. Stochastic Matrices

In [GWS-J48], Stewart proposes and analyzes an algorithm for computing the dominant eigenvector of irreducible non-negative matrices, and in particular, stochastic matrices.

A non-negative matrix $\mathbf{A}$ is *irreducible* if there is no permutation matrix $\mathbf{Q}$ so that $\mathbf{QAQ}^T$ is block triangular. Perron Frobenius theory implies that a non-negative irreducible matrix has a simple eigenvalue $\beta > 0$ that is equal to its spectral radius, and the associated left and right eigenvectors are positive, that is, $\mathbf{Ax} = \beta\mathbf{x}$ and $\mathbf{y}^T\mathbf{A} = \beta\mathbf{y}^T$. The eigenvalue β is called *Perron value*, and $\mathbf{x}$ and $\mathbf{y}$ are called *Perron vectors.*

A non-negative matrix $\mathbf{A}$ is *(row-)stochastic* if the elements in each row sum to one, $\mathbf{A1} = \mathbf{1}$, that is, the vector $\mathbf{1}$ of all ones is a right eigenvector of $\mathbf{A}$. If the stochastic matrix is irreducible, then the left eigenvector $\mathbf{y}$, normalized so that $\mathbf{y}^T\mathbf{1} = 1$, is the *stationary distribution* of $\mathbf{A}$.

Stewart considers nearly completely decomposable matrices. These are matrices that can be symmetrically permuted to a form

$$\mathbf{A} = \begin{bmatrix} \mathbf{A}_{11} & \mathbf{A}_{12} & \cdots & \mathbf{A}_{1k} \\ \mathbf{A}_{21} & \mathbf{A}_{22} & \cdots & \mathbf{A}_{2k} \\ \vdots & \vdots & \ddots & \vdots \\ \mathbf{A}_{k1} & \mathbf{A}_{k2} & \cdots & \mathbf{A}_{kk} \end{bmatrix},$$

where the off-diagonal matrices $\mathbf{A}_{ij}$ have small norm and the diagonal matrices $\mathbf{A}_{ii}$ are irreducible. Such matrices arise in economics and model systems whose states can be clustered into aggregates of tightly coupled states, with weak coupling among different aggregates. The classical 1961 paper by Simon and Ando [127] hinted toward computational procedures, and in the mid 1970s, Courtois [26, 27] developed and analyzed methods that exploit the dominance of the diagonal blocks in the computation of the stationary distribution. These are what we now call *aggregation–disaggregation methods* [97, 98], but at the time of [GWS-J48] they were just in their infancy. Pete Stewart's contribution was to make these methods accessible to the numerical community, and to present a simple approximation method combined with a computationally viable analysis.

The Algorithm

Stewart's method approximates the left Perron vector of $\mathbf{A}$ by computing the Perron vector of a smaller Rayleigh quotient and then projecting the small Perron vector back up. Let $\mathbf{X}$ be a nonsingular matrix, and

$$\mathbf{X} = \begin{bmatrix} \mathbf{X}_1 \ \mathbf{X}_2 \end{bmatrix}, \qquad \mathbf{X}^{-1} = \begin{bmatrix} \mathbf{Y}_1^{\mathrm{T}} \\ \mathbf{Y}_2^{\mathrm{T}} \end{bmatrix},$$

where $\mathbf{X}_1$ and $\mathbf{Y}_1$ have k columns. Using $\mathbf{X}$ in a similarity transformation of $\mathbf{A}$ gives

$$\begin{bmatrix} \mathbf{Y}_1^{\mathrm{T}} \\ \mathbf{Y}_2^{\mathrm{T}} \end{bmatrix} \mathbf{A} \begin{bmatrix} \mathbf{X}_1 \ \mathbf{X}_2 \end{bmatrix} = \begin{bmatrix} \mathbf{B} & * \\ * & * \end{bmatrix}.$$

The $k \times k$ matrix $\mathbf{B}$ is the above-mentioned Rayleigh quotient. Assuming that the eigenvalues of $\mathbf{B}$ contain the Perron value β of $\mathbf{A}$, Stewart computes the left Perron vector $\mathbf{v}$, i.e., $\mathbf{v}^{\mathrm{T}}\mathbf{B} = \beta \mathbf{v}^{\mathrm{T}}$, and chooses $\mathbf{v}^{\mathrm{T}}\mathbf{Y}_1^{\mathrm{T}}$ as the desired approximation to the left Perron vector $\mathbf{y}$ of the original matrix $\mathbf{A}$.

Stewart exploits the weakness of the off-diagonal blocks of $\mathbf{A}$ by constructing the matrices $\mathbf{X}_1$ and $\mathbf{Y}_1$ from the Perron vectors of the diagonal blocks,

$$\mathbf{X}_1 = \begin{bmatrix} \mathbf{x}_1 & & \\ & \ddots & \\ & & \mathbf{x}_k \end{bmatrix} \qquad \mathbf{Y}_1 = \begin{bmatrix} \mathbf{y}_1 & & \\ & \ddots & \\ & & \mathbf{y}_k \end{bmatrix},$$

where

$$\mathbf{A}_{ii}\mathbf{x}_i = \beta_i \mathbf{x}_i, \qquad \mathbf{y}_i^{\mathrm{T}}\mathbf{A}_{ii} = \beta_i \mathbf{y}_i^{\mathrm{T}}, \qquad 1 \le i \le k.$$

We see that the desired approximation can be constructed from multiples of the Perron vectors of the diagonal blocks,

$$\mathbf{v}^{\mathrm{T}}\mathbf{Y}_1^{\mathrm{T}} = \begin{bmatrix} v_1 \mathbf{y}_1^{\mathrm{T}} \cdots v_k \mathbf{y}_k^{\mathrm{T}} \end{bmatrix}.$$

Now Stewart employs the bounds from Sect. 7.1 to estimate the accuracy of this approximate left Perron vector. The generality of these bounds does away with the restriction from [26, 27] that $\mathbf{A}$ be diagonalizable.

A Caution

Like Wilkinson before him [157, Sect. 12], Stewart warns his readers not to put too much trust in normwise error bounds [GWS-J48, p. 273]:

> It is important not to expect too much of error bounds cast in terms of norms. In the first place, repeated use of inequalities such as the triangle inequality tends to make them pessimistic. In the second place, such a bound

can be difficult to interpret in terms of the components of the vector thus bounded. For example,[20] if $\|\mathbf{e}\| \leq \epsilon$ then any component of $\mathbf{e}$ can be as large as ϵ. But other things being equal, it is more likely that each component is of order $\epsilon/\sqrt{n}$.

7.4. Graded Matrices

In 1991, Stewart and Zhang investigated why the eigenvalues of a graded matrix tend to reflect the grading [GWS-J71], and Stewart takes up the subject again 10 years later [GWS-J109].

Grading

Informally, a matrix is "graded" if the magnitude of its elements decreases from left to right, or from top to bottom, or down along the diagonal. For instance, the matrices

$$\begin{bmatrix} 1 & 10^{-4} & 10^{-8} \\ 1 & 10^{-4} & 10^{-8} \\ 1 & 10^{-4} & 10^{-8} \end{bmatrix}, \quad \begin{bmatrix} 1 & 1 & 1 \\ 10^{-4} & 10^{-4} & 10^{-4} \\ 10^{-8} & 10^{-8} & 10^{-8} \end{bmatrix}, \quad \begin{bmatrix} 1 & 10^{-4} & 10^{-8} \\ 10^{-4} & 10^{-8} & 10^{-12} \\ 10^{-8} & 10^{-12} & 10^{-16} \end{bmatrix}$$

are graded by columns, by rows and diagonally, respectively.

The first paper [GWS-J71] appeared at a time when relative perturbation theory, and more generally, structured-perturbation theory, was just starting to flourish. Very informally, the purpose of this theory is to investigate whether the structure of a matrix can diminish the sensitivity of eigenvalues and eigenvectors to structured perturbations and to develop algorithms that exploit this structure. In [6] and a preprint of [36], Barlow, Demmel and Veselić consider the class of symmetric scaled-diagonally dominant matrices and show that small relative perturbations in the matrix elements cause only small relative changes in the eigenvalues. They also present algorithms that can compute the eigenvalues and eigenvectors to high relative accuracy.

In contrast, Stewart and Zhang are concerned with general, nonsymmetric matrices and absolute accuracy. Formally, they define an $n \times n$ matrix $\mathbf{A}$ to be *graded* if there is a diagonal matrix

$$\mathbf{D} = \begin{bmatrix} \delta_1 & & \\ & \ddots & \\ & & \delta_n \end{bmatrix}, \qquad \delta_1 \geq \cdots \geq \delta_n > 0,$$

so that $\mathbf{A} = \mathbf{BD}$, or $\mathbf{A} = \mathbf{DB}$, or $\mathbf{A} = \mathbf{DBD}$ and the matrix $\mathbf{B}$ is "well-behaved" in a sense to be defined later. The matrix $\mathbf{D}$ defines a downward grading. An upward grading can be expressed in an analogous fashion.

[20] Here $\mathbf{e}$ denotes a vector of n elements.

Dominant Eigenvalue

In their first result, Stewart and Zhang show that the dominant eigenvalue of a sufficiently strongly graded matrix $\mathbf{A} = \mathbf{DBD}$ can be approximated by the $(1,1)$ element of $\mathbf{A}$. Note that due to the downward grading, a_{11} has the potential to be an element of largest magnitude. They exploit this with Gershgorin's theorem, which says that the eigenvalues λ_i of $\mathbf{A}$ lie in the union of the disks

$$\left\{\lambda : |\lambda - a_{ii}| \le \sum_{j\neq i} |a_{ij}|\right\}.$$

Label the eigenvalues in the order of decreasing magnitude, $|\lambda_1| \ge \cdots \ge |\lambda_n|$. Stewart and Zhang's idea now is to confine the dominant eigenvalue λ_1 to the first disk by isolating it from all others. To this end, they denote by $b_{\max}$ the magnitude of a largest element of $\mathbf{B}$, that is, $b_{\max} = \max_{1\le i,j\le n} |b_{ij}|$. Then they apply the triangle inequality to $|\lambda - a_{ii}|$ and use $a_{ij} = \delta_i b_{ij} \delta_j$ to obtain a lower bound for eigenvalues in the first disk

$$|\lambda| \ge |a_{11}| - \sum_{j=2}^{n} |a_{1j}| \ge \delta_1^2 |b_{11}| - b_{\max}\, \delta_1 \sum_{j=2}^{n} \delta_j,$$

and an upper bound for eigenvalues in the other disks

$$|\lambda| \le |a_{ii}| + \sum_{j\neq i}^{n} |a_{ij}| \le b_{\max}\, \delta_i \sum_{j=1}^{n} \delta_j \le b_{\max}\, \delta_2 \sum_{j=1}^{n} \delta_j, \qquad 2 \le i \le n.$$

The first disk is isolated from all others if its lower bound is larger than the upper bounds of the other disks,

$$\delta_1^2 |b_{11}| - b_{\max}\, \delta_1 \sum_{j=2}^{n} \delta_j \ge b_{\max}\, \delta_2 \sum_{j=1}^{n} \delta_j.$$

Therefore, if

$$\frac{|b_{11}|}{b_{\max}} \ge \frac{\delta_1 \sum_{j=2}^{n} \delta_j + \delta_2 \sum_{j=1}^{n} \delta_j}{\delta_1^2}, \tag{7.10}$$

then the dominant eigenvalue λ_1 of $\mathbf{A}$ is simple and satisfies

$$|\lambda_1 - a_{11}| \le b_{\max}\, \delta_1 \sum_{j=2}^{n} \delta_j.$$

Stewart and Zhang present an example to illustrate the bound. They define the ratios of successive diagonal elements of $\mathbf{D}$ as *grading factors*

$$\gamma_i = \frac{\delta_{i+1}}{\delta_i}, \qquad 1 \le i \le n-1.$$

Because the diagonal elements are nonincreasing, the grading factors satisfy $\gamma_i \le 1$. The example is that of a uniformly graded matrix where all γ_i are identical, i.e., $\gamma_i = \gamma < 1$, because then $\delta_i/\delta_1 = \gamma^{i-1}$ and the right-hand side in (7.10) simplifies to

$$\frac{\delta_1 \sum_{j=2}^n \delta_j + \delta_2 \sum_{j=1}^n \delta_j}{\delta_1^2} = \gamma \sum_{j=0}^{n-2} \gamma^j + \gamma \sum_{j=0}^{n-1} \gamma^j \le \frac{2\gamma}{1-\gamma}.$$

Since (7.10) is satisfied if

$$\frac{|b_{11}|}{b_{\max}} \ge \frac{2\gamma}{1-\gamma}, \tag{7.11}$$

this condition implies an absolute error bound, which can also be expressed as the relative bound

$$\left|\frac{\lambda_1 - a_{11}}{a_{11}}\right| \le \frac{b_{\max}}{|b_{11}|} \frac{\gamma}{1-\gamma}.$$

Therefore, in the context of approximating the dominant eigenvalue of $\mathbf{A}$ by a_{11}, the matrix $\mathbf{B}$ is well-behaved if its $(1,1)$ element has large magnitude compared to all elements of $\mathbf{B}$. The above bound illustrates that the eigenvalue approximation improves as the strength of the grading increases, i.e., as γ becomes smaller.

The above deliberations also apply to one-sided gradings because $\mathbf{AD}$ and $\mathbf{DA}$ have the same eigenvalues as the diagonally graded matrix $\mathbf{D}^{1/2}\mathbf{A}\mathbf{D}^{1/2}$.

Subdominant Eigenvalues

Stewart and Zhang's idea is to approximate subdominant eigenvalues of $\mathbf{A}$ by dominant eigenvalues of certain submatrices, that is, to apply the condition (7.11) to a judiciously chosen submatrix. To produce such a submatrix they need to find a similarity transformation that reduces $\mathbf{A}$ to block triangular form

$$\begin{bmatrix} \mathbf{I} & \mathbf{0} \\ -\mathbf{P} & \mathbf{I} \end{bmatrix} \begin{bmatrix} \mathbf{A}_{11} & \mathbf{A}_{12} \\ \mathbf{A}_{21} & \mathbf{A}_{22} \end{bmatrix} \begin{bmatrix} \mathbf{I} & \mathbf{0} \\ \mathbf{P} & \mathbf{I} \end{bmatrix} = \begin{bmatrix} \mathbf{A}_{11} + \mathbf{A}_{12}\mathbf{P} & \mathbf{A}_{12} \\ \mathbf{0} & \mathbf{A}_{22} - \mathbf{P}\mathbf{A}_{12} \end{bmatrix}. \tag{7.12}$$

Finding the similarity transformation now amounts to finding a matrix $\mathbf{P}$ that satisfies (7.3). Partitioning $\mathbf{B}$ and $\mathbf{D}$ conformally,

$$\mathbf{B} = \begin{bmatrix} \mathbf{B}_{11} & \mathbf{B}_{12} \\ \mathbf{B}_{21} & \mathbf{B}_{22} \end{bmatrix}, \qquad \mathbf{D} = \begin{bmatrix} \mathbf{D}_1 & \mathbf{0} \\ \mathbf{0} & \mathbf{D}_2 \end{bmatrix},$$

and scaling $\mathbf{Q} = \mathbf{D}_2{}^{-1}\mathbf{P}\mathbf{D}_1$ in (7.3) gives

$$\mathbf{Q} = \mathbf{B}_{21}\mathbf{B}_{11}{}^{-1} + \mathbf{B}_{22}\mathbf{D}_2^2\,\mathbf{Q}\,\mathbf{D}_1^{-2}\mathbf{B}_{11}{}^{-1} - \mathbf{Q}\,\mathbf{B}_{12}\mathbf{D}_2^2\,\mathbf{Q}\,\mathbf{D}_1^{-2}\mathbf{B}_{11}{}^{-1}.$$

If the grading for this particular partition is strong enough so that $\gamma_k = \|\mathbf{D}_2\|_2\|\mathbf{D}_1{}^{-1}\|_2$ is sufficiently small, then $\mathbf{Q} \approx \mathbf{B}_{21}\mathbf{B}_{11}{}^{-1}$. Therefore the (2,2) block in (7.12) is

$$\mathbf{A}_{22} - \mathbf{P}\mathbf{A}_{12} \approx \mathbf{A}_{22} - \mathbf{A}_{21}\mathbf{A}_{11}{}^{-1}\mathbf{A}_{12},$$

which means that the eigenvalues of the Schur complement are approximations to eigenvalues of $\mathbf{A}$.

In [GWS-J109], Stewart returns to this subject and defines the *grading impediment* $\kappa_k = \|\mathbf{B}_{11}{}^{-1}\|_2\,\|\mathbf{B}\|_2$ to describe the accuracy of eigenvalue and eigenvector approximations for matrices with a one-sided grading.

Other Topics

In [GWS-J71], Stewart and Zhang discuss the condition number of a nondefective multiple eigenvalue in terms of the grading induced by the canonical angles between the associated left and right subspaces. They use the similarity transformation (7.12) to approximate eigenvectors, and then extend this approach to approximating singular values and singular vectors.

7.5. Rayleigh–Ritz Approximations

In the context of Lanczos methods, Saad had presented a $\sin\Theta$ theorem for a single Ritz vector of a Hermitian matrix [121, Theorem IV.4.6]. In [GWS-J108], Stewart extended this result to multidimensional Ritz spaces of non-Hermitian matrices by making use of the operator separation.

The situation is the following. Suppose we have a complex square matrix $\mathbf{A}$ with an invariant subspace $\hat{\mathcal{X}}_1$. We want to approximate this invariant subspace by vectors from a big space $\mathcal{X}$. Decompose $\mathcal{X} = \mathcal{X}_1 \oplus \mathcal{X}_2$, where the smaller space $\mathcal{X}_1$ has the same dimension as $\hat{\mathcal{X}}_1$. We approximate $\hat{\mathcal{X}}_1$ by $\mathcal{X}_1$, and want to bound the angle between the two spaces.

Stewart chooses a so-called orthogonal projection method. Let the columns of $\mathbf{X} = \begin{bmatrix}\mathbf{X}_1 & \mathbf{X}_2\end{bmatrix}$ be an orthonormal basis for $\mathcal{X}$ and the columns of $\mathbf{X}_1$ an orthonormal basis for $\mathcal{X}_1$. An orthogonal projection method requires that the Galerkin condition

$$\mathbf{A}\mathbf{X}_1 - \mathbf{X}_1\mathbf{B} \perp \mathcal{X}$$

be satisfied for some matrix $\mathbf{B}$. In terms of the orthonormal basis this means $\mathbf{X}^{\mathrm{H}}(\mathbf{A}\mathbf{X}_1 - \mathbf{X}_1\mathbf{B}) = \mathbf{0}$, which implies $\mathbf{B} = \mathbf{X}_1^{\mathrm{H}}\mathbf{A}\mathbf{X}_1$. The eigenvalues of $\mathbf{B}$ are Ritz values of $\mathbf{A}$.

Known Results for Hermitian Matrices

Let us suppose for the moment that $\mathbf{A}$ is Hermitian and that $\hat{\mathcal{X}}_1$ and $\mathcal{X}_1$ have dimension 1. The geometric intuition is as follows [121, Sect. IV.3.2]. We first project the unit vector $\widehat{\mathbf{x}}_1$ onto the big space $\mathcal{X}$ and then project the resulting vector onto the smaller subspace $\mathcal{X}_1$. With $\mathbf{P}_{\mathcal{X}} \equiv \mathbf{X}\mathbf{X}^{\mathrm{H}}$ being the orthogonal projector onto $\mathcal{X}$, the orthogonal projection of $\widehat{\mathbf{x}}_1$ onto $\mathcal{X}$ is

$$\mathbf{P}_{\mathcal{X}}\widehat{\mathbf{x}}_1 = \mathbf{y}\cos\theta,$$

where $\mathbf{y}$ is a unit vector in $\mathcal{X}$, and θ is the angle between between $\hat{\mathcal{X}}_1$ and $\mathcal{X}$. The orthogonal projection of $\mathbf{y}\cos\theta$ onto $\mathcal{X}_1$ is $\mathbf{x}_1 \cos\omega\cos\theta$, where $\mathbf{x}_1$ is a unit vector in $\mathcal{X}_1$, and ω is the angle between $\mathbf{y}$ and $\mathcal{X}_1$. The length of this projection is $\cos\omega\cos\theta$. This allows us to express the angle of interest θ_1, which is the angle between the invariant subspace $\hat{\mathcal{X}}_1$ and its approximation $\mathcal{X}_1$, as

$$\sin^2\theta_1 = 1 - \cos^2\omega\cos^2\theta = 1 - (1-\sin^2\omega)\ \cos^2\theta = \sin^2\theta + \sin^2\omega\cos^2\theta.$$

One can show that $\sin\omega\cos\theta \leq \sin\theta\ \|\mathbf{P}_{\mathcal{X}}\mathbf{A}(\mathbf{I}-\mathbf{P}_{\mathcal{X}})\|_2/\delta$, where δ is an eigenvalue separation. Putting everything together gives

$$\sin\theta_1 \leq \sin\theta\sqrt{1+\frac{\eta^2}{\delta^2}}, \qquad \text{where} \quad \eta \equiv \|\mathbf{P}_{\mathcal{X}}\mathbf{A}(\mathbf{I}-\mathbf{P}_{\mathcal{X}})\|_2.$$

Therefore, $\sin\theta_1$ can be bounded in terms of $\sin\theta$. This means that as soon as the invariant subspace is close to the bigger space $\mathcal{X}$, it is also automatically close to some small space $\mathcal{X}_1$ within $\mathcal{X}$. This was the Hermitian one-dimensional case.

Non-Hermitian Matrices

Now let us return to the general situation where $\mathbf{A}$ can be non-Hermitian, and $\hat{\mathcal{X}}_1$ and $\mathcal{X}_1$ can have larger dimension. Stewart replaces the eigenvalue separation by an operator separation. Furthermore, since the two-norm is invariant under multiplication by unitary matrices, he can express the operator separation (7.5) as a separation between subspaces,

$$\operatorname{sep}(\mathcal{X}_2,\ \hat{\mathcal{X}}_1) = \inf_{\|\mathbf{P}\|_2=1} \left\|\mathbf{P}\,(\mathbf{X}_2^{\mathrm{H}}\mathbf{A}\mathbf{X}_2) - (\widehat{\mathbf{X}}_1^{\mathrm{H}}\mathbf{A}\widehat{\mathbf{X}}_1)\,\mathbf{P}\right\|_2.$$

Then he derives a bound for the general case that exactly mirrors the bound in the Hermitian one-dimensional case,

$$\sin\theta_1 \leq \sin\theta\sqrt{1+\frac{\eta^2}{\operatorname{sep}(\mathcal{X}_2,\ \hat{\mathcal{X}}_1)^2}}.$$

This bound also holds in the Frobenius norm.

7.6. Powers of Matrices

In [GWS-J114], Stewart investigates what happens to powers of matrices when the factors are perturbed, and how the perturbations can affect the asymptotic behavior of the powers. Stewart extends three existing results for powers $\mathbf{A}^k$ to powers with perturbed factors,

$$\mathbf{S}_k = (\mathbf{A} + \mathbf{E}_k)(\mathbf{A} + \mathbf{E}_{k-1}) \cdots (\mathbf{A} + \mathbf{E}_1).$$

In the following discussion, let $\mathbf{A}$ be an $n \times n$ complex matrix with eigenvalues λ_i labeled in order of decreasing magnitude, $|\lambda_1| \geq \cdots \geq |\lambda_n|$. The *spectral radius* is $\rho(\mathbf{A}) \equiv |\lambda_1|$.

Convergence to Zero

Stewart first proves the classical result that a similarity transformation can bring the norm of a matrix arbitrarily close to its spectral radius.

To see why, consider a 3×3 matrix $\mathbf{A}$, and reduce it to Schur form

$$\mathbf{T} = \mathbf{V}^{\mathrm{H}}\mathbf{A}\mathbf{V} = \begin{bmatrix} \lambda_1 & t_{12} & t_{13} \\ & \lambda_2 & t_{23} \\ & & \lambda_3 \end{bmatrix}.$$

Define a diagonal matrix $\mathbf{D} = \operatorname{diag}(1, \alpha, \alpha^2)$, and use it in a similarity transformation of $\mathbf{T}$, which preserves the eigenvalues of $\mathbf{T}$,

$$\mathbf{D}^{-1}\mathbf{T}\mathbf{D} = \begin{bmatrix} \lambda_1 & \alpha\, t_{12} & \alpha^2\, t_{13} \\ & \lambda_2 & \alpha\, t_{23} \\ & & \lambda_3 \end{bmatrix}.$$

The off-diagonal elements approach zero as $\alpha \to 0$. With $\mathbf{X} = \mathbf{V}\mathbf{D}$, we obtain that $\|\mathbf{X}^{-1}\mathbf{A}\mathbf{X}\|_p \to \rho(\mathbf{A})$ as $\alpha \to 0$. This is essentially the proof of the classical result:

For any $\eta > 0$ there is a nonsingular matrix $\mathbf{X}$ such that

$$\|\mathbf{X}^{-1}\mathbf{A}\mathbf{X}\|_p \leq \rho(\mathbf{A}) + \eta,$$

hence

$$\|\mathbf{A}\|_p \leq \kappa_p(\mathbf{X})\ (\rho(\mathbf{A}) + \eta)\,, \tag{7.13}$$

where $\kappa_p(\mathbf{X}) = \|\mathbf{X}\|_p\, \|\mathbf{X}^{-1}\|_p$ is the condition number of $\mathbf{X}$ with respect to inversion.

For diagonalizable matrices $\mathbf{X}$, we have $\eta = 0$ and $\|\mathbf{A}\|_p \leq \kappa_p(\mathbf{X})\, \rho(\mathbf{A})$.

Now consider the power $\mathbf{A}^k$. Since $(\mathbf{X}^{-1}\mathbf{A}\mathbf{X})^k = \mathbf{X}^{-1}\mathbf{A}^k\mathbf{X}$, the bound (7.13) implies

$$\|\mathbf{A}^k\|_p \leq \kappa_p(\mathbf{X}) \ (\rho(\mathbf{A}) + \eta)^k ,$$

so that $\|\mathbf{A}^k\|_p^{1/k} \to \rho(\mathbf{A})$ as $k \to \infty$. One says that the *root convergence index* of $\mathbf{A}^k$ is $\rho(\mathbf{A})$. In particular, if $\rho(\mathbf{A}) < 1$, then $\lim_{k\to\infty} \mathbf{A}^k = \mathbf{0}$. The bound (7.13) shows that the asymptotic behavior of matrix powers is determined by the spectral radius of the matrix.

Now let us look at how the bound (7.13) can be extended to perturbed products $\mathbf{S}_k$ when $\rho(\mathbf{A}) < 1$. The behavior of the $\mathbf{S}_k$ had been investigated in detail by Higham and Knight [70] in the context of finite precision arithmetic, and even before that by Ostrowski [102]. Stewart derives a normwise version of a componentwise bound in [70, Sect. 3] as follows.

The bound (7.13) implies that there is a matrix $\mathbf{X}$ so that $\|\mathbf{X}^{-1}\mathbf{A}\mathbf{X}\|_p \leq \rho(\mathbf{A}) + \eta/2$ and

$$\|\mathbf{X}^{-1}(\mathbf{A} + \mathbf{E}_j)\mathbf{X}\|_p \leq \rho(\mathbf{A}) + \frac{\eta}{2} + \|\mathbf{X}^{-1}\mathbf{E}_j\mathbf{X}\|_p.$$

If $\kappa_p(\mathbf{X}) \ \|\mathbf{E}_j\|_p \leq \eta/2$ then

$$\|\mathbf{X}^{-1}(\mathbf{A} + \mathbf{E}_j)\mathbf{X}\|_p \leq \rho(\mathbf{A}) + \eta.$$

Thus, if we force the perturbations to be small enough so that

$$\|\mathbf{E}_j\|_p \leq \epsilon, \qquad \text{where} \qquad \epsilon \equiv \frac{\eta}{2\,\kappa(\mathbf{X})},$$

then $\|\mathbf{S}_k\|_p \leq \kappa_p(\mathbf{X}) \ (\rho(\mathbf{A}) + \eta)^k$. This gives Stewart's bound:

$$\limsup_k \|\mathbf{S}_k\|_p^{1/k} \leq \rho(\mathbf{A}) + \eta.$$

In particular, if also $\rho(\mathbf{A}) + \eta < 1$ then the products $\mathbf{S}_k$ converge to zero with root convergence index at most $\rho(\mathbf{A}) + \eta$.

Stewart points out that for a fixed error size ϵ, the best root convergence index one can expect is $\rho(\mathbf{A}) + \epsilon$. To see this, let $\mathbf{E}_j = \mathbf{E}$ where $\mathbf{E}$ is chosen so that $\rho(\mathbf{A} + \mathbf{E}) = \rho(\mathbf{A}) + \epsilon$.

A Simple Dominant Eigenvalue

The preceding results mainly concern normwise convergence. For matrices $\mathbf{A}$ with a simple dominant eigenvalue, $|\lambda_1| > |\lambda_2|$, Stewart also investigates componentwise convergence of the elements of $\mathbf{S}_k$.

Assume the matrix $\mathbf{A}$ has been scaled so that the dominant eigenvalue is equal to one, i.e., $\lambda_1 = 1$. Let $\mathbf{x}$ be a corresponding right eigenvector, $\mathbf{A}\mathbf{x} = \mathbf{x}$. For the exact powers $\mathbf{A}^k$ we know that if $\mathbf{y}$ is a left eigenvector associated with $\lambda_1 = 1$, then $\lim_{k\to\infty} \mathbf{A}^k = \mathbf{x}\mathbf{y}^{\mathrm{H}}$ with root convergence index $|\lambda_2|$.

However, to establish the convergence of perturbed products $\mathbf{S}_k$, Stewart requires a stringent condition on the norms of the perturbations. Since $\rho(\mathbf{A}) = 1$, the product $\prod_{j=1}^{\infty}(1 + \|\mathbf{E}_j\|_p)$ is finite if and only if the sum $\sum_{j=1}^{\infty}\|\mathbf{E}_j\|_p$ is finite. Under this condition, Stewart is able to bound the components of the vectors $\mathbf{s}_{k+1} = (\mathbf{A} + \mathbf{E}_{k+1})\mathbf{s}_k$, where $\mathbf{s}_0 \neq 0$, and to establish that the products $\mathbf{S}_k$ converge to a matrix whose columns are multiples of the right eigenvector $\mathbf{x}$. Formally:

If $\sum_{j=1}^{\infty}\|\mathbf{E}_j\|_p < \infty$, then there is a (possibly zero) vector $\mathbf{z}$ so that

$$\lim_{k\to\infty} \mathbf{S}_k = \mathbf{x}\mathbf{z}^{\mathrm{H}}.$$

The root convergence index is at most

$$\max\{|\lambda_2|, \limsup_k \|\mathbf{E}_k\|_p^{1/k}\}.$$

Stewart extends the proof of the above result to matrices with $\rho(\mathbf{A}) < 1$ and several dominant eigenvalues, and obtains: $\lim_{k\to\infty} \mathbf{S}_k = \mathbf{0}$.

Power Method

The power method generates a sequence of vectors that is to approximate a dominant eigenvector of a matrix. The power method iterates for $\mathbf{A}$ generated from a starting vector $\mathbf{x}_1 \neq \mathbf{0}$ are

$$\mathbf{x}_{k+1} = \nu_k\, \mathbf{A}\mathbf{x}_k = \nu_k \cdots \nu_1\, \mathbf{A}^k\mathbf{x}_1,$$

where ν_k are normalization factors often chosen so that $\|\mathbf{x}_{k+1}\|_p = 1$ or so that a particular component of $\mathbf{x}_{k+1}$ is equal to one. If $\mathbf{A}$ has a simple dominant eigenvalue, $|\lambda_1| < |\lambda_2|$, then for almost all starting vectors $\mathbf{x}_1$ the iterates $\mathbf{x}_k$ converge to a right eigenvector of λ_1 with rate $|\lambda_2|/|\lambda_1|$. This is in exact arithmetic.

In finite precision arithmetic, however, the computation of the power method iterates can be expressed as

$$\mathbf{x}_{k+1} = \nu_k\, (\mathbf{A} + \mathbf{E}_k)\mathbf{x}_k = \nu_k \cdots \nu_1\, \mathbf{S}_k\mathbf{x}_1,$$

where $\|\mathbf{E}_k\|_p/\|\mathbf{A}_k\|_p$ is on the order of roundoff error. Because the perturbations do not decrease in norm, one cannot apply the result of the preceding section to analyze the convergence.

As in the previous section, Stewart considers an idealized situation where we already know the eigenvector, that is,

$$\mathbf{A} = \begin{bmatrix} 1 & \mathbf{0} \\ \mathbf{0} & \mathbf{B} \end{bmatrix}, \qquad \text{where} \quad \|\mathbf{B}\|_p < 1,$$

and $\mathbf{x} = \begin{bmatrix} 1 & \mathbf{0} \end{bmatrix}^{\mathrm{H}}$ is a right eigenvector for the eigenvalue $\lambda_1 = 1$. Let us see what happens if we multiply the perturbed matrix $\mathbf{A} + \mathbf{E}$ by a vector that has a nonzero contribution in $\mathbf{x}$,

$$\begin{bmatrix} 1+e_{11} & \mathbf{e}_{12}^{\mathrm{H}} \\ \mathbf{e}_{21} & \mathbf{B}+\mathbf{E}_{22} \end{bmatrix} \begin{bmatrix} 1 \\ \mathbf{t} \end{bmatrix} = \begin{bmatrix} 1+e_{11}+\mathbf{e}_{12}^{\mathrm{H}}\mathbf{t} \\ \mathbf{e}_{21}+(\mathbf{B}+\mathbf{E}_{22})\mathbf{t} \end{bmatrix} = \nu \begin{bmatrix} 1 \\ \tilde{\mathbf{t}} \end{bmatrix},$$

where

$$\tilde{\mathbf{t}} = \frac{\mathbf{e}_{21}+(\mathbf{B}+\mathbf{E}_{22})\mathbf{t}}{1+e_{11}+\mathbf{e}_{12}^{\mathrm{H}}\mathbf{t}}, \qquad \nu = 1+e_{11}+\mathbf{e}_{12}^{\mathrm{H}}\mathbf{t}.$$

With $\|\mathbf{E}\|_p = \epsilon$, the norm of $\tilde{\mathbf{t}}$ can be bounded by

$$\|\tilde{\mathbf{t}}\|_p \le \frac{\|\mathbf{B}\|_p\|\mathbf{t}\|_p + \epsilon(1+\|\mathbf{t}\|_p)}{1-(1+\|\mathbf{t}\|_p)\epsilon}.$$

The question is, if we perform this matrix-vector multiplication repeatedly then what can we say about the trailing elements $\mathbf{t}_k$ of the iterates $\mathbf{x}_k^{\mathrm{H}} = \begin{bmatrix} 1 & \mathbf{t}_k^{\mathrm{H}} \end{bmatrix}$? Stewart answers this question by setting $t_k = \|\mathbf{t}_k\|_p$ and defining the iteration

$$t_{k+1} = \psi(t_k), \qquad \text{where} \quad \psi(t) = \frac{\|\mathbf{B}\|_p t + \epsilon(1+t)}{1-(1+t)\epsilon},$$

where $\|\mathbf{E}_k\|_p \le \epsilon$. He shows the following.

> If $4\epsilon < 1 - \|\mathbf{B}\|_p$, then the function ψ has a unique smallest fixed point
>
> $$t_* = \frac{2}{\tau + \sqrt{\tau^2 - 4}}, \qquad \text{where} \quad \tau \equiv \frac{1 - \|\mathbf{B}\|_p - 2\epsilon}{\epsilon}.$$
>
> Moreover, if $t_1 > t_*$ and t_1 is sufficiently close to t_*, then the iteration $t_{k+1} = \psi(t_k)$ converges to t_* from above.

Applied to the power method, this means that the backward errors should decrease with $1 - \|\mathbf{B}\|_p$. This is often not a problem since typically ϵ is on the order of roundoff error. Furthermore, as $\|\mathbf{B}\|_p$ approaches 1 the convergence is likely to slow down and the region of convergence tends to become smaller.

Stewart emphasizes that this is an idealized situation, and that in practice the errors must be multiplied by the condition number $\kappa_p(\mathbf{X})$ of the diagonalizing transformation. Moreover he shows that an ill-conditioned eigenvalue λ_1 leads to a similarity transformation $\mathbf{X}$ with a large condition number $\kappa_p(\mathbf{X})$. Therefore the accuracy of the power method tends to be lower for ill-conditioned dominant eigenvalues.

For the special case where the errors $\epsilon_k = \|\mathbf{E}_k\|_p$ do indeed decrease monotonically, Stewart shows:

> Let $0 < \|\mathbf{B}\|_p < 1$. For any t_1 there is an ϵ_1 so that if $\epsilon_1 \ge \epsilon_2 \ge \cdots$, then the iterates t_k converge monotonically to zero.

7.7. Impact

The concepts of subspace rotation and operator separation from [GWS-J15] and [GWS-J19] have proved to be versatile tools for a large variety of problems.

The operator separation is an extremely important concept for subspace sensitivity. Taking up Pete Stewart's formal definition, Jim Varah devotes an entire paper to the properties of sep [150]. In particular, he points out that the operator separation encompasses the *ϵ-pseudospectrum* of the matrix $\mathbf{B}$, that is, it is the set of all complex scalars λ for which $\text{sep}(\mathbf{B}, \lambda) \leq \epsilon$. The ϵ-pseudospectrum is a way to visualize the non-normality of a matrix and the sensitivity of its eigenvalues [140]. The extension of the pseudospectrum to two matrices is the so-called *spectrum separation*, that is, the size of the smallest perturbations $\mathbf{E}$ and $\mathbf{F}$ so that $\mathbf{B}+\mathbf{E}$ and $\mathbf{C}+\mathbf{F}$ have a common eigenvalue. Demmel [35] discusses the importance of the spectrum separation for the numerically stable computation of eigenvalue decompositions.

Perturbation bounds of the form (7.4), (7.7), and (7.8), as well as the operator separation, have also influenced subspace perturbation results of structured matrices and structured perturbations, such as Hamiltonian, skew-Hamiltonian, and block cyclic matrices [22, 84]. The bounds have also been applied to analyzing eigenvectors associated with polymer particles [52].

The algorithm for approximating dominant eigenvectors of non-negative matrices in [GWS-J48] has been applied to the analysis of transient behavior in the context of broadband networks [48].

The bounds for eigenvalues of Hermitian matrices in [GWS-J70] motivated the search for improved bounds with less stringent conditions [96].

Sun [136, 138] continued the work on condition numbers of multiple eigenvalues started by Stewart and Zhang in [GWS-J71].

The approach for analyzing the perturbed power method in [GWS-J114] was adopted in [44] to establish a corresponding result for nonlinear operators.

8

The SVD, Eigenproblem, and Invariant Subspaces: Algorithms

James W. Demmel

---

1. [GWS-J5] "Accelerating the Orthogonal Iteration for the Eigenvectors of a Hermitian Matrix," *Numerische Mathematik* **13** (1969) 362–376.
2. [GWS-J30] "Simultaneous Iteration for Computing Invariant Subspaces of Non-Hermitian Matrices," *Numerische Mathematik* **25** (1976) 123–136.
3. [GWS-J33] "Algorithm 506: HQR3 and EXCHNG: FORTRAN Subroutines for Calculating and Ordering the Eigenvalues of a Real Upper Hessenberg Matrix," *ACP Transactions on Mathematical Software* **2** (1976) 275–280.
4. [GWS-J37] (with C. A. Bavely) "An Algorithm for Computing Reducing Subspaces by Block Diagonalization," *SIAM Journal on Numerical Analysis* **16** (1979) 359–367.
5. [GWS-J75] (with R. Mathias) "A Block QR Algorithm and the Singular Value Decomposition," *Linear Algebra and its Applications* **182** (1993) 91–100.
6. [GWS-J102] "The QLP Approximation to the Singular Value Decomposition," *SIAM Journal on Scientific Computation* **20** (1999) 1336–1348.
7. [GWS-J107] (with Z. Jia) "An Analysis of the Rayleigh-Ritz Method for Approximating Eigenspaces," *Mathematics of Computation* **70** (2001) 637–647.

These papers form a sample of Stewart's work on algorithms for the eigenvalue problem, in particular computing selected invariant subspaces and corresponding eigenvalues and eigenvectors of both Hermitian and non-Hermitian matrices. We also include some contributions to the SVD. Algorithms for the generalized nonsymmetric eigenproblem, in particular the QZ algorithm, are discussed in Sect. 9.2.

M.E. Kilmer and D.P. O'Leary (eds.), *G.W. Stewart: Selected Works with Commentaries*, Contemporary Mathematicians, DOI 10.1007/978-0-8176-4968-5_8,

There are three themes we wish to highlight. The first theme is subspace iteration, for both Hermitian and non-Hermitian matrices. The emphasis here is on convergence analysis, which is closely related to Stewart's work on perturbation theory discussed in Chapt. 7 since one can view the difference between the current iterate and a fully converged version as a perturbation that is getting smaller. This first theme is covered in papers [GWS-J5,GWS-J30,GWS-J107].

The second theme is manipulating decompositions of non-Hermitian matrices, so as to extract information about desired invariant subspaces. Unlike Hermitian matrices, all of whose eigenvectors are orthogonal, eigenvectors of non-Hermitian matrices may point in arbitrarily close directions, and there may not even exist enough linearly independent eigenvectors to form a basis for $\mathbb{C}^n$ (e.g., when there are Jordan blocks of size 2×2 or larger). So computing eigenvectors (or more generally bases of invariant subspaces) involves more complicated tradeoffs in the non-Hermitian case than in the Hermitian case. Such tradeoffs are the subject of papers [GWS-J33,GWS-J37].

The third theme involves inexpensive approximations for the singular value decomposition (SVD). The SVD is the ultimate way to compute accurate bases for null-spaces and range-spaces of matrices, or to understand how sensitive these spaces are to perturbations in the matrix [GWS-J49]. But the SVD is just as expensive to update when the matrix changes slightly, say by adding a row, as it is to compute from scratch: either way the cost is $O(n^3)$. So the papers on this theme discuss inexpensive alternatives that approximate the SVD, and also how to refine them inexpensively (see also [GWS-J77] for motivation). Papers [GWS-J75, GWS-J102] fall into this category.

8.1. Who Invented Subspace Iteration?

Subspace iteration is a powerful and natural generalization of the power method for finding the "dominant" eigenvalue and eigenvector of a matrix, where dominant means of largest absolute value. In its simplest form, given an $n \times n$ matrix $\mathbf{A}$, the power method starts with a vector $\mathbf{x}_0$ and repeatedly forms $\mathbf{x}_i = \mathbf{A}\mathbf{x}_{i-1}$. If the eigenvalues λ_j of $\mathbf{A}$ satisfy $|\lambda_1| > |\lambda_2| \geq |\lambda_3| \geq \ldots \geq |\lambda_n|$, then $\mathbf{x}_i$ converges (in direction) to the eigenvector of λ_1, with error decreasing like $|\lambda_2/\lambda_1|^i$. It is common to "normalize" $\mathbf{x}_i$ by dividing by its norm, to simplify convergence testing.

The first and simplest generalization of the power method to an entire subspace is attributed to Bauer [11], who called it "stabilized simultaneous iteration," a variant of the "Treppeniteration." Bauer's algorithm is as follows. Instead of starting with a vector $\mathbf{x}_0$, we start with an $n \times p$ orthogonal matrix $\mathbf{Q}_0$ (with $p < n$), and repeat

(1) Form the product $\mathbf{Z}_i = \mathbf{A}\mathbf{Q}_{i-1}$.
(2) Form an orthogonal $\mathbf{Q}_i$ whose columns span the same space as $\mathbf{Z}_i$ (perhaps by computing the QR decomposition $\mathbf{Z}_i = \mathbf{Q}_i\mathbf{R}_i$).

If the $p+1$ largest eigenvalues of $\mathbf{A}$ have distinct absolute values, then Bauer showed that each column $j \le p$ of $\mathbf{Q}_i$ converges linearly to the eigenvector for λ_j, the error decreasing at every iteration by the factor

$$\max\left(\frac{|\lambda_j|}{|\lambda_{j-1}|}, \frac{|\lambda_{j+1}|}{|\lambda_j|}\right), \tag{8.1}$$

with eigenvalues converging at a similar rate.

Given a subspace spanned by the columns of $\mathbf{Q}_i$, it is natural to ask if any better approximations than merely its columns can be extracted from it. With hindsight, numerical analysts may all call out "use the Rayleigh quotient!" but it took until the late 1960s, when four researchers made this observation nearly simultaneously for the Hermitian case: Jennings [75], Rutishauser [119], Schwarz [126], and Stewart [GWS-J5]. In brief, the algorithm augments some or all steps of Treppeniteration by the following procedure:

(1) Form $\mathbf{Z}_i = \mathbf{A}\mathbf{Q}_{i-1}$.
(2) Form $\mathbf{B}_i = \mathbf{Q}_{i-1}^{\mathrm{H}}\mathbf{Z}_i$ (the Rayleigh-Quotient).
(3) Form the eigendecomposition $\mathbf{B}_i = \mathbf{U}_i\mathbf{\Lambda}_i\mathbf{U}_i^{\mathrm{H}}$ with the eigenvalues appearing on the diagonal of $\mathbf{\Lambda}_i$ in decreasing order of absolute value; these are the approximate eigenvalues of $\mathbf{A}$.
(4) Form $\mathbf{V}_i = \mathbf{Q}_{i-1}\mathbf{U}_i$; the columns of $\mathbf{V}_i$ are the approximate eigenvectors.

Note that it is still necessary to form $\mathbf{Q}_i$; this is computed as inexpensively as before, from the QR decomposition of $\mathbf{Z}_i$.

Both Stewart and Rutishauser independently showed the following much stronger convergence result holds for the above iteration than for Treppeniteration: Assuming $|\lambda_j| > |\lambda_{p+1}|$, the error in the jth eigenvalue decreases by a factor of

$$|\lambda_{p+1}/\lambda_j|^2 \tag{8.2}$$

at every iteration. This can be arbitrarily faster than the convergence rate (8.1) for the above Treppeniteration, both because the numerator is the first eigenvalue not in the subspace being computed and hence is smaller than before, and because the ratio is squared. The eigenvectors converge at the square root of this rate. Also, the convergence of an eigenvector depends on how well separated its eigenvalue (call it λ_j) is from the other eigenvalues; this is not surprising since as the difference $\min_{k\neq j}|\lambda_k - \lambda_j|$ shrinks, the eigenvector becomes more ill-conditioned. When this difference is zero, the eigenvector is not even uniquely defined, though the invariant subspace spanned by all the eigenvectors with eigenvalue λ_j is well defined. Given this potential difficulty, Stewart concludes the paper with an extension of his convergence theory to spaces spanned by eigenvectors of clusters of close eigenvalues.

It is of interest to recount how this idea came to be invented simultaneously by four researchers. This work constitutes half of Stewart's PhD dissertation [GWS-T1].[21] Stewart and Rutishauser do not refer to one another (Stewart

[21] The other half concerns Lehmer's method for finding the roots of a polynomial.

having submitted to the same journal a few months before Rutishauser, at roughly the same time he received his PhD, though his article appeared a few months after Rutishauser's). Only Rutishauser refers to Jennings (who only approximated Rayleigh–Ritz) and Schwarz (who did not analyze convergence). Although identical in exact arithmetic, Rutishauser's variant involved the eigendecomposition of $\mathbf{Z}_i^{\mathrm{T}}\mathbf{Z}_i$ instead of $\mathbf{B}_i$ as in Stewart's formulation; the former approach squares the eigenvalues and therefore can be less numerically stable if the eigenvalues span a wide range of values.

All of this was for the Hermitian case; in [GWS-J30] and later in [GWS-J107], Stewart takes on the harder non-Hermitian case. In [GWS-J30], Stewart considers how to extend the Rayleigh–Ritz procedure from the Hermitian to the non-Hermitian case. A natural idea is to replace the above step (3) with the eigendecomposition of $\mathbf{B}_i = \mathbf{U}_i\mathbf{\Lambda}_i\mathbf{U}_i^{-1}$, where $\mathbf{U}_i$ is of course no longer unitary. Variants of this were proposed by Bauer [11], Clint and Jennings [25], and later by Jia and Stewart in [GWS-J107]. However, $\mathbf{U}_i$ can be arbitrarily ill-conditioned, or not even exist if $\mathbf{B}_i$ has nontrivial Jordan blocks, so in [GWS-J30] Stewart proposes maintaining the orthogonality of $\mathbf{U}_i$ and settling for convergence to the Schur form instead. In other words, step (3) becomes

(3') $\mathbf{B}_i = \mathbf{U}_i\mathbf{T}_i\mathbf{U}_i^{\mathrm{H}}$,

where $\mathbf{U}_i$ is again unitary, but $\mathbf{T}_i$ is upper triangular, i.e., in Schur form. Stewart calls this "SRR iteration", where SRR is short for Schur–Rayleigh–Ritz. Thus, the columns of $\mathbf{V}_i$ in step (4) will converge to the Schur vectors, rather than the eigenvectors. In the Hermitian case, these are the same. Yousef Saad [122] remarks, "The main contribution of the paper was to show that one could — and should — use subspace iteration with the Schur vectors instead of eigenvectors. This was a novel idea and it became clear later that this was indeed a better approach."

In [GWS-J30], Stewart is able to extend the convergence theory of [GWS-J5] to the non-Hermitian case. The convergence ratio $|\lambda_{p+1}/\lambda_j|$ used in (8.2) is shown to apply in some sense to the leading $j\times j$ submatrix of the matrix $\mathbf{T}_i$ in line (3'), which is converging to Schur form. Specifically, Stewart proves under mild conditions that the subspace spanned by the leading j columns of $\mathbf{V}_i$ is converging at this rate to the invariant subspace spanned by the eigenvectors of the j eigenvalues largest in absolute value.

In [GWS-J107], instead of the Schur form, convergence to "eigenpairs" is considered, where $(\mathbf{N},\mathbf{X})$ is an ℓ-dimensional eigenpair of an $n\times n$ matrix $\mathbf{A}$ if $\mathbf{AX}=\mathbf{XN}$ and $\mathbf{X}^{\mathrm{H}}\mathbf{X}=\mathbf{I}_\ell$; in other words the columns of $\mathbf{X}$ form an orthonormal basis of an invariant subspace of $\mathbf{A}$, and if $\mathbf{A}$ were put into Schur form with the first ℓ Schur vectors spanning the same space as $\mathbf{X}$, then the leading $\ell\times\ell$ submatrix of the Schur form would be unitarily similar to $\mathbf{N}$. For this algorithm, line (3) in the subspace iteration algorithm above is replaced by

(3") Let $(\mathbf{N}_i,\mathbf{U}_i)$ be an ℓ-dimensional eigenpair of $\mathbf{B}_i$, where $\mathbf{N}_i$ is chosen to have eigenvalues closest to those of a desired eigenpair $(\mathbf{N},\mathbf{X})$.

For example, if $\mathbf{N}$ is supposed to contain the rightmost ℓ eigenvalues of $\mathbf{A}$ in the complex plane, $\mathbf{N}_i$ will contain the rightmost ℓ eigenvalues of $\mathbf{B}_i$.

Convergence of this iteration is studied as a function of $\epsilon = \sin(\mathbf{X}, \mathbf{Q}_i)$, the sine of the angle between the desired invariant subspace spanned by the columns of $\mathbf{X}$ and its approximation, spanned by the columns of $\mathbf{Q}_i$. Note that $\mathbf{Q}_i$ may have $p > \ell$ columns, i.e., more than $\mathbf{X}$, so that we must define

$$\sin(\mathbf{X}, \mathbf{Q}_i) = \max_{\mathbf{x} \in \text{range}(\mathbf{X})} \min_{\mathbf{q} \in \text{range}(\mathbf{Q}_i)} \sin(\mathbf{x}, \mathbf{q}).$$

In the Hermitian case, the convergence for the eigenvector of a single eigenvalue λ_j depended on the gap $\min_{k \neq j} |\lambda_k - \lambda_j|$ between λ_j and all the other eigenvalues; when this was small, convergence was slow. The analog of this simple gap for the non-Hermitian case uses the sep operator, defined earlier in (7.5), which measures the "separation" of the spectra of two square matrices. It is zero if and only if the matrices have a common eigenvalue. Suppose we put $\mathbf{Q}_i^{\mathrm{H}} \mathbf{A} \mathbf{Q}_i$ into Schur form

$$\begin{bmatrix} \mathbf{N}_i & \mathbf{E}_i \\ \mathbf{0} & \mathbf{C}_i \end{bmatrix},$$

where $\mathbf{N}_i$ is the $\ell \times \ell$ matrix whose eigenvalues are our desired approximations, and $\mathbf{C}_i$ is the $(p-\ell) \times (p-\ell)$ matrix of other eigenvalues. Then, to prove convergence, Jia and Stewart assume the *uniform separation condition*, that there is some constant $\alpha > 0$ such that for all i

$$\text{sep}(\mathbf{N}_i, \mathbf{C}_i) \geq \alpha.$$

This is the analog of assuming that the gap $\min_{k \neq j} |\lambda_k - \lambda_j|$ is bounded away from zero. Using this assumption they bound the error in the pair $(\mathbf{N}_i, \mathbf{X}_i)$ by an expression proportional to ϵ/α.

8.2. Extracting Invariant Subspaces

The subspace iteration algorithms of the last section are particularly well suited when it is computationally feasible to do matrix-vector products quickly, as is the case with sparse matrices. The algorithms from [GWS-J33] and [GWS-J37] discussed in this section compute invariant subspaces of matrices that are already in Schur form, and they require applying similarity transformations to the full matrix. Such matrices arise not just when starting with a dense nonsymmetric matrix whose invariant subspaces are desired, but in the course of nonsymmetric subspace iteration algorithms used above in [GWS-J30,GWS-J107] and in the Krylov–Schur algorithms of Sect. 10.1 and [GWS-J111,GWS-J113].

Suppose the matrix $\mathbf{A} = \mathbf{Q}\mathbf{T}\mathbf{Q}^{\mathrm{T}}$ has been reduced to real Schur form, where $\mathbf{Q}$ is orthogonal and

$$\mathbf{T} = \begin{bmatrix} \mathbf{T}_{11} & \mathbf{T}_{12} & \mathbf{T}_{13} \\ \mathbf{0} & \mathbf{T}_{22} & \mathbf{T}_{23} \\ \mathbf{0} & \mathbf{0} & \mathbf{T}_{33} \end{bmatrix}$$

is in real Schur form, $\mathbf{T}_{ii}$ is square of dimension n_i, and none of the diagonal blocks have any eigenvalues in common. Suppose one wants to compute an orthonormal basis of the n_1-dimensional invariant subspace associated with the eigenvalues of $\mathbf{T}_{11}$. Then, $\mathbf{AQ} = \mathbf{QT}$ implies that the first n_1 columns of $\mathbf{Q}$ form the desired basis. But what about a basis of the invariant subspace for the eigenvalues of $\mathbf{T}_{22}$ or $\mathbf{T}_{33}$? The Schur form is not unique since the eigenvalues can appear on the diagonal of $\mathbf{T}$ in any order. Stewart's approach in [GWS-J33] is to very inexpensively compute an orthogonal matrix to reorder the eigenvalues of $\mathbf{T}$ in any desired way. In other words, Stewart shows how to inexpensively compute an orthogonal $\mathbf{U}$ such that $\mathbf{T} = \mathbf{U}\hat{\mathbf{T}}\mathbf{U}^{\mathrm{T}}$ is also in Schur form, and the first n_2 eigenvalues on the diagonal of $\hat{\mathbf{T}}$ are the desired eigenvalues of $\mathbf{T}_{22}$. So from $\mathbf{A} = \mathbf{QU}\hat{\mathbf{T}}(\mathbf{QU})^{\mathrm{T}}$ we see that the first n_2 columns of $\mathbf{QU}$ span the desired invariant subspace.

Stewart's algorithm works analogously to bubble sort: if $\mathbf{T}_{ii}$ is an $m_i \times m_i$ diagonal block of $\mathbf{T}$ (so m_i is either 1 or 2) and $\mathbf{T}_{i+1,i+1}$ is an adjacent $m_{i+1} \times m_{i+1}$ diagonal block, then Stewart shows how to apply an $(m_i + m_{i+1}) \times (m_i + m_{i+1})$ orthogonal transformation (call it $\mathbf{U}_{i,i+1}$) just to the rows and columns of $\mathbf{T}$ containing $\mathbf{T}_{ii}$ and $\mathbf{T}_{i+1,i+1}$, to swap their locations on the diagonal. By such repeating swapping of adjacent blocks, any ordering can be achieved.

So how can this small (2×2, 3×3 or 4×4) transformation $\mathbf{U}_{i,i+1}$ be computed? Stewart uses the Hessenberg QR iteration algorithm itself, using the eigenvalues of the $\mathbf{T}_{ii}$ as shifts, so that they converge at the bottom right corner as desired. Numerical stability is guaranteed, since only orthogonal transformations are ever used.

But there is a question of convergence, since Hessenberg QR has no convergence guarantees [32], and if the eigenvalues of $\mathbf{T}_{ii}$ and $\mathbf{T}_{i+1,i+1}$ are sufficiently close, then convergence may not occur. (Stewart points out this possibility, and reasonably suggests that their order is therefore not numerically so well defined anyway.) Motivated by the desire to have a closed-form formulas with no convergence issues, several authors developed methods [23, 38, 112, 118]. But there were highly ill-conditioned examples where these closed-form formulas were numerically unstable, unlike Stewart's method.

So the ultimate goal was to combine the guaranteed timing of a closed-form formula with the guaranteed numerical stability of Stewart's approach using orthogonal transformations. The eventual solution used in LAPACK [1] managed to achieve both these goals by using a closed-form formula to compute an orthogonal transformation and explicitly testing to see if the swapping actually occurred stably [5]. It has yet to fail.

Stewart considers a related approach to this problem in [GWS-J37], in a paper co-authored by his student Connice Bavely. Suppose we can find a nonsingular matrix $\mathbf{X} = [\mathbf{X}_1, \ldots, \mathbf{X}_s]$, where $\mathbf{X}_i$ is $n \times n_i$, such that $\mathbf{X}^{-1}\mathbf{AX} = \mathbf{B} = \mathrm{diag}(\mathbf{B}_1, \mathbf{B}_2, \ldots, \mathbf{B}_s)$ where $\mathbf{B}_i$ is $n_i \times n_i$. Then we see that the columns of each $\mathbf{X}_i$ span an invariant subspace of $\mathbf{A}$ whose eigenvalues are those of $\mathbf{B}_i$. This simultaneous use of many invariant subspaces thus lets us block-diagonalize $\mathbf{A}$ and is useful for problems like computing functions of matrices, since, for example,

$$\mathbf{A}^k = (\mathbf{X}\mathbf{B}\mathbf{X}^{-1})^k = \mathbf{X}\mathbf{B}^k\mathbf{X}^{-1} = \mathbf{X}\,\mathrm{diag}(\mathbf{B}_1^k, \ldots, \mathbf{B}_s^k)\mathbf{X}^{-1} \tag{8.3}$$

reduces the problem of computing the kth power of a large matrix $\mathbf{A}$ to the same problem for smaller matrices $\mathbf{B}_i$.

We know, of course, that block-diagonalization is not always possible, for example, if $\mathbf{A}$ is a single Jordan block. But even if all the eigenvalues of $\mathbf{A}$ are distinct (imagine adding distinct tiny perturbations to the diagonal of a Jordan block), it might not be possible to block-diagonalize when $\mathbf{X}$ is so ill-conditioned that a computation like (8.3) that depends on $\mathbf{X}^{-1}$ is hopelessly inaccurate. For example, if $\mathbf{A} = \begin{bmatrix} 1 & 1 \\ 0 & 1+\epsilon \end{bmatrix}$, then $\mathbf{X}$ must be of the form $\begin{bmatrix} x & y/\epsilon \\ 0 & y \end{bmatrix}$ where x and y are arbitrary nonzero constants; no matter how x and y are chosen, the condition number of $\mathbf{X}$ grows at least as fast as $1/\epsilon$, and so is arbitrarily large when ϵ is small.

So the overall goal is twofold: (1) given a particular block diagonalization (i.e., partitioning of the spectrum of $\mathbf{A}$), find the best conditioned $\mathbf{X}$ that achieves it, and (2) find the finest possible partitioning of the spectrum that permits the condition number of $\mathbf{X}$ to be below some desired threshold τ.

In [GWS-J37], Stewart and Bavely address the second goal by using an interesting heuristic, which can be described as a greedy algorithm for finding the finest possible partitioning. Given a matrix $\mathbf{T}$ in Schur form, they first ask whether it is possible to divide the spectrum into two pieces, one with the eigenvalue at $\mathbf{T}$'s upper left (call it λ_1), and one with the rest. They do this by computing an explicit block-diagonalizing $\mathbf{X}$, with the help of the Bartels–Stewart algorithm from [GWS-J17] (see Sect. 4.4), comparing its condition number against the threshold τ. If the condition number of $\mathbf{X}$ is less than τ, they perform the partitioning and apply the same greedy algorithm to the trailing submatrix of $\mathbf{T}$. If this does not work, they look for another eigenvalue λ_j (or pair of eigenvalues) of $\mathbf{T}$ that is as close to λ_1 as possible, reorder it using Stewart's above technique from [GWS-J33] to be adjacent to λ_1, and again ask if the pair (λ_1, λ_j) can be split off from the rest of $\mathbf{T}$ with an acceptably conditioned $\mathbf{X}$. This repeats recursively, either splitting off a block at the upper left of $\mathbf{T}$, or finding an eigenvalue closest to an average of the eigenvalues from the upper left block and moving it to be adjacent to the upper left block, and trying again.

This is a useful heuristic that often works, as the numerical results show. Subsequent work clarified the difficulty of the problem and the optimal conditioning of $\mathbf{X}$, as described later.

8.3. Approximating the SVD

The two papers [GWS-J75] and [GWS-J102] were written 6 years apart; in retrospect they make an interesting pair to compare using a common notation. Both have the goal of taking an upper triangular matrix $\mathbf{R}_0 = \begin{bmatrix} \mathbf{S}_0 & \mathbf{H}_0 \\ \mathbf{0} & \mathbf{E}_0 \end{bmatrix}$ where the singular values of $\mathbf{S}_0$ are good approximations of the largest singular values of $\mathbf{R}_0$, and $\mathbf{H}_0$ and $\mathbf{E}_0$ are smaller, and producing a unitarily equivalent sequence $\mathbf{R}_i$ of upper triangular matrices that accentuate this property, so that the singular values of $\mathbf{S}_i$ converge to the largest singular values. The algorithm in [GWS-J75] is stated simply as follows (using MATLAB notation):

(1) $[\mathbf{Q}_0, \hat{\mathbf{R}}] = qr(\mathbf{R}_i^{\mathrm{T}})$.
(2) $[\mathbf{Q}_1, \mathbf{R}_{i+1}] = qr(\hat{\mathbf{R}}^{\mathrm{T}})$.

Here, the notation $qr(\mathbf{A})$ returns the $\mathbf{Q}$ and $\mathbf{R}$ factors from the QR decomposition of $\mathbf{A}$. Thus, the SVDs of $\mathbf{R}_i$ and $\mathbf{R}_{i+1} = \mathbf{Q}_1^{\mathrm{T}}\mathbf{R}_i\mathbf{Q}_0$ are simply related. Furthermore, since we can write the LQ decomposition of $\mathbf{R}_i^{\mathrm{T}}\mathbf{R}_i = (\mathbf{R}_i^{\mathrm{T}}\hat{\mathbf{R}}^{\mathrm{T}})\mathbf{Q}_0^{\mathrm{T}} \equiv \mathbf{LQ}$, then we see

$$\mathbf{QL} = \mathbf{Q}_0^{\mathrm{T}}\mathbf{R}_i^{\mathrm{T}}\hat{\mathbf{R}}^{\mathrm{T}} = \hat{\mathbf{R}}\hat{\mathbf{R}}^{\mathrm{T}} = \mathbf{R}_{i+1}^{\mathrm{T}}\mathbf{Q}_1^{\mathrm{T}}\mathbf{Q}_1\mathbf{R}_{i+1} = \mathbf{R}_{i+1}^{\mathrm{T}}\mathbf{R}_{i+1},$$

so that the algorithm is equivalent to one step of "LQ iteration" (an analog of QR iteration) on the symmetric positive definite matrix $\mathbf{R}_i^{\mathrm{T}}\mathbf{R}_i$. Thus Mathias and Stewart can apply convergence theory developed for the QR iteration. This insight also lets us rewrite the algorithm in the mathematically equivalent way

(1) $\hat{\mathbf{R}} = chol(\mathbf{R}_i\mathbf{R}_i^{\mathrm{T}})$.
(2) $\mathbf{R}_{i+1} = chol(\hat{\mathbf{R}}\hat{\mathbf{R}}^{\mathrm{T}})$.

where *chol* is the function that returns the Cholesky factor of its argument.

In [GWS-J102], Stewart makes the following observation: if QR (or Cholesky) is good for convergence, then QR (or Cholesky) with pivoting is probably even better. The pivoted QR version is as follows (where the MATLAB notation $[\mathbf{Q}, \mathbf{R}, \mathbf{P}] = qr(\mathbf{A})$ means that $\mathbf{A} = \mathbf{QRP}$ is the QR decomposition with column pivoting):

(1) $[\mathbf{Q}_0, \hat{\mathbf{R}}, \mathbf{P}_0] = qr(\mathbf{R}_i^{\mathrm{T}})$.
(2) $[\mathbf{Q}_1, \mathbf{R}_{i+1}, \mathbf{P}_1] = qr(\hat{\mathbf{R}}^{\mathrm{T}})$.

so that $\mathbf{R}_{i+1} = \mathbf{Q}_1^{\mathrm{T}}\mathbf{P}_0\mathbf{R}_i\mathbf{Q}_0\mathbf{P}_1^{\mathrm{T}}$ and $\mathbf{R}_i$ again have simply related SVDs. Letting $cholpiv(\mathbf{A})$ denote the upper triangular Cholesky factor of $\mathbf{A}$ obtained with complete (i.e., diagonal) pivoting, we may also write this iteration as:

(1) $\hat{\mathbf{R}} = cholpiv(\mathbf{R}_i\mathbf{R}_i^{\mathrm{T}})$.
(2) $\mathbf{R}_{i+1} = cholpiv(\hat{\mathbf{R}}\hat{\mathbf{R}}^{\mathrm{T}})$.

Once the algorithm has "settled" down so that the pivot matrices $\mathbf{P}_0$ and $\mathbf{P}_1$ are all the identity, the convergence analysis is identical to that of unshifted QR. The advantage is in the early steps, when the rank may be revealed more quickly than without pivoting. Indeed, Stewart makes the point that the intention is to run for just a few steps, as he illustrates with a number of interesting examples. For example, the well-known upper triangular Kahan matrix [78] has a tiny but unrevealed singular value, yet remains invariant under QR with column pivoting, whereas Stewart's algorithm reveals the rank after just one step.

8.4. Impact

Regarding the work of Stewart and others on subspace iteration, with hindsight we may well ask why not use (block) Lanczos or Arnoldi? Why throw away the information contained in all the previously computed approximate bases obtained from $\mathbf{Q}_{i-1}, \ldots, \mathbf{Q}_1$? At this time, Lanczos was not yet considered a reliable method because of the tendency of the allegedly orthogonal vectors it computed to not only just lose orthogonality, but also to become linearly dependent. Subspace iteration was the natural alternative. It was not until Paige's thesis [103,104] that this loss of orthogonality was analyzed and seen to be a friend (indicating convergence) instead of an enemy. It took even longer for the introduction of techniques like selective orthogonalization [113] to make reliable implementations of Lanczos available. But Parlett [114] points out reasons that subspace iteration remains useful even today. First, if no secondary storage is available, and fast storage can only hold a few n-vectors at a time, then it may be reasonable to discard previous vectors and keep new ones. Second, if the gap between the desired and undesired eigenvalues is sufficiently large, then very few steps will be needed for convergence. Finally, Rutishauser [119] shows that some of the convergence benefits of Lanczos can be retained (though he does not use this language) by multiplying $\mathbf{Q}_0$ not by $\mathbf{A}^k$, but by $p_k(\mathbf{A})$, a suitably chosen Chebyshev polynomial, which can of course be done efficiently with a three-term recurrence. Zhongxiao Jia [76] says of [GWS-J107], "Our collaboration was an memorable experience and produced an influential paper which has been cited many times, as Pete said and anticipated."

One of the best available implementations of subspace iteration for nonsymmetric matrices was written by Bai and Stewart [GWS-J76].

The problem of swapping diagonal blocks on the diagonal of a matrix in real Schur form is a fundamental subproblem of the nonsymmetric eigenproblem, and Stewart [GWS-J33] supplied the first practical, and guaranteed numerically stable, solution. It can fail to converge on rare, highly ill-conditioned problems, where the eigenvalues are so close that one should not try to reorder them anyway (since their values are interchangeably close). Combining guaranteed numerical stability with the guaranteed timing of a closed-form formula came later, but for many years Stewart's solution was the standard.

The difficulty of the problem of block-diagonalizing a matrix with a well-conditioned similarity transform was not completely understood until several years after Stewart's contributions in [GWS-J37]. In [33], it was determined how to produce a nearly optimally conditioned $\mathbf{X}$ that attains a given block-partitioning; not surprisingly each set of columns $\mathbf{X}_i$ spanning each invariant subspace should be chosen to be orthonormal. With this choice, the minimal condition number is roughly the largest norm of any spectral projector for any of the invariant subspaces. In contrast, in [GWS-J37] Stewart's choice of $\mathbf{X}$ could have the square of this minimal condition number. And the difficulty of the second goal, choosing the finest possible partitioning achievable with an $\mathbf{X}$ of condition number at most τ, was first completely explained by Gu in [60], who proved a conjecture of Demmel that the partitioning problem is NP-hard, even to divide it into just two pieces. This means that an efficient solution will have to be approximate, using heuristics like those of Stewart.

Stewart's work in [GWS-J75] and [GWS-J102] focused on locating gaps in the sequence of singular values and then obtaining good approximations to the large ones. He was motivated by applications requiring low-rank approximations to matrices, a strong theme throughout his work. In [GWS-J102], he distinguishes between what he calls *superior* and *inferior* singular subspaces, those corresponding to the largest and smallest singular values, respectively. Although he did not discuss the applications, such approximations are useful in the numerical solution of discretizations of ill-posed problems, where computing an approximate solution restricted to the right superior subspace can reduce the influence of noise in the measurements; see, for example, [63] for more information on these *truncated SVD methods* proposed in [56]. He also used these ideas in his work on subspace tracking (e.g., [GWS-J85]).

Stewart was aware of the efforts to compute reliable rank-revealing QR decompositions, and he remarks in [GWS-J102] that his algorithm "cleans up" the values produced by these algorithms. Further analysis on this was done by Huckaby and Chan [74].

A major theme in Stewart's work on perturbation in eigenproblems was the focus on appropriately-chosen invariant subspaces, rather than individual eigenvectors, when eigenvalues are clustered. Stewart viewed the perturbation of singular subspaces in a similar light. This point of view was not unique to him; along with Stewart, others including Davis and Kahan [31] and Varah [148] also advocated the analysis of perturbations of invariant subspaces in the late 1960s. However, it was through the clear exposition in Stewart's first textbook [GWS-B1] that this work became widely understood and used, and this, too, was an important contribution.

In Sect. 9.1, we will see how Stewart carried this same theme over to the generalized eigenproblem.

9

The Generalized Eigenproblem

Zhaojun Bai

1. [GWS-J16] "On the Sensitivity of the Eigenvalue Problem $Ax = \lambda Bx$," *SIAM Journal on Numerical Analysis* **9** (1972) 669–686.
2. [GWS-J18] (with C. B. Moler), "An Algorithm for Generalized Matrix Eigenvalue Problems," *SIAM Journal on Numerical Analysis* **10** (1973) 241–256.
3. [GWS-J27] "Gershgorin Theory for the Generalized Eigenvalue Problem $Ax = \lambda Bx$," *Mathematics of Computation* **29** (1975) 600–606.
4. [GWS-J38] "Perturbation Bounds for the Definite Generalized Eigenvalue Problem," *Linear Algebra and its Applications* **23** (1979) 69–85.

These papers describe Stewart's original and fundamental contributions on the generalized matrix eigenvalue problem. In these papers, Stewart systematically presented perturbation theory and sensitivity analysis for the problem, and (with Moler) introduced a landmark algorithm, namely the QZ algorithm, for computing eigenvalues and eigenvectors.

The generalized eigenvalue problem is that of finding the set of all $\boldsymbol{\lambda}$ for which the equation

$$\mathbf{A}\mathbf{x} = \lambda \mathbf{B}\mathbf{x}$$

has a nontrivial solution $\mathbf{x}$, where $\mathbf{A}$ and $\mathbf{B}$ are $n \times n$ with real or complex elements. The problem reduces to the ordinary eigenvalue problem when $\mathbf{B} = \mathbf{I}$. It is a natural formulation for many practical applications.

Although the generalization results from the simple replacement of an identity matrix by an arbitrary matrix $\mathbf{B}$, the problem has many features not shared with the ordinary eigenvalue problem: for example, a generalized eigenvalue problem

M.E. Kilmer and D.P. O'Leary (eds.), *G.W. Stewart: Selected Works with Commentaries*, Contemporary Mathematicians, DOI 10.1007/978-0-8176-4968-5_9,

can have infinite eigenvalues. Stewart was early to notice the importance of the generalized problem and to understand both the commonality and the essential differences between the two problems.

9.1. Perturbation Theory

In [GWS-J16], Stewart was among the first to study the sensitivity of the eigenvalues and the eigenvectors of the generalized eigenvalue problem to perturbations in **A** and **B**. In this paper, he presented a number of fundamental concepts and results for the problem.

- He introduced the notion of a *deflating subspace* and showed that a deflating subspace behaves much like an invariant subspace for the ordinary eigenvalue problem. The deflating subspace is later called an *eigenspace* in Stewart and Sun [GWS-B3].
- He introduced the *generalized Schur decomposition*, showing that there exist unitary matrices **U** and **V** so that $\mathbf{U}^*\mathbf{AV}$ and $\mathbf{U}^*\mathbf{BV}$ are both upper triangular. This is a generalization of the reduction of a matrix to triangular Schur form by a unitary similarity transformation. This reduction gives the theoretical foundation for the QZ algorithm that he and Moler designed.
- He introduced the "dif" function for deflating subspaces of $\mathbf{A} - \lambda\mathbf{B}$, analogous to the sep function for the conventional eigenvalue problem.
- He derived error bounds for approximate deflating subspaces. These bounds also provide information about the sensitivity of eigenvalues. The resulting perturbation bounds estimate realistically the sensitivity of the eigenvalues, even when **B** is singular or nearly singular.
- In the last section of the paper, he specialized the results to the important special case where **A** is Hermitian and **B** is positive definite, which is the subject of his further study in [GWS-J38].

This is a seminal work with an excellent development of just the right fundamental concepts, essential matrix decompositions, and elegant perturbation theory. It is written in a clear style, with good taste in its choice of results. It listed a number of open problems and inspired many other results on the perturbation analysis of the generalized eigenvalue problem, including, e.g., Bauer–Fike type theorems [45, 46, 88, 89], Hoffman–Wielandt type theorems [91, 132], stable eigendecompositions [34], and eigen-conditioning [67], and also inspired later work on definite pairs.

9.2. The QZ Algorithm

One approach to solving the generalized eigenvalue problem, when **B** is nonsingular, is to reformulate it as $\mathbf{B}^{-1}\mathbf{Ax} = \lambda\mathbf{x}$. This problem can be solved by applying the QR algorithm or the Arnoldi algorithm to the matrix $\mathbf{B}^{-1}\mathbf{A}$. This approach has major disadvantages: forming $\mathbf{B}^{-1}\mathbf{A}$ can be expensive for large problems, and, if it is not computed, then solving linear systems involving $\mathbf{B}$ must be done at each Arnoldi iteration. Moreover, any symmetry is lost, and serious stability issues arise if $\mathbf{B}$ is ill-conditioned. There was an important need for an efficient and stable algorithm, analogous to the QR algorithm, that worked directly with the data matrices $\mathbf{A}$ and $\mathbf{B}$.

Using the generalized Schur decomposition established by Stewart in [GWS-J16], Moler and Stewart [GWS-J18] cleverly discovered that it was possible to first simultaneously reduce $\mathbf{A}$ to Hessenberg form and $\mathbf{B}$ to triangular form using unitary transformations and then preserve this structure while implicitly applying the standard Hessenberg (implicit-shift) QR algorithm to $\mathbf{A}\mathbf{B}^{-1}$. Through meticulous attention to detail, they avoid many numerical pitfalls. For example, they do not compute the eigenvalues explicitly but instead represent them as ratios of diagonal entries of two matrices; by this device they avoid overflow for infinite eigenvalues and preserve more digits of accuracy. They present numerical results based on a FORTRAN implementation and note that the algorithm might be used to solve polynomial eigenvalue problems.

The paper concludes with acknowledgement to W. Kahan and J.H. Wilkinson for their helpful comments, reference to Linda Kaufman's analogous generalization of the LR algorithm [81] and to Charles Van Loan's error analysis of the QZ algorithm [143], and the presentation of a limerick.

The QZ algorithm is a backward-stable method for computing generalized eigenvalues and deflating subspaces of small- to medium-sized regular matrix pairs $(\mathbf{A}, \mathbf{B})$. It has undergone only a few modifications during the past 35 years, notably through the works by Ward [153], Kaufman [82], Kagstrom and Kressner [77], and others.

This is a very influential paper, with over 400 citations according to ISI Web of Science. Beresford Parlett says [115], "QZ is now an indispensable tool for matrix computations." Chris Paige [106] says,

> This is one of the very best papers in Numerical Linear Algebra. Before this there was no efficient, reliable and numerically stable algorithm for the general $Ax = \lambda Bx$ eigenproblem. Moler and Stewart's algorithm immediately became the gold standard, influencing us all with its theoretical beauty and practical applicability. The following anecdote will help put this in context. One of the great pleasures of working in London was the chance to listen to talks that James H. Wilkinson, FRS, would give at The National Physical Laboratory. He usually discussed work that he and his colleagues had done, but one time he gave a beautiful talk on this superb paper by Moler and Stewart. He finished by saying that this was one work that he himself would have been very proud to have produced.

The QZ algorithm is still the method of choice for solving the generalized non-Hermitian eigenvalue problem. It was first made widely available via EISPACK (extension, released on July 1975). Today, it is a fundamental algorithm in LAPACK [1]. The algorithm is widely used in applications in computational science and engineering, and even in the simulation of monetary models characterized by rational expectations and optimal monetary policy design [66].

9.3. Gershgorin's Theorem

In [GWS-J27], Stewart developed a generalization of Gershgorin's theorem for the generalized eigenvalue problem and applied it to obtain perturbation bounds for multiple eigenvalues.

- To avoid difficulties associated with infinite eigenvalues, and to symmetrize the roles of **A** and **B**, he proposed treating the generalized eigenvalue problem in the homogeneous form $\beta\mathbf{A}\mathbf{x} = \alpha\mathbf{B}\mathbf{x}$, where the eigenvalue is now represented by a pair of scalars $(\alpha, \beta) \neq (0, 0)$, corresponding to $\lambda = \alpha/\beta$. For example, when **B** is singular, a nonzero null vector of **B** is an eigenvector with $(\alpha, \beta) = (\alpha, 0)$. This formulation allows him to develop perturbation expansions for α and β individually and thus provide complete information about eigenvalue sensitivity.
- In a five-line proof, he established a remarkably simple and beautiful generalization of Gershgorin's theorem to the generalized eigenproblem, by localizing the set of possible α and β values in terms of the matrix elements, and even generalized the related result about the count of values in disjoint subregions.
- He introduced the use of the chordal metric in the perturbation theory. The results are interpreted in terms of the chordal metric on the Riemann sphere, which is especially convenient for treating infinite eigenvalues.
- He applied the Gershgorin theorem to produce a perturbation theory for the eigenvalue problem, analogous to results by Wilkinson [158] for the standard eigenproblem. These bounds were later simplified by Sun [135].

Stewart acknowledged a broad hint from Kahan in the paper and later in p. 156 of [GWS-B8]. The chordal metric introduced in this paper has become a standard tool in subsequent studies on the perturbation of other generalized eigenvalue problems, such as polynomial, periodic, and palindromic eigenvalue problems; see, for example [71].

9.4. Definite Pairs

For given $n \times n$ Hermitian matrices **A**, and **B**, the pair $(\mathbf{A}, \mathbf{B})$ is said to be a *definite pair* if

$$c(\mathbf{A}, \mathbf{B}) \equiv \inf_{\|\mathbf{x}\|=1} |\mathbf{x}^*(\mathbf{A} + i\mathbf{B})\mathbf{x}| > 0,$$

and otherwise *indefinite*. If $(\mathbf{A}, \mathbf{B})$ is definite, then both theoretical and computational advantages accrue: a definite problem has a complete system of eigenvectors and its eigenvalues are all real. The early study of definite pairs includes [2, 3, 28].

In [GWS-J38], Stewart showed that $(\mathbf{A}, \mathbf{B})$ is definite if and only if there exists a real number t such that the matrix $\mathbf{B}(t) \equiv \mathbf{A} \sin t + \mathbf{B} \cos t$ is positive definite. Subsequently, Stewart showed that under perturbations of $\mathbf{A}$ and $\mathbf{B}$, the eigenvalues behave much like the eigenvalues of a Hermitian matrix in the sense that there is a one-to-one pairing of the eigenvalues with the perturbed eigenvalues and a uniform bound for their differences (in the chordal metric). A sharper version of the theorem was obtained later in Sun [131]. In the paper, perturbation bounds were also developed for eigenvectors and eigenspaces. At the end of the article, Stewart proposed an open problem on whether certain results similar to the Davis–Kahan $\sin \Theta$ theorems [31] could be established. Sun [134] gave an positive answer to this question.

It is now taken for granted that the role played by definite pairs in the generalized eigenvalue problem is much the same as that of Hermitian matrices in the ordinary eigenvalue problems, owing much to this article. Many well-known results, including the Mirsky result [99], Davis–Kahan $\sin \Theta$ and $\sin 2\Theta$ theorems for the ordinary eigenvalue problem [31], eigenvalue approximations by Rayleigh quotients and approximate invariant subspaces, etc., were extended to definite pairs, too, in [14, 91, 92, 137]. The chordal metric was used throughout these extensions. Further related developments include results on the generalized singular value decomposition in [4, 90, 105, 109, 133, 144] and on the connection between definite pairs and matrix polynomials possessing appropriate analogs of definiteness [72].

10

Krylov Subspace Methods for the Eigenproblem

Howard C. Elman and Dianne P. O'Leary

---

1. [GWS-J111] "A Krylov-Schur Algorithm for Large Eigenproblems," *SIAM Journal on Matrix Analysis and Applications* **23** (2001) 601–614.
2. [GWS-J113] "Addendum to 'A Krylov-Schur Algorithm for Large Eigenproblems'," *SIAM Journal on Matrix Analysis and Applications* **24** (2002) 599–601.
3. [GWS-J110] "Backward Error Bounds for Approximate Krylov Subspaces," *Linear Algebra and its Applications* **340** (2002) 81–86.
4. [GWS-J112] "Adjusting the Rayleigh Quotient in Semiorthogonal Lanczos Methods," *SIAM Journal on Scientific Computing* **24** (2002) 201–207.

These papers comprise some of Stewart's recent contributions to the development and analysis of iterative algorithms based on Krylov subspace methods for computing eigenvalues. The task is to compute a few solutions (eigenvalues and eigenvectors) of the eigenvalue problem

$$\mathbf{A}\mathbf{v} = \lambda\mathbf{v},$$

where $\mathbf{A}$ is an $n \times n$ matrix, referenced through a matrix-vector product $\mathbf{y} \leftarrow \mathbf{A}\mathbf{x}$. The focus in this work is on generalizing existing methods to clarify their properties and enhance stability. Much of this work was later integrated in a uniform manner into his volume on eigensystems (Chap. 5 of [GWS-B8]).

10.1. A Krylov–Schur Algorithm

The standard techniques for computing eigenvalues of large matrices are variants of the Lanczos and Arnoldi methods [GWS-B8], [114,121]. The Lanczos method for

M.E. Kilmer and D.P. O'Leary (eds.), *G.W. Stewart: Selected Works with Commentaries*, Contemporary Mathematicians, DOI 10.1007/978-0-8176-4968-5_10,

Hermitian matrices constructs a set of orthonormal vectors that span the Krylov subspace $\mathcal{K}_k(\mathbf{A}, \mathbf{u}_1) = \text{span}\{\mathbf{u}_1, \mathbf{A}\mathbf{u}_1, \ldots, \mathbf{A}^{k-1}\mathbf{u}_1\}$ using a three-term recurrence of the form

$$\beta_k \mathbf{u}_{k+1} = \mathbf{A}\mathbf{u}_k - \alpha_k \mathbf{u}_k - \beta_{k-1}\mathbf{u}_{k-1}. \tag{10.1}$$

If

$$\mathbf{U}_k = [\mathbf{u}_1, \mathbf{u}_2, \ldots, \mathbf{u}_k]$$

is the matrix formed from the vectors generated by (10.1), then this process can be represented in terms of matrices as

$$\mathbf{A}\mathbf{U}_k = \mathbf{U}_k \mathbf{T}_k + \beta_k \mathbf{u}_{k+1} \mathbf{e}_k^{\mathrm{T}}, \tag{10.2}$$

where $\mathbf{e}_k$ is the unit vector of dimension k with 1 in its kth component. $\mathbf{T}_k$ is a $k \times k$ tridiagonal matrix that satisfies

$$\mathbf{T}_k = \mathbf{U}_k^{\mathrm{H}} \mathbf{A} \mathbf{U}_k.$$

Given an eigenvalue μ and eigenvector $\mathbf{w}$ of $\mathbf{T}_k$, estimates for those of $\mathbf{A}$ are μ and $\mathbf{v} = \mathbf{U}_k\mathbf{w}$, the so-called *Ritz value* and *Ritz vector*. It is known that as k increases, the Ritz values tend to eigenvalues of $\mathbf{A}$.

When $\mathbf{A}$ is non-Hermitian, there is an analogous *Arnoldi decomposition*

$$\mathbf{A}\mathbf{U}_k = \mathbf{U}_k \mathbf{H}_k + \beta_k \mathbf{u}_{k+1} \mathbf{e}_k^{\mathrm{T}}, \tag{10.3}$$

where the tridiagonal matrix is replaced by an upper-Hessenberg matrix $\mathbf{H}_k$. In this case, the computation analogous to (10.1) entails a k-term recurrence at step k, whose expense becomes large as k increases. To keep costs down, it is necessary to *restart* this procedure, for example, taking some suitable vector (such as $\mathbf{u}_k$) as a new starting vector to build a new Krylov subspace.

In [130], Sorensen proposed an efficient way to restart the Arnoldi process. It forces eigenvectors corresponding to desired eigenvalues (for example, the ones with rightmost real parts) to be more prominent in the system, and those corresponding to unwanted eigenvalues to be less prominent. In particular, given an Arnoldi decomposition

$$\mathbf{A}\mathbf{U}_m = \mathbf{U}_m \mathbf{H}_m + \beta_m \mathbf{u}_{m+1} \mathbf{e}_m^{\mathrm{T}}$$

and some $k < m$, choose $m-k$ shifts $\kappa_1, \kappa_2, \ldots, \kappa_{m-k}$ and apply $m-k$ steps of the implicitly shifted QR algorithm to $\mathbf{H}_m$. This gives a new upper-Hessenberg matrix $\widetilde{\mathbf{H}}_m = \mathbf{Q}^{\mathrm{H}} \mathbf{H}_m \mathbf{Q}$ and the modified decomposition

$$\mathbf{A}\widetilde{\mathbf{U}}_m = \widetilde{\mathbf{U}}_m \widetilde{\mathbf{H}}_m + \mathbf{u}_{m+1} \mathbf{c}_m^{\mathrm{H}}, \tag{10.4}$$

where $\widetilde{\mathbf{U}}_m = \mathbf{U}_m \mathbf{Q}$ and $\mathbf{c}_m = \beta_m \mathbf{Q}^{\mathrm{H}} \mathbf{e}_m$. It can be shown that the first $k-1$ components of $\mathbf{c}_m$ are zero [130]. Therefore, by restricting (10.4) to its leftmost k

columns, and letting $\widetilde{\mathbf{H}}_k$ be the leading $k \times k$ principal submatrix of $\widetilde{\mathbf{H}}_m$, one has a new Arnoldi decomposition of order k,

$$\mathbf{A}\widetilde{\mathbf{U}}_k = \widetilde{\mathbf{U}}_k\widetilde{\mathbf{H}}_k + \widetilde{\beta}_k\widetilde{\mathbf{u}}_{k+1}\mathbf{e}_k^{\mathrm{T}}. \tag{10.5}$$

A key observation concerning this construction is that $\widetilde{\mathbf{u}}_1$ is in the direction $p(\mathbf{A})\mathbf{u}_1$ where $p(t) = (t-\kappa_1)(t-\kappa_2)\cdots(t-\kappa_{m-k})$. That is, if the shifts can be chosen to be near a set of *unwanted* eigenvalues of $\mathbf{A}$, then the effect is that (10.5) corresponds to k completed steps of a(n) (implicitly) restarted Arnoldi computation, for which the starting vector $\widetilde{\mathbf{u}}_1$ is dominated by eigenvectors other than the $m-k$ undesired ones.

From this starting point, which is based on work of Sorensen and Lehoucq [87,130], Stewart shows in [GWS-J111] how to avoid two difficulties associated with the implicitly restarted Arnoldi (IRA) algorithm. In particular, it is known that the restarted Arnoldi process suffers from forward instability associated with the QR iteration [117]; i.e., the computed Hessenberg matrix $\widetilde{\mathbf{H}}_m$ of (10.4) may be far from the matrix that would be obtained in exact arithmetic. This causes difficulties for *purging* unwanted eigenvalues from the process, and *deflating* converged eigenvalues, since those actions require accurate identification of small subdiagonal entries of the Hessenberg matrix.

Stewart's improvements are derived from looking at a generic alternative to the Arnoldi decomposition, which he calls a *Krylov decomposition*

$$\mathbf{A}\mathbf{U}_k = \mathbf{U}_k\mathbf{B}_k + \mathbf{u}_{k+1}\mathbf{b}_{k+1}^{\mathrm{H}}. \tag{10.6}$$

Compared to (10.3), the only requirements on the components here are that the columns of $\mathbf{U}_k$ be linearly independent. Stewart then makes two observations. First, he shows that any Krylov decomposition is equivalent to an Arnoldi decomposition. That is, given the Krylov decomposition (10.6), there is a matrix $\widehat{\mathbf{U}}_k$ with orthonormal columns that span the same space as those of $\mathbf{U}_k$, and an upper Hessenberg matrix $\widehat{\mathbf{B}}_k$ such that

$$\mathbf{A}\widehat{\mathbf{U}}_k = \widehat{\mathbf{U}}_k\widehat{\mathbf{B}}_k + \beta_k\widehat{\mathbf{u}}_{k+1}\mathbf{e}_k^{\mathrm{T}}$$

is an Arnoldi decomposition. In particular, the eigenvalues of $\mathbf{B}_k$ are the same as those of $\widehat{\mathbf{B}}_k$.

The equivalence is established using the fact that subspaces of Krylov decompositions are invariant under two classes of operations, *similarity*, which changes $\mathbf{B}_k$, and *translation*, which changes $\mathbf{u}_{k+1}$. Since this invariance is the basis of the algorithm developed in this paper, we discuss how the equivalence is proved. Starting from (10.6), let $\mathbf{U}_k = \mathbf{Q}_k\mathbf{R}_k$ orthogonalize $\mathbf{U}_k$, so that $\mathbf{U}_k$ and $\mathbf{Q}_k$ span the same space. This leads, via similarity, to

$$\mathbf{A}\mathbf{Q}_k = \mathbf{Q}_k[\mathbf{R}_k\mathbf{B}_k\mathbf{R}_k^{-1}] + \mathbf{u}_{k+1}\widehat{\mathbf{b}}_{k+1}^{\mathrm{H}}, \quad \widehat{\mathbf{b}}_{k+1} = \mathbf{R}_k^{-\mathrm{H}}\mathbf{b}_{k+1}, \tag{10.7}$$

which is again a Krylov decomposition. Next, consider the translation $\mathbf{q}_{k+1} = \mathbf{u}_{k+1} + \mathbf{Q}_k\mathbf{c}_k$, where $\mathbf{c}_k$ denotes a vector of expansion coefficients to be determined momentarily. Then we have

$$\mathbf{A}\mathbf{Q}_k = \mathbf{Q}_k\,[\mathbf{R}_k\mathbf{B}_k\mathbf{R}_k^{-1} - \mathbf{c}_k\widehat{\mathbf{b}}_{k+1}^{\mathrm{H}}] + \mathbf{q}_{k+1}\widehat{\mathbf{b}}_{k+1}^{\mathrm{H}}.$$

We notice that the subspaces of dimension $k+1$ will also be the same, regardless of whether we use $\mathbf{q}_{k+1}$ or $\mathbf{u}_{k+1}$. The choice $\mathbf{c}_k = -\mathbf{Q}_k^{\mathrm{H}}\mathbf{u}_{k+1}$ makes $\mathbf{q}_{k+1}$ orthogonal to $\mathbf{Q}_k$. To complete the proof of equivalence to Arnoldi, we reduce $\widehat{\mathbf{b}}_{k+1}$ to a vector proportional to $\mathbf{e}_k$ using a unitary transformation $\mathbf{V}$, apply the similarity transformation defined by $\mathbf{V}$, and then we reduce the resulting matrix $\mathbf{V}^{\mathrm{H}}[\mathbf{R}_k\mathbf{B}_k\mathbf{R}_k^{-1} - \mathbf{c}_k\mathbf{b}_{k+1}^{\mathrm{H}}]\mathbf{V}$ to Hessenberg form using a unitary similarity transformation. It is straightforward to show that these operations leave the spaces unchanged and produce a decomposition with the structure of (10.1).

Stewart's second observation is that, at relatively little cost, the Arnoldi decomposition can be written in terms of an upper-triangular matrix instead of an upper-Hessenberg matrix, producing what he terms a *Krylov–Schur decomposition*

$$\mathbf{A}\mathbf{U}_k = \mathbf{U}_k\mathbf{S}_k + \mathbf{u}_{k+1}\mathbf{b}_{k+1}^{\mathrm{H}} \quad \text{or} \quad \mathbf{A}\mathbf{U}_k = \mathbf{U}_{k+1}\widehat{\mathbf{S}}_k\,, \tag{10.8}$$

where $\mathbf{S}_k$ is upper triangular and

$$\widehat{\mathbf{S}}_k = \begin{bmatrix} \mathbf{S}_k \\ \mathbf{b}_{k+1}^{\mathrm{H}} \end{bmatrix}. \tag{10.9}$$

The Ritz values are then the eigenvalues of $\mathbf{S}_k$, which can be obtained directly.

These observations are the basis for Stewart's *Krylov–Schur method* for computing eigenvalues. We outline the computations involved assuming, for simplicity, the diagonal Schur form.[22] Note that $\widehat{\mathbf{S}}_k$ contains the upper-triangular matrix $\mathbf{S}_k$ in its upper $k \times k$ block, together with a full $(k+1)$st row. Starting from a Krylov–Schur decomposition (10.8), the algorithm expands the Krylov subspace by first computing $\mathbf{A}\mathbf{u}_{k+1}$ and then orthogonalizing it against the previous space (spanned by the columns of $\mathbf{U}_k$). This expansion procedure can be performed multiple (say, $m-k$) times, leading to a decomposition of the form

$$\mathbf{A}\widehat{\mathbf{U}}_m = \widehat{\mathbf{U}}_{m+1}\widehat{\mathbf{S}}_m\,,$$

where the first k columns of $\widehat{\mathbf{U}}_{m+1}$ are the columns of $\mathbf{U}_k$ and $\widehat{\mathbf{S}}_m$ is an $(m+1)\times m$ matrix built from $\widehat{\mathbf{S}}_k$, with structure

$$\widehat{\mathbf{S}}_m = \left[\begin{array}{cccc} \mathbf{S}_k & * & \cdot & * \\ \mathbf{b}_{k+1}^{\mathrm{H}} & * & \cdot & * \\ & * & \cdot & * \\ & & * & * \\ \hline & & & * \end{array}\right].$$

[22] The case in which 2×2 blocks would be used to handle complex eigenvalues of real matrices is treated in Chap. 5 of [GWS-B8].

The $(k+1)$st row, inherited from the starting point (10.9), is full below the diagonal, and all subsequent rows are of Hessenberg structure (as in the Arnoldi computation). It is now straightforward and inexpensive to reduce the principal $m \times m$ submatrix of $\widehat{\mathbf{S}}_m$ to Schur form:

$$[\widehat{\mathbf{S}}_m]_{[1:m,1:m]} = \mathbf{Q}_m \mathbf{S}_m \mathbf{Q}_m^{\mathrm{H}}.$$

Thus, redefining $\mathbf{U}_m = \widehat{\mathbf{U}}_m \mathbf{Q}_m$, the result is an expanded Krylov–Schur decomposition

$$\mathbf{A}\mathbf{U}_m = \mathbf{U}_m \mathbf{S}_m + \mathbf{u}_{m+1} \mathbf{b}_{m+1}^{\mathrm{H}},$$

where once again, $\mathbf{S}_m$ is upper triangular.

A significant advantage of this procedure is the ease with which unwanted eigenvalue estimates can be purged and a restart performed. Since the Ritz values are immediately available from the diagonal of $\mathbf{S}_m$, unwanted ones can be moved to the "southeast" corner of the matrix. Assuming that this permutation has already been applied to the previous equation, this leads to the partitioned Krylov–Schur decomposition

$$\mathbf{A}[\mathbf{U}_k, \bar{\mathbf{U}}] = [\mathbf{U}_k, \bar{\mathbf{U}}] \begin{bmatrix} \mathbf{S}_{11} & \mathbf{S}_{12} \\ \mathbf{0} & \mathbf{S}_{22} \end{bmatrix} + \mathbf{u}_{k+1}[\mathbf{b}_1^{\mathrm{H}}, \mathbf{b}_2^{\mathrm{H}}],$$

where the diagonal of $\mathbf{S}_{22}$ contains the unwanted quantities. Then

$$\mathbf{A}\mathbf{U}_k = \mathbf{U}_k \mathbf{S}_{11} + \mathbf{u}_{k+1} \mathbf{b}_1^{\mathrm{H}}$$

is also a Krylov–Schur decomposition, which can be used to restart the process. Stewart proves that this decomposition is equivalent to that obtained from the implicitly restarted Arnoldi algorithm (10.5) when the same Ritz values are purged. He also presents a general discussion of options for deflating converged eigenvalues from the system and shows that the Krylov–Schur method adds minimal overhead relative to the implicitly restarted Arnoldi method.

Since, from (10.6), $\mathbf{B}_k = (\mathbf{U}_k^{\mathrm{H}} \mathbf{U}_k)^{-1} \mathbf{U}_k^{\mathrm{H}} \mathbf{A} \mathbf{U}_k$, the matrix $\mathbf{B}_k$ is a generalization of the scalar Rayleigh quotient $(\mathbf{u}_k^{\mathrm{H}} \mathbf{u}_k)^{-1} \mathbf{u}_k^{\mathrm{H}} \mathbf{A} \mathbf{u}_k$. See also [GWS-J19]. In a short addendum to this paper [GWS-J113], Stewart amplifies this observation, choosing any matrix $\mathbf{V}$ so that $\mathbf{V}^{\mathrm{H}}\mathbf{U}$ is nonsingular. Then if $\mathbf{U}_k \mathbf{w}$ is an eigenvector of $\mathbf{A}$ with eigenvalue λ, then $\mathbf{w}$ is an eigenvalue of $\widehat{\mathbf{B}} = (\mathbf{V}^{\mathrm{H}} \mathbf{U}_k)^{-1} \mathbf{V}^{\mathrm{H}} \mathbf{A} \mathbf{U}_k$ with the same eigenvalue. The choice $\mathbf{V} = (\mathbf{A} - \kappa \mathbf{I}) \mathbf{U}_k$ produces the so-called *harmonic Ritz values*, which are often superior for approximating interior eigenvalues. Using the translation property, Stewart then shows how to continue the Krylov–Schur method from $\mathbf{U}$ and $\widehat{\mathbf{B}}$.

10.2. Backward Error Analysis of Krylov Subspace Methods

In [GWS-J110], Stewart builds on the results above to characterize the backward error of Krylov subspace methods. He established the basis for these results in

Theorem 4.1 of [GWS-J111], where he showed that if we have an approximate Krylov decomposition $\mathbf{A}\mathbf{U}_k = \mathbf{U}_k\mathbf{B}_k + \mathbf{u}_{k+1}\mathbf{b}_{k+1}^{\mathrm{H}} + \mathbf{R}$, then

$$(\mathbf{A}+\mathbf{E})\mathbf{U}_k = \mathbf{U}_k\mathbf{B}_k + \mathbf{u}_{k+1}\mathbf{b}_{k+1}^{\mathrm{H}},$$

where

$$\frac{\|\mathbf{R}\|_2}{\|\mathbf{U}_k\|_2} \le \|\mathbf{E}\|_2 \le \|\mathbf{R}\|_2 \|\mathbf{U}_{k+1}^{\dagger}\|_2.$$

In [GWS-J110], he specializes to the case where $\mathbf{U}$ has orthonormal columns, allows a different choice of $\mathbf{B}_k$ in the computation of the residual (in particular, choosing it to be the Rayleigh quotient $\mathbf{U}_k^{\mathrm{H}}\mathbf{A}\mathbf{U}_k$), and extends the analysis to all unitarily invariant norms.[23]

Start with an approximate Krylov decomposition (10.6), written in generic form as

$$\mathbf{A}\mathbf{U}_k \approx \mathbf{U}_{k+1}\mathbf{G}. \tag{10.10}$$

This can be viewed as having come from Arnoldi's method or the method of the previous section, in floating-point arithmetic. The *Krylov residual* is defined to be

$$\mathbf{R} = \mathbf{A}\mathbf{U}_k - \mathbf{U}_{k+1}\mathbf{G}.$$

Stewart first shows that the choice of $\mathbf{G}$ that minimizes $\mathbf{R}$ in any unitarily invariant norm is the Rayleigh quotient $\mathbf{G} = \mathbf{U}_{k+1}^{\mathrm{H}}\mathbf{A}\mathbf{U}_k$.

Stewart then demonstrates how to specify an orthonormal matrix $\widehat{\mathbf{U}}_{k+1}$ such that $\|\mathbf{R}\|$ is minimal with respect to $\mathbf{U}_{k+1}$ (with $\mathbf{G}$ given as above). This is done by starting with an arbitrary orthonormal matrix $\mathbf{U}_{k+1}$, which can be viewed as an *approximate basis* for a Krylov decomposition of $\mathbf{A}$, and then finding a $(k+1)\times(k+1)$ unitary matrix $\mathbf{V}$ such that

$$\widehat{\mathbf{U}}_{k+1} = \mathbf{U}_{k+1}\mathbf{V}$$

makes the Krylov residual minimal. To achieve this, let

$$\mathbf{S} = \mathbf{A}\mathbf{U}_{k+1} - \mathbf{U}_{k+1}(\mathbf{U}_{k+1}^{\mathrm{H}}\mathbf{A}\mathbf{U}_{k+1}), \tag{10.11}$$

and explore the impact of $\mathbf{V} = [\mathbf{V}_1, \mathbf{v}_2]$ expressed in partitioned form. In particular, postmultiplying by $\mathbf{V}_1$ gives

$$\mathbf{S}\mathbf{V}_1 = \mathbf{A}(\mathbf{U}_{k+1}\mathbf{V}_1) - (\mathbf{U}_{k+1}\mathbf{V})[(\mathbf{U}_{k+1}\mathbf{V})^{\mathrm{H}}\mathbf{A}(\mathbf{U}_{k+1}\mathbf{V}_1)].$$

In light of the earlier discussion, it follows that $\mathbf{S}\mathbf{V}_1$ is the minimal Krylov residual for the basis $\widehat{\mathbf{U}}_{k+1} = \mathbf{U}_{k+1}\mathbf{V}$. We want to minimize this residual norm with respect to orthonormal matrices $\mathbf{V}_1$ of dimension $(k+1)\times k$. We use the fact that the

[23] A norm $\|\cdot\|$ is unitarily invariant if $\|\mathbf{U}\mathbf{A}\mathbf{V}\| = \|\mathbf{A}\|$ for all unitary matrices $\mathbf{U}$ and $\mathbf{V}$; examples used in this paper are the Euclidean and Frobenius norms.

singular values $\{\tau_i\}_{i=1}^{k}$ of $\mathbf{SV}_1$ are interleaved with the singular values $\{\sigma_j\}_{j=1}^{k+1}$ of $\mathbf{S}$, i.e.,

$$\sigma_1 \geq \tau_1 \geq \sigma_2 \geq \tau_2 \geq \cdots \geq \sigma_k \geq \tau_k \geq \sigma_{k+1}.$$

Stewart observes that the norm of $\mathbf{SV}_1$ will be minimized if the singular values of $\mathbf{SV}_1$ can be made as small as possible: $\tau_i = \sigma_{i+1}$. This can be achieved by choosing the columns of $\mathbf{V}_1$ to be the right singular vectors of $\mathbf{S}$ corresponding to $\sigma_2, \ldots, \sigma_{k+1}$, and choosing $\mathbf{v}_2$ to be the right singular vector associated with σ_1.

With this choice of $\widehat{\mathbf{U}}_{k+1}$, the residual $\mathbf{R} = \mathbf{A}\widehat{\mathbf{U}}_k - \widehat{\mathbf{U}}_{k+1}(\widehat{\mathbf{U}}_{k+1}^{\mathrm{H}}\mathbf{A}\widehat{\mathbf{U}}_k)$ is the (globally) optimal residual of the approximate Krylov decomposition with basis $\widehat{\mathbf{U}}_{k+1}$. It follows that with $\mathbf{E} = -\mathbf{R}\widehat{\mathbf{U}}_k^{\mathrm{H}}$, this approximate Krylov decomposition corresponds to an (exact) Krylov decomposition of a perturbed matrix:

$$(\mathbf{A} + \mathbf{E})\widehat{\mathbf{U}}_k = \widehat{\mathbf{U}}_{k+1}[\widehat{\mathbf{U}}_{k+1}^{\mathrm{H}}(\mathbf{A} + \mathbf{E})\widehat{\mathbf{U}}_k].$$

Moreover, by the construction of $\mathbf{V}$, the Euclidean and Frobenius norms of the Krylov residual are given by

$$\|\mathbf{R}\|_2 = \sigma_2, \quad \|\mathbf{R}\|_F = \sqrt{\sigma_2^2 + \cdots + \sigma_{k+1}^2}.$$

If the Krylov residual is zero (as for an exact decomposition), it follows that $\sigma_2 = 0$ and that the rank of $\mathbf{S}$ in (10.11) is at most one.

10.3. Adjusting the Rayleigh Quotient in Lanczos Methods

In [GWS-J112], Stewart resolves a difficulty associated with the Lanczos method for Hermitian problems, that of computing accurate estimates of eigenvectors in the presence of roundoff error. It is known that when the Lanczos decomposition (10.2) is constructed with floating-point arithmetic, the Lanczos vectors (columns of $\mathbf{U}_k$) tend to lose orthogonality. Despite this, under appropriate circumstances, the computed Ritz values, i.e., the eigenvalues of the tridiagonal matrix $\mathbf{T}_k$, are accurate approximations (to machine precision) of the true Ritz values. What is needed is for the Lanczos vectors to remain *semiorthogonal*, i.e., the off-diagonal elements of $\mathbf{U}_k^{\mathrm{H}}\mathbf{U}_k - \mathbf{I}$ should not be permitted to become larger than a multiple of $\sqrt{\epsilon}$, where ϵ is the machine precision [128].

However, because of loss of orthogonality, the computed Ritz vectors $\mathbf{U}_k\mathbf{w}$ (where $\mathbf{w}$ is an eigenvector of $\mathbf{T}_k$) might not be accurate estimates of the eigenvectors of $\mathbf{A}$. In exact arithmetic, we can assess how accurate the eigenvector approximation is: if $(\theta, \mathbf{z})$, with $\mathbf{z} = \mathbf{U}_k\mathbf{w}$, is a Ritz pair computed from $\mathbf{T}_k$, then the Euclidean norm of the eigenvalue residual satisfies

$$\|\mathbf{Az} - \theta\mathbf{z}\| = |\beta_k w_k|. \tag{10.12}$$

When matrix-vector products by $\mathbf{A}$ are expensive, the quantity on the right can be used to efficiently assess the accuracy of the estimates, but the equality fails to hold when loss of orthogonality occurs.

Stewart addresses these difficulties by adapting the (standard) methods used to ensure the computation of accurate Ritz values. Indeed, it is shown in [129] that loss of orthogonality can be monitored inexpensively, and full reorthogonalization of a new Lanczos vector against all old ones needs to be done only occasionally, when the requirement of semiorthogonality is violated. Stewart's crucial observation is that the computed semiorthogonal matrix $\mathbf{U}_k$ satisfies a recurrence of the form (10.3) where $\mathbf{H}_k$ is an upper-Hessenberg matrix whose nonzero entries outside its tridiagonal part are of magnitude $\sqrt{\epsilon}$. $\mathbf{H}_k$ could be computed explicitly; its nonzero entries above the first superdiagonal come from steps j in which the occasional full orthogonalization of $\mathbf{u}_j$ against $\{\mathbf{u}_1, \ldots, \mathbf{u}_{j-1}\}$ is performed to maintain semiorthogonality. Stewart suggests that Ritz vectors be calculated from $\mathbf{H}_k$ rather than $\mathbf{T}_k$.

If not, then $\mathbf{H}_k$ still has a use in assessing the accuracy of a Ritz vector computed from $\mathbf{T}_k$. Let $(\theta, \mathbf{w})$ be an eigenpair of $\mathbf{T}_k$, and let $\mathbf{z} = \mathbf{U}_k\mathbf{w}$. Taking $\mathbf{w}$ as an eigenvector estimate for $\mathbf{H}_k$, we have a residual $\mathbf{r} = \mathbf{H}_k\mathbf{w} - \theta\mathbf{w}$. Then

$$\mathbf{A}\mathbf{U}_k\mathbf{w} = \mathbf{U}_k\mathbf{H}_k\mathbf{w} + \beta_k\mathbf{u}_{k+1}\mathbf{e}_k^{\mathrm{T}}\mathbf{w},$$

which implies

$$\mathbf{A}\mathbf{z} - \theta\mathbf{z} = \mathbf{U}_k(\mathbf{H}_k\mathbf{w} - \theta\ \mathbf{w}) + \beta_k w_k \mathbf{u}_{k+1} = \mathbf{U}_{k+1}\begin{bmatrix} \mathbf{r} \\ \beta_k w_k \end{bmatrix}.$$

Taking Euclidean norms gives

$$\begin{aligned} \|\mathbf{A}\mathbf{z} - \theta\mathbf{z}\|^2 &= \left\| \mathbf{U}_{k+1}\begin{bmatrix} \mathbf{r} \\ \beta_k w_k \end{bmatrix} \right\|^2 \\ &\approx \left\| \begin{bmatrix} \mathbf{r} \\ \beta_k w_k \end{bmatrix} \right\|^2 \\ &= \|\mathbf{r}\|^2 + |\beta_k w_k|^2. \end{aligned} \tag{10.13}$$

Thus, an accurate approximation to the residual norm $\|\mathbf{A}\mathbf{z} - \theta\mathbf{z}\|$ is obtained from quantity used in (10.12) *plus* a term computed from the residual $\mathbf{r}$. Stewart then provides numerical evidence that estimates $(\theta, \mathbf{z})$ obtained by this means converge to eigenpairs of $\mathbf{A}$, and that the residual estimate of (10.13) is accurate.

10.4. Impact

In elegant style, these papers elucidate and improve upon the fundamental methods used for eigenvalue computations of large matrices. Despite 50 years of investigation into the Krylov methods, no one had made the observation that the Arnoldi and Lanczos decompositions fit into a broader and more general class of *Krylov decompositions*. Because of this, convergence analysis in inexact arithmetic had been very difficult. Stewart shows that a specific choice from this class, the Krylov–Schur decomposition, shares many of the advantages of the Arnoldi decomposition and yields much clearer derivations and implementations. Like the implicitly restarted Arnoldi method, it is amenable to restarts with starting vectors that have estimates of unwanted eigenvectors filtered out, and, at little cost, it avoids some of the difficulties associated with forward instability that have to be handled carefully in implementations of IRA. In addition, Stewart establishes the backward stability of these methods.

Finally, Stewart also takes on a long-standing problem associated with the Lanczos algorithm for Hermitian matrices, that of using it to accurately estimate eigenvectors without repeated orthogonalization of basis vectors. He demonstrates that this can be done at relatively little cost by working with the Hessenberg matrix associated with loss of orthogonality of the Lanczos vectors.

11

Other Contributions

Misha E. Kilmer and Dianne P. O'Leary

The preceding seven chapters of this commentary outlined some of Stewart's important contributions to matrix algorithms and matrix perturbation theory.

In fact, however, Stewart has a substantial body of research on other topics. He was a designer of a parallel computing architecture, the Maryland Crab, and the scheduling system for it [GWS-J61,GWS-J68,GWS-N16,GWS-N17,GWS-N18,GWS-N22]. In early work, he focused on solving nonlinear equations and finding roots of polynomials; see Sect. 5.1. Another singularity in his work is a paper on representing integers in noninteger bases [GWS-J63].

Stewart's most famous unpublished work is a 1976 technical report co-authored by Virginia Klema and Gene Golub [GWS-N6] on matrices $\mathbf{A}$ that are nearly rank-deficient. This is an important concept for matrices whose entries have some uncertainty. They consider the *numerical rank* of a matrix in a more precise way than any considered previously: $\mathbf{A}$ has numerical rank (δ, ϵ, r) if

$$r = \inf\{\text{rank}(\mathbf{B}) \, : \, \|\mathbf{A} - \mathbf{B}\| \leq \epsilon\}$$

and

$$\epsilon < \delta \leq \sup\{\eta \, : \, \|\mathbf{A} - \mathbf{B}\| \leq \eta \rightarrow \text{rank}(\mathbf{B}) \geq r\}.$$

They consider the implication of numerically rank-deficient matrices in the solution of least squares problems and describe pivoted QR and SVD algorithms for choosing a set of linearly independent columns that span the well-determined part of the range of $\mathbf{A}$.

Stewart is a true scholar, and his curiosity is notable. His work is marked by meticulous literature reviews, trying to go back to the origins of each idea. While teaching himself Latin during his lunch breaks, he decided to translate Karl Friedrich Gauss's work establishing the study of least squares problems [GWS-B4].

M.E. Kilmer and D.P. O'Leary (eds.), *G.W. Stewart: Selected Works with Commentaries*, Contemporary Mathematicians, DOI 10.1007/978-0-8176-4968-5_11,

This scholarship carries over to his editorial work. Richard Brualdi [20] says, "As an editor of *Linear Algebra and its Applications* (*LAA*), Pete Stewart demanded the same high standards that he demanded of himself, both in mathematical contribution and exposition. His vast knowledge and experience in both computational and numerical linear algebra have been an enormous resource for *LAA*. Now Pete is a Distinguished Editor of *LAA*, an honor reserved for very few people. We continue to make use of Pete's expertise."

No summary of his contributions would be complete without mention of the influence of his textbooks.

His *Introduction to Matrix Computations* [GWS-B1] gave a rather complete treatment of solution of dense linear systems, least squares problems, and eigenproblems, along with some perturbation theory. With his usual attention to clarity and precision, he defined a pseudolanguage called INFL for unambiguous presentation of algorithms.

His two *Afternotes* books [GWS-B5,GWS-B6] were his own transcriptions of lecture notes for numerical analysis courses that he taught at the University of Maryland. Informal in style, they present the student with an excellent selection of topics and results.

Finally, his two-volume work on matrix algorithms, decompositions and eigensystems [GWS-B7,GWS-B8], provides an encyclopedic survey of the literature and a concise explanation of issues and solutions revealing Pete's unique insight. Beresford Parlett [116] cites Theorem 1.8 in Volume II as having the "biggest direct influence on my work of Pete's deep understanding of Linear Algebra." Stewart's last student, Che-Rung (Roger) Lee [86] says of these two volumes:

> Computational linear algebra is a multidisciplinary subject. It involves numerical analysis, linear algebra, algorithm, and programming. Unlike other books that cover only one or two aspects, Pete's books treat each topic.... The books give enough elucidation on important subjects so that one can easily understand the derivation of algorithms, the analysis of errors, and the tricks of implementations. Meanwhile, they are not clogged with tedious details.... Although Pete once humbly said that there are already many good books for numerical linear algebra, we eagerly wait to see his next book, the next next one, and more. After all, Pete's style is one-of-a-kind.

His style is indeed one-of-a-kind. By his research, his clarity of presentation, and his influential textbooks, Stewart tremendously influenced computation and research in numerical linear algebra. His impact is aptly summarized by Yousef Saad [122]:

> Pete is someone who is deeply respected by his peers for his truly spectacular contributions to the field. But there are two other reasons why he inspires such deep respect. The first is that he is a humble individual who does his work quietly and meticulously in the tradition of previous generations of scholars. The second is that he is a very capable and effective hands-on

individual, someone who can produce several books, write software, and prove elegant theorems all at the same time.

Fueled by coffee, shoe leather (for walking the halls as his ideas percolate), and his insatiable curiosity, his scholarly activities will constitute an impressive legacy.

References

[1] E. Anderson, Z. Bai, C. Bischof, S. Blackford, J. Demmel, J. Dongarra, J. Du Croz, A. Greenbaum, S. Hammarling, A. McKenney, et al. *LAPACK Users' Guide*. SIAM, Philadelphia, PA, 3 edition, 1999. 29, 33, 100, 108

[2] Y.-H. Au-Yeung. Some theorems on the real pencil and simultaneous diagonalization of two Hermitian bilinear forms. *Proceedings of the American Mathematical Society*, 23:246–253, 1969. 109

[3] Y.-H. Au-Yeung. A theorem on a mapping from a sphere to the circle and the simultaneous diagonalization of two Hermitian matrices. *Proceedings of the American Mathematical Society*, 20:545–548, 1969. 109

[4] Z. Bai. The CSD, GSVD, their applications and computations. IMA Preprint Series 958, Institute of Mathematics and Applications, University of Minnesota, 1992. 109

[5] Z. Bai and J. W. Demmel. On swapping diagonal blocks in real Schur form. *Linear Algebra and its Applications*, 186:75–95, 1993. 100

[6] J. Barlow and J. W. Demmel. Computing accurate eigensystems of scaled diagonally dominant matrices. *SIAM Journal on Numerical Analysis*, 27(3):762–791, 1990. 84

[7] J. B. Barlow, M. M. Monahemi, and D. P. O'Leary. Constrained matrix Sylvester equations. *SIAM Journal on Matrix Analysis and Applications*, 13:1–9, 1992. 36

[8] J. L. Barlow, H. Erbay, and I. Slapnicar. An alternative algorithm for the refinement of ULV decompositions. *SIAM Journal on Matrix Analysis and Applications*, 27(1):198–211, 2006. 57

[9] J. G. P. Barnes. An algorithm for solving non-linear equations based on the secant method. *The Computer Journal*, 8(1):66–72, 1965. 49

[10] R. H. Bartels and G. H. Golub. The simplex method of linear programming using LU decomposition. *Communications of the ACM*, 12:266–268, 1969. 46

[11] F. L. Bauer. Das Verfahren der Treppeniteration und verwandte Verfahren zur Lösung algebraischer Eigenwertprobleme. *Z. Angew. Math. Phys.*, 8:214–235, 1957. 96, 98

[12] A. Ben-Israel. On error bounds for generalized inverses. *SIAM Journal on Numerical Analysis*, pages 585–592, 1966. 61

[13] M. W. Berry. Private communication, Aug., 2009. 41

[14] R. Bhatia and R.-C. Li. On perturbations of matrix pencils with real spectra. II. *Mathematics of Computation*, 65(214):637–645, 1996. 109

[15] Å. Björck. *Numerical Methods for Least Squares Problems.* SIAM, Philadelphia, PA, 1996. 50, 63, 68, 69

[16] Å. Björck and G. H. Golub. Numerical methods for computing angles between linear subspaces. *Mathematics of Computation*, 27:579–594, 1973. 64

[17] A. W. Bojanczyk, J. G. Nagy, and R. J. Plemmons. Block RLS using row Householder reflections. *Linear Algebra and its Applications*, 188:31–61, 1993. 43

[18] C. G. Broyden. A class of methods for solving nonlinear simultaneous equations. *Mathematics of Computation*, 19:577–593, 1965. 49

[19] C. G. Broyden. On the discovery of the 'good Broyden' method. *Mathematical Programming*, 87(2):209–213, 2000. 49

[20] R. A. Brualdi. Private communication, Aug. 17, 2009. 122

[21] J. R. Bunch. Private communication, Sep. 24, 2009. 28

[22] R. Byers and D. Kressner. Structured condition numbers for invariant subspaces. *SIAM Journal on Matrix Analysis and Applications*, 28(2):326–347, 2006. 93

[23] Z. H. Cao and F. G. Zhang. Direct methods for ordering eigenvalues of a real matrix. *Chinese University Journal of Computational Mathematics*, 1:27–36, 1981. In Chinese. 100

[24] X. W. Chang and C. C. Paige. On the sensitivity of the LU factorization. *BIT*, 38(3):486–501, 1998. 39

[25] M. Clint and A. Jennings. A simultaneous iteration method for the unsymmetric eigenvalue problem. *Journal of the Institute of Mathematics and its Applications*, 8:111–121, 1971. 98

[26] P.-J. Courtois. Error analysis in nearly-completely decomposable stochastic systems. *Econometrica*, 43(4):691–709, 1975. 82, 83

[27] P.-J. Courtois. *Decomposability.* Academic Press [Harcourt Brace Jovanovich Publishers], New York, 1977. Queueing and computer system applications, ACM Monograph Series. 82, 83

[28] C. R. Crawford. A stable generalized eigenvalue problem. *SIAM Journal on Numerical Analysis*, 13:854–860, 1976; 15:1070, 1978. 109

[29] G. B. Dantzig. *Linear Programming and Extensions.* Princeton University Press, Princeton, NJ, 1963. 45, 52

[30] C. Davis and W. M. Kahan. Some new bounds on perturbation of subspaces. *Bull. Amer. Math. Soc*, 75:863–868, 1969. 64, 72, 73

[31] C. Davis and W. M. Kahan. The rotation of eigenvectors by a perturbation. III. *SIAM Journal on Numerical Analysis*, 7:1–46, 1970. 64, 72, 73, 104, 109

[32] D. Day. How the QR algorithm fails to converge and how to fix it. Tech Report 96-0913J, Sandia National Laboratory, April 1996. 100

[33] J. Demmel. The condition number of equivalence transformations that block diagonalize matrix pencils. *SIAM Journal on Numerical Analysis*, 20(3):599–610, June 1983. 103

[34] J. W. Demmel and B. Kågström. Computing stable eigendecompositions of matrix pencils. *Linear Algebra and its Applications*, 88–89:139–186, 1987. 106

[35] J. W. Demmel. Computing stable eigendecompositions of matrices. *Linear Algebra and its Applications*, 79:163–193, 1986. 93

[36] J. W. Demmel and K. Veselić. Jacobi's method is more accurate than QR. *SIAM Journal on Matrix Analysis and Applications*, 13(4):1204–1245, 1992. 84

[37] I. I. Dikin. On convergence of an iterative process. *Upravlyaemye Systemy*, 12:54–60, 1974. In Russian. 68

[38] J. J. Dongarra, S. Hammarling, and J. Wilkinson. Numerical considerations in computing invariant subspaces. *SIAM Journal on Matrix Analysis and Applications*, 13:145–161, 1992. 100

[39] J. J. Dongarra, P. Luszczek, and A. Petitet. The LINPACK benchmark: Past, present and future. *Concurrency and Computation: Practice and Experience*, 15(9):803–820, 2003. 29

[40] P. Drineas, R. Kannan, and M. W. Mahoney. Fast Monte Carlo algorithms for matrices I: Approximating matrix multiplication. *SIAM Journal on Computing*, 36(1):132–157, 2007. 42

[41] P. Drineas, R. Kannan, and M. W. Mahoney. Fast Monte Carlo algorithms for matrices II: Computing a low-rank approximation to a matrix. *SIAM Journal on Computing*, 36(1):158–183, 2007.

[42] P. Drineas, R. Kannan, and M. W. Mahoney. Fast Monte Carlo algorithms for matrices III: Computing a compressed approximate matrix decomposition. *SIAM Journal on Computing*, 36(1):184–206, 2007.

[43] P. Drineas, M. W. Mahoney, and S. Muthukrishnan. Relative-error CUR matrix decompositions. *SIAM Journal on Matrix Analysis and Applications*, 30:844–881, 2008. 42

[44] S. Eastman and D. Estep. A power method for nonlinear operators. *Appl. Anal.*, 86(10):1303–1314, 2007. 93

[45] L. Elsner and P. Lancaster. The spectral variation of pencils of matrices. *J. Comp. Math.*, 3(3):262–274, 1985. 106

[46] L. Elsner and J.-G. Sun. Perturbation theorems for the generalized eigenvalue problem. *Linear Algebra and its Applications*, 48:341–357, 1982. 106

[47] H. Erbay, J. L. Barlow, and Z. Zhang. A modified Gram–Schmidt-based downdating technique for ULV decompositions with applications to recursive TLS problems. *Computational Statistics and Data Analysis*, 41(1):195–209, 2002. 55

[48] S. P. Evans. A quasi-stationary analysis of a virtual path in a B-ISDN network shared by services with very different characteristics. *Computer Networks and ISDN Systems*, 20(1-5):391–399, 1990. 93

[49] D. K. Faddeev, V. N. Kublanovskaya, and V. N. Faddeeva. Solution of linear algebraic systems with rectangular matrices. *Proc. Steklov Inst. Math*, 96:93–111, 1968. Reference provided by J. L. Barlow. 54

[50] H.-r. Fang and Y. Saad. Two classes of multisecant methods for nonlinear acceleration. *Numerical Linear Algebra with Applications*, 16(3):197–221, 2009. 49

[51] R. D. Fierro and J. R. Bunch. Bounding the subspaces from rank revealing two-sided orthogonal decompositions. *SIAM Journal on Matrix Analysis and Applications*, 16:743–759, 1995. 55

[52] K. Fukui, B. G. Sumpter, D. W. Noid, C. Yang, and R. E. Tuzun. Analysis of eigenvalues and eigenvectors of polymer particles: random normal modes. *Computational and Theoretical Polymer Science*, 11(3):191–196, 2001. 93

[53] B. S. Garbow, J. M. Boyle, J. J. Dongarra, and C. B. Moler. Matrix eigensystem routines EISPACK Guide extension, 1977. Volume 51 of Lecture Notes in Computer Science, Springer-Verlag, Berlin. 29

[54] P. E. Gill, G. H. Golub, W. Murray, and M.A. Saunders. Methods for modifying matrix factorizations. *Mathematics of Computation*, 28:505–535, 1974. 47, 49, 50, 52

[55] P. E. Gill, W. Murray, and M. H. Wright. *Practical Optimization.* Academic Press, New York, 1981. 47

[56] G. Golub and W. Kahan. Calculating the singular values and pseudo-inverse of a matrix. *J. Soc. Indust. Appl. Math.: Ser. B, Numer. Anal.*, 2:205–224, 1965. 104

[57] G. H. Golub, S. Nash, and C. Van Loan. A Hessenberg-Schur method for the problem $AX + XB = C$. *IEEE Trans. on Automatic Control*, AC-24:909–913, 1979. 36

[58] G. H. Golub and V. Pereyra. The differentiation of pseudo-inverses and nonlinear least squares problems whose variables separate. *SIAM Journal on Numerical Analysis*, 10:413–432, 1973. 65

[59] G. H. Golub and J. H. Wilkinson. Note on the iterative refinement of least squares solution. *Numerische Mathematik*, 9(2):139–148, 1966. 62

[60] M. Gu. Finding well-conditioned similarities to block-diagonalize nonsymmetric matrices is NP-hard. *Journal of Complexity*, 11:377–391, 1995. 104

[61] M. Gu and S. C. Eisenstat. Downdating the singular value decomposition. *SIAM Journal on Matrix Analysis and Applications*, 16(3):793–810, 1995. 54

[62] S. J. Hammarling. Numerical solution of the stable, non-negative definite Lyapunov equation. *IMA Journal of Numerical Analysis*, 2(3):303–323, 1982. 36

[63] P. C. Hansen. *Rank-Deficient and Discrete Ill-Posed Problems.* SIAM, Philadelphia, PA, 1998. 104

[64] P. C. Hansen, J. G. Nagy, and D. P. O'Leary. *Deblurring Images: Matrices, Spectra, and Filtering.* SIAM, Philadelphia, PA, 2006. 36

[65] R. J. Hanson and C. L. Lawson. Extensions and applications of the Householder algorithm for solving linear least squares problems. *Mathematics of Computation*, 23(108):787–812, 1969. 54

[66] F. Hespeler. Solution algorithm to a class of monetary rational equilibrium macromodels with optimal monetary policy design. *Comput. Econ.*, 31:207–223, 2008. 108

[67] D. J. Higham and N. J. Higham. Structured backward error and condition of generalized eigenvalue problems. *SIAM Journal on Matrix Analysis and Applications*, 20(2):493–512, 1998. 106

[68] N. J. Higham. A survey of condition number estimation for triangular matrices. *SIAM Review*, 29:575–596, 1987. 33, 55

[69] N. J. Higham. *Accuracy and Stability of Numerical Algorithms.* SIAM, Philadelphia, PA, 2002. 34

[70] N. J. Higham and P. A. Knight. Matrix powers in finite precision arithmetic. *SIAM Journal on Matrix Analysis and Applications*, 16(2):343–358, 1995. 90

[71] N. J. Higham, D. S. Mackey, and F. Tisseur. The conditioning of linearizations of matrix polynomials. *SIAM Journal on Matrix Analysis and Applications*, 28:1005–1028, 2006. 108

[72] N. J. Higham, D. S. Mackey, and F. Tisseur. Definite matrix polynomials and their linearization by definite pencils. *SIAM Journal on Matrix Analysis and Applications*, 31:478–502, 2009. 109

[73] A. S. Householder. *The Theory of Matrices in Numerical Analysis.* Blaisdell, New York, 1965. 28

[74] D. A. Huckaby and T. F. Chan. Stewart's pivoted QLP decomposition for low-rank matrices. *Numerical Linear Algebra with Applications*, 12:153–159, 2005. 104

[75] A. Jennings. A direct iteration method of obtaining latent roots and vectors of a symmetric matrix. *Proceedings of the Cambridge Philosphical Society*, 63:755–765, 1967. 97

[76] Z. Jia. Private communication, Aug. 22, 2009. 103
[77] B. Kågström and D. Kressner. Multishift variants of the QZ algorithm with aggressive early deflation. *SIAM Journal on Matrix Analysis and Applications*, 29(1):199–227, 2006. 107
[78] W. Kahan. Numerical linear algebra. *Canadian Math. Bulletin*, 9:757–801, 1966. 103
[79] W. Kahan. Spectra of nearly Hermitian matrices. *Proceedings of the American Mathematical Society*, 48(1):11–17, 1975. 81
[80] T. Kato. *Perturbation Theory for Linear Operators*. Springer-Verlag, Berlin, 1995. 72
[81] L. Kaufman. The LZ-algorithm to solve the generalized eigenvalue problem. *SIAM Journal on Numerical Analysis*, 11:997–1024, 1974. 107
[82] L. Kaufman. Some thoughts on the QZ algorithm for solving the generalized eigenvalue problem. *ACM Transactions on Mathematical Software*, 3(1):65–75, 1977. 107
[83] C. Kenney and A.J. Laub. Condition estimates for matrix functions. *SIAM Journal on Matrix Analysis and Applications*, 10:191–209, 1989. 33
[84] D. Kressner. Perturbation bounds for isotropic invariant subspaces of skew-Hamiltonian matrices. *SIAM Journal on Matrix Analysis and Applications*, 26(4):947–961, 2005. 93
[85] A. J. Laub and J. Xia. Fast condition estimation for a class of structured eigenvalue problems. *SIAM Journal on Matrix Analysis and Applications*, 30(4):1658–1676, 2009. 36
[86] C.-R. Lee. Private communication, Aug., 2009. 122
[87] R. B. Lehoucq and D. C. Sorensen. Deflation techniques for an implicity restarted Arnoldi iteration. *SIAM Journal on Matrix Analysis and Applications*, 17:789–821, 1996. 113
[88] R.-C. Li. A converse to the Bauer-Fike type theorem. *Linear Algebra and its Applications*, 109:167–178, 1988. 106
[89] R.-C. Li. On perturbation theorems for the generalized eigenvalues of regular matrix pencils. *Math. Numer. Sinica*, 11:10–19, 1989. In Chinese. Engl. trans. in *Chinese Journal of Numerical Mathematics and Applications* 11 (1989) 24-35. 106
[90] R.-C. Li. Bounds on perturbations of generalized singular values and of associated subspaces. *SIAM Journal on Matrix Analysis and Applications*, 14:195–234, 1993. 109
[91] R.-C. Li. On perturbations of matrix pencils with real spectra. *Mathematics of Computation*, 62:231–265, 1994. 106, 109
[92] R.-C. Li. On perturbations of matrix pencils with real spectra, a revisit. *Mathematics of Computation*, 72:715–728, 2003. 109
[93] K. J. R. Liu. Private communication, Sep. 1, 2009. 57
[94] F.T. Luk and S. Z. Qiao. A new matrix decomposition for signal processing. *Automatica (IFAC)*, 30:39–43, January 1994. 57
[95] H. M. Markowitz. The elimination form of the inverse and its application to linear programming. *Management Science*, 3(3):255–269, 1957. 46
[96] R. Mathias. Quadratic residual bounds for the Hermitian eigenvalue problem. *SIAM Journal on Matrix Analysis and Applications*, 19(2):541–550, 1998. 93
[97] C. D. Meyer. Stochastic complementation, uncoupling Markov chains, and the theory of nearly reducible systems. *SIAM Review*, 31(2):240–272, 1989. 82
[98] C. D. Meyer. Uncoupling the Perron eigenvector problem. *Linear Algebra and its Applications*, 114/115:69–94, 1989. 82

[99] L. Mirsky. Symmetric gauge functions and unitarily invariant norms. *Quart. J. Math.*, 11:50–59, 1960. 109

[100] D. P. O'Leary. On bounds for scaled projections and pseudoinverses. *Linear Algebra and its Applications*, 132:115–117, 1990. 68

[101] J. M. Ortega and W. C. Rheinboldt. *Iterative Solution of Nonlinear Equations in Several Variables.* SIAM, Philadelphia, PA, 2000. Vol. 30 of Classics in Applied Mathematics; Reprint of 1978 Academic Press book. 49

[102] A. M. Ostrowski. *Solution of equations in Euclidean and Banach spaces.* Academic Press, New York-London, 1973. Third edition of Solution of Equations and Systems of Equations. 90

[103] C. C. Paige. *The computation of eigenvalues and eigenvectors of very large sparse matrices.* PhD thesis, Inst. of Computer Science, University of London, 1971. 103

[104] C. C. Paige. Error analysis of the Lanczos algorithm for tridiagonalizing a symmetric matrix. *Journal of the Institute of Mathematics and its Applications*, 18:341–349, 1976. 103

[105] C. C. Paige. A note on a result of Sun Ji-Guang: sensitivity of the CS and GSV decompositions. *SIAM Journal on Numerical Analysis*, 21:186–191, 1984. 109

[106] C. C. Paige. Private communication, Aug. 14, 2009. 107

[107] C. C. Paige and M. A. Saunders. LSQR: An algorithm for sparse linear equations and sparse least squares. *ACM Transactions on Mathematical Software*, 8(1):43–71, 1982. 29

[108] C. C. Paige and M. Wei. History and generality of the CS decomposition. *Linear Algebra and its Applications*, 208/209:303–326, 1994. 64, 69

[109] C. C. Paige and M. A. Saunders. Towards a generalized singular value decomposition. *SIAM Journal on Numerical Analysis*, 18:398–405, 1981. 69, 109

[110] C.-T. Pan. A modification to the LINPACK downdating algorithm. *BIT*, 30:707–722, 1990. 52

[111] H. Park and L. Eldén. Downdating the rank-revealing URV decomposition. *SIAM Journal on Matrix Analysis and Applications*, 16(1):138–155, 1995. 57

[112] B. N. Parlett and K. C. Ng. Programs to swap diagonal blocks. CPAM Technical Report 381, University of California, Berkeley, CA, 1987. 100

[113] B. N. Parlett and D. S. Scott. The Lanczos algorithm with selective orthogonalization. *Mathematics of Computation*, 33:217–238, 1979. 103

[114] B. N. Parlett. *The Symmetric Eigenvalue Problem.* Prentice-Hall, Englewood Cliffs, NJ, 1980. 80, 103, 111

[115] B. N. Parlett. Private communication, Aug. 24, 2009. 107

[116] B. N. Parlett. Private communication, Jun. 9, 2009. 122

[117] B. N. Parlett and J. Le. Forward instability of tridiagonal QR. *SIAM Journal on Matrix Analysis and Applications*, 14:279–316, 1993. 113

[118] A. Ruhe. Perturbation bounds for means of eigenvalues and invariant subspaces. *BIT*, 10:343–354, 1970. 72, 100

[119] H. Rutishauser. Computational aspects of F. L. Bauer's simultaneous iteration method. *Numerische Mathematik*, 13(1):4–13, March 1969. 97, 103

[120] Y. Saad. Numerical solution of large Lyapunov equations. In *Signal Processing, Scattering and Operator Theory, and Numerical Methods, Proc. Int. Symp. MTNS-89*, volume III, pages 503–511. Birkhauser, Boston, 1990. 36

[121] Y. Saad. *Numerical Methods for Large Eigenvalue Problems.* Manchester University Press, Manchester, 1992. 87, 88, 111

[122] Y. Saad. Private communication, 2009. 98, 122

[123] M. A. Saunders. Large-scale linear programming using the Cholesky factorization. Technical Report STAN-CS-72-252, Computer Science Dept, Stanford University, 1972. 60 pp. 50, 52

[124] M. A. Saunders. Private communication, 2009. 29, 50, 52

[125] R. Schreiber and C. Van Loan. A storage-efficient WY representation for products of Householder transformations. *SIAM Journal on Scientific and Statistical Computing*, 10:53–57, 1989. 42

[126] H. R. Schwarz. *Numerik symmetrischer Matrizen.* Teubner, Stuttgart, 1968. 97

[127] H. Simon and A. Ando. Aggregation of variables in dynamic systems. *Econometrica*, 29(2):111–138, 1961. 82

[128] H. D. Simon. Analysis of the symmetric Lanczos algorithm with reorthogonalization methods. *Linear Algebra and its Applications*, 61:101–131, 1984. 117

[129] H. D. Simon. The Lanczos algorithm with partial reorthogonalization. *Mathematics of Computation*, 42:115–142, 1984. 118

[130] D. C. Sorensen. Implicit application of polynomial filters in a k-step Arnoldi method. *SIAM Journal on Matrix Analysis and Applications*, 13:357–385, 1992. 112, 113

[131] J.-G. Sun. A note on Stewart's theorem for definite matrix pairs. *Linear Algebra and its Applications*, 48:331–339, 1982. 109

[132] J.-G. Sun. The perturbation bounds of generalized eigenvalues of a class of matrix-pairs. *Math. Numer. Sinica*, 4:23–29, 1982. In Chinese. 106

[133] J.-G. Sun. Perturbation analysis for the generalized singular value decomposition. *SIAM Journal on Numerical Analysis*, 20:611–625, 1983. 109

[134] J.-G. Sun. The perturbation bounds for eigenspaces of a definite matrix pair. *Numerische Mathematik*, 41:321–343, 1983. 109

[135] J.-G. Sun. Gerschgorin type theorem and the perturbation of the eigenvalues of singular pencils. *Math. Numer. Sinica*, 7:253–264, 1985. In Chinese. Engl. trans. in *Chinese Journal of Numerical Mathematics and Applications* 10 (1988) 1-13. 108

[136] J.-G. Sun. On condition numbers of a nondefective multiple eigenvalue. *Numerische Mathematik*, 61(2):265–275, 1992. 93

[137] J.-G. Sun. Backward perturbation analysis of certian characteristic subspaces. *Numerische Mathematik*, 65:357–382, 1993. 109

[138] J.-G. Sun. On worst-case condition numbers of a nondefective multiple eigenvalue. *Numerische Mathematik*, 69(3):373–382, 1995. 93

[139] V. L. Syrmos. Disturbance decoupling using constrained Sylvester equations. *IEEE Trans. on Automatic Control*, 39:797–803, 1994. 36

[140] L. N. Trefethen and M. Embree. *Spectra and pseudospectra.* Princeton University Press, Princeton, NJ, 2005. 93

[141] S. Van Criekingen. A linear algebraic analysis of diffusion synthetic acceleration for three-dimensional transport equations. *SIAM Journal on Numerical Analysis*, 43:2034–2059, 2005. 68

[142] C. Van Loan and J. Speiser. Computation of the C-S decomposition, with application to signal processing. In *SPIE Conference on Advanced Algorithms and Architectures for Signal Processing, 19-20 Aug. 1986*, pages 71–78. San Diego, CA; (USA), 1987. 68

[143] C. F. Van Loan. A general matrix eigenvalue algorithm. *SIAM Journal on Numerical Analysis*, 12:819–834, 1975. 107

[144] C. F. Van Loan. Generalizing the singular value decomposition. *SIAM Journal on Numerical Analysis*, 13:76–83, 1976. 69, 109

[145] C. F. Van Loan. Computing the CS and the generalized singular value decompositions. *Numerische Mathematik*, 46(4):479–491, 1985. 69

[146] C. F. Van Loan. On estimating the condition of eigenvalues and eigenvectors. *Linear Algebra and its Applications*, 88-89:715–732, 1987. 33

[147] C. F. Van Loan. The ubiquitous Kronecker product. *J. Computational and Applied Mathematics*, 123:85–100, 2000. 36

[148] J. M. Varah. *The Computation of Bounds for the Invariant Subspaces of a General Matrix Operator*. PhD thesis, Stanford University, Department of Computer Science, Stanford, CA, 1967. 104

[149] J. M. Varah. Invariant subspace perturbations for a non-normal matrix. In *Information processing 71 (Proc. IFIP Congress, Ljubljana, 1971), Vol. 2: Applications*, pages 1251–1253. North-Holland, Amsterdam, 1972. 72

[150] J. M. Varah. On the separation of two matrices. *SIAM Journal on Numerical Analysis*, 16(2):216–222, 1979. 76, 93

[151] J. J. Vartiainen. *Unitary Transformations for Quantum Computing*. PhD thesis, Helsinki University of Technology, Department of Engineering Physics and Mathematics, 2005. 69

[152] Z. Z. Wang and F. H. Qi. Analysis of multiframe super-resolution reconstruction for image anti-aliasing and aeblurring. *Image and Vision Computing*, 23:393–404, 2005. 68

[153] R. C. Ward. The combination shift QZ algorithm. *SIAM Journal on Numerical Analysis*, 12:835–853, 1975. 107

[154] P-Ä. Wedin. On pseudo-inverses of perturbed matrices. Technical Report, Department of Computer Science, Lund University, Sweden, 1969. 68

[155] P-Ä. Wedin. Perturbation theorey for pseudo-inverses. *BIT*, 13:217–232, 1973. 65, 68

[156] J. H. Wilkinson. *Rounding Errors in Algebraic Processes*. York House, London, 1963. reprinted by Dover, 1994. 36

[157] J. H. Wilkinson. Modern error analysis. *SIAM Review*, 13(4):548–568, 1971. 83

[158] J. H. Wilkinson. *The Algebraic Eigenvalue Problem*. Oxford University Press, Oxford, 1965. 30, 36, 61, 108

[159] P. Wolfe. The secant method for simultaneous nonlinear equations. *Communications of the ACM*, 2(12):12–13, 1959. 49

[160] Z. Zlatev, J. Wasniewski, and K. Schaumburg. Condition number estimators in a sparse matrix software. *SIAM Journal on Scientific and Statistical Computing*, 7(4):1175–1189, 1986. 33

Index

[GWS-B1]: 28, 104, 122
[GWS-B2]: 27, 28, 33, 50, 135, 137
[GWS-B3]: 28, 68, 106
[GWS-B4]: 60, 121
[GWS-B5]: 122
[GWS-B6]: 122
[GWS-B7]: 45, 122
[GWS-B8]: 108, 111, 114, 122
[GWS-J102]: 95, 96, 101, 102, 104, 519, 604
[GWS-J103]: 27, 41, 42, 135, 230
[GWS-J107]: 95, 96, 98, 99, 103, 519, 618
[GWS-J108]: 71, 87, 389, 475
[GWS-J109]: 71, 84, 87, 389, 481
[GWS-J110]: 111, 115, 116, 693, 713
[GWS-J111]: 99, 111, 113, 116, 693, 694
[GWS-J112]: 111, 117, 693, 720
[GWS-J113]: 99, 111, 115, 693, 720
[GWS-J114]: 71, 89, 93, 389, 504
[GWS-J117]: 42
[GWS-J118]: 28, 41, 42, 136, 242
[GWS-J119]: 59
[GWS-J11]: 46
[GWS-J12]: 46
[GWS-J13]: 46
[GWS-J14]: 46
[GWS-J15]: 71, 72, 93, 389, 390
[GWS-J16]: 57, 105–107, 630, 631
[GWS-J17]: 27, 34, 101, 135, 151
[GWS-J18]: 105, 107, 630, 650
[GWS-J19]: 37, 57, 71, 72, 77, 78, 79, 80, 93, 115, 389, 404
[GWS-J1]: 46
[GWS-J21]: 46
[GWS-J22]: 46
[GWS-J23]: 59
[GWS-J27]: 105, 108, 630, 667
[GWS-J28]: 59, 66
[GWS-J29]: 31, 45, 46, 54, 57, 261, 262
[GWS-J30]: 95, 96, 98, 99, 519, 536
[GWS-J31]: 45, 49, 57, 99, 261, 278
[GWS-J32]: 27, 30, 135, 159
[GWS-J33]: 95, 96, 99–101, 103, 104, 519
[GWS-J34]: 27, 37, 57, 75, 135, 162
[GWS-J35]: 57, 59, 63, 68, 73, 338, 353
[GWS-J37]: 96, 95, 99–101, 103, 104, 519, 558
[GWS-J38]: 57, 105, 106, 109, 630, 675
[GWS-J39]: 57
[GWS-J40]: 45, 50–52, 57, 261, 303
[GWS-J42]: 27, 31, 55, 57, 135, 173
[GWS-J44]: 46
[GWS-J47]: 64, 69
[GWS-J48, p. 273]: 83
[GWS-J48]: 71, 82, 93, 389, 443
[GWS-J49]: 27, 39, 96, 135, 182
[GWS-J4]: 59–62, 338, 339
[GWS-J56]: 36, 55
[GWS-J5]: 96–98, 519, 520
[GWS-J60]: 59
[GWS-J61]: 121
[GWS-J63]: 121
[GWS-J65]: 59, 67, 68, 338, 383

[GWS-J68]: 121
[GWS-J6]: 46
[GWS-J70]: 71, 80, 93, 389, 459
[GWS-J71]: 71, 84, 87, 93, 389, 464
[GWS-J73]: 45, 54, 56, 57, 261, 315
[GWS-J75]: 95, 96, 101, 102, 104, 519, 593
[GWS-J76]: 103
[GWS-J77]: 45, 56, 57, 96, 261, 323
[GWS-J78]: 27, 38, 39, 52, 135, 194
[GWS-J7]: 46
[GWS-J82]: 57
[GWS-J83]: 46
[GWS-J85]: 57, 104
[GWS-J87]: 45, 52, 58, 261, 330
[GWS-J89]: 27, 43, 135, 200
[GWS-J8]: 46
[GWS-J92]: 27, 38, 135, 212
[GWS-J94]: 27, 33, 135, 219
[GWS-J97]: 59
[GWS-N16]: 121
[GWS-N17]: 121
[GWS-N18]: 121
[GWS-N22]: 121
[GWS-N24]: 59
[GWS-N29]: 57
[GWS-N6]: 54, 59, 121
[GWS-N9]: 29
[GWS-T1]: 97

Part III

Reprints

12

Papers on Matrix Decompositions

1. [GWS-B2] Introduction from LINPACK *Users Guide,* (with J. J. Dongarra, J. R. Bunch, and C. B. Moler), SIAM, Philadelphia (1979).
2. [GWS-J17] (with R. H. Bartels), "Algorithm 432: Solution of the Matrix Equation $AX + XB = C$," *Communications of the ACM* **15** (1972) 820–826.
3. [GWS-J32] "The Economical Storage of Plane Rotations," *Numerische Mathematik* **25** (1976) 137–138.
4. [GWS-J34] "Perturbation Bounds for the QR Factorization of a Matrix," *SIAM Journal on Numerical Analysis* **14** (1977) 509–518.
5. [GWS-J42] (with A. K. Cline, C. B. Moler, and J. H. Wilkinson), "An Estimate for the Condition Number of a Matrix," *SIAM Journal on Numerical Analysis* **16** (1979) 368–375.
6. [GWS-J49] "Rank Degeneracy," *SIAM Journal on Scientific and Statistical Computing* **5** (1984) 403–413.
7. [GWS-J78] "On the Perturbation of LU, Cholesky, and QR Factorizations," *SIAM Journal on Matrix Analysis and Applications,* **14** (1993) 1141–1145.
8. [GWS-J89] "On Graded QR Decompositions of Products of Matrices," *Electronic Transactions in Numerical Analysis* **3** (1995) 39–49.
9. [GWS-J92] "On the Perturbation of LU and Cholesky Factors," *IMA Journal of Numerical Analysis,* **17** (1997) 1–6.
10. [GWS-J94] "The Triangular Matrices of Gaussian Elimination and Related Decompositions," *IMA Journal of Numerical Analysis,* **17** (1997) 7–16.
11. [GWS-J103] "Four Algorithms for the the (sic) Efficient Computation of Truncated Pivoted QR Approximations to a Sparse Matrix," *Numerische Mathematik,* **83** (1999) 313–323.

12. [GWS-J118] (with M. W. Berry and S. A. Pulatova) "Algorithm 844: Computing Sparse Reduced-Rank Approximations to Sparse Matrices," *ACM Transactions on Mathematical Software (TOMS)* **31** (2005) 252–269.

12.1. [GWS-B2] (with J. J. Dongarra, J. R. Bunch, and C. B. Moler) Introduction from LINPACK *Users Guide*

[GWS-B2] (with J. J. Dongarra, J. R. Bunch, and C. B. Moler), Introduction from LINPACK *Users Guide*, SIAM, Philadelphia (1979).

LINPACK
INPACK
NPACK
PACK
ACK
CK
K

USERS' GUIDE

J.J. Dongarra C.B. Moler
J.R. Bunch G.W. Stewart

Introduction

1. Overview

LINPACK is a collection of Fortran subroutines which analyze and solve various systems of simultaneous linear algebraic equations. The subroutines are designed to be completely machine independent, fully portable, and to run at near optimum efficiency in most operating environments.

Many of the subroutines deal with square coefficient matrices, where there are as many equations as unknowns. Some of the subroutines process rectangular coefficient matrices, where the system may be over- or underdetermined. Such systems are frequently encountered in least squares problems and other statistical calculations. Different subroutines are intended to take advantage of different special properties of the matrices and thereby save computer time and storage.

The entire coefficient matrix will usually be stored in the computer memory, although there are provisions for band matrices and for processing large rectangular matrices row by row. This means that on most contemporary computers, LINPACK will handle full matrices of order less than a few hundred and band matrices of order less than several thousand. There are no subroutines for general sparse matrices or for iterative methods for very large problems.

Most linear equation problems will require the use of two LINPACK subroutines, one to process the coefficient matrix and one to process a particular right hand side. This division of labor results in significant savings of computer time when there is a sequence of problems involving the same matrix, but different right hand sides. This situation is so common and the savings so important that no provision has been made for solving a single system with just one subroutine.

We make a somewhat vague distinction between a matrix "factorization" and a matrix "decomposition". For either, a given matrix is expressed as the product of two or three matrices of various special forms which subsequently allow rapid solution of linear systems and easy calculation of quantities useful in analyzing such systems. With a factorization, the user is rarely concerned about the details of the factors; they are simply passed from one subroutine to another. With a decomposition, the user will often be interested in accessing and manipulating the individual factors themselves. For the most part, factorizations are associated with standard problems involving square matrices and decompositions with more esoteric problems involving both square and non-square matrices.

A subroutine naming convention is employed in which each subroutine name is a coded specification of the computation done by that subroutine. All names consist of five letters in the form TXXYY . The first letter, T , indicates the matrix data type. Standard Fortran allows the use of three such types:

S REAL
D DOUBLE PRECISION
C COMPLEX

In addition, some Fortran systems allow a double precision complex type:

Z COMPLEX*16

The next two letters, XX , indicate the form of the matrix or its decomposition:

GE General
GB General band
PO Positive definite
PP Positive definite packed
PB Positive definite band
SI Symmetric indefinite
SP Symmetric indefinite packed
HI Hermitian indefinite
HP Hermitian indefinite packed
TR Triangular
GT General tridiagonal
PT Positive definite tridiagonal
CH Cholesky decomposition
QR Orthogonal-triangular decomposition
SV Singular value decomposition

The final two letters, YY , indicate the computation done by a particular subroutine:

FA Factor
CO Factor and estimate condition
SL Solve
DI Determinant and/or inverse and/or inertia
DC Decompose
UD Update
DD Downdate
EX Exchange

The following chart shows all the LINPACK subroutines. The initial S in the names may be replaced by D , C or Z and the initial C in the complex-only names may be replaced by a Z .

	CO	FA	SL	DI
SGE	✓	✓	✓	✓
SGB	✓	✓	✓	✓
SPO	✓	✓	✓	✓
SPP	✓	✓	✓	✓
SPB	✓	✓	✓	✓
SSI	✓	✓	✓	✓
SSP	✓	✓	✓	✓
CHI	✓	✓	✓	✓
CHP	✓	✓	✓	✓
STR	✓		✓	✓
SGT			✓	
SPT			✓	

	DC	SL	UD	DD	EX
SCH	✓		✓	✓	✓
SQR	✓	✓			
SSV	✓				

The remaining sections of this Introduction cover some software design and numerical analysis topics which apply to the entire package. Each of the chapters 1 through 11 describes a particular group of subroutines, ordered roughly as indicated by the preceding chart. Each chapter includes Overview, Usage and Examples sections which are intended for all users. In addition many chapters include additional sections on Algorithms, Programming Details and Performance which are intended for users requiring more specific information. In order to make each chapter fairly self-contained, some material is repeated in several related chapters.

2. Software Design

The overall design of LINPACK has been strongly influenced by TAMPR and by the BLAS. TAMPR is a software development system created by Boyle and Dritz (1974). It manipulates and formats Fortran programs to clarify their structure. It also generates variants of programs. The "master versions" of all the LINPACK subroutines are those which use complex arithmetic; versions which use single precision, double precision, and double precision complex arithmetic have been produced automatically by TAMPR. A user may thus convert from one type of arithmetic to another by simply changing the declarations in his program and changing the first letter of the LINPACK subroutines being used.

Anyone reading the Fortran source code for LINPACK subroutines should find the loops and logical structures clearly delineated by the indentation generated by TAMPR.

The BLAS are the Basic Linear Algebra Subprograms designed by Lawson, Hanson, Kincaid and Krogh (1978). They contribute to the speed as well as to the modularity and clarity of the LINPACK subroutines. LINPACK is distributed with versions of the BLAS written in standard Fortran which are intended to provide reasonably efficient execution in most operating environments. However, a particular computing installation may substitute machine language versions of the BLAS and thereby perhaps improve efficiency.

LINPACK is designed to be completely machine independent. There are no machine dependent constants, no input/output statements, no character manipulation, no COMMON or EQUIVALENCE statements, and no mixed-mode arithmetic. All the subroutines (except those whose names begin with Z) use the portable subset of Fortran defined by the PFORT verifier of Ryder (1974).

There is no need for machine dependent constants because there is very little need to check for "small" numbers. For example, candidates for pivots in Gaussian elimination are checked against an exact zero rather than against some small quantity. The test for singularity is made instead by estimating the condition of the matrix; this is not only machine independent, but also far more reliable. The convergence of the iteration in the singular value decomposition is tested in a machine independent manner by statements of the form

```
TEST1 = something not small
TEST2 = TEST1 + something possibly small
IF (TEST1 .EQ. TEST2) ...
```

The absence of mixed-mode arithmetic implies that the single precision subroutines do not use any double precision arithmetic and hence that the double precision subroutines do not require any kind of extended precision. It also implies that LINPACK does not include a subroutine for iterative improvement; however, an example in Chapter 1 indicates how such a subroutine could be added by anyone with easy access to mixed-mode arithmetic. (Some of the BLAS involve mixed-mode arithmetic, but they are not used by LINPACK.)

Floating point underflows and overflows may occur in some of the LINPACK subroutines. Any underflows which occur are harmless. We hope that the operating system sets underflowed quantities to zero and continues operation without producing any error messages. With some operating systems, it may be necessary to insert control cards or call special system subroutines to achieve this type of underflow handling.

Overflows, if they occur, are much more serious. They must be regarded as error situations resulting from improper use of the subroutines or from unusual scaling. Many precautions against overflow have been taken in LINPACK, but it is impossible to absolutely prevent overflow without seriously degrading performance on reasonably scaled problems. It is expected that overflows will cause the operating system to terminate the computation and that the user will have to correct the program or rescale the problem before continuing.

Fortran stores matrices by columns and so programs in which the inner loop goes up or down a column, such as

```
      DO 20 J = 1, N
         DO 10 I = 1, N
            A(I,J) = ...
   10    CONTINUE
   20 CONTINUE
```

generate sequential access to memory. Programs in which the inner loop goes across a row cause non-sequential access. Sequential access is preferable on operating systems which employ virtual memory or other forms of paging. LINPACK is consequentially "column oriented". Almost all the inner loops occur within the BLAS and, although the BLAS allow a matrix to be accessed by rows, this provision is never used by LINPACK. The column orientation requires revision of some conventional algorithms, but results in significant improvement in performance on operating systems with paging and cache memory.

All square matrices which are parameters of LINPACK subroutines are specified in the calling sequences by three arguments, for example

```
      CALL SGEFA(A,LDA,N,...)
```

Here A is the name of a two-dimensional Fortran array, LDA is the leading dimension of that array, and N is the order of the matrix stored in the array or in a portion of the array. The two parameters LDA and N have different meanings and need not have the same value. The amount of storage reserved for the array A is determined by a declaration in the user's program and LDA refers to the leading, or first, dimension as specified in this declaration. For example, the declaration

```
REAL A(50,50)
```

or

```
DIMENSION A(50,50)
```

should be accompanied by the initialization

```
DATA LDA/50/
```

or the statement

```
LDA = 50
```

The value of LDA should not be changed unless the declaration is changed. The order N of a particular coefficient matrix may be any value not exceeding the leading dimension of the array, that is $N \leq LDA$. The value of N may be changed by the user's program as systems of different orders are processed.

Rectangular matrices require a fourth argument, for example

```
CALL SQRDC(X,LDX,N,P,...)
```

Here the matrix is called X to adhere to the notation common in statistics, LDX is the leading dimension of the two-dimensional array, N is the number of rows in the matrix, and P is the number of columns. Note that the default Fortran typing conventions must be overridden by declaring P to be an integer. This conforms to usual statistical notation and is the only argument of a LINPACK subroutine which does not have the default type.

Many of the LINPACK subroutines have one or two arguments with the names JOB and INFO . JOB is always an input parameter. It is set by the user, often by simply including an integer constant in the call, to specify which of several possible computations are to be carried out. For example, SGESL solves a system of equations involving either the factored matrix or its transpose, and JOB should be zero or nonzero accordingly.

INFO is always an output parameter. It is used to return various kinds of diagnostic information from LINPACK routines. In some situations, INFO may be regarded as an error parameter. For example, in SPOFA , it is used to indicate that the matrix is not positive definite. In other situations, INFO may be one of the primary output quantities. For example, in SCHDC , it is an indication of the rank of a semi-definite matrix.

A few LINPACK subroutines require more space for storage of intermediate results than is provided by the primary parameters. These subroutines have a parameter WORK which is a one-dimensional array whose length is usually the number of rows or columns of the matrix being processed. The user will rarely be interested in the contents of WORK and so must merely provide the appropriate declaration.

Most of the LINPACK subroutines do not call any other LINPACK subroutine. The only set of exceptions involves the condition estimator subroutines, with names ending in CO, each of which calls the corresponding FA routine to factor the matrix. However, almost all the LINPACK subroutines call one or more of the BLAS. To facilitate construction of libraries, the source code for each LINPACK subroutine includes comments which list all of the BLAS and Fortran-supplied functions required by that subroutine.

3. General Numerical Properties

The purpose of this section is to give an informal discussion of a group of closely-related topics -- errors, detection of singularity, accuracy of computed results, and especially scaling. By scaling we mean the multiplication of the rows and columns of a matrix A by nonzero scaling factors. This amounts to replacing A by D_rAD_c, where D_r and D_c are diagonal matrices consisting respectively of the row and column scaling factors. Many matrix problems have mathematically equivalent scaled versions. For example, the linear system

$$Ax = b \tag{3.1}$$

is equivalent to the system

$$(D_rAD_c)(D_c^{-1}x) = (D_rb) \tag{3.2}$$

and this latter system can be solved by applying, say, SGEFA to D_rAD_c and then SGESL to D_rb to give $D_c^{-1}x$.

Scaling is important for two reasons. First, even if two formulations of a problem are mathematically equivalent, it does not follow that their numerical solution in the presence of rounding error will give identical results. For example, it is easy to concoct cases where the use of LINPACK routines to solve (3.1) and (3.2) will give accurate answers for one and inaccurate answers for the other. The second reason is that some of the LINPACK routines provide the user with numbers from which the accuracy of his solutions can be estimated. However, the numbers and their interpretation depend very much on how the problem has been scaled. It is thus important for the LINPACK user to have some idea of what is a good scaling.

For simplicity we shall confine ourselves to the case of a square matrix A of order n, although much of what we have to say applies to the general rectangular matrices treated by the QR and SV routines. The discussion is informal, and the reader is referred to other parts of this guide and to the literature for details.

Scaling problems arise as a consequence of errors in the matrix A . These errors may have a number of sources. If the elements of A have been measured they will be inaccurate owing to the limitations of the measuring instruments. If they have been computed, truncation error or rounding error will contaminate them. Even a matrix that is known exactly may not be representable within the finite word length of a digital computer. For example, the binary expansion of 1/10 is nonterminating and must be rounded.

Subsequent computations performed on A itself may sometimes be regarded as another of initial error. In particular, most of the LINPACK routines have the property that the computed solutions are the exact solutions of a problem involving a slightly perturbed matrix $A+E'$. We can then lump E' with the other errors in A to get a total error matrix E that accounts for initial errors and all the rounding errors made by the LINPACK routine in question. It will often happen that E' will be insignificant compared with the other errors, in which case we may assert that for practical purposes rounding error has played no role in the computation.

The presence of errors in A raises two problems. In the first place the error matrix E may be so large that $A+E$ can be singular, in which case many of the computations for which LINPACK was designed become meaningless. In the second place, even if E is small enough so that A is nonsingular, the solution of a problem involving $A+E$ may differ greatly from the one involving A . Thus we are led to ask two questions: How near is A to a singular matrix, and how greatly does an error E cause the solution to change? The answer to the second question depends on the problem being solved. For definiteness we shall suppose that we are trying to solve the linear system (3.1).

An answer to the above questions can be phrased in terms of a condition number associated with the matrix A ; however, we must first make precise what we mean by such vague phrases as "near", "differ greatly", etc. What is needed is a notion of size for a matrix, so that we can compare matrices unambiguously. The measure of size that we shall adopt in this introductory section is the number $\nu(A)$, which is defined to be the largest of the absolute values of the elements of A ; i.e.

$$\nu(A) = \max\{|a_{ij}|: \ i,j = 1,2,\ldots,n\} .$$

We can now say that E is smaller than A if $\nu(E) < \nu(A)$. In particular, the ratio

$\nu(E)/\nu(A)$ can be regarded as a relative indication of how much smaller E is than A. For example, if all the elements of A are roughly the same size and $\nu(E)/\nu(A) = 10^{-6}$, then the elements of A and $A+E$ agree to about six significant figures.

The LINPACK routines allow the user to estimate a condition number $\kappa(A) \geq 1$ with the following properties. The smallest matrix E for which $A+E$ is singular satisfies

(3.3) $$\frac{1}{f_n\kappa(A)} \leq \frac{\nu(E)}{\nu(A)} \leq \frac{f_n}{\kappa(A)} .$$

Moreover, if $x+h$ denotes the solution of the linear system

$$(A+E)(x+h) = b ,$$

then for all sufficiently small E

(3.4) $$\frac{\nu(h)}{\nu(x)} \leq g_n\kappa(A) \frac{\nu(E)}{\nu(A)} .$$

The coefficients f_n and g_n depend on how κ was estimated, and they tend to make the bounds conservative.

It is instructive to examine the bounds (3.3) and (3.4) informally. If we assume that f_n is about one, then (3.3) says that a relative perturbation of order κ^{-1} can make A singular, but no smaller perturbation can. For example, if the elements of A are roughly the same size and $\kappa(A) = 10^6$, then perturbations in the sixth place of the elements of A can make A singular while perturbations in the seventh will not. If A is known only to six figures, it must then be considered effectively singular.

The second bound (3.4) says that the relative error in x due to the perturbation E is $\kappa(A)$ times the relative error that E induces in A. In the above example, if E is 10^{-10}, then $\nu(h)/\nu(x)$ is 10^{-4}, and we should not expect the components of x to be accurate to more than four figures.

Both (3.3) and (3.4) suggest that one may encounter serious problems in dealing with matrices for which κ is large. Such matrices are said to be <u>ill-conditioned with respect to inversion</u>. For properly scaled problems, κ provides a quantitative measure of difficulty we can expect. For example, the bound (3.4) gives the following rule of thumb: if $\kappa(A)$ is 10^k then the solution of a linear system computed in t-digit (decimal) arithmetic will have no more than $t-k$ accurate figures. We stress again that A must be properly scaled for the rule of thumb to apply, a point which we shall return to in a moment.

Estimates for a condition number κ may be obtained from all LINPACK routines with the suffixes CO and DC (for the CHDC and QRDC routines, the pivoting option must be taken with JPVT initialized to zero). The details are as follows.

In CO routines take

$$\kappa \cong 1/\text{RCOND}, \quad (f_n = n,\ g_n = n^2).$$

In SVDC routines take

$$\kappa \cong S(1)/S(N), \quad (f_n = \sqrt{n},\ g_n = n).$$

In CHDC routines take

$$\kappa \cong (A(1,1)/A(N,N))**2 \quad (f_n = \sqrt{n},\ g_n = n),$$

where A is the output array.

In QRDC routines take

$$\kappa \cong A(1,1)/A(N,N), \quad (f_n = \sqrt{n},\ g_n = n),$$

where A is the output array.

For further details and sharper bounds in terms of norms, see the documentation for the individual routines.

A weak point in the above discussion is that the measure $\nu(E)$ attempts to compact information about the size of the elements of E into a single number and therefore cannot take into account special structure in E . If some of the elements of E are constrained to be much smaller than $\nu(E)$, the bounds (3.3) and (3.4) can be misleading. As an example, consider

$$A = \begin{bmatrix} 1 & 0 \\ 0 & 10^{-6} \end{bmatrix} .$$

It is easily seen that $\kappa(A) = 10^6$, and in fact that bound (3.3) is realistic in the sense that if

$$E = \begin{bmatrix} 0 & 0 \\ 0 & -10^{-6} \end{bmatrix}$$

then $\nu(E) = 1/\kappa$ and A+E is singular. On the other hand, if we attempt to solve the system (3.1) in 8-digit arithmetic, we shall obtain full 8-digit accuracy in the solution -- contrary to our rule of thumb.

What has gone wrong? The answer is that in solving (3.1), we have in effect solved a perturbed system $A+\tilde{E}$, where the elements of $\tilde{E}$ have the following orders of magnitude:

$$\begin{bmatrix} 10^{-8} & 0 \\ 0 & 10^{-14} \end{bmatrix} .$$

Now the estimate $\nu(\tilde{E}) = 10^{-8}$ is unrealistic in (3.4) since it cannot account for the critical fact that $|e_{22}|$ is not greater than 10^{-14} .

The above example suggests that the condition number will not be meaningful unless the elements of the error matrix E are all about the same size. Any scaling D_rAD_c of A automatically induces the same scaling D_rED_c of E . Consequently, we recommend the following scaling strategy.

> Estimate the absolute size of the errors in the matrix A and then scale so that as nearly as possible all the estimates are equal.

The rest of this section is devoted to elaborating the practical implications of this strategy.

In the first place, it is important to realize that the errors to be estimated include all initial errors -- measurement, computational, and rounding -- and in many applications rounding error will be the least of these. An exception is when the elements of A are known or computed to high accuracy and are rounded to accommodate the floating-point word of the computer in question. In this case the error in an element is proportional to the size of the element, and the strategy reduces to scaling the original matrix so that all its elements are roughly equal. This is a frequently recommended strategy, but it is applicable only when the matrix E represents a uniform relative error in the elements of A . It should be noted, however, that "equilibration" routines intended to scale a matrix A so that its elements are roughly equal in size can be applied instead to the matrix of error estimates to obtain the proper scaling factors.

It may be impossible to equilibrate the matrix of error estimates by row and column scaling, in which case the condition number may not be a reliable indicator of near singularity. This possibility is even more likely in special applications, where the class of permissible scaling operations may be restricted. For example, in dealing with symmetric matrices one must restrict oneself to scaling of the form DAD in order to preserve symmetry. As another example, row scaling a least squares problem amounts to solving a weighted problem, which may be impermissible in statistical applications.

Finally, although we have justified our scaling strategy by saying that it makes the condition number meaningful, the same scaling can have a beneficial effect on the numerical behavior of a number of LINPACK routines, in particular all routines based on Gaussian elimination with pivoting (i.e. GE and SI routines). However, the numerical properties of other routines are essentially unaffected by some types of scaling. For example, scaling of the form DAD will not change to any great extent the accuracy of solutions computed from PO routines. Likewise, QRDC is unaffected by column scaling, as also are the GE and GB routines.

12.2. [GWS-J17] (with R. H. Bartels), "Algorithm 432: Solution of the Matrix Equation $AX + XB = C$"

[GWS-J17] (with R. H. Bartels), "Algorithm 432: Solution of the Matrix Equation $AX + XB = C$," *Communications of the ACM* **15** (1972) 820–826.
`http://doi.acm.org/10.1145/361573.361582`

```
      SUBROUTINE PIVOT (N)
C PURPOSE - TO PERFORM THE PIVOT OPERATION BY UPDATING THE
C           INVERSE OF THE BASIS AND Q VECTOR.
C
      COMMON AM,Q,L1,B,NL1,NL2,A,NE1,NE2,IR,MBASIS,W,Z
      DIMENSION AM(50,50), Q(50), B(50,50), A(50)
      DIMENSION W(50), Z(50), MBASIS(100)
      DO 1 I=1,N
    1   B(IR,I)=B(IR,I)/A(IR)
      Q(IR)=Q(IR)/A(IR)
      DO 3 I=1,N
        IF (I.EQ.IR) GO TO 3
        Q(I)=Q(I)-Q(IR)*A(I)
        DO 2 J=1,N
          B(I,J)=B(I,J)-B(IR,J)*A(I)
    2     CONTINUE
    3   CONTINUE
C UPDATE THE INDICATOR VECTOR OF BASIC VARIABLES
      NL1=MBASIS(IR)
      L=N+IR
      NL2=MBASIS(L)
      MBASIS(IR)=NE1
      MBASIS(L)=NE2
      L1=L1+1
      RETURN
      END
      SUBROUTINE PPRINT (N)
C PURPOSE - TO PRINT THE CURRENT SOLUTION TO COMPLEMENTARY
C           PROBLEM AND THE ITERATION NUMBER.
C
      COMMON AM,Q,L1,B,NL1,NL2,A,NE1,NE2,IR,MBASIS,W,Z
      DIMENSION AM(50,50), Q(50), B(50,50), A(50)
      DIMENSION W(50), Z(50), MBASIS(100)
      WRITE(6,1) L1
    1 FORMAT (10X,13HITERATION NO.,I4)
      I=N+1
      J=1
    2 K1=MBASIS(I)
      K2=MBASIS(J)
      IF (Q(J).GE.0.0) GO TO 3
      Q(J)=0.0
    3 IF (K2.EQ.1) GO TO 5
      WRITE(6,4) K1,Q(J)
    4 FORMAT (10X,2HZ(,I4,2H)=,F15.5)
      GO TO 7
    5 WRITE(6,6) K1,Q(J)
    6 FORMAT (10X,2HW(,I4,2H)=,F15.5)
    7 I=I+1
      J=J+1
      IF (J.LE.N) GO TO 2
      RETURN
      END
```

Editor's note: *Algorithm 432 described here is available on magnetic tape from the Department of Computer Science, University of Colorado, Boulder, CO 80302. The cost for the tape is \$16.00 (U.S. and Canada) or \$18.00 (elsewhere). If the user sends a small tape (wt. less than 1 lb.) the algorithm will be copied on it and returned to him at a charge of \$10.00 (U.S. only). All orders are to be prepaid with checks payable to ACM Algorithms. The algorithm is re corded as one file of BCD 80 character card images at 556 B.P.I., even parity, on seven track tape. We will supply the algorithm at a density of 800 B.P.I. if requested. The cards for the algorithm are sequenced starting at 10 and incremented by 10. The sequence number is right justified in colums 80. Although we will make every attempt to insure that the algorithm conforms to the description printed here, we cannot guarantee it, nor can we guarantee that the algorithm is correct.—L.D.F.*

Algorithm 432

Solution of the Matrix Equation AX + XB = C [F4]

R.H. Bartels and G.W. Stewart [Recd. 21 Oct. 1970 and 7 March 1971]
Center for Numerical Analysis, The University of Texas at Austin, Austin, TX 78712

Key Words and Phrases: linear algebra, matrices, linear equations
CR Categories: 5.14
Language: Fortran

Description

The following programs are a collection of Fortran IV subroutines to solve the matrix equation

$$AX + XB = C \tag{1}$$

where A, B, and C are real matrices of dimensions $m \times m$, $n \times n$, and $m \times n$, respectively. Additional subroutines permit the efficient solution of the equation

$$A^T X + XA = C, \tag{2}$$

where C is symmetric. Equation (1) has applications to the direct solution of discrete Poisson equations [2].

It is well known that (1) has a unique solution if and only if the eigenvalues $\alpha_1, \alpha_2, \ldots, \alpha_m$ of A and $\beta_1, \beta_2, \ldots, \beta_n$ of B satisfy

$$\alpha_i + \beta_j \neq 0 \quad (i = 1, 2, \ldots, m; j = 1, 2, \ldots, n).$$

One proof of the result amounts to constructing the solution from complete systems of eigenvalues and eigenvectors of A and B, when they exist. This technique has been proposed as a computational method (e.g. see [1]); however, it is unstable when the eigensystem is ill conditioned. The method proposed here is based on the Schur reduction to triangular form by orthogonal similarity transformations.

Equation (1) is solved as follows. The matrix A is reduced to lower real Schur form A' by an orthogonal similarity transformation U; that is A is reduced to the real, block lower triangular form.

$$A' = U^T A U = \begin{bmatrix} A'_{11} & & & \bigcirc \\ A'_{21} & A'_{22} & & \\ \cdot & \cdot & \cdot & \\ \cdot & \cdot & & \cdot \\ \cdot & \cdot & & \cdot \\ A'_{p1} & A'_{p2} & \cdots & A'_{pp} \end{bmatrix},$$

where each matrix A'_{ii} is of order at most two. Similarly B is reduced to upper real Schur form by the orthogonal matrix V:

$$B' = V^T B V = \begin{bmatrix} B'_{11} & B'_{12} & \cdots & B'_{1q} \\ & B'_{22} & \cdots & B'_{2q} \\ & & \cdot & \cdot \\ & & \cdot & \cdot \\ & \bigcirc & & \cdot \\ & & & B'_{qq} \end{bmatrix},$$

This research was supported in part by Grant DA-ARO(D)-31-124-G721, Army Research Office, Durham, and by National Science Foundation Grant GP-5253 awarded to The University of Texas at Austin.

where again each B'_{ii} is of order at most two. If

$$C' = U^TCV = \begin{bmatrix} C'_{11} & \cdots & C'_{1q} \\ \cdot & & \\ \cdot & & \\ C'_{p1} & \cdots & C'_{pq} \end{bmatrix}$$

and

$$X' = U^TXV = \begin{bmatrix} X'_{11} & \cdots & X'_{1q} \\ \cdot & & \\ \cdot & & \\ X'_{p1} & \cdots & X'_{pq} \end{bmatrix},$$

then eq. (1) is equivalent to

$$A'X' + X'B' = C'.$$

If the partitions of A', B', C', and X' are conformal, then

$$A'_{kk}X'_{kl} + X'_{kl}B'_{ll} = C'_{kl} - \sum_{j=1}^{k-1} A'_{kj}X'_{jl} - \sum_{i=1}^{l-1} X'_{ki}B'_{il} \qquad (k = 1, 2, \cdots, p; \; l = 1, 2, \cdots, q). \tag{3}$$

These equations may be solved successively for $X'_{11}, X'_{21}, \ldots, X'_{p1}, X'_{12}, X'_{22}, \ldots$ The solution of (1) is then given by $X = UX'V^T$.

The reduction of A and B to real Schur form is accomplished by standard techniques. The matrix B is reduced to upper Hessenberg form by Householder's method [4, p. 34], and the upper Hessenberg matrix is in turn reduced to real Schur form by the QR algorithm [3]. The product of the transformations used in the reductions is accumulated to form the matrix V. The reduction of A to lower real Schur form is accomplished by reducing the transpose of A to upper real Schur form and transposing back.

Since the QR algorithm is an iterative method that, as used here, reduces the subdiagonal elements of an upper Hessenberg matrix to zero, some criterion must be adopted for determining when an element is negligible. In these programs an element of H is considered negligible if it is less than or equal to $\epsilon_H \| H \|_\infty$ where ϵ_H is a constant supplied by the user. This criterion is appropriate if the elements of H are all of roughly the same size. A different criterion may be required if the elements vary widely and the small elements are significant, as when the elements decrease greatly in size as one passes from the upper left to the lower right corners of A (see, for example, the criterion in [3]).

The solution for X'_{kl} in (3) still requires the solution of a matrix equation of the form (1). However, in this case the matrices A'_{kk} and B'_{ll} are of order at most two; hence the solution of (3) can be obtained by solving a linear system of order at most four. For example, if A'_{kk} and B'_{ll} are both of order two, then

$$\begin{bmatrix} a'_{11} + b'_{11} & a'_{12} & b'_{21} & 0 \\ a'_{21} & a'_{22} + a'_{11} & 0 & b'_{21} \\ b'_{12} & 0 & a'_{11} + b'_{22} & a'_{12} \\ 0 & b'_{12} & a'_{21} & a'_{22} + b'_{22} \end{bmatrix} \begin{bmatrix} x'_{11} \\ x'_{21} \\ x'_{12} \\ x'_{22} \end{bmatrix} = \begin{bmatrix} d_{11} \\ d_{21} \\ d_{12} \\ d_{22} \end{bmatrix},$$

where a'_{ij}, b'_{ij}, and x'_{ij} denote the elements of A'_{kk}, B'_{ll}, and X'_{lk} and d_{ij} denotes the elements of the right-hand side of (3). The systems arising from (3) are solved using the Crout reduction. Once calculated, the solution X'_{kl} may be stored in the locations occupied by C_{kl}, which is no longer needed.

The programs contain provisions for skipping the reduction of A to real Schur form, so that once A' and U have been calculated they may be used to solve new systems with different matrices B and C. Likewise, the reduction of B may be skipped. These provisions may be used to advantage in the iterative refinement of the computed solution X_1 of (1). Namely, let the residual matrix $R_1 = C - AX_1 - X_1B$ be computed in double precision and rounded to single precision (on many computers this may be done with single-precision multiplications and double-precision additions). Use the programs to solve the system $AY_1 + Y_1B = R_1$. Then $X_2 = X_1 + Y_1$ will in general be a more accurate approximate solution. This process may be iterated, no step after the computation of X_1 requiring reductions of A and B. This iteration is perfectly analogous to the iterative refinement of approximate solutions of linear systems described by Wilkinson [4, p. 255].

The following trick enables one to use an upper rather than a lower real Schur form of A in the solution of (1). Let D be the matrix with ones on the secondary diagonal and zeros elsewhere. Then

$$(DAD)DX + DXB = DC. \tag{4}$$

Moreover, if $A' = U^TAU$ is an upper real Schur form for A, then $DA'D = (DUD)^T(DAD)(DUD)$ is a lower real Schur form for DAD. Hence to calculate DX, which is X with its rows written in reverse order, one may use the above algorithm with $DA'D$ and DUD to solve (4). A similar trick enables one to use a lower real Schur form for B.

In principle, the algorithm described above can be used to solve the symmetric problem (2). However, it is possible to take advantage of the symmetry. Let U be orthogonal and $A' = U^TAU$ be in upper real Schur form. Partition A', $C' = U^TCU$, and $X' = U^TXU$ in the form

$$A' = \begin{bmatrix} A'_{11} & A'_{12} \\ 0 & A'_{22} \end{bmatrix}, \quad X' = \begin{bmatrix} X'_{11} & X'^T_{21} \\ X'_{21} & X'_{22} \end{bmatrix}, \quad C' = \begin{bmatrix} C'_{11} & C'^T_{21} \\ C'_{21} & C'_{22} \end{bmatrix},$$

where A'_{11}, X'_{11}, and C'_{11} are at most of order 2. Then from the equation $A'^TX' + X'A' = C'$, it follows that

$$A'^T_{22}X'_{22} + X'_{22}A'_{22} = C'_{22} - X'_{21}A'_{12} - A'^T_{12}X'_{21}.$$

Hence, once X'_{11} and X'_{21} have been calculated, the size of the problem can be reduced.

The matrix X'_{21} is computed as described above for the general case. The matrix X'_{11} satisfies the symmetric equation

$$A'^T_{11}X'_{11} + X'_{11}A'_{11} = C'_{11}, \tag{5}$$

whose solution is trivial when A'_{11} is of order unity. When A'_{11} is of order two, equation (5) gives a new linear system of order three for the three distinct elements of X'_{11}.

A mild saving in operations may be realized in the computation of $C' = U^TCU$ and $X = UX'U^T$. Let $C = T + T^T$, where T is upper triangular. Then

$$C' = U^TCU = U^TTU + (U^TTU)^T.$$

Thus one need calculate only U^TTU, and, since T is upper triangular, the product TU can be computed with about half the operations required for the computation of CU.

The number of multiplications required for the solution of (1) is probably overestimated by

$$(2 + 4\sigma)(m^3 + n^3) + \frac{5}{2}(mn^2 + nm^2)$$

where σ is the average number of QR steps required to make a subdiagonal element negligible. The first term is due to the reduction of A and B to real Schur form. A like estimate for the solution of (2) is given by

$$(2 + 4\sigma)n^3 + \frac{7}{2}n^3;$$

the first term is again due to the reduction of A to real Schur form.

To solve the nonsymmetric problem, the user must furnish $2m^2 + 2n^2 + mn$ storage locations to hold the matrices A, U, B, V, and C. If A, B, and C are required for later use, they must be stored elsewhere, since the programs overwrite A and B with their real Schur forms and C with the solution. The symmetric problem requires $3n^2$ locations to hold A, U, and C.

In assessing the effects of rounding error on the algorithm, we should consider the algorithm stable if the computed solution were near a matrix $\overline{X}$ that satisfied

$$(A+E)\overline{X} + \overline{X}(B+F) = C + G$$

for some small E, F, and G. We are unable to establish such a result. However, an elementary rounding error analysis, combined with the known properties of the other algorithms used in the method, shows that the residual matrix is small compared with the larger of $\| A \| \| X \|$ and $\| B \| \| X \|$.

Here follows a brief description of the programs listed below. Detailed information on their use will be found in the program listings themselves. The casual user need only familiarize himself with the programs *AXPXB* and *ATXPXA*, which coordinate the other programs for the solutions of (1) and (2), respectively.

AXPXB. The coordinating program for the solution of (1). Given A, B and C the program overwrites C with the solution X. The lower real Schur form of A overwrites A, and the upper real Schur form of B overwrites B. The user may furnish the real Schur forms and skip the reductions. The subroutine requires the subroutines *HSHLDR, BCKMLT, SCHUR, SHRSLV*, and *SYSSLV*.

ATXPXA. The coordinating program for the solution of (2). Given A and C the program overwrites C with the solution X. The upper real Schur form of A overwrites A. The user may furnish the real Schur form and skip the reduction. The subroutine requires the subroutines *HSHLDR, BCKMLT, SCHUR, SYMSLV*, and *SYSSLV*.

HSHLDR. Reduces a matrix A to upper Hessenberg form. The upper Hessenberg form and a history of the transformations overwrite A.

BCKMLT. Takes the output A of *HSHLDR* and computes the orthogonal matrix U that reduces the original matrix A to upper Hessenberg form. At the user's option the elements of U can overwrite A.

SCHUR. Computes an upper real Schur form of an upper Hessenberg matrix A. *SCHUR* is an adaptation of thel Agol procedure *hqr* by Martin, Peters, and Wilkinson [1]. The product of the transformations used in the reduction is accumulated. *SCHUR* leaves undisturbed the elements below the third subdiagonal of the array containing A. (N.b. The modifications made in *hqr* to find a real Schur form make *SCHUR* an inefficient program for calculating the eigenvalues of an upper Hessenberg matrix.)

SHRSLV. Solves an equation of the form (1), where A is in lower real Schur form and B is in upper real Schur form.

SYMSLV. Solves an equation of the form (2), where A is in upper real Schur form.

SYSSLV. Solves a system of linear equations.

When $m \geq n$, *AXPXB* can be modified so that the real Schur forms of A and B share the storage originally allocated to A and the matrix V occupies the locations occupied by B. The modifications are as follows. Replace the section labeled *"IF REQUIRED, REDUCE B TO UPPER REAL SCHUR FORM"* with

```
35 IF(EPSB .LT. 0.) GO TO 45
      CALL HSHLDR (B, N, NB)
      DO 40 I = 1, N
        IF (I .NE. 1) A(I, I+4) = B(I-1, N1)
        DO 40 J = I, N
          A(I, J+5) = B(I, J)
40 CONTINUE
      CALL BCKMLT(B,B,N,NB,NB)
      CALL SCHURA(1,6),B,N,NA,NB,EPSB,FAIL)
      FAIL = -FAIL
      IF(FAIL .NE. 0) RETURN
```

In the sections labeled *"TRANSFORM C"* and *"TRANSFORM C BACK TO THE SOLUTION"* replace all occurrences of the variable V with B and all references to $A(I,M1)$ with $A(M1,I)$. Change the call to *SHRSLV* to

CALL SHRSLV(A,A(1,6),C,M,N,NA,NA,NC).

Note that in this modification the reduction of B to real Schur form cannot be skipped without also skipping the reduction of A.

When $m \leq n$ a similar modification can be made to store the Schur form of A and B together in B.

References

1. Bickley, W.G. and McNamee, J. Matrix and other direct methods for the solution of systems of linear difference equations. *Philos. Trans. Roy. Soc.* (London) Ser. A, *252* (1960), 69–131.
2. Dorr, Fred W. The direct solution of the discrete Poisson equation on a rectangle. *SIAM Rev. 12* (1970), 248–263.
3. Martin, R.S., Peters, G., and Wilkinson, J.H. The QR algorithm for real Hessenberg matrices. (Handbook series linear algebra.) *Numer. Math. 14* (1970), 219–231.
4. Wilkinson, J.H. *The Algebraic Eigenvalue Problem*. Clarendon, Oxford, 1965.

Algorithm

```
      SUBRØUTINE AXPXB(A,U,M,NA,NU,B,V,N,NB,NV,C,NC,EPSA,
     1EPSB,FAIL)
C
C AXPXB IS A FØRTRAN IV SUBRØUTINE TØ SØLVE THE REAL MATRIX
C EQUATIØN AX + XB - C.  THE MATRICES A AND B ARE TRANS-
C FØRMED INTØ REAL SCHUR FØRM, AND THE TRANSFØRMED SYSTEM IS
C SØLVED BY BACK SUBSTITUTIØN.  THE PRØGRAM REQUIRES THE
C AUXILIARY SUBRØUTINES HSHLDR, BCKMLT, SCHUR, AND SHRSLV.
C THE PARAMETERS IN THE ARGUMENT LIST ARE
C
C            A       A DØUBLY SUBSCRIPTED ARRAY CØNTAINING THE
C                    MATRIX A.  ØN RETURN, THE LØWER TRIANGLE
C                    AND SUPERDIAGØNAL ØF THE ARRAY A CØNTAIN
C                    A LØWER REAL SCHUR FØRM ØF A.  THE ARRAY
C                    A MUST BE DIMENSIØNED AT LEAST M+1 BY
C                    M+1.
C            U       A DØUBLY SUBSCRIPTED ARRAY THAT, ØN
C                    RETURN, CØNTAINS THE ØRTHØGØNAL MATRIX
C                    THAT REDUCES A TØ REAL SCHUR FØRM.
C            M       THE ØRDER ØF THE MATRIX A.
C            NA      THE FIRST DIMENSIØN ØF THE ARRAY A.
C            NU      THE FIRST DIMENSIØN ØF THE ARRAY U.
C            B       A DØUBLY SUBSCRIPTED ARRAY CØNTAINING THE
C                    MATRIX B.  ØN RETURN, THE UPPER TRIANGLE
C                    AND SUBDIAGØNAL ØF THE ARRAY B CØNTAIN AN
C                    UPPER REAL SCHUR FØRM ØF B.  THE ARRAY B
C                    MUST BE DIMENSIØNED AT LEAST M+1 BY M+1.
C            V       A DØUBLY SUBSCRIPTED ARRAY THAT, ØN
C                    RETURN, CØNTAINS THE ØRTHØGØNAL MATRIX
C                    THAT REDUCES B TØ REAL SCHUR FØRM.
C            N       THE ØRDER ØF THE MATRIX B.
C            NB      THE FIRST DIMENSIØN ØF THE ARRAY B.
C            NV      THE FIRST DIMENSIØN ØF THE ARRAY V.
C            C       A DØUBLY SUBSCRIPTED ARRAY CØNTAINING THE
C                    MATRIX C.  ØN RETURN, C CØNTAINS THE
C                    SØLUTIØN MATRIX X.
C            NC      THE FIRST DIMENSIØN ØF THE ARRAY C.
C            EPSA    A CØNVERGENCE CRITERIØN FØR THE REDUCTIØN
C                    ØF A TØ SCHUR FØRM.  EPSA SHØULD BE SET
C                    SLIGHTLY SMALLER THAN 10.**(-N), WHERE N
C                    IS THE NUMBER ØF SIGNIFICANT DIGITS IN
C                    THE ELEMENTS ØF THE MATRIX A.
C            EPSB    A CØNVERGENCE CRITERIØN FØR THE REDUCTIØN
C                    ØF B TØ REAL SCHUR FØRM.
C            FAIL    AN INTEGER VARIABLE THAT, ØN RETURN,
C                    CØNTAINS AN ERRØR SIGNAL.  IF FAIL IS
C                    PØSITIVE (NEGATIVE) THEN THE PRØGRAM WAS
C                    UNABLE TØ REDUCE A (B) TØ REAL SCHUR
C                    FØRM.  IF FAIL IS ZERØ, THE REDUCTIØNS
C                    PRØCEEDED WITHØUT MISHAP.
C
C WHEN EPSA IS NEGATIVE THE REDUCTIØN ØF A TØ REAL SCHUR
C FØRM IS SKIPPED AND THE ARRAYS A AND U ARE ASSUMED TØ
C CØNTAIN THE SCHUR FØRM AND ACCØMPANYING ØRTHØGØNAL MATRIX.
C THIS PERMITS THE EFFICIENT SØLUTIØN ØF SEVERAL EQUATIØNS
C ØF THE FØRM AX + BX = C WHEN A DØES NØT CHANGE.  LIKEWISE,
C IF EPSB IS NEGATIVE, THE REDUCTIØN ØF B TØ REAL SCHUR FØRM
C IS SKIPPED.
C
      REAL
     1A(NA,1),U(NU,1),B(NB,1),V(NV,1),C(NC,1),EPSA,EPSB,TEMP
      INTEGER
     1M,NA,NU,N,NB,NV,NC,FAIL,M1,MM1,N1,NM1,I,J,K
      M1 = M+1
      MM1 = M-1
      N1 = N+1
      NM1 = N-1
C
C IF REQUIRED, REDUCE A TØ UPPER REAL SCHUR FØRM.
C
      IF(EPSA .LT. 0.) GØ TØ 35
      DØ 10 I=1,M
        DØ 10 J=I,M
          TEMP = A(I,J)
          A(I,J) = A(J,I)
          A(J,I) = TEMP
   10 CØNTINUE
      CALL HSHLDR(A,M,NA)
      CALL BCKMLT(A,U,M,NA,NU)
      IF(MM1 .EQ. 0) GØ TØ 25
      DØ 20 I=1,MM1
        A(I+1,I) = A(I,M1)
   20 CØNTINUE
      CALL SCHUR(A,U,M,NA,NU,EPSA,FAIL)
```

```
      IF(FAIL .NE. 0) RETURN
   25 DØ 30 I=1,M
        DØ 30 J=I,M
          TEMP = A(I,J)
          A(I,J) = A(J,I)
          A(J,I) = TEMP
   30 CØNTINUE
C
C IF REQUIRED, REDUCE B TØ UPPER REAL SCHUR FØRM.
C
   35 IF(EPSB .LT. 0.) GØ TØ 45
      CALL HSHLDR(B,N,NB)
      CALL BCKMLT(B,V,N,NB,NV)
      IF(NM1 .EQ. 0) GØ TØ 45
      DØ 40 I=1,NM1
        B(I+1,I) = B(I,N1)
   40 CØNTINUE
      CALL SCHUR(B,V,N,NB,NV,EPSB,FAIL)
      FAIL = -FAIL
      IF(FAIL .NE. 0) RETURN
C
C TRANSFØRM C.
C
   45 DØ 60 J=1,N
        DØ 50 I=1,M
          A(I,M1) = 0.
          DØ 50 K=1,M
            A(I,M1) = A(I,M1) + U(K,I)*C(K,J)
   50   CØNTINUE
        DØ 60 I=1,M
          C(I,J) = A(I,M1)
   60 CØNTINUE
      DØ 80 I=1,M
        DØ 70 J=1,N
          B(N1,J) = 0.
          DØ 70 K=1,N
            B(N1,J) = B(N1,J) + C(I,K)*V(K,J)
   70   CØNTINUE
        DØ 80 J=1,N
          C(I,J) = B(N1,J)
   80 CØNTINUE
C
C SØLVE THE TRANSFØRMED SYSTEM.
C
      CALL SHRSLV(A,B,C,M,N,NA,NB,NC)
C
C TRANSFØRM C BACK TØ THE SØLUTIØN.
C
      DØ 100 J=1,N
        DØ 90 I=1,M
          A(I,M1) = 0.
          DØ 90 K=1,M
            A(I,M1) = A(I,M1) + U(I,K)*C(K,J)
   90   CØNTINUE
        DØ 100 I=1,M
          C(I,J) = A(I,M1)
  100 CØNTINUE
      DØ 120 I=1,M
        DØ 110 J=1,N
          B(N1,J) = 0.
          DØ 110 K=1,N
            B(N1,J) = B(N1,J) + C(I,K)*V(J,K)
  110   CØNTINUE
        DØ 120 J=1,N
          C(I,J) = B(N1,J)
  120 CØNTINUE
      RETURN
      END
      SUBRØUTINE SHRSLV(A,B,C,M,N,NA,NB,NC)
C SHRSLV IS A FØRTRAN IV SUBRØUTINE TØ SØLVE THE REAL MATRIX
C EQATIØN AX + XB = C, WHERE A IS IN LØWER REAL SCHUR FØRM
C AND B IS IN UPPER REAL SCHUR FØRM.  SHRSLV USES THE AUX-
C ILIARY SUBRØUTINE SYSSLV, WHICH IT CØMMUNICATES WITH
C THRØUGH THE CØMMØN BLØCK SLVBLK.  THE PARAMETERS IN THE
C ARGUMENT LIST ARE
C            A       A DØUBLY SUBSCRIPTED ARRAY CØNTAINING THE
C                    MATRIX A IN LØWER REAL SCHUR FØRM.
C            B       A DØUBLY SUBSCRIPTED ARRAY CØNTAINING THE
C                    MATRIX B IN UPPER REAL SCHUR FØRM.
C            C       A DØUBLY SUBSCRIPTED ARRAY CØNTAINING THE
C                    MATRIX C.
C            M       THE ØRDER ØF THE MATRIX A.
C            N       THE ØRDER ØF THE MATRIX B.
C            NA      THE FIRST DIMENSIØN ØF THE ARRAY A.
C            NB      THE FIRST DIMENSIØN ØF THE ARRAY B.
C            NC      THE FIRST DIMENSIØN ØF THE ARRAY C.
C
       REAL
     1A(NA,1),B(NB,1),C(NC,1),T,P
       INTEGER
     1M,N,NA,NB,NC,K,KM1,DK,KK,L,LM1,DL,LL,I,IB,J,JA,NSYS
       CØMMØN/SLVBLK/T(5,5),P(5),NSYS
      L = 1
   10   LM1 = L-1
        DL = 1
        IF(L .EQ. N) GØ TØ 15
        IF(B(L+1,L) .NE. 0.) DL = 2
   15   LL = L+DL-1
        IF(L .EQ. 1) GØ TØ 30
        DØ 20 J=L,LL
          DØ 20 I=1,M
            DØ 20 IB=1,LM1
              C(I,J) = C(I,J) - C(I,IB)*B(IB,J)
   20   CØNTINUE
   30   K = 1
   40     KM1 = K-1
          DK = 1
          IF(K .EQ. M) GØ TØ 45
          IF(A(K,K+1) .NE. 0.) DK = 2
   45     KK = K+DK-1
          IF(K .EQ. 1) GØ TØ 60
          DØ 50 I=K,KK
            DØ 50 J=L,LL
              DØ 50 JA=1,KM1
                C(I,J) = C(I,J) - A(I,JA)*C(JA,J)
   50     CØNTINUE
   60     IF(DL .EQ. 2) GØ TØ 80
          IF(DK .EQ. 2) GØ TØ 70
          T(1,1) = A(K,K) + B(L,L)
          IF(T(1,1) .EQ. 0.) STØP
          C(K,L) = C(K,L)/T(1,1)
          GØ TØ 100
   70     T(1,1) = A(K,K) + B(L,L)
          T(1,2) = A(K,KK)
          T(2,1) = A(KK,K)
          T(2,2) = A(KK,KK) + B(L,L)
          P(1) = C(K,L)
          P(2) = C(KK,L)
          NSYS = 2
          CALL SYSSLV
          C(K,L) = P(1)
          C(KK,L) = P(2)
          GØ TØ 100
   80     IF(DK .EQ. 2) GØ TØ 90
          T(1,1) = A(K,K) + B(L,L)
          T(1,2) = B(LL,L)
          T(2,1) = B(L,LL)
          T(2,2) = A(K,K) + B(LL,LL)
          P(1) = C(K,L)
          P(2) = C(K,LL)
          NSYS = 2
          CALL SYSSLV
          C(K,L) = P(1)
          C(K,LL) = P(2)
          GØ TØ 100
   90     T(1,1) = A(K,K) + B(L,L)
          T(1,2) = A(K,KK)
          T(1,3) = B(LL,L)
          T(1,4) = 0.
          T(2,1) = A(KK,K)
          T(2,2) = A(KK,KK) + B(L,L)
          T(2,3) = 0.
          T(2,4) = T(1,3)
          T(3,1) = B(L,LL)
          T(3,2) = 0.
          T(3,3) = A(K,K) + B(LL,LL)
          T(3,4) = T(1,2)
          T(4,1) = 0.
          T(4,2) = T(3,1)
          T(4,3) = T(2,1)
          T(4,4) = A(KK,KK) + B(LL,LL)
          P(1) = C(K,L)
          P(2) = C(KK,L)
          P(3) = C(K,LL)
          P(4) = C(KK,LL)
          NSYS = 4
          CALL SYSSLV
          C(K,L) = P(1)
          C(KK,L) = P(2)
          C(K,LL) = P(3)
          C(KK,LL) = P(4)
  100   K = K + DK
        IF(K .LE. M) GØ TØ 40
      L = L + DL
      IF(L .LE. N) GØ TØ 10
      RETURN
      END
      SUBRØUTINE ATXPXA(A,U,C,N,NA,NU,NC,EPS,FAIL)
C
C ATXPXA IS A FØRTRAN IV SUBRØUTINE TØ SØLVE THE REAL MATRIX
C EQUATIØN TRANS(A)*X + X*A = C, WHERE C IS SYMMETRIC AND
C TRANS(A) DENØTES THE TRANSPØSE ØF A.  THE EQUATIØN IS
C TRANSFØRMED SØ THAT A IS IN UPPER REAL SCHUR FØRM, AND THE
C TRANSFØRMED EQUATIØN IS SØLVED BY A RECURSIVE PRØCEDURE.
C THE PRØGRAM REQUIRES THE AUXILIARY SUBRØUTINES HSHLDR,
C BCKMLT, SCHUR, AND SYMSLV.  THE PARAMETERS IN THE ARGUMENT
C LIST ARE
C            A       A DØUBLY SUBSCRIPTED ARRAY CØNTAINING THE
C                    MATRIX A.  ØN RETURN, THE UPPER TRIANGLE
C                    AND THE FIRST SUBDIAGØNAL ØF THE ARRAY A
C                    CØNTAIN AN UPPER REAL SCHUR FØRM ØF A.
C                    THE ARRAY A MUST BE DIMENSIØNED AT LEAST
C                    N+1 BY N+1.
C            U       A DØUBLY SUBSCRIPTED ARRAY THAT, ØN
C                    RETURN, CØNTAINS THE ØRTHØGØNAL MATRIX
C                    THAT REDUCES A TØ UPPER REAL SCHUR FØRM.
C            C       A DØUBLY SUBSCRIPTED ARRAY CØNTAINING THE
C                    MATRIX C.  ØN RETURN, C CØNTAINS THE
C                    SØLUTIØN MATRIX X.
C            N       THE ØRDER ØF THE MATRIX A.
C            NA      THE FIRST DIMENSIØN ØF THE ARRAY A.
C            NU      THE FIRST DIMENSIØN ØF THE ARRAY U.
C            NC      THE FIRST DIMENSIØN ØF THE ARRAY C.
C            EPS     A CØNVERGENCE CRITERIØN FØR THE REDUCTIØN
C                    ØF A TØ REAL SCHUR FØRM.  EPS SHØULD BE
C                    SET SLIGHTLY SMALLER THAN 10.**(-N),
C                    WHERE N IS THE NUMBER ØF SIGNIFICANT
C                    DIGITS IN THE ELEMENTS ØF THE MATRIX A.
C            FAIL    AN INTEGER VARIABLE THAT, ØN RETURN,
C                    CØNTAINS AN ERRØR SIGNAL.  IF FAIL IS
C                    NØNZERØ, THEN THE PRØGRAM WAS UNABLE TØ
C                    REDUCE A TØ REAL SCHUR FØRM.  IF FAIL IS
C                    ZERØ, THE REDUCTIØN PRØCEEDED WITHØUT
C                    MISHAP.
C
C WHEN EPS IS NEGATIVE, THE REDUCTIØN ØF A TØ REAL SCHUR
C FØRM IS SKIPPED AND THE ARRAYS A AND U ARE ASSUMED TØ
C CØNTAIN THE SCHUR FØRM AND ACCØMPANYING ØRTHØGØNAL MATRIX.
C THIS PERMITS THE EFFICIENT SØLUTIØN ØF SEVERAL EQUATIØNS
C WITH DIFFERENT RIGHT HAND SIDES.
C
       REAL
     1A(NA,1),U(NU,1),C(NC,1),EPS
       INTEGER
     1N,NA,NU,NC,FAIL,N1,NM1,I,J,K
       N1 = N+1
       NM1 = N-1
C
C IF REQUIRED, REDUCE A TØ LØWER REAL SCHUR FØRM.
C
```

```
      IF(EPS .LT. 0.) GØ TØ 15
      CALL HSHLDR(A,N,NA)
      CALL BCKMLT(A,U,N,NA,NU)
      DØ 10 I=1,NM1
        A(I+1,I) = A(I,N1)
   10 CØNTINUE
      CALL SCHUR(A,U,N,NA,NU,EPS,FAIL)
      IF(FAIL .NE. 0) RETURN
C
C TRANSFØRM C.
C
   15 DØ 20 I=1,N
        C(I,I) = C(I,I)/2.
   20 CØNTINUE
      DØ 40 I=1,N
        DØ 30 J=1,N
          A(N1,J) = 0.
          DØ 30 K=I,N
            A(N1,J) = A(N1,J) + C(I,K)*U(K,J)
   30   CØNTINUE
        DØ 40 J=1,N
          C(I,J) = A(N1,J)
   40 CØNTINUE
      DØ 60 J=1,N
        DØ 50 I=1,N
          A(I,N1) = 0.
          DØ 50 K=1,N
            A(I,N1) = A(I,N1) + U(K,I)*C(K,J)
   50   CØNTINUE
        DØ 60 I=1,N
          C(I,J) = A(I,N1)
   60 CØNTINUE
      DØ 70 I=1,N
        DØ 70 J=I,N
          C(I,J) = C(I,J) + C(J,I)
          C(J,I) = C(I,J)
   70 CØNTINUE
C
C SØLVE THE TRANSFØRMED SYSTEM.
C
      CALL SYMSLV(A,C,N,NA,NC)
C
C TRANSFØRM C BACK TØ THE SØLUTIØN.
C
      DØ 80 I=1,N
        C(I,I) = C(I,I)/2.
   80 CØNTINUE
      DØ 100 I=1,N
        DØ 90 J=1,N
          A(N1,J) = 0.
          DØ 90 K=I,N
            A(N1,J) = A(N1,J) + C(I,K)*U(J,K)
   90   CØNTINUE
        DØ 100 J=1,N
          C(I,J) = A(N1,J)
  100 CØNTINUE
      DØ 120 J=1,N
        DØ 110 I=1,N
          A(I,N1) = 0.
          DØ 110 K=1,N
            A(I,N1) = A(I,N1) + U(I,K)*C(K,J)
  110   CØNTINUE
        DØ 120 I=1,N
          C(I,J) = A(I,N1)
  120 CØNTINUE
      DØ 130 I=1,N
        DØ 130 J=I,N
          C(I,J) = C(I,J) + C(J,I)
          C(J,I) = C(I,J)
  130 CØNTINUE
      RETURN
      END
      SUBRØUTINE SYMSLV(A,C,N,NA,NC)
C
C SYMSLV IS A FØRTRAN IV SUBRØUTINE TØ SØLVE THE REAL MATRIX
C EQUATIØN TRANS(A)*X + X*A = C, WHERE C IS SYMMETRIC, A IS
C IN UPPER REAL SCHUR FØRM, AND TRANS(A) DENØTES THE TRANS-
C PØSE ØF A.  SYMSLV USES THE AUXILIARY SUBRØUTINE SYSSLV,
C WHICH IT CØMMUNICATES WITH THRØUGH THE CØMMØN BLØCK
C SLVBLK.  THE PARAMETERS IN THE ARGUMENT LIST ARE
C            A       A DØUBLY SUBSCRIPTED ARRAY CØNTAINING THE
C                    MATRIX A IN UPPER REAL SCHUR FØRM.
C            C       A DØUBLY SUBSCRIPTED ARRAY CØNTAINING THE
C                    MATRIX C.
C            N       THE ØRDER ØF THE MATRIX A.
C            NA      THE FIRST DIMENSIØN ØF THE ARRAY A.
C            NC      THE FIRST DIMENSIØN ØF THE ARRAY C.
C
      REAL
     1A(NA,1),C(NC,1),T,P
      INTEGER
     1N,NA,NC,K,KK,DK,KM1,L,LL,DL,LDL,I,IA,J,NSYS
      CØMMØN/SLVBLK/T(5,5),P(5),NSYS
      L = 1
   10   DL = 1
        IF(L .EQ. N) GØ TØ 20
        IF(A(L+1,L) .NE. 0.) DL = 2
   20   LL = L+DL-1
        K = L
   30     KM1 = K-1
          DK = 1
          IF(K .EQ. N) GØ TØ 35
          IF(A(K+1,K) .NE. 0.) DK = 2
   35     KK = K+DK-1
          IF(K .EQ. L) GØ TØ 45
          DØ 40 I=K,KK
            DØ 40 J=L,LL
              DØ 40 IA=L,KM1
                C(I,J) = C(I,J) - A(IA,I)*C(IA,J)
   40     CØNTINUE
   45     IF(DL .EQ. 2) GØ TØ 60
          IF(DK .EQ. 2 ) GØ TØ 50
          T(1,1) = A(K,K) + A(L,L)
          IF(T(1,1) .EQ. 0.) STØP
          C(K,L) = C(K,L)/T(1,1)
          GØ TØ 90
   50     T(1,1) = A(K,K) + A(L,L)
          T(1,2) = A(KK,K)
          T(2,1) = A(K,KK)
          T(2,2) = A(KK,KK) + A(L,L)
          P(1) = C(K,L)
          P(2) = C(KK,L)
          NSYS = 2
          CALL SYSSLV
          C(K,L) = P(1)
          C(KK,L) = P(2)
          GØ TØ 90
   60     IF(DK .EQ. 2) GØ TØ 70
          T(1,1) = A(K,K) + A(L,L)
          T(1,2) = A(LL,L)
          T(2,1) = A(L,LL)
          T(2,2) = A(K,K) + A(LL,LL)
          P(1) = C(K,L)
          P(2) = C(K,LL)
          NSYS = 2
          CALL SYSSLV
          C(K,L) = P(1)
          C(K,LL) = P(2)
          GØ TØ 90
   70     IF(K .NE. L) GØ TØ 80
          T(1,1) = A(L,L)
          T(1,2) = A(LL,L)
          T(1,3) = 0.
          T(2,1) = A(L,LL)
          T(2,2) = A(L,L) + A(LL,LL)
          T(2,3) = T(1,2)
          T(3,1) = 0.
          T(3,2) = T(2,1)
          T(3,3) = A(LL,LL)
          P(1) = C(L,L)/2.
          P(2) = C(LL,L)
          P(3) = C(LL,LL)/2.
          NSYS = 3
          CALL SYSSLV
          C(L,L) = P(1)
          C(LL,L) = P(2)
          C(L,LL) = P(2)
          C(LL,LL) = P(3)
          GØ TØ 90
   80     T(1,1) = A(K,K) + A(L,L)
          T(1,2) = A(KK,K)
          T(1,3) = A(LL,L)
          T(1,4) = 0.
          T(2,1) = A(K,KK)
          T(2,2) = A(KK,KK) + A(L,L)
          T(2,3) = 0.
          T(2,4) = T(1,3)
          T(3,1) = A(L,LL)
          T(3,2) = 0.
          T(3,3) = A(K,K) + A(LL,LL)
          T(3,4) = T(1,2)
          T(4,1) = 0.
          T(4,2) = T(3,1)
          T(4,3) = T(2,1)
          T(4,4) = A(KK,KK) + A(LL,LL)
          P(1) = C(K,L)
          P(2) = C(KK,L)
          P(3) = C(K,LL)
          P(4) = C(KK,LL)
          NSYS = 4
          CALL SYSSLV
          C(K,L) = P(1)
          C(KK,L) = P(2)
          C(K,LL) = P(3)
          C(KK,LL) = P(4)
   90   K = K + DK
        IF(K .LE. N) GØ TØ 30
        LDL = L + DL
        IF(LDL .GT. N) RETURN
        DØ 120 J=LDL,N
          DØ 100 I=L,LL
            C(I,J) = C(J,I)
  100     CØNTINUE
          DØ 120 I=J,N
            DØ 110 K=L,LL
              C(I,J) = C(I,J) - C(I,K)*A(K,J) - A(K,I)*C(K,J)
  110       CØNTINUE
            C(J,I) = C(I,J)
  120   CØNTINUE
      L = LDL
      GØ TØ 10
      END
      SUBRØUTINE HSHLDR(A,N,NA)
C
C HSHLDR IS A FØRTRAN IV SUBRØUTINE TØ REDUCE A MATRIX TØ
C UPPER HESSENBERG FØRM BY ELEMENTARY HERMITIAN TRANSFØRMA-
C TIØNS (THE METHØD ØF HØUSEHØLDER).  THE PARAMETERS IN THE
C ARGUMENT LIST ARE
C            A       A DØUBLY SUBSCRIPTED ARRAY CØNTAINING THE
C                    MATRIX A.  ØN RETURN, THE UPPER TRIANGLE
C                    ØF THE ARRAY A MATRIX AND THE (N+1)-TH
C                    CØLUMN CØNTAIN THE SUBDIAGØNAL ELEMENTS
C                    ØF THE TRANSFØRMED MATRIX.  ØN RETURN,
C                    THE LØWER TRIANGLE AND THE (N+1)-TH RØW
C                    ØF THE ARRAY A CØNTAIN A HISTØRY ØF THE
C                    TRANSFØRMATIØNS.
C            N       THE ØRDER ØF THE MATRIX A.
C            NA      THE FIRST DIMENSIØN ØF THE ARRAY A.
C
      REAL
     1A(NA,1),MAX,SUM,S,P
      INTEGER
     1N,NA,NM2,N1,L,L1,I,J
      NM2 = N-2
      N1 = N+1
      IF(N .EQ. 1) RETURN
      IF(N .GT. 2) GØ TØ 5
      A(1,N1) = A(2,1)
      RETURN
    5 DØ 80 L=1,NM2
```

```
      L1 = L+1
      MAX = 0.
      DØ 10 I=L1,N
        MAX = AMAX1(MAX,ABS(A(I,L)))
 10   CØNTINUE
      IF(MAX .NE. 0.) GØ TØ 20
      A(L,N1) = 0.
      A(N1,L) = 0.
      GØ TØ 80
 20   SUM = 0.
      DØ 30 I=L1,N
        A(I,L) = A(I,L)/MAX
        SUM = SUM + A(I,L)**2
 30   CØNTINUE
      S = SIGN(SQRT(SUM),A(L1,L))
      A(L,N1) = -MAX*S
      A(L1,L) = S + A(L1,L)
      A(N1,L) = S*A(L1,L)
      DØ 50 J=L1,N
        SUM = 0.
        DØ 40 I=L1,N
          SUM = SUM + A(I,L)*A(I,J)
 40     CØNTINUE
        P = SUM/A(N1,L)
        DØ 50 I=L1,N
          A(I,J) = A(I,J) - A(I,L)*P
 50   CØNTINUE
      DØ 70 I=1,N
        SUM = 0.
        DØ 60 J=L1,N
          SUM = SUM + A(I,J)*A(J,L)
 60     CØNTINUE
        P = SUM/A(N1,L)
        DØ 70 J=L1,N
          A(I,J) = A(I,J) - P*A(J,L)
 70   CØNTINUE
 80 CØNTINUE
    A(N-1,N1) = A(N,N-1)
    RETURN
    END
    SUBRØUTINE BCKMLT(A,U,N,NA,NU)
C
C BCKMLT IS A FØRTRAN IV SUBRØUTINE THAT, GIVEN THE ØUTPUT
C ØF THE SUBRØUTINE HSHLDR, CØMPUTES THE ØRTHØGØNAL MATRIX
C THAT REDUCES A TØ UPPER HESSENBERG FØRM.  THE PARAMETERS
C IN THE ARGUMENT LIST ARE
C            A      A DØUBLY SUBSCRIPTED ARRAY CØNTAINING THE
C                   ØUTPUT FRØM HSHLDR.
C            U      A DØUBLY SUBSCRIPTED ARRAY THAT, ØN
C                   RETURN, CØNTAINS THE ØRTHØGØNAL MATRIX.
C            N      THE ØRDER ØF THE MATRIX A IN HSHLDR.
C            NA     THE FIRST DIMENSIØN ØF THE ARRAY A.
C            NU     THE FIRST DIMENSIØN ØF THE ARRAY U.
C
C THE ARRAYS A AND U MAY BE IDENTIFIED IN THE CALLING
C SEQUENCE.  IF THIS IS DØNE, THE ELEMENTS ØF THE ØRTHØGØNAL
C MATRIX WILL ØVERWRITE THE ØUTPUT ØF HSHLDR.
C
      REAL
     1A(NA,1),U(NU,1),SUM,P
      INTEGER
     1N,NA,N1,NM1,NM2,LL,L,L1,I,J
      N1 = N+1
      NM1 = N-1
      NM2 = N-2
      U(N,N) = 1.
      IF(NM1 .EQ. 0) RETURN
      U(NM1,N) = 0.
      U(N,NM1) = 0.
      U(NM1,NM1) = 1.
      IF(NM2 .EQ. 0) RETURN
      DØ 40 LL=1,NM2
        L = NM2-LL+1
        L1 = L+1
        IF(A(N1,L) .EQ. 0.) GØ TØ 25
        DØ 20 J=L1,N
          SUM = 0.
          DØ 10 I=L1,N
            SUM = SUM + A(I,L)*U(I,J)
 10       CØNTINUE
          P = SUM/A(N1,L)
          DØ 20 I=L1,N
            U(I,J) = U(I,J) - A(I,L)*P
 20     CØNTINUE
 25     DØ 30 I=L1,N
            U(I,L) = 0.
            U(L,I) = 0.
 30     CØNTINUE
        U(L,L) = 1.
 40 CØNTINUE
    RETURN
    END
    SUBRØUTINE SCHUR(H,U,NN,NH,NU,EPS,FAIL)
C
C SCHUR IS A FØRTRAN IV SUBRØUTINE TØ REDUCE AN UPPER
C HESSENBERG MATRIX TØ REAL SCHUR FØRM BY THE QR METHØD WITH
C IMPLICIT ØRIGIN SHIFTS.  THE PRØDUCT ØF THE TRANSFØRMA-
C TIØNS USED IN THE REDUCTIØN IS ACCUMULATED.  SCHUR IS AN
C ADAPTATIØN ØF THE ALGØL PRØGRAM HQR BY MARTIN, PETERS, AND
C WILKINSØN (NUMER. MATH. 14 (1970) 219-231).  THE PARA-
C METERS IN THE ARGUMENT LIST ARE
C            H      A DØUBLY SUBSCRIPTED ARRAY CØNTAINING THE
C                   UPPER HESSENBERG MATRIX H.  ØN RETURN, H
C                   CØNTAINS AN UPPER REAL SCHUR FØRM ØF H.
C                   THE ELEMENTS ØF THE ARRAY H BELØW THE
C                   THIRD SUBDIAGØNAL ARE UNDISTURBED.
C            U      A DØUBLY SUBSCRIPTED ARRAY CØNTAINING ANY
C                   MATRIX.  ØN RETURN, U CØNTAINS THE MATRIX
C                   U*R(1)*R(2)..., WHERE R(I) ARE THE TRANS-
C                   FØRMATIØNS USED IN THE REDUCTIØN ØF H.
C            NN     THE ØRDER ØF THE MATRICES H AND U.
C            NH     THE FIRST DIMENSIØN ØF THE ARRAY H.
C            NU     THE FIRST DIMENSIØN ØF THE ARRAY U.
C            EPS    A NUMBER USED IN DETERMINING WHEN AN
C                   ELEMENT ØF H IS NEGLIGIBLE.  H(I,J) IS
C                   NEGLIGIBLE IF ABS(H(I,J)) IS LESS THAN ØR
C                   EQUAL TØ EPS TIMES THE INFINITY NØRM ØF
C                   H.
C            FAIL   AN INTEGER VARIABLE THAT, ØN RETURN,
C                   CØNTAINS AN ERRØR SIGNAL. IF FAIL IS
C                   PØSITIVE, THEN THE PRØGRAM FAILED TØ MAKE
C                   THE FAIL-1 ØR FAIL-2 SUBDIAGØNAL ELEMENT
C                   NEGLIGIBLE AFTER 30 ITERATIØNS.
C
      REAL
     1H(NH,1),U(NU,1),EPS,HN,RSUM,TEST,P,Q,R,S,W,X,Y,Z
      INTEGER
     1NN,NA,NH,FAIL,I,ITS,J,JL,K,L,LL,M,MM,M2,M3,N,NA
      LØGICAL
     1LAST
      N = NN
      HN = 0.
      DØ 20 I=1,N
        JL = MAX0(1,I-1)
        RSUM = 0.
        DØ 10 J=JL,N
          RSUM = RSUM + ABS(H(I,J))
 10     CØNTINUE
        HN = AMAX1(HN,RSUM)
 20 CØNTINUE
    TEST = EPS*HN
    IF(HN .EQ. 0.) GØ TØ 230
 30 IF(N .LE. 1) GØ TØ 230
    ITS = 0
    NA = N-1
    NM2 = N-2
 40 DØ 50 LL=2,N
    L = N-LL+2
      IF(ABS(H(L,L-1)) .LE. TEST) GØ TØ 60
 50 CØNTINUE
    L = 1
    GØ TØ 70
 60 H(L,L-1) = 0.
 70 IF(L .LT. NA) GØ TØ 72
    N = L-1
    GØ TØ 30
 72 X = H(N,N)/HN
    Y = H(NA,NA)/HN
    R = (H(N,NA)/HN)*(H(NA,N)/HN)
    IF(ITS .LT. 30) GØ TØ 75
    FAIL = N
    RETURN
 75 IF(ITS.EQ.10 .ØR. ITS.EQ.20) GØ TØ 80
    S = X + Y
    Y = X*Y - R
    GØ TØ 90
 80 Y = (ABS(H(N,NA)) + ABS(H(NA,NM2)))/HN
    S = 1.5*Y
    Y = Y**2
 90 ITS = ITS + 1
    DØ 100 MM=L,NM2
      M = NM2-MM+L
      X = H(M,M)/HN
      R = H(M+1,M)/HN
      Z = H(M+1,M+1)/HN
      P = X*(X-S) + Y + R*(H(M,M+1)/HN)
      Q = R*(X+Z-S)
      R = R*(H(M+2,M+1)/HN)
      W = ABS(P) + ABS(Q) + ABS(R)
      P = P/W
      Q = Q/W
      R = R/W
      IF(M .EQ. L) GØ TØ 110
    IF(ABS(H(M,M-1))*(ABS(Q)+ABS(R)) .LE. ABS(P)*TEST)
   1GØ TØ 110
100 CØNTINUE
110 M2 = M+2
    M3 = M+3
    DØ 120 I=M2,N
      H(I,I-2) = 0.
120 CØNTINUE
    IF(M3 .GT. N) GØ TØ 140
    DØ 130 I=M3,N
      H(I,I-3) = 0.
130 CØNTINUE
140 DØ 220 K=M,NA
      LAST = K.EQ.NA
      IF(K .EQ. M) GØ TØ 150
      P = H(K,K-1)
      Q = H(K+1,K-1)
      R = 0.
      IF(.NØT.LAST) R = H(K+2,K-1)
      X = ABS(P) + ABS(Q) + ABS(R)
      IF(X .EQ. 0.) GØ TØ 220
      P = P/X
      Q = Q/X
      R = R/X
150   S = SQRT(P**2 + Q**2 + R**2)
      IF(P .LT. 0.) S = -S
      IF(K .NE. M) H(K,K-1) = -S*X
      IF(K.EQ.M .AND. L.NE.M) H(K,K-1) = -H(K,K-1)
      P = P + S
      X = P/S
      Y = Q/S
      Z = R/S
      Q = Q/P
      R = R/P
      DØ 170 J=K,NN
        P = H(K,J) + Q*H(K+1,J)
        IF(LAST) GØ TØ 160
        P = P + R*H(K+2,J)
        H(K+2,J) = H(K+2,J) - P*Z
160     H(K+1,J) = H(K+1,J) - P*Y
        H(K,J) = H(K,J) - P*X
170   CØNTINUE
      J = MIN0(K+3,N)
      DØ 190 I=1,J
        P = X*H(I,K) + Y*H(I,K+1)
        IF(LAST) GØ TØ 180
        P = P + Z*H(I,K+2)
        H(I,K+2) = H(I,K+2) - P*R
```

```
 180      H(I,K+1) = H(I,K+1) - P*Q
          H(I,K) = H(I,K) - P
 190    CØNTINUE
        DØ 210 I=1,NN
          P = X*U(I,K) + Y*U(I,K+1)
          IF(LAST) GØ TØ 200
          P = P + Z*U(I,K+2)
          U(I,K+2) = U(I,K+2) - P*R
 200      U(I,K+1) = U(I,K+1) - P*Q
          U(I,K) = U(I,K) - P
 210    CØNTINUE
 220  CØNTINUE
      GØ TØ 40
 230  FAIL = 0
      RETURN
      END
      SUBRØUTINE SYSSLV
C
C SYSSLV IS A FØRTRAN IV SUBRØUTINE THAT SØLVES THE LINEAR
C SYSTEM AX = B ØF ØRDER N LESS THAN 5 BY CRØUT REDUCTIØN
C FØLLØWED BY BACK SUBSTITUTIØN.  THE MATRIX A, THE VECTØR
C B, AND THE ØRDER N ARE CØNTAINED IN THE ARRAYS A,B, AND
C THE VARIABLE N ØF THE CØMMØN BLØCK SLVBLK.  THE SØLUTIØN
C IS RETURNED IN THE ARRAY B.
C
      CØMMØN/SLVBLK/A(5,5),B(5),N
      REAL MAX
    1 NM1 = N-1
      N1 = N+1
C
C CØMPUTE THE LU FACTØRIZATIØN ØF A.
C
      DØ 80 K=1,N
        KM1 = K-1
        IF(K.EQ.1) GØ TØ 20
        DØ 10 I=K,N
          DØ 10 J=1,KM1
            A(I,K) = A(I,K) - A(I,J)*A(J,K)
  10    CØNTINUE
  20    IF(K.EQ.N) GØ TØ 100
        KP1 = K+1
      MAX = ABS(A(K,K))
        INTR = K
        DØ 30 I=KP1,N
          AA = ABS(A(I,K))
          IF(AA .LE. MAX) GØ TØ 30
          MAX = AA
          INTR = I
  30    CØNTINUE
        IF(MAX .EQ. 0.) STØP
        A(N1,K) = INTR
        IF(INTR .EQ. K) GØ TØ 50
        DØ 40 J=1,N
          TEMP = A(K,J)
          A(K,J) = A(INTR,J)
          A(INTR,J) = TEMP
  40    CØNTINUE
  50    DØ 80 J=KP1,N
          IF(K.EQ.1) GØ TØ 70
          DØ 60 I=1,KM1
            A(K,J) = A(K,J) - A(K,I)*A(I,J)
  60      CØNTINUE
  70      A(K,J) = A(K,J)/A(K,K)
  80  CØNTINUE
C
C INTERCHANGE THE CØMPØNENTS ØF B.
C
 100  DØ 110 J=1,NM1
        INTR = A(N1,J)
        IF(INTR .EQ. J) GØ TØ 110
        TEMP = B(J)
        B(J) = B(INTR)
        B(INTR) = TEMP
 110  CØNTINUE
C
C SØLVE LX = B.
C
 200  B(1) = B(1)/A(1,1)
      DØ 220 I=2,N
        IM1 = I-1
        DØ 210 J=1,IM1
          B(I) = B(I) - A(I,J)*B(J)
 210    CØNTINUE
        B(I) = B(I)/A(I,I)
 220  CØNTINUE
C
C SØLVE UX = B.
C
 300  DØ 310 II=1,NM1
        I = NM1-II+1
        I1 = I+1
        DØ 310 J=I1,N
          B(I) = B(I) - A(I,J)*B(J)
 310  CØNTINUE
      RETURN
      END
```

12.3. [GWS-J32] "The Economical Storage of Plane Rotations"

[GWS-J32] "The Economical Storage of Plane Rotations," *Numerische Mathematik* **25** (1976) 137–138.
http://dx.doi.org/10.1007/BF01462266

Numer. Math. 25, 137—138 (1976)

The Economical Storage of Plane Rotations

G. W. Stewart*

Received February 3, 1975

Summary. Plane rotations, which have long been used in matrix computations to annihilate selected elements of a matrix, have the drawback that their definition requires two numbers. In this note it is shown how this information may be stably compacted into a single number.

The object of this note is to describe how plane rotations can be economically represented on a computer. Plane rotations have long been used as a very stable computational device for introducing zeros into a matrix (see, e.g., [3]). Since a general plane rotation differs from the identity matrix in only a 2×2 principal submatrix, it is sufficient for our purposes to consider the 2×2 rotation

$$R=\begin{pmatrix}\gamma & \sigma\\ -\sigma & \gamma\end{pmatrix},$$

where

$$\gamma^2+\sigma^2=1. \tag{1}$$

If $(\alpha,\beta)^T\neq 0$ and γ and σ are defined by

$$\gamma=\frac{\alpha}{\sqrt{\alpha^2+\beta^2}},\qquad \sigma=\frac{\beta}{\sqrt{\alpha^2+\beta^2}}, \tag{2}$$

then

$$R\begin{pmatrix}\alpha\\ \beta\end{pmatrix}=\begin{pmatrix}\sqrt{\alpha^2+\beta^2}\\ 0\end{pmatrix}.$$

When plane rotations are used in matrix computations, it is often necessary to save them for later use. Since a plane rotation can in general introduce only a single zero element into a matrix, it is important to represent it by a single number which can be stored in the place of the newly introduced zero. We propose that this number be the number ϱ which is calculated as follows:

$$\begin{aligned}&\text{if } \gamma=0 && \text{then } \varrho=1,\\ &\text{if } |\sigma|<|\gamma| && \text{then } \varrho=\tfrac{1}{2}\operatorname{sign}(\gamma)\sigma,\\ &\text{if } |\gamma|\leqq|\sigma| && \text{then } \varrho=2\operatorname{sign}(\sigma)/\gamma.\end{aligned}$$

* This work was supported in part by the Office of Naval Research under Contract No. N00014-67-A-0239-0037.

The numbers γ and σ (up to a common factor of ± 1) can be recovered from ϱ as follows:

$$\begin{aligned} &\text{if } \varrho = 1 && \text{then } \gamma = 0, && \sigma = 1, \\ &\text{if } |\varrho| < 1 && \text{then } \sigma = 2\varrho, && \gamma = \sqrt{1-\sigma^2}, \\ &\text{if } |\varrho| > 1 && \text{then } \gamma = 2/\varrho, && \sigma = \sqrt{1-\gamma^2}. \end{aligned}$$

In other words, we save $\frac{1}{2}\sigma$ or $\frac{1}{2}\gamma$, whichever is smaller; however, to distinguish between the two numbers we store the reciprocal of $\frac{1}{2}\gamma$, which is always greater than unity. Since it is the smaller of γ or σ that is used to recover the other via (1), the process is numerically stable. The reason for saving $\frac{1}{2}\gamma$ or $\frac{1}{2}\sigma$ instead of γ or σ is to allow for the case $\gamma = 0$ where γ^{-1} is undefined. This is signaled by setting $\varrho = 1$, a value that cannot otherwise occur. The factor $\frac{1}{2}$ has been chosen to avoid rounding errors on a binary machine. On a hexadecimal machine a factor of $\frac{1}{16}$ would be more appropriate.

The formulas given above determine R up to a factor ± 1. Since either of R or $-R$ can introduce a zero, this is not a drawback, provided one must take care to work with the same rotation throughout.

This technique for representing rotations can be extended to the various scaled rotations (fast Givens transformations) that have been recently introduced [1, 2]. For example, if we scale the rotation determined by (2) so that the larger of γ and σ is unity, then we may store it as follows:

$$\begin{aligned} &\text{if } \alpha = 0 && \text{then } \varrho = 1, \\ &\text{if } \alpha > \beta && \text{then } \varrho = \tfrac{1}{2}\beta/\alpha, \\ &\text{if } \alpha \leqq \beta && \text{then } \varrho = 2\beta/\alpha, \end{aligned} \tag{3}$$

The scaled transformation can then be recovered as follows:

$$\begin{aligned} &\text{if } \varrho = 1 && \text{then } \gamma = 0, && \sigma = 1, \\ &\text{if } \varrho < 1 && \text{then } \gamma = 1, && \varrho = 2\varrho, \\ &\text{if } \varrho > 1 && \text{then } \gamma = 2/\varrho, && \sigma = 1. \end{aligned} \tag{4}$$

References

1. Gentleman, W. M.: Least squares computations by Givens transformations without square roots. University of Waterloo, Department of Applied Analysis and Computer Science, Research Report CSRR-2062 (1972)
2. Hammerling, S.: A note on modifications of the Givens method. To appear J. Inst. Math. Appl.
3. Wilkinson, J. H.: The algebraic eigenvalue problem. London and New York: Oxford University Press (Clarendon) 1965

G. W. Stewart
Dept. of Computer Science
University of Maryland
College Park
Maryland 20742
USA

12.4. [GWS-J34] "Perturbation Bounds for the QR Factorization of a Matrix"

[GWS-J34] "Perturbation Bounds for the QR Factorization of a Matrix," *SIAM Journal on Numerical Analysis* **14** (1977) 509–518.
http://dx.doi.org/10.1137/0714030

SIAM J. NUMER. ANAL.
Vol. 14, No. 3, June 1977

PERTURBATION BOUNDS FOR THE *QR* FACTORIZATION OF A MATRIX*

G. W. STEWART†

Abstract. Let A be an $m \times n$ matrix of rank n. The QR factorization of A decomposes A into the product of an $m \times n$ matrix Q with orthonormal columns and a nonsingular upper triangular matrix R. The decomposition is essentially unique, Q being determined up to the signs of its columns and R up to the signs of its rows. If E is an $m \times n$ matrix such that $A+E$ is of rank n, then $A+E$ has an essentially unique factorization $(Q+W)(R+F)$. In this paper bounds on $\|W\|$ and $\|F\|$ in terms of $\|E\|$ are given. In addition perturbation bounds are given for the closely related Cholesky factorization of a positive definite matrix B into the product R^TR of a lower triangular matrix and its transpose.

1. Introduction. Let A be a real $m \times n$ matrix ($A \in \mathbb{R}^{m \times n}$) whose columns are linearly independent. In many applications it is required to have an orthonormal basis for the space spanned by the columns of $A\,[\mathcal{R}(A)]$. This amounts to knowing a matrix $Q \in \mathbb{R}^{m \times n}$ with orthonormal columns such that $\mathcal{R}(Q) = \mathcal{R}(A)$. Since each column of A can be expressed as a linear combination of the columns of Q, there is a nonsingular matrix $R \in \mathbb{R}^{n \times n}$ such that

$$A = QR, \tag{1.1}$$

and since $Q^TQ = I$, we have

$$R = Q^TA.$$

Of course the decomposition (1.1) is not unique. If $U \in \mathbb{R}^{n \times n}$ is orthogonal, then $A = (QU)(U^TR)$ is another such decomposition.

It may happen that the columns of A are perturbed by the addition of a matrix $E \in \mathbb{R}^{m \times n}$, in which case the orthonormal basis for $\mathcal{R}(A)$ must be recomputed. It is of interest to know if there is a factorization of $A+E$ that is near the factorization (1.1). Conditions under which this is true are given in the following theorem.

THEOREM 1.1. *Let $A = QR$, where $A \in \mathbb{R}^{m \times n}$ has rank n and $Q^TQ = I$. Let $E \in \mathbb{R}^{m \times n}$ satisfy*

$$\|A^\dagger\|\,\|E\| < 1.$$

Then there are matrices $W \in \mathbb{R}^{m \times n}$ and $F \in \mathbb{R}^{n \times n}$ such that

$$A + E = (Q+W)(R+F), \tag{1.2a}$$

$$(Q+W)^T(Q+W) = I, \tag{1.2b}$$

$$\frac{\|F - Q^TE\|}{\|R\|} \leq \left[\frac{\kappa(A)\dfrac{\|E\|}{\|A\|}}{1-\kappa(A)\dfrac{\|E\|}{\|A\|}}\right] \equiv \mu^2, \tag{1.2c}$$

* Received by the editors March 15, 1975, and in revised form February 23, 1976.

† Computer Science Department, University of Maryland, College Park, Maryland 20742. This work was supported in part by the Office of Naval Research under Contract N00014-67-A-0128-0018.

(1.2d) $$\|W\| \leq \mu \left[1 + \mu \frac{\kappa(A)}{1 - \kappa(A)\frac{\|E\|}{\|A\|}} \right],$$

(1.2e) $$\|Q^T W\| \leq \frac{\kappa(A)}{1 - \kappa(A)\frac{\|E\|}{\|A\|}} \mu^2.$$

Here $\|\cdot\|$ denotes the Frobenius norm defined by $\|A\|^2 = \text{trace}\,(A^T A)$. The matrix $A^\dagger$ is the pseudo-inverse of A, defined by $A^\dagger = (A^T A)^{-1} A^T$, and

$$\kappa(A) = \|A\|\,\|A^\dagger\|.$$

For the sake of continuity, the proof of Theorem 1.1 will be deferred to the end of this section.

Equations (1.2a) and (1.2b) in Theorem 1.1 simply state that the columns of $Q + W$ form an orthonormal basis for $\mathcal{R}(A+E)$. The inequality (1.2c) states that up to terms of second order in $\|E\|$, the matrix F may be taken to be $Q^T E$. Inequality (1.2d) gives a bound on $\|W\|$, in which the relative error $\|E\|/\|A\|$ is effectively magnified by $\kappa(A)$. Inequality (1.2e) says that the projection of W onto $\mathcal{R}(A)$ is of second order in $\|E\|$. If the bound (1.2d) is nearly an equality, this implies that for E sufficiently small the columns of W will be almost orthogonal to those of Q.

Although Theorem 1.1 is sufficient for many applications, it is not directly applicable to perturbations of the QR factorization of the matrix A. In this factorization the matrix R in (1.1) is taken to be upper triangular. The QR factorization is essentially unique, Q being determined up to the signs of its columns and R up to the signs of its rows. The columns of Q are just the vectors that would be obtained by applying the Gram–Schmidt orthogonalization to the columns of A in their natural order. Since $A^T A = R^T R$, the matrix R is the upper triangular factor in the Cholesky decomposition of the positive definite matrix $A^T A$ [5].

The difficulty of Theorem 1.1 in connection with the QR factorization is that the matrix F is not upper triangular so that the perturbed factorization (1.2a) is no longer a QR factorization. Since the QR factorization arises in a number of numerical applications, it is important to be able to assess its sensitivity to perturbations. For example, the standard convergence proofs for the QR algorithm require perturbation theory for the QR factorization [6].

Another reason for working with the QR factorization is that the upper triangularity of R represents an uncoupling of the first columns of A from the later ones. Specifically, if $A^{|k}$ is the matrix consisting of the first column of A and $R^{|k}$ is the leading principal submatrix of R of order k, then

$$A^{|k} = Q^{|k} R^{|\overline{k}}.$$

It follows that the QR factorization of $A^{|k} + E^{|k}$ is independent of the remaining columns of E (whose matrix is denoted by $E^{n-k|}$). If $\|E^{|k}\|$ is significantly less than $\|E^{n-k|}\|$, then the perturbation bounds obtained for the first part of the QR

factorization will be smaller than those for the remaining part. Such oblique perturbations arise in the analysis of methods of simultaneous iteration for finding eigenvalues of a matrix [2], [5].

The object of this paper is to develop a perturbation theory for the QR factorization of a matrix of full rank. Throughout the paper we shall assume the hypotheses and notation of Theorem 1.1. We note for future reference that since $\|\cdot\|$ is invariant under orthogonal transformations, we have

$$\|A\| = \|R\|, \qquad \|A^{\dagger}\| = \|R^{-1}\|,$$

and

$$\kappa(A) = \kappa(R).$$

In § 2 we shall derive first order perturbation equations for the QR factorization of A. It will turn out that the equations will involve a linear operator from the space of upper triangular matrices into the space of symmetric matrices, and the remainder of § 2 will be devoted to computing bounds on the norm of the inverse of the operator. In § 3 the perturbation bounds will be derived, and in § 4, perturbation bounds for the Cholesky factorization of a positive definite matrix will be given.

It should be noted that all the results of this paper hold when $\kappa(A)$ is replaced by $\kappa_2(A) = \|A\|_2\|A^{\dagger}\|_2$ and terms of the form $\|E\|/\|A\|$ are replaced by $\|E\|/\|A\|_2$. Here $\|\cdot\|_2$ denotes the spectral norm [4].

Another extension of these results is that they hold over the complex field. However, in this case it is necessary to restrict R and F so that their diagonal elements are real.

Proof of Theorem 1.1. So far as satisfying (1.2a) and (1.2b) are concerned, we may take for $R+F$ any matrix S such that the columns of $(A+E)S^{-1}$ are orthonormal. If U is any orthogonal matrix, then $U^T(A+E)S^{-1}$ has orthonormal columns if and only if $(A+E)S^{-1}$ does. Take $U = (Q, \hat{Q})$, where $\mathcal{R}(\hat{Q})$ is the orthogonal complement of $\mathcal{R}(A)$. Then $\hat{Q}^TA = 0$. Moreover from the above observations, we may take for S any matrix such that S^{-1} orthogonalizes the columns of

$$B = U^T(A+E) = \begin{pmatrix} Q^TA + Q^TE \\ \hat{Q}^TA + \hat{Q}^TE \end{pmatrix} = \begin{pmatrix} R + Q^TE \\ \hat{Q}^TE \end{pmatrix} = \begin{pmatrix} R+G \\ \hat{G} \end{pmatrix}.$$

Now since $\|R^{-1}\|\,\|G\| \leq \|R^{-1}\|\,\|E\| < 1$, $R+G$ is nonsingular. Set

$$B(R+G)^{-1} = \begin{pmatrix} I \\ \hat{G}(R+G)^{-1} \end{pmatrix} = \begin{pmatrix} I \\ H \end{pmatrix}.$$

Then from standard perturbation theory on inverses of matrices [4], we have

$$\|H\| \leq \frac{\|\hat{G}\|\,\|R^{-1}\|}{1 - \|G\|\,\|R^{-1}\|} \leq \mu,$$

where μ is defined in (1.2c). The matrix $I + H^TH$ is positive definite and hence has a unique positive definite square root. It is easily verified that

$$B(R+G)^{-1}(I+H^TH)^{-1/2}$$

has orthonormal columns. Hence we may take

$$R+F=(I+H^TH)^{1/2}(R+G), \tag{1.3}$$

thus defining F.

To obtain the bound (1.2c) write

$$(I+H^TH)^{1/2}=I+K. \tag{1.4}$$

By diagonalizing the symmetric matrix H^TH by an orthogonal similarity, one gets

$$\|K\|_2 \leqq \tfrac{1}{2}\|H^TH\|_2 \leqq \tfrac{1}{2}\mu^2. \tag{1.5}$$

Now from (1.3) and (1.4)

$$F=G+K(R+G),$$

and (1.2c) follows from (1.5) and the facts that $G=Q^TE$ and $1+\|G\|/\|R\|<2$.

To obtain the bound (1.2d), we first obtain a bound on $\|(R+F)^{-1}\|_2$. In view of (1.2a) and (1.2b), $A+E$ and $R+F$ have the same nonzero singular values [4]. Let $\sigma_{\min}(A)$ denote the smallest singular value of A. Then $\sigma_{\min}(R+F) \geqq \sigma_{\min}(A)-\|E\|_2$. But

$$\begin{aligned}\|(R+F)^{-1}\|_2 &= \frac{1}{\sigma_{\min}(R+F)} \leqq \frac{1}{\sigma_{\min}(A)-\|E\|_2} \\ &= \frac{\sigma_{\min}^{-1}(A)}{1-\sigma_{\min}^{-1}(A)\|E\|_2} = \frac{\|A^\dagger\|_2}{1-\|A^\dagger\|_2\|E\|_2}.\end{aligned} \tag{1.6}$$

Now from (1.2a),

$$W=(E-QF)(R+F)^{-1},$$

whence

$$U^TW=\begin{pmatrix} G-F \\ \hat{G} \end{pmatrix}(R+F)^{-1}. \tag{1.7}$$

Hence

$$\|W\|=\|U^TW\| \leqq \|(R+F)^{-1}\|_2(\|G-F\|+\|\hat{G}\|),$$

and (1.2d) follows from (1.2c), (1.6) and the fact that $\|\hat{G}\| \leqq \|E\|$.

Finally from (1.2a), $Q^TW=(G-F)(R+F)^{-1}$, from which the bound (1.2e) follows easily. □

2. The operator $F \mapsto R^TF+F^TR$. In this section we shall be concerned with deriving perturbation equations for the upper triangular part of the QR factorization and with describing the properties of an operator associated with these equations. Since by hypothesis $A+E$ has full rank n, $A+E$ has an essentially unique QR factorization which we may write in the form

$$A+E=(Q+W)(R+F),$$

where $A=QR$ is the QR factorization of A. Since $(Q+W)^T(Q+W)=I$,

$$(R+F)^T(R+F)=(A+E)^T(A+E),$$

and since $R^TR = A^TA$,

$$R^TF + F^TR = A^TE + E^TA + E^TE - F^TF.$$

Introducing the operator $\hat{\mathbf{T}} = F \mapsto R^TF + F^TR$, we can write this in the form

$$\hat{\mathbf{T}}F = A^TE + E^TA + E^TE - F^TF. \tag{2.1}$$

The operator $\hat{\mathbf{T}}$ is a linear operator that maps $\mathbb{R}^{n\times n}$ into the space of symmetric matrices, and as such it must be singular. However, in our applications we require F to be upper triangular, and consequently, if we denote by $\mathbf{T}$ the restriction of $\hat{\mathbf{T}}$ to the space of upper triangular matrices, then F must satisfy

$$\mathbf{T}F = A^TE + E^TA + E^TE - F^TF. \tag{2.2}$$

If $\mathbf{T}$ is nonsingular and E and F are small enough, we may ignore the last two terms on the right-hand side of (2.2) and write

$$F \cong \mathbf{T}^{-1}(A^TE + E^TA). \tag{2.3}$$

Any bound on $\|F\|$ will thus have to depend on a bound on $\|\mathbf{T}^{-1}\|$. The following theorem furnishes such a bound.

THEOREM 2.1. *Let $R \in \mathbb{R}^{n\times n}$ be upper triangular and define the operator $\mathbf{T}$ that maps the space of upper triangular matrices into the space of symmetric matrices by*

$$\mathbf{T}F = R^TF + F^TR.$$

Then $\mathbf{T}$ is nonsingular if and only if R is nonsingular. Moreover,

$$\|\mathbf{T}^{-1}B\| \leqq \tau\|B\|, \tag{2.4a}$$

where

$$\tau = n\|R^{-1}\|\left[1 + \frac{1}{\sqrt{2}}\kappa(R)\right]. \tag{2.4b}$$

Proof. The theorem is clearly true for $n = 1$. Assume that the theorem is true for a fixed value of $n \geqq 1$ and let $R \in \mathbb{R}^{n\times n}$ be upper triangular and nonsingular. We shall show that for any $\rho \neq 0$ the augmented system

$$\begin{pmatrix} R^T & 0 \\ r^T & \rho \end{pmatrix}\begin{pmatrix} F & f \\ 0 & \varphi \end{pmatrix} + \begin{pmatrix} F^T & 0 \\ f^T & \varphi \end{pmatrix}\begin{pmatrix} R & r \\ 0 & \rho \end{pmatrix} = \begin{pmatrix} B & b \\ b^T & \beta \end{pmatrix} \tag{2.5}$$

has a solution, and we shall give bounds on its norm. (The augmented matrices in (2.5) will be denoted by $\tilde{R}$, $\tilde{F}$, and $\tilde{B}$.)

If equation (2.5) is expanded according to its partitioning, there result the following three equations:

$$R^TF + F^TR = B, \tag{2.6}$$

$$R^Tf + F^Tr = b, \tag{2.7}$$

$$2r^Tf + 2\rho\varphi = \beta. \tag{2.8}$$

By the induction hypothesis (2.6) has a solution F that satisfies (2.4).

We shall need a special bound on $R^{-T}F^T$. To obtain it we make the following observation: if L is lower triangular, then

$$\sqrt{2}\|L\| \leqq \|L+L^T\|.$$

To see this simply note that

$$\|L+L^T\|^2 = 2\sum_{i>j} l_{ij}^2 + 4\sum_i l_{ii}^2$$

$$\geqq 2\sum_{i\geqq j} l_{ij}^2 = 2\|L\|^2.$$

Now from (2.6),

$$FR^{-1} + R^{-T}F^T = R^{-T}BR^{-1}, \tag{2.9}$$

and since $R^{-T}F^T$ is lower triangular,

$$\|R^{-T}F^T\| \leqq \frac{1}{\sqrt{2}}\|R^{-T}BR^{-1}\| \leqq \frac{1}{\sqrt{2}}\|R^{-1}\|^2\|B\|. \tag{2.10}$$

Returning to the problem of bounding $\|\tilde{F}\|$, we note that equations (2.7) and (2.8) can be written in the form

$$\tilde{R}^T \tilde{f} = \begin{pmatrix} b \\ \beta/2 \end{pmatrix} - \begin{pmatrix} F^T r \\ 0 \end{pmatrix},$$

where $\tilde{f}$ is the last column of $\tilde{F}$. Hence from (2.10),

$$\|\tilde{f}\| \leqq \|\tilde{R}^{-1}\| \left\| \begin{pmatrix} b \\ \beta/2 \end{pmatrix} \right\| + \frac{1}{\sqrt{2}}\|R^{-1}\|^2\|B\|\,\|r\| \leqq \|\tilde{R}^{-1}\|\left[1+\frac{1}{\sqrt{2}}\kappa(R)\right]\|\hat{B}\|.$$

Hence

$$\|\tilde{F}\| \leqq \left[\tau + \|\tilde{R}^{-1}\|\left(1+\frac{1}{\sqrt{2}}\kappa(R)\right)\right]\|\tilde{B}\|,$$

from which the bound (2.4) follows.

It remains to show that if R is singular then $\mathbf{T}$ is singular. We shall do this by exhibiting an upper triangular F for which $\mathbf{T}F=0$. If the last diagonal element of R is zero, take $F=e_ne_n^T$. Otherwise partition R in the form

$$R = \begin{pmatrix} R_{11} & R_{12} \\ 0 & R_{22} \end{pmatrix},$$

where R_{22} is nonsingular. Then R_{11} is singular and there is a nonzero vector x such that $R_{11}^T x = 0$. Set

$$F = \begin{pmatrix} 0 & xe_1^T \\ 0 & F_{22} \end{pmatrix}$$

where F_{22} is to be determined. Then

$$\mathbf{T}F = \begin{pmatrix} 0 & 0 \\ 0 & R_{12}^T x e_1^T + e_1 x^T R_{12} + R_{22}^T F_{22} + F_{22}^T R_{22} \end{pmatrix}. \tag{2.11}$$

Since R_{22} is nonsingular, we can solve the equation $R_{22}^TF_{22}+F_{22}^TR_{22}=-R_{12}^Txx^T-xx^TR_{12}$ for a matrix F_{22} that makes the (2,2)-element in the partition (2.11) equal to zero.

Theorem 2.1 is not quite delicate enough for our purposes since it fails to take into account the special nature of the right-hand side of (2.3). To see that this is indeed a special case, let $U=(Q,\hat{Q})$ be orthogonal. Then with $G=Q^TE$ and $\hat{G}=\hat{Q}^TE$,

$$A^TE=A^TUU^TE=(A^TQ,A^T\hat{Q})\binom{Q^TE}{\hat{Q}^TE}$$

$$=(R^T,0)\binom{G}{\hat{G}}=R^TG.$$

Hence (2.3) can be written in the form

$$F\cong\mathbf{T}^{-1}(R^TG+G^TR)=\mathbf{T}^{-1}\hat{\mathbf{T}}G,$$

and what we require is a bound on $\|\mathbf{T}^{-1}\hat{\mathbf{T}}\|$. Such a bound is given in the following theorem. The proof is a simple variation of the proof of Theorem 2.1 and will not be reproduced here.

THEOREM 2.2. *In Theorem* 2.1 *let* $\hat{\mathbf{T}}$ *denote the natural extension of* $\mathbf{T}$ *to* $\mathbb{R}^{n\times n}$. *Then if* $F=\mathbf{T}^{-1}\hat{\mathbf{T}}G$, *we have*

(2.12a) $$\|F\|\leqq\sigma\|G\|,$$

where

(2.12b) $$\sigma=n(2+\sqrt{2})\kappa(R).$$

The bound in Theorem 2.2 represents a considerable improvement over the bound in Theorem 2.1, for σ is proportional to $\kappa(R)$, whereas if R is scaled so that $\|R\|=1$, then τ is proportional to $\kappa^2(R)$, potentially a much larger number.

It is important to realize that we have given much away in deriving σ and τ, and they are likely to overestimate $\|\mathbf{T}^{-1}\hat{\mathbf{T}}\|$ and $\|\mathbf{T}^{-1}\|$ in specific cases. For example, if $R=I$, we may take $\tau=1$ and $\sigma=2$.

As was mentioned in § 1, the results of this paper can be extended to the complex field. The difficulties in doing this involve the operator $\mathbf{T}$ which must now be defined as $F\mapsto R^HF+F^HR$. The first difficulty is that $\mathbf{T}^{-1}$ is no longer well defined, for (2.8) becomes

(2.13) $$\text{real}\,(r^Hf)+\text{real}\,(\rho\varphi)=\beta,$$

which does not completely determine φ. However, if we restrict R and F to have real diagonal elements, which is no restriction at all in our applications, then φ in (2.13) and hence $\mathbf{T}^{-1}$ is uniquely determined.

A second difficulty is that the bound

(2.14) $$\sqrt{2}\|L\|\leqq\|L+L^H\|$$

is not true for general, triangular, complex matrices (for a counter example, let $L=iI$). However, if in addition L has real diagonal elements, then (2.14) holds. In our applications $L=R^{-H}F^H$ does indeed have real diagonal elements.

3. Perturbation bounds. We turn now to the problem of obtaining perturbation bounds for the QR factorization. From the introductory material in the last section, the perturbation of R will be the solution of the nonlinear equation (2.2), or what is equivalent, the solution of the equation

$$\mathbf{T}F = R^T(Q^TE) + (E^TQ)R + E^TE - F^TF. \tag{3.1}$$

Nonlinear equations of this form have been investigated by the author in [3] where a similar equation arises in connection with the eigenvalue problem. The techniques developed there can be modified slightly to prove the first part of the following theorem.

THEOREM 3.1. *Let A, Q, and R be as in Theorem* 1.1. *Let E satisfy*

$$\|A^\dagger\|\,\|E\| < \tfrac{1}{2}, \tag{3.2}$$

and let τ and σ be defined by (2.4b) *and* (2.12b). *Let*

$$F_1 = \mathbf{T}^{-1}[R^T(Q^TE) + (E^TQ)R].$$

If

$$\tau\|F_1\| < \tfrac{1}{4}, \tag{3.3}$$

then there is a unique solution of (3.1) *that satisfies*

$$\|F\| < 2\|F_1\| \leqq 2\sigma\|E\|. \tag{3.4}$$

Moreover $A + E = (Q + W)(R + F)$, where $Q + W$ has orthonormal columns

$$\|W\| \leqq \frac{3\kappa(A)\dfrac{\|E\|}{\|A\|}}{1 - 2\kappa(A)\dfrac{\|E\|}{\|A\|}}, \tag{3.5}$$

and

$$\frac{\|F\|}{\|A\|} \leqq \frac{\|E\|}{\|A\|} + \|W\|\left(1 + \frac{\|E\|}{\|A\|}\right). \tag{3.6}$$

Proof. As was mentioned above, the techniques developed in [3] can be used to establish the existence of an F satisfying (3.1) and (3.4). From (3.3) it follows that $\|F\|\,\|R^{-1}\| < 1$ and hence that $R + F$ is nonsingular. Since $(A + E)^T(A + E) = (R + F)^T(R + F)$, the matrix $Q + W$ defined by

$$Q + W = (A + E)(R + F)^{-1} \tag{3.7}$$

has orthonormal columns.

To get the bounds (3.5) and (3.6), we first must find a bound on $\|FR^{-1}\|$. Since (3.5) and (3.6) clearly hold for $n = 1$, we may assume that $n \geqq 2$. Then (3.3) and (3.4) imply that

$$\|F\|\,\|R^{-1}\| < \tfrac{1}{6}. \tag{3.8}$$

From (3.1) and the definition of $\mathbf{T}$ we have

$$FR^{-1} + R^{-T}F^{-T} = GR^{-1} + R^{-T}G^{-T} + R^{-T}E^TER^{-1} + R^{-T}F^TFR^{-1}$$

where $G = Q^T E$. Hence

$$\sqrt{2}\|FR^{-1}\| - \|FR^{-1}\|^2 \leqq 2\|GR^{-1}\| + \|ER^{-1}\|^2$$
$$\leqq \|E\| \|A^\dagger\|(2 + \|A^\dagger\| \|E\|).$$

In view of (3.2) and (3.8),

$$\|FR^{-1}\| \leqq 2\|A^\dagger\| \|E\| < 1. \tag{3.9}$$

Now from (3.7),

$$W = E(R+F)^{-1} - QFR^{-1}(I+FR^{-1})^{-1}. \tag{3.10}$$

From (3.9) it follows that

$$\|FR^{-1}\| \|(I+FR^{-1})^{-1}\| \leqq \frac{2\|A^\dagger\| \|E\|}{1-2\|A^\dagger\| \|E\|}. \tag{3.11}$$

The bound (3.5) then follows from (1.6), (3.10), and (3.11).

Finally for the bound (3.6), note that

$$F = Q^T E + Q^T W(R+F).$$

Since $\|R+F\| = \|A+E\| \leqq \|A\| + \|E\|$, the bound follows. □

It is instructive to compare the bound on F in Theorem 3.1 with that of Theorem 1.1. For small E, the bound in Theorem 1.1 is of the order $\|E\|$, while in Theorem 3.1 it is of order $\kappa(A)\|E\|$ (cf. (3.5) and (3.6)). Unfortunately the factor $\kappa(A)$ is the price we must pay for insisting on the upper triangularity of F. That we cannot avoid paying it is shown by the following example.

Example 3.2. The well-known matrix, illustrated here for $n = 5$,

$$\begin{pmatrix} 1 & -1 & -1 & -1 & -1 \\ 0 & 1 & -1 & -1 & -1 \\ 0 & 0 & 1 & -1 & -1 \\ 0 & 0 & 0 & 1 & -1 \\ 0 & 0 & 0 & 0 & 1 \end{pmatrix}$$

has a condition number of about 2^n [1]. If we perturb its last row by adding ε to each element, we get up to terms of order ε^2 the QR factorization (again illustrated for $n = 5$).

$$\begin{pmatrix} 1 & -1 & -1 & -1 & -1 \\ 0 & 1 & -1 & -1 & -1 \\ 0 & 0 & 1 & -1 & -1 \\ 0 & 0 & 0 & 1 & -1 \\ \varepsilon & \varepsilon & \varepsilon & \varepsilon & 1+\varepsilon \end{pmatrix}$$

$$\cong \begin{pmatrix} 1 & 0 & 0 & 0 & -\varepsilon \\ 0 & 1 & 0 & 0 & -2\varepsilon \\ 0 & 0 & 1 & 0 & -4\varepsilon \\ 0 & 0 & 0 & 1 & -8\varepsilon \\ \varepsilon & 2\varepsilon & 4\varepsilon & 8\varepsilon & 1 \end{pmatrix} \begin{pmatrix} 1 & -1 & -1 & -1 & \varepsilon-1 \\ 0 & 1 & -1 & -1 & 2\varepsilon-1 \\ 0 & 0 & 1 & -1 & 4\varepsilon-1 \\ 0 & 0 & 0 & 1 & 8\varepsilon-1 \\ 0 & 0 & 0 & 0 & 16\varepsilon+1 \end{pmatrix}$$

The perturbations in both Q and R are of order 2^n.

4. Perturbation bounds for the Cholesky factorization. If B is a positive definite matrix, then B can be written in the form

$$B = R^T R,$$

where R is upper triangular. If H is a symmetric matrix such that $\|B^{-1}\|\,\|H\| < 1$, then $B + H$ is also positive definite and has a Cholesky factorization

$$B + H = (R + F)^T (R + F).$$

Clearly F satisfies the equation

$$\mathbf{T}F = H - F^T F. \tag{4.1}$$

Applying the techniques of [3] to this equation, we obtain the following theorem.

THEOREM 4.1. *Let B be positive definite and $B = R^T R$, where R is upper triangular. Let H be symmetric with $\|B^{-1}\|\,\|H\| < 1$. Let $F_1 = \mathbf{T}^{-1}H$, where $\mathbf{T}$ is the operator defined in § 2. Then if*

$$\tau \|F_1\| < \tfrac{1}{4},$$

there is a unique upper triangular matrix F satisfying

$$\|F\| < 2\|F_1\| \leqq 2\tau \|H\|$$

such that $B + H = (R + F)^T (R + F)$.

REFERENCES

[1] G. H. GOLUB AND W. KAHAN, *Calculating the singular values and pseudo-inverse of a matrix*, this Journal, 2 (1965), pp. 202–224.
[2] H. RUTISHAUSER, *Computational aspects of F. L. Bauer's simultaneous iteration method*, Numer. Math., 13 (1969), pp. 4–13.
[3] G. W. STEWART, *Error and perturbation bounds for subspaces associated with certain eigenvalue problems*, SIAM Rev., 15 (1973), pp. 727–764.
[4] ———, *Introduction to Matrix Computations*, Academic Press, New York, 1973.
[5] ———, *Simultaneous iteration for computing invariant subspaces of a non-Hermitian matrix*, Numer. Math., 25 (1976), pp. 123–136.
[6] J. H. WILKINSON, *The Algebraic Eigenvalue Problem*, Clarendon, Oxford, 1965.

12.5. [GWS-J42] (with A. K. Cline, C. B. Moler, and J. H. Wilkinson), "An Estimate for the Condition Number of a Matrix"

[GWS-J42] (with A. K. Cline, C. B. Moler, and J. H. Wilkinson), "An Estimate for the Condition Number of a Matrix," *SIAM Journal on Numerical Analysis* **16** (1979) 368–375.
`http://dx.doi.org/10.1137/0716029`

SIAM J. NUMER. ANAL.
Vol. 16, No. 2, April 1979

0036-1429/79/1602-0017 $01.00/0

AN ESTIMATE FOR THE CONDITION NUMBER OF A MATRIX*

A. K. CLINE,† C. B. MOLER,‡ G. W. STEWART¶ AND J. H. WILKINSON§

Abstract. It is important in practice when solving linear systems to have an economical method for estimating the condition number $\kappa(A)$ of the matrix of coefficients. An algorithm involving $O(n^2)$ arithmetic operations is described; it gives a reliable indication of the order of magnitude of $\kappa(A)$.

1. Introduction. The sensitivity of the solution of a nonsingular system of linear equations

$$Ax = b \tag{1.1}$$

with respect to perturbation E in A is directly related to the *condition number* $\kappa(A)$ of A with respect to inversion. If

$$(A+E)y = b \tag{1.2}$$

then

$$\|y-x\|/\|y\| \leqq \|A^{-1}E\|. \tag{1.3}$$

Let

$$\varepsilon = \|E\|/\|A\| \tag{1.4}$$

be the relative error in A. Then the error in x, measured relative to the perturbed solution y, satisfies

$$\|y-x\|/\|y\| \leqq \varepsilon \|A\| \, \|A^{-1}\|. \tag{1.5}$$

The condition number of A with respect to inversion is defined by

$$\kappa(A) = \|A\| \, \|A^{-1}\|. \tag{1.6}$$

Obviously $\kappa(A)$ is a function of the norm which is used. We shall be interested in l_1, l_2 and l_∞ norms and the variation of κ with the norm will be its least interesting aspect.

We observe that in deriving (1.3) we have majorized $\|A^{-1}Ey\|$ by $\|A^{-1}E\| \, \|y\|$ and for some y and E this will be an overestimate. When $Ey = 0$, for example, we have $y = x$. Further, in deriving (1.5) from (1.3), we have majorized $\|A^{-1}E\|$ by $\|A^{-1}\| \, \|E\|$. When $\kappa(A)$ is large this will be a severe overestimate for some E; for example, if $E = \varepsilon A$ we have $\|A^{-1}E\| = \varepsilon$ which is independent of κ.

In spite of these remarks the inequality (1.3) will be quite realistic for almost all perturbations E.

The sensitivity to perturbations e in b is also related to $\kappa(A)$. If

$$Ay = b + e, \qquad \|e\|/\|b\| \leqq \varepsilon, \tag{1.7}$$

we have

$$\|y-x\| \leqq \|A^{-1}e\| \leqq \varepsilon \|A^{-1}\| \, \|b\|, \tag{1.8}$$

* Received by the editors December 5, 1977. This work was performed under the auspices of the U.S. Energy Research and Development Administration and the U.S. National Science Foundation.

† Computer Science Department, University of Texas at Austin, Austin, Texas 78712.

‡ Department of Mathematics and Statistics, University of New Mexico, Albuquerque, New Mexico 87131.

¶ Computer Science Department, University of Maryland, College Park, Maryland 20742.

§ National Physical Laboratory, Teddington, England.

while from (1.1)

$$\|b\| \leqq \|A\| \, \|x\|. \tag{1.9}$$

Hence

$$\|y-x\|/\|x\| \leqq \varepsilon\kappa(A), \tag{1.10}$$

which is essentially the same bound as in (1.5). However, this last result is very deceptive. Although the bound is attainable the probability that (1.10) is realistic is rather low. Its derivation is dependent on (1.9) and this relation is very weak for almost all b when $\kappa(A)$ is large. Such probability considerations are of vital importance in what follows.

2. Analysis in terms of SVD of A. Deeper insight into the effect of perturbations may be gained via the singular value decomposition (SVD) of A. We write

$$A = U\Sigma V^T \tag{2.1}$$

where U and V are orthogonal and $\Sigma = \operatorname{diag}(\sigma_i)$, the σ_i being the singular values of A. From (2.1)

$$Av_i = \sigma_i u_i, \qquad A^T u_i = \sigma_i v_i \tag{2.2}$$

and since $\|A\|_2 = \sigma_1$ and $\|A^{-1}\|_2 = \sigma_n^{-1}$ we have $\|A\|_2\|A^{-1}\|_2 = \sigma_1/\sigma_n$.

We may expand both b and e in terms of the orthogonal vectors u_i in the form

$$b = \|b\|\Sigma\alpha_i u_i \qquad (\Sigma\alpha_i^2 = 1), \tag{2.3}$$

$$e = \varepsilon\|b\|\Sigma\beta_i u_i \qquad (\Sigma\beta_i^2 = 1), \tag{2.4}$$

and from (2.2)

$$x = \|b\|\Sigma(\alpha_i/\sigma_i)v_i, \qquad y - x = \varepsilon\|b\|\Sigma(\beta_i/\sigma_i)v_i. \tag{2.5}$$

It is clear that

$$\|y-x\|/\|x\| = \varepsilon\sigma_1/\sigma_n = \varepsilon\kappa(A) \tag{2.6}$$

only in the extreme case when

$$b = \|b\|u_1, \qquad e = \varepsilon\|b\|u_n. \tag{2.7}$$

When σ_1/σ_n is large $\|y-x\|/\varepsilon\|x\|$ will be of the order of unity for any e satisfying (1.7) unless α_n, which gives the exponent of b in the direction of u_n, is exceptionally small. Also since

$$\|x\|/\|b\| = [\Sigma(\alpha_i/\sigma_i)^2]^{1/2} \tag{2.8}$$

this ratio will be of order σ_n^{-1} (i.e. $\|A^{-1}\|_2$) unless α_n is exceptionally small.

3. Estimates for $\kappa(A)$. It is important in practice when solving linear systems to have some estimate of $\kappa(A)$ which will give at least a reliable indication of its order of magnitude. When a linear system $Ax = b$ has been solved by a direct method one has some factorization of A and it is natural to make use of this in determining the estimate of $\kappa(A)$. For the moment we assume that we have exact factors of A and comment later on the effect of the rounding errors made when deriving the factorization.

The problem is perhaps simplest when we have a factorization of the form

$$A = QR \tag{3.1}$$

where Q is orthogonal and R is upper triangular. In this case

(3.2) $$\|A\|_2 = \|R\|_2, \quad \|A^{-1}\|_2 = \|R^{-1}\|_2, \quad \kappa_2(A) = \kappa_2(R),$$

where $\kappa_2(\cdot)$ denotes the condition number corresponding to the l_2 norm. However, we concentrate on an estimate of $\kappa_1(R)$, since the l_1 norm can be computed with less expense. Certainly $\|R\|_1$ is immediately available, while $\|R^{-1}\|_1$ may be obtained by computing the columns of R^{-1} one by one. There is no need to store R^{-1} since $\|R^{-1}\|_1$ is merely the maximum of the l_1 norms of its columns. However, the computation of R^{-1} involves $n^3/6$ multiplication and one would like to derive an estimate of $\|R^{-1}\|$ by a process which involves only $O(n^2)$ operations and at most $O(n)$ additional storage locations.

Now we know that if

(3.3) $$Rx = b$$

then $\max(\|x\|_2/\|b\|_2) = \sigma_n^{-1} = \|R^{-1}\|_2$, this bound being attained when $b = u_n$ where $U\Sigma V^T$ is now the SVD of R. Note that

(3.4) $$A = QR = (QU)\Sigma V^T$$

so that the SVD's of A and R are closely related. Is it possible to choose a right-hand side b in some ad hoc manner so that $\|x\|_2$ has a near-maximum norm? From equation (2.8) we see that when $\kappa(A)$ is large the probability that $\|x\|_2/\|b\|_2$ will give a good estimate of σ_n^{-1} is quite high. Our object is to choose b in such a way as to reinforce this natural probability.

It is tempting to suggest that one takes a random b, computes the corresponding x and then uses this x as a new right-hand side. Indeed this is effectively what is done when one wishes to find the eigenvector of R of smallest modulus by inverse iteration. However, it is well known that a second step of inverse iteration is often extremely unsatisfactory. The problem is that in order to obtain a 'large' solution to the system $Rx = b$ one requires a right-hand side in which u_n is substantially represented. The x derived at the end of the first step will usually have v_n as its dominant component; if v_n is almost orthogonal to u_n the latter will be poorly represented in x. This is a surprisingly common situation.

Discussion in terms of the SVD, however, suggests the use of the two-step process

(3.5) $$R^T x = b, \qquad Ry = x.$$

If then

(3.6) $$b = \Sigma \alpha_i v_i$$

we have

(3.7) $$x = \Sigma(\alpha_i/\sigma_i)u_i, \qquad y = \Sigma(\alpha_i/\sigma_i^2)v_i.$$

Provided b has a component of v_n which is not unduly small the vector y is likely to be completely dominated by its component of v_n. Unlike the situation with two steps of inverse iteration we now have the full benefit of the factor σ_i^{-2}. From (3.7)

(3.8) $$\|y\|_2/\|x\|_2 = [\Sigma(\alpha_i/\sigma_i^2)^2/\Sigma(\alpha_i/\sigma_i)^2]^{1/2}$$

and we shall have a good estimate of σ_n^{-1} provided α_n/σ_n is not very small relative to the other α_i/σ_i. This is a much healthier situation, the more so in the very important case when $\sigma_n \ll \sigma_1$, i.e. when A is ill-conditioned.

If we start then from a *random* vector b the probability of obtaining a good estimate for $\|R^{-1}\|$ and hence for $\|A^{-1}\|$ is high. We have the possibility of enhancing this probability by a judicious choice of b. Our object is to choose b so that the solution x of $R^Tx = b$ is such that $\|x\|/\|b\|$ is as large as possible. The following simple strategy suggests itself.

Let us take $b_s = \pm 1$, the sign being determined at the stage when x_s is computed via the relation

$$r_{ss}x_s = b_s - (r_{1s}x_1 + \cdots + r_{s-1,s}x_{s-1}). \tag{3.9}$$

The sign of b_s is chosen to be the opposite of that of the inner-product in parentheses on the right-hand side of (3.9). This gives $|x_s|$ the larger of the two possible values.

This strategy was very successful in practice, but the following matrix of order 4 revealed its weakness. Let

$$R^T = \left[\begin{array}{c|c} I & 0 \\ \hline kE & I \end{array}\right], \qquad E = \begin{bmatrix} 1 & -1 \\ -1 & 1 \end{bmatrix}. \tag{3.10}$$

Then

$$R^{-T} = \left[\begin{array}{c|c} I & 0 \\ \hline -kE & I \end{array}\right], \qquad \|R^T\|_1 = \|R^{-T}\|_1 = 1 + 2k. \tag{3.11}$$

If k is large, both $\|R^T\|_1$ and $\|R^{-T}\|_1$ are large. Without loss of generality we can assume $b_1 = +1$, giving $x_1 = 1$. Either choice of b_2 gives $|x_2| = 1$. If we take $b_2 = +1$ then both $|x_3|$ and $|x_4|$ are unity and we shall not have a 'large' x. Note that if the ambiguity in choice is resolved by taking $b_2 = -1$ we have a similar failure with

$$E = \begin{bmatrix} 1 & 1 \\ 1 & 1 \end{bmatrix}. \tag{3.12}$$

Returning to the E of (3.10) and assuming that we take $b_s = +1$ on each occasion we have

$$R^Tb = b, \qquad Rb = b \tag{3.13}$$

and the process fails to indicate the ill-condition of R.

Clearly the weakness in the strategy is that the decision on the sign of b_s takes place on a purely local criterion. The value of such x_s affects all later values x_i and we need some 'look-ahead' feature in our strategy. This leads naturally to the following improved strategy.

Each x_i $(i > s)$ is determined by the relation

$$r_{ii}x_i = (-r_{1i}x_1 - \cdots - r_{s-1,i}x_{s-1}) + (-r_{si}x_s - \cdots - r_{i-1,i}x_{i-1} + b_i) \tag{3.14}$$

where the right-hand side has been split into two parts. The first part is determined irrevocably once $x_1, \cdots, x_{s-1}$ have been assigned. We denote this part by $p_i^{(s-1)}$. At the stage when we are about to assign b_s and compute x_s let us assume that we have already computed and stored the quantities $p_i^{(s-1)}$ $(i = s, \cdots, n)$. Now the two possible values of x_s are given by

$$r_{ss}x_s = -p_s^{(s-1)} + b_s = -p_s^{(s-1)} \pm 1. \tag{3.15}$$

We denote these two possible values of x_s by x_s^+ and x_s^-. If we compute both of these for the moment we can use these to give two different sets of updates of the

$p_i^{(s-1)}$ to $p_i^{(s)}$. We may write

$$p_i^{(s)+} = p_i^{(s-1)} + r_{si}x_s^+, \qquad p_i^{(s)-} = p_i^{(s-1)} + r_{si}x_s^- \tag{3.16}$$

and our decision should depend on the size of these $|p_i^{(s)}|$ as well as on the size of x_s itself. As a reasonable criterion we could take $b_s = \pm 1$ according as

$$|-p_s^{(s-1)} + 1| + \sum_{i=s+1}^{n} |p_i^{(s)+}| \geqq \text{ or } < |-p_s^{(s-1)} - 1| + \sum_{i=s+1}^{n} |p_i^{(s)-}|. \tag{3.17}$$

There is about twice as much work in the solution of $R^T x = b$ with this improved strategy as with the original but since we have subsequently to solve $Ry = x$ this reduces the overall factor to 1.5. It will be seen that this modification immediately deals with the example of (3.10). Indeed if

$$E = \begin{bmatrix} \pm 1 & \pm 1 \\ \pm 1 & \pm 1 \end{bmatrix} \tag{3.18}$$

then with all possible combinations of signs an x is produced such that $\|x\|_1 > k$.

Two comments may be made on this second strategy.

(i) Since it is the size of x which really interests us it might be advisable to replaced (3.17) by the criterion

$$\frac{|-p_s^{(s-1)} + 1|}{|r_{ss}|} + \sum_{i=s+1}^{n} \frac{|p_i^{(s)+}|}{|r_{ii}|} \geqq \text{ or } < \frac{|-p_s^{(s-1)} - 1|}{|r_{ss}|} + \sum_{i=s+1}^{n} \frac{|p_i^{(s)-}|}{|r_{ii}|}. \tag{3.19}$$

However, this modification increases the volume of computation appreciably.

(ii) Matrices which arise in practice can in no sense be said to have random elements. Many of them will turn out to be very special and choosing a vector b with elements ± 1 must increase the probability of an accidental failure. If one takes $b_i = \pm \theta_i$ with the sign chosen as in our second strategy but with each θ_i a random number between $\frac{1}{2}$ and 1 the chances of accidental failure must be diminished. Indeed if b_i is chosen in this way even the *first* strategy is virtually certain to succeed with the example of (3.10).

4. Estimate from *LU* decomposition. In practice dense systems of linear equations are most commonly solved by Gaussian elimination with some form of pivoting. This provides permutation matrices P and Q, a unit lower-triangular matrix L and an upper-triangular matrix U such that

$$PAQ = LU. \tag{4.1}$$

The matrices P and Q cover interchanges resulting from pivoting. With partial pivoting $Q = I$ and with either partial pivoting or complete pivoting

$$|l_{ij}| \leqq 1. \tag{4.2}$$

For simplicity we shall write A in place of PAQ from now on. Our previous discussion suggests that we now use the two-step process

$$(LU)^T x = b, \qquad LUy = x \tag{4.3}$$

and use $\|y\|/\|x\|$ as our estimate of $\|A^{-1}\|$. The first step then is the solution of $U^T L^T x = b$ which is done in the two stages

$$U^T z = b, \qquad L^T x = z. \tag{4.4}$$

We can use either of the strategies discussed in the previous section to attempt to maximize z.

That the first strategy described in § 3 may fail is shown by the example

$$U^T = \left[\begin{array}{c|c} I & 0 \\ \hline kE & I \end{array}\right], \quad L^T = \left[\begin{array}{c|c} I & E \\ \hline U & I \end{array}\right] \tag{4.5}$$

with E as defined in (3.10). This leads to the choice $b_s = +1$ $(s = 1, \cdots, 4)$ and

$$U^T b = b, \qquad L^T b = b, \tag{4.6}$$

$$Lb = b, \qquad Ub = b. \tag{4.7}$$

No indication is given of the ill-condition of A. The second strategy immediately overcomes this difficulty.

There remains the potential weakness that any advantage we may have gained by a good choice of b when computing z may be vitiated when we compute x. (It should be appreciated that when solving the pair $R^T x = b$, $Ry = x$ this danger does not exist; moderate success with $R^T x = b$ ensures almost complete success when solving $Ry = x$.) However, the danger of this would appear to be slight. Pivoting usually ensures that any ill-condition in A is reflected in a corresponding ill-condition in U; L is usually quite well-conditioned. The ill-condition of U is to be expected since when A is exactly singular some u_{ii}, usually u_{nn}, must be zero. The most common situation with an ill-conditioned A is that u_{nn} is small. It must be emphasized though that A *can* be almost singular without *any* u_{ii} being unduly small.

That L can be ill-conditioned in spite of pivoting is illustrated by the example

$$l_{ii} = 1, \qquad l_{ij} = -1 \qquad (i > j), \tag{4.8}$$

for which the maximum element of L^{-1} is 2^{n-2}; indeed L is "almost singular" even for quite modest values of n. If such an L is associated with a well-conditioned U then decisions should be made when working with L and not with U. The danger is exemplified by the case when U is diagonal with $u_{nn} = 1$, $u_{ii} = -1$ $(i \neq n)$. Then the solution of $U^T z = b$ is $(-1, -1, \cdots, -1, 1)$ with either of our two strategies and the solution of $L^T x = z$ is $(0, 0, \cdots, 0, 1)$. In solving $LUy = x$ there is no further amplification. Examples of this kind are extremely special and the danger in practice is likely to be greatly diminished by taking $b_i = \pm\theta_i$ as discussed under (ii) in § 3.

5. Rounding errors in the factors. In practice the factors will not be exact but if the factorization algorithm is stable the computed factors will be exact for some matrix $(A + E)$ where $\|E\|/\|A\|$ is some fairly modest multiple of the machine precision. The errors made in solving the triangular systems are unlikely to be of much importance in practice particularly as we are trying to use right-hand sides such that the solution truly reflects the ill-condition, if any. However, it must be faced that at best we can obtain only $\kappa(A + E)$. Since $A + E$ may be singular when A is not and vice-versa, there is naturally a limitation on the information that can be derived from the computed factors.

This limitation becomes serious only when A is very close to singularity. For if

$$\|E\|/\|A\| = \varepsilon \tag{5.1}$$

we have

$$(1 - \varepsilon)\|A\| \leqq \|A + E\| \leqq (1 + \varepsilon)\|A\| \tag{5.2}$$

while

$$\frac{\|A^{-1}\|(1-2\varepsilon\kappa)}{1-\varepsilon\kappa} \leqq \|(A+E)^{-1}\| \leqq \frac{\|A^{-1}\|}{1-\varepsilon\kappa}. \tag{5.3}$$

Combining (5.2) and (5.3), we obtain

$$\frac{(1-\varepsilon)(1-2\varepsilon\kappa)}{1-\varepsilon\kappa} \leqq \frac{\kappa(A+E)}{\kappa(A)} \leqq \frac{(1+\varepsilon)}{1-\varepsilon\kappa}. \tag{5.4}$$

Results are not likely to be of interest unless, say, $\varepsilon\kappa(A) \leqq 0.1$ and we have then

$$\frac{8}{9}(1-\varepsilon) \leqq \frac{\kappa(A+E)}{\kappa(A)} \leqq \frac{10}{9}(1+\varepsilon).$$

Since we are primarily concerned with the order of magnitude of $\kappa(A)$, the implicit use of $A+E$ in place of A is of little importance.

6. Implementation. This technique is implemented in LINPACK, a collection of FORTRAN subroutines for solving various forms of linear equations that is being developed by Argonne National Laboratory and three universities. The actual details of the implementation are described in the LINPACK documentation [1].

Since the technique is designed to cause growth in the size of the elements of various vectors, there is a definite possibility of floating point arithmetic overflow. In the Gaussian elimination subroutines, for example, the most crucial point is the division by the diagonal elements of U in the solution of $U^Tz = b$. The elements of b are in principle ± 1, but since only the direction of z is important, b can be rescaled as necessary to avoid any dangerous divisions. Consider the 3×3 example

$$\begin{pmatrix} u_{11} & 0 & 0 \\ u_{12} & \eta & 0 \\ u_{13} & u_{23} & u_{33} \end{pmatrix} \begin{pmatrix} z_1 \\ z_2 \\ z_3 \end{pmatrix} = \begin{pmatrix} \pm 1 \\ \pm 1 \\ \pm 1 \end{pmatrix}$$

where η is very small. The first equation gives $z_1 = 1/u_{11}$ with no difficulty. Then the η in the second equation is noted. The right-hand side and the solution developed so far can both be scaled by η, giving

$$z_1 = \eta/u_{11}, \qquad z_2 = \pm 1 - u_{12}/u_{11}.$$

Finally,

$$z_3 = (\pm\eta - u_{13}z_1 - u_{23}z_2)/u_{33}.$$

No division by η is required. In the extreme case when $\eta = 0$, the right-hand side becomes the zero vector and z becomes a null vector of U^T. The subsequent steps produce a null vector of A, which is exactly what is desired.

The LINPACK subroutines are designed with the assumption that overflows are fatal errors, but that underflows are quietly replaced by zeros. Consequently, a quantity RCOND which estimates $1/\kappa(A)$ is computed. If $\kappa(A)$ is very large, this may underflow to zero. If an exact zero diagonal element occurs in the triangular factors, the rescaling strategy automatically results in a zero RCOND. Consequently, there is no necessity to make exact singularity a special case.

The programs used in testing LINPACK evaluate the condition estimator by also inverting the matrices and computing the actual $\kappa_1(A)$. The ratio of the estimated condition to the actual condition is printed out as one of the testing diagnostics. This ratio must be less than 1.0 if the programs are working correctly. The ratio is regarded as "suspicious" if it is less than the arbitrary value 0.1. We have no justification for picking 0.1, except that the estimator very rarely produces a suspicious estimate.

In one test, J. T. Goodman of the University of New Mexico generated 1350 matrices with elements chosen randomly from various distributions. The orders of the matrices ranged from $n = 10$ to $n = 50$.

Of 550 matrices generated with elements normally distributed with mean zero and variance one, only one matrix was noted in which the ratio was less than 0.1. The matrix was of order 30 and the ratio was 0.077.

Of 300 matrices generated with elements uniformly distributed on the interval $(-1, 1)$, all ratios were greater than 0.1 with the majority of the ratios between 0.55 and 0.65. No ratios higher than 0.8 were observed.

Of 400 matrices generated with elements -1, 0, or 1 with equal probability, two matrices exhibited ratios less than 0.1. One matrix was of order 20 with a ratio of 0.062, the other a matrix of order 30 with a ratio of 0.092. The majority of the ratios were between 0.45 and 0.55. No ratio higher than 0.8 was observed.

Of 100 matrices of order 10 generated from Householder reflections, all ratios were greater than 0.1. Indeed, the condition estimator performed better on these matrices of order 10 than on any of the other three types of matrices of order 10. Four ratios were observed in the 0.95 to 1.0 range. The majority of the ratios were in the 0.50 to 0.55 range.

None of these matrices were particularly badly conditioned and so ratios near 1.0 are not expected.

In a second test, several dozen matrices were generated which had condition numbers in the range from 10^3 to 10^{12}. For these matrices, the estimate is quite accurate and ratios between 0.99 and 1.00 were fairly common, although this is of no great importance.

REFERENCE

[1] J. J. Dongarra, J. R. Bunch, C. B. Moler, and G. W. Stewart, *LINPACK Users' Guide*, Society for Industrial and Applied Mathematics, Philadelphia, 1979.

12.6. [GWS-J49] "Rank Degeneracy"

[GWS-J49] "Rank Degeneracy," *SIAM Journal on Scientific and Statistical Computing* **5** (1984) 403–413.
http://dx.doi.org/10.1137/0905030

SIAM J. SCI. STAT. COMPUT.
Vol. 5, No. 2, June 1984

013

RANK DEGENERACY*

G. W. STEWART†

Abstract. At some point in many applications a decision must be made about the rank of a matrix. The decision is frequently complicated by the fact that the matrix in question is contaminated with errors. This paper surveys some of the more commonly used methods for approximating rank, with particular attention being paid to the effects of errors.

Key words. rank, QR decomposition, singular value decomposition, inverse, cross product, matrix perturbation theory

1. Introduction. The problem treated in this paper is the following. Let X be an $n \times p$ matrix that satisfies

$$k \equiv \operatorname{rank}(X) \leqq p \leqq n.$$

Suppose further that X itself cannot be observed; rather we are given a perturbed matrix

$$\tilde{X} = X + E, \tag{1.1}$$

where E represents a matrix of errors. From the matrix $\tilde{X}$ we wish to

1. determine k

and

2. find an approximation $\hat{X}$ to X that is of rank k.

There are at least two reasons for considering this problem. First, if $k < p$ the attempt to use $\tilde{X}$ instead of X in a computational procedure may result in violently inaccurate results. For example, regression coefficients calculated from $\tilde{X}$ may be inaccurate owing to the discontinuity of the pseudo-inverse around a rank deficient matrix [13]. If, however, one can determine k, the use of a rank deficient approximation $\hat{X}$ may restore the lost accuracy. Thus one reason for solving the problem is to obtain a regularization procedure for certain ill-posed problems.

The second application occurs in areas, such as bifurcation theory, where the matrix X depends on a parameter λ. In computations for such problems, the normal state of affairs is that X is of full rank p. However, at certain critical values of λ the rank drops below p, and special computational action must be taken, action that usually involves, at least implicitly, a rank degenerate approximation $\hat{X}$ (e.g. see [10]).

It is important to realize that the problem as stated may be unsolvable, even when a fair amount is known about the matrix E. For example, suppose that the elements of E are known to be bounded by 10^{-2} and consider the following two decompositions of the form (1.1):

$$\begin{bmatrix} 1.010 & -0.005 \\ 0.005 & 0.005 \end{bmatrix} = \begin{bmatrix} 1 & 0 \\ 0 & 0 \end{bmatrix} + \begin{bmatrix} 0.010 & -0.005 \\ 0.005 & 0.005 \end{bmatrix},$$

$$\begin{bmatrix} 1.010 & -0.005 \\ 0.005 & 0.005 \end{bmatrix} = \begin{bmatrix} 1 & 0 \\ 0 & 0.01 \end{bmatrix} + \begin{bmatrix} 0.010 & -0.005 \\ 0.005 & -0.005 \end{bmatrix}.$$

In both cases the matrix $\tilde{X}$ is the same and the elements of E are bounded by 10^{-2}.

* Received by the editors June 4, 1982, and in revised form April 6, 1983.

† Computer Science Department, University of Maryland, College Park, Maryland 20742.

However, in the first case $k = 1$, whereas in the second $k = 2$. On the other hand if

$$\tilde{X} = \begin{bmatrix} 1.010 & 0.005 \\ 0.005 & 0.100 \end{bmatrix},$$

then under the assumption that the elements of E are bounded by 10^{-2} we can say definitely that $k = 2$.

There are two conclusions to be drawn from these examples. First, we must know something about E to say anything at all. Second, unless we have very precise information about E, we can only obtain lower bounds on k, usually by showing that $\tilde{X}$ could have come from a matrix X whose rank is equal to the lower bound. *The decision to accept this lower bound as a "numerical rank" must be made by considering its consequences for the specific problem being solved.* The failure to appreciate this simple fact has, in the author's opinion, resulted in many misguided attempts to determine rank automatically, perhaps the most notorious examples being pseudo-inverse programs with built-in tolerances that are invisible to the user.

Because of the difficulties introduced by the matrix E, the next four sections of this paper will be devoted to describing how to compute indicators of possible rank deficiency. Throughout these sections, we shall use the terms like "sufficiently small" without specifying precisely what we mean. Only in the last section will an attempt be made to say how small is small.

Throughout this paper we shall let $\|\cdot\|_2$ denote the ordinary Euclidean vector norm defined by

$$\|x\|_2^2 = x^T x$$

or the spectral matrix norm defined by

$$\|X\|_2 = \max_{\|x\|_2 = 1} \|Xx\|_2.$$

The symbol $\|\cdot\|_F$ will denote the Frobenius matrix norm defined by

$$\|X\|_F^2 = \sum x_{ij}^2 = \text{trace}\,(X^T X).$$

For more on matrix and vector norms see [12].

2. The singular value decomposition. Perhaps the most widely recommended tool for detecting rank degeneracies is the singular value decomposition. It can be shown (e.g. see [12]) that there are orthogonal matrices U and V such that[1]

$$U^T X V = \begin{bmatrix} \Psi \\ 0 \end{bmatrix}, \tag{2.1a}$$

where

$$\Psi = \text{diag}\,(\psi_1, \psi_2, \cdots, \psi_p) \tag{2.1b}$$

with

$$\psi_1 \geqq \psi_2 \geqq \cdots \geqq \psi_p \geqq 0. \tag{2.1c}$$

It is sometimes more convenient to write the decomposition in the factored form

$$X = U_X \Psi V^T, \tag{2.2}$$

[1] Most numerical analysts would write Σ where we have written Ψ. However, this usage is impossible for statisticians, to whom Σ means a variance. Since Ψ is little used in either camp, we have adopted it here. As a mnemonic take ψ to stand for "psingular" value.

where U_X comes from the partition

$$U = (U_X \quad U_P) \tag{2.3}$$

with U_X having p columns.

The singular value decomposition is closely related to the spectral decomposition of the *cross-product matrix*

$$A = X^T X. \tag{2.4}$$

Specifically, from (2.2) and the fact that $U_X^T U_X = I$ it follows that

$$A = V\Psi^2 V^T. \tag{2.5}$$

Since A is symmetric and V is orthogonal, the squares of the singular values of X are the eigenvalues of A. The eigenvectors are the corresponding columns of V.

From (2.1) and (2.2) it follows that rank $(X) =$ rank (Ψ). Consequently if ψ_{k+1}, $\psi_{k+2}, \cdots, \psi_p$ are sufficiently small and ψ_k is sufficiently large, it is natural to accept a rank k approximation to X. The best such approximation may be obtained as follows. Let

$$\hat{\Psi} = \operatorname{diag}(\psi_1, \psi_2, \cdots, \psi_k, 0, \cdots, 0),$$

and in analogy with (2.2) let

$$\hat{X} = U_X \Psi \hat{V}^T. \tag{2.6}$$

Then a classical result of Eckart and Young [8] states that

$$\psi_{k+1}^2 + \cdots + \psi_p^2 = \|X - \hat{X}\|_F^2 = \min_{\operatorname{rank}(\bar{X}) \leq k} \|X - \bar{X}\|_F^2.$$

Mirsky [9] has proved the corresponding result for the spectral norm:

$$\psi_{k+1} = \|X - \hat{X}\|_2 = \min_{\operatorname{rank}(\bar{X}) \leq k} \|X - \bar{X}\|.$$

In applications one must compute the singular value decomposition of $\tilde{X}$ not X. Even if X is exactly rank degenerate, $\tilde{X}$ will not be, and the sizes of the singular values of $\tilde{X}$ corresponding to the zero singular values of X will depend on the properties of E (e.g., none of them can be larger than $\|E\|_2$). Thus something must be known about E before we can decide which singular values of $\tilde{X}$ to ignore.

Once a decision has been made about the rank of X, that is, once one has decided on a value of k, one can work with the approximation $\hat{X}$ in (2.6). In practice it is important to remember that $\hat{X}$ will be the best approximation to $\tilde{X}$, from which it has been computed, not to X, which is unobserved. Nonetheless, this approximation may be good enough in many cases.

In most problems it will be unnecessary to compute $\hat{X}$ explicitly; rather it can be manipulated in the factored form (2.6), with considerable savings in computations.

3. The QR decomposition. A second very effective tool for detecting rank degeneracy is the QR decomposition. Specifically, it can be shown [12] that given any permutation matrix J, there is an orthogonal matrix Q such that

$$Q^T X J = \begin{bmatrix} R \\ 0 \end{bmatrix},$$

where R is an upper triangular matrix with nonnegative diagonal elements. Moreover,

J may be chosen by the method of *column pivoting* [3] so that

$$r_{jj}^2 \geqq \sum_{i+1}^{k} r_{ik}^2 \qquad (k=j, j+1, \cdots, p). \tag{3.1}$$

As with the singular decomposition, the QR decomposition can be written in a factored form

$$XJ = Q_X R, \tag{3.2}$$

where Q_X is taken from the partition

$$Q = (Q_X \quad Q_P)$$

with Q_X having the same dimensions as X.

The QR decomposition can also be related to the cross-product matrix A in (2.4). Specifically, from (3.2), it follows that

$$J^T A J = R^T R. \tag{3.3}$$

Since R is upper triangular, $R^T R$ is the *Cholesky factorization* [12] of A with its rows and columns symmetrically permuted according to J.

The application of the QR decomposition to the problem of detecting rank degeneracy goes as follows. Let R be partitioned in the form

$$R = \begin{bmatrix} R_{11} & R_{12} \\ 0 & R_{22} \end{bmatrix},$$

where R_{11} is $k \times k$. If $r_{k+1,k+1}$ is negligible, then (3.1) assures us that each column of R_{22} also has negligible norm; indeed

$$\rho \equiv \|R_{22}\|_F \leqq (p-k)^{1/2} r_{k+1,k+1}. \tag{3.4}$$

Moreover, if $r_{kk} \neq 0$ and we set

$$\hat{R} = \begin{bmatrix} R_{11} & R_{12} \\ 0 & 0 \end{bmatrix}, \tag{3.5a}$$

then

$$\hat{X} = Q_X \hat{R} J^T \tag{3.5b}$$

is an approximation to X that has the following properties [15]:

1. $\hat{X}$ is of rank k.
2. $\|X - \hat{X}\|_F = \|R_{22}\|_F$.
3. XJ and $\hat{X}J$ differ only in their last $p-k$ columns.
4. If $\bar{X}$ is any matrix satisfying 1 and 3, then $\|X - \bar{X}\|_F \geqq \|R_{22}\|_F$.

According to the fourth property, which is an analogue of the Eckart–Young–Mirsky result, the rank degenerate approximation (3.5) is the best that can be obtained by altering only the last $p-k$ columns of X. When $k = p-1$, the number ρ in (3.4) has a particularly nice interpretation: it is the norm of the smallest perturbation of the last column of XJ that will make X exactly degenerate. By computing $\rho = r_{pp}$ for a sequence of permutations J that moves each column of X to the last, one obtains a set of numbers

$$\{\rho_1, \rho_2, \cdots, \rho_p\}, \tag{3.6}$$

corresponding to the columns of X, which tell how much each column must be altered to make X degenerate. These numbers can be efficiently computed from the upper triangular factor of the QR factorization of X by means of the LINPACK subroutine SCHEX [3].

It is unfortunate that there seems to be no simple, sharp relation between the number ρ in (3.4) for the QR decomposition with column pivoting and the last $p-k$ singular values of $\tilde{X}$. When $k=p-1$, there is empirical evidence that will be of the same order of magnitude as ψ_p [14]. Moreover, it can be shown that

$$\psi_p \leqq \min \rho_i \leqq p\psi_p, \tag{3.7}$$

where the ρ_i are the numbers appearing in (3.6). The general folklore says that the QR decomposition with column pivoting is about as good as the singular value decomposition.

The practical considerations that arise in using the QR decomposition are similar to the ones discussed in the section on the singular value decomposition. There is, however, one additional point. It frequently happens that once a value of k for which ρ is negligible has been determined, the problem can be recast entirely in terms of the first k columns of XJ, with a potentially large saving in work.

4. The inverse cross-product matrix. We have seen [cf. (2.5) and (3.3)] that the singular value decomposition and the QR decomposition are intimately related to the cross-product matrix $A = X^TX$. Since many procedures for solving problems involving X form A^{-1} explicitly (e.g. the solution of least squares problems by means of inverting the normal equations [11]), it is natural to ask what can be learned about rank degeneracy from A^{-1}.

The principal result is that

$$a_{jj}^{(-1)} = \rho_j^{-2}, \tag{4.1}$$

where $a_{jj}^{(-1)}$ denotes the jth diagonal of A^{-1} and ρ_j is the jth member of (3.6). To see this, suppose that J has been chosen so that the jth column of X is the last column of XJ. If R is partitioned in the form

$$R = \begin{bmatrix} R_* & r \\ 0 & \rho_j \end{bmatrix},$$

then from (3.3)

$$(J^TAJ)^{-1} = \begin{bmatrix} R_* & r \\ 0 & \rho_j \end{bmatrix}^{-1} \begin{bmatrix} R_* & r \\ 0 & \rho_j \end{bmatrix}^{-T}.$$

From this it is easily seen that the (p,p)-element of $(J^TAJ)^{-1}$, which is just $a_{jj}^{(-1)}$, is equal to ρ_j^{-2}.

In view of (3.7) and (4.1) it is possible to detect rank degeneracy by examining the diagonal elements of A^{-1}. However, one cannot determine a numerical rank without further calculation. One possibility is to determine an index j for which ρ_j is minimal and then use a Gauss–Jordan SWEEP operator [1], [6], to compute the inverse cross-product matrix for a matrix corresponding to X with its jth column deleted. Iterating this process leads to a sort of reverse pivoting algorithm, whose properties are not well understood at this time. (For anyone who wishes to pursue this matter, we note that this algorithm is essentially equivalent to calculating the Cholesky decomposition of A^{-1} with pivoting. We also note that there is an equivalent, but more stable algorithm that works with R instead of A^{-1}.)

5. Computational properties. There are two aspects to the algorithmic realization of a general procedure: the amount of work required and the effects of rounding error. We shall consider each in turn.

Most algorithms for computing the singular value decomposition are based on a preliminary reduction to bidiagonal form followed by an iteration for the singular values [3], [5]. Although the amount of work required is $O(np^2)$, the order constant is rather large, so that, in the author's opinion, the singular value decomposition should not be calculated explicitly unless there is a specific need for it.

In fact, the QR decomposition can often be used in place of the singular value decomposition. There are essentially three methods for computing a QR decomposition: the Gram–Schmidt algorithm with reorthogonalization [7], the Golub–Householder method [4], and a method based on plane rotations [3], [4]. The first two require that X be maintained in high-speed memory. The last permits the formation of R by bringing in X a row at a time; however, column pivoting is not possible, at least directly, and the storage of the rotations requires as much memory as the storage of X. All the methods require $O(np^2)$ work, but the constant is much smaller than the one for the singular value decomposition.

An important composite algorithm for the singular value decomposition can considerably reduce the amount of work required when $n \gg p$. Specifically, it follows from (2.5) and (3.3) (with $J = I$) that the singular value decomposition of R is given by

$$W^T R V = \Psi,$$

where V and Ψ are as in (2.1). It further follows from (3.2) that

$$X = (Q_X W)\Psi V^T$$

is the singular value factorization of X. This suggests that to get the singular value factorization of X one first compute the QR factorization of X and then the singular value decomposition of the small $p \times p$ matrix R. A program implementing this approach is given in [2].

A similar approach can be used to compute the QR factorization *with column pivoting* of X when n is so large that X must be brought into memory by rows (or groups of rows). Namely, the R-factor of the QR factorization without column pivoting is first computed, say by plane rotations. Then the QR decomposition of R is computed with pivoting; i.e., $W^T RJ = R'$. It then follows that

$$XJ = Q_X W R'$$

is the QR factorization, with pivoting, of X.

The computation of the cross-product matrix A requires less work than the computation of either the QR or the singular value decompositions. Moreover, it can be computed in the form

$$A = \sum_{i=1}^{n} x_i x_i^T, \tag{5.1}$$

which allows the rows x_i^T of X to be brought into main storage one at a time. Great savings can be effected when X is sparse, since if x_{ij} is zero, the row $x_{ij} x_i^T$ in the sum (5.1) need not be accumulated in A.

Once A has been computed, its spectral decomposition may be computed to get the singular values Ψ and the singular vectors V [cf. (2.5)]. Alternatively, the Cholesky factorization of A may be computed to get the R-factor of X (cf. (3.3)). Finally, A

may be inverted. Of the three alternatives, the last is the most common; however, the author has a predilection for the first or second for reasons that will be given shortly.

In discussing the effects of rounding error on the decompositions, we assume that the calculations are carried out in floating-point arithmetic and that no overflows or destructive underflows occur. Using standard techniques from rounding-error analysis [17], it can be shown that the *computed* QR decomposition of X satisfies

$$Q^T(X+F)J=\begin{bmatrix} R \\ 0 \end{bmatrix}, \tag{5.2a}$$

where

$$\|F\|_F \leqq \phi(n,p)\varepsilon_M. \tag{5.2b}$$

In (5.2b) the number ε_M is the rounding unit for the arithmetic in question; e.g., if thirteen decimal digits are carried in the calculation, then ε_M will be about 10^{-13}. The function ϕ is a slowly growing function of n and p. A similar result holds for the singular value decomposition.

The kind of result embodied in (5.2) has two important implications. First, it may be possible to choose ε_M so that F is much smaller than the error matrix E. If this is done, then the entire effects of rounding error can be regarded as coming from an insignificant perturbation of E. Since any reasonable procedure must be insensitive to minor changes in E, about which not much is known, the effects of rounding error can be ignored.

The second implication is that when X is known exactly, one can tell a true rank degeneracy from a spurious one by increasing the precision. For example, suppose that the singular values of X are computed at precision ε_M and ψ_p is of order ε_M. Then there is some question as to whether ψ_p is nonzero or if the value observed is due to rounding error. If computation is reperformed in double precision ($\varepsilon'_M \approx \varepsilon_M^2$) and the resulting singular value ψ'_p is of order ε_M, then X cannot have been degenerate. On the other hand, if ψ'_p is of order ε'_M, then there is strong reason to believe that the true singular value is zero and that the computed nonzero values are due to rounding error. It is very important to remember that in applying this technique the elements of X must be computed to the same or greater precision as that in which the singular values are computed; otherwise errors made in computing X will comprise the larger part of the errors in the singular values.

An example will make this point clear. Consider the matrix

$$X=\begin{bmatrix} 3.142 & 6.285 \\ 2.718 & 5.436 \end{bmatrix}, \tag{5.3}$$

which is assumed to be known exactly. If ρ from the QR decomposition of X is computed in four decimal digit, floating-point arithmetic, the computed value is $\rho = 10^{-3}$. This is near enough $\varepsilon_M \approx 10^{-4}$ so that we cannot tell whether it is zero or not. If the calculations are repeated with $\varepsilon'_M \approx 10^{-8}$, the results are $\rho' = 6.542 \cdot 10^{-4}$. This is nowhere nearly as small as ε'_M and hence X is of full rank.

On the other hand, let X in (5.3) be regarded as a four-digit approximation to the matrix

$$\begin{bmatrix} \pi & 2\cdot\pi \\ e & 2\cdot e \end{bmatrix}. \tag{5.4}$$

When the procedure is applied in four digit arithmetic, the result is the same as before.

However, when the procedure is repeated in eight digit arithmetic, we must work with the matrix

$$\begin{bmatrix} 3.1415927 & 6.2831853 \\ 2.7182818 & 5.4365637 \end{bmatrix},$$

which is an eight digit approximation to (5.4). The result is $\rho' = 10^{-8}$, which is convincing, although not rigorous evidence that the rank of the matrix (5.4) is one—not two.

Rounding error has more serious effects on the cross-product matrix than on the singular value decomposition or the QR decomposition. To see why this should be so, assume that all the elements of X are roughly the same size, and consider the effects on the singular values of rounding the elements of X with rounding unit ε_M. The resulting matrix $X' = X + F$ will have an error matrix F with $\|F\|_2 \approx \psi_1 \varepsilon_M$, and this will introduce a perturbation of the same size in ψ_p [12]. Thus as long as

$$\frac{\psi_p}{\psi_1} > \varepsilon_M, \tag{5.5}$$

rounding errors will leave some accuracy in ψ_p.

Now suppose that $A = X^T X$ is rounded at the same precision ε_M, giving $A' = A + G$, where $\|G\|_2 \approx \psi_1^2 \varepsilon_M$ (since $\|A\|_2 = \psi_1^2$). The corresponding perturbation in the smallest eigenvalue ψ_p^2 of A will be of order $\|G\|_2$. Hence to retain any accuracy in ψ_p, as computed from A', we must have

$$\left|\frac{\psi_p}{\psi_1}\right|^2 > \varepsilon_M. \tag{5.6}$$

A comparison of (5.5) and (5.6) shows that the second condition will be violated before the first. For example, if $\varepsilon_M = 10^{-4}$, then the condition (5.5) requires that $\psi_p/\psi_1 > 10^{-4}$, whereas condition (5.6) requires that $\psi_p/\psi_1 > 10^{-2}$. The general conclusion to be derived from this is that it requires twice the precision to accommodate the same range of singular values when one works with A instead of X.

The matrix X of (5.3) furnishes a nice example. Its smallest singular value is about $3 \cdot 10^{-4}$. On the other hand, the smallest eigenvalue of the matrix

$$\bar{A} = \begin{bmatrix} 17.26 & 34.52 \\ 34.52 & 69.05 \end{bmatrix}, \tag{5.7}$$

which is $X^T X$ rounded to four digits, is $2 \cdot 10^{-3}$. This gives the unacceptably large value of $4.5 \cdot 10^{-2}$ as an approximation to ψ_2.

If passing from X to A can loose some of the information about X, passing from A to A^{-1} can loose all of it. For example, the inverse, rounded to four digits, of the matrix $\bar{A}$ in (5.7) is

$$B = \begin{bmatrix} 400.1 & -200.0 \\ -200.0 & 100.0 \end{bmatrix}.$$

The exact inverse of B is

$$B^{-1} = \begin{bmatrix} 10 & 20 \\ 20 & 40.01 \end{bmatrix},$$

which bears only a passing resemblance to $\bar{A}$, which it is supposed to approximate. Phenomena like this will occur whenever (5.6) comes near to being violated. In general, one should use the inverse cross-product matrix only if one is prepared to compute in high precision, and even then only with safeguards.

6. How small is small. In this section we shall treat the problem of determining when a singular value is negligible. The principal problem is that we must work with the singular values $\tilde{\psi}_1, \tilde{\psi}_2, \cdots, \tilde{\psi}_p$ of $\tilde{X}$, not those of X. In particular, we should like to know what values of $\tilde{\psi}_p$ are consistent with the hypothesis that $\psi_p = 0$.

The basic result that will be used is a characterization of $\tilde{\psi}_p$ by means of a perturbation expansion. Suppose that ψ_{p-1} and ψ_p are well separated relative to E, say $\psi_{p-1} - \psi_p > 5\|E\|_2$. Then [16]

(6.1) $$\tilde{\psi}_p^2 = (\psi_p + u_p^T E v_p)^2 + \|U_P^T E v_p\|_2^2 + O(\psi_p \|E\|_2^2),$$

where U_P is the matrix in the partition (2.3).

In order to apply this result, something must be known about E. We shall examine a simple but revealing model in which the elements of E are assumed to be independent random variables with mean zero and standard deviation σ. It is also assumed that a rough estimate of the size of σ is available.

The expected value of the second term in (6.1) is

$$E(\|U_P^T E v_p\|_2^2) = (n-p)\sigma^2.$$

Hence if ψ_p^2 is significantly greater than $(n-p)\sigma^2$, then

$$\tilde{\psi}_p \approx \psi_p + u_p^T E v_p.$$

The term $u_p^T E v_p$ has standard deviation σ, which is a fortiori small compared to ψ_p. Thus in this case, ψ_p and $\tilde{\psi}_p$ are not likely to differ by much, and we are justified in concluding that $\psi_p \neq 0$ if we observe a value of $\tilde{\psi}_p^2$ that is significantly greater than $(n-p)\sigma^2$.

On the other hand, if $\psi_p = 0$, then ignoring the $O(\psi_p\|E\|_2^2)$ term in (6.1), we have

$$E(\tilde{\psi}_p^2) = (n-p+1)\sigma^2.$$

Thus values of $\tilde{\psi}_p^2$ near $n\sigma^2$ are not inconsistent with ψ_p being zero. However, any decision to treat ψ_p as if it were zero must be made in the context of the problem being solved—as we were at pains to point out in the introduction.

The assumption of a common standard deviation in the above analysis has important implications for the scaling of $\tilde{X}$. Specifically, *if the elements of E are assumed to be independent, then the rows and columns of X must be scaled so that all the elements of E are roughly equal.* The failure to observe this dictum can result in spurious indications of degeneracy, a phenomenon sometimes referred to as artificial ill conditioning.

To see how artificial ill conditioning arises, consider the behavior of the matrix

$$X_t = (X_* \, tx)$$

as t approaches zero. Let

$$R_t = \begin{bmatrix} R_* & tr \\ 0 & t\rho \end{bmatrix}$$

be the R-factor of X_t. Then it can be shown that the smallest singular value $\psi_p^{(t)}$ of X_t is asymptotic to $t\rho$:

(6.2) $$\psi_p^{(t)} \sim t\rho.$$

Moreover the right singular vector $v_p^{(t)}$ satisfies

$$v_p^t \sim \begin{bmatrix} -tR_*^{-1}r \\ 1 \end{bmatrix}, \tag{6.3}$$

and consequently the left singular vector $u_p^{(t)} = X_t v_p^{(t)}/\|X_t v_p^{(t)}\|_2$ satisfies

$$\lim_{t\to 0} u_p^{(t)} = \frac{x - X_* R_*^{-1} r}{\|x - X_* R_*^{-1} r\|_2} \equiv u_p^{(0)}.$$

Now the error matrix E_t associated with X_t inherits the scaling of X; that is, E_t may be written in the form

$$E_t = (E_* \; te). \tag{6.4}$$

Assuming $E_* \neq 0$, we have

$$\lim_{t\to 0} \|E_t\|_2 = \|E_*\|_2 > 0. \tag{6.5}$$

It follows from (6.2) and (6.5) that by taking t small enough we can make $\psi_p^{(t)}$ arbitrarily smaller than $\|E_t\|_2$. The same kind of analysis shows that $\tilde{\psi}_p^{(t)}$ will also be small compared with $\|E_t\|_2$, from which an inexperienced person might conclude that X_t is degenerate.

The fallacy in so concluding may be exposed by considering the perturbation expansion (6.1). From (6.3) and (6.4) it follows that

$$E_t v_p^{(t)} = t(e - E_* R_*^{-1} r) \equiv t\bar{e}.$$

Consequently,

$$\psi_p^{(t)2} \sim (\psi_p^{(t)} + u_p^{(t)T} E_t v_p^{(t)})^2 + \|U^T E_t v_p^{(t)}\|_2^2 = t^2[(\rho + u_p^{(0)T}\bar{e})^2 + \|U_p^T \bar{e}\|_2^2].$$

Now the components of $\bar{e}$ are independent with variance $\sigma^2(1 + \|R_*^{-1} r\|_2^2)$. Hence if

$$\rho \gg [(n-p)(1 + \|R_*^{-1} r\|_2^2)]^{1/2},$$

then the perturbation introduced in $\psi_p^{(t)}$ by E_t is small relative to $\psi_p^{(t)}$, and it is not reasonable to conclude from a small value of $\tilde{\psi}_p^{(t)}$ that $\psi_p^{(t)}$ could have been zero.

We conclude this section with a caveat. Although the analysis given here is quite successful as far as it goes, there are many problems to which it does not apply. The sticky point is the assumption of the independence of the components of E, which is patently false in many applications. Consider, for example, a polynomial regression problem in which a polynomial of degree $p-1$ is to be fit to ordinates $y_1, y_2, \cdots, y_n$ observed at distinct abscissas $t_1, t_2, \cdots, t_n$. The resulting regression matrix X has rows of the form

$$x_i^T = (1, t_i, t_i^2, \cdots, t_i^{p-1}) \qquad (i = 1, 2, \cdots, n).$$

Now suppose the t_i are determined with errors e_i. The resulting regression matrix will have rows of the form

$$\tilde{x}_i^T = [1, \; t_i + e_i, \; (t_i + e_i)^2, \; \cdots, \; (t_i + e_i)^{p-1}].$$

Clearly the errors in the ith row, depending as they do solely on e_i, are correlated, and the analysis given above does not apply. This can be verified independently from the fact that, no matter how small the singular values of X are, the matrix $\tilde{X}$ cannot become degenerate unless the errors e_i are large enough to allow the t_i to cluster into a set of fewer than p distinct points. The problem of how to make decisions about rank in the presence of highly structured errors is an open question that needs further research.

REFERENCES

[1] F. L. BAUER AND C. REINSCH, *Inversion of positive definite matrices by the Gauss–Jordan method*, in Handbook for Automatic Computation II. Linear Algebra, J. H. Wilkinson and C. Reinsch, eds., Springer, New York, 1971, pp. 45–49.

[2] TONY CHAN, *An improved algorithm for computing the singular value decomposition*, ACM Trans. Math. Software, 8 (1982), pp. 72–83.

[3] J. J. DONGARRA, J. R. BUNCH, C. B. MOLER AND G. W. STEWART, *The* LINPACK *Users' Guide*, Society for Industrial and Applied Mathematics, Philadelphia, 1979.

[4] G. H. GOLUB, *Numerical methods for solving least squares problems*, Numer. Math., 7 (1965), pp. 206–216.

[5] G. H. GOLUB AND C. REINSCH, *Singular value decomposition and least squares solution*, Numer. Math., 14 (1970), pp. 403–420.

[6] J. H. GOODNIGHT, *A tutorial on the* SWEEP *operator*, Amer. Statistician, 33 (1979), pp. 149–158.

[7] J. W. DANIEL, W. B. GRAGG, L. KAUFMAN AND G. W. STEWART, *Reorthogonalization and stable algorithms for updating the Gram–Schmidt* QR *factorization*, Math. Comp., 30 (1976), pp. 772–795.

[8] C. ECKART AND G. YOUNG, *The approximation of one matrix by another of lower rank*, Psychometrica, 1 (1936), pp. 211–218.

[9] L. MIRSKY, *Symmetric gauge functions and unitarily invariant norms*, Quart. J. Math., 11 (1960), pp. 50–59.

[10] W. C. RHEINBOLDT, *Numerical methods for a class of finite dimensional bifurcation problems*, SIAM J. Numer. Anal., 15 (1978), pp. 1–11.

[11] G. A. F. SEBER, *Linear Regression Analysis*, John Wiley, New York, 1977.

[12] G. W. STEWART, *Introduction to Matrix Computations*, Academic Press, New York, 1974.

[13] ———, *On the perturbation of pseudo-inverses, projections, and linear least squares problems*, SIAM Rev., 19 (1977), pp. 634–662.

[14] ———, *The efficient generation of random orthogonal matrices with an application to condition estimators*, SIAM J. Numer. Anal., 17 (1980), pp. 403–409.

[15] ———, *Assessing the effects of variable error in linear regression*, University of Maryland Computer Science Technical Report No. 818, 1979.

[16] ———, *A second order perturbation expansion for small singular values*, University of Maryland Computer Science Technical Report No. 1241, 1982.

[17] J. H. WILKINSON, *The Algebraic Eigenvalue Problem*, Oxford University Press, Oxford, 1965.

12.7. [GWS-J78] "On the Perturbation of LU, Cholesky, and QR Factorizations"

[GWS-J78] "On the Perturbation of LU, Cholesky, and QR Factorizations," *SIAM Journal on Matrix Analysis and Applications,* **14** (1993) 1141–1145.
http://dx.doi.org/10.1137/0614078

SIAM J. MATRIX ANAL. APPL.
Vol. 14, No. 4, pp. 1141–1145, October 1993

ON THE PERTURBATION OF LU, CHOLESKY, AND QR FACTORIZATIONS*

G. W. STEWART†

Abstract. In this paper error bounds are derived for a first-order expansion of the LU factorization of a perturbation of the identity. The results are applied to obtain perturbation expansions of the LU, Cholesky, and QR factorizations.

Key words. perturbation theory, LU factorization, QR factorization, Cholesky factorization

AMS subject classification. 15A18

1. Introduction. Let A be of order n, and suppose that the leading principal submatrices of A are nonsingular. Then A has an LU factorization

$$A = LU, \tag{1.1}$$

where L is lower triangular and U is upper triangular. The factorization is not unique; however, any other LU factorization must have the form

$$A = (LD)(D^{-1}U),$$

where D is a nonsingular diagonal matrix. Thus, if the diagonal elements of L (or U) are specified, the factorization is uniquely determined.

The purpose of this note is to establish a first-order perturbation expansion for the LU factorization of A along with bounds on the second-order terms. At least three authors have considered the perturbation of LU, Cholesky, and QR factorizations [1], [2], [4]. The chief difference between their papers and this one is that the former treat perturbations bounds for the decompositions in question, while here we treat the accuracy of a perturbation expansion.

Throughout this note $\|\cdot\|$ will denote a family of absolute, consistent matrix norms; i.e.,

$$|A| \le |B| \implies \|A\| \le \|B\|$$

and

$$\|AB\| \le \|A\|\|B\|$$

whenever the product AB is defined. Thus the bounds of this paper will hold for the Frobenius norm, the 1-norm, and the ∞-norm, but not for the 2-norm (for more on these norms see [3]).

2. Perturbation of the identity. The heart of this note is the observation that the LU factorization of the matrix $I + F$, where F is small, has a simple perturbation expansion. Specifically, write

$$F = F_{\mathrm{L}} + F_{\mathrm{U}},$$

* Received by the editors February 28, 1992; accepted for publication (in revised form) March 9, 1992. This work was supported in part by Air Force Office of Sponsored Research contract AFOSR-87-0188.

† Department of Computer Science and Institute for Advanced Computer Studies, University of Maryland, College Park, Maryland 20742 (stewart@cs.umd.edu).

where F_L is strictly lower triangular and F_U is upper triangular. Then

$$(I + F_L)(I + F_U) = I + F_L + F_U + F_L F_U = I + F + O(\|F\|^2), \tag{2.1}$$

and the product of the unit lower triangular matrix $I + F_L$ and the upper triangular matrix $I + F_U$ reproduces $I + F$ up to terms of order $\|F\|^2$. The following theorem shows that we can move these lower-order terms to the right-hand side of (2.1) to get an LU factorization of $I + F$.

THEOREM 2.1. *If*

$$\|F\| \le \frac{1}{4},$$

then there is a strictly lower triangular matrix G_L *and an upper triangular matrix* G_U *satisfying*

$$\|G_L + G_U\| \le \frac{\|F\|^2}{1 - 2\|F\| + \sqrt{1 - 4\|F\|}}$$

such that

$$(I + F_L + G_L)(I + F_U + G_U) = I + F. \tag{2.2}$$

Proof. From (2.2) it follows that the perturbations G_L and G_U must satisfy

$$G_L + G_U = -(F_L F_U + F_L G_U + G_L F_U + G_L G_U).$$

Starting with $G_L^0 = 0$ and $G_U^0 = 0$, generate strictly lower triangular and upper triangular iterates according to the formula

$$G_L^{k+1} + G_U^{k+1} = -(F_L F_U + F_L G_L^k + G_U^k F_U + G_L^k G_U^k). \tag{2.3}$$

Because $\|\cdot\|$ is absolute,

$$\|G_L^k\|, \|G_U^k\| \le \|G_L^k + G_U^k\|.$$

Hence if we set $\phi = \|F\|$, $\gamma_0 = 0$, and define the sequence $\{\gamma_k\}$ by

$$\gamma_{k+1} = \phi^2 + 2\phi\gamma_k + \gamma_k^2, \qquad k = 0, 1, \ldots, \tag{2.4}$$

then $\|G_L^k + G_U^k\| \le \gamma_k$.

Now by graphing the right-hand side of (2.4), it is easy to see that if $\phi \le \frac{1}{4}$, then the sequence γ_k converges monotonically to

$$\gamma_* = \frac{\phi^2}{1 - 2\phi + \sqrt{1 - 4\phi}},$$

which is, therefore, an upper bound on $\|G_L^k + G_U^k\|$ for all k. It remains only to show that the sequence $G_L^k + G_U^k$ converges.

From (2.3) it follows that

$$\begin{aligned}(G_L^{k+1} + G_U^{k+1}) - (G_L^k + G_U^k) = {} & F_L(G_U^{k-1} - G_U^k) + (G_L^{k-1} - G_L^k)F_U \\ & + (G_L^{k-1} - G_L^k)G_U^{k-1} + G_L^k(G_U^{k-1} - G_U^k).\end{aligned}$$

Hence

$$\|(G_{\mathrm{L}}^{k+1}+G_{\mathrm{U}}^{k+1})-(G_{\mathrm{L}}^{k}+G_{\mathrm{U}}^{k})\|\le 2(\phi+\gamma_*)\|(G_{\mathrm{L}}^{k}+G_{\mathrm{U}}^{k})-(G_{\mathrm{L}}^{k-1}+G_{\mathrm{U}}^{k-1})\|.$$

If $2(\phi+\gamma_*)<1$, which is certainly true if $\phi\le\frac{1}{4}$, then the series of differences is majorized by a geometric series, and the sequence converges.

There are some comments to be made on this theorem. In the first place the first-order expansion is particularly simple: Split F into its lower and upper triangular parts. We will take advantage of this simplicity in the next section, where we will derive perturbation expansions and asymptotic bounds for the LU, Cholesky, and QR factorizations.

The condition that $\|F\|\le\frac{1}{4}$ is perhaps too constraining, since the LU factorization of $I+F$ exists provided that $\|F\|<1$. However, as $\|F\|$ approaches one, it is possible for the factors in the decomposition to grow arbitrarily, in which case the bounds on the second-order terms must also grow. Thus the more restrictive condition can be seen as the price we pay for bounds that do not explode.

As $\|F\|$ goes to zero, the bound quickly assumes the asymptotic form

$$\|G_{\mathrm{L}}+G_{\mathrm{U}}\|\lesssim\|F\|^2;$$

i.e., the order constant for the second-order terms is essentially one. If we write this in the form

$$\frac{\|G_{\mathrm{L}}+G_{\mathrm{U}}\|}{\|F\|}\lesssim\|F\|,$$

we see that the *relative* error in the first-order expansion is of the same order as the perturbation itself, with order constant one.

Finally, Theorem 2.1 treats an LU decomposition of $I+F$ in which L is unit lower triangular. In analyzing symmetric permutations, we may want to take $L=U^{\mathrm{T}}$. In this case, we may work with slightly different matrices, illustrated below for $n=3$:

$$\hat{F}_{\mathrm{L}}=\begin{pmatrix}\frac{1}{2}f_{11} & 0 & 0\\ f_{21} & \frac{1}{2}f_{22} & 0\\ f_{31} & f_{32} & \frac{1}{2}f_{33}\end{pmatrix}\quad\text{and}\quad \hat{F}_{\mathrm{U}}=\begin{pmatrix}\frac{1}{2}f_{11} & f_{12} & f_{13}\\ 0 & \frac{1}{2}f_{22} & f_{23}\\ 0 & 0 & \frac{1}{2}f_{22}\end{pmatrix}. \tag{2.5}$$

If $\hat{G}_{\mathrm{L}}$ and $\hat{G}_{\mathrm{U}}$ are defined analogously, the proof of Theorem 2.1 goes through *mutatis mutandis*. For these matrices a useful inequality is

$$\|\hat{F}_{\mathrm{L}}\|_{\mathrm{F}},\,\|\hat{F}_{\mathrm{U}}\|_{\mathrm{F}}\le\frac{1}{\sqrt{2}}\|F\|_{\mathrm{F}}, \tag{2.6}$$

where $\|\cdot\|_{\mathrm{F}}$ denotes the Frobenius norm.

3. Applications. In this section we will apply the results of the previous section to get perturbation expansions for the LU, Cholesky, and QR decompositions. We will present only first-order terms, since bounds for the second-order terms can be derived from Theorem 2.1, and since the rate of convergence of these bounds to zero suggests that the first-order expansions will be satisfactory for all but the most delicate work. We will also derive asymptotic bounds for the first-order terms.

Our first application is to the problem we began with: the perturbation of the LU decomposition. Let A have the LU decomposition (1.1) and let $\tilde{A}=A+E$. Then

$$L^{-1}\tilde{A}U^{-1}=I+L^{-1}EU^{-1}\equiv I+F.$$

Let F_{L} and F_{U} be as in the last section. Then $I+F \cong (I+F_{\mathrm{L}})(I+F_{\mathrm{U}})$ is the first-order approximation to the LU factorization of $I+F$. It follows that

$$\tilde{A} \cong L(I+F_{\mathrm{L}})(I+F_{\mathrm{U}})U$$

is the first-order approximation to the LU factorization of $\tilde{A}$. Note that because F_{L} is unit lower triangular, this expansion preserves the scaling of the diagonal elements of L.

By taking norms we can derive the following asymptotic perturbation bound

$$\frac{\|\tilde{L}-L\|}{\|L\|} \lesssim \|L^{-1}\|\|U^{-1}\|\|A\|\frac{\|E\|}{\|A\|} \equiv \kappa_{\mathrm{LU}}(A)\frac{\|E\|}{\|A\|}. \tag{3.1}$$

Thus $\kappa_{\mathrm{LU}}(A) = \|L^{-1}\|\|U^{-1}\|\|A\|$ serves as a condition number for the LU decomposition of A. When A is square, this number is never less than the usual condition number $\kappa(A) = \|A\|\|A^{-1}\|$ and can be much larger. Bounds on the U factor can be derived similarly.

An unhappy aspect of the bound (3.1) is that it overestimates the perturbation of the leading part of the LU factorization. Specifically, if we partition

$$\begin{pmatrix} A_{11} & A_{12} \\ A_{21} & A_{22} \end{pmatrix} = \begin{pmatrix} L_{11} & 0 \\ L_{21} & L_{22} \end{pmatrix}\begin{pmatrix} U_{11} & U_{12} \\ 0 & U_{22} \end{pmatrix},$$

then $A_{11} = L_{11}U_{11}$, and the condition number for this part of the factorization is $\kappa_{\mathrm{LU}}(A_{11})$, which is in general smaller than $\kappa_{\mathrm{LU}}(A)$. The perturbation in L_{21} can then be estimated from the equation $L_{21} = A_{21}U_{11}^{-1}$.[1]

If A is symmetric and positive definite, then A has the Cholesky factorization

$$A = R^{\mathrm{T}}R,$$

where R is upper triangular. Let $\tilde{A} = A + E$, where E is symmetric. Setting $F = R^{-\mathrm{T}}ER^{-1}$ and defining $\hat{F}_{\mathrm{U}}$ as in (2.5), we have

$$\tilde{R} \cong (I+\hat{F}_{\mathrm{U}})R.$$

By (2.6) and the consistency of the 2-norm with the Frobenius norm, we have

$$\frac{\|\tilde{R}-R\|_{\mathrm{F}}}{\|R\|_2} \lesssim \frac{1}{\sqrt{2}}\|R^{-\mathrm{T}}\|_2\|R^{-1}\|_2\|E\|_{\mathrm{F}} = \frac{\kappa_2(A)}{\sqrt{2}}\frac{\|E\|_{\mathrm{F}}}{\|A\|_2},$$

where $\kappa_2(A) = \|A\|_2\|A^{-1}\|_2$ is the usual condition number in the 2-norm.

Finally, let A, now rectangular, be of full-column rank, and consider the QR factorization

$$A = QR,$$

[1] An alternative approach is to set

$$\check{L} = \begin{pmatrix} L_{11} & 0 \\ L_{21} & I \end{pmatrix},$$

so that

$$\check{L}^{-1}\begin{pmatrix} A_{11} \\ A_{21} \end{pmatrix}U_{11}^{-1} = \begin{pmatrix} I \\ 0 \end{pmatrix} \equiv J.$$

The proof of Theorem 2.1 can easily be adapted to give a bound on the perturbation of the LU factorization of a perturbation of J and hence on the perturbation of the LU factorization of $\binom{A_{11}}{A_{21}}$.

where Q has orthonormal columns and R is upper triangular with positive diagonal elements. The key to the derivation of the bounds is the equation

$$A^{\mathrm{T}}A = R^{\mathrm{T}}R;$$

i.e., R is the Cholesky factor of $A^{\mathrm{T}}A$.

As usual, let $\tilde{A} = A + E$, and let E_A be the orthogonal projection of E onto the column space of A. Then $A^{\mathrm{T}}E = A^{\mathrm{T}}E_A$. It follows that

$$\tilde{A}^{\mathrm{T}}\tilde{A} \cong A^{\mathrm{T}}A + A^{\mathrm{T}}E_A + E_A^{\mathrm{T}}A \equiv A^{\mathrm{T}}A + F.$$

Hence, with $\hat{F}_{\mathrm{U}}$ as above, we have

$$\tilde{R} \cong (I + \hat{F}_{\mathrm{U}})R.$$

In particular,

$$\frac{\|\tilde{R} - R\|_{\mathrm{F}}}{\|R\|_2} \lesssim \sqrt{2}\kappa_2(A)\frac{\|E_A\|_{\mathrm{F}}}{\|A\|_2},$$

where $\kappa_2(A) = \|R\|_2\|R^{-1}\|_2$. Since $\tilde{Q} = \tilde{A}\tilde{R}^{-1}$, we have

$$\tilde{Q} \cong Q(I - \hat{F}_{\mathrm{U}}) + ER^{-1},$$

from which it follows that

$$\|\tilde{Q} - Q\|_{\mathrm{F}} \lesssim \kappa_2(A)\frac{\sqrt{2}\|E_A\|_{\mathrm{F}} + \|E\|_{\mathrm{F}}}{\|A\|_2}. \tag{3.2}$$

Asymptotically, the bounds derived in this section agree with the bounds in [1], [4], with the exception of (3.2), which is a little sharper owing to the presence of $\|E_A\|_{\mathrm{F}}$.

Acknowledgment. This paper has been much improved by the suggestions of Jim Demmel and Nick Higham.

REFERENCES

[1] A. Barrland, *Perturbation bounds for the LDL^{H} and the LU factorizations*, BIT, 31 (1991), pp. 358–363.
[2] G. W. Stewart, *Perturbation bounds for the QR factorization of a matrix*, SIAM J. Numer. Anal., 14 (1977), pp. 509–518.
[3] G. W. Stewart and J.-G. Sun, *Matrix Perturbation Theory*, Academic Press, Boston, 1990.
[4] J.-G. Sun, *Perturbation bounds for the Cholesky and QR factorizations*, BIT, 31 (1991), pp. 341–352.

12.8. [GWS-J89] "On Graded QR Decompositions of Products of Matrices"

[GWS-J89] "On Graded QR Decompositions of Products of Matrices," *Electronic Transactions in Numerical Analysis* **3** (1995) 39–49.

Electronic Transactions on Numerical Analysis.
Volume 3, pp. 39-49, March 1995.

ISSN 1068-9613.

ETNA
Kent State University
etna@mcs.kent.edu

ON GRADED QR DECOMPOSITIONS OF PRODUCTS OF MATRICES *

G. W. STEWART†

Abstract. This paper is concerned with the singular values and vectors of a product $M_m = A_1A_2\cdots A_m$ of matrices of order n. The chief difficulty with computing them directly from M_m is that with increasing m the ratio of the small to the large singular values of M_m may fall below the rounding unit, so that the former are computed inaccurately. The solution proposed here is to compute recursively the factorization $M_m = QRP^{\mathrm{T}}$, where Q is orthogonal, R is a graded upper triangular, and P^{T} is a permutation.

Key words. QR decomposition, singular value decomposition, graded matrix, matrix product.

AMS subject classification. 65F30.

1. Introduction. This paper is concerned with calculating the singular values of a product $M_m = A_1A_2\cdots A_m$ of matrices of order n. The chief difficulty with the natural algorithm—compute M_m and then compute its singular value decomposition—is that as m increases, the singular values will tend to spread out and the smaller singular values will be inaccurately computed. The rule of thumb is that singular values of M_m whose ratio to the largest singular value is less than the rounding unit will have no accuracy.

The product singular value decomposition (PSVD) is one way of circumventing this problem [1]. Here the product M_m is decomposed in the form

$$M_m = (UT_1Q_1^{\mathrm{T}})(Q_1T_2Q_2^{\mathrm{T}})\cdots(Q_{m-1}T_mV^{\mathrm{T}}), \tag{1.1}$$

where U, V, and the Q_k are orthogonal, the T_k are triangular, and

$$\Sigma = T_1T_2\cdots T_m$$

is diagonal. Thus the product (1.1) collapses into the singular value decomposition $U\Sigma V^{\mathrm{T}}$ of M_m.

The chief drawback of the PSVD is that it is expressed in terms of all the factors of M_m. This means that as m increases the storage required for the decomposition increases. Moreover, there seems to be no easy way to pass from the PSVD of M_m to that of M_{m+1}; the multiplication by A_{m+1} changes the entire decomposition, so that the work required to append a factor also increases with m.

An alternative is an algorithm for computing the singular value decomposition of a product of two matrices [4]. Given the singular value decomposition of M_m, it can be used to calculate the singular value decomposition of $M_{m+1} = M_mA_{m+1}$. However, it is not clear how the increasing spread of the singular values affects this algorithm.

In this paper we take a different tack and work with decompositions of the form $A = QRP^{\mathrm{T}}$, where Q is orthogonal, R is upper triangular, and P is a permutation. The algorithm works to insure that R is a graded; that is, $R = D\hat{R}$, where the

* Submitted April 29, 1994. Accepted for publication March 21, 1995. Communicated by J. W. Demmel.

† Department of Computer Science and Institute for Advanced Computer Studies, University of Maryland, College Park, MD 20742. This work was supported by The National Science Foundation under Grant CCR 9115568.

elements in the upper half of $\hat{R}$ are of order one and D is a diagonal matrix whose diagonal elements decrease in magnitude. The rule of thumb mentioned above—that the small singular values of a matrix are calculated inaccurately—need not hold for graded triangular matrices, so that very small singular values of M_m can usually be calculated from its graded factor R_m. Moreover, the product of graded triangular matrices is graded, so that in a decomposition analogous to (1.1), we can multiply out the triangular factors. Thus the passage from a graded QRP decomposition of M_m to one of M_{m+1} does not involve all the previous decompositions.

In the next section, we will show how to compute a graded QRP decomposition of a product AB from a graded QRP decomposition of A. In Section 3 we argue informally that the algorithm preserves grading. Here we also point out that good estimates of the singular values can be obtained with a small amount of additional work (the mathematical justification is given in Appendix A). In Section 4 we give numerical examples to show that this procedure can be applied recursively to compute singular values of a product of matrices whose singular values have ratios far below the rounding unit. Matlab code for the procedure is given in an appendix to this paper.

A final point. Although we have chosen to work with QRP decompositions in which P is a permutation, it will be clear that P could equally well be an orthogonal matrix. Thus the approach taken here also applies to two-sided orthogonal decompositions such as the URV decomposition [6].

2. The Algorithm. In this section we will describe the algorithm for updating graded QRP decompositions. The description will be in two stages: first an overview at the matrix level, then a detailed description. The latter is required to understand why the algorithm preserves grading. We will assume that the reader is familiar with the pivoted QR decomposition and plane rotations.

The input to the algorithm is a graded QRP decomposition

$$A = Q_A R_A P_A^{\mathrm{T}}$$

of a matrix A and an unfactored matrix B. The output is a graded QRP decomposition of

$$C \equiv AB = Q_C R_C P_C^{\mathrm{T}}. \tag{2.1}$$

The steps are as follows.

1. Compute the pivoted QR decomposition $P_A^{\mathrm{T}} B = Q_B R_B P_C^{\mathrm{T}}$.
2. Compute an orthogonal matrix U such that $\hat{R}_A = U^{\mathrm{T}} R_A Q_B$ is upper triangular.
3. Set $Q_C = Q_A U$.
4. Set $R_C = \hat{R}_A R_B$.

It is easy to verify that the quantities so computed satisfy (2.1).

In some applications it is necessary to work with products of the form AB^{-1}. The above algorithm can be adapted to compute a graded QRP decomposition of AB^{-1}. The trick is to compute a graded PRQ decomposition of B, so that when the decomposition is inverted, it becomes a graded QRP decomposition of B^{-1}. This leads to the following algorithm.

1. Compute the pivoted QR decomposition $BP_A = P_C R_B Q_B^{\mathrm{T}}$.
2. Compute an orthogonal matrix U such that $\hat{R}_A = U^{\mathrm{T}} R_A Q_B$ is upper triangular.

3. Set $Q_C = Q_A U$.
4. Set $R_C = \hat{R}_A R_B^{-1}$.

In applications in which it is desired to compute powers of a single matrix A, we may wish to proceed by matrix squaring; that is, by computing the sequence A, A^2, $A^4, \ldots$. Thus given a graded QRP decomposition of A^k we wish to compute a graded QRP decomposition of $A^k A^k$. More generally, given graded QRP decompositions of A and B it is possible to compute a graded decomposition of their product, i.e., of $(Q_A R_A P_A^{\mathrm{T}})(Q_B R_B P_B^{\mathrm{T}})$. The algorithm goes as follows.

1. Compute the QR-decomposition $P_A^{\mathrm{T}} Q_B = VD$. Note that D will be diagonal because $P_A^{\mathrm{T}} Q_B$ is orthogonal.
2. Compute an orthogonal matrix U such that $\hat{R}_A = U^{\mathrm{T}} R_A V$ is upper triangular.
3. Set $P_C = P_B$, $Q_C = Q_A U$, and $R_C = \hat{R}_A D R_B$.

We will now consider the particulars. Since the three algorithms sketched above are variants of one another, we will treat only the first. The reader may find it helpful to consult the matlab program in Appendix B.

The first step in the above algorithm is accomplished with plane rotations. Specifically, rotations in the $(i, i+1)$-plane eliminate elements in $P^{\mathrm{T}} B$ in the order indicated below:

$$\begin{pmatrix} b & b & b & b & b \\ b^4 & b & b & b & b \\ b^3 & b^7 & b & b & b \\ b^2 & b^6 & b^9 & b & b \\ b^1 & b^5 & b^8 & b^{10} & b \end{pmatrix}.$$

Before the kth column is processed, the column in the trailing principal submatrix whose 2-norm is largest is located, and the *entire* column is swapped with the kth. (The matlab code in the appendix inefficiently computes the norms *ab initio*. In practice, the norms should be downdated as in [2].)

In the second step, the rotations generated in the first are applied to R_A. When a rotation in the $(i, i+1)$-plane is postmultiplied into R_A, it creates a nonzero element in the $(i+1, i)$-position. This element is eliminated by premultiplying by a rotation in the $(i, i+1)$-plane. The process is illustrated in Figure 2.1. Here the arrows indicate the plane of the rotation to be applied, and a hat indicates an element destined to be annihilated.

The rotations from the first step can be applied to R_A as soon as they are generated, which saves having to store them. Similarly the rotations that make up U can be accumulated in Q_A, so that at the end Q_A will be transformed into Q_C. Thus in the implementation steps one, two and three are completely interleaved.

When a product of the form $A_1 A_2 \cdots A_m$ is to be processed, the procedure can be started by setting $A = Q_A = P_A = I$ and $B = A_1$. In this case step two should be skipped and the rotations from step one accumulated in Q_A.

The algorithm is quite inexpensive. An operation count shows that it requires $5\frac{1}{6}n^3$ additions and $10\frac{1}{6}n^3$ multiplications. When n is large enough, scaled rotations [3, §5.1.13] can be used to reduce the number of multiplications to $5\frac{1}{6}n^3$. In this case, the algorithm can be compared with five matrix multiplications.

3. Grading. Grading enters into the algorithm in three places: in step one, where it is established, and in steps two and three, where it must be maintained. We will consider each step in turn.

$$
\begin{array}{cccc} & & \downarrow & \downarrow \\ r & r & r & r \\ 0 & r & r & r \\ 0 & 0 & r & r \\ 0 & 0 & 0 & r \end{array}
\Longrightarrow
\begin{array}{ccccc} & & & & \\ & r & r & r & r \\ & 0 & r & r & r \\ \rightarrow & 0 & 0 & r & r \\ \rightarrow & 0 & 0 & \hat{r} & r \end{array}
\Longrightarrow
\begin{array}{cccc} & \downarrow & \downarrow & \\ r & r & r & r \\ 0 & r & r & r \\ 0 & 0 & r & r \\ 0 & 0 & 0 & r \end{array}
\Longrightarrow
$$

$$
\begin{array}{ccccc} & & & & \\ & r & r & r & r \\ \rightarrow & 0 & r & r & r \\ \rightarrow & 0 & \hat{r} & r & r \\ & 0 & 0 & 0 & r \end{array}
\Longrightarrow
\begin{array}{cccc} \downarrow & \downarrow & & \\ r & r & r & r \\ 0 & r & r & r \\ 0 & 0 & r & r \\ 0 & 0 & 0 & r \end{array}
\Longrightarrow
\begin{array}{ccccc} & & & & \\ \rightarrow & r & r & r & r \\ \rightarrow & \hat{r} & r & r & r \\ & 0 & 0 & r & r \\ & 0 & 0 & 0 & r \end{array}
\Longrightarrow
\begin{array}{cccc} r & r & r & r \\ 0 & r & r & r \\ 0 & 0 & r & r \\ 0 & 0 & 0 & r \end{array}
$$

FIG. 2.1. *Reduction of* R_A

◇

In step one we have used pivoting on column norms to establish a grading. This is the simplest method in an arsenal of techniques designed to reveal the rank of a matrix. Although there is a famous counter-example for which the method fails [3, §5.5.7], in practice it works well. However, as we mentioned in the introduction, other rank revealing decompositions may be substituted for the QR decomposition obtained by pivoting on column norms.

Turning now to step two of the algorithm, let us see how grading in R_A can be lost. The transformation of the trailing 2×2 matrix is typical. Let it be written

$$\begin{pmatrix} \alpha & \beta \\ 0 & \delta \end{pmatrix}.$$

For definiteness we will suppose that $\sqrt{\alpha^2 + \beta^2} = 1$ and that δ is small, so that the matrix has a marked grading. After postmultiplying by the rotation from the reduction of $P^{\mathrm{T}}B$, we have

$$\begin{pmatrix} \bar{\alpha} & \bar{\beta} \\ \bar{\gamma} & \bar{\delta} \end{pmatrix},$$

where $\sqrt{\bar{\alpha}^2 + \bar{\beta}^2} = 1$ and $\sqrt{\bar{\gamma}^2 + \bar{\delta}^2} = |\delta|$. Thus the postmultiplication leaves the grading unaffected, at least normwise.

We must now premultiply by a plane rotation to reduce $\bar{\gamma}$ to zero. This rotation has the form

$$\frac{1}{\nu}\begin{pmatrix} \bar{\alpha} & \bar{\gamma} \\ -\bar{\gamma} & \bar{\alpha} \end{pmatrix},$$

where

$$\nu = \sqrt{\bar{\alpha}^2 + \bar{\gamma}^2}.$$

It follows that the $(2,2)$-element is

$$\hat{\epsilon} = \frac{\bar{\alpha}\bar{\delta} - \bar{\beta}\bar{\gamma}}{\nu},$$

whence

$$|\hat{\epsilon}| \leq \sqrt{2}\frac{\epsilon}{\nu}. \tag{3.1}$$

Now the way grading is lost is for a large element in one row to overwhelm a small element in another row. The inequality (3.1) shows that this can happen only if ν is small or equivalently $\bar{\alpha}$ is small. But

$$\bar{\alpha} = c\alpha + s\beta,$$

where c and s are the cosine and sine from the reduction of $P_A^{\mathrm{T}}B$. Since these numbers are unrelated to α and β, it is extremely unlikely that α will be very much less than one.

Finally, consider the grading of the product $\hat{R}_A R_B$. Let

$$\hat{R}_A = \mathrm{diag}(\epsilon_1, \ldots, \epsilon_n)\hat{R} \quad \text{and} \quad R_B = \mathrm{diag}(\delta_1, \ldots, \delta_n)R,$$

where the elements of $\hat{R}$ and R are of order one in magnitude and the ϵ's and δ's are decreasing. Then the (i,j)-element of $\hat{R}_A R_B$ is

$$\delta_i\epsilon_i\big(\hat{r}_{ii}r_{ij} + \sum_{k=i+1}^{j}\frac{\delta_k}{\delta_i}\hat{r}_{ik}r_{kj}\big).$$

Since $\delta_k/\delta_i < 1$, we can expect the ith row of R_C to be about $\delta_i\epsilon_i$ in size.

It is hardly necessary to point out that these arguments are informal. Even the definition of a graded triangular matrix is too stringent: in practice we would expect the rows of a graded matrix to have occasional elements that are uncharacteristically small. Nonetheless, experience, folklore, and limited analytical results all suggest than the ills that can potentially beset graded matrices do not often occur in practice.

If the grading is sharp enough, we can get a cheap estimate of the singular values of a graded triangular matrix R. Let δ_1, δ_2, ..., δ_n be the grading factors. Suppose that we have determined an orthogonal matrix V such that RV is lower triangular. Since postmultiplication by V does not change the norms of the rows of R, the elements of RV will be of the sizes indicated below for $n = 5$:

$$RV \cong \begin{pmatrix} \delta_1 & & & & \\ \delta_2 & \delta_2 & & & \\ \delta_3 & \delta_3 & \delta_3 & & \\ \delta_4 & \delta_4 & \delta_4 & \delta_4 & \\ \delta_5 & \delta_5 & \delta_5 & \delta_5 & \delta_5 \end{pmatrix}. \tag{3.2}$$

Let $\rho_1 = 0$, $\rho_i = \delta_i/\delta_{i-1}$ $(i = 2, \ldots, n)$, and $\rho_{n+1} = 0$. Then if ρ_i and ρ_{i+1} are not too large, the ith diagonal element of RV approximates a singular value of R with relative error that is approximately bounded by $\frac{1}{2}(\rho_i^2 + \rho_{i+1}^2)$. (For more details see Appendix A.) Thus for about a third again the work (that is, the cost of computing RV) we can obtain estimates of the singular values, estimates that are often very accurate.

4. Numerical Results. The main problem with testing the algorithm is constructing test cases whose answers can be easily recognized. The solution taken here is an extension of an idea in [4].

The matrices U and V of left and right singular vectors are calculated from a random normal matrix, and a diagonal matrix Σ is chosen. Two matrices A and B are defined by

$$A = U\Sigma V^{\mathrm{T}} \quad \text{and} \quad B = V\Sigma U^{\mathrm{T}}.$$

For a given m, a QRP decomposition of the product

$$M_{2m+1} = A\overbrace{BABA\cdots BA}^{m}$$

is calculated. The singular value decomposition of M_{2m+1} is

$$M_{2m+1} = U\Sigma^{2m+1}V^{\mathrm{T}},$$

so that a correct answer can easily be recognized. Note that B and A are accumulated individually to avoid working with the positive definite product BA, which might create a bias in favor of the algorithm.

The first test, in which $\Sigma = \mathrm{diag}(1, 0.1, 0.01, 0.001, 0.0001)$, is designed to push the singular values toward the underflow point. The following table contains the singular values of M_{2m+1} and their relative errors for three values of m.

```
m=5
    1.0e+00   1.0e-11   1.0e-22   1.0e-33   1.0e-44
    3.9e-15   1.1e-14   1.1e-14   4.0e-14   6.3e-13
m=10
    1.0e+00   1.0e-21   1.0e-42   1.0e-63   1.0e-84
    7.4e-15   2.0e-14   2.1e-14   6.2e-14   1.3e-12
m=20
    1.0e+00   1.0e-41   1.0e-82   1.0-123   1.0-164
    1.4e-14   3.9e-14   4.1e-14   1.0e-13   2.6e-12
```

For $m = 20$ the smallest singular value decreases by 160 orders of magnitude, yet it is computed with almost full accuracy. Incidentally, most of the inaccuracy in this singular value is not due to the algorithm at all, but to the fact that the singular value 0.0001 is represented inaccurately in the original matrix.

The second test takes longer run down a gentler grade. For this case

$$\Sigma = \mathrm{diag}(1, 0.99, 0.8, 0.7, 0.6).$$

The first two rows of the following results are as above.

```
m=20
    1.0e+00   6.6e-01   1.3e-02   1.0e-04   4.4e-07
    1.3e-14   4.6e-15   1.8e-14   4.0e-15   6.5e-15
    9.1e-01   7.2e-01   1.3e-02   1.0e-04   4.4e-07
    9.3e-02   8.5e-02   7.7e-05   5.0e-06   4.4e-07
m=40
    1.0e+00   4.4e-01   1.9e-04   1.4e-08   2.8e-13
    2.5e-14   8.6e-15   3.8e-14   7.0e-15   1.3e-14
    9.3e-01   4.7e-01   1.9e-04   1.4e-08   2.8e-13
    6.8e-02   6.4e-02   3.6e-08   4.3e-10   1.0e-11
m=80
    1.0e+00   1.9e-01   4.2e-08   2.4e-16   1.1e-25
```

4.8e-14	1.8e-14	7.1e-14	1.5e-14	2.7e-14
9.8e-01	2.0e-01	4.2e-08	2.4e-16	1.1e-25
1.7e-02	1.7e-02	7.9e-14	1.5e-14	2.7e-14

Again the algorithm computes the singular values with almost full accuracy. Rounding error accumulates, but very slowly.

The third row contains the diagonal elements of RV [see (3.2)], which approximate the singular values of the product, and the fourth row contains their relative errors. For $m = 20$, the last three singular values are well approximated (note how it is the square of the grading that governs the accuracy). There is even useful information about the magnitudes of the first two singular values. The approximations become more accurate as m increases.

To see how the algorithm performs on a matrix of moderate size with a natural distribution of singular values, A was taken to be a random normal matrix of order 50 with B generated as described above. The algorithm was run for $m = 2$. The results are too voluminous to present in their entirity, but the data for the smallest six singular values illustrates what is happening.

2.9e+00	6.7e-01	1.5e-01	3.7e-02	2.7e-04	1.2e-07
7.4e-15	5.0e-15	1.0e-15	1.1e-14	1.2e-14	1.0e-15
3.0e+00	5.4e-01	1.8e-01	3.8e-02	2.7e-04	1.2e-07
5.9e-02	2.3e-01	1.8e-01	4.7e-03	2.5e-08	7.0e-08

The singular values of the computed product are quite accurate. The approximations give one or two figures — more when the local grading is strong.

In a completely different example, we tested the algorithm for updating the product of two graded QR decompositions by using it to compute A^{2^k}. The initial matrix was $A = X^{-1}\text{diag}(1, 0.8, 0.7, 0.5)X$, where X is a random normal matrix. The initial graded QRP decomposition of A was computed from the usual algorithm. Then the algorithm for updating the product of graded QRP decompositions was used to compute the graded QRP decompositions of A^2, A^4, ...

The smallest eigenvalue of $C_k = A^{2^k}$ is 0.5^{2^k}. Even for modest values of k this eigenvalue is too small to be computed from C_k itself. However, since the R factor of the QRP decomposition of C_k is graded, we can accurately compute its inverse and hence C_k^{-1}. From C_k^{-1} we can compute the inverse of the smallest eigenvalue of C_k. For $k = 8$, this eigenvalue is on the order of 10^{-77}. It was represented in the QRP decomposition of C_8 to about thirteen decimal digits.

5. Conclusions. The main recommendation for this algorithm is its simplicity and efficiency. The matlab code attests its simplicity. Its efficiency is attested by the operation counts. The only question is: does it work? Here we have only been able to give an informal analysis and limited examples, either of which alone might be considered insufficient. Together, I believe, they make a strong case for the algorithm.

REFERENCES

[1] A. Bojanczyk, J. G. Nagy, and R. J. Plemmons, *Block RLS using row Householder reflections*, Linear Algebra Appl., 188–189 (1993), pp. 31–62.

[2] J. J. Dongarra, J. R. Bunch, C. B. Moler, and G. W. Stewart, *LINPACK User's Guide*, SIAM, Philadelphia, 1979.

[3] G. H. Golub and C. F. Van Loan, *Matrix Computations*, Johns Hopkins University Press, Baltimore, Maryland, 2nd ed., 1989.

[4] M. T. Heath, A. J. Laub, C. C. Paige, and R. C. Ward, *Computing the SVD of a product of two matrices*, SIAM J. Sci. Statist. Comput., 7 (1986), pp. 1147–1159.

[5] R. Mathias and G. W. Stewart, *A block qr algorithm and the singular value decomposition*, Linear Algebra Appl., 182 (1993), pp. 91–100.
[6] G. W. Stewart, *An updating algorithm for subspace tracking*, IEEE Trans. Signal Processing, 40 (1992), pp. 1535–1541.

Appendix A: On the Singular Values of Graded Matrices. In this appendix we show that the singular values of a matrix with the structure (3.2) are approximated by the diagonal elements of the matrix. We will use the following theorem of Mathias and Stewart [5] (which we have weakened slightly for the sake of simplicity). Here $\|A\|$ denotes the spectral norm of A and $\inf(A)$ is the smallest singular value of A.

Let

$$T = \begin{matrix} & k & n-k \\ k & P & H \\ n-k & 0 & Q \end{matrix}$$

be a block triangular matrix, and suppose that $\|Q\|/\inf(P), \|H\|/\inf(P) < 1$. *Let* $\sigma_1 \geq \cdots \geq \sigma_n$ *be the singular values of* T *and* $\hat{\sigma}_1 \geq \cdots \geq \hat{\sigma}_n$ *be the singular values of* $\mathrm{diag}(P, Q)$. *Then*

$$1 \geq \frac{\hat{\sigma}_i}{\sigma_i} \geq (1-\tau^2)^{\frac{1}{2}}, \qquad i = 1, 2, \ldots, k,$$

and

$$1 \geq \frac{\sigma_i}{\hat{\sigma}_i} \geq (1-\tau^2)^{\frac{1}{2}}, \qquad i = k+1, k+2, \ldots, n,$$

where

$$\tau^2 = \frac{(\|H\|/\inf(P))^2}{1 - (\|Q\|/\inf(P))^2}.$$

The theorem states that approximations of the singular values of T with high relative accuracy can be found in the diagonal blocks of P and Q.

To apply the theorem, let the matrix $L = RV$ of (3.2) be written in the form

$$L = \mathrm{diag}(\delta_1, \delta_2, \ldots, \delta_n)\hat{L},$$

where the rows of $\hat{L}$ have norm one. Partition L in the form

$$\begin{matrix} & k-1 & 1 & n-k \\ k-1 & L_{11} & 0 & 0 \\ 1 & \ell_{21}^{\mathrm{T}} & \lambda_{22} & 0 \\ n-k & L_{31} & \ell_{32} & L_{33} \end{matrix}.$$

We wish to assess the accuracy of λ_{22} as a singular value of L. The approach is to apply the above theorem twice.

In the first application we set $P = L_{11}$, which amounts to throwing out ℓ_{21}^{T} and L_{31}. In this case

$$\inf(P) = \inf(L_{11}) = \inf(\mathrm{diag}(\delta_1, \ldots, \delta_{k-1})\hat{L}_{11}) \geq \delta_{k-1}\inf(L_{11}).$$

Moreover, since the rows of $\hat{L}$ have norm one,

$$\|H\|, \|Q\| \leq \delta_k \sqrt{1 + (\delta_{k+1}/\delta_k)^2 + \cdots (\delta_n/\delta_k)^2}.$$

Consequently if we set

$$\text{(A.1)} \qquad \begin{aligned} \rho_{1k} &= \frac{\delta_k \sqrt{1 + (\delta_{k+1}/\delta_k)^2 + \cdots (\delta_n/\delta_k)^2}}{\delta_{k-1} \inf(L_{11})} \\ &\equiv \rho_k \frac{\sqrt{1 + (\delta_{k+1}/\delta_k)^2 + \cdots (\delta_n/\delta_k)^2}}{\inf(L_{11})}, \end{aligned}$$

then

$$\tau_1^2 = \frac{\rho_{1k}^2}{1 - \rho_{1k}^2}$$

bounds the perturbation in the singular values.

The second step is to take $P = \lambda_{22}$ in the matrix

$$\begin{pmatrix} \lambda_{22} & 0 \\ \ell_{32} & L_{33}, \end{pmatrix},$$

which amounts to setting ℓ_{32} to zero, leaving λ_{22} as a singular value of the perturbed matrix. Here $\inf(P) = |\lambda_{22}|$. Moreover,

$$\|H\|, \|Q\| \leq \delta_{k+1} \sqrt{1 + (\delta_{k+2}/\delta_{k+1})^2 + \cdots (\delta_n/\delta_{k+1})^2}.$$

Consequently, if we set

$$\rho_{2,k+1} = \rho_{k+1} \frac{\sqrt{1 + (\delta_{k+2}/\delta_{k+1})^2 + \cdots (\delta_n/\delta_{k+1})^2}}{|\lambda_{kk}|/\delta_k},$$

then

$$\tau_2^2 = \frac{\rho_{2,k+1}^2}{1 - \rho_{2,k+1}^2}$$

bounds the effect of the second perturbation.

For small ρ_{1k} and $\rho_{2,k+1}$, the total bound on the relative perturbation is approximately $(\rho_{1k}^2 + \rho_{2,k+1}^2)/2$. Note that in (A.1) we have written ρ_{1k} as the product of ρ_k with a magnification factor that in general will be greater than one. If the grading is strong the denominator $\sqrt{1 + (\delta_{k+1}/\delta_k)^2 + \cdots (\delta_n/\delta_k)^2}$ will be very near one. In our application, the pivoting in the algorithm tends to keep $\inf(L_{11})$ from becoming very small. Hence it is the ratio $\rho_k = \delta_k/\delta_{k-1}$ that controls the accuracy of the kth diagonal element as an approximate singular value. Similar comments can be made about $\rho_{2,k+1}$.

Appendix B: Program Listings.

```
function [rc, qc, pc] = prodqrp(ra, qa, pa, b, first)
%
% Given a qrp factorization of a matrix a and a matrix b
% prodqrp computes a qrp factorization of a*b.
% To start the factorization set invoke
%
%    prodqrp(eye(n), eye(n), 1:n, a, 1)
%
% Designed and coded by G. W. Stewart, April 8, 1994.
%
  n = size(ra, 1);
  qc = qa;
%
% Accumulate the permutation pa in b.
%
  for i=1:n-1
     b([i,pa(i)],:) = b([pa(i),i],:);
  end
%
% Compute a qrp factorization of b and update.
%
  for k=1:n-1
     for j=k:n, nrm(j)=norm(b(k:n,j)); end
     [xx, pc(k)] = max(nrm(k:n));
     pc(k) = pc(k)+k-1;
     b(1:n, [k,pc(k)]) = b(1:n, [pc(k),k]);
     for i=n-1:-1:k
        [c, s, nu] = rotgen(b(i,k), b(i+1,k));
        b(i,k) = nu; b(i+1,k) = 0.;
        b(i:i+1, k+1:n) = [c, s; -s, c]*b(i:i+1, k+1:n);
        if (first == 0)
           ra(1:i+1, i:i+1) = ra(1:i+1, i:i+1)*[c, -s; s, c];
           [c, s, nu] = rotgen(ra(i,i),ra(i+1,i));
           ra(i,i) = nu; ra(i+1,i) = 0;
           ra(i:i+1, i+1:n) = [c , s; -s, c]*ra(i:i+1, i+1:n);
        end
        qc(i:i+1, 1:n) = [c, s; -s, c]*qc(i:i+1, 1:n);
     end
  end
  rc = ra*b;
```

```
function [c, s, norm] = rotgen(a,b)
%
% This function generates a plane rotation.
%
   scale = abs(a) + abs(b);
   if scale == 0
      c = 1; s = 0; norm = 0;
   else
      norm = scale*sqrt((a/scale)^2+(b/scale)^2);
      c = a/norm;
      s = b/norm;
   end
```

12.9. [GWS-J92] "On the Perturbation of LU and Cholesky Factors"

[GWS-J92] "On the Perturbation of LU and Cholesky Factors," *IMA Journal of Numerical Analysis*, **17** (1997) 1–6.
`http://dx.doi.org/10.1093/imanum/17.1.1`

IMA Journal of Numerical Analysis (1997) **17**, 1–6

On the perturbation of *LU* and Cholesky factors*

G. W. STEWART

Department of Computer Science and Institute for Advanced Computer Studies, University of Maryland, College Park, MD 20742, USA

[Received 27 October 1995 and in revised form 8 January 1996]

In a recent paper, Chang and Paige have shown that the usual perturbation bounds for Cholesky factors can systematically overestimate the errors. In this note we sharpen their results and extend them to the factors of the *LU* decomposition. The results are based on a new formula for the first-order terms of the error in the factors.

1. Introduction

Let A be a positive definite matrix of order n. Then A has a unique Cholesky factorization of the form $A = R^{\mathrm{T}}R$, where R is upper triangular with positive diagonal elements.

Let $\tilde{A} = A + E$ be a perturbation of A in which E is symmetric. If E is sufficiently small, then $\tilde{A}$ also has a Cholesky factorization:

$$A + E = (R + F_R)^{\mathrm{T}}(R + F_R).$$

Several workers (Barrlund (1991), Stewart (1977, 1993), Sun (1992)) have given bounds on the matrix F_R. The common result is essentially that

$$\frac{\|F_R\|_{\mathrm{F}}}{\|R\|_2} \leqslant \frac{1}{\sqrt{2}}\kappa_2(A)\frac{\|E\|_{\mathrm{F}}}{\|A\|_2} + \mathrm{O}(\|E\|_2^2). \tag{1.1}$$

Recently Chang and Paige (Chang *et al* (1996)) have shown that (1.1) can consistently overestimate the error in the Cholesky factor and have proposed new bounds. Following Stewart (1977), they note that

$$E = R^{\mathrm{T}}F_R + F_R^{\mathrm{T}}R + \mathrm{O}(\|F_R\|^2). \tag{1.2}$$

Consequently if one defines the linear operator $\mathbf{T}_R$ on the space of upper triangular matrices by

$$\mathbf{T}_R(F) = R^{\mathrm{T}}F + F^{\mathrm{T}}R,$$

then

$$F_R \cong \check{F}_R \equiv \mathbf{T}_R^{-1}(E),$$

and

$$\|F_R\| \lesssim \|\mathbf{T}_R^{-1}\|\ \|E\|,$$

*This report is available by anonymous ftp from thales.cs.umd.edu in the directory pub/reports.

where $\|\cdot\|$ denotes a suitably chosen norm. By examining the matrix representation of $\mathbf{T}_R$, Chang and Paige were able to show that bounds based on $\|\mathbf{T}_R^{1}\|$ are sharper than the conventional bounds. They also derive a lower bound for $\|\mathbf{T}_R^{1}\|$, and show by example that the failure of (1.1) is somehow connected with pivoting in the computation of the decomposition.

The purpose of this note is to generalize the results of Chang and Paige by a different approach. We will exhibit an explicit matrix representation of $\check{F}_R$. The representation is invariant under a certain kind of diagonal scaling, and by adjusting the scaling we can improve the usual bound. This results in good, informative bounds—though not as good as those of Chang and Paige, which are by definition optimal. Since the approach works for the more general LU decomposition, we will treat that case first and then specialize to the Cholesky decomposition.†

Throughout this note $\|\cdot\|$ will denote an absolute norm, such as the 1-norm, the ∞-norm, or the Frobenius norm, which will also be denoted by $\|\cdot\|_F$. The matrix 2-norm, which is not absolute, will be denoted by $\|\cdot\|_2$. For any nonsingular matrix X we will define

$$\kappa_\nu(X) = \|X\|_\nu \, \|X^{-1}\|_\nu.$$

For more on norms see Stewart & Sun (1990).

2. The *LU* decomposition

Let A be a matrix of order n whose leading principal submatrices are nonsingular. Then A can be written in the form

$$A = LU,$$

where L is lower triangular and U is upper triangular. The decomposition is not unique, but it can be made so by specifying the diagonal elements of L. (The conventional choice is to require them to be one.)

If E is sufficiently small, $A + E$ has an LU factorization:

$$A + E = (L + F_L)(U + F_U). \tag{2.1}$$

Again, the factorization is not unique, but it can be made so, say by requiring that the diagonals of L remain unaltered.

Multiplying out the right-hand side of (2.1) and ignoring higher-order terms, we obtain a linear matrix equation for first-order approximations $\check{F}_L$ and $\check{F}_U$ to F_L and F_U:

$$L\check{F}_U + \check{F}_L U = E.$$

We shall show how to solve this equation in terms of two matrix operators.

† Chang and Paige have incorporated some of these results in their own paper and kindly added me as a coauthor. But they should be given sole credit for their approach and its execution.

Let $0 \leqslant p \leqslant 1$, and define $\mathcal{L}_p$ and $\mathcal{U}_p$ as illustrated below for a 3×3 matrix:

$$\mathcal{L}_p(X) = \begin{pmatrix} px_{11} & 0 & 0 \\ x_{21} & px_{22} & 0 \\ x_{31} & x_{32} & px_{33} \end{pmatrix} \quad \text{and} \quad \mathcal{U}_p(X) = \begin{pmatrix} px_{11} & x_{12} & x_{13} \\ 0 & px_{22} & x_{23} \\ 0 & 0 & px_{33} \end{pmatrix}.$$

It then follows that for any matrix X,

$$X = \mathcal{L}_p(X) + \mathcal{U}_{1-p}(X), \tag{2.2}$$

and

$$\|\mathcal{L}_p(X)\|,\ \|\mathcal{U}_p(X)\| \leqslant \|X\|.$$

Finally, if X is symmetric

$$\|\mathcal{U}_{1/2}(X)\|_F \leqslant \frac{1}{\sqrt{2}}\|X\|_F. \tag{2.3}$$

Our basic result is the following:

$$\check{F}_L = L\mathcal{L}_p(L^{-1}EU^{-1}) \quad \text{and} \quad \check{F}_U = \mathcal{U}_{1-p}(L^{-1}EU^{-1})U.$$

To see this, write

$$\begin{aligned} L[\mathcal{U}_{1-p}&(L^{-1}EU^{-1})U] + [L\mathcal{L}_p(L^{-1}EU^{-1})]U \\ &= L[\mathcal{U}_{1-p}(L^{-1}EU^{-1}) + \mathcal{L}_p(L^{-1}EU^{-1})]U \\ &= L[L^{-1}EU^{-1})U \qquad \text{by (2.2)} \\ &= E. \end{aligned}$$

The number p is a normalizing parameter, controlling how much of the perturbation is attached to the diagonals of L and U. If $p = 0$, the diagonal elements of L do not change. If $p = 1$, the diagonal elements of U do not change.

We can take norms in the expressions $\check{F}_L$ and $\check{F}_U$ to get first-order perturbation bounds for the LU decomposition, but it is possible to introduce degrees of freedom in the expressions that can later be used to reduce the bounds. Specifically, for any nonsingular diagonal matrix D_L, we have

$$\check{F}_L = LD_L\mathcal{L}_p(D_L^{-1}L^{-1}EU^{-1}) \equiv \hat{L}\mathcal{L}_p(\hat{L}^{-1}EU^{-1}).$$

Consequently

$$\|\check{F}_L\| \leqslant \|\hat{L}\|\ \|\hat{L}^{-1}\|\ \|U^{-1}\|\ \|E\|,$$

or

$$\frac{\|\check{F}_L\|}{\|L\|} \leqslant \frac{\kappa(\hat{L})\kappa(U)\ \|E\|}{\|L\|\ \|U\|}. \tag{2.4}$$

Since $\|A\| \leqslant \|L\|\ \|U\|$, we have

$$\frac{\|\check{F}_L\|}{\|L\|} \leqslant \kappa(\hat{L})\kappa(U)\frac{\|E\|}{\|A\|}. \tag{2.5}$$

Similarly, if D_U is a nonsingular diagonal matrix and we set

$$\hat{U} = D_U U,$$

then

$$\frac{\|\check{F}_U\|}{\|U\|} \leq \kappa(L)\kappa(\hat{U})\frac{\|E\|}{\|A\|}. \tag{2.6}$$

The bounds (2.5) and (2.6) differ from the usual bounds (see, e.g., Stewart (1993)) by the substitution of $\hat{L}$ or $\hat{U}$ for L or U. However, if the diagonal matrices D_L and D_U are chosen appropriately, $\kappa(\hat{L})$ and $\kappa(\hat{U})$ can be far less than $\kappa(L)$ or $\kappa(U)$. For example, if

$$U = \begin{pmatrix} 1 & \varepsilon \\ 0 & \varepsilon \end{pmatrix}, \tag{2.7}$$

then $\kappa_1(U) \cong 1/\varepsilon$. But if we set $D_U = \operatorname{diag}(1, 1/\varepsilon)$, then $\kappa(\hat{U}) \cong 1$.

Poorly scaled but essentially well-conditioned matrices like U in (2.7) occur naturally. If A is ill-conditioned and the LU decomposition of A is computed with pivoting, the ill-conditioning of A will usually reveal itself in the diagonal elements of U. In Stewart (1995), the author has shown that such upper triangular matrices are artificially ill-conditioned in the sense that they can be made well-conditioned by scaling their rows.

If $\kappa(\hat{L}) = 1$ (it cannot be less), then the bound (2.4) reduces to

$$\|\check{F}_L\| \leq \|U^{-1}\| \, \|E\|. \tag{2.8}$$

It is reasonable to ask if there are problems for which we can replace $\|U^{-1}\|$ by an even smaller number and still have inequality for all E. The answer depends on p. For example, suppose that $\|\cdot\|$ is the ∞-norm. Let $\mathbf{e}_i$ be the unit coordinate vectors, and let k be such that $\|\mathbf{e}_k^{\mathrm{T}} U^{-1}\| = \|U^{-1}\|$. Let $E = \mathbf{e}_n \mathbf{e}_k^{\mathrm{T}}$, so that $\|E\| = 1$. Then it is easy to see that

$$\|\hat{L}\mathscr{L}_p(\hat{L}^{-1}EU^{-1})\| = \|\mathscr{L}_p(\mathbf{e}_n\mathbf{e}_k^{\mathrm{T}}U^{-1})\|.$$

Hence

$$\|U^{-1}\| \, \|E\| \geq \|\check{F}_L\| \geq p \, \|U^{-1}\| \, \|E\|. \tag{2.9}$$

Consequently, if p is near one, (2.8) is essentially the smallest bound that holds uniformly for all E.

The reason for the appearance of the factor p in (2.9) is that the error may concentrate in the last column of $L^{-1}EU^{-1}$, in which case it is reduced by a factor of at least p by the operator $\mathscr{L}_p$. This can happen, for example, when $L = I$ and $U = \operatorname{diag}(I_{n-1}, \varepsilon)$ for ε small. However, if p is small, the perturbation will show up in $\check{F}_U$, for which the factor is $1 - p$.

The bounds (2.5) and (2.6) suggest a strategy for estimating the condition of the LU factorizations. Van der Sluis (1969) has shown that in the 2-norm, the condition number is approximately minimized when the rows or columns of the matrix are scaled to have norm one. Thus the strategy is to so scale $\hat{L}$ and $\hat{U}$ and use a condition estimator to estimate the condition of L, $\hat{L}$, U, $\hat{U}$.

In Stewart (1993) it is shown how to obtain rigorous bounds for the errors in the first-order approximations $\check{F}_L$ and $\check{F}_U$. Since the second-order terms decay rapidly, the error bounds are less important than the condition that insures their existence: namely

$$\|L^{-1}\| \, \|U^{-1}\| \, \|E\| \leqslant \tfrac{1}{4}.$$

3. The Cholesky decomposition

We now return to the Cholesky decomposition. In analyzing the perturbation of the Cholesky factor R it is natural to take $p = \frac{1}{2}$ so that symmetry is preserved. In this case the solution of the perturbation equation becomes

$$\check{F}_R = \mathcal{U}_{1/2}(R^{-\mathrm{T}}ER^{-1})R.$$

Hence if $\hat{R}$ is defined in analogy with $\hat{L}$ and $\hat{U}$, it follows from (2.3) that

$$\frac{\|\check{F}_R\|_{\mathrm{F}}}{\|R\|_2} \leqslant \frac{1}{\sqrt{2}} \kappa_2(\hat{R})\kappa_2(R) \frac{\|E\|_{\mathrm{F}}}{\|A\|_2}.$$

Moreover, by a variant of the argument that led to (2.9), for any A, there is an E such that

$$\|R^{-1}\|_2 \, \|E\|_{\mathrm{F}} \geqslant \|\check{F}_R\|_{\mathrm{F}} \geqslant \tfrac{1}{2} \, \|R^{-1}\|_2 \, \|E\|_{\mathrm{F}}$$

which shows that we cannot reduce the constant ρ in the bound

$$\|\check{F}_R\|_{\mathrm{F}} \leqslant \rho \, \|E\|_{\mathrm{F}}$$

to less than $\frac{1}{2} \, \|R^{-1}\|$.

Acknowledgement

This work was supported in part by the National Science Foundation under grant CCR 95503126.

References

Barrlund, A. 1991 Perturbation bounds for the LDL^{H} and the LU factorizations. *BIT* **31,** 358–63.

Chang, X.-W., Paige, C. C., & Stewart, G. W. 1996 New perturbation analyses for the Cholesky factorization. *IMA J. Numer. Anal.* **16,** 457–484.

Stewart, G. W. 1977 Perturbation bounds for the *QR* factorization of a matrix. *SIAM J. Numer. Anal.* **1977,** 509–18.

Stewart, G. W. 1993 On the perturbation of *LU*, Cholesky, and *QR* factorizations. *SIAM J. Matrix Anal. Applic.* **14,** 1141–6.

Stewart, G. W. 1995 The triangular matrices of Gaussian elimination and related decompositions. *Technical Report CS-TR-3533 UMIACS-TR-95-91,* University of Maryland, Department of Computer Science.

STEWART, G. W., & SUN, J.-G. 1990 *Matrix Perturbation Theory.* (Boston: Academic).
SUN, J.-G. 1992 Rounding-error and perturbation bounds for the Cholesky and LDL^T factorizations. *Lin. Algebra Applic.* **173,** 77–98.
VAN DER SLUIS, A. 1969 Condition numbers and equilibration of matrices. *Numerische Mathematik* **14,** 14–23.

12.10. [GWS-J94] "The Triangular Matrices of Gaussian Elimination and Related Decompositions"

[GWS-J94] "The Triangular Matrices of Gaussian Elimination and Related Decompositions," *IMA Journal of Numerical Analysis*, **17** (1997) 7–16.
http://dx.doi.org/10.1093/imanum/17.1.7

IMA Journal of Numerical Analysis (1997) **17**, 7–16

The triangular matrices of Gaussian elimination and related decompositions*

G. W. STEWART

Department of Computer Science and Institute for Advanced Computer Studies, University of Maryland, College Park, MD 20742, USA

[Received 27 October 1995 and in revised form 8 January 1996]

It has become commonplace that triangular systems are solved to higher accuracy than their condition would warrant. This observation is not true in general, and counterexamples are easy to construct. However, it is often true of the triangular matrices from pivoted LU or QR decompositions. It is shown that this fact is closely connected with the rank-revealing character of these decompositions.

1. Introduction

In 1961, Wilkinson (1961) published a ground-breaking error analysis of Gaussian elimination. In the course of the paper he observed that triangular systems are frequently solved more accurately than their condition would warrant. In support of this observation he offered some examples and suggestive analyses, but no general theorems.

Wilkinson's observation has stood the test of time. But it has a touch of mystery about it. No general results can be proved, because it is easy to find innocuous looking matrices that are quite ill behaved—for example, an upper triangular matrix of standard normal deviates row scaled so that its diagonals are one. Thus any general bounds have to be weak. The weakness usually manifests itself by the appearance of a factor of 2^n in the bounds.

However, the matrices Wilkinson was chiefly concerned with were the unit lower triangular matrix L and the upper triangular matrix U that are produced by pivoted Gaussian elimination applied to a matrix A. It is the purpose of this paper to show that these matrices have special properties that derive from the rank revealing character of Gaussian elimination.

Specifically, Wilkinson (1965, pp 213–4) has noted that partial or complete pivoting tends to reveal ill-conditioning in the matrix in the sense that the diagonal elements of U show a steady decrease in size. As we shall see, to the extent that these diagonals are 'ball-park' estimates of the smallest singular values of the corresponding leading principal submatrices of A, we have the following two consequences.

1. Any ill-conditioning of U is artificial in the sense that if U is row scaled so that its diagonals are one, then its smallest singular value is near one.
2. The matrix L is well-conditioned.

* This report is available by anonymous ftp from thales.cs.umd.edu in the directory pub/reports.

The first of these consequences implies that systems involving U will be solved accurately, since the best bounds on the accuracy of the solution do not depend on row scaling (Higham (1996)). The second implies *ipso facto* that systems involving L will be solved accurately.

It might be objected that we have traded one mystery for another—the other being that Gaussian elimination tends to be rank revealing. The proper response is that it is not very mysterious. If A is exactly defective in rank, then Gaussian elimination must produce an exact zero on a diagonal of U, and by continuity the element will tend to remain small when A is perturbed slightly. In fact, most people are surprised to learn that Gaussian elimination can fail to detect a near degeneracy. Moreover, the examples on which Gaussian elimination does fail are invariably constructed by forming the product of carefully chosen triangular factors. The mystery would be if Nature, who is ignorant of LU factorizations, should contrive to produce such matrices.

In the next section we will establish the basic results on rank revealing triangular matrices. In the following section we will apply the results to Gaussian elimination, and in §4 to the QR and Cholesky factorizations. The paper concludes with a brief recapitulation.

In the following, $\|\cdot\|$ denotes the 2-norm. Since the basic results are stated in terms of singular values, we will assume the reader is familiar with their elementary properties. For details see Golub & Van Loan (1989), Horn & Johnson (1991), Stewart & Sun (1990).

2. Lower bounds for singular values

The purpose of this section is to show that a triangular matrix whose principal minors reveal their rank becomes well-conditioned when its rows are equilibrated. The heart of the development is a technical lemma that relates the smallest singular value of a triangular matrix to those of its largest leading principal submatrix.

LEMMA 2.1 Let

$$\hat{R} = \begin{pmatrix} R & r \\ 0 & 1 \end{pmatrix}$$

be upper triangular. Let ρ be the smallest singular value of R and $\hat{\rho}$ be the smallest singular value of $\hat{R}$. If for some $\beta, \delta \in (0, 1]$ the smallest singular value σ of the matrix

$$\begin{pmatrix} R & r \\ 0 & \delta \end{pmatrix}$$

satisfies

$$\sigma \geqslant \beta\delta, \tag{2.1}$$

then

$$\hat{\rho} \geqslant \frac{\beta\rho}{\sqrt{\beta^2+\rho^2}}. \tag{2.2}$$

Proof. Let

$$\begin{pmatrix} R & r \\ 0 & 1 \end{pmatrix}\begin{pmatrix} x \\ \xi \end{pmatrix} = \hat{\rho}\begin{pmatrix} y \\ \eta \end{pmatrix},$$

where the vectors $(x^{\mathrm{T}} \ \xi)$ and $(y^{\mathrm{T}} \ \eta)$ have norm one. Set

$$e = -\hat{\rho}R^{-1}y \qquad \text{and} \qquad \tilde{x} = x + e,$$

so that

$$\begin{pmatrix} R & r \\ 0 & 1 \end{pmatrix}\begin{pmatrix} \tilde{x} \\ \xi \end{pmatrix} = \hat{\rho}\begin{pmatrix} 0 \\ \eta \end{pmatrix}.$$

Let

$$\nu = \left\| \begin{pmatrix} \tilde{x} \\ \xi \end{pmatrix} \right\|.$$

Then it follows from the relation

$$\frac{1}{\nu}\begin{pmatrix} R & r \\ 0 & \delta \end{pmatrix}\begin{pmatrix} \tilde{x} \\ \xi \end{pmatrix} = \begin{pmatrix} 0 \\ \nu^{-1}\hat{\rho}\delta\eta \end{pmatrix}$$

that $\nu^{-1}\hat{\rho}\delta\eta \geqslant \sigma$. Hence by (2.1),

$$\frac{\hat{\rho}\eta\delta}{\nu} \geqslant \beta\delta,$$

or

$$\hat{\rho} \geqslant \frac{\beta\nu}{\eta}. \tag{2.3}$$

Now

$$\|e\| = \|\hat{\rho}R^{-1}y\| \leqslant \rho^{-1}\hat{\rho}\ \|y\| = \rho^{-1}\hat{\rho}\sqrt{1-\eta^2}.$$

Hence

$$\nu \geqslant 1 - \rho^{-1}\hat{\rho}\sqrt{1-\eta^2},$$

and it follows from (2.3) that

$$\hat{\rho} \geqslant \frac{\beta\rho}{\rho\eta + \beta\sqrt{1-\eta^2}}. \tag{2.4}$$

The maximum of the denominator of this expression occurs when

$$\eta^2 = \frac{\rho^2}{\beta^2 + \rho^2},$$

and its value is $\sqrt{\beta^2 + \rho^2}$. The inequality (2.2) follows on substituting this value in (2.4). □

We wish to apply this lemma to get lower bounds on the singular values of a row-equilibrated upper triangular matrix. We will denote the original matrix by U and assume without loss of generality that its diagonal elements δ_i are positive. We will also assume that the δ_i are not greater than one, a situation that can always be achieved by multiplying U by a scalar. Let

$$D = \operatorname{diag}(\delta_1, \delta_2, \ldots, \delta_n),$$

so that the equilibrated matrix is $D^{-1}U$.

To pin down what we mean for a triangular matrix to be rank revealing, we make the following definition.

Definition 2.2 Let U be upper triangular of order n and let σ be the smallest singular value of U. We say that U is *rank revealing of quality* β if

$$\sigma = \beta\, |u_{nn}|.$$

The quality factor β in this definition is always less than or equal to one, since $\sigma \leq |u_{nn}|$. The nearer β is to one, the better $|u_{nn}|$ estimates σ. It should be stressed that this definition has been tailored to the requirements of this paper and is not meant to preempt other definitions of what it means to reveal rank.

We are going to use our lemma to get a recursion for a lower bound on the singular values of the leading principal submatrices of $D^{-1}U$. Let U_k and D_k denote the leading principal submatrices of order k of U and D. Let σ_k be the smallest singular value of U_k and, suppose U_k reveals σ_k with quality β_k, so that

$$\sigma_k = \beta_k \delta_k.$$

Since the diagonal elements of D_{k-1}^{-1} are all greater than one, the smallest singular value σ of

$$\begin{pmatrix} D_{k-1}^{-1} U_{k-1} & D_{k-1}^{-1} u_k \\ 0 & \delta_k \end{pmatrix}$$

is not less than σ_k. (This statement follows from the fact that the smallest singular value of a matrix A is $\inf_{\|x\|=1} \|Ax\|$.) Thus

$$\sigma \geq \beta_k \delta_k,$$

which is just the hypothesis (2.1) of Lemma 2.1. Applying the lemma with

$$\hat{R} = \begin{pmatrix} D_{k-1}^{-1}U_{k-1} & D_{k-1}^{-1}u_k \\ 0 & 1 \end{pmatrix},$$

we find that if ρ_k is the smallest singular value of $D_k^{-1}U_k$ (i.e., the equilibrated principal minor) then

$$\rho_k \geqslant \frac{\beta_k \rho_{k-1}}{\sqrt{\beta_k^2 + \rho_{k-1}^2}}. \tag{2.5}$$

Since the quantities in the inequality (2.5) do not change when U is multiplied by a scalar, we may drop the assumption that the diagonals of U are not greater than one. Hence we have the following theorem.

THEOREM 2.3 If for $k = 2, 3, \ldots, n$ the leading principal minor U_k of U reveals its rank with quality β_k and ρ_k is the smallest singular value of U_k row-scaled so that its diagonal elements are one, then the ρ_k satisfy the recursion (2.5).

There is a corresponding theorem for lower triangular matrices; the only difference is that the scaling is by columns rather than rows.

The recursion, starting with $\rho_1 = 1$, allows us to compute lower bounds on the smallest singular value of the leading principal minors of the equilibrated submatrices. In general if the β_k are uniformly bounded away from 1, the recursion converges to zero. However, if the β_k are uniformly bounded away from zero, the convergence is sublinear in the sense that

$$\frac{\rho_{k+1}}{\rho_k} \to 1.$$

Typically, a sequence converging sublinearly to zero shows an initial sharp decrease followed by an increasingly slow approach to zero. Our recursion is no exception. Table 2.1 exhibits values of the lower bounds for ρ_k when β_k is held constant. In fact it can be shown that in the limit the iterates approach $\beta/\sqrt{k}$. (For a general theory of sublinear convergence see Stewart (1995a).)

The price we pay for these slowly decreasing bounds is the requirement that U satisfy a strong rank-revealing condition. The diagonals of U must not just approximate the corresponding singular values of U but instead must

TABLE 2.1

	β		
k	0·50	0·10	0·01
5	2·4E − 1	5·0E − 2	5·0E − 3
10	1·6E − 1	3·3E − 2	3·3E − 3
100	5·0E − 2	1·0E − 2	1·0E − 3
1000	1·6E − 2	3·2E − 2	3·2E − 3

TABLE 2.2
Some experiments with triangular matrices

1	2	3	4
2·6E − 10	3·6E − 01	3·1E − 02	1·1E − 10
1·0E − 10	1·4E − 01	2·1E − 02	6·7E − 11
5·6E − 07	3·4E − 01	1·9E − 02	1·1E − 06
3·0E − 07	1·3E − 01	1·1E − 02	2·8E − 07
1·8E − 08	3·6E − 01	3·6E − 02	1·7E − 08
1·6E − 08	1·4E − 01	2·6E − 02	1·5E − 08
2·1E − 11	3·4E − 01	5·3E − 02	3·2E − 11
1·3E − 11	1·3E − 01	3·2E − 02	8·1E − 12
8·2E − 07	3·6E − 01	1·8E − 02	4·8E − 07
5·4E − 07	1·3E − 01	1·5E − 02	2·7E − 07

1. U consisting of standard normal deviates.
2. R from the pivoted QR decomposition of U in 1.
3. The upper triangular part of the LU decomposition of UQ, U from 1 and Q a random orthogonal matrix.
4. The upper triangular part of the LU decomposition of QU, U from 1 and Q a random orthogonal matrix.

approximate the smallest singular value of the corresponding leading principal submatrix. By the interlacing theorem for singular values (Horn & Johnson (1991, p 149)), the latter can be smaller than the former.

To illustrate the bounds, let us consider an upper triangular matrix whose elements are standard normal deviates and also some triangular matrices that can be obtained from them by computing factorizations. Each column of Table 2.2 contains five replications of an experiment involving an upper triangular matrix of order 25. The source of these triangular matrices is explained in the legend. Each double entry consists of the smallest singular value of the equilibrated matrix and the lower bound computed from the recursion (2.5).

The bounds are remarkably sharp, which suggests that we gave little away in their derivation. If we had used the minimum of the ratios $\beta = \sigma_k/\delta_k$ for the β_k, the bounds would not have been much worse. Owing to the sublinear convergence of the recursion, the bound would quickly drop to a little below β and then stagnate.

Turning now to the kinds of matrix used in the experiments, we note from the first column that a random normal matrix U is not good at revealing its rank, as evidenced by the small lower bound. Since the bound is nearly attained, we observe a small singular value in $D^{-1}U$.

The numbers in the second column come from the R factor in a pivoted QR decomposition of the U in column one. Such a decomposition is generally an excellent rank revealer (Stewart (1980)), and indeed we observe that the bounds and the smallest singular value are near one.

We will discuss the experiments of the third and fourth columns after we apply our results to Gaussian elimination.

3. Gaussian elimination

In applying our results to Gaussian elimination we will have to make use of some empirical facts about the growth of elements in the course of the algorithm. For experiments and analyses concerning this important topic, see the paper by Trefethen & Schreiber (1990).

Let the matrix A of order n be decomposed by Gaussian elimination with pivoting, so that

$$P^{\mathrm{T}}AQ = LDU,$$

where P and Q are permutation matrices, L is a unit lower triangular matrix, and U is unit upper triangular. When partial pivoting is used, $Q = I$ and the elements of L are less than one in magnitude. When complete pivoting is used, both L and U have elements less than one in magnitude.

We will assume that $\|L\|$ and $\|U\|$ are slowly growing functions of n. For complete pivoting, we have the bound $\|L\|$, $\|U\| \leqslant n$, which is often an overestimate. For partial pivoting, the bound continues to hold for $\|L\|$. The fact that $\|U\|$ grows slowly is related to the slow growth of elements in Gaussian elimination.

Let A_k, L_k, D_k, and U_k be the leading principal submatrices of A, L, D, and U. Let σ_k denote the smallest singular value of $D_k U_k$, and τ_k the smallest singular value of A_k. Define

$$\beta_k = \frac{\sigma_k}{|\delta_k|} \qquad \text{and} \qquad \gamma_k = \frac{\tau_k}{|\delta_k|}.$$

Thus γ_k is the quality of $|\delta_k|$ as a revealer of the rank of A.

Now since $A_k = L_k(D_k U_k)$, we have $\tau_k \leqslant \sigma_k \|L_k\|$, or

$$\beta_k \geqslant \frac{\gamma_k}{\|L_k\|}.$$

By our hypothesis on the size of $\|L\|$, if U_k reveals the rank of the A_k in the sense that γ_k is near one, the matrix $D_k U_k$ also reveals its own rank. Since these statements are true for all k, it follows from the considerations of the last section that the smallest singular value of U is near one.

The well-conditioning of L can be deduced by the same argument applied to A^{T}.

The last two columns in Table 2.2 show two aspects of these results. In the third, the columns of the matrix U of standard normal deviates was scrambled by postmultiplication by a random orthogonal matrix Q, and Gaussian elimination with partial pivoting was applied to obtain a new upper triangular matrix. This new matrix is rank-revealing and consequently scaling its rows makes its smallest singular value near one. The bounds also show that Gaussian elimination with partial pivoting is not as good at revealing rank as pivoted QR.

In the fourth column, the matrix U is replaced by QU and subjected to Gaussian elimination with partial pivoting. Here the new upper triangular matrix completely fails to reveal the rank of the original, and when it is scaled it has small singular value. The reason is easy to see. If we let $Q = LR$ be a partially pivoted LU decomposition of Q, then RU is the upper triangular matrix computed from QU. But as it turns out Q, L, and R are well-conditioned, hence RU remains rank concealing.

These results have implications for the perturbation theory of the LU decomposition. Let $A + E$ have the LU decomposition $(L + F_L)(U + F_U)$ and let S be an arbitrary nonsingular diagonal matrix. The author has shown (Stewart (1995b)) that for any absolute norm $\|\cdot\|$

$$\frac{\|F_U\|}{\|U\|} \leqslant \kappa(L)\kappa(SU)\frac{\|E\|}{\|A\|},$$

where as usual $\kappa(X) = \|X\|\,\|X^{-1}\|$. The results derived here say that if U reveals the rank of A, the factor $\kappa(L)\kappa(SU)$ can be made near one (i.e., U is insensitive to perturbations in A). However, it is important to keep in mind that the insensitivity is normwise. Small elements of U are generally quite sensitive, as common sense would dictate.

4. Related decompositions

Stronger results can be obtained for the triangular matrices produced by orthogonal decompositions. For example, let A be an $m \times n$ matrix with $m \geqslant n$. The pivoted QR decomposition factors A in the form

$$AP = QR,$$

where P is a permutation matrix, Q is an $m \times n$ matrix with orthonormal columns and R is upper triangular with positive diagonal elements. The pivoting insures that

$$r_{kk}^2 \geqslant \sum_{i=k}^{j} r_{ij}^2, \qquad j = k+1,\dots,n.$$

The pivoted QR decomposition is known empirically to reveal the rank of A in the following sense (Stewart (1980)). If A_k denotes the matrix consisting of the first k columns of A, then r_{kk} is an approximation to the smallest singular value of A_k. Since the singular values of the leading principal matrix R_k of R are the same as the singular values of A_k, the matrix R_k will also be rank revealing. Consequently, the smallest singular value of the matrix obtained by scaling so its diagonal elements are one will be near one. Thus the R factor from a pivoted QR factorization is another source of triangular systems that tend to be solved accurately.

These results possibly explain an observation of Golub on the use of Householder transformations to solve least-squares problems (Golub (1965)). He noted that column pivoting slightly improved the accuracy of the computed solutions. From the point of view taken here; pivoting would make the R more rank revealing and hence the QR equations for the least-squares solution would be solved more accurately.

Since the R factor of AP is the Cholesky factor of P^TA^TAP, pivoted Cholesky factorization with diagonal pivoting of positive definite matrices also give rise to systems that can be solved accurately. The same can be said of the systems resulting from two-sided orthogonal decompositions like the URV and ULV decompositions.

5. Conclusions

We do not claim to have solved the mysteries of Gaussian elimination in this paper. The basic result is that if Gaussian elimination produces a rank-revealing LDU factorization with L and U of modest norm, then L and U, suitably scaled, must be well-conditioned, and systems involving them will be solved accurately. That pivoted Gaussian elimination should be rank revealing is not in itself surprising, and all the counterexamples I am aware of are obtained by starting with rank-concealing triangular matrices (cf. column four in Table 2.2). The fact (critical to our analysis) that L and U are of modest size is guaranteed for complete pivoting, but for partial pivoting, what inhibits growth of the elements of U is imperfectly understood (Trefethen & Schreiber 1990)).

For the pivoted QR and Cholesky factorizations we need no auxiliary hypothesis about the sizes of the triangular factor. All we need to believe is that the factorizations reveal rank.

Perhaps the most unusual feature of the analysis is the nature of the recursion (2.5). It is a great leveler, eager to reduce ρs that are greater than β and reluctant to reduce them much further. For example, if $\rho = 2\beta$, then ρ is reduced by a factor of 0·45, whereas if $\rho = \frac{1}{2}\beta$ it is reduced by a factor of only 0·89. Thus, even if a triangular matrix consistently overestimates the rank of its principal submatrices, the overestimates have only a one-time effect and do not propagate exponentially in the bounds. May we have more bounds of this nature!

Acknowledgement

This work was supported in part by the National Science Foundation under grant CCR 95503126.

References

Golub, G. H. 1965 Numerical methods for solving least squares problems. *Numerische Mathematik* **7**, 206–16.

Golub, G. H., & Van Loan, C. F. 1989 *Matrix Computations* 2nd edition. Baltimore, MD: Johns Hopkins University Press.

Higham, N. J. 1996 *Accuracy and Stability of Numerical Algorithms.* Philadelphia, PA: SIAM.

Horn, R. A., & Johnson, C. R. 1991 *Topics in Matrix Analysis.* Cambridge: Cambridge University Press.

Stewart, G. W. 1980 The efficient generation of random orthogonal matrices with an application to condition estimators. *SIAM J. Numer. Anal.* **17**, 403–4.

Stewart, G. W. 1995a On sublinear convergence. *Technical Report CS-TR-3534 UMIACS-TR-95-92*, University of Maryland, Department of Computer Science.

STEWART, G. W. 1995b On the perturbation of LU and Cholesky factors. *Technical Report CS-TR-3535 UMIACS-TR-95-93*, University of Maryland, Department of Computer Science.

STEWART, G. W., & SUN, J.-G. 1990 *Matrix Perturbation Theory*. Boston: Academic.

TREFETHEN, L. N., & SCHREIBER, R. S. 1990 Average-case stability of Gaussian elimination. *SIAM J. Matrix Anal. Applic.* **11,** 335–60.

WILKINSON, J. H. 1961 Error analysis of direct methods of matrix inversion. *J. ACM* **8,** 281–330.

WILKINSON, J. H. 1965 *The Algebraic Eigenvalue Problem*. Oxford: Clarendon.

12.11. [GWS-J103] "Four Algorithms for the the (sic) Efficient Computation of Truncated Pivoted QR Approximations to a Sparse Matrix"

[GWS-J103] "Four Algorithms for the the (sic) Efficient Computation of Truncated Pivoted QR Approximations to a Sparse Matrix," *Numerische Mathematik,* **83** (1999) 313–323.
http://dx.doi.org/10.1007/s002110050451

Numer. Math. (1999) 83: 313–323

Numerische
Mathematik

Four algorithms for the the efficient computation of truncated pivoted QR approximations to a sparse matrix*, **

G.W. Stewart

Department of Computer Science and Institute for Advanced Computer Studies, University of Maryland, College Park, MD 20742, USA

Received February 23, 1998 / Revised version received April 16, 1998

Dedicated to Olof Widlund on his 60th birthday

Summary. In this paper we propose four algorithms to compute truncated pivoted QR approximations to a sparse matrix. Three are based on the Gram–Schmidt algorithm and the other on Householder triangularization. All four algorithms leave the original matrix unchanged, and the only additional storage requirements are arrays to contain the factorization itself. Thus, the algorithms are particularly suited to determining low-rank approximations to a sparse matrix.

Mathematics Subject Classification (1991): 65F20, 65F50

1. Introduction

Let X be an $n \times p$ matrix with $n > p$. This paper concerns the approximation of X by a matrix $\hat{X}$ of rank $k < p$. The most elegant solution to this problem is the truncated singular value approximation which gives an $\hat{X}$ that deviates minimally from X in any unitarily invariant norm — in particular in both the spectral and Frobenius norms.[1]

---

* This report is available by anonymous ftp from `thales.cs.umd.edu` in the directory `pub/reports` or on the web at `http://www.cs.umd.edu/~stewart/`

** This work was supported by the National Science Foundation under grant CCR-95-03126

[1] Throughout this paper, $\|\cdot\|$ will denote the Frobenius norm defined by $\|X\|^2 = \sum_{i,j} x_{ij}^2$. The algorithms in this paper are based on two standard algorithms — the Gram–Schmidt

A different approximation may be had from the pivoted QR decomposition. This decomposition is a factorization of the form

$$X = Q \begin{pmatrix} R \\ 0 \end{pmatrix} \Pi^{\mathrm{T}},$$

where Q is orthogonal, R is upper triangular, and Π is a permutation matrix that corresponds to column interchanges in X (i.e., pivoting). If we ignore pivoting for the moment and partition

$$(1) \qquad X \equiv (X_1^{(k)} \; X_2^{(k)})(Q_1^{(k)} \; Q_2^{(k)} \; Q_\perp) \begin{pmatrix} R_{11}^{(k)} & R_{12}^{(k)} \\ 0 & R_{22}^{(k)} \\ 0 & 0 \end{pmatrix},$$

where $R_{11}^{(k)}$ is of order k, then

$$(2) \qquad \hat{X} = Q_1^{(k)}(R_{11}^{(k)} \; R_{12}^{(k)})$$

is a rank-k approximation to X with the following properties.

1. $X_1^{(k)} = Q_1^{(k)} R_{11}^{(k)}$.
2. $\|X_2^{(k)} - Q_1^{(k)} R_{12}^{(k)}\| = \|R_{22}^{(k)}\|$.
3. If $U(S_1 \; S_2)$ is another rank k approximation to X with $X_1^{(k)} = US_1$, then in any unitarily invariant norm

$$\|X_2^{(k)} - Q_1^{(k)} R_{12}^{(k)}\| \leq \|X_2^{(k)} - US_2\|.$$

In other words, the approximation (2) reproduces the first k columns of X exactly, and among all such approximations it reproduces the last $p-k$ columns optimally. In particular $R_{22}^{(k)}$ is sufficiently small, the approximation $\hat{X}$ solves our problem. We will call (2) the truncated pivoted QR approximation to X.

The singular value and pivoted QR decompositions loose some of their their appeal when X is large and sparse. The problem is that the conventional algorithms for computing these decomposition proceed by transformations that quickly destroy the sparsity of X. For the singular value decomposition, recourse must be had to iterative approximations. The purpose of this paper is to show that the pivoted QR decomposition can be implemented in such a way that the sparsity of X is not compromised.

We actually give four algorithms, three based on Gram–Schmidt orthogonalization and the other on Householder triangularization. The key idea is

algorithm and Householder triangularization. We assume that the reader is familiar with the basic facts about these algorithms. A good reference is the excellent book by Å. Björck [1].

to build up the Q-factor column by column and the R-factor row by row, leaving the original matrix X unchanged. Only when a column of X is needed to form a column of Q is it transformed.[2] A key feature of the algorithms is the downdating of column norms to control pivoting.

The first Gram–Schmidt variant computes the explicit factorization and is conceptually simpler. The second is a new variant of the Gram–Schmidt method, here called the quasi-Gram–Schmidt algorithm. It is based on the fact that since $Q_1^{(k)} = X_1^{(k)} R_{11}^{(k)-1}$, we can store the sparse matrix $X_1^{(k)}$ instead of the (usually) dense matrix $Q_1^{(k)}$. The third algorithm expands on this idea to produce an approximation of the form YTZ^{T} where Y and Z consist respectively of columns and rows of X.

The algorithm based on Householder triangularization provides an implicit basis for the orthogonal complement of the column space of the decomposition. It is closely related to an algorithm of Quintana-Ortí, Sun, and Bischof [2] for blocking the triangularization.

2. The Gram–Schmidt variant

Let $X = (x_1 \ \cdots \ x_p)$ be partitioned by columns, and as above let

$$X_1^{(k)} = (x_1 \ \cdots \ x_k), \quad \text{and} \quad X_2^{(k)} = (x_{k+1} \ \cdots \ x_p),$$

Suppose we have computed a QR factorization.

$$X_1^{(k-1)} = Q_1^{(k-1)} R_{11}^{(k-1)},$$

where $R_{11}^{(k-1)}$ is upper triangular of order $k-1$. Then the corresponding QR approximation is

$$\hat{X}^{(k-1)} = Q_1^{(k-1)} (R_{11}^{(k-1)} \ \ R_{12}^{(k-1)}),$$

where

$$R_{12}^{(k-1)} = Q_1^{(k-1)\mathrm{T}} X_2^{(k-1)}.$$

Now suppose we want to update this approximation to include x_k. This can be accomplished by one step of the Gram–Schmidt algorithm:

$$\begin{array}{ll} 1. & r_{1k} = Q_1^{(k-1)\mathrm{T}} x_k \\ 2. & q_k = x_k - Q_1^{(k-1)} r_{1k} \\ 3. & r_{kk} = \|q_k\| \\ 4. & q_k = q_k / r_{kk} \end{array} \tag{3}$$

[2] The approach is analogous to the Crout variant of Gaussian elimination.

Then

$$X_1^{(k)} = (X_1^{(k-1)} \;\; x_k) = (Q_1^{(k-1)} \;\; q_k) \begin{pmatrix} R_{11}^{(k-1)} & r_{1k} \\ 0 & r_{kk} \end{pmatrix}$$

is the updated QR factorization. To compute the rest of the approximation, we need only compute the kth row of R, which has the form

$$r_{k,k+1}^{\mathrm{T}} = q_k^{\mathrm{T}} X_2^{(k)}. \tag{4}$$

This updating procedure is very close to what we promised above. Each step generates a new column of Q and a new row of R. The only part of X that is modified is the column that eventually becomes the new column of Q. The only other use made of X is a vector-matrix product in (4) to fill out the kth row of R. However, there remain two problems.

The first problem is that to get good approximations we must incorporate column pivoting into our algorithm. Specifically, at the beginning of the kth stage of the algorithm, we must select the column x_j of $(x_k \;\cdots\; x_p)$ for which $\|(I - Q_1^{(k)} Q_1^{(k)\mathrm{T}})x_j\|$ is maximal and interchange it with x_k. The computation of $(I - Q_1^{(k)} Q_1^{(k)\mathrm{T}})x_j$ for all $j \geq k$ effectively transforms X into a dense matrix.

The answer to this problem lies in the following well-known formula:

$$\|(I - Q_1^{(k)} Q_1^{(k)\mathrm{T}})x_j\|^2 = \|x_j\|^2 - \sum_{i=1}^{k} r_{ij}^2. \tag{5}$$

This formula allows us to downdate the column norms of X as we compute rows of R to give the norms of the projected columns.

The second problem is that the Gram–Schmidt algorithms as implemented above is not guaranteed to produce a matrix Q_k whose columns are orthogonal to working accuracy. The cure for this problem is a process in which q_k is reorthogonalized against the previous q_j. The details of this process, which also alters the kth columns of R have been well treated in [1], and we omit them.

Figure 1 contains an algorithm implementing this method. The algorithm continues until the projected columns norms, which correspond to the norms of the columns of $R_{22}^{(k)}$ in (1), become sufficiently small, at which point the process terminates with the rank k approximation $\hat{X}^{(k)} = Q[:, 1{:}k]*R[1{:}k, 1{:}p]$. To simplify the program, we have made explicit interchanges in X. The only unusual feature is the omission of the statement

$$R[1{:}k{-}1, k] = Q[:, 1{:}k{-}1]^{\mathrm{T}}*X[:, k] \tag{6}$$

which corresponds to statement 3 in (3). The reason is that these numbers have already been computed as we have added rows to R.

Given an $n \times p$ $(n \geq p)$ matrix X this algorithm returns a truncated pivoted QR decomposition of X. Initially, the matrices Q and R are void.

```
ν_j = ||X[:, j]||²,    j = 1, ..., p
Determine an index p_1 such that ν_{p_1} is maximal
for k = 1 to p
    X[:, k] ↔ X[:, p_k]
    R[1:k−1, k] ↔ R[1:k−1, p_k]
    Q[:, k] = X[:, k] − Q[:, 1:k−1]*R[1:k−1, k]
    R[k, k] = ||Q[:, k]||
    Q[:, k] = Q[:, k]/R[k, k]
    If necessary reorthogonalize Q[:, k] and adjust R[1:k+1, k]
    R[k, k+1:p] = Q[:, k]^T*X[:, k+1:p]
    ν_j = ν_j − R[k, j]²,    j = k+1, ..., p
    Determine an index p_{k+1} ≥ k+1 such that ν_{p_{k+1}} is maximal
    if (ν_{p_{k+1}} is sufficiently small) leave k fi
end for k.
```

Fig. 1. Truncated pivoted QR factorization: Gram–Schmidt version

It is important to recognize that this algorithm and the ones to follow are limited by the accuracy attainable from the formula (5). Specifically, if $\|x_j\| = 1$ and $\|(I - Q_1^{(k)} Q_1^{(k)\mathrm{T}})x_j\|^2$ is less than the rounding unit (denoted by ϵ_{M}), its computed value will be completely inaccurate. Since the norm of the projected X is the square root of this quantity, our method cannot distinguish among vectors whose projected norms are less than $\sqrt{\epsilon_{\mathrm{M}}}$. In IEEE double precision, for example, any vector x whose norm falls below about $10^{-7}\|x\|$ can no longer be a candidate for selection. Unfortunately, the standard fix of recomputing the norms of the projected vectors is not available to our methods. Fortunately, when one is seeking a low-rank approximation, this difficulty is unlikely to be a serious limitation.

3. The quasi-Gram–Schmidt method

If the original matrix X is very large and sparse, it may be impossible to store the factors $Q_1^{(k)}$ and $(R_{11}^{(k)}\ R_{12}^{(k)})$, which are generally dense. If one requires an explicit low-rank approximation of X, then nothing much can be done about this situation. But if one only needs $Q_1^{(k)}$, say to compute projections, then one can store $X_1^{(k)}$ and $R_{11}^{(k)}$ and use the relation

$$Q_1^{(k)} = X_1^{(k)} R_{11}^{(k)-1}$$

to recover the action of Q. More precisely, If we require to compute, say, $w = Q_1^{(k)} v$, we can use the following algorithm.

Given an $n{\times}p$ ($n \geq p$) matrix X this algorithm the the factor $R_{11}^{(k)}$ from the truncated pivoted QR decompostion. Initially, the matrix R is void.

1. $\nu_j = \|X[:, j]\|^2, \qquad j = 1, \ldots, p$
2. Determine an index p_1 such that ν_{p_1} is maximal
3. **for** $k = 1$ **to** p
4. $\quad X[:, k] \leftrightarrow X[:, p_k]$
5. $\quad v = X[:, 1{:}k{-}1]^{\mathrm{T}} {*} X[:, k]$
6. $\quad$ Solve the system $R[1{:}k{-}1, 1{:}k{-}1]^{\mathrm{T}} {*} R[1{:}k{-}1, k] = v$
7. $\quad$ Solve the system $R[1{:}k{-}1, 1{:}k{-}1] {*} w = R[1{:}k{-}1, k]$
8. $\quad q = X[:, k] - X[:, 1{:}k{-}1] {*} w$
9. $\quad v = X[:, 1{:}k{-}1]^{\mathrm{T}} {*} q$
10. $\quad$ Solve the system $R[1{:}k{-}1, 1{:}k{-}1]^{\mathrm{T}} {*} r = v$
11. $\quad$ Solve the system $R[1{:}k{-}1, 1{:}k{-}1] {*} w = r$
12. $\quad q = q - X[:, 1{:}k{-}1] {*} w$
13. $\quad R[1{:}k{-}1, k] = R[1{:}k{-}1, k] {+} r$
14. $\quad R[k, k] = \|q\|$
15. $\quad q = q / R[k, k]$
16. $\quad r[k{+}1{:}p] = q^{\mathrm{T}} {*} X[:, k{+}1{:}p]$
17. $\quad \nu_j = \nu_j - r[j]^2, \qquad j = k{+}1, \ldots, p$
18. $\quad$ Determine an index $p_{k+1} \geq k{+}1$ such that $\nu_{p_{k+1}}$ is maximal
19. $\quad$ **if** ($\nu_{p_{k+1}}$ is sufficiently small) **leave** k **fi**
20. **end for** k.

Fig. 2. Truncated pivoted QR factorization: quasi-Gram–Schmidt version

$$
\begin{aligned}
&1. \quad \text{Solve the system } R_{11}^{(k)} z = v \\
&2. \quad w = X_1^{(k)} z
\end{aligned}
\tag{7}
$$

This involves only the solution of a triangular system of order k and a sparse matrix-vector multiplication.[3]

The algorithm in Fig. 1 can be adapted to this storage scheme. Since we do not have $Q_1^{(k-1)}$ explicitly, we must simulate its action by (7) and its variant for computing $Q_1^{(k-1)\mathrm{T}} v$. If we are only updating $R_{11}^{(k-1)}$, we must also add the equivalent of (6). These considerations lead to the algorithm in Fig. 2.

To downdate the column norms we must form the entire kth row of R (statement 3), which for economy of storage is discarded. However, if p is small, so that the storage of $R_{12}^{(k)}$ is not too burdensome, one can retain it as in the first algorithm.

Statements 10–13 are a "reorthogonalization" of q, and it has a profound effect on the algorithm, as the following example shows.

---

[3] This method is closely related to Saunder's method of seminormal equations. See [1, p. 70].

ℓ	no reorth	reorth
1	9.0e−15	2.2e−15
2	1.5e−13	3.7e−15
3	2.2e−11	3.6e−14
4	3.0e−09	1.3e−13
5	3.1e−06	1.6e−12
6	4.1e−05	1.2e−11
7	7.1e−04	1.0e−10
8	1.3e−01	1.1e−09
9	1.6e+00	2.6e−08
10	2.0e+00	1.1e−06

Fig. 3. Decay of orthogonality

A sequence 100×10 matrices X was generated having singular values of

$$10^{-1},\ldots,10^{-\ell},\qquad \ell=1,\ldots,10$$

in equally spaced logarithmic decrements. The above algorithm (without pivoting) was run on each X, once without reorthogonalization and once with. The matrices $Q_1^{(10)} = X_1^{(10)}R_{11}^{(10)-1}$ were tested for orthogonality by computing $\|I - Q_1^{(10)\mathrm{T}}Q_1^{(10)}\|_2$. The results are displayed in Fig. 3. The first column shows a strong loss of orthogonality — the kind we expect from the classical Gram–Schmidt algorithm. The second column shows a slower loss of orthogonality that is proportional to the smallest singular value of X. Note that this loss is the same we would observe if we computed $Q_1^{(10)} = X_1^{(10)}R_{11}^{(10)-1}$ from the exact values of $X_1^{(10)}$ and $R_{11}^{(10)}$.

These numbers suggest that with reorthogonalization the quasi-Gram–Schmidt algorithm is quite satisfactory. The same method without reorthogonalization may also be satisfactory provided the matrices $X_1^{(k)}$ are not too ill conditioned. This is precisely what we would expect in low-rank approximation, where ill-conditioning indicates that we have achieved a satisfactory approximation.

4. A sparse approximation

The quasi-Gram–Schmidt method provides us with a set of columns of X that approximate its column space and a triangular matrix R that orthogonalizes those columns. But it does not provide a sparse factorization of X itself. It this section we will show how such an approximation can be obtained from two applications of the quasi-Gram–Schmidt method.

Let us suppose that we have applied the quasi-Gram–Schmidt algorithm with pivoting to X to get a matrix Y of columns of X whose QR factorization is $Y = PR$. Note that the quasi-Gram–Schmidt algorithm also provides R but not Q. Now apply the algorithm to X^{T} to get a matrix Z^{T} of rows of X with QR factorization $Z = QS$. To get an approximation to X, we will determine a matrix T such that

$$\|YTZ^{\mathrm{T}} - X\|^2 = \min. \tag{8}$$

Let $(P\ P_\perp)$ and $(Q\ Q_\perp)$ be orthogonal. Since the Frobenius norm is unitarily invariant, the least squares problem (8) is equivalent to the problem

$$\left\| \begin{pmatrix} P^{\mathrm{T}} \\ P_\perp^{\mathrm{T}} \end{pmatrix} (YTZ^{\mathrm{T}} - X)(Q\ Q_\perp) \right\|^2 = \min.$$

Equivalently

$$\left\| \begin{pmatrix} RTS^{\mathrm{T}} & 0 \\ 0 & 0 \end{pmatrix} - \begin{pmatrix} P^{\mathrm{T}}XQ & P^{\mathrm{T}}XQ_\perp \\ P_\perp^{\mathrm{T}}XQ & P_\perp^{\mathrm{T}}XQ_\perp \end{pmatrix} \right\|^2 = \min. \tag{9}$$

Thus T must satisfy

$$T = R^{-1}P^{\mathrm{T}}XQS^{-\mathrm{T}}.$$

But $P^{\mathrm{T}}XQ = R^{-\mathrm{T}}Y^{\mathrm{T}}XZS^{-1}$. Hence

$$T = (R^{\mathrm{T}}R)^{-1}Y^{\mathrm{T}}XZ(S^{\mathrm{T}}S)^{-1}.$$

If X is sparse, the computation of $Y^{\mathrm{T}}XZ$ involves only sparse matrix-vector operations. If the approximation is low-rank, then manipulating the dense matrices R, S and T present no problems. Note that Y and Z need not have the same dimensions.

It is worth noting that we can also write the approximation in the form

$$YTZ^{\mathrm{T}} = (PP^{\mathrm{T}})X(QQ^{\mathrm{T}}).$$

In other words the approximation is obtained by projecting the columns and rows of X onto the spaces spanned by Y and Z.

We can easily bound the error in the approximation. We have $\|P_\perp^{\mathrm{T}}X\| = \|R_{22}^{(k)}\|$, where $R_{22}^{(k)}$ is the part of the R-factor of X we have ignored. The square of its Frobenius norm can be computed as the sum of squares of the norms of the projected vectors not included in the decomposition—precisely the quantities we compute to control the pivoting. Similarly $\|XQ_\perp\| = \|S_{22}^{(\ell)}\|$, where $S_{22}^{(\ell)}$ is the part of the R-factor of X^{T} that we have ignored. It follows from (9) that

$$\|YTZ^{\mathrm{T}} - X\| \le \sqrt{\|R_{22}^{(k)}\|^2 + \|S_{22}^{(\ell)}\|^2}.$$

For low-rank approximations this bound is likely to be an overestimate by a factor of around $\sqrt{2}$, since the term $\|P_\perp XQ_\perp\|^2$ implicitly enters twice.

5. Householder triangularization

We now turn to using orthogonal triangularization to compute truncated pivoted QR approximations. We begin with a brief description of how the pivoted QR decomposition is computed via Householder transformations. Before the kth stage of the reduction, we have determined Householder transformations $H_1, \ldots, H_{k-1}$ and a permutation matrix Π_{k-1} such that

$$H_{k-1} \cdots H_1 X \Pi_{k-1} = \begin{pmatrix} R_{11}^{(k-1)} & \bar{R}_{12}^{(k-1)} \\ 0 & \bar{X}_{22}^{(k-1)} \end{pmatrix}, \tag{10}$$

where $R_{11}^{(k-1)}$ is triangular of order $k-1$. One then determines the column of largest 2-norm of $\bar{X}_{22}^{(k-1)}$ and swaps it with the first, along with the corresponding columns of $R_{12}^{(k-1)}$, to get

$$H_{k-1} \cdots H_1 X \Pi_{k-1} \bar{\Pi}_k = \begin{pmatrix} R_{11}^{(k-1)} & \tilde{R}_{12}^{(k-1)} \\ 0 & \tilde{X}_{22}^{(k-1)} \end{pmatrix}.$$

Here $\bar{\Pi}_k$ is a permutation matrix representing the interchange of columns. Now a Householder transformation $\tilde{H}_k = I - \tilde{u}_k \tilde{u}_k^{\mathrm{T}}$ is determined so that

$$\tilde{H}_k \tilde{X}_{22}^{(k-1)} = \begin{pmatrix} r_{kk} & \bar{r}_{k,k+1}^{\mathrm{T}} \\ 0 & \bar{X}_{22}^{(k)} \end{pmatrix}.$$

If we set $u_k^{\mathrm{T}} = (0\ \tilde{u}_k^{\mathrm{T}})$, $H_k = I - u_k u_k^{\mathrm{T}}$, and $\Pi_k = \Pi_{k-k} \bar{\Pi}_k$, then (10) holds with k advanced by one.

If after the kth stage the largest column norm of $\bar{X}_{22}^{(K+1)}$ is sufficiently small, we may terminate the procedure with a truncated approximation to X of rank k, in which the matrix Q^{T} is represented as the product $H_k \cdots H_1$.

In these terms, the pivoting strategy introduced above amounts to pivoting on the column of X for which the corresponding column norm of $\bar{X}_{22}^{(k-1)}$ is maximal. As above, the squares of the norms are (ignoring interchanges) given by

$$\|x_j^{(k-1)}\|^2 = \|x_j\|^2 - \sum_{i=1}^{k-1} r_{ij}^2, \qquad j = k+1, \ldots, p$$

If we know the first $k-1$ rows of R, we can downdate the column norms of X to give the column norms of $\bar{X}_{kk}$.

Moreover, to compute the Householder transformation at the kth stage we need only one column from $\bar{X}_{22}^{(k-1)}$ to compute u_k. This column can be computed on the spot from the corresponding column of X, used to generate u_k, and discarded. Thus there is no need to transform the entire matrix X.

These ideas are best implemented using an elegant representation for the product of Householder transformations due to Schreiber and Van Loan [3]. Specifically, if we set

$$U^{(k)} = (u_1 \ \cdots \ u_k),$$

then

$$(I - u_k u_k^{\mathrm{T}}) \cdots (I - u_1 u_1^{\mathrm{T}}) = U^{(k)} T^{(k)\mathrm{T}} U^{(k)\mathrm{T}}, \tag{11}$$

where the unit lower triangular matrix T_k can be generated by the following recurrence:

$$T^{(k)} = \begin{pmatrix} T^{(k-1)} & -T^{(k-1)} U^{(k-1)\mathrm{T}} u_k \\ 0 & 1 \end{pmatrix}. \tag{12}$$

We will call (11) a UTU representation of the product on the left-hand side.

If we maintain the pivoted Q-factor of X in UTU form, we can compute the truncated pivoted factorization without modifying X. Specifically, suppose that just before the kth step, we have $U^{(k-1)}$, $T^{(k-1)}$, and an auxiliary matrix $S^{(k-1)} = U^{(k-1)\mathrm{T}} X$, along with the current downdated column norms of X. Then the pivot column can be determined from the downdated norms. If x denotes the value of this column in the original matrix X, the transformed value can be calculated as $\bar{x} = x - U^{(k-1)} T^{(k-1)\mathrm{T}} U^{(k-1)\mathrm{T}} x$. From this the kth Householder vector u_k can be computed, and $U^{(k-1)}$, $T^{(k-1)}$, and $S^{(k-1)}$ can be updated. Finally, if $u_{(k)}^{\mathrm{T}}$ denotes the kth row of $U^{(k)}$ and $x_{(k)}^{\mathrm{T}}$ denotes the kth row of X, then the kth row of R can be computed in the form $x_{(k)}^{\mathrm{T}} - u_{(k)}^{\mathrm{T}} T^{(k)} S^{(k)}$. From this the new column norms can be computed and the process repeated.

The code in Fig. 4 implements this sketch. As with the Gram–Schmidt version, the algorithm is quite efficient in terms of both operations and storage. Excepting interchanges, which we have incorporated explicitly to simplify the exposition, the array X is never modified. One column of it is used at each stage to compute the current Householder transformation, and the vector-matrix product $u^{\mathrm{T}} X$ must be formed to update the auxiliary matrix S. If the additional storage for S is too burdensome, the rows of R may be dropped after they have been used to downdate the column norms, and R can be reconstituted in S when the algorithm terminates. At this time one may go on to compute the first k columns of Q explicitly. Alternatively, one can leave the entire matrix Q in UTU form.

Given an $n{\times}p$ $(n \geq p)$ matrix X this algorithm returns a truncated pivoted QR decomposition of X. The Q-factor is returned in UTU form. Initially the matrices U, T, R, and S are void.

$\nu_j = \|X[:,j]\|^2, \quad j = 1,\ldots,p$
Determine an index p_1 such that ν_{p_1} is maximal
for $k = 1$ **to** p
    $X[:,k] \leftrightarrow X[:,p_k]$
    $R[1{:}k{-}1,k] \leftrightarrow R[1{:}k{-}1,p_k]$
    $S[1{:}k{-}1,k] \leftrightarrow S[1{:}k{-}1,p_k]$
    $\bar{x} = X[:,k] - U[:,1{:}k{-}1]{*}T^{\mathrm{T}}{*}U[:,1{:}k{-}1]^{\mathrm{T}}{*}X[:,k]$
    Determine u from $\bar{x}$
    $T = \begin{pmatrix} T & -T{*}U^{\mathrm{T}}{*}u \\ 0 & 1 \end{pmatrix}$
    $U = (U \;\; u)$
    $S = \begin{pmatrix} S \\ u^{\mathrm{T}}X \end{pmatrix}$
    $R = \begin{pmatrix} R \\ X[k,:] - U[k,:]{*}T^{\mathrm{T}}{*}S \end{pmatrix}$
    $\nu_j = \nu_j - R[k,j]^2, \quad j = k{+}1,\ldots,n$
    Determine an index $p_{k+1} \geq k{+}1$ such that $\nu_{p_{k+1}}$ is maximal
    if ($\nu_{p_{k+1}}$ is sufficiently small) **leave** k **fi**
end for k

Fig. 4. Truncated pivoted QR decomposition: UTU version

References

1. Björck, Å (1996): Num. Methods Least Squares Problems. SIAM, Philadelphia
2. Quintana-Ortí, G., Sun, X., Bischof, C.H. (1998): A BLAS-3 version of the QR factorization with column pivoting. SIAM J. Scientific Comput. **19**, 1486–1494
3. Schreiber, R., Van Loan, C.F. (1989): A storage efficient WY representation for products of Householder transformations. SIAM J. Scientific Stat. Comput. **10**, 53–57

12.12. [GWS-J118] (with M. W. Berry and S. A. Pulatova) "Algorithm 844: Computing Sparse Reduced-Rank Approximations to Sparse Matrices"

[GWS-J118] (with M. W. Berry and S. A. Pulatova) "Algorithm 844: Computing Sparse Reduced-Rank Approximations to Sparse Matrices," *ACM Transactions on Mathematical Software (TOMS)* **31** (2005) 252–269.
http://doi.acm.org/10.1145/1067967.1067972

Algorithm 844: Computing Sparse Reduced-Rank Approximations to Sparse Matrices

MICHAEL W. BERRY and SHAKHINA A. PULATOVA
University of Tennessee, Knoxville
and
G. W. STEWART
University of Maryland, College Park

In many applications—latent semantic indexing, for example—it is required to obtain a reduced rank approximation to a sparse matrix A. Unfortunately, the approximations based on traditional decompositions, like the singular value and QR decompositions, are not in general sparse. Stewart [(1999), 313–323] has shown how to use a variant of the classical Gram–Schmidt algorithm, called the quasi–Gram-Schmidt–algorithm, to obtain two kinds of low-rank approximations. The first, the SPQR, approximation, is a pivoted, Q-less QR approximation of the form $(XR_{11}^{-1})(R_{11}\ R_{12})$, where X consists of columns of A. The second, the SCR approximation, is of the form the form $A \cong XTY^{\mathrm{T}}$, where X and Y consist of columns and rows A and T, is small. In this article we treat the computational details of these algorithms and describe a MATLAB implementation.

Categories and Subject Descriptors: G.1.3 [**Numerical Analysis**]: Numerical Linear Algebra—*Sparse, structured, and very large systems (direct and iterative method)*

General Terms: Algorithms

Additional Key Words and Phrases: Sparse approximations, Gram–Schmidt algorithm, MATLAB

1. INTRODUCTION

In a number of applications [Berry et al. 1999; Jiang and Berry 2000; Stuart and Berry 2003; Berry and Martin 2004] one is given a large matrix A and wishes

The research of M.W. Berry and S. A. Pulatova was supported in part by the National Science Foundation under grant CISE-EIA-99-72889. The research of G. W. Stewart was supported in part by the National Science Foundation under grant CCR0204084. Part of this work was performed as faculty appointee at the Mathematical and Computational Sciences Division of the National Institute for Standards and Technology.
Authors' addresses: M. W. Berry and S. A. Pulatova, Department of Computer Science, 203 Claxton Complex, University of Tennessee, Knoxville, TN 37996; email: {berry, pulatova}@cs.utk.edu; G. W. Stewart, Department of Computer Science and Institute for Advanced Computer Studies, University of Maryland, College Park, MD 20742; email: stewart@cs.umd.edu.

ACM Transactions on Mathematical Software, Vol. 31, No. 2, June 2005, Pages 252–269.

to find a reduced-rank approximation to A. This approximation is invariably expressed in the form

$$A \cong XTY^{\mathrm{T}} \tag{1}$$

where X and Y are full-rank matrices and T is nonsingular (T may be the identity matrix). When A is $m \times n$ and T is of order k, this approximation requires $(m+n+k)k$ words to store, as opposed to mn for the full A. Moreover, the matrix-vector product Ax requires $(m+n+k)k$ additions and multiplications to compute, as opposed, again, to mn additions and multiplications for the full A. Clearly, if k is small, great savings are to be had by using the reduced-rank approximation (1).

A widely used reduced-rank approximation is the truncated singular value decomposition, which is known to be optimal in the sense that the Frobenius norm $\|A - XTY^{\mathrm{T}}\|$ is minimized. There are stable direct methods for its computation; however, these methods compute the full decomposition and are not suitable for very large matrices. Fortunately, there are iterative methods that produce the approximation (1) without having to compute the full SVD. These methods require only the formation of matrix-vector products and do not alter A.

An alternative is the pivoted QR decomposition, which generally gives results comparable to the SVD (see, e.g., Stewart [1980] or Stewart [1998, §5.2]). For large A, the Gram–Schmidt algorithm can be adapted to compute this decomposition. Again, A is not altered, and the principle operations are matrix-vector multiplications. This article is concerned with elaborations of this approach to reduced-rank approximations.

When A is large and sparse the situation is not as simple. For A, the storage and operation counts given above become proportional to the number of nonzero elements in A. Since the factors X, T, and Y are generally not sparse, the storage and operation counts for the approximation remain the same. Thus as k increases, we will reach a point where it becomes necessary to abandon the factored form. Note that we do not have the ability to choose k, since the accuracy required of the approximation, which depends on k, is governed by the application.

In this article we are going to describe two approximations based on the pivoted QR decomposition that produce approximations in which X or both X and Y are sparse. The first approximation is called *sparse pivoted QR approximation* (SPQR). It is computed by an algorithm, called the *quasi-Gram–Schmidt* algorithm, that produces a factorization in which X consists of a selection of columns of A. In the second approximation, called the *sparse column-row (SCR) approximation,* X consists of a selection of the columns of A, and Y consists of a selection of the rows of A, so that when A is sparse so are both X and Y. These methods were first described by Stewart [1999], and the quasi-Gram–Schmidt method has been analyzed in Stewart [2004].

At this point we should mention the sparse low-rank approximation of Zhang, Zha, and Simon [Zhang et al. 2002]. The idea is to approximate the dominant singular vectors by sparse vectors x and y and form $\hat{A} = A - \delta xy^{\mathrm{T}}$, where δ is chosen to minimize the Frobenius of $\hat{A}$. The process is then repreated recursively on $\hat{A}$, until a satisfactorily accurate approximation is obtained. Experimental results are promising; and the algorithm should be considered a serious alternative to the approximations given here.

The purpose of this article is to give the computational details leading to the accompanying MATLAB functions for the SPQR and the SCR approximations. In the next section we introduce the pivoted QR decomposition. In Section 3 we derive the quasi-Gram–Schmidt method and apply it to the computation of the sparse pivoted QR approximation. In Section 4 we show how to compute the SCR approximation. We also show how it can be applied to an information retrieval process known as latent semantic indexing. The implementation details for our algorithms are described in Section 5. In Section 6 we compare the SPQR approximation with the singular value decomposition. Finally, in Section 7 we discuss some sparsity issues that arise in producing implementations in compiled programming languages like C or Fortran. The MATLAB programs are listed in appendices.

Throughout this article, $\|\cdot\|$ will denote the Frobenius norm defined by

$$\|A\|^2 = \sum_{i,j} a_{ij}^2,$$

and $\|\cdot\|_2$ the spectral norm defined by

$$\|A\|_2 = \max_{\|x\|=1} \|Ax\|.$$

2. THE PIVOTED QR FACTORIZATION

As above, let A be an $m \times n$ matrix, not necessarily sparse. A pivoted QR (PQR) factorization has the form

$$AP = QR, \tag{2}$$

where P is a permutation matrix, Q is orthonormal, and R is upper triangular. The exact form depends on the the sizes of m and n. If $m \geq n$, then Q is $m \times n$ and R is $n \times n$. If $m < n$, then Q is $m \times m$ and R is $m \times n$. Although our algorithms apply to both cases, for ease of exposition we will assume that $m \geq n$ in what follows.

A rank k approximation to A can be obtained by partitioning the factorization (2). Let $B = AP$ and write

$$\begin{pmatrix} B_1^{(k)} & B_2^{(k)} \end{pmatrix} = \begin{pmatrix} Q_1^{(k)} & Q_2^{(k)} \end{pmatrix} \begin{pmatrix} R_{11}^{(k)} & R_{12}^{(k)} \\ 0 & R_{22}^{(k)} \end{pmatrix}, \tag{3}$$

where $B_1^{(k)}$ has k columns. Then our approximation is

$$\tilde{B}^{(k)} = Q_1^{(k)} \begin{pmatrix} R_{11}^{(k)} & R_{12}^{(k)} \end{pmatrix}. \tag{4}$$

Note that

$$B - \tilde{B}^{(k)} = Q_2^{(k)} \begin{pmatrix} 0 & R_{22}^{(k)} \end{pmatrix}.$$

Since $Q_2^{(k)}$ is orthonormal, the error in $\tilde{B}^{(k)}$ as an approximation to B is

$$\left\|B - \tilde{B}^{(k)}\right\| = \left\|R_{22}^{(k)}\right\|. \tag{5}$$

We will not compute the entire decomposition (3). Rather we will bring in columns of A one at a time and use each to compute an additional column of Q and row of R. Thus at the end of the kth step of this algorithm we will have computed the approximation (4).

The process of selecting columns is called *column pivoting*, or for short simply *pivoting*. The order in which the columns are selected determines the permutation P. Equation (5) suggests that at the beginning of the kth step we should choose the column of A in such a way as to make $\|R_{22}^{(k)}\|$ small. The classical choice is to bring in the column of A that corresponds to the column of $R_{22}^{(k-1)}$ of largest norm. (For more on this choice see Björck [1996]; Stewart [1998].)

Surprisingly, we can implement this strategy without computing $R_{22}^{(k-1)}$ itself. Consider the partition (3), in which the superscripts (k) are replaced by $(k-1)$. Let b_j and r_j denote the jth columns of B and R. Because Q is orthonormal, we have

$$\|b_j\| = \|r_j\|. \tag{6}$$

Now for $j \geq k$ partition

$$r_j = \begin{pmatrix} r_1^{(j)} \\ r_2^{(j)} \end{pmatrix}$$

where $r_1^{(j)}$ has $k-1$-components. Thus $r_1^{(j)}$ is the jth column of $R_{12}^{(k-1)}$, and $r_2^{(j)}$ is the jth column of $R_{22}^{(k-1)}$. It then follows that

$$\left\|r_2^{(j)}\right\|^2 = \|b_j\|^2 - \left\|r_1^{(j)}\right\|^2 = \|b_j\|^2 - r_{1j}^2 - r_{2j}^2 - \cdots - r_{k-1,j}^2. \tag{7}$$

Thus at each stage we can compute the squares of the norms of the columns of $R_{22}^{(k-1)}$. Moreover, the sum of these numbers is $\|R_{22}^{(k-1)}\|^2$, so that by (5) we get, almost for free, the value of the norm of the error in our reduced-rank approximation. This number can be used to determine when we have a satisfactorily accurate reduced-rank approximation.

Unfortunately, the expression (7) has a dark side. If $\|r_2^{(j)}\|^2$ is small compared to $\|b_j\|^2$, there will be cancellation in the computation of $\|r_2^{(j)}\|^2$. In particular, in IEEE double-precision arithmetic, if $\|r_2^{(j)}\|^2 \leq 10^{-16}\|b_j\|^2$ we can expect no accuracy in the computed value. On taking square roots we find that we can use the formula (7) only when

$$\left\|r_2^{(j)}\right\| > 10^{-8}\|b_j\|.$$

This means that if all the columns of A have norm one, we cannot reliably compute reduced-rank approximation that is more accurate than 10^{-8}. However, this accuracy is usually more than enough.

3. THE QUASI-GRAM–SCHMIDT METHOD

In this section we will describe the quasi-Gram–Schmidt method. We will begin with a description of the classical Gram–Schmidt method.

Suppose we have a QR factorization

$$B = QR \tag{8}$$

of B and wish to compute a QR factorization

$$(B\ a) = (Q\ q)\begin{pmatrix} R & r \\ 0 & \rho \end{pmatrix}$$

of $(B\ a)$. The second column of this equality gives us the relation

$$a = Qr + \rho q.$$

Since $Q^{\mathrm{T}}Q = I$ and $Q^{\mathrm{T}}q = 0$, we have

$$r = Q^{\mathrm{T}}a. \tag{9}$$

Since $\|q\| = 1$, we have

$$\rho = \|a - Qr\| \tag{10}$$

and

$$q = \rho^{-1}(a - Qr). \tag{11}$$

Equations (9), (10), and (11) are effectively an algorithm for extending our original QR factorization.

Unfortunately, cancellation in the formation of $a - Qr$ can cause the computed q to be far from orthogonal to the columns of Q. The cure for this problem is *reorthogonalization*, in which the process is repeated on $a - Qr$. Specifically, we have the following algorithm, in which we use MATLAB notation.

```
1. r = Q'*a
2. q = a - Q*r
3. s = Q'*q
4. r = r + s                                  (12)
5. q = q - Q*s
6. rho = norm(q)
7. q = q/rho
```

Typically, this algorithm produces a q that is orthogonal to the columns of Q to working accuracy.[1]

We can use this algorithm to compute a PQR factorization of A simply by selecting columns of A and updating the QR factorization of B. To start the process off, one sets $R_{11}^{(1)} = \rho = \|a\|$, and $Q_1^{(1)} = \rho^{-1}a$, where a is the first column

[1]In extremely unlikely cases the algorithm may produce a zero q at step 2 or in which case special action must be taken (see Stewart [1998, §5.1.4]). In practice, most programs that use Gram–Schmidt orthogonalization ignore this problem.

selected from the columns of A. In this way we compute the decompositions

$$B_1^{(k)} = Q_1^{(k)} R_{11}^{(k)},$$

where $R_{11}^{(k)}$ is $k \times k$.

A problem with this algorithm is that it only computes the factor $R_{11}^{(k)}$ in (3). However it is easy to see that row k of $R_{12}^{(k)}$ is simply $q_k^{\mathrm{T}}\hat{A}$, where q_k is the kth column of $Q_1^{(k)}$ and $\hat{A}$ consists of the $n-k$ columns of A that are not in $B_1^{(k)}$. If n is large, this computation may be the most expensive part of the algorithm. Note that even if we do not want the $R_{12}^{(k)}$ we must still form (and discard) the product $q_k^{\mathrm{T}}\hat{A}$ in order to compute the column norms of $R_{22}^{(k)}$ as described in the last section.

Returning now to the Gram–Schmidt algorithm, we note that even if A is sparse, Q is in general not sparse. If m is very large, we may be unable to store Q. To circumvent this problem, we observe that it follows from (8) that

$$Q = BR^{-1}.$$

Consequently, we can form the product `Q'*a` in (12) by the following algorithm.

```
d = a'*B
r = (d/R)'
```

Similarly we can form the product `Q*r` by

```
p = R\r
q = B*p
```

This leads to the following *quasi-Gram–Schmidt step* (in which we have put the code for the classical Gram-Schmidt step on the right).

```
d = a'*B                 r = Q'*a
r = (d/R)'               q = a - Q*r
p = R\r
q = a - B*p
d = q'*B                 s = Q'*q
s = (d/R)'                                  (13)
r = r + s                r = r + s
p = R\s                  q = q - Q*s
q = q - B*s
rho = norm(q)            rho = norm(q)
```

This code computes only `r` and `rho`—the quantities needed to update `R`. It does not compute `q` in the form `q/rho`, as does the classical Gram–Schmidt algorithm. Instead, `q` is defined by the relation

$$q = (B \ a)\begin{pmatrix} R & r \\ 0 & \rho \end{pmatrix}^{-1} = \rho^{-1}(a - BR^{-1}r), \tag{14}$$

and is so computed in our algorithms.

The quasi-Gram–Schmidt step can be applied successively to columns of A, as described above for the classical Gram–Schmidt algorithm, to produce a pivoted, Q-less PQR factorization, which we will call a *sparse-PQR (SPQR) factorization.*

We will call the corresponding approximation $(B_1^{(k)} R_{11}^{(k)-1})(R_{11}^{(k)}\ R_{12}^{(k)})$ the *SPQR approximation*. The algorithm not only dispenses with the storage for Q, but it replaces dense products involving Q with sparse products involving columns of A. The only strictly dense operations involve R_{11} and R_{12}. But since the order of $(R_{11}\ R_{12})$ is $k \times n$, if $m \gg n$ these operations account for little of the total work.

Once again there is a dark side—there may be a progressive loss of orthogonality in the matrix BR^{-1}. However, an analysis of the quasi-Gram–Schmidt algorithm [Stewart 2004] shows that the loss of orthogonality is proportional to the condition number $\|R\|\|R^{-1}\|$ of R, which is usually good enough.

A nice feature of the SPQR approximation (and QR, approximations in general) is that having computed an approximation of order k, one has immediately all the approximations of order $\ell < k$. Simply, work with the first ℓ columns of B and rows of R.

4. SPARSE COLUMN-ROW APPROXIMATIONS

When $m \gg n$, the SPQR approximation is satisfactory. But when m and n are nearly equal, the storage of R becomes a problem. We can circumvent this problem at the cost of performing another factorization.

Specifically, first apply the quasi-Gram–Schmidt algorithm to the columns of A to get a representative set of columns X of A and an upper triangular matrix R corresponding to R_{11}. Let the error in the corresponding reduced-rank decomposition be ϵ_{col}. Now apply the same algorithm to A^{T} to get a representative set Y^{T} of rows and another upper triangular matrix S. Let the error be ϵ_{row}. We then seek a matrix T such that

$$\|A - XTY^{\text{T}}\|^2 = \min.$$

In Stewart [1999] it is shown that the minimizer is

$$T = R^{-1}R^{-\text{T}}(X^{\text{T}}AY)S^{-1}S^{-\text{T}}.$$

Moreover,

$$\|A - XTY^{\text{T}}\|^2 \leq \epsilon_{\text{col}}^2 + \epsilon_{\text{row}}^2. \tag{15}$$

We will call this approximation a sparse column-row approximation, or for short an SCR approximation. Such approximations are economical to use. For example, to compute $y = XTY^{\text{T}}x$ we compute $r = Y^{\text{T}}x$, $s = Tr$, and $y = Xs$. This requires two sparse matrix-vector multiplications and one dense matrix-vector multiplication in which the matrix is small.

It may happen that X and Y do not have the same number of columns, in which case T will not be square. This causes no problems in matrix-vector multiplications.

Some care must be taken in computing the matrix T. The crux of the matter is to form $X^{\text{T}}AY$ correctly. If, for example, m is large and we first calculate AY, we end up with a large, potentially full matrix. The cure for this problem is to partition Y by columns, writing

$$X^{\text{T}}A(y_1\ y_2\ \cdots\ y_k).$$

We can then calculate $X^{\mathrm{T}}AY$ column by column as follows.

```
1. T = [];
2. for j=1:k
3.     T = [T, X'*(A*Y(:,j)];
4. end
```

A variant of this decomposition may be useful in latent semantic indexing (LSI), a device for retrieving documents from a query vector of terms [Berry and Browne 1999; Berry et al. 1995, 1999; Deerwester et al. 1990]. Briefly (here we follow Deerwester et al. [1990]), one starts with a term-document matrix A whose (i, j)-element is the number of times term i occurs in document j. One then calculates the singular value approximation

$$A = U_k \Sigma_k V_k^{\mathrm{T}}. \tag{16}$$

In the parlance of LSI, the columns of U_k are called term vectors and columns of V_k^{T} are called document vectors. Given a query vector q of terms, one computes a corresponding document vector by the formula

$$d = \Sigma_k^{-1} U_k^{\mathrm{T}} q$$

and compares it with the columns of V_k to determine which columns are related to the the query vector q. For example, one might compute the cosines of the angles between d and the columns of V_k and choose the columns corresponding to the larger ones.

We can rewrite the XTY^{T} in the form of (16). Specifically,

$$XTY^{\mathrm{T}} = (XR^{-1})(R^{-\mathrm{T}}X^{\mathrm{T}}AYS^{-1})(S^{-\mathrm{T}}Y^{\mathrm{T}}) \equiv PWQ^{\mathrm{T}}.$$

Now mathematically, P and Q are orthogonal. Consequently, if we compute the singular value decomposition $W = M\Sigma N^{\mathrm{T}}$ of W and set

$$U = PM \quad \text{and} \quad V = QN,$$

then U and V are orthonormal, and

$$XTY^{\mathrm{T}} = U\Sigma V^{\mathrm{T}}. \tag{17}$$

We can use this decomposition as described above to perform LSI. Of course we do not explicitly form U and V; rather we keep and apply them in factored form:

$$U = XR^{-1}M \quad \text{and} \quad V = YS^{-1}N.$$

It should be stressed that the relation between (17) and (16) is purely formal. It is an open question whether LSI performed using the former will share the good properties of ordinary LSI. Theorem 6.1 below encourages us to conjecture that it will.

5. A MATLAB IMPLEMENTATION

In this section we will describe a MATLAB function spqr to compute SPQR approximations. The basic algorithm is simple and the chief implementation problem is how to package it. We begin by looking at the input parameters.

The essential input is the matrix A and a tolerance `tol` to tell when to stop the factorization. Although the algorithm always terminates after a finite number of steps, if `tol` is too small, `spqr` may be committed to performing an unacceptably large number of operations. For this reason, a third parameter `maxcol` puts an upper bound on the number of columns of A to be used in the approximation. Of course, one can always set `maxcol` greater than or equal to n, it which case it has no effect. The program terminates at the first step k for which $\|R_{22}^{(k)}\| < \texttt{tol}$. Consequently, if `tol` is zero, `spqr` is forced to include `maxcol` columns of A.

Thus the basic calling sequence for `spqr` is

```
spqr(A, tol, maxcol)
```

Although A is presumed to be a MATLAB sparse matrix, `spqr` also works when A is dense. We will now turn to the output parameters.

When `spqr` finishes we need to know four things.

(1) The number columns of A involved in the approximation.
(2) The matrix $(R_{11} \;\; R_{12})$.
(3) The relation of the columns of $(R_{11} \;\; R_{12})$ to those of A.
(4) The error in the approximation.

The first is returned in the output parameter `ncols`. The second in the output matrix R.

The third and fourth items are connected with the way `spqr` implements pivoting. The function begins with two arrays: `colx`, which is initialized to `1:n`, and `norms`, which is initialized so that `norms(j)` is the norm of the `j`th column of A. At step `k`, `spqr` determines the first index `j` $\geq$ `k` for which `norms(j)` is maximal and swaps components `j` and `k` of both `colx` and `norms`, along with the corresponding columns of R. The column `A(:,colx(j))` is then used to advance the approximation. After the quasi-Gram–Schmidt orthogonalization has been computed the elements of the `k`th row of $R_{22}^{(k)}$ are computed and used to downdate `norms(k+1:n)`. The error in the current approximation is also computed and stored in `norms(k)`.

From this description it follows that on return

$$B = A(:, \texttt{colx}),$$

which provides the relation between R and A. Moreover, the error in the approximation is `norms(ncols)`. However, we get a little more. For `j` $\leq$ `ncols`, the arrays `colx` and `R(1:j,:)` contain the SPQR approximation associated with `A(:,colx(1:j))`, and by construction its error is `norms(j)`. Thus by setting `tol` to zero, we can track the quality of all the approximations from 1 to `maxcols`.

The function `spqr` has three optional input arguments used to fine tune the decomposition. The first `fullR` has a default value of 1 (`true`). If it is present and 0 (`false`), then only $R_{11}^{(\texttt{ncols})}$ is computed. This is useful when the primary concern is with the space spanned by the columns `A(:,colx(1:ncols))`. The second optional parameter `pivot` has the default value 1. If it is present and 0 pivoting is suppressed—the columns of A are processed in their natural order. Finally, the optional parameter `cn` (for compute norms) has a default value of 1.

If it is present and `fullR | pivot | cr` is zero, then the computation of `norms` is suppressed and on return `norms = []`.

Thus the final calling sequence is

```
[ncols, R, colx, norms] = spqr(A, tol, maxcol, fullR, pivot, cn)
```

in which `fullR`, `pivot`, and `cn` are optional. For a concise summary see the prologue to `spqr`.

6. COMPARISON WITH THE SVD

In this section we will make some theoretical and timing comparisons between the quasi-QR and the SVD reduced-rank approximations. SVD approximations are rightly regarded as the ones to beat. Gaps in the singular values reveal numerical rank with great reliability, and the reduced rank-approximations it produces are optimal. However, as we shall see, it can be expensive to compute.

Since the SVD is so highly regarded, it is sometimes objected that other approximations may not reproduce the row and column spaces from the SVD to sufficient accuracy. This is particularly important in applications where we are not interested in the approximations themselves but in the subspaces they define. We are now going to show that if any reduced-rank approximation is accurate then it contains good approximations to the singular vectors corresponding to large singular values.

THEOREM 6.1. *Let $A = XY^{\mathrm{T}} + E$. Let $\mathcal{X}$ be the space spanned by the columns of X, and $\mathcal{Y}$ be the space spanned by the columns of Y. Let $\sigma > 0$ be a singular value of A with normalized left and right singular vectors u and v, so that $Av = \sigma u$ and $u^{\mathrm{T}}A = \sigma v^{\mathrm{T}}$. Then*

$$\sin\angle(u, \mathcal{X}),\ \sin\angle(v, \mathcal{Y}) \leq \frac{\|E\|_2}{\sigma}. \tag{18}$$

PROOF. We will establish the first inequality, the second being established similarly. Let $X_\perp$ be an orthonormal basis for the orthogonal complement of $\mathcal{X}$. Then $\|X_\perp^{\mathrm{T}}u\|$ is the sine of the angle between u and $\mathcal{X}$ [Stewart 2001, §4.2.a]. Now

$$X_\perp^{\mathrm{T}}Av = \sigma X_\perp^{\mathrm{T}}u,$$

and

$$X_\perp^{\mathrm{T}}Av = X_\perp^{\mathrm{T}}XY^{\mathrm{T}}v + X_\perp^{\mathrm{T}}Ev = X_\perp^{\mathrm{T}}Ev,$$

since by construction $X_\perp^{\mathrm{T}}X = 0$. It then follows that $\sigma X_\perp^{\mathrm{T}}u = X_\perp^{\mathrm{T}}Ev$, whence $\|X_\perp^{\mathrm{T}}u\|_2 \leq \|E\|_2/\sigma$, which is just the first inequality in (18). □

This theorem says that if a reduced-rank approximation is accurate, then its column space must contain accurate approximations to the left singular vectors corresponding to singular values that are large compared to $\|E\|$. An analogous statement is true of the row space and the right singular vectors.

Turning now to timing examples, we will use matrices A of order $n = 10{,}000$ generated by the MATLAB function `sprandn`, which produces a "random" sparse matrix with a given distribution of singular values (for more, issue the command edit sprandn in MATLAB). The first distribution we consider is given by the MATLAB statement

```
s = logspace(0, -6, n)
```

Thus the common logarithms of the singular values are equally spaced between 0 and −6. For ncr=10:5:40 we timed the call

```
[ncr, cx, nr, rx, T, rsd] = cra(A, 1e-5, ncr);
```

which will produce an approximation of rank ncr. We also timed the MATLAB function

```
[U, S, V] = svds(A, ncr);
```

which produces the matrices required to compute an SVD approximation of rank ncr.

The results are summarized in the following table, in which the time is reported in seconds.

ncr	SPQR	SVD
10	2.6	42.4
15	3.0	35.7
20	3.4	52.6
25	3.7	57.3
30	4.1	70.5
35	4.4	91.4
40	4.8	120.0

It is seen that the SVD times are worse by factors ranging from 16 for ncr $=$ 10 to 25 for ncr $=$ 40. Regarding storage, the SVD requires $(n+m)k$ floating-point words, whereas the SQR requires only k^2 words.

In the above example the singular values of the test matrix had no gaps, and consequently the reduced-rank approximations are not very good—either for the SVD or the SPQR approximations. In a different experiment, we generated singular values by the statements

```
s = logspace(0, -4, n);
s(20:n) = 1e-6*s(20:n);
```

This places a multiplicative gap of about 10^{-6} between the 19th and 20th singular values. For nc $=$ 19, 20 we timed the call

```
spqr(A, 1.e-2, nc, 1)
```

and the above call to svds. The results were

nc	SPQR	SVD
19	1.8	4.4
20	1.8	323.6

The improved performance for the SVD when nc $=$ 19 is explained by the fact that svds is being asked to find a singular subspace whose singular values are well separated from the remaining singular values. Under such circumstances

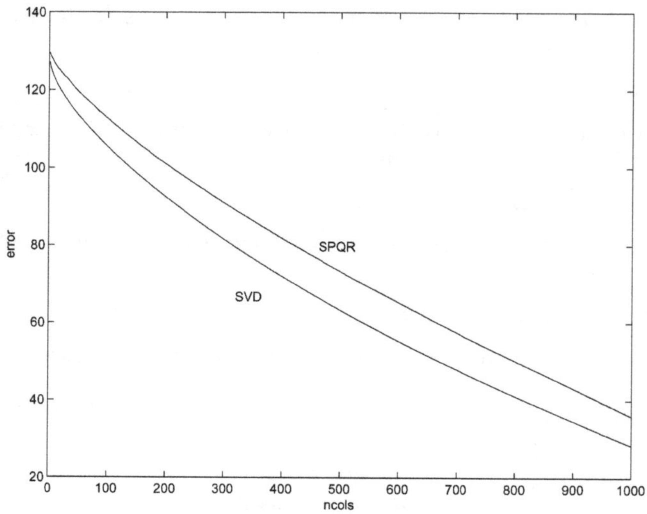

Fig. 1. Error in SVD and SPQR approximation for CRAN.

iterative methods for the SVD converge rapidly. The dismal performance of the SVD for nc = 20 is harder to explain. The function svds is being asked to find the 20th singular value, which is small compared with $\|A\|$ and is poorly separated from the other small singular values. Experience has shown this to be a difficult task. Be that as it may, the 20th singular value has to be computed to estimate the error in the approximation with nc = 19.

As a final example we consider a term-document matrix from an LSI application.[2] The matrix (or rather its transpose) is 4,612 by 1,398. Figure 1 graphs the error (Frobenius norm) in the SVD and SPQR approximations for ncols ranging from 1–1000. The SVD approximation is better, as it must be, but the SPQR approximation tracks it nicely. More specifically from $k = 200$ to $k = 1000$ the ratios of the error in SPQR to those in SVD vary almost linearly between 1.09 and 1.26. The slope of the least squares linear approximation is $2.1 \cdot 10^{-4}$, which implies that the discrepancy between the SVD and SPQR approximations grows very slowly with k.

The SVD approximations were not actually computed: that would have taken too long. Instead the norms of the SVD approximation were computed from the singular values of A, which were computed via the statement;

```
R = qr(A);
sig = svd(full(R));
```

The total time was about 4 minutes. By contrast the time to compute the entire SPQR approximation with 1000 columns (and hence all the decompositions with fewer columns) was about 2.5 minutes.

[2]Specifically the matrix CRAN generated from the Cranfield collection and available at http://www.cs.utk.edu/~lsi/.

```
  1 2 3 4
1 a     b
2   c
3 d
4     e
5 f     g
6     h


val       : a d f c e h b g   (floating-point nnz)
col_start : 1 4 5 7 9         (integer n+1)
rx        : 1 3 5 2 4 6 1 5   (integer nnz)
```

Fig. 2. Compressed column representation of a 6×4 matrix.

7. SPARSITY CONSIDERATIONS

The timings of the last section show that the MATLAB implementation of the SCR approximation is considerably faster than computing the SVD to obtain an equivalent approximation. The code is simple because MATLAB hides the implementation of the sparse matrix-vector multiplications that are at the heart of the algorithms. It is therefore natural to try to improve on the performance of the MATLAB implementation by writing the algorithm—sparse operations and all—in a compiled language like C or Fortran. This section is devoted to sparsity issues that must inform such an attempt.

For definiteness we will consider the problem of computing a pivoted SPQR approximation for a matrix A of order n where n is large. We will assume that the number of columns nc in the approximation is small compared to n. Finally, we will assume that A is represented in compressed column (CC) form, which we will now briefly describe. For definiteness, we will assume 1-based indexing and use MATLAB statements in the examples.

An example of CC representation is given in Figure 2. The nonzero values of the elements of the sparse matrix A are stored in column major order in an array `val`. The length of the array is nnz—the number of nonzero elements of A. An integer array `rx` of length nnz contains the row indices of the corresponding elements in `val`. Another integer array, `col_start`, of length $n + 1$ tells where the columns start in `val` and `rx`. Specifically, the first element in column j is `val[col_start[j]]`. The value of `col_start[n+1]` is set to `nnz+1`. This means that the length of column j is `col_start[j+1]-col_start[j]`.

Formation of the matrix-vector products Ax and $x^{\mathrm{T}}A$ are easy in this representation. The following code computes $y = Ax$

```
y(1:n) = 0;
for j=1:n
   for ii = col_start(j):col_start(j+1)-1;
      i = rx(ii);                                  (19)
      y(i) = y(i) + val(ii)*x(j);
   end
end
```

Similarly, we can compute $y = x^{T}A$ as follows.

```
for j=1:n
   y(j) = 0;
   for ii = col_start(j):col_start(j+1)-1;
      i = rx(ii);
      y(j)  = y(j) + x(i)*val(ii);
   end
end
```
(20)

Both algorithms require nnz additions and multiplications. Both traverse the array val in its natural order, which makes for good cache usage. The first traverses x in its natural order, but its references to y jump around; the reverse is true for the second algorithm.

Now if we examine the quasi-Gram–Schmidt algorithm, we find we must perform the following operations involving the matrix A.

(1) Extract the pivot column from A.
(2) Calculate matrix-vector products of the form `x'*A(:,colx(1:k-1))`.
(3) Calculate matrix-vector products of the form `A(:,colx(:,1:k-1))*x`.
(4) Calculate matrix-vector products of the form `x'*A(:,colx(:,k+1:n))`. (As we have mentioned, when n is large, this calculation is the most expensive part of the algorithm.)

CC storage is ideal for performing all these operations. For example, to calculate `A(:,colx(1:k-1))*x`, we need only replace the outer `for` statement in (19) by

```
for j=colx(1:k-1)
```

Again, to compute `x'*A(:,colx(:,k+1:n))` we change the outer `for` statement in (20) with

```
for j=colx(k+1:n)
```

The algorithms no longer access val sequentially, but access within an individual column is concentrated in the contiguous part of val where its elements lie. Thus the CC representation goes hand-in-glove with the computation of the SPQR approximation.

The situation is different when we must compute the SPQR approximation of A^{T}, as is required when we wish to compute an SCR approximation. There are two major alternatives. We can work with A^{T}, or we can write a row-oriented version of spqr that works directly with A.

Regarding the first alternative, the SPARSKIT package by Saad [1994] gives an algorithm for transposing a matrix in condensed format in place.[3] The algorithm requires an additional working integer array of size nnz and O(nnz) operations. A disadvantage is that the elements of each column, though they remain contiguous in val, no longer occur in their natural order. This makes no difference for our algorithms for forming matrix-vector products.

[3]The algorithm assumes condensed row format, but it can easily be adapted to CC format.

The second alternative is to write a row-oriented version of spqr to compute a decomposition of the form

$$\hat{P}^{\mathrm{T}} A = \hat{R}^{\mathrm{T}} \hat{Q}^{\mathrm{T}},$$

where $\hat{P}$ is a permutation, $\hat{R}$ is upper trapezoidal, and $\hat{Q}$ is orthogonal. The algorithm is completely analogous to the SPQR; however, it pivots on rows of A producing a pivot array rowx corresponding to colx in SPQR. The operations we must perform are essentially the transposes of the operations listed above for SPQR.

(1) Extract the pivot row from A.
(2) Calculate matrix-vector products of the form A(rowx(1:k-1),:)*x.
(3) Calculate matrix-vector products of the form x'*A(rowx(1:k-1),:).
(4) Calculate matrix-vector products of the form A(rowx(k+1:n,:))*x.

The natural way to implement the row-oriented algorithm is to transform A into compressed row format. Once again, SPARSKIT provides an algorithm. The advantage of this approach is that the translation from spqr to the row-oriented version is purely mechanical. The disadvantage is that the storage requirements are doubled.

An alternative is to work with the CC format, perhaps augmented by additional arrays. However, this creates difficulties in implementing the row-oriented algorithm. Specifically, consider the adaptation of (19) to compute A(rowx(1:k-1),:)*x.

```
y(1:n) = 0;
for j=1:n
   for ii = col_start(j):col_start(j+1)-1;
      i = rx(ii);
      if i in rowx(1:k-1)                              (21)
         y(i) = y(i) + val(ii)*x(j);
      end
   end
end
```

There are two problems with this algorithm—one easily solved, the other more difficult.

The first problem is that with each iteration of the inner loop rowx(1:k-1) must be searched to determine if it contains i as an entry. The cure is to negate the indices of rx corresponding to row i when row i is brought into the factorization. Then we may replace the the conditional part of the inner loop with

```
if rx(i) < 0
   y(-i) = y(-i) + val(ii)*x(j);
end
```

The second problem is that by our assumptions k $\ll$ n. Now the loops in (21) traverse all the nnz elements in the matrix A. But we actually work with only those few elements in the rows indexed by rowx(1:k-1). In other words, for most of the time, the body of the double loop does nothing. The cure for

```
   1  2  3  4
1  a        b
2     c
3  d
4        e
5  f        g
6        h
```

```
val        :  a d f c e h b g   (floating-point nnz)
col_start  :  1 4 5 7 9         (integer n)
row_index  :  1 3 5 2 4 6 1 5   (integer nnz)
row_start  :  1 3 4 5 6 8 9     (integer m+1)
row_elp    :  1 7 4 2 5 3 8 6   (integer nnz) (elp = element pointer)
col_index  :  1 1 1 2 3 3 4 4   (integer nnz)
```

Fig. 3. Compressed-column representation of a 6×4 matrix with row links.

this problem is to store a copy of the matrix `A(rowx(1:k-1)` in compressed-row format. Because $k \ll n$, the extra storage is insignificant. Moreover, it is then easy to perform the operations `x'*A(rowx(1:k-1),:)` and `A(rowx(1:k-1),:)*x`. We choose the compressed row-form because it is easy to add additional rows to it as k increases.

Now consider the product `A(rowx(k+1:n,:))*x`. Assuming that we have negated the elements of `rx` corresponding to rows `rowx(1:k)`, we can perform this multiplication by modifying the body of the loop (21) as follows.

```
if rx(i) > 0
   y(i) = y(i) + val(ii)*x(j);
end
```

Since $k \ll n$, the body of the loop is performing useful work most of the time.

Surprisingly, the problem of extracting the pivot row from a compressed column form is also difficult. For definiteness, let the index of that row be `ipvt`. The following algorithm does the job.

```
for j=1:n
   for ii=col_start(j):col_start(j+1)-1
      if rx(ii) > ipvt, break, end
      if rx(ii) == ipvt
         % A(ipvt, j) = val(ii) is in row ipvt;
         break;
      end
   end
end
```

Unfortunately, if $\text{ipvt} = n$, we must traverse the entire matrix just to extract the pivot row. Thus the use of this algorithm has the potential to add $O(\text{nzz})$ work at each step of the algorithm.

A solution is to augment the compressed column format to allow access to the rows. Figure 3 shows one such scheme, which we will call compressed-column, linked-row representation (CCLR representation). With it we can access the row `ipvt` as follows.

```
for jjj = row_start(ipvt):row_start(ipvt+1)-1
   jj = row_elp(jjj);
   j = cx(jj);
   % A(ipvt, j) = val(jj) is in row ipvt
end
```

This ability to traverse rows allows one to implement the row-oriented algorithm in exactly the same manner as the column oriented algorithm. However, there are two differences that may affect efficiency. First there are two levels of indirection from `jjj` to `jj` to `j`. Second, row traversals do not access the elements of `val` sequentially. Thus it may still pay to maintain a copy of `A(rowx(1:k-1),:)` and to compute `A(rowx(1:k-1),:)*x` directly from the column oriented form.

To sum up, if we assume that we have 4-byte integers and 8-byte floating-point words, then compressed column storage requires 12 nnz + 4 n bytes of memory. To compute the SPQR approximation of A^T we have the following options.

(1) Transpose A in place. Storage: 16 nnz + 4 n (4 nnz of which is temporary and can be allocated as an automatic variable). Additional work: O(nnz) for the initial transpose.
(2) Copy A to compressed row format. Storage: 24 nnz + 8 n. Additional work: O(nnz) for the conversion.
(3) Use CC representation, and copy `A(:,colx(1:k-1))`. Storage 12 nnz + 4 n. Additional work: up to O(nnz) per step to extract rows.
(4) Use CCLR representation, copy `A(:,colx(1:k-1))`, and use the row links only to extract the pivot row. Storage: 20 nnz + 8 n. Additional work $O(1)$ per step.
(5) Use CCLR representation: 20 nnz + 8 n. Additional work: O(nnz) per step from extra overhead in processing rows.

Items 1, 2, and 4 emerge as the strongest options, playing off storage, work, and ease of programming against each other. Item 1 is attractive because of its low storage requirements and the fact that one does not have to code a row-oriented version of `spqr`. Item 2 doubles the storage, but makes the coding of the row-oriented version trivial. Item 4 almost doubles the storage, and the copying complicates the row-orient algorithm. But it is attractive when additional row operations involving A are anticipated.

It should be stressed that the above analysis was done under a number of special hypotheses—e.g., nc $\ll$ n. Change the hypotheses and the the results may change. Moreover, the nature of the problem may make other storage schemes preferable. However, the analysis illustrates the questions that should be asked by someone implementing the MATLAB algorithms in a language where sparseness must be explicitly taken into account.

REFERENCES

BERRY, M. AND BROWNE, M. 1999. *Understanding Search Engines: Mathematical Modeling and Text Retrieval*. SIAM, Philadelphia, PA.

BERRY, M. W., DRMAČ, Z., AND JESSUP, E. 1999. Matrices, vector spaces, and information retrieval. *SIAM Rev. 41*, 335–362.

BERRY, M. W., DUMAIS, S. T., AND O'BRIEN, G. W. 1995. Using linear algebra for intelligent information retrieval. *SIAM Rev. 37*, 573–595.

BERRY, M. W. AND MARTIN, D. I. 2004. Principal component analysis for information retrieval. In *Handbook of Parallel Computing and Statistics*. Marcel Dekker, New York. To appear.

BJÖRCK, Å. 1996. *Numerical Methods for Least Squares Problems*. SIAM, Philadelphia.

DEERWESTER, S., DUMAIS, S., FURNAS, G., LANDAUER, T., AND HARSHMAN, R. 1990. Indexing by latent semantic analysis. *J. Amer. Soc. Info. Sci. 41*, 391–407.

JIANG, P. AND BERRY, M. W. 2000. Solving total least squares problems in information retrieval. *Lin. Alg. Appl. 316*, 137–156.

SAAD, Y. 1994. SPARSEKIT: A basic tool kit for sparse matrix computations. Available at `www-users.cs.umn.edu/~saad/software/SPARSKIT/sparskit.html`.

STEWART, G. W. 1980. The efficient generation of random orthogonal matrices with an application to condition estimators. *SIAM J. Num. Anal. 17*, 403–409.

STEWART, G. W. 1998. *Matrix Algorithms I: Basic Decompositions*. SIAM, Philadelphia.

STEWART, G. W. 1999. Four algorithms for the the efficient computation of truncated pivoted qr approximations to a sparse matrix. *Numerische Mathematik 83*, 313–323.

STEWART, G. W. 2001. *Matrix Algorithms II: Eigensystems*. SIAM, Philadelphia.

STEWART, G. W. 2004. Error analysis of the quasi-Gram–Schmidt algorithm. Tech. Rep. CMSC TR-4572, Department of Computer Science, University of Maryland.

STUART, G. W. AND BERRY, M. W. 2003. A comprehensive whole genome bacterial phylogeny using correlated peptide motifs defined in a high dimensional vector space. *J. Bioinformatics and Computational Bio. 1*, 475–493.

ZHANG, Z., ZHA, H., AND SIMON, H. 2002. Low-rank approximations with sparse factors I: Basic algorithms and error analysis. *SIAM J. Matrix Anal. Appl. 23*, 706–727.

Received June 2004; revised December 2004; accepted January 2005

13

Papers on Updating and Downdating Matrix Decompositions

1. [GWS-J29] (with W. B. Gragg), "A Stable Variant of the Secant Method for Solving Nonlinear Equations," *SIAM Journal on Numerical Analysis* **13** (1976) 889–903.
2. [GWS-J31] (with J. W. Daniel, W. B. Gragg, L. Kaufman), "Reorthogonalization and Stable Algorithms for Updating the Gram-Schmidt QR Factorization," *Mathematics of Computation* **30** (1976) 772–795.
3. [GWS-J40] "The Effects of Rounding Error on an Algorithm for Downdating a Cholesky Factorization," *Journal of the Institute for Mathematics and its Applications (IMA), Applied Mathematics* **23** (1979) 203–213.
4. [GWS-J73] "An Updating Algorithm for Subspace Tracking," *IEEE Transactions on Signal Processing* **40** (1992) 1535–1541.
5. [GWS-J77] "Updating a Rank-Revealing ULV Decomposition," *SIAM Journal on Matrix Analysis and Applications* **14** (1993) 494–499.
6. [GWS-J87] "On the Stability of Sequential Updates and Downdates," *IEEE Transactions on Signal Processing* **43** (1995) 2642–2648.

13.1. [GWS-J29] (with W. B. Gragg), "A Stable Variant of the Secant Method for Solving Nonlinear Equations"

[GWS-J29] (with W. B. Gragg), "A Stable Variant of the Secant Method for Solving Nonlinear Equations," *SIAM Journal on Numerical Analysis* **13** (1976) 889–903.
http://dx.doi.org/10.1137/0713070

SIAM J. NUMER. ANAL.
Vol. 13, No. 6, December 1976

A STABLE VARIANT OF THE SECANT METHOD FOR SOLVING NONLINEAR EQUATIONS*

W. B. GRAGG† AND G. W. STEWART‡

Abstract. The usual successive secant method for solving systems of nonlinear equations suffers from two kinds of instabilities. First the formulas used to update the current approximation to the inverse Jacobian are numerically unstable. Second, the directions of search for a solution may collapse into a proper affine subspace, resulting at best in slowed convergence and at worst in complete failure of the algorithm. In this report it is shown how the numerical instabilities can be avoided by working with factorizations of matrices appearing in the algorithm. Moreover, these factorizations can be used to detect and remedy degeneracies among the directions.

1. Introduction. In this paper we shall be concerned with the successive secant method for solving the system of nonlinear equations

$$f(x)=0, \tag{1.1}$$

where f is a mapping from some domain in real n-space into real n-space ($f\colon D\subset \mathbb{R}^n \to \mathbb{R}^n$). Given approximations $x_1, x_2, \cdots, x_{n+1}$ to a solution of (1.1), a new approximation x_* is generated as follows. Let $l\colon \mathbb{R}^n \to \mathbb{R}^n$ be the affine function that interpolates f at $x_1, x_2, \cdots, , x_{n+1}$; that is,

$$f_i := f(x_i) = l(x_i), \qquad i=1,2,\cdots,n+1. \tag{1.2}$$

Then x_* is taken to be the zero of the function l. If the points $x_1, x_2, \cdots x_{n+1}$ are affinely independent, then l is uniquely defined. The approximation x_* will be uniquely defined provided the vectors $f_1, f_2, \cdots, f_{n+1}$ are affinely independent (cf. (1.4) below). The method derives its name from the fact that the ith coordinate function of l represents the secant hyperplane interpolating the ith coordinate function of f.

Various formulas can be written for the approximation x_* (see [2] for the detailed discussion of secant methods and their convergence theory). We shall use the following representation. Let X be the $n\times(n+1)$ matrix ($X\in\mathbb{R}^{n\times(n+1)}$) defined by

$$X := (x_1, x_2, \cdots, x_{n+1}),$$

and let

$$F := (f_1, f_2, \cdots, f_{n+1}).$$

Define the operator Δ by

$$\Delta X = (x_2-x_1, x_3-x_1, \cdots, x_{n+1}-x_1).$$

* Received by the editors August 19, 1974 and in revised form October 1, 1975.

† Department of Mathematics, University of California at San Diego, La Jolla, California 92037. The research of this author was supported by the Air Force Office of Scientific Research, Air Force Systems Command, USAF, under Grant AFOSR 71-2006.

‡ Department of Computer Science, University of Maryland, College Park, Maryland 20742. The research of this author was supported in part by the Office of Naval Research under Contract N00014-67-A-0128-0018.

Then it is easily verified that the function l defined by

$$l(x)=f_1+\Delta F(\Delta X)^{-1}(x-x_1) \tag{1.3}$$

satisfies (1.2). It follows from solving the equation $l(x)=0$ that

$$x_*=x_1-\Delta X(\Delta F)^{-1}f_1. \tag{1.4}$$

The existence of the inverses in (1.3) and (1.4) is guaranteed by the affine independence of the columns of X and F.

The new approximation x_* will not in general be an exact zero of f, and the process must be repeated iteratively. This may be done in several ways. We shall be concerned with the successive variant in which x_* replaces one of the points x_i. Conventionally this is done in one of two ways. Either x_* replaces x_{n+1}, or x_* replaces that column of X for which the corresponding column of F has largest norm. In any case the iterative process generates sequences of matrices $X_1, X_2, \cdots$ and a corresponding sequence $F_1, F_2, \cdots$ with X_{k+1} differing from X_k in only a single column (in practice it may be necessary to permute the columns of X_k before inserting $x_*^{(k)}$; see § 4.2 below).

When f is differentiable, the matrix $\Delta F(\Delta X)^{-1}$ in (1.4) may be regarded as an approximation to the Jacobian f' of f. Thus the secant formula (1.4) is a discretization of Newton's method, a method that under appropriate conditions converges quadratically to a zero of f. The convergence theory for the successive secant method suggests that if the matrices ΔX_k remain uniformly nonsingular, the n steps of the secant method will be roughly comparable to one step of Newton's method (see [2] and [3]). This has important computational consequences. The ab initio calculation of $(\Delta F)^{-1}f_1$ requires $O(n^3)$ operations (see, e.g., [5]), and therefore n steps of the secant method will require $O(n^4)$ operations, which may be prohibitively large. The usual cure for this problem is to calculate $(\Delta F_{k+1})^{-1}$ directly from $(\Delta F_k)^{-1}$ (actually the inverses of slightly different matrices are calculated). Since F_k and F_{k+1} are simply related, this can be done in $O(n^2)$ operations, giving a satisfactory $O(n^3)$ operation count for n steps of the successive secant method (for the first such implementation see [4]).

The method outlined above has two serious defects. First the scheme for updating $(\Delta F)^{-1}$ is numerically unstable. Second, the columns of the matrices X_k may tend to collapse into proper affine subspaces of $\mathbb{R}^n$, resulting in the prediction of wild points, or at least in slowed convergence. The first problem arises whenever ΔF_k is ill-conditioned. In this case $(\Delta F_k)^{-1}$ is computed inaccurately and these inaccuracies transmit themselves to subsequent inverses, even though the corresponding ΔF's are well-conditioned. The same problem occurs in linear programming (see, e.g., [1]), and one could adopt the usual solution of periodically reinverting ΔF. However, this entails extra work for the reinversion and extra storage to hold the matrix F. Moreover, one must face the tricky problem of deciding when to reinvert.

The problem of degeneracy among the columns of X arises, among other occasions, when one of the component functions of f is linear. Then the linear component and the corresponding component of l, call it l_i, are identical. It follows that x^* lies in the proper affine subspace defined by $l_i(x)=0$. Ultimately all the columns of some X_k must lie in this subspace, and ΔX_k will be singular. The matrix ΔF_k may not be singular, but it will almost certainly be ill-conditioned, and the prediction $x_*^{(k)}$ will be spurious. Moreover, as noted above, the inaccuracies in $(\Delta F_k)^{-1}$ will propagate themselves via the update formulas.

The purpose of this paper is to show how the two problems mentioned above can be resolved by generating and updating QR factorizations of the matrices X_k and F_k. The factorization of F permits the $O(n^2)$ solution of the equation $\Delta Fz = f_1$, which is equivalent to forming $(\Delta F)^{-1}f_1$. The factorization of X enables one to detect degeneracies in the columns of X. Moreover, the factorization can be used to alter a column of X in such a way as to reduce or remove the degeneracy. The factorizations of X_{k+1} and F_{k+1} can be obtained from those of X_k and F_k in $O(n^2)$ operations.

In the next section we shall introduce the factorizations, show how they may be used to execute a step of the secant method, and show how they may be updated. We shall also show that the updating method is numerically stable. In § 3, we shall show how the factorization can be used to detect and remove degeneracies in X. In § 4 some comments on the practicalities of implementing these methods are given, and in § 5 some numerical examples.

2. Factorization. In this section we shall be concerned with the stable implementation of a single secant step. Suppose that at step k we are given nonsingular matrices P_k and Q_k such that the matrices Y_k and G_k defined by

$$X_k = P_k^T Y_k \tag{2.1}$$

and

$$G_k = Q_k F_k \tag{2.2}$$

are upper trapezoidal, i.e. zero below the diagonal. (Numerically the matrices P_k and Q_k will be very nearly orthogonal, but we need not assume so.) Because premultiplication by a matrix acts column by column on the multiplicand, we have

$$\Delta X_k = P_k^T(\Delta Y_k)$$

and

$$\Delta G_k = Q_k(\Delta F_k).$$

Moreover, the matrices ΔY_k and ΔG_k are upper Hessenberg, i.e. zero below the first subdiagonal.

Now let $x_*^{(k)}$ be the vector obtained from a single secant step:

$$x_*^{(k)} = x_1^{(k)} - \Delta X_k(\Delta F_k)^{-1} f_1^{(k)}. \tag{2.3}$$

If we set $y_*^{(k)} = P_k^{-T} x_*^{(k)}$, then (2.3) can be written in the form

$$y_*^{(k)} = y_1^{(k)} - \Delta Y_k(\Delta G_k)^{-1} g_1^{(k)}, \tag{2.4}$$

where $y_1^{(k)}$ and $g_1^{(k)}$ are the first columns of Y_k and G_k. Equation (2.4) suggests the following algorithm.

(2.5)
1. Solve the system $\Delta G_k z = g_1^{(k)}$
2. $y_*^{(k)} = y_1^{(k)} - \Delta Y_k z$
3. $x_*^{(k)} = P_k^T y_*^{(k)}$
4. $f_*^{(k)} = f(x_*^{(k)})$
5. $g_*^{(k)} = Q_k f_*^{(k)}$

This algorithm produces not only the secant approximation $x_*^{(k)}$ but also the function value $f_*^{(k)}$ and its Q-transform $g_*^{(k)}$. Excepting step 4, the bulk of the work done by the algorithm is concentrated in step 1. Since ΔG_k is an upper Hessenberg matrix, step 1 can be accomplished by standard techniques in $O(n^2)$ operations [5, p. 218]. Thus a knowledge of the factorizations (2.1) and (2.2) allows us to compute a secant approximation in $O(n^2)$ operations.

Of course $x_*^{(k)}$ must replace a column of X_k and $f_*^{(k)}$ replace the corresponding column of F_k. This amounts to replacing the same columns of Y_k and G_k by $y_*^{(k)}$ and $g_*^{(k)}$ to give new matrices Y_k^* and G_k^*. In principle, algorithm (2.5) can be applied to these new matrices to give another approximation. In practice, however, G_k^* will no longer be upper trapezoidal and step 1 of (2.5) cannot be effected in $O(n^2)$ operations. To circumvent this difficulty we shall show how to construct orthogonal matrices R_k and S_k such that

$$Y_{k+1} := R_k Y_k^*$$

and

$$G_{k+1} := S_k G_k^*$$

are upper trapezoidal. If we then set

$$P_{k+1} := R_k P_k$$

and

$$Q_{k+1} := S_k Q_k,$$

then the relations (2.1) and (2.2) will be satisfied with k replaced by $k+1$, and algorithm (2.5) may be efficiently reapplied.

For definiteness we shall deal with the computation of R_k and illustrate the general procedure by a specific example. For numerical reasons that will be discussed in § 4, the order of the columns of Y and G cannot be assigned arbitrarily. This means that although $y_*^{(k)}$ may replace, say, column l of Y, it may have to be inserted at some other position, say in column m. In the specific case where $n = 7$, $l = 1$, and $m = 3$, we shift column 2 into column 1, shift column 3 into column 2 and overwrite column 3 with $y_*^{(k)}$. This gives a matrix Y_k^* whose nonzero elements have the distribution

$$
(2.6)\qquad
\begin{array}{cccccccc}
x & x & x & x & x & x & x & x \\
x & x & x & x & x & x & x & x \\
0 & x & x & x & x & x & x & x \\
0 & 0 & x & x & x & x & x & x \\
0 & 0 & x^3 & 0^3 & x & x & x & x \\
0 & 0 & x^2 & 0 & 0^2 & x & x & x \\
0 & 0 & x^1 & 0 & 0 & 0^1 & x & x
\end{array} .
$$

The matrix R_k is computed as the product of 9 plane rotations or Householder transformations: $R_k = H_9 H_8 \cdots H_2 H_1$. In the first stage, the transformations H_1, H_2, and H_3 are chosen in the usual way (see [5, p. 47]) to introduce zeros into the elements of the "stalactite" in column 3. These transformations will enter nonzero elements in the zero positions labeled 1, 2, and 3, so that the matrix will be in Hessenberg form:

$$
\begin{array}{cccccccc}
x & x & x & x & x & x & x & x \\
x^4 & x & x & x & x & x & x & x \\
0 & x^5 & x & x & x & x & x & x \\
0 & 0 & x^6 & x & x & x & x & x \\
0 & 0 & 0 & x^7 & x & x & x & x \\
0 & 0 & 0 & 0 & x^8 & x & x & x \\
0 & 0 & 0 & 0 & 0 & x^9 & x & x
\end{array} .
$$

Now the transformations $H_4, \cdots, H_9$ are chosen to introduce zeros in the elements labeled $4, \cdots, 9$, bringing the matrix to trapezoidal form. The matrix $P_{k+1} = H_9 \cdots H_1 P_k$ can be formed directly by multiplying the transformations into P_k as they are generated. The matrix G_k^* also has the form (2.6) and is updated similarly.

The procedure sketched above is perfectly general. If column l is to be deleted and a vector inserted in column m, the vectors between column l (exclusive) and m (inclusive) are shifted one column toward column l and the new vector is inserted. The matrix is then reduced to triangular form as illustrated above. From the standpoint of operations, the case $l = m = 1$ is the worst, requiring the introduction of $2n - 3$ zeros. In all cases the operation count for the updating is $O(n^2)$.

The method is extremely stable in the sense that there are small matrices Z_k and H_k such that $P_k^T Y_k = X_k + Z_k$ and $Q_k(F_k + H_k) = G_k$. This implies that if no further rounding errors are made in algorithm (2.5), the value of $x_*^{(k)}$ is the value that would have been obtained by taking a secant step with the slightly perturbed matrices $X_k + Z_k$ and $F_k + H_k$.

The derivation of H_k is typical. The errors for each column are independent of one another, and it is sufficient to follow the history of a single column from its insertion as $g_*^{(k)}$. Now $g_*^{(k)}$ is computed according to (2.5). It follows from standard rounding error assumptions [5] that the computed $g_*^{(k)}$ satisfies

$$g_*^{(k)} = Q_k f_*^{(k)} + e_*^{(k)},$$

where

$$\|e_*^{(k)}\| \leqq n^{3/2} \|Q_k\| \|f_*^{(k)}\| \varepsilon.$$

Here $\|\cdot\|$ denotes the spectral norm [5, p. 57] and ε is a small constant that depends on the arithmetic used to compute $g_*^{(k)}$. It follows that

$$g_*^{(k)} = Q_k(f_*^{(k)} + h_*^{(k)}),$$

where

$$\|h_*^{(k)}\| = \|Q_k^{-1} e_*^{(k)}\| \leqq n^{3/2} \|Q_k\| \|Q_k^{-1}\| \|f_*^{(k)}\| \varepsilon. \tag{2.7}$$

Now the matrices Q_k are computed as the product of orthogonal matrices (see § 4.4 below) and will themselves be very nearly orthogonal (for detailed error analyses of orthogonal transformations see [5]). It follows that certainly

$$\|h_*^{(k)}\| \leqq 2n^{3/2} \|f_*^{(k)}\| \varepsilon. \tag{2.8}$$

Thus when $g_*^{(k)}$ is inserted in G_k, the error bound for the corresponding column of H_k^* is satisfactorily small.

As the matrix G_k^* and the subsequent G's are updated, the column of H corresponding to the inserted $g_*^{(k)}$ will grow, but very slowly as an elementary error analysis will show. Even this slow growth might be intolerable over a large number of iterations, but after about n iterations the column is discarded (this may be forced if necessary), and its replacement is born anew with little error. It is true that the matrices P_k and Q_k will slowly deviate from orthogonality, but orthogonality is not required in the above analysis. All that is needed is that P_k and Q_k be well-conditioned so that in the case of Q_k we may pass from (2.7) to (2.8). Since P_k and Q_k are computed as products of orthogonal matrices, their condition cannot deteriorate in any reasonable number of iterations.

Two points in the above analysis bear stressing. First the matrices Z_k and H_z are uniformly bounded, provided no column is retained longer than a fixed number of iterations and the matrices P_k and Q_k remain well-conditioned. In effect we can use and update the factorizations as long as we like. This is especially important in parameterized problems in which the factorizations from the solution of one problem are used to start the solution of a nearby problem (cf. § 4.5). The second point is that the analysis implies that the error in any column will be small compared with the norm of that column. Even if the columns vary widely in size (in the matrix G they will), the error associated with a large column cannot overwhelm a small column.

3. Detecting and correcting degeneracy. As was pointed out in § 1, the columns of X will be affinely dependent whenever ΔX is singular. In this section we shall show how the factorization of X introduced in the last section can be used to tell when ΔX is singular and if necessary remove the singularity by altering a column of X. The method to be used cannot be justified with complete rigor, although a suggestive theorem can be proved.

Actually we shall work with the matrices Y and ΔY, which are the ones that are at hand. There is some ambiguity in speaking of the singularity of ΔY, since its columns may vary widely in size. For the sake of uniformity we shall instead examine the matrix A obtained from ΔY by scaling its columns so they have 2-norm unity:

$$A := \left(\frac{y_2 - y_1}{\|y_2 - y_1\|}, \frac{y_3 - y_1}{\|y_3 - y_1\|}, \cdots, \frac{y_{n+1} - y_1}{\|y_{n+1} - y_1\|}\right). \tag{3.1}$$

There is more than just convention in this choice. The convergence proofs for the secant method require a uniform upper bound on the condition of the matrices A generated by the iteration.

The method for correcting degeneracies may be justified heuristically as follows. If A is nearly singular, then it has approximate left and right null vectors; that is there are vectors u and v with $\|u\| = \|v\| = 1$ such that $\|Au\|$ and $\|v^TA\|$ are small; say they are less than some fixed tolerance α. Now to say that $\|v^TA\|$ is small is to say that v is almost orthogonal to each column of A. Thus the condition of A may be improved by replacing some column with the vector v. However, it is important that v not replace a column that is already independent of the other columns of A. The vector u may be used to find a suitable column. Let u_ν be the component of u that is largest in absolute value: $|u_\nu| \geqq |u_i|$ $(i = 1, 2, \cdots, n)$. Then the νth column of A is given by

$$a_\nu = \frac{Au}{u_\nu} - \sum_{i \neq \nu} \frac{u_i}{u_\nu} a_i. \tag{3.2}$$

Since $|u_\nu| \geqq n^{-1/2}$, the vector Au/u_ν is negligible, and (3.2) effectively expresses a_ν as a linear combination of the other columns of A. Thus v should replace a_ν to give a new matrix A_1.

If A_1 is nearly singular, the process may be reapplied to give a matrix A_2, and so on. The following theorem shows that if α is not too large, then the sequence of matrices A_k so generated must terminate. We establish the result for rectangular matrices with an eye to applications to least squares problems.

THEOREM 3.1. *Let $A_0 \in \mathbb{R}^{m \times n}$ $(m \geqq n)$ have columns of norm unity. Given $\alpha > 0$, generate a sequence $A_0, A_1, \cdots$ of matrices as follows. Let A_k be given and suppose that there are vectors u_k and v_k satisfying*

$$\|u_k\| = \|v_k\| = 1 \tag{3.3}$$

and

$$\|A_k u_k\|, \|A_k^T v_k\| \leqq \alpha. \tag{3.4}$$

Let $u_\nu^{(k)}$ be a maximal component of u_k: $|u_\nu^{(k)}| \geqq |u_i^{(k)}|$ $(i = 1, 2, \cdots, n)$. The matrix A_{k+1} is then the matrix obtained by replacing the ν-th column of A_k by v_k. If there are *no vectors u_k and v_k satisfying* (3.3) *and* (3.4), *end the sequence with A_k. Then if*

(3.5)
$$\alpha < \frac{1}{\sqrt{n}(1+\sqrt{n})},$$

the sequence terminates with some A_k where $k < n$.

Proof. We shall show that in passing from A_k to A_{k+1}, the column that was thrown out must be a column of A_0. This is clearly true for the matrix A_0 itself. Assuming its truth for $A_0, A_1, \cdots, A_{k-1}$, we can, by rearranging the columns of A_k write A_k in the form

$$A_k = (v_0, v_1, \cdots, v_{k-1}, a_{k+1}^{(k)}, \cdots, a_n^{(k)}),$$

where $a_{k+1}^{(k)}, \cdots, a_n^{(k)}$ are columns of A_0. Thus we must show that $u_i^{(k)} (i = 1, 2, \cdots, k)$ cannot be maximal.

The case $i = 1$ is typical. Write A_k in the form $A_k = (v_0, A_2^{(k)})$. Then it follows from (3.4) that

$$\|v_0^T A_2^{(k)}\| \leqq \sqrt{n-1}\,\alpha.$$

But if we write $u_k = (u_1^{(k)}, w_k^T)^T$, then

$$\begin{aligned} \alpha &\geqq |v_0^T A u_k| = |v_0^T v_0 u_1^{(k)} + v_0^T A_2^{(k)} w_k| \\ &\geqq |u_1^{(k)}| - \|v_o^T A_2^{(k)}\| \|w_k\| \\ &\geqq |u_1^{(k)}| - \sqrt{n-1}\,\alpha. \end{aligned}$$

The inequality (3.5) then implies that $|u_1^{(k)}| < n^{-1/2}$ and $u_1^{(k)}$ cannot be maximal.

Now either the sequence terminates before $k = n-1$, or we must arrive at the matrix A_{n-1}. Since at this point all the columns of A_0 but one have been replaced, the matrix A_{n-1} satisfies $A_{n-1}^T A_{n-1} = (I + E)$, where $|e_{ij}| \leqq \alpha$. Thus

$$\|E\| \leqq n\alpha.$$

For any vector u with $\|u\| = 1$, we have

$$\begin{aligned} \|A_{n-1}u\|^2 &= |u^T A_{n-1}^T A_{n-1} u| = |u^T (I+E) u| \\ &\geqq 1 - |u^T E u| \geqq 1 - n\alpha > \alpha > \alpha^2 \end{aligned}$$

and the sequence terminates with A_{n-1}. □

So far as the secant method is concerned, the main problem is to compute the vectors u and v associated with the matrix A defined by (3.1). Since A is upper Hessenberg, this can be done efficiently by a variant of the inverse power method. The motivation for the method is that if A is nearly singular, then A^{-1} will be large. Unless the elements of A^{-1} are specially distributed, the vector $u' = A^{-1}e$ will be large for almost any choice of e with $\|e\| = 1$. If we set $u = u'/\|u'\|$, then $\|Au\| = \|e\|/\|u'\| = 1/\|u'\|$ is small.

Because A is upper Hessenberg, it can be reduced by orthogonal transformations to triangular form in $O(n^2)$ operations; that is we can cheaply compute an orthogonal matrix R such that

$$B = RA$$

is upper triangular. We then solve the system $Bu' = e$. Since $\|Au'\| = \|R^T Bu'\| = \|R^T e\| = \|e\|$, we can work with the vector $u' = B^{-1}e$ rather than $A^{-1}e$. The components of e are taken to be $\pm 1/\sqrt{n}$, where the signs are chosen to enhance the size of the solution. Specifically,

(3.6)
1. $u'_n = n^{-1/2}/b_{nn}$
2. For $i = n-1, n-2, \cdots, 1$
    1. $\sigma = -\sum_{j=i+1}^{n} b_{ij}u'_j$
    2. $u'_i = [\sigma + \operatorname{sgn}(\sigma) n^{-1/2}]/b_{ii}$

The vector v is obtained by solving the system $B^T w = e$ in a manner analogous to (3.6) and setting $v = R^T w/\|R^T w\|$.

If $\|u'\|$ is large then a column of A, say the νth, must be replaced. From the definition of A, this amounts to replacing the $(\nu+1)$st column of Y by $y_1 + \lambda v$, where λ is arbitrary. We are now in a position to describe our overall algorithm for detecting and removing degeneracies.

(3.7)
1. Form A according to (3.1)
2. Calculate u' as described above
3. If $\|u'\| \geqq tol$
    1. Find ν so that $|u_\nu| \geqq |u_i| (i = 1, 2, \cdots, n)$
    2. Calculate v as described above
    3. $y^* = y_1 + \min\{\|y_i - y_1\| : i = 2, \cdots, n+1\} v$
    4. Insert y^* in Y, throwing out column $v+1$
    5. Go to 1
4. $\cdots$

As we mentioned at the beginning of this section, the above algorithm cannot be justified with complete rigor. Here we summarize the difficulties.

Statement 1. In the formation of A, the vector y_1 has been given a special role as a pivot. If another column of Y is used as a pivot, a different matrix A will be obtained. For example, if y_1, y_2, and y_3 are situated as shown

and y_1 is the pivot, then the vectors may well be judged to be affinely dependent. On the other hand if y_2 is the pivot, they will definitely be judged independent, since $y_1 - y_2$ and $y_3 - y_2$ are orthogonal. We have chosen y_1 as a pivot because the ordering imposed on the columns of Y and G creates the presumption that $x_1 = P^T y_1$ is nearer the zero of f than are the other columns of X (see § 4.2).

Statement 3. If $\|u'\|$ is large, then A is certainly nearly singular. However it is conceivable that A could be nearly singular and the algorithm for computing u' fail to give a large vector. We feel that this is extremely unlikely (it is equivalent to the failure of the widely used inverse power method for finding eigenvectors [5, p. 619]).

The value of *tol* should not be too large, otherwise slow convergence or wild predictions may result. On the other hand, Theorem 3.1 below suggests that it should not be too small. We have used a value of 100 in our numerical experiments (for $n = 100$, the bound (3.5) gives $\alpha^{-1} > 110$).

Statement 3.3. The form of y^* shows that our method for removing degeneracies amounts to taking a "side step" from y_1 along the direction v. The length of the side step is arbitrary. We have chosen the distance between y_1 and y_2 as the length, since x_1 and x_2 are presumed to be the points nearest the zero of f.

Statement 3.5. With *tol* suitably chosen, the only way this statement could cause an infinite loop is for $\|Av\|$ to be repeatedly smaller than *tol*. This is unlikely; however, the fastidious user might place an upper bound on the number of attempts to remove the degeneracy in A. Alternatively he can replace only previously untouched vectors.

4. Practical details. In this section we shall consider some of the practical problems that will arise when the method is implemented.

4.1. Economics. Since the matrices X and F are never used by the algorithm, it is necessary to store only the matrices Y, P, G, and Q. The number of nonzero elements in these matrices is about $3n^2$; however, if they are stored conventionally as separate arrays, they will require about $4n^2$ locations. Since the lower part of the array in which G (*or* Y) is stored is zero, this part of the array can be used as a workspace in which ΔG and ΔY are formed and manipulated.

In assessing the amount of work involved, we assume that plane rotations are used for all reductions. We shall count the number of rotations and the number of multiplications, which multiplications correspond roughly to the number of data accesses. The results are summarized below, where only the leading term of the count is given.

a. *Secant step*

$$\text{rot} = n - 1, \qquad \text{mult} = 3n^2.$$

b. *Function evaluation*

$$\text{rot} = 0, \qquad \text{mult} = 2n^2.$$

c. *Insertion and updating* (worst case in which y^* is inserted in the first column replacing y_1)

$$\text{rot} = n - 1, \qquad \text{mult} = 12n^2.$$

d. *Insertion and updating* (typical case in which y^* is inserted in the first column replacing y_{n+1})

$$\text{rot} = n - 1, \qquad \text{mult} = 6n^2.$$

e. *Checking degeneracy* (computation of u)

$$\text{rot} = n - 1, \qquad \text{mult} = 2.5n^2.$$

f. *Fixing degeneracy* (computation of v, evaluation of g^*, insertion of $y =$ and g^* [typical case])

$$\text{rot} = 2n - 2, \qquad \text{mult} = 14.5n^2.$$

Thus a typical iteration without degeneracy will consist of $a+b+2d+e$, or $3n-3$ rotations and $19.5n^2$ multiplications. With degeneracy, a typical iteration will require $5n-5$ rotations and $34n^2$ multiplications.

4.2. Order of the columns of Y and G. In forming ΔG preliminary to the computation of g^*, the vector g_1 is subtracted from the other columns of G. If $\|g_1\|$ is much larger than $\|g_i\|$, then the vector g_i will be overwhelmed by g_1. To avoid this we order the columns of G so that $\|g_1\| \leqq \|g_2\| \leqq \cdots \leqq \|g_{n+1}\|$. The matrix Y inherits this order, and since $\|f_i\| = \|g_i\|$, it may be presumed that when the process is converging, the vector x_i is nearer the solution than x_{i+1}. The ordering has the advantage that it gives a favorable operation count for the updates in the case when y^* replaces the column for which the norm of g is largest.

4.3. Communication with the user. The user must of course furnish code to evaluate the function f, which is customarily done in a subprogram provided by the user. After the secant prediction y^* has been calculated, the user must decide whether the process has converged. If is has not, he must decide whether the predicted point is acceptable and if not what to do about it. Since no single strategy is likely to be effective in all cases, we have left a blank section in our implementation of the algorithm where the user may code his own decisions.

4.4. Obtaining initial factorizations. The updating algorithm can be used to obtain the factorizations (2.1) and (2.2) at the start of the algorithm. The user of course must furnish $n+1$ vectors $x_1, x_2, \cdots, x_{n+1}$ in the matrix X. At the kth $(k=0, 1, \cdots, n)$ step of the initialization procedure, assume that the factorizations of the matrices $X^{|k} = (x_1, \cdots, x_k)$ and $F^{|k} = (f_1, f_2, \cdots, f_k)$ are known; i.e.,

$$X^{|k} = P^T Y^{|k}, \qquad G^{|k} = QF^{|k},$$

where $Y^{|k} = (y_1, \cdots, y_k)$ and $G^{|k} = (g_1, \cdots, g_k)$ are upper trapezoidal. Calculate the vectors $y_{k+1} = Px_{k+1}$ and $g_{k+1} = Qf_{k+1}$. Append a column to $Y^{|k}$ and $G^{|k}$ and insert y_{k+1} and g_{k+1}, making sure that the columns just appended are the ones to be discarded, and update as usual. After the nth step all the vectors in X and F will have been incorporated into the factorization.

4.5. Using an old Jacobian. When a sequence of closely related problems is being solved, the solution of one may be a good approximation to that of the next. Moreover the approximation to the old Jacobian implicitly contained in the matrices Y, P, G, and Q may also be a good approximation to the new Jacobian. Unfortunately the new iteration cannot simply be started with the old matrices Y, P, G, and Q, as the following hypothetical example shows.

Consider the case illustrated below in which the numbers associated with the points give the norms of the function values.

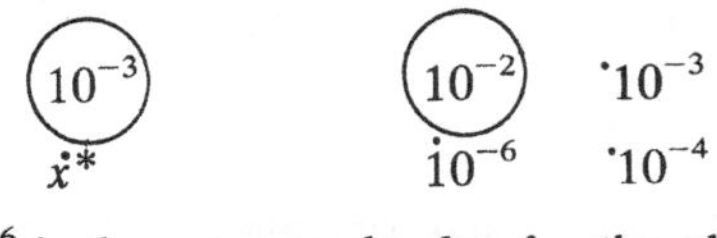

The point labeled 10^{-6} is the converged value for the old iteration. When the process is restarted with the new function, the point will have a much higher function value, say the circled 10^{-2}. Consequently the prediction x^* will be far removed from the original points, and when y^* is inserted into Y, the array will be judged to be degenerate. Moreover the function value at x^* will have a norm (10^{-3} in the example) which is out of scale with the old values. Thus both the G and Y arrays must be rescaled before they can be used with the new function.

Our method of scaling consists of two steps. First the columns of ΔY are scaled so that their norms are equal to $\|y^*-y_1\|$. The modification is extended to G by linearity. Then, with g_1' denoting the new g value at y_1, the columns of G are increased by $g_1'-g_1$. This scaling technique is described below. The notation Insert (g, i, j) means insert g into column i of G, throwing out column j, then update as usual.

1. Calculate the new value g_1' corresponding to y_1
2. $y^* = y_1 - \Delta Y(\Delta G)^{-1} g_1'$
3. For $i = 2, 3, \cdots, n+1$
   1. $\omega_i = \|y^* - y_1\| / \|y_i - y_1\|$
   2. $y_i \leftarrow y_i + \omega_i(y_i - y_1)$
   3. $g_i \leftarrow g_i + \omega_i(g_i - g_1)$
4. Insert $(g_1' - g_1, 1, 1)$, multiplying the update transformations into g_1
5. $g_i = g_i + (g_1' - g_1)$, $(i = 2, 3, \cdots, n+1)$
6. Insert $(g_1', 1, 1)$

It should be noted that statements 3.2 and 3.3 do not destroy the upper triangularity of the matrices Y and G, since only the first elements if y_1 and g_1 are nonzero. Statements 4, 5, and 6 are a circumlocution designed to avoid excessive updating. Statement 4 transforms the system so that $g_1' - g_1$ is nonzero in only its first component, after which G may be altered without destroying its upper triangularity (statement 5). Statement 6 places g_1' in its rightful position.

The y^* predicted by the scaled Y and G will be the same as the y^* of statement 1. The columns of G need no longer be in order of increasing norm; but since all but the first represent old data, they should be discarded as soon as possible.

4.6. Incorporating linearities. As was mentioned in § 1, degeneracies are certain to develop when some of the component functions are linear. Since the procedure for removing degeneracies is about as expensive as a secant step, it is important to be able to deal directly with such linearities. This may be done as follows.

Assume that $f: \mathbb{R}^{n+l} \to \mathbb{R}^n$, and that the equation $f(x) = 0$ is supplemented by l linear equations of the form

$$Ax = b, \tag{4.1}$$

where $A \in \mathbb{R}^{l \times (n+l)}$ is of full rank. Suppose that we are given a unitary matrix U such that

$$AU = (0 \quad T), \tag{4.2}$$

where T is square. Set $\hat{x} = U^T x$ and partition $\hat{x}$ in the form $\hat{x} = (\hat{x}_1^T, \hat{x}_2^T)^T$, where $\hat{x}_2 \in \mathbb{R}^l$. Then from (4.1) and (4.2),

$$T\hat{x}_2 = b. \tag{4.3}$$

Since A is of full rank, T is nonsingular and any solution of the system (4.1) must have $\hat{x}_2 = T^{-1}b$.

Define the function $f: \mathbb{R}^n \to \mathbb{R}^n$ by

$$\hat{f}(\hat{x}_1) = f\left[U\begin{pmatrix}\hat{x}_1\\ T^{-1}b\end{pmatrix}\right].$$

Then $\hat{f}(\hat{x}_1) = 0$ if and only if

$$x = U\begin{pmatrix}\hat{x}_1\\ T^{-1}b\end{pmatrix}$$

satisfies $f(x) = 0$ and $Ax = b$. The secant method may now be applied to $\hat{f}$.

The matrix U required by this process may be obtained in the usual way as the product of Householder transformations [5]. When this is done, the matrix T will be triangular, which makes (4.3) easy to solve.

5. Numerical examples and conclusions. The algorithm described in the above sections has been tried on a variety of problems. Here we summarize the results of three tests that exhibit the typical behavior of the algorithm.

The first example involves the function whose ith component is given by

$$f_i(x) = i - \sum_{j=1}^{i} x_i + q_i \sum_{j=i}^{n} (1 - x_i)^2.$$

This function has a solution at $\hat{x} = (1, 1, \cdots, 1)^T$. At the solution its Jacobian is the lower triangular matrix whose nonzero elements are all -1, a nicely conditioned matrix. The numbers q_i may be chosen ad libitum to make the function more or less nonlinear. Table 1 summarizes the results of applying the above algorithm to

TABLE 1

$\|e\|$	$\|f\|$	$\|u\|$
7.7×10^{-1}	9.0×10^{-1}	2.7×10^{0}
1.3×10^{-1}	3.1×10^{-1}	1.2×10^{2}
7.5×10^{-1}	2.8×10^{1}	1.4×10^{1}
1.2×10^{-2}	1.3×10^{-2}	5.7×10^{1}
2.9×10^{-3}	4.7×10^{-3}	6.2×10^{2}
9.8×10^{-3}	4.3×10^{-1}	1.3×10^{1}
2.4×10^{-4}	2.8×10^{-4}	1.5×10^{2}
3.0×10^{-3}	1.0×10^{-2}	1.2×10^{1}
1.1×10^{-5}	3.3×10^{-5}	2.4×10^{1}
1.6×10^{-6}	4.6×10^{-6}	4.3×10^{1}
4.3×10^{-7}	1.5×10^{-6}	2.5×10^{1}
1.2×10^{-7}	4.2×10^{-7}	2.8×10^{1}

this function with $n = 15$ and $q_i = 0.3$ $(i = 1, 2, \cdots, n)$. The initial estimate was the point $(0.8, 1.2, 0.8, 1.2, \cdots, 0.8^T$. The remaining 15 points required by the algorithm were obtained by adding alternately ± 0.05 to the successive components of the initial estimate. The results are summarized in Table 1, where $\|e\|$ denotes the Euclidean norm of the error in the current iterate, $\|f\|$ denotes the

TABLE 2

$\|e\|$	$\|f\|$	$\|u\|$
4.5×10^{-1}	4.5×10^{-1}	1.6×10^{0}
7.9×10^{-2}	1.1×10^{-1}	2.6×10^{1}
1.0×10^{-2}	8.2×10^{-3}	2.5×10^{1}
3.6×10^{-3}	4.1×10^{-3}	7.5×10^{1}
3.2×10^{-4}	2.6×10^{-4}	7.2×10^{1}
1.0×10^{-4}	1.3×10^{-4}	1.2×10^{1}
2.9×10^{-6}	2.3×10^{-6}	4.6×10^{3}
1.0×10^{-4}	3.4×10^{-4}	5.0×10^{0}
5.4×10^{-8}	1.1×10^{-7}	4.7×10^{0}

TABLE 3

$\|e\|$	$\|f\|$	$\|u\|$
4.5×10^{-1}	4.1×10^{-1}	1.6×10^{0}
4.3×10^{-2}	6.5×10^{-2}	2.6×10^{1}
4.7×10^{-3}	3.1×10^{-3}	2.3×10^{1}
1.4×10^{-3}	1.2×10^{-3}	1.7×10^{2}
3.7×10^{-3}	1.1×10^{-2}	4.5×10^{0}
2.9×10^{-5}	3.9×10^{-5}	6.9×10^{0}
2.8×10^{-6}	3.8×10^{-6}	4.2×10^{0}
7.0×10^{-8}	9.0×10^{-8}	5.6×10^{0}
4.5×10^{-1}	1.5×10^{0}	1.0×10^{1}
6.6×10^{-2}	9.1×10^{-2}	6.4×10^{9}
2.5×10^{-3}	2.1×10^{-3}	9.6×10^{1}
9.7×10^{-4}	1.0×10^{-3}	1.5×10^{1}
2.5×10^{-5}	2.3×10^{-5}	1.1×10^{2}
1.0×10^{-3}	8.1×10^{-4}	2.8×10^{2}
9.9×10^{-4}	8.1×10^{-4}	2.4×10^{0}
2.7×10^{-7}	4.2×10^{-7}	1.0×10^{1}
4.5×10^{-1}	1.5×10^{0}	1.0×10^{1}
5.1×10^{-2}	6.7×10^{-2}	3.3×10^{0}
2.3×10^{-3}	2.5×10^{-3}	7.1×10^{0}
1.7×10^{-4}	1.7×10^{-4}	1.9×10^{1}
1.0×10^{-6}	7.2×10^{-7}	6.8×10^{1}
4.5×10^{-1}	1.5×10^{0}	1.4×10^{2}
6.7×10^{-1}	1.5×10^{0}	9.7×10^{0}
1.2×10^{-7}	1.5×10^{-7}	1.5×10^{1}
4.5×10^{-1}	1.5×10^{0}	1.5×10^{1}
1.1×10^{-7}	2.0×10^{-7}	2.8×10^{1}
4.5×10^{-1}	1.5×10^{0}	2.8×10^{1}
8.0×10^{-8}	1.5×10^{-7}	5.6×10^{1}

Euclidean norm of the current function value, and $\|u\|$ denotes the norm of the vector u used to check degeneracies. Of the starting values only the central one is reported. At three points it was necessary to rectify a degeneracy; otherwise the convergence is routine (the iteration was terminated when $\|f\| \leqq 10^{-6}$).

The second example uses the same function with $n = 5, q_1 = q_2 = q_3 = q_4 = 0.5$ and $q_5 = 0$. The starting points are generated in the same way as for the first example. Since the fifth component of the function is linear, degeneracy can be expected in the iteration. It occurs at the seventh step ($\|u\| = 4.6 \cdot 10^3$) and is handled easily. (See Table 2).

The third example tests the algorithm for reusing old information. The function depends on a parameter s and is defined by

$$f_i(x) = i \cdot s - \sum_{j=1}^{i} x_i + q_i \sum_{j=1}^{n} (s - x_i)^2.$$

With $n = 5$ and $q_i = 0.3$ the zero $(s, s, s, s, s)^T$ was found for $s = 1.0, 1.2, 1.4, 1.6, 1.8, 2.0$. The information from one solution was used to start the next. The results are summarized in Table 3. The last three solutions are atypical in that they require effectively only a single iteration to converge. This is because the error vectors and the function values were the same at each new starting point, and this information had been preserved from the last solution.

These examples are given principally to illustrate the behavior of the algorithm. Additional experiments suggest that the local behavior of the method is quite good. Indeed if one believes that the algorithm for fixing degeneracies will work, one can apply the theory in [3] to give local convergence proofs. However, we believe it is too early to make general claims about the algorithm. For example, we do not know if damping techniques can be used to make it effective on problems where it otherwise would not work.

REFERENCES

[1] R. H. BARTELS, J. STOER AND CH. ZENGER, *A realization of the simplex method based on triangular decomposition*, Handbook for Automatic Computation II. Linear Algebra, J. H. Wilkinson and C. Reinsch, eds., Springer, New York, 1971, pp. 152–190.

[2] R. P. BRENT, *On maximizing the efficiency of algorithms for solving systems of nonlinear equations*, IBM Research RC 3725, Yorktown Heights, New York, 1972.

[3] J. M. ORTEGA AND W. C. RHEINBOLDT, *Iterative Solution of Nonlinear Equations in Several Variables*, Academic Press, New York, 1970.

[4] P. WOLFE, *The secant method for simultaneous nonlinear equations*, Comm. ACM, 2 (1959), pp. 12–13.

[5] J. H. WILKINSON, *The Algebraic Eigenvalue Problem*, Clarendon Press, Oxford, 1965.

13.2. [GWS-J31] (with J. W. Daniel, W. B. Gragg, L. Kaufman), "Reorthogonalization and Stable Algorithms for Updating the Gram-Schmidt QR Factorization"

[GWS-J31] (with J. W. Daniel, W. B. Gragg, L. Kaufman), "Reorthogonalization and Stable Algorithms for Updating the Gram-Schmidt QR Factorization," *Mathematics of Computation* 30 (1976) 772–795.
http://www.jstor.org/stable/2005398

MATHEMATICS OF COMPUTATION, VOLUME 30, NUMBER 136
OCTOBER 1976, PAGES 772–795

Reorthogonalization and Stable Algorithms for Updating the Gram-Schmidt QR Factorization

By J. W. Daniel, W. B. Gragg, L. Kaufman and G. W. Stewart*

Abstract. Numerically stable algorithms are given for updating the Gram-Schmidt QR factorization of an $m \times n$ matrix A $(m \geq n)$ when A is modified by a matrix of rank one, or when a row or column is inserted or deleted. The algorithms require $O(mn)$ operations per update, and are based on the use of elementary two-by-two reflection matrices and the Gram-Schmidt process with reorthogonalization. An error analysis of the reorthogonalization process provides rigorous justification for the corresponding ALGOL procedures.

1. Introduction. In many applications, most notably to linear least squares problems, it is important to have the QR factorization of a real $m \times n$ matrix A $(m \geq n)$ into the product of an $m \times n$ matrix Q with orthonormal columns and an $n \times n$ upper triangular matrix R. When A has full rank n the QR factorization is classically computed, in $O(mn^2)$ multiplications and additions, by the Gram-Schmidt process; the diagonal elements of R may then be taken positive and, with this normalization, the factorization is unique. In cases when the rank of A is nearly deficient the columns of Q, computed by the Gram-Schmidt process in the presence of rounding error, can deviate arbitrarily far from orthonormality.

The purpose of this paper is to provide numerically stable and relatively efficient algorithms for updating the Gram-Schmidt QR factorization of A when a row or column is inserted or deleted, or when A is modified by a matrix of rank one: $A \leftarrow \bar{A} = A + vu^T$, where u and v are (column) vectors. The paper may thus be considered supplementary to the important survey [2] of Gill, Golub, Murray and Saunders. It is emphasized that a principal aim of [2] was to update the *complete* orthogonal decomposition of A. This requires storage of an $m \times m$ orthogonal matrix and $O(m^2)$ arithmetic operations per update. The algorithms presented here arose from the desire to efficiently extend a stable modification of the secant method [4] to nonlinear least squares problems. The knowledge of an $m \times n$ Q then suffices, and we normally have $m \gg n$. The storage is thus reduced to $O(mn)$, and we shall show that the same is true of the operation counts.

The principal tools which we shall use are the Gram-Schmidt process, *with*

Received March 17, 1975.

AMS (MOS) subject classifications (1970). Primary 65F05; Secondary 15–04, 15A06, 62–04, 62J05, 65F20, 65F25, 65G05, 90C05, 90C30.

* This research was supported in part by the Office of Naval Research under Contracts N00014-67-A-0126-0015 and N00014-67-A-0314-0018, by the Air Force Office of Scientific Research under Grant AFOSR 71-2006, and by NSF MCS 75-23333.

reorthogonalization, and elementary two-by-two reflectors (or Givens matrices). The Gram-Schmidt process is in essence an algorithm for appending columns. Reorthogonalization is used to insure numerical stability, that is to preserve (near) orthogonality of the columns of the computed Q. There are actually two distinct algorithms for the general rank one update, and each has implications for the special updates. One algorithm, which will not be described in detail, uses Givens matrices to obtain an intermediate problem of appending a column; after the Gram-Schmidt process is applied the initial transformations must be undone. We shall present a slightly more efficient algorithm which uses the Gram-Schmidt process first, and then Givens matrices, to reduce the problem to that of appending a row. We then observe that the algorithm given in [2] for appending a row applies also to the Gram-Schmidt QR factorization. The algorithm for the stable deletion of rows is essentially the reverse of that for appending a row, but the arguments involved seem rather subtle.

In the next section we give a heuristic discussion of the reorthogonalization process. This is followed in Section 3 by a description of the updating algorithms. Section 4 is devoted to an error analysis of the reorthogonalization process. This allows us to set certain parameters in the ALGOL codes of Section 5, where some numerical results are also presented.

We shall use the Householder notational conventions [5], with the addition that $\mathbf{R}^{m \times n}$ denotes the set of real $m \times n$ matrices. Also, $\|\cdot\|$ refers to the Euclidean vector norm, as well as to its induced matrix norm: $\|A\| = \max\{\|Ax\|: \|x\| = 1\}$ for $A \in \mathbf{R}^{m \times n}$.

2. The Gram-Schmidt Process, With Reorthogonalization. We first recall the basic step of the Gram-Schmidt process. Let

$$Q = (q_1, q_2, \ldots, q_n) \in \mathbf{R}^{m \times n} \qquad (m \geqslant n)$$

have orthonormal columns, so that $Q^TQ = I_n$, and let $v \in \mathbf{R}^m$. We seek vectors $q \in \mathbf{R}^m$, $r \in \mathbf{R}^n$ and a scalar ρ so that

$$(Q, v) = (Q, q)\begin{pmatrix} I & r \\ 0 & \rho \end{pmatrix} \quad \text{and} \quad Q^Tq = 0.$$

The last column is $v = Qr + q\rho$ and multiplication by Q^T gives $r = Q^Tv$. Setting $v' \equiv q\rho$, we now have

$$v' = v - Qr = (I - QQ^T)v.$$

If $m > n$, we also insist that $\|q\| = 1$ which gives $\rho = \|v'\|$, and then

$$q = v'/\rho \quad \text{if } \rho \neq 0.$$

The process fails if $\rho = 0$, in which case $v = Qr$ is in the range of Q and $(Q, v) = Q(I, r)$ has rank n. In particular, when $m = n$ the matrix Q is orthogonal so we must have

$$q \equiv 0 \quad \text{and} \quad \rho \equiv 0.$$

The process, without normalization, uses $2mn$ multiplications and about the same number of additions.

If $m > n$ and the process fails, then, theoretically, any unit vector orthogonal to the range of Q could be substituted for q. The corresponding numerical problem is more subtle. If the process is carried out in the presence of rounding error, it is unlikely that ρ would vanish exactly, but it could be quite small. The process is designed to force $Q^T v'$ to be small relative to $\|v\|$, and indeed our error analysis will show this to be true. But even if $\|Q^T v'\| = \epsilon\|v\|$, with ϵ small, if ρ is extremely small, then the normalized vector $q = v'/\rho$ would satisfy only $\|Q^T q\| = \epsilon\|v\|/\rho$; and the error relative to $\|v\|$ could be very large. Thus, there could be a catastrophic loss of orthogonality in the computed q.

To rectify this situation one reasons as follows. If $\|v'\|/\|v\|$ is small, then numerical cancellation has occurred in forming v'. Thus, v', as well as q, are likely to be inaccurate relative to their lengths. If one attempts to correct v' by reorthogonalizing it, that is by applying the process again with v replaced by v', then one gets (approximately)

$$s = Q^T v' \quad \text{and} \quad v'' = v' - Qs = v - Q(r + s).$$

Comparing this with the desired result, $v' = v - Qr$, one sees that v' should be replaced by v'' and r by $r + s$. If $\|v''\|/\|v'\|$ is not too small, then v'' may be safely scaled to give a satisfactory q. Otherwise the process is repeated.

Thus, if η is a parameter satisfying $0 \ll \eta < 1$, for instance $\eta = 1/\sqrt{2}$, we have the tentative algorithm:

$$r^0 = 0, \qquad v^0 = v,$$

$$\text{for } k = 1, 2, 3, \ldots \text{ until } \|v^k\| > \eta\|v^{k-1}\|$$

$$\left| \begin{array}{ll} s^k = Q^T v^{k-1}, & r^k = r^{k-1} + s^k, \\ u^k = Qs^k, & v^k = v^{k-1} - u^k, \end{array} \right.$$

$$r = r^k, \qquad \rho = \|v^k\|, \qquad q = v^k/\rho.$$

It is unlikely that this iterative reorthogonalization process would fail to terminate, for ultimately rounding errors would force some iterate v^k to have substantial components (relative to $\|v^{k-1}\|$) orthogonal to the range of Q. However, this is only a "probabilistic" argument and so in Section 4 we shall give a completely rigorous alternative.

3. The Updating Algorithms.

Givens Matrices. A Givens matrix is a matrix of the form

$$G = \begin{pmatrix} \gamma & \sigma \\ \sigma & -\gamma \end{pmatrix}: \gamma = \cos\theta, \qquad \sigma = \sin\theta;$$

G is orthogonal and symmetric and $\det G = -1$. If $x = (\xi_1, \xi_2)^T$, then Gx is the reflection of x in the line which meets the axis $\xi_1 \geqslant 0$ in the angle $\theta/2$. The angle θ can be chosen so that

$$Gx = \tau e_1 = (\tau, 0)^T, \qquad \tau = \pm\|x\|.$$

If $\xi_2 = 0$, we take $\theta = 0$ so that $\gamma = 1$ and $\sigma = 0$. Otherwise, we compute

$$\mu = \max\{|\xi_1|, |\xi_2|\}, \qquad |\tau| = \mu \operatorname{sqrt}[(\xi_1/\mu)^2 + (\xi_2/\mu)^2],$$

$$\tau = \pm|\tau|, \qquad \gamma = \xi_1/\tau \quad \text{and} \quad \sigma = \xi_2/\tau.$$

This computation of $|\tau| = \|x\|$ avoids artificial problems of overflow and underflow; a corresponding device will be used to compute the length of any vector. The sign of τ remains unspecified.

The computation of $z = Gy$, $y = (\eta_1, \eta_2)^T$, may be done rather efficiently as follows. First compute $\nu = \sigma/(1 + \gamma)$, and then

$$\begin{pmatrix} \zeta_1 \\ \zeta_2 \end{pmatrix} = \begin{pmatrix} \gamma & \sigma \\ \sigma & -\gamma \end{pmatrix}\begin{pmatrix} \eta_1 \\ \eta_2 \end{pmatrix} = \begin{pmatrix} \gamma\eta_1 + \sigma\eta_2 \\ \nu(\eta_1 + \zeta_1) - \eta_2 \end{pmatrix}.$$

If G is applied to a $2 \times n$ matrix in this way, the cost is $3n$ multiplications and additions, instead of the usual $4n$ multiplications and $2n$ additions. Finally, the sign of τ is chosen so no cancellation occurs in the formation of ν:

$$\tau = |\tau| \operatorname{sign} \xi_1, \qquad \operatorname{sign} \xi \equiv \begin{cases} 1, & \xi \geqslant 0, \\ -1, & \xi < 0. \end{cases}$$

Of course, by a trivial modification, G can be chosen so Gx is a scalar multiple of the second axis vector $e_2 = (0, 1)^T$. Also, we shall actually use $n \times n$ Givens matrices $G_{i,j}$ which deviate from the identity only in the submatrix G formed from rows and columns i and j.

General Rank One Updates. Let

$$A = QR \in \mathbf{R}^{m \times n} \quad (m > n), \quad u \in \mathbf{R}^n \quad \text{and} \quad v \in \mathbf{R}^m.$$

Observe that

$$\bar{A} \equiv A + vu^T = (Q, v)\begin{pmatrix} R \\ u^T \end{pmatrix}.$$

Step 1. Apply the Gram-Schmidt process (with reorthogonalization) to obtain

$$(Q, v) = (Q, q)\begin{pmatrix} I & r \\ 0 & \rho \end{pmatrix}, \qquad Q^Tq = 0, \qquad \|q\| = 1.$$

We then have

$$\bar{A} = (Q, q)\left[\begin{pmatrix} R \\ 0^T \end{pmatrix} + \begin{pmatrix} r \\ \rho \end{pmatrix} u^T\right] \equiv \widetilde{Q}\widetilde{R},$$

and $\widetilde{Q}$ has orthonormal columns.

Step 2. Choose Givens matrices $G_{n,n+1}, G_{n-1,n}, \ldots, G_{1,2}$ so that

$$G\begin{pmatrix} r \\ \rho \end{pmatrix} \equiv G_{1,2} \cdots G_{n,n+1}\begin{pmatrix} r \\ \rho \end{pmatrix} \equiv \tau e_1.$$

That is, choose the $G_{i,i+1}$ $(i = n, n-1, \ldots, 1)$ to successively introduce zeros into the vector from the bottom element through the second. The matrix G is orthogonal. The $(n+1) \times n$ matrix

$$G\begin{pmatrix} R \\ 0^T \end{pmatrix} \equiv G_{1,2} \cdots G_{n,n+1}\begin{pmatrix} R \\ 0^T \end{pmatrix} \equiv R'$$

is upper Hessenberg (= almost triangular), and so is

$$G\widetilde{R} = R' + \tau e_1 u^T \equiv \hat{R}.$$

Moreover, by the orthogonality of G, the matrix $\widetilde{Q}G^T = \widetilde{Q}G_{n,n+1} \cdots G_{1,2} \equiv \hat{Q}$ has orthonormal columns and $\bar{A} = \hat{Q}\hat{R}$.

Step 3. Choose Givens matrices $H_{1,2}, H_{2,3}, \ldots, H_{n,n+1}$ to successively annihilate the subdiagonal elements of $\hat{R}$, giving

$$H\hat{R} \equiv H_{n,n+1} \cdots H_{1,2}\hat{R} \equiv \begin{pmatrix} \bar{R} \\ 0^T \end{pmatrix}$$

with $\bar{R}$ upper triangular. Then

$$\hat{Q}H^T = \hat{Q}H_{1,2} \cdots H_{n,n+1} \equiv (\bar{Q}, \bar{q})$$

has orthonormal columns and

$$\bar{A} = (\bar{Q}, \bar{q})\begin{pmatrix} \bar{R} \\ 0^T \end{pmatrix} = \bar{Q}\bar{R},$$

as required.

This algorithm uses approximately $2(l+3)mn + 3n^2$ multiplications and additions, where l is the number of orthogonalization steps ($l-1$ *re*orthogonalizations). The algorithm appears not to be valid for $m = n$, but in this case it actually simplifies. For then Q is orthogonal, and so

$$\bar{A} = Q(R + ru^T), \qquad r \equiv Q^T v.$$

Steps 2 and 3 apply, with one fewer Givens transformation each, to achieve the result.

Deleting a Column. Let

$$A = QR \in \mathbf{R}^{m \times n} \qquad (m \geqslant n)$$

and let a be the kth column of A, so that

$$A \equiv (A_1, a, A_2) \equiv Q(R_1, r, R_2).$$

Then

$$a = Qr \quad \text{and} \quad \bar{A} \equiv (A_1, A_2) = Q(R_1, R_2) \equiv Q\hat{R}.$$

The matrix $\hat{R}$ is upper Hessenberg. For instance, when $n = 6$ and $k = 3$, we have

$$\hat{R} = \begin{pmatrix} x & x & x & x & x \\ & x & x & x & x \\ & & x & x & x \\ & & x & x & x \\ & & & x & x \\ & & & & x \end{pmatrix},$$

where x denotes a possibly nonnull element and the elements not indicated are null. Thus, only Step 3 is needed, and this simplifies. We choose Givens matrices $H_{k,k+1}$, $H_{k+1,k+2}, \ldots, H_{n-1,n}$ so that

$$H\hat{R} = H_{n-1,n} \cdots H_{k,k+1}\hat{R} = \begin{pmatrix} \bar{R} \\ 0^T \end{pmatrix}$$

with $\bar{R}$ upper triangular. Then

$$QH^T = QH_{k,k+1} \cdots H_{n-1,n} \equiv (\bar{Q}, \bar{q})$$

has orthonormal columns and $\bar{A} = \bar{Q}\bar{R}$, as required.

This algorithm uses about $3[m + (n-k)/2](n-k)$ multiplications and additions together with mk more if, as will be done in our ALGOL code, the deleted column vector $a = Qr$ is retrieved.

Inserting a Column. Let

$$A \equiv (A_1, A_2) = Q(R_1, R_2) \equiv QR \in \mathbf{R}^{m\times(n-1)} \qquad (m \geqslant n)$$

with $A_1 \in \mathbf{R}^{m\times(k-1)}$ and $A_2 \in \mathbf{R}^{m\times(n-k)}$. If the vector $a \in \mathbf{R}^m$ is inserted "between" A_1 and A_2, it becomes the kth column of

$$\bar{A} \equiv (A_1, a, A_2) = (Q, a)\begin{pmatrix} R_1 & 0 & R_2 \\ 0^T & 1 & 0^T \end{pmatrix}.$$

We apply Step 1 to get

$$(Q, a) = (Q, q)\begin{pmatrix} I & r \\ 0 & \rho \end{pmatrix}, \qquad Q^Tq = 0, \qquad \|q\| = 1.$$

Then

$$\bar{A} = (Q, q)\begin{pmatrix} R_1 & r & R_2 \\ 0^T & \rho & 0^T \end{pmatrix} \equiv \tilde{Q}\tilde{R},$$

where $\tilde{R}$ is of the form ($n = 7, k = 3$)

$$\tilde{R} = \begin{pmatrix} x & x & x & x & x & x & x \\ & x & x & x & x & x & x \\ & & x & x & x & x & x \\ & & x & & x & x & x \\ & & x & & & x & x \\ & & x & & & & x \\ & & x & & & & \end{pmatrix}.$$

Step 2 is consequently simplified and Step 3 is not needed. We choose Givens matrices $G_{n-1,n}, G_{n-2,n-1}, \ldots, G_{k,k+1}$ so that

$$G\widetilde{R} \equiv G_{k,k+1} \cdots G_{n-1,n}\widetilde{R} \equiv \overline{R}$$

is upper triangular. This fills only the diagonal positions in columns $n, n-1, \ldots, k+1$. Then

$$\widetilde{Q}G^T = \widetilde{Q}G_{n-1,n} \cdots G_{k,k+1} \equiv \overline{Q}$$

has orthonormal columns and $\overline{A} = \overline{Q}\overline{R}$, as required.

This algorithm uses approximately $2lmn + 3[m + (n-k)/2](n-k)$ multiplications and additions.

Inserting a Row. Let

$$A = QR \in \mathbf{R}^{(m-1)\times n} \qquad (m > n) \quad \text{and} \quad a \in \mathbf{R}^n.$$

Without loss of generality we may *append* a^T to A. Then

$$\overline{A} \equiv \begin{pmatrix} A \\ a^T \end{pmatrix} = \begin{pmatrix} Q & 0 \\ 0^T & 1 \end{pmatrix}\begin{pmatrix} R \\ a^T \end{pmatrix} \equiv \widetilde{Q}\widetilde{R}.$$

Here $\widetilde{Q}$ already has orthonormal columns so Step 1 can be avoided. So can Step 2 provided Step 3 is altered only slightly. We choose Givens matrices $H_{1,n+1}, H_{2,n+1}, \ldots, H_{n,n+1}$ so that

$$H\widetilde{R} \equiv H_{n,n+1} \cdots H_{1,n+1}\widetilde{R} \equiv \begin{pmatrix} \overline{R} \\ 0^T \end{pmatrix}$$

with $\overline{R}$ upper triangular. Then

$$\widetilde{Q}H^T = \widetilde{Q}H_{1,n+1} \cdots H_{n,n+1} \equiv (\overline{Q}, \overline{q})$$

has orthonormal columns and $\overline{A} = \overline{Q}\overline{R}$, as required.

This algorithm uses approximately $3(m + n/2)n$ multiplications and additions.

Deleting a Row. Again, we may delete the last row. Let

$$A \equiv \begin{pmatrix} \overline{A} \\ a^T \end{pmatrix} \equiv \begin{pmatrix} Q \\ q^T \end{pmatrix} R \in \mathbf{R}^{m\times n} \qquad (m > n)$$

with

$$(Q^T, q)\begin{pmatrix} Q \\ q^T \end{pmatrix} = Q^TQ + qq^T = I_n.$$

Now also,

$$\begin{pmatrix} \overline{A} \\ a^T \end{pmatrix} = \begin{pmatrix} Q & 0 \\ q^T & 1 \end{pmatrix}\begin{pmatrix} R \\ 0^T \end{pmatrix}.$$

Apply Step 1, the Gram-Schmidt process (with reorthogonalization), to obtain

$$\begin{pmatrix} Q & 0 \\ q^T & 1 \end{pmatrix} = \begin{pmatrix} Q & \widetilde{q} \\ q^T & \sigma \end{pmatrix}\begin{pmatrix} I & r \\ 0 & \rho \end{pmatrix}.$$

The first iteration simplifies, since

$$r = (Q^T, q)\begin{pmatrix} 0 \\ 1 \end{pmatrix} = q,$$

and then

$$v' = \begin{pmatrix} \widetilde{q} \\ \sigma \end{pmatrix}\rho = \begin{pmatrix} 0 \\ 1 \end{pmatrix} - \begin{pmatrix} Q \\ q^T \end{pmatrix} q = \begin{pmatrix} -Qq \\ 1 - q^T q \end{pmatrix}.$$

Since $\|\widetilde{q}\|^2 + \sigma^2 = 1$, we have

$$\begin{aligned} \rho^2 &= \|Qq\|^2 + (1 - q^Tq)^2 = q^T(I - qq^T)q + (1 - q^Tq)^2 \\ &= 1 - q^Tq = \sigma\rho. \end{aligned}$$

If $\rho \neq 0$, then $\sigma = \rho$. If $\rho = 0$, then $q^Tq = 1$; and since $q^Tq + \sigma^2 \leqslant 1$, we have $\sigma = \rho$ in any case. Step 1 thus provides

$$\begin{pmatrix} Q & 0 \\ q^T & 1 \end{pmatrix} = \begin{pmatrix} Q & \widetilde{q} \\ q^T & \rho \end{pmatrix}\begin{pmatrix} I & q \\ 0 & \rho \end{pmatrix} \quad \text{with } \widetilde{q}\rho = -Qq.$$

Hence, we have

$$\begin{pmatrix} \bar{A} \\ a^T \end{pmatrix} = \begin{pmatrix} Q & \widetilde{q} \\ q^T & \rho \end{pmatrix}\begin{pmatrix} R \\ 0^T \end{pmatrix}.$$

We now choose Givens matrices $H_{n,n+1}, H_{n-1,n+1}, \ldots, H_{1,n+1}$ so that

$$(q^T, \rho)H^T \equiv (q^T, \rho)H_{n,n+1} \cdots H_{1,n+1} \equiv (0^T, \tau).$$

Moreover, $\tau = \pm 1$ since orthogonal transformations preserve length. Now the matrix

$$\begin{pmatrix} Q & \widetilde{q} \\ q^T & \rho \end{pmatrix} H^T \equiv \begin{pmatrix} \bar{Q} & \bar{q} \\ 0^T & \tau \end{pmatrix}$$

has orthonormal columns and so $\bar{q} = 0$. Finally, we have

$$H\begin{pmatrix} R \\ 0^T \end{pmatrix} = H_{1,n+1} \cdots H_{n,n+1}\begin{pmatrix} R \\ 0^T \end{pmatrix} \equiv \begin{pmatrix} \bar{R} \\ \pm \bar{r}^T \end{pmatrix},$$

with $\bar{R}$ upper triangular, and

$$\begin{pmatrix} \bar{A} \\ a^T \end{pmatrix} = \begin{pmatrix} \bar{Q} & 0 \\ 0^T & \pm 1 \end{pmatrix}\begin{pmatrix} \bar{R} \\ \pm \bar{r}^T \end{pmatrix} = \begin{pmatrix} \bar{Q}\bar{R} \\ \bar{r}^T \end{pmatrix},$$

as required.

This algorithm uses about $2(l + 1)mn + 3n^2/2$ multiplications and additions, where l is the number of orthogonalization iterations.

4. Construction of the Orthogonalization Code.

LEMMA 4.1 (ON MATRIX BY VECTOR MULTIPLICATION). *Let $A \in \mathbf{R}^{m\times n}$, $x \in \mathbf{R}^n$ and let $y \in \mathbf{R}^m$ be the result of the algorithm*

$$\begin{aligned} &y_0 = 0,\\ &\text{for } k = 1, 2, \ldots, n\\ &\quad \big\lfloor\, y_k = y_{k-1} + a_k\xi_k + e_k,\\ &y = y_n, \end{aligned}$$

in which the (error) vectors $\{e_k\}_1^n$ satisfy

$$\|e_k\| \leqslant \alpha\|y_{k-1}\| + \beta\|a_k\|\,|\xi_k|.$$

Then $y = Ax + e$ with

$$\|e\| \leqslant [(n-1)\alpha + \min\{m^{1/2}, n^{1/2}\}\beta](1+\alpha)^{n-1}\|A\|\,\|x\|.$$

Proof. By induction on k,

$$y_k = Ax_k + \sum_1^k e_j, \qquad x_k \equiv (\xi_1, \ldots, \xi_k, 0, \ldots, 0)^T, \qquad \|y_k\| \leqslant \|Ax_k\| + \sum_1^k \|e_j\|,$$

$$\|e_{k+1}\| \leqslant \alpha\|Ax_k\| + \beta\|a_{k+1}\|\,|\xi_{k+1}| + \alpha\sum_1^k \|e_j\|$$

and

$$\begin{aligned} \sum_1^{k+1} \|e_j\| &\leqslant \alpha\|Ax_k\| + \beta\|a_{k+1}\|\,|\xi_{k+1}| + (1+\alpha)\sum_1^k \|e_j\| \\ &\leqslant \alpha\sum_1^k (1+\alpha)^{k-j}\|Ax_j\| + \beta\sum_1^{k+1} (1+\alpha)^{k+1-j}\|a_j\|\,|\xi_j| \\ &\leqslant (1+\alpha)^k\left(\alpha\sum_1^k \|Ax_j\| + \beta\sum_1^{k+1} \|a_j\|\,|\xi_j|\right). \end{aligned}$$

The result now follows from

$$e = \sum_1^n e_j, \qquad \|Ax_j\| \leqslant \|A\|\,\|x_j\| \leqslant \|A\|\,\|x\|$$

and

$$\sum_1^n \|a_j\|\,|\xi_j| \leqslant \|A\|_F\|x\|, \qquad \|A\|_F \equiv \left(\sum_1^n \|a_j\|^2\right)^{1/2} \leqslant \min\{m^{1/2}, n^{1/2}\}\|A\|.$$

Applications. If y, a and ξ are floating point vectors and scalar, respectively, and $y' = y + a\xi + e'$ is computed in floating point arithmetic [10], [1], [6] then a typical element satisfies

$$\eta' = \eta + \alpha\xi + \epsilon' = \eta(1 + \delta'') + \alpha\xi(1 + \delta)(1 + \delta'),$$

where δ, δ' and δ'' are rounding errors; thus

$$\epsilon' = \eta\delta'' + \alpha\xi(\delta + \delta' + \delta\delta').$$

We assume rounded arithmetic operations and denote the basic machine unit by δ_0 $(=2^{-t})$. Quantities which are in practice only slightly larger than δ_0 will be denoted by $\delta_1, \delta_2, \delta_3, \ldots$; in the same vein we put $\delta_{-1} \equiv \delta_0/(1 + \delta_0)$. When $\xi = 1$ (vector addition), we have $\delta = 0$.

a. (*Weak*) *Single Precision* (possibly large relative error in addition). Here we have $|\delta| \leqslant \delta_{-1}$, $|\delta'| \leqslant 3\delta_0/2$ and $|\delta''| \leqslant 3\delta_0/2$. Thus,

$$|\epsilon'| \leqslant \frac{3}{2}\delta_0|\eta| + \frac{5}{2}\delta_0'|\alpha\xi|, \qquad \delta_0' \equiv \delta_0\left(1 + \frac{1}{5}\delta_{-1}\right).$$

Hence, in Lemma 4.1 we can take

$$\|e_k\| \leqslant \frac{3}{2}\delta_0\|y_{k-1}\| + \frac{5}{2}\delta_0'\|a_k\|\,|\xi_k|$$

to obtain $y = Ax + e$ with

$$\|e\| \leqslant \frac{1}{2}[3(n-1) + 5\min\{m^{1/2}, n^{1/2}\}]\delta_1\|A\|\,\|x\|$$

and

$$\delta_1 \equiv \delta_0\left(1 + \frac{3}{2}\delta_0\right)^n.$$

b. *Inner Products Computed in Double Precision* (followed by rounding to single precision). If y is the unrounded vector, then Lemma 4.1 applies, with

$$\|e_k\| \leqslant \frac{3}{2}\delta_0^2(\|y_{k-1}\| + \|a_k\|\,|\xi_k|),$$

to provide a bound for $\|y - Ax\|$. The double precision vector y is then rounded to yield the single precision vector z for which $\|z - y\| \leqslant \delta_0\|y\|$. From the triangle inequality it follows that $z = Ax + f$ with

$$\|f\| \leqslant \delta_0\|Ax\| + \frac{3}{2}(n + \min\{m^{1/2}, n^{1/2}\})\delta_2^2\|A\|\,\|x\|$$

and

$$\delta_2^2 \equiv \delta_0^2(1 + \delta_0)\left(1 + \frac{3}{2}\delta_0^2\right)^n.$$

These bounds, which do not appear in [10], [11], [1], [9] for instance, are basic to our further analysis.

In the following the Wilkinson symbols fl and fl_2 indicate the use of (weak) single precision and accumulated inner products, respectively.

THEOREM 4.1. *Let* $Q \in \mathbf{R}^{m\times n}$ $(m > n)$ and $v \in \mathbf{R}^m$ *have floating point entries, and let the vectors* $\{v^k\}_0^\infty$ *be computed from the algorithm:*

$$
\begin{aligned}
&v^0 = v,\\
&\textit{for } k = 1, 2, 3, \ldots\\
&\left\lfloor
\begin{aligned}
s^k &= \mathrm{fl}_2(Q^T v^{k-1}),\\
u^k &= \mathrm{fl}(Qs^k),\\
v^k &= v^{k-1} - u^k.
\end{aligned}\right.
\end{aligned}
$$

If

$$Q^TQ \equiv I + E, \qquad \|E\| \leqslant \epsilon, \qquad \gamma \equiv (1 + \epsilon)^{1/2},$$

$$\alpha \geqslant \frac{3}{2}\gamma\delta \quad \textit{and} \quad \beta \equiv \epsilon + \frac{3}{2}(n^{1/2} + 1)^2\gamma^2\delta,$$

where δ (in practice only slightly larger than the basic machine unit δ_0) is defined below, then the following inequalities hold:

1. $\|Q^T v^{k+1}\| \leqslant \alpha\|v^k\| + \beta\|Q^T v^k\|$;
2. $\|v^{k+1}\| \leqslant [\|v^k\|^2 - (1 - \epsilon)\|Q^T v^k\|^2]^{1/2} + \alpha\|v^k\| + \beta\|Q^T v^k\|$;
3. $\|v^{k+1}\| \geqslant [\|v^k\|^2 - (1 + \epsilon)\|Q^T v^k\|^2]^{1/2} - \alpha\|v^k\| - \beta\|Q^T v^k\|$ *provided* $\gamma\|Q^T v^k\| \leqslant \|v^k\|$.

Likewise, if fl_2 *is replaced by* fl, *and*

$$\alpha \geqslant \frac{3}{2}(m^{1/2} + 1)^2\gamma^2\delta.$$

Proof. We may suppose $k = 1$. We elaborate only on the fl_2 case, putting $r \equiv s'$ and $u \equiv u'$. From the applications of Lemma 4.1 we have

$$r = Q^T v + c, \qquad u = Qr + e, \qquad v' = v - u + f,$$

with

$$\|c\| \leqslant \delta_0\|Q^T v\| + \frac{3}{2}(m + n^{1/2})\delta_3^2\|Q\|\,\|v\|,$$

$$\delta_3^2 \equiv \delta_0^2(1 + \delta_0)\left(1 + \frac{3}{2}\delta_0^2\right)^m, \qquad \|e\| \leqslant \frac{1}{2}(3n + 5n^{1/2} - 3)\delta_1\|Q\|\,\|r\|$$

and

$$\|f\| \leqslant \frac{1}{2}(\|u\| \leqslant \|v\|).$$

Elimination of u from the above equalities and inequalities gives

$$r = Q^T v + c, \qquad v' = v - Qr + g,$$

with $g \equiv e - f$ and

$$\|g\| \leqslant \frac{3}{2}\delta_0\|v\| + \frac{3n + 5n^{1/2}}{2}\delta_1\|Q\|\,\|r\|.$$

(We have actually used a slightly sharper bound for $\|e\|$ to avoid introducing δ_4.)

Eliminating r in a similar manner, we find $v' = (I - QQ^T)v - h$ with $h \equiv Qc + g$,

$$\|h\| \leqslant \frac{3}{2}\delta_4\|v\| + \frac{3}{2}(n^{1/2}+1)^2\delta_5\|Q\|\,\|Q^Tv\|,$$

$$\delta_4 \equiv \delta_0 + (m + n^{1/2})\delta_3^2\left(1 + \frac{3n + 5n^{1/2}}{2}\delta_1\right)(1+\epsilon)$$

and

$$\delta_5 \equiv \delta_0(1+\delta_0)\left(1 + \frac{3}{2}\delta_0\right)^n.$$

In the single precision case a corresponding rather precise bound is

$$\|h\| \leqslant \frac{3}{2}\delta_6\|Q\|\,[(m^{1/2}+1)^2\|Q\|\,\|v\| + (n^{1/2}+1)^2\|Q^Tv\|]$$

with

$$\delta_6 \equiv \delta_0\left(1 + \frac{3}{2}\delta_0\right)^m\left(1 + \frac{3n + 5n^{1/2}}{2}\delta_1\right),$$

and we now define

$$\delta \equiv \max\{\delta_4, \delta_5, \delta_6\}.$$

Since

$$Q^Tv' = Q^Tv - (I+E)Q^Tv - Q^Th = -EQ^Tv - Q^Th,$$

we obtain the first bound by taking norms and using $\|Q\| \leqslant \gamma$. The remaining bounds follow from

$$|\|v'\| - \|(I - QQ^T)v\|| \leqslant \|h\|,$$

$$\|(I - QQ^T)v\|^2 = \|v\|^2 - \|Q^Tv\|^2 + (Q^Tv)^TEQ^Tv$$

and

$$|(Q^Tv)^TEQ^Tv| \leqslant \|E\|\,\|Q^Tv\|^2.$$

This completes the proof.

The quantities δ_k/δ_0 $(k \geqslant 0)$ are nondecreasing functions of δ_0, m, n and ϵ. For instance, if $\delta_0 \leqslant 5 \cdot 10^{-7}$, $m \leqslant 10^5$, $n \leqslant 10^5$ and $\epsilon \leqslant 1$ we have $\delta < 1.11\delta_0$.

We shall now apply Theorem 4.1 to construct the orthogonalization code. We shall assume that the numbers α, β and ϵ are not extremely large. Restrictions on their size will be expressed in the form of certain "τ-conditions" which will hold for a wide range of practical cases. We shall see that the (provable) attainable limiting precision of the reorthogonalization process is determined by the (minimal) value of α. In practice this is about $3\delta_0/2$ when accumulated inner products are used and $3m\delta_0/2$ otherwise.

If $v \neq 0$ and $\|Q^Tv\|/\|v\| \leqslant \xi \leqslant 1/\gamma$, then Theorem 4.1.3 implies

$$\|v'\|/\|v\| \geqslant [1 - (\gamma\xi)^2]^{1/2} - \alpha - \beta\xi.$$

The right side is positive if and only if

$$\psi(\xi) \equiv (\beta^2 + \gamma^2)^{1/2}\xi^2 + 2\alpha\beta\xi - (1 - \alpha^2) < 0.$$

Equivalently, by Descartes' rule, we must have $\alpha < 1$ and

$$\xi < \bar{\xi} \equiv \frac{1 - \alpha^2}{[(1 - \alpha^2)\gamma^2 + \beta^2]^{1/2} + \alpha\beta},$$

the positive zero of ψ. Hence, by Theorem 4.1.1, $\alpha < 1$ and $\xi < \bar{\xi}$ imply

$$\frac{\|Q^T v'\|}{\|v'\|} \leqslant \xi_1 \equiv \varphi(\xi) \equiv \frac{\alpha + \beta\xi}{[1 - (\gamma\xi)^2]^{1/2} - \alpha - \beta\xi}.$$

The function

$$\varphi\colon\ [0, \bar{\xi}) \longrightarrow [\underline{\xi}, +\infty), \qquad \underline{\xi} \equiv \alpha/(1 - \alpha),$$

is strictly increasing with reciprocal inverse function

$$\frac{1}{\varphi^{(-1)}(\xi)} = \frac{\alpha\beta(1 + \xi)^2 + \xi\{\beta^2(1 + \xi)^2 + \gamma^2[\xi^2 - \alpha^2(1 + \xi)^2]\}^{1/2}}{\xi^2 - \alpha^2(1 + \xi)^2}.$$

The fixed points of φ satisfy

$$\pi(\xi) \equiv \xi^2[(\gamma\xi)^2 - 1] + (1 + \xi)^2(\alpha + \beta\xi)^2 \equiv \xi^2\psi(\xi) + \chi(\xi) = 0,$$

and the polynomial χ has positive coefficients. By Descartes' rule π can have at most two positive zeros. Now $\pi(0) > 0$ and $\pi(\bar{\xi}) > 0$. For the limiting values $\alpha = \beta = 0$ and $\gamma = 1$ we have $\pi(\xi) = \xi^2(\xi^2 - 1)$ and the derivative $\pi'(\xi) = 2\xi(2\xi^2 - 1)$; hence, $\pi'(1/\sqrt{2}) = 0$ and $\pi(1/\sqrt{2}) = -\tfrac{1}{4} < 0$. For "general" values of α, β and γ we insist that $\pi(1/\sqrt{2}) < 0$, that is

$$\tau_1 \equiv \epsilon + (3 + 2\sqrt{2})(\sqrt{2}\alpha + \beta)^2 < 1.$$

This τ-condition implies that φ has exactly two fixed points ξ^* and ξ^{**} for which $\underline{\xi} < \xi^* < \xi^{**} < \bar{\xi}$. If the algorithm of Theorem 4.1 is started with any vector $v^0 \neq 0$ satisfying

$$\|Q^T v^0\|/\|v^0\| \leqslant \xi_0 < \xi^{**},$$

then it follows from the monotonicity and continuity of φ that

$$\|Q^T v^k\|/\|v^k\| \leqslant \xi_k \equiv \varphi(\xi_{k-1}) \longrightarrow \xi^* \qquad (k \longrightarrow \infty).$$

For practical values of α, β and ϵ the difference equation $\xi_{k+1} = \varphi(\xi_k)$ is extremely "stiff".

To set a sharp termination criterion we shall ultimately need a rather precise upper bound ξ^+ for ξ^*. For now suppose that

$$\xi^+ \equiv \theta\alpha > \xi^* \quad \text{with } \theta > 1$$

as yet unspecified. From the above we may terminate the iteration when

$$[\alpha\|v^{k-1}\| + \beta\|Q^T v^{k-1}\|]/\|v^k\| < \xi^+,$$

since the left side is at most ξ_k. Equivalently, we may terminate when

$$\|v^{k-1}\| + \omega\|Q^T v^{k-1}\| < \theta\|v^k\|, \qquad \omega \equiv \beta/\alpha.$$

The *termination parameters* (ω, θ) will be specified by the user. Increasing α corresponds to decreasing ω, or ϵ, and a weaker accuracy requirement.

We now investigate the possibility of nontermination. If $\|Q^T v^k\|/\|v^k\| \geqslant \xi^+$ then

$$\|Q^T v^{k-l}\|/\|v^{k-l}\| \geqslant \varphi^{(-l)}(\xi^+) \qquad (0 \leqslant l \leqslant n),$$

where $\varphi^{(-l)}$ is the lth compositional power of $\varphi^{(-1)}$. In other words

$$\|v^{k-l}\| \leqslant \|Q^T v^{k-l}\|/\varphi^{(-l)}(\xi^+) \qquad (0 \leqslant l \leqslant n).$$

Then, by induction using Theorem 4.1.1,

$$\xi^+\|v^k\| \leqslant \|Q^T v^0\| \prod_1^k [\beta + \alpha/\varphi^{(-l)}(\xi^+)].$$

In particular

$$\xi^+\|v^k\|/\|v^0\| \leqslant \gamma\rho^k, \qquad \rho \equiv \beta + \alpha/\varphi^{(-1)}(\xi^+).$$

Hence, if $\rho < 1$ and the termination criterion continually fails to be satisfied, then $v^k \longrightarrow 0$.

Our explicit expression for $\varphi^{(-1)}$ provides the means for studying ρ. The apparent difficulty in guaranteeing that $\rho < 1$ lies in the fact that if ξ^+ is extremely close to ξ^* $(\doteq\alpha)$, then the denominator

$$(\xi^+)^2 - \alpha^2(1 + \xi^+)^2 = \alpha^2[\theta^2 - (1 + \theta\alpha)^2]$$

can become extremely small. We now choose θ conveniently to make the term in brackets equal to unity. This gives

$$\theta = ((2 - \alpha^2)^{1/2} + \alpha)/(1 - \alpha^2)$$

which is about $\sqrt{2}$ for small α. Some simple estimates then show that $\rho < 1$ provided

$$\tau_2 \equiv \theta[(1 + \epsilon/2)\alpha + (\theta + 1)(1 + \alpha)\beta] < 1.$$

In fact $\rho < \tau_2$, so in practice ρ is substantially smaller than unity. Finally, the condition that $\xi^+ > \xi^*$ is implied by $\pi(\theta\alpha) < 0$, which reduces to

$$\tau_3 \equiv (1 + \epsilon)\theta^4\alpha^2 + \theta(2 + \theta\alpha)(1 + \theta\alpha)^2\beta < 1.$$

The practical reorthogonalization process will thus either terminate quickly or else we shall soon have $\|v^k\| \leqslant \sigma\|v^0\|$, where $\sigma > 0$ is a parameter somewhat smaller than the basic machine unit δ_0 (for instance $\sigma = \delta_0/10$). In this case v^k is certainly indistinguishable from rounding error; and if $v^k \neq 0$, we can legitimately replace v^k by $\|v^k\|e_l$, where the axis vector e_l is chosen so the lth row $e_l^TQ = (Q^Te_l)^T$ of Q has minimal length. (If $v^k = 0$, we replace v^k by e_l, but put $r = r^k$ and $\rho = 0$.) Our final, and most stringent, τ-conditions will guarantee the convergence of this alternative procedure.

First, we have

$$\|Q\|_F^2 = \sum_1^m \|Q^T e_i\|^2 \geqslant m\|Q^T e_l\|^2;$$

and then,

$$m^{1/2}\|Q^T e_l\| \leqslant \|Q\|_F \leqslant n^{1/2}\|Q\| \leqslant \gamma n^{1/2}.$$

That is

$$\|Q^T e_l\|/\|e_l\| \leqslant \gamma(n/m)^{1/2}.$$

We now obtain a lower bound, ξ^-, for ξ^{**}. Since $\pi(\xi^{**}) = 0$, we have, from the quadratic formula,

$$(\xi^{**})^2 = \frac{1 + \{1 - 4[(\alpha + \beta\xi^{**})\gamma(1 + \xi^{**})]^2\}^{1/2}}{2\gamma^2}.$$

Since $\xi^{**} < \bar{\xi} < 1/\gamma < 1$, $\gamma < 1 + \epsilon/2$ and $\sqrt{1-\eta} > 1 - \eta$ for $0 < \eta < 1$, we find

$$\xi^{**} > \left[\frac{1 - 4(2 + \epsilon)(\alpha + \beta)^2}{1 + \epsilon}\right]^{1/2} \equiv \xi^-.$$

The alternative procedure thus converges if $\gamma(n/m)^{1/2} \leqslant \xi^-$, or equivalently

$$\tau_4' \equiv (1 + \epsilon)^2 n/m + 4(2 + \epsilon)(\alpha + \beta)^2 \leqslant 1.$$

For practically small values of α, β and ϵ this τ-condition is surely satisfied if m is sufficiently greater than n. However, when reorthogonalization is applied, we *always* have $m > n$ and some trivial rearrangement shows that the alternative procedure converges for *all* such m if

$$\tau_4 \equiv 2(n + 1)[\epsilon + 4(\alpha + \beta)^2] \leqslant 1.$$

For practical values of α and β this is roughly equivalent with $2(n + 1)\epsilon < 1$.

For minimal α ($= 3\gamma\delta/2$ or $3(m^{1/2} + 1)^2\gamma^2\delta/2$) the numbers τ_k are increasing functions of δ_0, m, n and ϵ. The τ-conditions are all satisfied if, for instance, $\delta_0 \leqslant 5 \cdot 10^{-7}$, $n \leqslant 10^3$, $n < m \leqslant 10^4$ and $\epsilon < 10^{-4}$. Also, we can choose $\epsilon = n \max|\epsilon_{ij}|$, so in these cases we have $\epsilon \leqslant 10^{-4}$ provided $\max|\epsilon_{ij}| \leqslant n\delta_0$.

Although we shall not pursue the matter in detail, we wish to show how Theorem 4.1.2 can be used to obtain an upper bound for $\|Q^T v\|/\|v\|$ from a lower bound for $\|v'\|/\|v\|$. Thus, assume α, β and ϵ are sufficiently small, fix η so that $\alpha + \beta\gamma \leqslant \eta \leqslant 1$, and suppose

$$\eta \leqslant \|v'\|/\|v\|, \quad \tilde{\xi} \equiv \|Q^T v\|/\|v\|.$$

Then

$$\eta \leqslant [1 - (1 - \epsilon)\tilde{\xi}^2]^{1/2} + \alpha + \beta\tilde{\xi},$$

or equivalently

$$(1 - \epsilon + \beta^2)\tilde{\xi}^2 - (\eta - \alpha)\beta\tilde{\xi} - [1 - (\eta - \alpha)^2] \leqslant 0.$$

Hence,

$$\frac{\|Q^T v\|}{\|v\|} \leqslant \frac{(\eta - \alpha)\beta + \{(1 - \epsilon)[1 - (\eta - \alpha)^2] + \beta^2\}^{1/2}}{1 - \epsilon + \beta^2} \equiv \xi.$$

As α, β and ϵ tend to zero we have $\xi \longrightarrow (1 - \eta^2)^{1/2}$. Theorem 4.1.1 then yields the bound

$$\frac{\|Q^T v'\|}{\|v'\|} \leqslant \frac{\alpha + \beta\xi}{\eta} \sim \frac{\alpha + (1 - \eta^2)^{1/2}\beta}{\eta}.$$

This *indicates* that our termination criterion with

$$\omega = 0, \quad \theta = 1/\eta, \quad 0 \ll \eta < 1,$$

is not unreasonable, especially when α and β are of comparable size. On the other hand, when accumulated inner products are used, we may have $\alpha \ll \beta$ for large n and we would have to take $\eta \doteq 1$ to *guarantee* a limiting precision of $\xi^+ \doteq \sqrt{2}\alpha$.

5. ALGOL Procedures and Numerical Results. The principal practical results of this paper are summarized in the following package of ALGOL procedures.

```
comment: ALGOL procedures for updating Gram-Schmidt QR factorizations;
begin
  integer base; real lnbase;
  real omega, theta, sigma; label fail;
  comment: These are global entities. base and lnbase, the base of the machine arith-
    metic and its natural logarithm, are used in the procedure length. The others are
    relevant to the procedure orthogonalize. omega and theta are used to specify the
    termination criterion and sigma is used to test for restarting. See Section 4. The
    error exit fail is taken if termination is not obtained in a reasonable number of
    iterations;

real procedure length(n, x);
value n; integer n; real array x;
comment: Computes the accumulated Euclidean length of x[1 : n].
  Can be coded in machine language for greater efficiency;
begin
  integer k; real s, t; double ss, tt;
  ss := 0; t := 0;
  for k := 1 step 1 until n do t := max(t, abs(x[k]));
  if t > 0 then
  begin
    t := base ↑ entier(ln(t)/lnbase);
    for k := 1 step 1 until n do
    begin
```

```
        tt := x[k]/t; ss := ss + tt↑2
        end k
    end;
    s := ss; length := t × sqrt(s)
end length;

procedure orthogonalize(m, n, Q, v, r, rho);
value m, n; integer m, n; real rho; real array Q, v, r;
comment: Assuming Q[1 : m, 1 : n] (m ≥ n) has (nearly) orthonormal columns this
    procedure orthogonalizes v[1 : m] to the columns of Q, and normalizes the result if
    m > n. r[1 : n] is the array of "Fourier coefficients", and rho is the distance from
    v to the range of Q. r and its corrections are computed in double precision. For
    more detail see Sections 2 and 4;
begin
    Boolean restart, null; integer i, j, k; real rho 0, rho 1, t;
    double ss, qq, vv; real array u[1 : m], s[1 : n];
    label again, standardexit;
    restart := null := false;
    for j := 1 step 1 until n do r[j] := 0;
    rho := rho 0 := length(m, v);
    k := 0;
again:
    comment: Take a Gram-Schmidt iteration, ignoring r on later steps
        if previous v was null;
    for i := 1 step 1 until m do u[i] := 0;
    for j := 1 step 1 until n do
    begin
        ss := 0;
        for i := 1 step 1 until m do
        begin
            qq := Q[i, j]; vv := v[i]; ss := ss + qq × vv
        end i;
        s[j] := t := ss;
        for i := 1 step 1 until m do u[i] := u[i] + Q[i, j] × t
    end j;
    if ¬ null then
    for j := 1 step 1 until n do r[j] := r[j] + s[j];
    for i := 1 step 1 until m do v[i] := v[i] − u[i];
    rho 1 := length(m, v); t := length(n, s);
    k := k + 1;
    comment: Treat the special case m = n if necessary;
    if m = n then
    begin
        for i := 1 step 1 until m do v[i] := 0;
```

```
        rho := 0; go to standardexit
    end;
    comment: Test for nontermination;
    if rho 0 + omega × t ≥ theta × rho 1 then
    begin
        comment: Exit to fail if too many iterations;
        if k > 4 then go to fail;
        comment: Restart if necessary;
        if ¬ restart ∧ rho 1 ≤ rho × sigma then
        begin
            restart := true;
            comment: Find first row of minimal length of Q;
            for i := 1 step 1 until m do u[i] := 0;
            for j := 1 step 1 until n do
            for i := 1 step 1 until m do u[i] := u[i] + Q[i, j]↑2
            t := 2;
            for i := 1 step 1 until m do
            if u[i] < t then begin k := i; t := u[k] end;
            comment: Take correct action if v is null;
            if rho 1 = 0 then begin null := true; rho 1 := 1 end;
            comment: Reinitialize v and k;
            for i := 1 step 1 until m do v[i] := 0;
            v[k] := rho 1; k := 0
        end;
        comment: Take another iteration;
        rho 0 := rho 1; go to again
    end;
    comment: Normalize v and take the standard exit;
    for i := 1 step 1 until m do v[i] := v[i]/rho 1;
    if ¬ null then rho := rho 1 else rho := 0;
standardexit:
end orthogonalize;

procedure computereflector(x, y, c, s);
real x, y, c, s;
comment: Computes parameters for the Givens matrix G for which
    (x, y)G = (z, 0). Replaces (x, y) by (z, 0);
begin
    real t, u, v, mu;
    u := x; v := y;
    if v = 0 then begin c := 1; s := 0 end
    else
    begin
```

```
      mu := max(abs(u), abs(v));
      t := mu × sqrt((u/mu)↑2 + (v/mu)↑2);
      if u < 0 then t := -t;
      c := u/t; s := v/t; x := t; y := 0
   end
end computereflector;

procedure applyreflector(c, s, k, l, x, y, j);
value c, s, k, l; integer k, l, j; real c, s, x, y;
comment: When called with x := x[j] and y := y[j], this procedure replaces the two
   column matrix (x[k:l], y[k:l]) by (x[k:l], y[k:l])G, where G is the Givens matrix
   determined by c and s. Uses the Jensen device [8];
begin
   real t, u, v, nu;
   nu := s/(1 + c);
   for j := k step 1 until l do
   begin
      u := x; v := y; x := t := u × c + v × s; y := (t + u) × nu − v
   end j
end applyreflector;

procedure rankoneupdate(m, n, Q, R, u, v);
value m, n; integer m, n; real array Q, R, u, v;
comment: Updates the factorization A = Q[1 : m, 1 : n]R[1 : n, 1 : n] (m ⩾ n)
   when the outer product of v[1 : m] and u[1 : n] is added to A;
begin
   integer i, j, k; real c, s, rho; real array t[1 : n];
   orthogonalize(m, n, Q, v, t, rho);
   computereflector(t[n], rho, c, s);
   applyreflector(c, s, n, n, R[n, n], rho, j);
   applyreflector(c, s, 1, m, Q[i, n], v[i], i);
   for k := n − 1 step − 1 until 1 do
   begin
      computereflector(t[k], t[k + 1], c, s);
      applyreflector(c, s, k, n, R[k, j], R[k + 1, j], j);
      applyreflector(c, s, 1, rr 2[i, k], Q[i, k + 1], i)
   end k;
   for j := 1 step 1 until n do R[1, j] := R[1, j] + t[1] × u[j];
   for k := 1 step 1 until n − 1 do
   begin
      computereflector(R[k, k], R[k + 1, k], c, s);
      applyreflector(c, s, k + 1, n, R[k, j], R[k + 1, j], j);
      applyreflector(c, s, 1, m, Q[i, k], Q[i, k + 1], i)
   end k;
```

```
    computereflector(R[n, n], rho, c, s);
    applyreflector(c, s, 1, m, Q[i, n], v[i], i);
end rankoneupdate;

procedure deletecolumn(m, n, Q, R, k, v);
value m, n, k; integer m, n, k; real array Q, R, v;
comment: Updates the factorization A = Q[1 : m, 1 : n]R[1 : n, 1 : n] (m ≥ n) when
    the kth column of A is deleted. Returns the deleted column in v[1 : m];
begin
    integer i, j, l; real c, s, t;
    for i := 1 step 1 until m do v[i] := 0;
    for l := 1 step 1 until k do
    begin
        t := R[l, k];
        for i := 1 step 1 until m do v[i] := v[i] + Q[i, l] × t
    end l;
    for l := k step 1 until n − 1 do
    begin
        computereflector(R[l, l + 1], R[l + 1, l + 1], c, s);
        applyreflector(c, s, l + 2, n, R[l, j], R[l + 1, j], j);
        applyreflector(c, s, 1, m, Q[i, l], Q[i, l + 1], i)
    end l;
    for j := k step 1 until n − 1 do
    for i := 1 step 1 until j do R[i, j] := R[i, j + 1];
    for i := 1 step 1 until n do R[i, n] := 0;
    for i := 1 step 1 until m do Q[i, n] := 0
end deletecolumn;

procedure insertcolumn(m, n, Q, R, k, v);
value m, n, k; integer m, n, k; real array Q, R, v;
comment: Updates the factorization A = Q[1 : m, 1 : n − 1]R[1 : n − 1, 1 : n − 1]
    (m ≥ n) when the vector v[1 : m] is inserted between columns k − 1 and k of A;
begin
    integer i, j, l; real c, s; real array u[1 : n];
    for j := n − 1 step −1 until k do
    for i := 1 step 1 until j do R[i, j + 1] := R[i, j];
    for j := k + 1 step 1 until n do R[j, j] := 0;
    orthogonalize(m, n − 1, Q, v, u, u[n]);
    for i := 1 step 1 until m do Q[i, n] := v[i];
    for l := n − 1 step −1 until k do
    begin
        computereflector(u[l], u[l + 1], c, s);
        applyreflector(c, s, l + 1, n, R[l, j], R[l + 1, j], j);
        applyreflector(c, s, 1, m, Q[i, l], Q[i, l + 1], i)
```

```
    end l;
    for i := 1 step 1 until k do R[i, k] := u[i]
end insertcolumn;

procedure insertrow(m, n, Q, R, k, u);
value m, n, k; integer m, n, k; real array Q, R, u;
comment: Updates the factorization A = Q[1 : m - 1, 1 : n]R[1 : n, 1 : n]
    (m > n) when the vector u[1 : n] is inserted between rows k - 1
    and k of A;
begin
    integer i, j, l; real c, s; real array v[1 : m];
    for i := 1 step 1 until m do v[i] := 0; v[k] := 1;
    for l := 1 step 1 until n do
    begin
        for i := m - 1 step -1 until k do Q[i + 1, l] := Q[i, l]; Q[k, l] := 0;
        computereflector(R[l, l], u[l], c, s);
        applyreflector(c, s, l + 1, n, R[l, j], u[j], j);
        applyreflector(c, s, 1, m, Q[i, l], v[i], i)
    end l
end insertrow;

procedure deleterow(m, n, Q, R, k, u);
value m, n, k; integer m, n, k; real array Q, R, u;
comment: Updates the factorization A = Q[1 : m, 1 : n]R[1 : n, 1 : n] (m > n)
    when the kth row of A is deleted. Returns the deleted row in u[1 : n];
begin
    integer i, j, l; real c, s, t; real array v[1 : m];
    for i := 1 step 1 until m do v[i] := 0; v[k] := 1;
    orthogonalize(m, n, Q, v, u, t);
    for i := k step 1 until m - 1 do v[i] := v[i + 1];
    for l := n step -1 until 1 do
    begin
        for i := k step 1 until m - 1 do Q[i, l] := Q[i + 1, l]
        computereflector(t, u[l], c, s);
        applyreflector(c, s, l, n, u[j], R[l, j], j);
        applyreflector(c, s, 1, m - 1, v[i], Q[i, l], i);
        Q[m, l] := 0
    end l;
    for j := 1 step 1 until n do u[j] := t × u[j]
end deleterow;

procedure QRfactor(m, n, A, Q, R);
value m, n; integer m, n; real array A, Q, R;
comment: Computes a Gram-Schmidt QR factorization, Q[1 : m, 1 : n]R[1 : n, 1 : n],
    of A[1 : m, 1 : n] (m ≥ n);
```

```
begin
   integer i, k; real array v[1 : m];
   for k := 1 step 1 until n do
   begin
      for i := 1 step 1 until m do v[i] := A [i, k]; insertcolumn(m, k, Q, R, k, v)
   end k

end QRfactor;
fail:
end Gram-Schmidt QR updating procedures;
```

Extensive tests with small matrices have been made to check the logic of our codes. We report in detail only the following larger tests with the view of obtaining numerical experience with the inevitable problem of error propagation.

Experiment 1. To test the numerical stability of the procedure orthogonalize, as well as its dependence on the termination criterion, we have constructed numerical Gram-Schmidt QR factorizations, Q_nR_n, of the Hilbert sections

$$H_n \equiv (1/(i+j-1)) \in \mathbf{R}^{100\times n}, \qquad n = 1, 2, \ldots, 100.$$

Since this is done by successively appending columns, no Givens transformations are used and the propagation of rounding errors is due solely to the orthogonalization process. Moreover, the (double precision) Frobenius norms

$$\|Q_nR_n - H_n\|_F \quad \text{and} \quad \|Q_n^TQ_n - I_n\|_F$$

are easily updated. A slight and trivial extension of our error analysis shows that, if the recommended termination pair (ω, θ) is used and the final scale factor ρ is computed using an accumulated inner product, then we have

$$\|Q_n^TQ_n - I_n\|_F \leqslant \kappa n^{1/2}\delta_0$$

with κ essentially independent of m and n. Our experiments were done on a Burroughs 6700 computer, for which $\delta_0 = 0.5/8^{12}$. Table 1 gives an indicative selection of the quantities

$$\bar{d}_n^k \equiv \|Q_nR_n - H_n\|_F/n^{1/2}\delta_0, \qquad \bar{e}_n^k \equiv \|Q_n^TQ_n - I\|_F/n^{1/2}\delta_0,$$

for three different termination criteria ($k = 1, 2, 3$): (1) (ω, θ) as prescribed in Section 4 with minimal α and $\epsilon = \|Q_n^TQ_n - I_n\|_F$, (2) $(\omega, \theta) \equiv (0, \sqrt{2})$ and (3) $(\omega, \theta) \equiv (0, 10)$.

TABLE 1

n	$\bar{d}_n^1$	$\bar{e}_n^1$	$\bar{d}_n^2$	$\bar{e}_n^2$	$\bar{d}_n^3$	$\bar{e}_n^3$
20	0.26	0.90	0.25	1.08	0.25	1.7
40	0.27	1.03	0.26	1.68	0.27	54.0
60	0.26	0.93	0.26	1.42	0.27	44.1
80	0.23	0.90	0.23	1.28	0.24	38.2
100	0.21	0.95	0.21	1.25	0.22	34.2

In cases one and two, for $n \geqslant 2$ and with only one exception, the number of orthogonalization iterations was constant at two for small values of n and three for large n. The jumps from two to three iterations occurred at $n = 12$ and $n = 38$, respectively. Case three was similar except that only one iteration was used for $n = 2$, and the jump then occurred earlier, at $n = 26$. A fourth run was made restricting the number of iterations to two (one reorthogonalization). For this we had

$$\bar{e}_n^4 > 10^{10} \quad \text{for } n \geqslant 45,$$

that is all orthonormality in the computed Q_n had been lost.

In earlier runs we had set the restart parameter σ *equal* to the basic unit δ_0. From about fifty calls to the procedure orthogonalize two restarts were observed to occur. Since restarting is expensive we have thus recommended a somewhat smaller value of σ, in order to give the probabalistic heuristic of Section 2 a fair chance. No restarts were observed in our experience with the code as given ($\sigma = \delta_0/10$).

Experiment 2. We now consider updating the numerical QR factorizations, $Q_m R_m$, of the Hilbert sections

$$H_m \equiv (1/(i + j - 1)) \in \mathbf{R}^{m \times 10}$$

by appending and dropping rows. After obtaining the initial factorization of H_{10} as above, that is with the procedure *QRfactor*, we used *insertrow* to append rows up to $m = 50$ and then *deleterow* to successively drop the last row and return to $m = 10$. The recommended termination parameters were used consistently, with minimal α and $\epsilon = \|Q_m^T Q_m - I_{10}\|_F$. Table 2 lists typical values of the magnified error norms

$$d_m \equiv \|Q_m R_m - H_m\|_F/\delta_0, \qquad e_m \equiv \|Q_m^T Q_m - I_{10}\|_F/\delta_0,$$

which no longer can be computed recursively. For ascending m we have also given the quantities

$$\tilde{d}_m \equiv \ln(d_m)/\ln(m), \qquad \tilde{e}_m \equiv \ln(e_m)/\ln(m).$$

TABLE 2

m	d_m	$\tilde{d}_m$	e_m	$\tilde{e}_m$
10	0.7	−0.13	3	0.54
20	10.4	0.78	37	1.20
30	18.9	0.86	65	1.23
40	32.1	0.94	88	1.21
50	51.9	1.01	123	1.23
40	52.0		122	
30	51.6		120	
20	50.6		118	
10	47.6		106	

For ascending m these results indicate error growth of roughly m in d_m and $m^{5/4}$ in e_m. For descending m both errors are moderately decreasing.

Departments of Computer Science and Mathematics
University of Texas at Austin
Austin, Texas 78712

Department of Mathematics
University of California, San Diego
La Jolla, California 92093

Department of Computer Science
University of Colorado
Boulder, Colorado 80302

Department of Computer Science
University of Maryland
College Park, Maryland 20742

1. GEORGE E. FORSYTHE & CLEVE B. MOLER, *Computer Solution of Linear Algebraic Systems,* Prentice-Hall, Englewood Cliffs, N. J., 1967. MR 36 #2306.

2. P. E. GILL, G. H. GOLUB, W. MURRAY & M. A. SAUNDERS, "Methods for modifying matrix factorizations," *Math. Comp.,* v. 28, 1974, pp. 505–535. MR 49 #8299.

3. PHILIP E. GILL, WALTER MURRAY & MICHAEL A. SAUNDERS, "Methods for computing and modifying the *LDV* factors of a matrix," *Math. Comp.,* v. 29, 1975, pp. 1051–1077.

4. W. B. GRAGG & G. W. STEWART, "A stable variant of the secant method for solving nonlinear equations," *SIAM J. Numer. Anal.,* v. 13, 1976.

5. ALSTON S. HOUSEHOLDER, *The Theory of Matrices in Numerical Analysis,* Blaisdell, New York, 1964. MR 30 #5475.

6. DONALD E. KNUTH, *The Art of Computer Programming.* Vol. 2: *Seminumerical Algorithms,* Addison-Wesley, Reading, Mass., 1969. MR 44 #3531.

7. CHARLES L. LAWSON & RICHARD J. HANSON, *Solving Least Squares Problems,* Prentice-Hall Ser. in Automatic Computation, Prentice-Hall, Englewood Cliffs, N. J., 1974. MR 51 #2270.

8. HEINZ RUTISHAUSER, *Handbook for Automatic Computation.* Vol. 1/Part a: *Description of ALGOL* 60, Die Grundlehren der math. Wissenschaften, Band 135, Springer-Verlag, New York, 1967.

9. G. W. STEWART, *Introduction to Matrix Computations,* Academic Press, New York, 1973.

10. J. H. WILKINSON, *Rounding Errors in Algebraic Processes,* Prentice-Hall, Englewood Cliffs, N. J., 1963. MR 28 #4661.

11. J. H. WILKINSON, *The Algebraic Eigenvalue Problem,* Clarendon Press, Oxford, 1965. MR 32 #1894.

13.3. [GWS-J40] "The Effects of Rounding Error on an Algorithm for Downdating a Cholesky Factorization"

[GWS-J40] "The Effects of Rounding Error on an Algorithm for Downdating a Cholesky Factorization," *Journal of the Institute for Mathematics and its Applications (IMA), Applied Mathematics* **23** (1979) 203–213.

J. Inst. Maths Applics (1979) **23**, 203–213

The Effects of Rounding Error on an Algorithm for Downdating a Cholesky Factorization

G. W. STEWART

Department of Computer Science, University of Maryland

[Received 29 November 1977 and in revised form 31 August 1978]

Let the positive definite matrix A have a Cholesky factorization $A = R^T R$. For a given vector x suppose that $\tilde{A} = A - xx^T$ has a Cholesky factorization $\tilde{A} = \tilde{R}^T\tilde{R}$. This paper considers an algorithm for computing $\tilde{R}$ from R and x and an extension for removing a row from the QR factorization of a regression problem. It is shown that the algorithm is stable in the presence of rounding errors. However, it is also shown that the matrix $\tilde{R}$ can be a very ill-conditioned function of R and x.

1. Introduction

LET A be a positive definite matrix of order p. Then A can be factorized in the form

$$A = R^T R$$

where R is upper triangular. This "Cholesky factorization" of A is unique up to the signs of the rows of R (e.g. see Stewart, 1973).

In this paper we shall be concerned with the following problem. Given a p-vector x and the matrix R find the Cholesky factorization of the matrix

$$\tilde{A} = A - xx^T, \tag{1.1}$$

where it is assumed that x is such that $\tilde{A}$ is positive definite. We shall refer to this problem as the *downdating* problem.

An important application of downdating is the removal of an observation from a linear regression problem that is being solved by means of the QR factorization. Specifically, consider the problem of choosing β to minimize

$$\rho^2 = \|X\beta - y\|^2,$$

where X is an $n \times p$ matrix of rank p and $\|\cdot\|$ denotes the usual Euclidean vector norm defined by $\|x\|^2 = x^T x$. It is well known that X can be factorized in the form

$$X = QR,$$

where R is upper triangular and Q has orthonormal columns, i.e. $Q^T Q = I$. If z is defined by

$$z = Q^T r,$$

This research was supported in part by the Office of Naval Research under Contract No. N00014-76-C-0391.

then the solution of the regression problem is given by

$$\beta = R^{-1}z$$

and the residual sum of squares by

$$\rho^2 = \|y\|^2 - \|z\|^2.$$

When n is large, it may be impossible to retain the elements of the $n \times p$ matrix Q in the main memory of the computer performing the calculation. In this case one may compute R, z, and ρ without explicitly forming Q (Golub, 1965; Lawson & Hanson, 1974). Although this suffices for the computation of β, one is left with the problem of performing a variety of statistical computations when one knows only R, z, and ρ. (It is interesting to note that aficionados of the normal equations have the same problem; they cannot retain X in main memory and must work instead with X^TX, X^Ty, and $\|y\|^2$ (Dempster, 1969).)

One frequently occurring requirement is to remove an observation from the regression, that is to remove a row x^T from X and the corresponding component η from y. Without loss of generality we may suppose that x^T is the last row of X, so that X can be written in the form

$$X = \begin{pmatrix} \tilde{X} \\ x \end{pmatrix}.$$

It follows that

$$\tilde{X}^T\tilde{X} = X^TX - xx^T. \tag{1.2}$$

Now

$$X^TX = R^TQ^TQR = R^TR,$$

which shows that the triangular part of the QR factorization of X is the Cholesky factor of X^TX. Likewise $\tilde{R}$, the triangular part of the QR factorization of $\tilde{X}$, is the Cholesky factor of $\tilde{X}^T\tilde{X}$. Comparing (1.1) and (1.2), we see that the problem is one of downdating the Cholesky factorization of X^TX. There is, of course, more to it than this, for we must also compute the downdated vector $\tilde{z}$ and residual sum of squares $\tilde{\rho}^2$.

In this paper we shall give a rounding error analysis of an algorithm for computing $\tilde{R}$, $\tilde{z}$, and $\tilde{\rho}$. Our conclusions are that the algorithm is remarkably stable; however, this stability does not guarantee that the results are accurate, for the downdating problem can be quite ill conditioned. We begin with a discussion of this ill-conditioning, before going on to a description of the algorithm and the subsequent error analysis.

2. The Condition of the Downdating Problem

Let A, $\tilde{A}$, and x be as in the previous section with A having a Cholesky factorization R^TR. Let $\tilde{A}$ have the Cholesky factorization $\tilde{R}^T\tilde{R}$. In applications we will of course not know A and B. Rather we will be given R and x and be required to compute $\tilde{R}$. Consequently, we are interested in assessing the effects of perturbations in R and x on $\tilde{R}$.

We shall first consider a perturbation E in R. Assume that the matrix $(R+E)^T(R+E)-xx^T$ has a Cholesky factor $\bar{R}$. We wish to assess the size $\|\tilde{R}-\bar{R}\|$, where here $\|\cdot\|$ denotes the spectral matrix norm (Stewart, 1973; Wilkinson, 1965). We begin by comparing the singular values (Stewart, 1973) of $\tilde{R}$ and $\bar{R}$, which we denote by $\tilde{\sigma}_1 \geqslant \tilde{\sigma}_2 \geqslant \ldots \geqslant \tilde{\sigma}_p$ and $\bar{\sigma}_1 \geqslant \bar{\sigma}_2 \geqslant \ldots \geqslant \bar{\sigma}_p$. Now

$$\begin{aligned}\bar{R}^T\bar{R} &= (R+E)^T(R+E)-xx^T \\ &= R^TR-xx^T+R^TE+E^TR+E^TE \\ &= \tilde{R}^T\tilde{R}+R^TE+E^TR+E^TE.\end{aligned}$$

Since $\tilde{\sigma}_1^2, \tilde{\sigma}_2^2, \ldots, \tilde{\sigma}_p^2$ are the eigenvalues of $\tilde{R}^T\tilde{R}$ and likewise $\bar{\sigma}_1^2, \bar{\sigma}_2^2, \ldots, \bar{\sigma}_p^2$ are the eigenvalues of $\bar{R}^T\bar{R}$, it follows from the classical perturbation theory for eigenvalues of symmetric matrices (Stewart, 1973; Wilkinson, 1965) that for $i = 1, 2, \ldots, p$

$$|\tilde{\sigma}_i^2-\bar{\sigma}_i^2| \leqslant \|R^TE+E^TR+E^TE\| \leqslant 2\sigma_1\varepsilon+\varepsilon^2,$$

where $\sigma_1 = \|R\|$ and $\varepsilon = \|E\|$. In particular

$$1-\frac{2\sigma_1\varepsilon+\varepsilon^2}{\tilde{\sigma}_i^2} \leqslant \left(\frac{\bar{\sigma}_i}{\tilde{\sigma}_i}\right)^2 \leqslant 1+\frac{2\sigma_1\varepsilon+\varepsilon^2}{\tilde{\sigma}_i^2},$$

and it follows from the inequality $|1-\sqrt{1+x}| \leqslant |x|$ that

$$\tilde{\sigma}_i-\frac{2\sigma_1\varepsilon+\varepsilon^2}{\tilde{\sigma}_i} \leqslant \bar{\sigma}_i \leqslant \tilde{\sigma}_i+\frac{2\sigma_1\varepsilon+\varepsilon^2}{\tilde{\sigma}_i}. \tag{2.1}$$

Now

$$\|\tilde{R}-\bar{R}\| \geqslant \max|\tilde{\sigma}_i-\bar{\sigma}_i|;$$

hence (2.1) has the disturbing implication that $\|\tilde{R}-\bar{R}\|$ can be as large as $(2\sigma_1\varepsilon+\varepsilon^2)/\tilde{\sigma}_p$. In particular if $\tilde{\sigma}_p \leqslant \sqrt{2\sigma_1\varepsilon+\varepsilon^2}$, we cannot guarantee that $\tilde{\sigma}_p$ and $\bar{\sigma}_p$ agree in any significant figures.

Casting the results in terms of relative errors (e.g. rounding errors) may make this clearer. Suppose that the original matrix R has nonzero elements all of about the same size, and these elements are perturbed by a relative error of order ε_M. Then in the above, $\varepsilon \cong \varepsilon_M\|R\| = \varepsilon_M\sigma_1$, so that if

$$\frac{\sigma_1}{\tilde{\sigma}_p} \geqslant \sqrt{\varepsilon_M} \tag{2.2}$$

$\tilde{\sigma}_p$ may be obliterated by the error in R. The square root has the implication that in downdating one cannot tolerate spread of singular values of half the computational precision without losing all precision in the smallest singular value.

The significance of these results depends on the application. If one is concerned principally with A, then they are not very disturbing. In particular, if $\bar{A} = \bar{R}^T\bar{R}$ and $\tilde{A} = A-xx^T$, then it follows that

$$\|\tilde{A}-\bar{A}\| = \|R^TE-E^TR+E^TE\| \leqslant 2\sigma_1^2\varepsilon_M(1+\varepsilon_M).$$

Consequently, since $\sigma_1^2 = \|A\|$

$$\frac{\|\tilde{A}-\bar{A}\|}{\|\bar{A}\|} \leqslant \varepsilon_M(1-\varepsilon_M)\frac{\|A\|}{\|\bar{A}\|}$$

and it follows that $\bar{A}$ has high relative error only if $\|\bar{A}\|$ is appreciably smaller than $\|A\|$.

On the other hand, if one is concerned with X, the results imply that there may be no $\bar{X}$ near $\tilde{X}$ such that $\bar{X}$ has the QR factorization $\bar{X} = \bar{Q}\bar{R}$. This follows from the fact that the singular values of $\tilde{X}$ (resp. $\bar{X}$) and $\tilde{R}(\bar{R})$ are the same. Hence

$$\|\tilde{X}-\bar{X}\| \geqslant \max |\tilde{\rho}_i - \bar{\sigma}_i|,$$

which can be as large as $\sqrt{\varepsilon_M}\|X\| \geqslant \sqrt{\varepsilon_M}\|\tilde{X}\|$. It is interesting to note that no such phenomenon occurs when one appends a row to X; in this case $\|\tilde{X}-\bar{X}\|$ is always of order $\varepsilon_M\|\tilde{X}\|$ or less.

Perturbations in x have much the same effect. If

$$\bar{R}^T\bar{R} = R^TR-(x+f)(x+f)^T,$$

then a repetition of the above argument shows that

$$|\sigma_i - \bar{\sigma}_i| \leqslant \frac{2\|x\|\,\|f\|+\|f\|^2}{\bar{\sigma}_i}.$$

That the bound (2.1) is realistic can be seen by considering the scalar case $p = 1$. This case actually arises in practice; for in the regression problem mentioned in Section 1, X becomes an n-vector, x becomes a component of X, and R becomes $\|X\|$. Thus the downdating problem becomes: given the norm of a vector find the new norm after a component has been removed from the vector. The results of this section have the following implications for downdating norms in t-digit arithmetic. If ever a sequence of downdates reduces the norm by a factor greater than $10^{t/2}$, the results can be expected to be completely spurious.

3. The Algorithm

The algorithm described in this section is an extension of one due to Saunders (1972) (see also Gill *et al.*, 1974, Golub & Styan, 1974, or Lawson & Hanson, 1974). We shall use the notation introduced in Section 1. We assume that the reader is familiar with computations with plane rotations (for details see Stewart (1973) or Wilkinson (1965)). In order that the several computational steps of the algorithm will not be lost in the derivations, we proceed immediately to a description of the entire algorithm.

1. Solve the system $a^TR = x^T$.
2. If $\|a\| \geqslant 1$, report R^TR-xx^T indefinite and stop.
3. Compute $\alpha = \sqrt{1-\|a\|^2}$.
4. For $i = p, p-1, \ldots, 1$ determine plane rotations U_i in the $(i, p+1)$ plane such that

$$U_1 \ldots U_{p-1} U_p \begin{pmatrix} a \\ \alpha \end{pmatrix} = \begin{pmatrix} 0 \\ 1 \end{pmatrix}.$$

5. Calculate

$$\begin{pmatrix} \tilde{R} \\ x^T \end{pmatrix} = U_1 \ldots U_{p-1} U_p \begin{pmatrix} R \\ 0 \end{pmatrix}.$$

To justify this algorithm, we first show that the condition $\|a\| < 1$ is necessary and sufficient for $R^T R - xx^T$ to be positive definite. In fact

$$(R^T R - xx^T) = R^T (I - aa^T) R. \tag{3.1}$$

Now the eigenvalues of $I - aa^T$ are $1 - \|a\|^2$ of multiplicity unity and 1 of multiplicity $p-1$. It follows that $I - aa^T$, and hence $\tilde{R}^T \tilde{R}$, is positive definite if and only if $\|a\|^2 < 1$.

Let $Q = U_1 \ldots U_{p-1} U_p$. Then

$$Q \begin{pmatrix} a & R \\ \alpha & 0 \end{pmatrix} = \begin{pmatrix} 0 & \tilde{R} \\ \beta & b^T \end{pmatrix}, \tag{3.2}$$

in which we must verify that $\beta = 1$ and $b = x$. Since $Q^T Q = I$, if each side of (3.2) is multiplied by its transpose, the result is

$$\begin{pmatrix} 1 & a^T R \\ R^T a & R^T R \end{pmatrix} = \begin{pmatrix} \beta^2 & \beta b^T \\ \beta b & \tilde{R}^T \tilde{R} + bb^T \end{pmatrix}.$$

It follows immediately that $\beta = 1$, $b = x$, and

$$R^T R = \tilde{R}^T \tilde{R} + xx^T.$$

But it is easily seen from the form of the plane rotations U_i that $\tilde{R}$ is upper triangular. Hence $\tilde{R}$ is the downdated Cholesky factor.

In applications to regression problems, it is necessary to compute $\tilde{z}$ and $\tilde{\rho}$. One way of approaching this is to observe that the Cholesky factor of $(X, y)^T (X, y)$ is

$$\begin{pmatrix} R & z \\ 0 & \rho \end{pmatrix}. \tag{3.3}$$

Thus the new decomposition can be determined by removing $(x^T \eta)$ from (3.3). However we prefer to use a different algorithm for two reasons. First if one has several vectors y, the algorithm must be repeated for each one, with considerable diseconomies in time and storage. Second, the augmented downdating may fail, even though R by itself can be downdated, and it is desirable not to confound these sources of failure.

In the description of our proposed algorithm below, c_i and s_i are the cosines and sines defining the plane rotations U_i.

1. Set $\tilde{\eta}_0 = \eta$.
2. For $i = 1, 2, \ldots, p$ compute

$$\begin{aligned} \tilde{z}_i &= (z_i + s_i \tilde{\eta}_{i-1})/c_i, \\ \eta_i &= s_i \tilde{z}_i + c_i \tilde{\eta}_{i-1}. \end{aligned} \tag{3.4}$$

3. If $\tilde{\eta}_p > \rho$ stop.
4. $\tilde{\rho} = \sqrt{\rho^2 - \tilde{\eta}_p^2}$.

To show that these formulas indeed produce the required $\tilde{z}$ and $\tilde{\rho}$, we first observe that they are well defined, since $\alpha \neq 0$ implies that no c_i can be zero. Now the two relations in step two of (3.4) are equivalent to

$$\begin{pmatrix} c_i & -s_i \\ s_i & c_i \end{pmatrix}\begin{pmatrix} \tilde{z}_i \\ \tilde{\eta}_{i-1} \end{pmatrix} = \begin{pmatrix} z_i \\ \tilde{\eta}_i \end{pmatrix}. \tag{3.5}$$

It follows that

$$\begin{pmatrix} z \\ \tilde{\eta}_p \end{pmatrix} = U_p^T \ldots U_2^T U_1^T \begin{pmatrix} \tilde{z} \\ \eta \end{pmatrix} = Q^T \begin{pmatrix} \tilde{z} \\ \eta \end{pmatrix},$$

whence

$$Q\begin{pmatrix} R & z \\ 0 & \tilde{\eta}_p \end{pmatrix} = \begin{pmatrix} \tilde{R} & \tilde{z} \\ x^T & \eta \end{pmatrix}.$$

It then follows that

$$\begin{pmatrix} R & z \\ 0 & \rho \end{pmatrix}^T \begin{pmatrix} R & z \\ 0 & \rho \end{pmatrix} = \begin{pmatrix} \tilde{R}^T\tilde{R} + xx^T & \tilde{R}^T\tilde{z} + \eta x \\ \tilde{z}^T\tilde{R} + \eta x^T & z^Tz + \rho^2 \end{pmatrix}.$$

But $z^Tz + \rho^2 = \tilde{z}^T\tilde{z} + \tilde{\rho}^2 + \eta^2$; hence

$$\begin{pmatrix} R & z \\ 0 & \rho \end{pmatrix}^T \begin{pmatrix} R & z \\ 0 & \rho \end{pmatrix} - \begin{pmatrix} x \\ \eta \end{pmatrix}(x\eta)^T = \begin{pmatrix} \tilde{R} & \tilde{z} \\ 0 & \tilde{\rho} \end{pmatrix}^T \begin{pmatrix} \tilde{R} & \tilde{z} \\ 0 & \tilde{\rho} \end{pmatrix},$$

and $\tilde{z}$ and $\tilde{\rho}$ comprise the last column of the Cholesky factor of the downdated augmented system.

We note that if $\eta_p > \rho$, then ρ is not large enough to accommodate the decrease in the residual due to the deletion of $(x^T\eta)$, and the algorithm should be stopped.

4. The Effects of Rounding Error

In this section we shall adopt the conventions and assumptions usual in floating-point rounding-error analyses. If e is an arithmetic expression with a specified order of evaluation, $fl(e)$ will denote the result of evaluating e in floating-point arithmetic. We shall assume that floating-point multiplication and division satisfy

$$fl(a \circ b) = a \circ b(1+\varepsilon), \quad \circ = \times, \div,$$

where

$$|\varepsilon| \leq \varepsilon_M.$$

Here ε_M is the rounding unit of the computer in question (i.e. ε_M is approximately the largest number ε for which $fl(1+\varepsilon) = 1$). We assume addition and subtraction satisfy

$$fl(a \pm b) = a(1+\varepsilon_1) \pm b(1+\varepsilon_2)$$

where $|\varepsilon_1|, |\varepsilon_2| \leq \varepsilon_M$. Finally we assume that

$$fl(\sqrt{a}) = (1+\varepsilon)\sqrt{a},$$

where again $|\varepsilon| \leqslant \varepsilon_M$. As is customary, we ignore problems of overflow and underflow.

In order to simplify our bounds we shall freely discard higher order terms in ε_M. For example $(1+\varepsilon_M)(1+\varepsilon_M)$ will be approximated by $1+2\varepsilon_M$. Although our results will no longer have the status of theorems, their derivation will be considerably less cluttered. Moreover, if p is sufficiently small, say $p\varepsilon_M < 0{\cdot}01$, then the bounds can be made rigorous by multiplying by a factor near unity.

We begin with the computation of a. Here, and in what follows, all quantities stand for their computed, not their true, values. The solution of triangular systems has been analysed elsewhere (Stewart, 1973; Wilkinson, 1965), and we merely quote the results. The vector a satisfies

$$a^T(R+F) = x^T \tag{4.1}$$

where

$$|f_{ij}| \leqslant (j+2)|r_{ij}|\varepsilon_M. \tag{4.2}$$

It follows that if r_j and f_j denote the jth columns of R and F, then

$$\|f_j\| \lesssim \sqrt{j}(j+2)\|r_j\|\varepsilon_M. \tag{4.3}$$

We turn now to the computation of α. If we compute α^2 in the order $1-(a_1^2+a_2^2+\ldots+a_p^2)$ we have

$$\alpha^2 = (1+\varepsilon_0)-a_1^2(1+\varepsilon_1)-a_2^2(1+\varepsilon_2)-\ldots-a_p^2(1+\varepsilon_p)$$

where $|\varepsilon_0| \leqslant \varepsilon_M$ and $|\varepsilon_i| \leqslant (p-i+3)\varepsilon_M (i \geqslant 1)$. Hence, since $\|a\|^2 < 1$,

$$\alpha = (1+\tau_1)\sqrt{1-\|a\|^2+\tau_2} \tag{4.4a}$$

where

$$|\tau_1| \lesssim \varepsilon_M, \qquad |\tau_2| \lesssim (p+3)\varepsilon_M.$$

The computation and application of plane rotations has been analysed in detail by Wilkinson (1965), where he shows that there are exact rotations $\hat{U}_1, \hat{U}_2, \ldots, \hat{U}_p$ such that for any vector v

$$fl(U_1 \ldots U_{p-1}U_pv) = \hat{U}_1 \ldots \hat{U}_{p-1}\hat{U}_pv+g$$

where

$$\|g\| \lesssim 6p\|v\|\varepsilon_M. \tag{4.4b}$$

Here we have suppressed some second order terms that account for the slow growth in a bound on $\|U_1 \ldots U_{p-1}U_pv\|$.

Let

$$\hat{Q} = \hat{U}_1 \ldots \hat{U}_{p-1}\hat{U}_p. \tag{4.5}$$

We first consider the application of $\hat{Q}$ to the vector $(a^T, \alpha)^T$. From the results quoted above

$$\hat{Q}\begin{pmatrix} a \\ \alpha \end{pmatrix} = \begin{pmatrix} g_0 \\ \beta \end{pmatrix} \tag{4.6}$$

where

$$\|g_0\| \lesssim 6p\varepsilon_M(\|a\|^2+\alpha^2)^{1/2}.$$

Now from (4.4a)

$$\|a\|^2+\alpha^2 = \|a\|^2+(1+\tau_1)^2(1-\|a\|^2+\tau_2)$$
$$\cong 1+2\|a\|\tau_1+\tau_2.$$

Hence

$$\|g_0\| \lesssim 6p\varepsilon_M \tag{4.7}$$

and

$$\begin{aligned}\beta &= (\|a\|^2+\alpha^2-\|g_0\|^2)^{1/2}\\ &= 1+\sigma_0,\end{aligned} \tag{4.8}$$

where

$$|\sigma_0| \lesssim \frac{p+5}{2}\,\varepsilon_M. \tag{4.9}$$

We next consider the application of $\hat{Q}$ to $(r_j^T, 0)^T$. We have

$$\hat{Q}\begin{pmatrix} r_j \\ 0 \end{pmatrix} = \begin{pmatrix} \tilde{r}_j+g_j \\ \xi_j+\gamma_j \end{pmatrix}, \tag{4.10}$$

where

$$\|g_j\|,\ |\gamma_j| \lesssim 6p\|r_j\|\varepsilon_M. \tag{4.11}$$

Here ξ_j is the computed value. We wish to find a bound on $|x_j-\xi_j|$. Since $\hat{Q}$ is orthogonal, we have from (4.6), (4.8) and (4.10) that

$$\begin{aligned} a^T r_j &= (g_0^T, \beta)\begin{pmatrix} \tilde{r}_j+g_j \\ \xi_j+\gamma_j \end{pmatrix} \\ &\cong \beta\xi_j+\gamma_j+\tilde{g}_0^T r_j \\ &= \xi_j+\sigma_0\xi_j+\gamma_j+g_0^T\tilde{r}_j. \end{aligned}$$

But from (4.1)

$$a^T r_j = x_j + a^T f_j.$$

Hence

$$x_j-\xi_j \cong \sigma_0\xi_j+\gamma_j+g_0^T\tilde{r}_j-a^T f_j.$$

Since up to terms of order ε_M, $\|r_j\| = \|(\tilde{r}_j^T, \xi_j)\|$, we have from (4.3), (4.7), (4.9), and (4.11) that

$$x_j-\xi_j = \sigma_j$$

where

$$|\sigma_j| \lesssim \left[\frac{13p+5}{2}+\sqrt{j}(j+2)\right]\|r_j\|\varepsilon_M. \tag{4.12}$$

To summarize we have shown that there is an orthogonal matrix $\hat{Q}$ such that

$$\hat{Q}\begin{pmatrix} R \\ 0 \end{pmatrix} = \begin{pmatrix} \tilde{R}+G \\ x^T+s^T \end{pmatrix},$$

where G and s satisfy (4.11) and (4.12). In other words, the computed downdated Cholesky factor $\tilde{R}$ is very near the factor obtained by downdating with a slightly perturbed vector x. The error G in $\tilde{R}$ is unimportant, except as it may affect subsequent downdates; however, the results of Section 2 show that the error s in x may seriously affect the accuracy of $\tilde{R}$.

Two other points. First, the higher order term in (4.12) is due to the solution of the triangular system $a^T R = x^T$. The factor $(j+2)$ can be removed from this term by accumulating inner products in double precision; however, in practice this is unnecessary, since the term does not dominate its companion (and this only in column p) until $p = 40$, and it is not yet double when $p = 150$.

Second, the bounds are given column by column and hence are independent of column scaling. This is not surprising, since the computations in each column are independent of one another.

We turn now to the analysis of the errors involved in downdating z. Define

$$w_i = (z_1, z_2, \ldots, z_{i-1}, \tilde{z}_i, \ldots, \tilde{z}_p, \tilde{\eta}_i)^T,$$

so that from (3.5)

$$w_i = fl(U_i^T w_{i-1}), \quad i = 1, 2, \ldots, p.$$

However, the evaluation of $fl(U^T w_{i-1})$ is not the straightforward one implied by (3.5); rather it is the indirect one implied by the formulas in step 2 of (3.4), which we now analyse. We have

$$\tilde{z}_i = \frac{[z_i(1+\varepsilon_1)+s_i\tilde{\eta}_{i-1}(1+\varepsilon_2)(1+\varepsilon_3)]}{c_i}(1+\varepsilon_4)$$

where $|\varepsilon_i| \leqslant \varepsilon_M$. Thus

$$z_i = c_i\tilde{z}_i(1+\varepsilon_1)^{-1}(1+\varepsilon_4)^{-1} - s_i\tilde{\eta}_{i-1}(1+\varepsilon_2)(1+\varepsilon_3)(1+\varepsilon_1)^{-1},$$

and it follows that

$$z_i = c_i\tilde{z}_i(1+\varepsilon_5) - s_i\tilde{\eta}_{i-1}(1+\varepsilon_6)$$

where $|\varepsilon_5| \lesssim 2\varepsilon_M$ and $|\varepsilon_6| \lesssim 3\varepsilon_M$. Likewise

$$\tilde{\eta}_i = s_i\tilde{z}_i(1+\varepsilon_7) + c_i\tilde{\eta}_{i-1}(1+\varepsilon_8),$$

where $|\varepsilon_7|, |\varepsilon_8| \lesssim 2\varepsilon_M$.

These results show that as far as rounding errors are concerned, the formulas in (3.4) are equivalent to the direct application of (3.5), with the exception that the term ε_6 has the bound $3\varepsilon_M$ instead of $2\varepsilon_M$. This means that the error analysis of Wilkinson cited above goes through *mutatis mutandis*, with the result that the right-hand side of the bound (4.4b) becomes $7p\|v\|\varepsilon_M$. Hence

$$w_p = \hat{Q}^T w_0 + g,$$

where $\|g\| \lesssim 7p\|w_0\|\varepsilon_M$. Since $\hat{Q}$ is orthogonal,

$$\hat{Q}\begin{pmatrix} z \\ \tilde{\eta} \end{pmatrix} = \begin{pmatrix} \tilde{z}+h \\ \eta+\tau \end{pmatrix},$$

where

$$\|h\|, |\tau| \lesssim 7p(\|\tilde{z}\|^2+\eta^2)\varepsilon_M.$$

Thus the computed $\tilde{z}$ is very near the vector that would be obtained by downdating z with a slightly perturbed η. It should be noted that the transformation $\hat{Q}$ is the same as the one defined by (4.5) in the previous analysis.

It goes without saying that these bounds are an extreme over-estimate of the errors that would be encountered in practice. None the less they suffice to demonstrate the exceptional stability of the algorithm. Any inaccuracies observed in the results cannot be attributed to the algorithm; they must instead be due to the ill-conditioning of the problem. This raises the question: does the algorithm provide some way of detecting ill conditioning? We shall answer this question in the next section.

5. The Meaning of $\|a\|$

It is a consequence of the results of Section 2 that ill-conditioning in the downdating problem is associated with small singular values in $\tilde{R}$. In Section 3 it was shown that if $\|a\| = 1$, then $\tilde{R}$ is singular, i.e. $\tilde{\sigma}_p = 0$. It is therefore reasonable to conjecture that values of a, near unity will be associated with ill-conditioned problems and vice versa. However, just as the determinant is a poor indicator of the condition of a matrix, the value of $\|a\|$ may be a poor indicator of the condition of the downdating problem. In this section we shall show that the value of $\|a\|$ will reliably signal trouble.

We first show that the value of $\|a\|$ cannot cry wolf; if it is near unity, then the problem must be ill-conditioned. It follows from (3.1) and the fact that the smallest eigenvalue of $I - aa^T$ is $1 - \|a\|^2$ that the smallest eigenvalue of $\tilde{R}^T\tilde{R}$ is

$$\lambda_{\min}(\tilde{R}^T\tilde{R}) \leqslant \|R\|_2^2(1 - \|a\|^2).$$

Since $\tilde{\sigma}_p^2 = \lambda_{\min}(\tilde{R}^T\tilde{R})$,

$$\frac{\tilde{\sigma}_p}{\sigma_1} \leqslant \sqrt{1 - \|a\|^2}.$$

It follows from the discussion surrounding (2.2) that if $1 - \|a\|^2 = O(\varepsilon_M)$ then $\tilde{R}$ can be expected to lose about half its accuracy.

We cannot show that a small value of $\tilde{\sigma}_p$ implies that $\|a\|$ is near unity. However we can show that if any singular value of R is reduced in the downdating by a significant factor, then $\|a\|$ must be near unity. We start by developing an expression for $\|a\|^2$. First

$$\begin{aligned}\|a\|^2 &= x^T R^{-1} R^{-T} x = x^T (R^T R)^{-1} x \\ &= x^T(\tilde{R}^T\tilde{R} + xx^T)^{-1} x \\ &= x^T \tilde{R}^{-1}(I + \tilde{R}^{-T} x x^T \tilde{R}^{-1})\tilde{R}^{-T} x.\end{aligned}$$

Set

$$b = \tilde{R}^{-T} x,$$

so that

$$\|a\|^2 = b^T(I + bb^T)^{-1} b.$$

Since b is an eigenvector of $I + bb^T$ corresponding to the eigenvalue $1 + \|b\|^2$, it follows that

$$\|a\|^2 = \frac{\|b\|^2}{1 + \|b\|^2}. \tag{5.1}$$

We next obtain a lower bound on $\|b\|^2$. Let $\tilde{v}_i$ be the right singular vector corresponding to $\tilde{\sigma}_i$ and let $\tilde{V}_i = (\tilde{v}_i, \tilde{v}_{i+1}, \ldots, \tilde{v}_p)$. Then if $\tilde{V}_i^T x \neq 0$

$$\|b\| = \|R^{-T}x\| \geq \frac{\|\tilde{V}_i^T x\|}{\tilde{\sigma}_i}. \tag{5.2}$$

But from the minimax theorems (Stewart, 1973; Wilkinson, 1965)

$$\sigma_i^2 \leq \|\tilde{V}_i^T R^T R \tilde{V}_i\| \leq \|\tilde{V}_i^T \tilde{R}^T \tilde{R} \tilde{V}_i\| + \|\tilde{V}_i^T x x^T \tilde{V}_i\|$$
$$= \tilde{\sigma}_i^2 + \|V_i^T x\|^2.$$

Hence

$$\|\tilde{V}_i^T x\|^2 \geq \sigma_i^2 - \tilde{\sigma}_i^2. \tag{5.3}$$

Combining (5.1), (5.2), and (5.3) gives

$$\|a\|^2 \geq \frac{(\sigma_i/\tilde{\sigma}_i)^2 - 1}{(\sigma_i/\tilde{\sigma}_i)^2 + 1}.$$

Thus a large value of $(\sigma_i/\tilde{\sigma}_i)^2$ will be reflected by the nearness of $\|a\|^2$ to unity.

I would like to thank Dr Michael Saunders, whose good advice introduced me to the algorithm and inspired me to analyse it.

References

DEMPSTER, A. P. 1969 *Elements of Continuous Multivariate Analysis*. Reading, Massachusetts: Addison-Wesley.

GILL, P. E., GOLUB, G. H., MURRAY, W. & SAUNDERS, M. A. 1974 Methods for modifying matrix factorizations. *Maths Comput.* **28**, 505–535.

GOLUB, G. H. 1965 Numerical methods for solving least squares problems. *Num. Math.* **7**, 206–216.

GOLUB, G. H. & STYAN, G. P. 1974 Numerical computations for univariate linear models, *J. Stat. Comput. Simul.* **2**, 253–274.

LAWSON, C. L. & HANSON, R. J. 1974 *Solving Least Squares Problems*. Englewood Cliffs, N.J.: Prentice-Hall.

SAUNDERS, M. A. 1972 Large scale linear programming using the Cholesky factorization. Stanford University report STAN–CS–72–252.

STEWART, G. W. 1973 *Introduction to Matrix Computations*. New York: Academic Press.

WILKINSON, J. H. 1965 *The Algebraic Eigenvalue Problem*. Oxford: Clarendon Press.

13.4. [GWS-J73] "An Updating Algorithm for Subspace Tracking"

[GWS-J73] "An Updating Algorithm for Subspace Tracking," *IEEE Transactions on Signal Processing* **40** (1992) 1535–1541.
http://dx.doi.org/10.1109/78.139256

An Updating Algorithm for Subspace Tracking

G. W. Stewart

***Abstract*—In certain signal processing applications it is required to compute the null space of a matrix whose rows are samples of a signal with p components. The usual tool for doing this is the singular value decomposition. However, the singular value decomposition has the drawback that it requires $O(p^3)$ operations to recompute when a new sample arrives. In this paper, we show that a different decomposition, called the URV decomposition, is equally effective in exhibiting the null space and can be updated in $O(p^2)$ time. The updating technique can be run on a linear array of p processors in $O(p)$ time.**

I. Introduction

MANY problems in digital signal processing require the computation of an approximate null space of an $n \times p$ matrix A whose rows represent samples of a signal (see [9] for examples and references). Specifically, we must find an orthogonal matrix $V = (V_1 V_2)$ such that 1) AV_1 has no small singular values; 2) AV_2 is small.

In this case we say that A has approximate rank k, where k is the number of columns in V_1. In applications V_1 corresponds to the signal while V_2 corresponds to noise. We will call the space spanned by the columns of V_2 the error space.

As the signal changes, so does its error space. Since the *ab initio* computation of an error space is expensive, it is desirable to use the previously computed error space adaptively to approximate the new error space—a process that is generally called *updating*. Our specific updating problem can be described as follows. Given the error space of a matrix A compute the error space of the matrix

$$A_z = \begin{pmatrix} \beta A \\ z^H \end{pmatrix}$$

where z is a new sample and $\omega \leq 1$ is a "forgetting factor" that damps out the effect of the previous samples. To simplify the exposition, we will take $\beta = 1$ in this paper (however, see the end of Section II, where the problem of tolerances is treated).

The usual approach to computing error spaces has been via the singular value decomposition [4], [8]. Specifically, there are orthogonal matrices U and V such that

$$U^H AV = \begin{pmatrix} \Sigma \\ 0 \end{pmatrix}$$

where

$$\Sigma = \text{diag}(\sigma_1, \cdots, \sigma_p)$$

with

$$\sigma_1 \geq \cdots \geq \sigma_p.$$

The procedure for computing error spaces is to determine an integer k such that σ_k is above the noise level, while σ_{k+1} is below it. The columns of V corresponding to $\sigma_{k+1}, \cdots, \sigma_n$ then span the error space.

Although the singular value decomposition furnishes an elegant solution to the problem of calculating error spaces, it has two disadvantages: it is expensive to compute and it is difficult to update. The initial cost of computing a singular value decomposition would not be an objection, if the decomposition could be cheaply updated; however, all known updating schemes require on the order of p^3 operations (e.g., see [2]). Recently, abridged updating schemes that produce an approximate singular value decomposition have been proposed [7]. However, the effectiveness of this approach has not yet been demonstrated.

The difficulties in working with the singular value decomposition have sparked an interest in rank revealing QR decompositions, which decompose the matrix into the product of an orthogonal matrix, an upper triangular matrix, and a permutation matrix in such a way that the effective rank of the matrix is obvious [3]. However, a QR decomposition—even a rank-revealing one—does not provide an explicit basis for the error space. In this paper, we will consider an intermediary between the singular value decomposition and the QR decomposition, a two-sided orthogonal decomposition that we will call the URV decomposition, that has some of the virtues of both.

In the next section we will introduce the URV decomposition and its rank revealing variant. This section also contains a discussion of how to determine rank in the presence of errors. Since the updating will be accomplished by plane rotations, we give a brief review of their properties in Section III. In the following section we will show how to compute a rank revealing URV decomposition of a triangular matrix. This special case will be used in Section V where we show how to update a rank revealing URV decomposition in such a way that it remains rank revealing. In Section VI we will show that the updating algorithm can be implemented on a linear array of processors in such a way that it runs in $O(p)$ time. Finally, in the last sections we will make some general observations on the updating algorithm.

Manuscript received July 19, 1990; revised March 23, 1991. This work was supported in part by the Air Force Office of Scientific Research under Contract AFOSR-87-0188.

The author is with the Department of Computer Science and Institute for Advanced Computer Studies, University of Maryland, College Park, MD 20742.

IEEE Log Number 9107657.

1053-587X/92$03.00 © 1992 IEEE

Throughout this paper $\|\cdot\|$ will denote the Euclidean vector norm and the Frobenius matrix norm defined by

$$\|A\|^2 = \sum_{i,j} |a_{ij}|^2.$$

The smallest singular value of a matrix A will be written inf (A).

II. URV Decompositions

Suppose for the moment that A has rank k. Then there are orthogonal matrices U and V such that

$$A = U \begin{pmatrix} R & 0 \\ 0 & 0 \end{pmatrix} V^H \tag{2.1}$$

where R is an upper triangular matrix of order k. We will call this decomposition a URV decomposition. Unlike the singular value decomposition the URV decomposition is not unique; in fact, the singular value decomposition is itself a URV decomposition. However, we will be concerned with the case where R is not diagonal but fully triangular.

Now suppose that A is nearly of rank k in the sense that its singular values satisfy

$$\sigma_1 \geq \cdots \geq \sigma_k > \sigma_{k+1} \geq \cdots \geq \sigma_p$$

where σ_k is large compared to σ_{k+1}. It can be shown that there is a URV decomposition of A of the form

$$A = U \begin{pmatrix} R & F \\ 0 & G \end{pmatrix} V^H \tag{2.2}$$

where

1) R and G are upper triangular,
2) inf $(R) \cong \sigma_k$,
3) $\sqrt{\|F\|^2 + \|G\|^2} \cong \sqrt{\sigma_{k+1}^2 + \cdots + \sigma_p^2}$.

The singular value decomposition itself is an example of such a decomposition, but there are many others. We will call any such decomposition a *rank revealing* URV decomposition. From such a decomposition we can extract the error subspace, just as we did with the singular value decomposition. However, as we shall see, rank revealing URV decompositions are easier to compute and update than the singular value decomposition.

In practice the small singular values of A will come from noise, and the user must furnish a tolerance to distinguish them from the singular values associated with the signal. Unfortunately, the relation between noise and the small singular values is not simple, and the mathematical form of the tolerance is a matter of some delicacy.

We will adopt a very simple model. Suppose that A has the form

$$A = \hat{A} + E$$

where $\hat{A}$ has rank exactly k. We will assume that the errors are roughly of the same size, say ϵ, so that when the forgetting factor is taken into account, E has the form

$$E = \begin{pmatrix} \beta^{n-1} e_1^H \\ \beta^{n-2} e_2^H \\ \vdots \\ e_n^H \end{pmatrix}$$

where the components of the e_i are approximately ϵ in size. Let the columns of V_2 form an orthonormal basis for the error space of $\hat{A}$. Then our tolerance should approximate the norm of

$$AV_2 = EV_2$$

(remember $\hat{A}V_2 = 0$). Now the ith row of EV_2 consists of $p - k$ elements of size roughly $\beta^{n-1}\epsilon$. Consequently,

$$\|EV_2\|^2 \cong (p-k)\epsilon^2 \sum_{i=1}^{n} \beta^{2(n-i)} < \frac{(p-k)\epsilon}{1-\beta^2}.$$

Consequently the tolerance, we call it tol, should be chosen so that

$$\text{tol} \geq \sqrt{\frac{p-k}{1-\beta^2}}\,\epsilon. \tag{2.3}$$

Note that it is better to choose tol a little too large than too small. In the latter case, the dimension of the error space will be underestimated. On the other hand, if the tolerance is a little too large and there is a good signal-to-noise ratio, the tolerance will insinuate itself between the signal and the noise, and the dimension of the error space will be correctly estimated.

III. Plane Rotations

The chief computational tool of this paper is the plane rotation, which will be used to introduce zeros selectively into matrices to be updated. Since treatments of plane rotations are widely available (e.g., see [4]), we will not go into the numerical details here. Instead we will sketch the few basic facts needed to understand the updating algorithm and introduce some conventions for describing reductions based on plane rotations.

Fig. 1 shows two rows of a matrix before and after the application of a plane rotation. The X's represent nonzero elements, the 0's represent zero elements, and the E's represent small elements. The plane rotation has been chosen to introduce a zero into the position occupied by the checked X in column 2. When the rotation is applied the following rules hold:

1) A pair of X's remains a pair of X's (columns 1 and 3).

2) An X and an O are replaced by a pair of X's (column 4).

3) A pair of O's remains a pair of O's (column 5).

4) An X and an E are replaced by a pair of X's (column 6).

5) A pair of E's remains a pair of E's (column 7).

BEFORE

1	2	3	4	5	6	7
X	X	X	O	O	X	E
X	$\check{X}$	X	X	O	E	E

AFTER

1	2	3	4	5	6	7
X	X	X	X	O	X	E
X	O	X	X	O	X	E

Fig. 1. Application of a plane rotation.

$$\begin{array}{ccccc} \rightarrow & r & r & r & r \\ & 0 & r & r & r \\ & 0 & 0 & r & r \\ & 0 & 0 & 0 & r \\ \rightarrow & \check{x} & x & x & x \end{array} \Longrightarrow \begin{array}{ccccc} & r & r & r & r \\ \rightarrow & 0 & r & r & r \\ & 0 & 0 & r & r \\ & 0 & 0 & 0 & r \\ \rightarrow & 0 & \check{x} & x & x \end{array} \Longrightarrow \begin{array}{ccccc} & r & r & r & r \\ & 0 & r & r & r \\ \rightarrow & 0 & 0 & r & r \\ & 0 & 0 & 0 & r \\ \rightarrow & 0 & 0 & \check{x} & x \end{array} \Longrightarrow$$

$$\begin{array}{ccccc} & r & r & r & r \\ & 0 & r & r & r \\ & 0 & 0 & r & r \\ \rightarrow & 0 & 0 & 0 & r \\ \rightarrow & 0 & 0 & 0 & \check{x} \end{array} \Longrightarrow \begin{array}{cccc} r & r & r & r \\ 0 & r & r & r \\ 0 & 0 & r & r \\ 0 & 0 & 0 & r \\ 0 & 0 & 0 & 0 \end{array}$$

Fig. 2. Reduction to triangular form.

The fact that a pair of small elements remains small (column 7) follows from the fact that a plane rotation is orthogonal and cannot change the norm of any vector to which it is applied. This is one of the key observations of this paper since the point of the updating algorithm is to keep small elements small.

It requires about $4p$ multiplications and $2p$ additions to apply a plane rotation to two rows of length p. The multiplication count can be reduced by using so-called fast rotations; however, in either case the work involved is $O(p)$.

Premultiplication by a plane rotation operates on the rows of the matrix. We will call such rotations *left rotations*. Postmultiplication by *right rotations* operates on the columns. Analogous rules hold for the application of a right rotation to two columns of a matrix.

When rotations are used to update a URV decomposition, the right rotations must be multiplied into V. To get a complete update of the decomposition, we must also multiply the left rotations into U. However, in most signal processing applications U is not needed, and this step can be omitted.

Algorithms that use plane rotations are best described by pictures. To fix our conventions, we will show how the matrix

$$\begin{pmatrix} R \\ x^H \end{pmatrix}$$

can be reduced to upper triangular form by left rotations. Here we assume the R is itself upper triangular.

The reduction is illustrated in Fig. 2. The elements of R and x^H are represented generically by r's and x's. The first step in the reduction is to eliminate the first element of x^H by a rotation that acts on x^H and the first row of R. The element to be eliminated has a check over it and the two rows that are being combined are indicated by the arrows to the left of the array.

According to this notation, the second step combines the second row of R with x^H to eliminate the second element of the latter. Note that r_{21}, which is zero, forms a pair of zeros with the first component of x^H, so that the zero we introduced in the first step is not destroyed in the second step. The third and fourth steps of the reduction are similar.

For column operations with right rotations we will use an analogous notation. The main difference is that the arrows will point down to the columns being combined.

IV. Deflation and Refinement

In applications we can expect occasional changes in the rank of the matrix A. An increase in rank usually makes itself felt in an obvious way. On the other hand, a decrease in rank can hide itself in the matrix R of the URV decomposition. Thus any updating algorithm must be able to detect rank degeneracy in R and act accordingly. In this section we will show how to compute a rank revealing URV decomposition of a $k \times k$ upper triangular matrix R.

The first step is to determine if R is defective in rank, that is, if inf (R) is less than a prescribed tolerance (see the discussion at the end of Section II). This problem has been extensively studied under the rubric of condition estimation (see [6] for a survey), and there exist reliable algorithms that, given a triangular matrix R, produce a vector w of norm one such that with

$$b = Rw$$

we have

$$\eta \equiv \|b\| \cong \inf(R). \tag{4.1}$$

The algorithms generally require a single solution of a triangular system whose matrix is R and therefore require only $O(k^2)$ work.

The next step is to determine a sequence $V_1^H, V_2^H, \cdots, V_{k-1}^H$ of rotations that eliminate the first $k - 1$ component of w, so that w is zero except for its last component, which is one. The reduction is illustrated in Fig. 3. Let $Q^H = V_{k-1}^H V_{k-2}^H \cdots V_1^H$ denote the product of the rotations obtained from this step.

Next we determine an orthogonal matrix P such that $P^H RQ$ is upper triangular. This may be done by applying $V_1, V_2, \cdots, V_{k-1}$ from the right to R as shown in Fig. 4. The result of applying a rotation V_i to two columns of R is to place a nonzero element below the diagonal of R. A left rotation then eliminates this element and restores triangularity. The matrix P^H is the product of the left rotations. The entire process requires only $O(k^2)$ time.

The appearance of the quantities e in the last array of Fig. 4 is meant to indicate that r_{kk} must be small on completion of the process. In fact, the norm of the last column

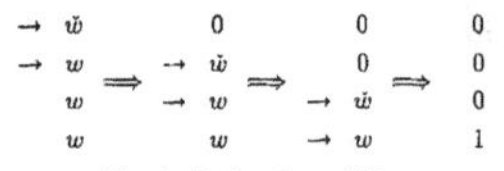

Fig. 3. Reduction of W.

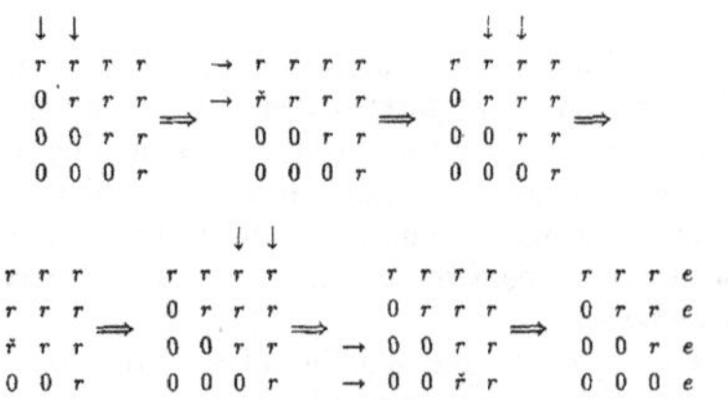

Fig. 4. Triangularization of RQ.

is the number η defined by (4.1). To see that this is true, note that from (4.1)

$$b' \equiv P^H b = (P^H RQ)(Q^H w) \equiv R'w'.$$

Since the last component of w' is one and all its other components are zero, we see that the last column of R is b'. Since the norm of a vector is unchanged when it is multiplied by a unitary matrix, it follows that

$$\|b'\| = \|b\| = \eta.$$

Since $\eta \cong \inf(R)$, we have produced a URV decomposition that reveals that R has a small singular value. We can continue the process with the leading principle matrix of R' of order $k - 1$.

Although this procedure solves the problem of deflating a rank deficient triangular matrix, it is possible to refine the decomposition, bringing it nearer to diagonality. The procedure begins by reducing the first $k - 1$ elements in the last column of R to zero. This is done by a sequence of right rotations. The process is illustrated in Fig. 5. Again only $O(k^2)$ time is required. However, the right rotations must be accumulated in V.

The second step in the refinement is to reduce R to upper triangular form by a sequence of left rotations as illustrated in Fig. 6.

This second step also requires $O(k^2)$ time, and the final matrix is clearly in URV form. However, more can be said: the norm of the last column is less than the absolute value of the (k, k)-element before the refinement was begun. To see this, note that in the first step of the refinement procedure, the (k, k)-element does not increase. Thus, at the end of the first step the norm of the last column is less than the absolute value of the (k, k)-element before the refinement was begun. The same is true at the end of the second step, since this step uses only right rotations and cannot change the norm of a column.

In practice, the refinement step can reduce the size of the part last column lying above the diagonal so that it is insignificant compared to the diagonal element. The effect of this is to polish the approximation to the error space.

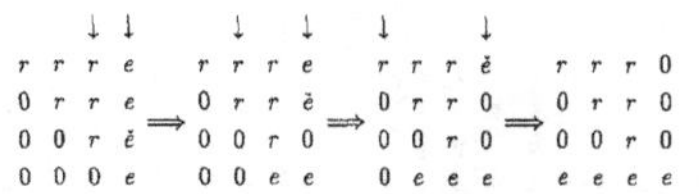

Fig. 5. Reducing the last column.

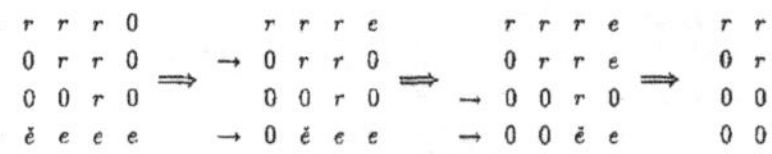

Fig. 6. Reducing the last row.

Whether the refinement is worth its cost will have to be determined experimentally. Preliminary runs indicate that the refined version is preferable to the unrefined version when it is used with the MUSIC algorithm [5].

V. Updating a URV Factorization

In this section we will show how to update a rank revealing URV decomposition of A when a row z^H is appended. Specifically, we will suppose that A has the URV decomposition (2.2), where V is known. To decide what is small we will suppose we have a user supplied tolerance, tol, and that

$$\nu \overset{\text{def}}{=} \sqrt{\|F\|^2 + \|G\|^2} \leq \text{tol}.$$

As is implied by (2.3), the tolerance may depend on dimensions of A and R and the size of the forgetting factor.

The first step is to compute

$$(x^H y^H) = z^H V$$

where x is of dimension k, i.e., the order of R. Our problem then becomes one of updating the matrix

$$\hat{A} = \begin{pmatrix} R & F \\ 0 & G \\ x^H & y^H \end{pmatrix}.$$

There are two cases to consider. The first, and simplest occurs when

$$\sqrt{\nu^2 + \|y\|^2} \leq \text{tol}. \tag{5.1}$$

In this case we reduce $\hat{A}$ to triangular form by a sequence of left rotations as in Fig. 2. Since the new value of ν will be given by the left-hand side of (5.1), we are assured that within our tolerance the rank cannot increase. However, it is possible for the rank to decrease. Hence we must check and possibly reduce R as described Section IV. The time required for this case is $O(p^2)$.

If (5.1) is not satisfied, there is a possibility that there is an increase in rank. Since the increase in rank can be at most one, the problem is to transform the matrix to upper triangular form without destroying all the small values in F and G.

The first step is to reduce y^H so that it has only one nonzero component and G remains upper triangular. The

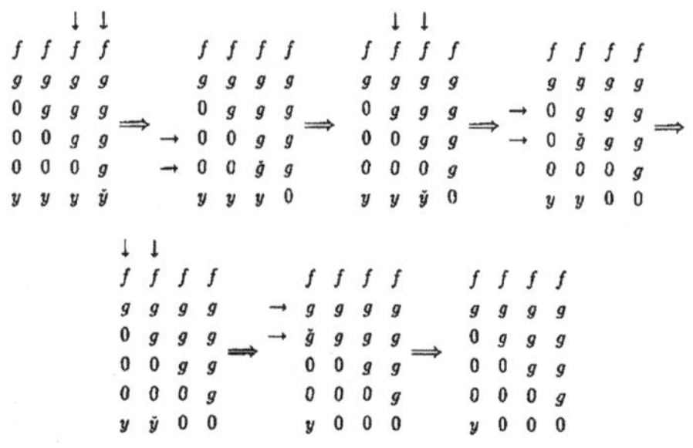

Fig. 7. Reduction of y_H.

reduction is illustrated in Fig. 7. Since R and x^H are not involved in this part of the reduction, we show only F, G, and y^H (n.b., the f's in the figure represent entire columns of F).

Finally, the entire matrix

$$\begin{matrix} R & f & f & f & f \\ 0 & g & g & g & g \\ 0 & 0 & g & g & g \\ 0 & 0 & 0 & g & g \\ 0 & 0 & 0 & 0 & g \\ x^H & y & 0 & 0 & 0 \end{matrix}$$

is reduced to triangular form in the usual way to give a matrix of the form

$$\begin{matrix} R & y & f & f & f \\ 0 & y & g & g & g \\ 0 & 0 & g & g & g \\ 0 & 0 & 0 & g & g \\ 0 & 0 & 0 & 0 & g \\ 0 & 0 & 0 & 0 & 0. \end{matrix} \tag{5.2}$$

Then k is increased by one, and the new R is checked for degeneracy and if necessary reduced as described in Section II. The result is the updated URV decomposition.

VI. Parallelization

In this section we will show that the updating algorithm can be implemented on an array of p processors to yield an $O(p)$ algorithm. To simplify matters, we will consider shared memory implementation; however, it will be clear that the algorithms can be implemented on a linear array of distributed memory processors, provided fine-grain communication is sufficiently fast, e.g., on a linear systolic array.

Since we are concerned with the existence of a parallel algorithm, rather than in the details of a particular implementation, we will use the method of *precedence diagrams*. The idea is to write down an order in which operations can be performed consistently and then assign operations to processors in such a way that no two simultaneous operations are performed by the same processor. In our case, the assignments will amount to making each processor responsible for a row of the matrix and its neighbors.

Fig. 8. Precedence diagram for $VP_1 \cdots P_{p-1}$.

We begin with the updating of V by rotations. Specifically, we desire to compute $VP_1 \cdots P_{p-1}$, where P_i combines columns i and $i+1$ of V. Fig. 8 shows a precedence diagram for this computation for $p = 5$. The circles represent elements of V. The numbers between any two circles represent the time at which a rotation can be applied to the two corresponding elements. We will call these numbers *ticks*.

The increasing ticks in the first row reflect the fact that P_i must be applied before P_{i+1}. From an arithmetic point of view, the ticks need not increase as we go down a column, since a right rotation can be applied simultaneously to all the elements of a pair of columns. We have allowed the ticks to increase to reflect the realities of communication: on a shared memory system, there will be contention along a column as the processors attempt to access the same rotation; in a distributed memory system the rotation must be passed down the column. In general, the method of precedence diagrams does not require one to write down the best diagram—only a correct one.

Fig. 9 shows the assignment of operations to processors. Operations between two horizontal lines are performed by the same processor, which, as we have observed, amounts to an assignment of the elements of V by rows to the processors. The diagram makes it obvious that the updating of V can be performed with p processors in about $2p$ ticks.

Let us now consider a more complicated case: the reduction of R in Fig. 4. The construction of an appropriate precedence diagram is simplified by the fact that the right rotations can be applied in their entirety before the left rotations are generated and applied. Thus we can construct the diagram for the right rotations and then fill in the left rotations in a suitable manner. The result is shown in Fig. 10. A number between two circles in a row represents a right rotation; between two circles in a column, a left rotation. From this diagram it is seen that the reduction can be carried out in about $3p$ ticks.

Finally, we consider the most difficult task of all: the construction of a precedence diagram for the reduction of y and G in Fig. 7. The difficulty is that the right and left rotations must be interspersed; for if they are not, the right rotations will fill out the bottom half of G. The precedence diagram is for the matrix G, the matrix F being in

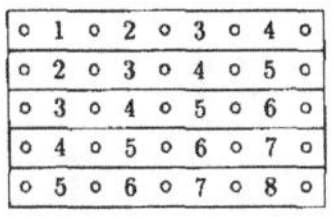

Fig. 9. Assignment of $VP_1 \cdots P_{p-1}$.

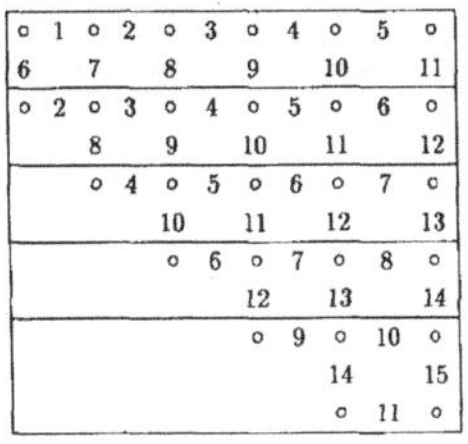

Fig. 10. Reduction of w and R.

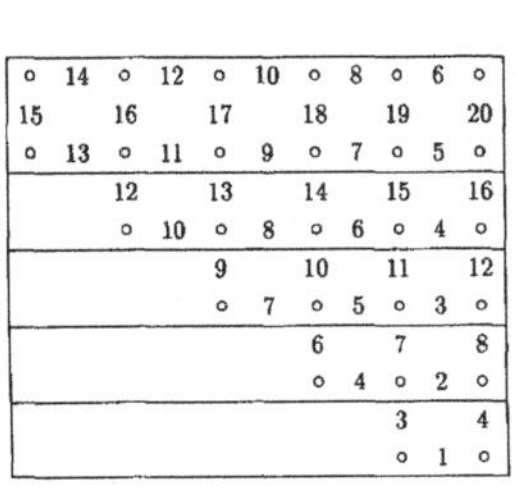

Fig. 11. Reduction of y and G.

Fig. 11 handled independently by other processors. This reduction requires about $4(p - k)$ ticks plus k ticks to apply the rotations to F.

Inspection of these diagrams shows that at each tick information passes only between neighboring processors. This implies that the algorithms are suitable for implementation on a linear systolic array.

The other parts of the updating algorithm can be analyzed similarly. In all cases a row oriented assignment of elements to processors results in parallel implementations that take $O(p)$ ticks.

VII. Comments

We have shown how to update a rank revealing URV factorization of a matrix using plane rotations. In this concluding section we will try to put our contribution in perspective.

Initialization. An attractive feature of the algorithm is that it requires no initial calculation of a decomposition. Instead one starts with a degenerate URV decomposition, in which $k = 0$, $F = 0$, and $V = I$, and applies the updating algorithm as the rows of A enter. This is an important economy when it comes to implementing the algorithm with special purpose hardware.

Efficiency. The total operation count for one update is a multiple of p^2. This should be contrasted with a cost of $O(p^3)$ for a complete update of a singular value decomposition. In addition, the updating algorithm is rich in left rotations, which need not be accumulated in some applications. In fact, if rank changes are rare, the algorithm will seldom use any right rotations.

Reliability. The singular value decomposition is generally regarded as the most reliable technique for computing null spaces. However, the algorithm presented here is almost as reliable. The crux of the matter is the reliability of the condition estimator that produces the approximate null vector in (4.1). Although counterexamples exist for most condition estimators, a wealth of experiments and experience has shown them to be very reliable in real-life applications [6].

Effect of Rounding Errors. The algorithms proposed here are completely stable. Standard-rounding error analysis shows that the updated matrix is orthogonally equivalent to an original matrix that is perturbed by quantities proportional to the rounding unit times the norm of the matrix [10]. Moreover, the matrix V deviates only very slowly from orthogonality, the more so since V changes only when a change in rank is suspected.

Parallelization. We have seen that the algorithm can be implemented on a linear array of p processors so that it runs in order p time. On the other hand, the singular value decomposition requires p^2 processors to achieve the same updating time.

Numerical Results. In [1], the algorithm has been used with the MUSIC algorithm to estimate directions of arrival for simulated data. The algorithm performs well, estimating the rank correctly and providing a sufficiently accurate error subspace to obtain the directions of arrival.

Availability. An experimental Fortran version of the algorithm is available by anonymous ftp at thales.cs.umd.edu in the file pub/reports/uast.f.

Acknowledgment

This work has its origins in a workshop on the singular value decomposition in signal processing, organized by R. J. Vaccaro. The author is indebted to F. Luk for mentioning complete orthogonal decompositions as an alternative to QR decomposition.

References

[1] G. Adams, M. F. Griffin, and G. W. Stewart, "Direction-of-arrival estimation using the rank-revealing URV decomposition," in *Proc. IEEE Int. Conf. Acoust., Speech, Signal Processing*, 1991.

[2] J. R. Bunch and C. P. Nielsen, "Updating the singular value decomposition," *Numer. Math.*, vol. 31, pp. 111-129, 1978.

[3] T. F. Chan, "Rank revealing QR factorizations," *Linear Alg. Its Appl.*, vol. 88/89, pp. 67-82, 1987.

[4] G. H. Golub and C. F. Van Loan, *Matrix Computations*, 2nd ed. Baltimore, MD: Johns Hopkins University Press, 1989.
[5] M. F. Griffin, personal communication, 1990.
[6] N. J. Higham, "A survey of condition number estimation for triangular matrices," *SIAM Rev.*, vol. 29, pp. 575-596, 1987.
[7] M. Moonen, P. Van Dooren, and J. Vandewalle, "Combined Jacobi-type algorithms in signal processing," in *Proc. 2nd Int. Workshop SVD, Signal Processing*, 1990, pp. 83-88.
[8] G. W. Stewart, *Introduction to Matrix Computations*. New York: Academic, 1974.
[9] R. J. Vaccaro, Ed., *Proc. 2nd Int. Workshop SVD Signal Processing*, to be published.
[10] J. H. Wilkinson, *The Algebraic Eigenvalue Problem*. Oxford, England: Clarendon, 1965.

G. W. Stewart received the Ph.D. degree in mathematics (numerical analysis) from the University of Tennessee in 1968.

He has been with the Department of Computer Science of the University of Texas (1968-1972), Carnegie-Mellon University (1972-1974), and the University of Maryland where he is also a Research Professor in the Institute for Advanced Computer Studies. His interests are matrix theory, numerical linear algebra, and the numerical treatment of Markov chains. He is the author or coauthor of *Introduction to Matrix Computations*, LINPACK, and *Matrix Perturbation Theory*.

13.5. [GWS-J77] "Updating a Rank-Revealing ULV Decomposition"

[GWS-J77] "Updating a Rank-Revealing ULV Decomposition," *SIAM Journal on Matrix Analysis and Applications* **14** (1993) 494–499.
`http://dx.doi.org/10.1137/0614034`

SIAM J. MATRIX ANAL. APPL.
Vol. 14, No. 2, pp. 494–499, April 1993

UPDATING A RANK-REVEALING ULV DECOMPOSITION*

G. W. STEWART†

Abstract. A ULV decomposition of a matrix A of order n is a decomposition of the form $A = ULV^{\mathrm{H}}$, where U and V are orthogonal matrices and L is a lower triangular matrix. When A is approximately of rank k, the decomposition is rank revealing if the last $n-k$ rows of L are small. This paper presents algorithms for updating a rank-revealing ULV decomposition. The algorithms run in $O(n^2)$ time, and can be implemented on a linear array of processors to run in $O(n)$ time.

Key words. ULV decomposition, URV decomposition, rank revealing, updating

AMS(MOS) subject classifications. 65F25, 65F25, 65F35

1. Introduction. Let A be an $m \times n$ matrix. A rank-revealing URV decomposition of A is a reduction of A by unitary transformations to a triangular matrix of the form

$$U^{\mathrm{H}}AV = \begin{pmatrix} R & H \\ 0 & E \\ 0 & 0 \end{pmatrix}. \tag{1.1}$$

The decomposition is rank revealing in the sense that the matrices H and E are smaller than some prespecified tolerance, and the smallest singular value of R is greater than that tolerance. Such decompositions—they are not unique—are useful in solving rank-deficient systems. Moreover, if $V = (V_1\ V_2)$ is partitioned conformally, then in the spectral norm

$$\|AV_2\| = \left\| \begin{pmatrix} H \\ E \end{pmatrix} \right\|, \tag{1.2}$$

so that the columns of V_2 form an orthonormal basis for an approximate null space of A, something required in signal processing applications like direction of arrival estimation.

The advantage of the URV decomposition over the more familiar singular value decomposition is that it can be updated when a row is added to A (in many applications, A is first multiplied by a constant less than one to damp out old data, a process known as exponential windowing). The updating procedure, which is described in [7], requires $O(n^2)$ operations and preserves the rank-revealing character of the decomposition. Moreover, it can be implemented on a linear array of n processors to run in $O(n)$ time.

Although the URV decomposition is fully satisfactory for applications like recursive least-squares, it is less satisfactory for applications in which an approximate null space is needed. The reason is the presence of H in (1.2). To see that it should not be there, let $(U_1\ U_2)$ be a partition of U conformal with (1.1). Then it is easily seen that $\|U_2^{\mathrm{H}}A\| = \|E\|$. Consequently the last $n-k$ singular values of A are less than or equal to $\|E\|$, and the corresponding left singular vectors form an approximate null

* Received by the editors April 17, 1991; accepted for publication (in revised form) August 23, 1991.

† Department of Computer Science and Institute for Advanced Computer Studies, University of Maryland, College Park, Maryland 20742 (stewart@cs.umd.edu). This work was supported in part by Air Force Office of Scientific Research contract AFOSR-87-0188.

$$\begin{array}{ccccccccccc}
\rightarrow & \check{w} & & & 0 & & & 0 & & 0 \\
\rightarrow & w & & \rightarrow & \check{w} & & & 0 & & 0 \\
 & w & \Longrightarrow & \rightarrow & w & \Longrightarrow & \rightarrow & \check{w} & \Longrightarrow & 0 \\
 & w & & & w & & \rightarrow & w & & 1
\end{array}$$

FIG. 2.1. *Reduction of W.*

space whose residual has norm less than or equal to $\|E\|$. Thus V_2 is not the best available approximate null space.

In [7] this problem was circumvented by including a refinement step that reduces the size of H. The properties of this refinement have been analyzed in [8]. In experiments with the MUSIC algorithm for direction of arrival estimation, the refinement was found to improve the results [1]. However, it adds extra work, and aesthetically it has the appearance of a stopgap. The purpose of this paper is to present an alternative.

Specifically, we will describe how to update a lower triangular decomposition of the form

$$U^{\mathrm{H}}AV = \begin{pmatrix} L & 0 \\ H & E \\ 0 & 0 \end{pmatrix}, \tag{1.3}$$

where U and V are orthogonal, L is well conditioned, and H and E are small. We will call such a decomposition a rank-revealing ULV decomposition. The updating algorithm consists of two parts: an algorithm to bring a lower triangular matrix into rank-revealing ULV form and the updating algorithm proper. We will present the former in the next section and the latter in §3. The paper concludes with some general observations on the algorithms.

2. Deflation. Although adding a row to a matrix cannot decrease its rank, in many applications the matrix is first multiplied by a constant less than one to damp out old information. Under such circumstances, it is possible for the matrix L in (1.3) to become effectively rank deficient. Here we present an algorithm to calculate a rank-revealing ULV decomposition of a lower triangular matrix L.

The first step is to determine a vector w of norm one such that $\omega = \|w^{\mathrm{H}}L\|$ approximates the smallest singular value of L. This can be done by using any of a number of reliable condition estimators [4].[1] If ω is greater than a prescribed tolerance, then there is nothing to be done. Otherwise, we must modify L by unitary transformations so that its last row becomes small. We will use plane rotations to accomplish this reduction. The reader is assumed to be familiar with plane rotations, which are discussed in most texts on numerical linear algebra (e.g., see [3], [6], and [9]).

We begin by reducing w the nth unit vector $\mathbf{e}_n$. The reduction is illustrated in Fig. 2.1. The two arrows represent the plane of the rotation, and it annihilates the component of w with a check over it. Denote the product of rotations by

$$P = P_1P_2\cdots P_{n-1}.$$

[1] It is worth noting that these estimators succeed, where QR with pivoting fails, in revealing the rank of a widely cited matrix of Kahan [5].

$$
\begin{array}{c}\rightarrow\\ \rightarrow\\ \\ \\ \end{array}
\begin{array}{cccc} l & 0 & 0 & 0\\ l & l & 0 & 0\\ l & l & l & 0\\ l & l & l & l \end{array}
\Longrightarrow
\begin{array}{cccc} \downarrow & \downarrow & & \\ l & \check{l} & 0 & 0\\ l & l & 0 & 0\\ l & l & l & 0\\ l & l & l & l \end{array}
\Longrightarrow
\begin{array}{c}\\ \rightarrow\\ \rightarrow\\ \\ \end{array}
\begin{array}{cccc} l & 0 & 0 & 0\\ l & l & 0 & 0\\ l & l & l & 0\\ l & l & l & l \end{array}
\Longrightarrow
$$

$$
\begin{array}{cccc} & \downarrow & \downarrow & \\ l & 0 & 0 & 0\\ l & l & \check{l} & 0\\ l & l & l & 0\\ l & l & l & l \end{array}
\Longrightarrow
\begin{array}{c}\\ \\ \rightarrow\\ \rightarrow \end{array}
\begin{array}{cccc} l & 0 & 0 & 0\\ l & l & 0 & 0\\ l & l & l & 0\\ l & l & l & l \end{array}
\Longrightarrow
\begin{array}{cccc} & & \downarrow & \downarrow \\ l & 0 & 0 & 0\\ l & l & 0 & 0\\ l & l & l & \check{l}\\ l & l & l & l \end{array}
\Longrightarrow
\begin{array}{cccc} l & 0 & 0 & 0\\ l & l & 0 & 0\\ l & l & l & 0\\ h & h & h & e \end{array}
$$

FIG. 2.2. *Reduction of PL.*

Next we apply the rotations P_i to L from the left, as is shown in Fig. 2.2. The application of P_i produces a nonzero element in the $(i, i+1)$-element of L. This element is removed by postmultiplying by a plane rotation Q_i. Let $Q = Q_1 Q_2 \cdots Q_{n-1}$ be the product of these rotations.

The appearance of h's and e in the last matrix of Fig. 2.2 is meant to indicate that the last row of PLQ is small. To see that this is true, write

$$\omega = \|w^{\mathrm{H}} L\| = \|w^{\mathrm{H}} P^{\mathrm{H}} PLQ\| = \|\mathbf{e}_n^{\mathrm{T}} (PLQ)\|.$$

Thus the last row of PLQ has norm ω.

If the $(n-1) \times (n-1)$ leading principle submatrix of L is sufficiently well conditioned, we have our rank-revealing ULV decomposition. If not, we can repeat the deflation procedure until a sufficiently well conditioned matrix is found in the upper right-hand corner.

3. Updating. We now turn to the updating step. Specifically, we suppose that we are given an additional row z^{H} and wish to determine a rank-revealing ULV decomposition of

$$\begin{pmatrix} A \\ z^{\mathrm{H}} \end{pmatrix}.$$

We begin by forming $z^{\mathrm{H}} V$ to bring the row into the coordinate system of the current decomposition. Thus the problem of updating reduces to the problem of updating

$$\begin{pmatrix} L & 0 \\ H & E \\ x^{\mathrm{H}} & y^{\mathrm{H}} \end{pmatrix},$$

where we have partitioned $z^{\mathrm{H}} V = (x^{\mathrm{H}} \; y^{\mathrm{H}})$.

We now reduce y^{H} to $\|y\| \mathbf{e}_1^{\mathrm{T}}$, as is shown in Fig. 3.1, where only the parts cor-

$$
\begin{array}{cccc} & & \downarrow & \downarrow \\ e&0&0&0\\ e&e&0&0\\ e&e&e&0\\ e&e&e&e\\ y&y&y&\breve{y}\end{array}
\Longrightarrow
\begin{array}{ccccc} & & & & \\ &e&0&0&0\\ &e&e&0&0\\ \rightarrow&e&e&e&\breve{e}\\ \rightarrow&e&e&e&e\\ &y&y&y&0\end{array}
\Longrightarrow
\begin{array}{cccc} & \downarrow & \downarrow & \\ e&0&0&0\\ e&e&0&0\\ e&e&e&0\\ e&e&e&e\\ y&y&\breve{y}&0\end{array}
\Longrightarrow
$$

$$
\begin{array}{ccccc} & & & & \\ &e&0&0&0\\ \rightarrow&e&e&\breve{e}&0\\ \rightarrow&e&e&e&0\\ &e&e&e&e\\ &y&y&0&0\end{array}
\Longrightarrow
\begin{array}{cccc} \downarrow & \downarrow & & \\ e&0&0&0\\ e&e&0&0\\ e&e&e&0\\ e&e&e&e\\ y&\breve{y}&0&0\end{array}
\Longrightarrow
\begin{array}{ccccc} & & & & \\ \rightarrow&e&\breve{e}&0&0\\ \rightarrow&e&e&0&0\\ &e&e&e&0\\ &e&e&e&e\\ &y&0&0&0\end{array}
\Longrightarrow
\begin{array}{cccc} & & & \\ e&0&0&0\\ e&e&0&0\\ e&e&e&0\\ e&e&e&e\\ y&0&0&0\end{array}
$$

FIG. 3.1. *Reducing* y.

$$
\begin{array}{cccccc} &l&0&0&0&0\\ &l&l&0&0&0\\ &l&l&l&0&0\\ \rightarrow&h&h&h&e&0\\ &h&h&h&e&e\\ \rightarrow&x&x&x&\breve{y}&0\end{array}
\Longrightarrow
\begin{array}{cccccc} &l&0&0&0&0\\ &l&l&0&0&0\\ \rightarrow&l&l&l&0&0\\ &x&x&x&y&0\\ &h&h&h&e&e\\ \rightarrow&x&x&\breve{x}&0&0\end{array}
\Longrightarrow
$$

$$
\begin{array}{cccccc} &l&0&0&0&0\\ \rightarrow&l&l&0&0&0\\ &l&l&l&0&0\\ &x&x&x&y&0\\ &h&h&h&e&e\\ \rightarrow&x&\breve{x}&0&0&0\end{array}
\Longrightarrow
\begin{array}{cccccc} \rightarrow&l&0&0&0&0\\ &l&l&0&0&0\\ &l&l&l&0&0\\ &x&x&x&y&0\\ &h&h&h&e&e\\ \rightarrow&\breve{x}&0&0&0&0\end{array}
\Longrightarrow
\begin{array}{ccccc} l&0&0&0&0\\ l&l&0&0&0\\ l&l&l&0&0\\ x&x&x&y&0\\ h&h&h&e&e\\ 0&0&0&0&0\end{array}
$$

FIG. 3.2. *Triangularization step.*

responding to y^{H} and E are shown. At the end of the reduction the matrix has the form

$$
\begin{array}{ccccc} l&0&0&0&0\\ l&l&0&0&0\\ l&l&l&0&0\\ h&h&h&e&0\\ h&h&h&e&e\\ x&x&x&y&0\end{array}
$$

One way to finish the update is to continue as in Fig. 3.2 to incorporate x and what is left of y into the decomposition. This process moves the presumably large elements of x^{H} into the first row of H. If there has been no increase in rank, this destroys the rank-revealing character of the matrix. However, in that case the deflation procedure

$$
\begin{array}{ccccc}
\downarrow & & & \downarrow & \\
l & 0 & 0 & \check{y} & 0 \\
l & l & 0 & y & 0 \\
l & l & l & y & 0 \\
h & h & h & e & 0 \\
h & h & h & e & e \\
0 & 0 & 0 & y & 0
\end{array}
\Longrightarrow
\begin{array}{ccccc}
 & \downarrow & & \downarrow & \\
l & 0 & 0 & 0 & 0 \\
l & l & 0 & \check{y} & 0 \\
l & l & l & y & 0 \\
h & h & h & e & 0 \\
h & h & h & e & e \\
y & 0 & 0 & y & 0
\end{array}
\Longrightarrow
\begin{array}{ccccc}
 & & \downarrow & \downarrow & \\
l & 0 & 0 & 0 & 0 \\
l & l & 0 & 0 & 0 \\
l & l & l & \check{y} & 0 \\
h & h & h & e & 0 \\
h & h & h & e & e \\
y & y & 0 & y & 0
\end{array}
\Longrightarrow
\begin{array}{ccccc}
 & & & & \\
l & 0 & 0 & 0 & 0 \\
l & l & 0 & 0 & 0 \\
l & l & l & 0 & 0 \\
h & h & h & e & 0 \\
h & h & h & e & e \\
y & y & y & y & 0
\end{array}
$$

FIG. 3.3. *Eliminating the tower of y's.*

will restore the small elements.

If y is small enough, then the rank cannot increase. In that case there is another way of proceeding. Perform the reduction of Fig. 3.2 but skip the first step. This will give a matrix having the form of the first matrix in Fig. 3.3 with a tower of contributions from the scalar y at the bottom. Rotations are used as shown in the figure to reduce the tower. This fills up the bottom row again, but now the elements are of the same size as y. Since y is small, the algorithm of Fig. 3.2 can be used to complete the decomposition without destroying the rank-revealing structure.

It is worth noting that if the y's in the tower are small compared with the diagonal elements of L, the y's along the last row will actually be of order $\|y\|^2$. If only an approximate decomposition is required, it may be possible to neglect them.

4. Comments. We mentioned in the introduction that a ULV decomposition can be expected to give a higher quality approximate null space than a URV decomposition. However, there are trade-offs. It costs more to deflate the URV decomposition if we insist on refinement steps. On the other hand the updating algorithm for the URV decomposition is much simpler. In fact, when there is no change of rank, it amounts to the usual LINPACK updating algorithm SCHUD [2]. Only experience with real-life problems will tell us under what circumstances one decomposition is to be preferred to the other.

Both sets of algorithms are stable and reliable. They are stable because they use orthogonal transformations straightforwardly with no additional implicit relations. They are as reliable as their underlying condition estimators.

In [7] we showed that the algorithms for the URV decomposition could, in principle, be parallelized on a linear array of processors. The same is true of the algorithms for updating a ULV decomposition. Since the techniques to show that the algorithms have parallel implementations are the same for both decompositions, we do not give the details here.

REFERENCES

[1] G. ADAMS, M. F. GRIFFIN, AND G. W. STEWART, *Direction-of-arrival estimation using the rank-revealing* URV *decomposition*, in Proc. IEEE Internat. Conf. Acoustics, Speech, and Signal Processing, Washington, DC, 1991, to appear.

[2] J. J. DONGARRA, J. R. BUNCH, C. B. MOLER, AND G. W. STEWART, LINPACK *User's Guide*, Society for Industrial and Applied Mathematics, Philadelphia, PA, 1979.

[3] G. H. GOLUB AND C. F. VAN LOAN, *Matrix Computations*, 2nd ed., The Johns Hopkins University Press, Baltimore, MD, 1989.

[4] N. J. HIGHAM, *A survey of condition number estimation for triangular matrices*, SIAM Rev., 29 (1987), pp. 575–596.

[5] W. KAHAN, *Numerical linear algebra*, Canad. Math. Bull., 9 (1966), pp. 757–801.
[6] G. W. STEWART, *Modifying pivot elements in Gaussian elimination*, Math. Comp., 28 (1974), p. 1974.
[7] ———, *An updating algorithm for subspace tracking*, Tech. Rep. CS-TR 2494, Dept. of Computer Science, Univ. of Maryland, College Park, MD, 1990; IEEE Trans. Signal Processing, 40 (1992), pp. 1535–1541.
[8] ———, *On an algorithm for refining a rank-revealing* URV *decomposition and a perturbation theorem for singular values*, Tech. Rep. CS-TR 2626, Dept. of Computer Science, Univ. of Maryland, College Park, MD, 1991.
[9] D. S. WATKINS, *Fundamentals of Matrix Computations*, John Wiley, New York, 1991.

13.6. [GWS-J87] "On the Stability of Sequential Updates and Downdates"

[GWS-J87] "On the Stability of Sequential Updates and Downdates," *IEEE Transactions on Signal Processing* **43** (1995) 2642–2648.
http://dx.doi.org/10.1109/78.482114

On the Stability of Sequential Updates and Downdates

G. W. Stewart

Abstract—**The updating and downdating of Cholesky decompositions has important applications in a number of areas. There is essentially one standard updating algorithm, based on plane rotations, which is backward stable. Three downdating algorithms have been treated in the literature: the LINPACK algorithm, the method of hyperbolic transformations, and Chambers' algorithm. Although none of these algorithms is backward stable, the first and third satisfy a relational stability condition. In this paper, it is shown that relational stability extends to a sequence of updates and downdates. In consequence, other things being equal, if the final decomposition in the sequence is well conditioned, it will be accurately computed, even though intermediate decompositions may be almost completely inaccurate. These results are also applied to the two-sided orthogonal decompositions, such as the URV decomposition.**

I. Introduction

LET A be a positive definite matrix of order p. Then, A can be written in the form $A = R^{\mathrm{T}}R$, where R is an upper triangular matrix with positive diagonal elements. This factorization is called the *Cholesky decomposition* A, and the matrix R is called its *Cholesky factor.*

In some applications—recursive least squares, for example—it is required to compute the Cholesky decomposition $S^{\mathrm{T}}S$ of $B = A + xx^{\mathrm{T}}$, where x is a given p-vector. Although the *ab initio* calculation of S requires $O(p^3)$ arithmetic operations, it turns out that S can be computed from R and x in $O(p^2)$ operations, a process that is known as *updating.* The usual updating algorithm is numerically stable.

The inverse process of computing the Cholesky decomposition $R^{\mathrm{T}}R$ of $A = B - x^{\mathrm{T}}x$ from that of B is called *downdating.* Three algorithms for downdating have appeared in the literature: Chambers' algorithm [1], the LINPACK algorithm [2] (due to Michael Saunders and first published in [3]), and the method of plane hyperbolic transformations (which in another guise is due to Golub [4]).[1] Although Chambers' algorithm and the LINPACK algorithms are not stable in the usual backward sense, it has been shown [8], [9] to have an important property, which we will call *relational stability.* Specifically, the mathematical relations that hold between the true quantities continue to hold for the computed quantities provided they are perturbed slightly. We will show that relational stability is preserved in a sequence of updates and downdates. This, combined with the block perturbation theory of Eldén and Park [10] (see also [11]), implies that if the final result of sequence of updates and downdates is well conditioned then it will be computed accurately.

The method of plane hyperbolic transformations is not relationally stable. Consequently, as we will show by an example, it can introduce unnecessary errors in the course of a sequence of updates and downdates.

Rank degenerate problems usually require a decomposition that reveals the rank and provides a basis for the null space of the matrix in question. Two-sided orthogonal-triangular, such as the URV and ULV decompositions [12], [13], perform these functions and in addition can be efficiently updated and downdated. We will show that the relational stability of the updating and downdating algorithms extends to these algorithms. In particular, if the nondegenerate part of the matrix is well conditioned, then the basis for the null space is accurately computed.

This paper is organized as follows. In the next section we sketch the results of rounding error analyses for the various algorithms. In Section III, we review the perturbation theory for the Cholesky decomposition. In Section IV, we establish the relational stability of a sequence of updates and downdates and derive error bounds for the results. Section V is devoted to an example illustrating the results of the previous section. In Section VI, we derive bounds for URV updating. The paper concludes with some observations on downdating and exponential windowing.

Throughout the paper, $\|A\|$ will denote the Frobenius norm of the matrix A, which is defined by

$$\|A\|^2 = \sum_{i,j} a_{ij}^2.$$

The quantity $\|x\|$ is the ordinary Euclidean norm of the vector x. For more on norms see [14].

II. Rounding Error Analyses

In this section, we will review the rounding error analyses of updating by plane rotations and downdating by Chambers' and the LINPACK algorithms.

Manuscript received May 25, 1994; revised March 14, 1995. This work was supported by the National Science Foundation under Grant CCR 9115568. The associate editor coordinating the review of this paper and approving it for publication was Dr. Andreas Spanias.

The author is with the Department of Computer Science and Institute for Advanced Computer Studies, University of Maryland, College Park, MD 20742 USA.

IEEE Log Number 9415075.

[1] A close reading of the papers involved suggests that Chambers thought he was merely rederiving Golub's hyperbolic algorithm, since he explicitly attributes the algorithm to Golub. However, his derivation resulted in a different formula for one of the downdated quantities—in effect a different algorithm with different properties. It should be noted that methods that carry along additional data, such as the rows of the original matrix [5] or its Q-factor [6], [7], have superior numerical properties. However, they cost more in both time and storage.

1053-587X/95$04.00 © 1995 IEEE

The updating algorithm in general use is due to Bogert and Burris [15] and Golub [16]. The idea behind the algorithm is to compute an orthogonal matrix Q such that

$$Q^{\mathrm{T}}\begin{pmatrix} R \\ x^{\mathrm{T}} \end{pmatrix} = \begin{pmatrix} S \\ 0 \end{pmatrix}$$

where S is upper triangular. It then follows from the orthogonality of Q that

$$R^{\mathrm{T}}R + xx^{\mathrm{T}} = S^{\mathrm{T}}S$$

so that S is the Cholesky factor of $R^{\mathrm{T}}R + xx^{\mathrm{T}} = A + xx^{\mathrm{T}}$.

The algorithm is stable in the backward sense. The very general rounding-error of plane rotations by Wilkinson [17, p.131] applies to give the following result. If we let S denote the *computed* matrix, then there is an orthogonal matrix Q and a $(p+1) \times p$ matrix F satisfying

$$\|F\| \le K\|S\|\epsilon_{\mathrm{M}}$$

such that

$$Q^{\mathrm{T}}\left[\begin{pmatrix} R \\ x^{\mathrm{T}} \end{pmatrix} + F\right] = \begin{pmatrix} S \\ 0 \end{pmatrix}. \tag{1}$$

Here, ϵ_{M} is the rounding unit for the machine in question and K is a constant that depends on p and the details of the computer arithmetic. Thus, the computed result, however inaccurate, comes from a slightly perturbed problem.

In exact arithmetic, both Chambers' algorithm and the LINPACK algorithm produce an orthogonal Q matrix such that

$$Q^{\mathrm{T}}\begin{pmatrix} S \\ 0 \end{pmatrix} = \begin{pmatrix} R \\ x^{\mathrm{T}} \end{pmatrix}.$$

It follows that

$$R^{\mathrm{T}}R = S^{\mathrm{T}}S - xx^{\mathrm{T}}.$$

Thus, R is the Cholesky factor of the matrix $S^{\mathrm{T}}S - xx^{\mathrm{T}} = B - xx^{\mathrm{T}}$.

For both downdating algorithms it has been shown [8], [9] that if R denotes the computed matrix then there is an orthogonal matrix Q and a $(p+1) \times p$ matrix E satisfying

$$\|E\| \le K\|S\|\epsilon_{\mathrm{M}} \tag{2}$$

such that

$$Q^{\mathrm{T}}\begin{pmatrix} S \\ 0 \end{pmatrix} = \begin{pmatrix} R \\ x^{\mathrm{T}} \end{pmatrix} + E. \tag{3}$$

This result is not backward stability, since it is not possible to concentrate the entire error in the matrix S and the vector x^{T}. Instead, we will call it *relational stability* because the defining mathematical relation between the true quantities continues to be satisfied up to a small error by the computed quantities. We will see later that relational stability has important consequences for the accuracy of the computed results.

Note that (1) can be brought into the form (3) by defining $E = -Q^{\mathrm{T}}F$. It is this common form that we will use to treat sequential updates and downdates.

The method of hyperbolic transformations is neither backward or relationally stable. The unhappy consequences of this fact will be seen in §V.

III. Perturbation Theory

The error analyses of updating and downdating reveal that the true result can be obtained from the computed result by perturbing its cross-product matrix slightly and computing the Cholesky factor. To find out how accurate the result actually is, we must call on perturbation theory.

The perturbation theory for Cholesky decompositions has been studied in a number of places. Since here we are concerned with small perturbations, we will give an asymptotic result that is sharp up to second-order terms in the error [18].

Theorem 1: Let A be positive definite, and let $\tilde{A} = A + H$, where H is symmetric. Then, for all sufficiently small H, $\tilde{A}$ is positive definite. If R is the Cholesky factor of A and $\tilde{R}$ is the Cholesky factor of $\tilde{A}$, then

$$\frac{\|\tilde{R} - R\|}{\|R\|} \lesssim \frac{\|R^{-1}\|^2}{\sqrt{2}}\|H\|. \tag{4}$$

Note that this result puts an inherent limit on the accuracy we can expect in a computed Cholesky factor. For example, if we merely round the elements of A, then

$$\|H\| \le \|A\|\epsilon_{\mathrm{M}} \le \|R\|^2\epsilon_{\mathrm{M}}$$

where ϵ_{M} is the rounding unit. It follows that

$$\frac{\|\tilde{R} - R\|}{\|R\|} \lesssim \frac{\kappa^2(R)}{\sqrt{2}}\epsilon_{\mathrm{M}} \tag{5}$$

where $\kappa(R) = \|R\|\|R^{-1}\|$ is the condition number of R. It would be unfair to expect an algorithm to produce a result more accurate than the right hand side of (5).

If A and R are partitioned in the forms

$$A = \begin{pmatrix} A_{11} & A_{12} \\ A_{21} & A_{22} \end{pmatrix}$$

and

$$R = \begin{pmatrix} R_{11} & R_{12} \\ 0 & R_{22} \end{pmatrix}$$

where A_{11} and R_{11} are of order k, then the Cholesky factor of A_{11} is R_{11}. The perturbation analysis above shows that the accuracy of $\tilde{R}_{11}$ depends not on the condition of R but on the condition of R_{11}. Thus, the Cholesky factor of a well-conditioned leading principal submatrix of A will by insensitive to perturbations, even though A as a whole may be ill conditioned: the large errors end up in the terminal columns of R. We will use this fact in analyzing URV decompositions.

IV. Sequential Updating

In this section, we will show that a sequence of relationally stable updates and downdates is relationally stable. We will begin by considering a single downdate followed by an update.

Let R_0 be the matrix to be downdated and let x_0 be the vector to be removed. Let the computed result be R_1. Similarly, let R_1 be updated by the vector x_1 to give R_2. Then, by the rounding error analyses just cited, there is an orthogonal matrix Q_0 and a small matrix E_1 such that

$$Q_0\begin{pmatrix} R_0 \\ 0 \\ x_1^{\mathrm{T}} \end{pmatrix} = \begin{pmatrix} R_1 \\ x_0^{\mathrm{T}} \\ x_1^{\mathrm{T}} \end{pmatrix} + E_1.$$

Similarly there is an orthogonal matrix Q_1 and a small matrix E_2 such that

$$Q_1^{\mathrm{T}}\begin{pmatrix} R_1 \\ x_0^{\mathrm{T}} \\ x_1^{\mathrm{T}} \end{pmatrix} = \begin{pmatrix} R_2 \\ x_0^{\mathrm{T}} \\ 0 \end{pmatrix} + E_2.$$

If we set

$$Q^{\mathrm{T}} = Q_1^{\mathrm{T}} Q_0$$

and

$$E = Q_1^{\mathrm{T}} E_1 + E_2$$

then

$$Q^{\mathrm{T}}\begin{pmatrix} R_0 \\ 0 \\ x_1^{\mathrm{T}} \end{pmatrix} = \begin{pmatrix} R_2 \\ x_0^{\mathrm{T}} \\ 0 \end{pmatrix} + E.$$

Thus, a downdate followed by an update is stable and the norm of the error

$$\|E\| \le \|E_0\| + \|E_1\|$$

is bounded by the sum of the norms of the errors in the individual steps.

This analysis clearly extends to any sequence of n updates and downdates. Specifically, collect the vectors appearing in updates in the matrix $X_{\mathrm{u}}^{\mathrm{T}}$ and the vectors appearing in downdates in the $X_{\mathrm{d}}^{\mathrm{T}}$. Then, there is an orthogonal transformation Q and a matrix E such that

$$Q\begin{pmatrix} R_0 \\ 0 \\ X_{\mathrm{u}}^{\mathrm{T}} \end{pmatrix} = \begin{pmatrix} R_n \\ X_{\mathrm{d}}^{\mathrm{T}} \\ 0 \end{pmatrix} + E. \tag{6}$$

The norm of the error E is bounded by the sum of of the norms of the backward errors in the individual updates and downdates.

To derive a specific bound for the error, we note that the error bound (2) for updating and downdating involve computed Cholesky factors. Consequently, if we let

$$\rho = \max\{\|R_i\| : i = 1, \cdots, k,\}$$

then a common bound on all the errors E_k is $K\rho\epsilon_{\mathrm{M}}$. It follows that the error in (6) is bounded by

$$\|E\| \le nK\rho\epsilon_{\mathrm{M}}. \tag{7}$$

To assess the accuracy of R_n, we do a block perturbation analysis in the spirit of Eldén and Park [10]. Specifically, from (6) it follows that

$$R_n^{\mathrm{T}} R_n = R_0^{\mathrm{T}} R_0 + X_{\mathrm{u}} X_{\mathrm{u}}^{\mathrm{T}} - X_{\mathrm{d}} X_{\mathrm{d}}^{\mathrm{T}} + H.$$

If we set

$$\hat{\rho} = \max\left\{\rho, \left\|\begin{pmatrix} R_0 \\ X_{\mathrm{d}}^{\mathrm{T}} \end{pmatrix}\right\|\right\}.$$

and assume that

$$nK\epsilon_{\mathrm{M}} < 1$$

then

$$\|H\| = \|E\|(2 + \|E\|)\left\|\begin{pmatrix} R_0 \\ X_{\mathrm{d}}^{\mathrm{T}} \end{pmatrix}\right\| \le 3nK\epsilon_{\mathrm{M}}\hat{\rho}^2. \tag{8}$$

It now follows from (4) that if S_n is the Cholesky factor of $R_0^{\mathrm{T}} R_0 + X_{\mathrm{u}} X_{\mathrm{u}}^{\mathrm{T}} - X_{\mathrm{d}} X_{\mathrm{d}}^{\mathrm{T}}$ (i.e., the true Cholesky factor) then

$$\frac{\|S_n - R_n\|}{\|R_n\|} \lesssim \sqrt{4.5}nK\hat{\rho}^2\|R_n^{-1}\|^2\epsilon_{\mathrm{M}}. \tag{9}$$

This bound is quite crude and no doubt can be refined. However, it already tells us that if $\hat{\rho}^2\|R_n^{-1}\|^2$ is not large, the computed Cholesky factor will be a good approximation to the true one, *no matter how inaccurate the intermediate quantities may be.* The factor R_n^{-1} will be large when R is ill-conditioned. The factor $\hat{\rho}$ is essentially the norm of the matrix one would get if all the updates but none of the downdates were performed. If all the rows of $X_{\mathrm{u}}^{\mathrm{T}}$ are of a size, then $\hat{\rho}$ can be expected to grow like $\sqrt{n}$. However, if even one row is very much larger than the others, the bound tells us to expect a persisting inaccuracy in the subsequent computed Cholesky factors. This phenomena has been observed in [5].

V. A Numerical Example

To illustrate the the above results we will give a numerical example in which a downdate from a well-conditioned matrix R_0 to an ill-conditioned matrix R_1 is followed by an update to a well conditioned matrix R_2. The calculations were performed in *matlab* with a rounding unit of $2 \cdot 10^{-16}$.

The following is a description of the experiment. The idea is generate an ill-conditioned matrix R_1 and create R_0 and R_1 by updating it.

1) Let R_1 be the R-factor from the QR-factorization of a matrix of independent normal random variables with mean zero and variance one. This will produce a well conditioned matrix.
2) Set the $(2,2)$-element of R_1 to 10^{-7} to produce an ill-conditioned R-factor.
3) Let x be a random normal vector and update R_1 and x to get the matrix R_0.
4) Let y be a random normal vector and update R_1 and y to get the matrix R_2.
5) Let R_{l1} be the result of using the LINPACK algorithm to downdate R_0 and x. Let R_{l2} be the result of updating R_{l1} and y.
6) Let R_{c1} be the result of using Chambers' algorithm to downdate R_0 and x. Let R_{c2} be the result of updating R_{c1} and y.
7) Let R_{h1} be the result of using plane hyperbolic transformations to downdate R_0 and x . Let R_{h2} be the result of updating R_{h1} and y.

Table I gives the result of twenty repetitions (steps 1–6 above) of this procedure for $p = 5$. The asterisks indicated cases where the hyperbolic downdating could not be carried out.

The results are entirely consistent with theory. Since R_1 is ill conditioned, any attempt to compute it by downdating a

TABLE I
A DOWNDATE FOLLOWED BY AN UPDATE

$$\rho_{pi} = \frac{\|R_{pi} - R_i\|}{\|R_i\|}$$

$\kappa(R_1)$	$\kappa(R_2)$	ρ_{l1}	ρ_{l2}	ρ_{c1}	ρ_{c2}	ρ_{h1}	ρ_{h2}
1e+08	1e+01	4e−02	3e−15	1e−02	1e−15	2e−02	2e−10
2e+08	1e+02	8e−02	1e−13	1e−01	5e−13	*	*
2e+08	5e+00	1e−03	1e−16	3e−03	1e−16	7e−05	1e−10
4e+07	3e+00	7e−03	2e−16	8e−03	1e−16	3e−05	2e−10
1e+08	5e+00	3e−04	2e−16	2e−04	1e−16	9e−04	1e−11
7e+07	5e+00	4e−03	1e−16	7e−03	1e−16	8e−03	1e−10
6e+08	4e+00	7e−02	3e−16	1e−01	4e−16	*	*
1e+08	4e+01	3e−03	2e−16	1e−03	1e−16	5e−03	1e−10
1e+08	4e+00	6e−02	1e−16	1e−02	1e−16	1e−02	2e−10
4e+08	1e+01	4e−01	9e−16	3e−01	3e−16	*	*
1e+09	1e+01	1e−02	2e−16	1e−02	1e−16	*	*
7e+07	1e+01	1e−03	8e−16	8e−03	4e−16	2e−03	7e−10
6e+07	1e+01	5e−03	3e−16	7e−03	4e−16	8e−03	6e−10
6e+07	4e+01	1e−02	4e−15	4e−04	4e−16	1e−02	2e−10
5e+07	1e+01	5e−04	1e−16	6e−04	1e−16	5e−04	3e−11
1e+08	1e+01	1e−02	3e−16	9e−03	2e−16	6e−03	1e−09
3e+08	8e+00	3e−02	3e−16	7e−03	3e−16	2e−02	7e−11
2e+08	1e+01	8e−02	6e−16	6e−02	3e−16	8e−02	2e−10
1e+10	4e+02	2e−02	1e−14	1e−02	2e−15	3e−02	2e−10
6e+07	8e+01	9e−04	4e−16	1e−03	6e−16	1e−05	1e−11

well-condition matrix must result in inaccuracies proportional to the square of the condition number. All the algorithms exhibit these inaccuracies. The difference between the algorithms becomes apparent when we examine the errors in the approximations to R_2. Here, the two relationally stable algorithms restore almost full accuracy, while the hyperbolic algorithm loses several figures. However, not all of the error in R_{h1} is carried forward to R_{h2}: presumably some component of the error introduced by the hyperbolic rotations can be accounted for by relational perturbations, a point that deserves further study.

In three cases, the hyperbolic downdate fails when a quantity that should be positive turns out negative. In all cases the other algorithms go through to completion. However, this comparison is a little unfair to the hyperbolic approach. The condition numbers of the matrices R_1 in Table I are on the order of 10^8, close to the point where the perturbation theory predicts no accuracy for the computed results. If we make R_2 even a little more ill conditioned, Chambers' algorithm begins to fail.[2] Decrease the condition number a little, and all algorithms go through to completion.

[2] The LINPACK algorithm continues to perform well, but this is an artifact of the simplicity of the example and special properties of the algorithm. In more realistic settings, the LINPACK algorithm would also fail.

VI. URV DECOMPOSITIONS

In this section, we will apply our results to sequential updates and downdates of URV decompositions. A URV decomposition of a matrix X is a decomposition of the form

$$U^{\mathrm{T}}XV = \begin{pmatrix} R \\ 0 \end{pmatrix}$$

where U and V are orthogonal and R is upper triangular. Any matrix has infinitely many URV decompositions. One of them, the singular value decomposition (R diagonal), is widely used because it exhibits approximate rank degeneracies in X and provides an orthonormal basis for an approximate null space of the matrix. However, it cannot be efficiently updated or downdated.

Rank-revealing URV decompositions overcome the computational deficiencies of the singular value decomposition. Suppose that X has been obtained from a matrix of exactly rank k by perturbing it by some noise. (We use the term "noise" rather than "error" to distinguish the perturbation from effects due to rounding error.) Then, there is a URV decomposition in which R takes the form

$$R = \begin{pmatrix} T & F \\ 0 & G \end{pmatrix}$$

where T is a well-conditioned conditioned matrix of order k and F and G are the same size as the noise (F may actually be much smaller, even zero). The virtues of a rank-revealing URV decomposition are that it can be updated and downdated. Moreover, if V is partitioned in the form

$$V = (V_1 V_2)$$

then V_1 and V_2 provide orthonormal bases for approximate row and null space of R. In general, U, which can be arbitrarily large, is not saved.

Although the updating and downdating algorithms are quite complicated—they involve decisions about rank and procedures for keeping the small part of the decomposition small—nonetheless they fall within the purview of the analyses discussed above. Specifically, if the LINPACK or Chambers' algorithm is used to perform downdates, there are orthogonal matrices U and V such that the computed R_n satisfies

$$U^{\mathrm{T}} \begin{pmatrix} R_0 \\ 0 \\ X_{\mathrm{u}}^{\mathrm{T}} \end{pmatrix} V = \begin{pmatrix} R_n \\ X_{\mathrm{d}}^{\mathrm{T}} V \\ 0 \end{pmatrix} + E \qquad (10)$$

where as above $\|E\| \leq nK\rho\epsilon_{\mathrm{M}}$ (cf. (7)).

In interpreting this bound, there are two questions we can ask. One question is, "How accurate is V?" Actually, this question is not well posed, since there is no unique URV decomposition associated with the data. We can, however, show that the V-factor of any URV decomposition satisfying a relation such as (10) must produce approximate null spaces that lie near that produced by V (see the appendix to this paper).

But there is a simpler alternative. For *any* V, there is a unique URV decomposition of R_0 that is obtained by computing the Cholesky decomposition of $V^{\mathrm{T}}(R_0^{\mathrm{T}} R_0 + X_{\mathrm{u}} X_{\mathrm{u}}^{\mathrm{T}} -$

$X_dX_d^T)V$. Now, the URV algorithm does not compute the Cholesky decomposition of this matrix; instead it computes the Cholesky decomposition of $V^T(R_0^TR_0 + X_uX_u^T - X_dX_d^T + H)V$, where H satisfies (8), and it is from this decomposition that we deduce that that we have revealed the rank. Thus, if this decomposition is accurate, V truly furnishes a basis for an approximate null space. Thus the second question is, "How accurate is R_n ?"

Here, we are on familiar territory. If the matrix T_n is well conditioned, by the comments at the end of Section III it will be accurately computed. The matrices G_n and F_n, which consist of noise, will be less accurately computed. However, R_n^{-1} will be approximated by G_n^{-1}, so that the factor ρR_n^{-1} in (9) can be regarded as a signal-to-noise ratio. If this ratio is substantially above $\sqrt{\epsilon_M}$, then F and G will be computed with reasonable accuracy. Specific bounds may be obtained as above.

It should not be thought that V is near the matrix that would have been obtained by exact computation. The algorithm for determining rank involves discrete decisions, and if rounding error causes a change in any of these decisions, the computed decomposition will diverge sharply from the exact one. Nonetheless, by the analysis sketched above, we will have computed a rank-revealing URV decomposition.

One final point. The matrix V in (10) is defined as the exact product of the rotations computed in the course of the sequential updates and downdates. The computed V, being contaminated with rounding error, will diverge from the original. However, this divergence will be very slow and corresponds to the factor n in (8).

VII. Conclusion

Downdating has had bad press in some circles. Part of it is no doubt due to unfortunate experiences with bad algorithms, such as hyperbolic downdating. However, a great deal of it is the result of not understanding the limitations of both updating and downdating.

An extremely simple example will illustrate the problems. Let R be the scalar 1, and suppose that in ten-digit decimal floating-point arithmetic we wish to incorporate $x = 5 \cdot 10^{-6}$; that is we wish to update

$$\begin{pmatrix} 1 \\ 5 \cdot 10^{-6} \end{pmatrix}.$$

The exact update is $1 + 2.5 \cdot 10^{-11}$. The computed update will be 1. There is no trace of the number $5 \cdot 10^{-6}$; it has been swallowed by the update, and a subsequent downdate cannot recover it. Thus, downdating is sometimes blamed for inaccuracies that are implicit in the updating procedure.

However, downdating has limitations of its own. If for example, the computed update is perturbed (as in real life it might be by rounding error) to become $1.000\,000\,001$, then the computed downdate will be about $3.2 \cdot 10^{-5}$. This is inaccurate, as we would expect; but if a relationally stable algorithm is used the inaccuracy will go away on subsequent updates. Something worse happens when the problem is perturbed to become $0.999\,999\,999\,9$. Now, the downdating process fails completely, and there is no chance to regain accuracy in a subsequent update.

The lesson is that when the condition numbers of the triangular factors approach $1/\sqrt{\epsilon_M}$, both updating and downdating become problematical, but increase the conditional number a little, and relationally stable algorithms will perform well. When inaccuracies are inherent in the problem, they will, of course, produce inaccurate answers; but well-conditioned R-factors will be computed accurately.

In some applications, exponential windowing is an alternative to downdating. In this method, the matrix R is multiplied by a factor $\beta < 1$ before each update, which damps the influence of older updates. Now, when the sequence of vectors x_i^T represents a stationary process, exponential windowing is to be preferred to downdating. It is simpler and has better numerical properties [19], [20]. However, in nonstationary situations, the two techniques will produce different R-factors, so that they are not just different numerical algorithms computing the same thing. In this case, the decision between the two must depend on their behavior in the application in question. An important contribution of this paper, then, is to show when numerical considerations need not enter into this decision.

Appendix

Recall that a computed URV decomposition satisfies

$$U^T \begin{pmatrix} R_0 \\ 0 \\ X_u^T \end{pmatrix} V = \begin{pmatrix} R_n \\ X_d^T V \\ 0 \end{pmatrix} + E \tag{11}$$

where U and V are orthogonal, and E satisfies the bound $\|E\| \le nK\rho\epsilon_M$. The matrix R_n will be rank revealing if it has the form

$$\begin{pmatrix} T_n & F_n \\ 0 & G_n \end{pmatrix}$$

where

$$\|T_n^{-1}\|\|G_n\|, \|T_n^{-1}\|\|G_n\| \ll 1. \tag{12}$$

Now, let

$$\bar{U}^T \begin{pmatrix} R_0 \\ 0 \\ X_u^T \end{pmatrix} \bar{V} = \begin{pmatrix} R_n \\ X_d^T \bar{V} \\ 0 \end{pmatrix} + \bar{E} \tag{13}$$

where $\bar{U}$ and $\bar{V}$ are orthogonal and $\bar{E}$ satisfies the same bound. We are going to show that if

$$V = (V_1 V_2) \quad \text{and} \quad \bar{V} = (\bar{V}_1 \bar{V}_2)$$

and we set

$$W = \bar{V}^T V = \begin{pmatrix} \bar{V}_1^T V_1 & \bar{V}_1^T V_2 \\ \bar{V}_2^T V_1 & \bar{V}_2^T V_2 \end{pmatrix} = \begin{pmatrix} W_{11} & W_{12} \\ W_{21} & W_{22} \end{pmatrix}$$

then W_{12} is small. Note that this implies that the space $\mathcal{R}(\bar{V}_1)$ spanned by the columns of $\bar{V}_1$ is almost orthogonal $\mathcal{R}(V_2)$. Since $\mathcal{R}(V_1)$ is exactly orthogonal $\mathcal{R}(V_2)$, it follows that $\mathcal{R}(V_1)$ and $\mathcal{R}(\bar{V}_1)$ are in some sense near each other.

More precisely, we will show that $\|W_{12}\|_2$ is small, where $\|W_{12}\|_2$ is the spectral norm of W_{12}—the largest singular value of W_{12}. This number is also the sine of the largest canonical angle be between $\mathcal{R}(V_1)$ and $\mathcal{R}(V_2)$ (see [22, ch. 11]).

We begin with a lemma.

Lemma 1: Let

$$A = \begin{pmatrix} A_{11} & A_{12} \\ A_{12}^{\mathrm{T}} & A_{22} \end{pmatrix}$$

be positive definite, and let

$$W = \begin{pmatrix} W_{11} & W_{12} \\ W_{21} & W_{22} \end{pmatrix}$$

be orthogonal. Suppose that

$$\|A - W^{\mathrm{T}}AW\|_2 \le \epsilon$$

and that

$$\eta \equiv \|A_{11}^{-1}\|(\epsilon + 2\|A_{12}\| + \|A_{22}\|) < \frac{1}{2}. \tag{14}$$

Then

$$\|W_{12}\|_2^2 \le \frac{1 - \sqrt{1 - 4\eta^2}}{2} < 2\eta^2. \tag{15}$$

Proof: The $(1,2)$ block of $W^{\mathrm{T}}AW$ is $W_{11}^{\mathrm{T}}A_{11}W_{12} + W_{11}^{\mathrm{T}}A_{12}W_{22} + W_{21}^{\mathrm{T}}A_{12}^{\mathrm{T}}W_{12} + W_{21}^{\mathrm{T}}A_{22}W_{22}$. Hence

$$\begin{aligned}\epsilon \ge \|W_{11}^{\mathrm{T}}A_{11}W_{12}\|_2 - \|W_{11}^{\mathrm{T}}A_{12}W_{21}\|_2 \\ - \|W_{21}^{\mathrm{T}}A_{12}^{\mathrm{T}}W_{12}\|_2 - \|W_{21}^{\mathrm{T}}A_{22}W_{22}\|_2.\end{aligned}$$

Since $\|W_{ij}\|_2 \le 1$

$$\|W_{11}^{\mathrm{T}}A_{11}W_{12}\|_2 \le \epsilon + 2\|A_{12}\|_2 + \|A_{22}\|_2$$

and since

$$\|W_{11}^{\mathrm{T}}A_{11}W_{12}\|_2 \ge \frac{\|W_{12}\|_2}{\|A_{11}^{-1}\|_2\|W_{11}^{-1}\|_2}$$

$$\frac{\|W_{12}\|_2}{\|W_{11}^{-1}\|_2} \le \eta.$$

By the orthogonality of W, we have

$$W_{11}^{\mathrm{T}}W_{11} + W_{12}^{\mathrm{T}}W_{12} = I.$$

Thus, if $\omega = \|W_{12}\|_2$ is the largest singular value of W_{12}, then $\sqrt{1-\omega^2}$ is the smallest singular value of W_{11}. Hence, $\|W_{11}^{-1}\|_2^{-1} = \sqrt{1-\omega^2}$, and

$$\omega^2(1 - \omega_2) \le \eta^2.$$

Thus, ω^2 is the smallest root of the quadratic equation $\omega^4 - \omega^2 + \eta = 0$, which gives (15).

Now, from (11) and (13)

$$\begin{aligned}R_0^{\mathrm{T}}R_0 + X_{\mathrm{u}}^{\mathrm{T}}X_{\mathrm{u}} - X_{\mathrm{d}}^{\mathrm{T}} &= VR_n^{\mathrm{T}}R_nV^{\mathrm{T}} + VE^{\mathrm{T}}EV^{\mathrm{T}} \\ R_0^{\mathrm{T}}R_0 + X_{\mathrm{u}}^{\mathrm{T}}X_{\mathrm{u}} - X_{\mathrm{d}}^{\mathrm{T}} &= \bar{V}R_n^{\mathrm{T}}R_n\bar{V}^{\mathrm{T}} + \bar{V}\bar{E}^{\mathrm{T}}\bar{E}\bar{V}^{\mathrm{T}}.\end{aligned}$$

Set

$$A = R_n^{\mathrm{T}}R_n = \begin{pmatrix} T_n^{\mathrm{T}}T_n & T_n^{\mathrm{T}}F_n \\ 0 & F_n^{\mathrm{T}}F_n + G_n^{\mathrm{T}}G_n \end{pmatrix}$$

$H = VE^{\mathrm{T}}EV^{\mathrm{T}}$, $\bar{H} = \bar{V}\bar{E}^{\mathrm{T}}\bar{E}\bar{V}^{\mathrm{T}}$, and $W = \bar{V}^{\mathrm{T}}V$. Then it follows that

$$W^{\mathrm{T}}AW - A = W^{\mathrm{T}}\bar{H}W - H.$$

If

$$\epsilon = \|H\| + \|\bar{H}\|$$

and η, defined by (14) is less than $\frac{1}{2}$, then

$$\|\bar{V}_1^{\mathrm{T}}V_2\|_2 < 2\eta.$$

It is instructive to bound η. Since $\|H\|$ and $\|\bar{H}\|$ satisfy (8), we have

$$\eta \le (6nK\hat{\rho}^2\epsilon_{\mathrm{M}} + 2\|F_n\|\|\|T_n\| + \|F_n^{\mathrm{T}}F_n + G_n^{\mathrm{T}}G_n\|)\|T_n^{-1}\|^2.$$

The first term $6nK\hat{\rho}^2\|T_n^{-1}\|^2\epsilon_{\mathrm{M}}$, which represents the contribution of rounding error, is precisely the term that must be small for R_n to be computed accurately. The term $\|F_n^{\mathrm{T}}F_n + G_n^{\mathrm{T}}G_n\|\|T_n^{-1}\|^2$ is small by virtue of (12). If we write the middle term in the form

$$2\kappa(T_n)\|T_n^{-1}\|\|F_n\|$$

we see that, (12) notwithstanding, this term is potentially larger than the others. Now the algorithm for updating URV decompositions contains a refinement step that is specifically designed to make F_n small. The above analysis suggests that such a step is fully justified.

ACKNOWLEDGEMENTS

Parts of this paper were inspired by the block perturbation analysis of Park and Eldén [21], which showed how to avoid intermediate quantities in assessing the accuracy of the final result. And many thanks to H. Park for her useful comments at every stage of this research.

REFERENCES

[1] J. M. Chambers, "Regression updating," *J. Am. Statist. Assoc.*, vol. 66, pp. 744–748, 1971.
[2] J. J. Dongarra, J. R. Bunch, C. B. Moler, and G. W. Stewart, *LINPACK User's Guide*. Philadelphia, PA: SIAM, 1979.
[3] P. E. Gill, G. H. Golub, W. Murray, and M. A. Saunders, "Methods of modifying matrix factorizations," *Mathematics of Computation*, vol. 28, pp. 505–535, 1974.
[4] G. H. Golub, "Matrix decompositions and statistical computation," in *Statistical Computation*, R.C. Milton and J. A. Nelder, Eds. New York: Academic, 1969, pp. 365–397.
[5] Å. Björck, H. Park, and L. Eldén, "Accurate downdating of least squares solutions," *SIAM J. Matrix Anal. Applicat.*, vol. 15, pp. 549–568, 1992.
[6] J. Daniel, W. B. Gragg, L. Kaufman, and G. W. Stewart, "Reorthogonalization and stable algorithms for updating the Gram-Schmidt QR factorization," *Mathemat. Computation*, vol. 30, pp. 772–795, 1976.
[7] C. C. Paige, "Error analysis of some techniques for updating orthogonal decompositions," *Mathemat. Computation*, vol. 34, pp. 465–471, 1980.
[8] G. W. Stewart, "The effects of rounding error on an algorithm for downdating a Cholesky factorization," *J. Inst. Mathemat. Applicat.*, vol. 23, pp. 203–213, 1979.
[9] A. Bojanczyk, R. P. Brent, P. Van Dooren, and F. de Hoog, "A note on downdating the Cholesky factorization" *SIAM J. Scientific Statist. Computing*, vol. 8, pp. 210–221, 1987.
[10] L. Eldén and H. Park, "Block downdating of least squares solutions," *SIAM J. Matrix Anal. Applicat.*, vol. 15, pp. 1018–1034, 1992.
[11] S. J. Olszanskyj, J. M. Lebak, and A. W. Bojanczyk, "Rank-k modificaton methods for recursive least squares problems," *Numerical Algorithms*, vol. 7, pp. 325–354, 1994.

[12] G. W. Stewart, "An updating algorithm for subspace tracking," *IEEE Trans. Signal Processing*, vol. 40, pp. 1535–1541, 1992.
[13] ——, "Updating a rank-revealing ULV decomposition," *SIAM J. Matrix Anal. Applicat.*, vol. 14, pp. 494–499, 1993.
[14] G. H. Golub and C. F. Van Loan, *Matrix Computations*. Baltimore, MD: Johns Hopkins University Pr., 1989, 2nd ed.
[15] D. Bogert and W. R. Burris, "Comparison of least squares algorithms," Neutron Physics Division, Oak Ridge National Laboratory, Report ORNL-3499, vol. 1, sec. 5.5, 1963.
[16] G. H. Golub, "Numerical methods for solving least squares problems," *Numerische Mathematik*, vol. 7, pp. 206–216, 1965.
[17] J. H. Wilkinson, *The Algebraic Eigenvalue Problem*. Oxford, England: Clarendon Pr., 1965.
[18] G. W. Stewart, "On the perturbation of LU, Cholesky, and QR factorizations," *SIAM J. Matrix Anal. Applicat.*, vol. 14, pp. 1141–1146, 1993.
[19] M. Moonen, "Jacobi-Type updating algorithms for signal processing, systems identification and control," Ph.D. thesis, Katholieke Universiteit Leuven, Leuven, Belgium, 1990.
[20] G. W. Stewart, "Error analysis of QR updating with exponential windowing," *Mathemat. Computation*, vol. 59, pp. 135–140, 1992.
[21] L. Eldén and H. Park, "Perturbation analysis for block downdating of a Cholesky decomposition," Department of Mathematics, Linköping University, Technical Report LiTH-MAT-R-93-26, 1993.
[22] G. W. Stewart and J.-G. Sun, *Matrix Perturbation Theory*. Boston, MA: Academic, 1990.

G. W. Stewart received the degree in mathematics in 1969 under the late Alston Householder.

He is a professor in the Computer Science Department and Research Professor in the Institute for Advanced Comuter Studies at the University of Maryland, College Park. He is the author of many papers on various aspects of numerical analysis and matrix computation, with applications in statistical computing, signal processing, and stochastic processes. His books include *Introduction to Matrix Computation, Matrix Perturbation Theory* (with J. G. Sun), and *Afternotes on Numerical Analysis* (forthcoming). He is a coauthor of the LINPACK package for linear algebra and has translated Gauss's later works on least squares.

14

Papers on Least Squares, Projections, and Generalized Inverses

1. [GWS-J4] "On the Continuity of the Generalized Inverse," *SIAM Journal on Applied Mathematics* **17** (1969) 33–45.
2. [GWS-J35] "On the Perturbation of Pseudo-inverses, Projections and Linear Least Squares Problems," *SIAM Review* **19** (1977) 634–662.
3. [GWS-J65] "On Scaled Projections and Pseudoinverses," *Linear Algebra and its Applications* **112** (1989) 189–193.

14.1. [GWS-J4] "On the Continuity of the Generalized Inverse"

[GWS-J4] "On the Continuity of the Generalized Inverse," *SIAM Journal on Applied Mathematics* **17** (1969) 33–45.
http://dx.doi.org/10.1137/0117004

SIAM J. APPL. MATH.
Vol. 17, No. 1, January 1969

ON THE CONTINUITY OF THE GENERALIZED INVERSE*

G. W. STEWART†

1. Statement of the problem. The generalized inverse $A^\dagger$ of a matrix A was first introduced by Moore [9]. It was rediscovered by Penrose [10] who characterized it as the unique matrix X satisfying

$$(XA)^H = XA, \tag{1.1}$$

$$(AX)^H = AX, \tag{1.2}$$

$$AXA = A, \tag{1.3}$$

$$XAX = X. \tag{1.4}$$

When A is square and nonsingular, the Moore-Penrose generalized inverse is identical with the usual inverse, A^{-1}, of A. In this case it is well known (e.g., see [12]) that, for any matrix norm $\|\cdot\|$ with $\|I\| = 1$, if

$$\|A^{-1}\| \, \|E\| < 1,$$

then

$$\frac{\|(A + E)^{-1} - A^{-1}\|}{\|A^{-1}\|} \leqq \frac{\kappa(A)\|E\|/\|A\|}{1 - \kappa(A)\|E\|/\|A\|}, \tag{1.5}$$

where

$$\kappa(A) = \|A\| \, \|A^{-1}\|. \tag{1.6}$$

This implies that the inverse of a nonsingular matrix is a continuous function of the elements of the matrix; that is, for any nonsingular A,

$$\lim_{E \to 0} (A + E)^{-1} = A^{-1}.$$

The number $\kappa(A)$ appearing in (1.6) is called a condition number with respect to inversion for the matrix A. It reflects how perturbations in the matrix may be magnified in its inverse.

The Moore-Penrose generalized inverse of a matrix is not necessarily a continuous function of the elements of the matrix. For example, if

$$A = \operatorname{diag}(1, 0)$$

and

$$E = \operatorname{diag}(0, 1),$$

---

* Received by the editors April 26, 1968.

† Mathematics Division, Oak Ridge National Laboratory, and Computing Technology Center, Oak Ridge Gaseous Diffusion Plant, Oak Ridge, Tennessee 37831. This research was sponsored by the United States Atomic Energy Commission under contract with Union Carbide Corporation.

then for $\varepsilon \neq 0$,

$$(A + \varepsilon E)^\dagger = (A + \varepsilon E)^{-1} = \operatorname{diag}(1, \varepsilon^{-1}),$$

and $(A + \varepsilon E)^\dagger$ has no limit as ε approaches zero.

Ben-Israel [3] has shown that, for a sequence of matrices E_n satisfying

$$\|A^\dagger\| \, \|E_n\| < 1 \tag{1.7}$$

and

$$\lim_{n\to\infty} E_n = 0, \tag{1.8}$$

a sufficient condition that

$$\lim_{n\to\infty} (A + E_n)^\dagger = A^\dagger \tag{1.9}$$

is that, for all sufficiently large n, the rows and columns of E_n lie in the row and column spaces of A. He also gives a perturbation bound quite analogous to (1.5). The object of this paper is to show that *under* (1.7) *and* (1.8) *a necessary and sufficient condition for* (1.9) *is that, for all sufficiently large n,*

$$\operatorname{rank}(A + E_n) = \operatorname{rank}(A).$$

This will be done by exhibiting explicit bounds for

$$\|(A + E)^\dagger - A^\dagger\|.$$

However, because the columns (or rows) of A and $A + E$ may span different spaces, this bound is considerably more complicated than the right-hand side of (1.5).

In the next section some results from matrix theory that will be needed later are stated. In § 3 a slightly different proof of Ben-Israel's theorem is given. In § 4–§ 5 the main theorem is proved, first for matrices of full rank and then for general matrices. Finally in § 6 the results are applied to the linear least squares problem.

2. Preliminaries. The generalized inverse of a matrix A is closely related to the projectors P_A and R_A which project a vector onto the column and row spaces of A: namely,

$$P_A = AA^\dagger$$

and

$$R_A = A^\dagger A.$$

P_A and R_A are Hermitian idempotents:

$$P_A^H = P_A^2 = P_A$$

and

$$R_A^H = R_A^2 = R_A.$$

Conversely any Hermitian idempotent is a projector. If P projects onto a given space, then $I - P$ projects onto the orthogonal complement of that space.

For any matrix A,

$$(A^H)^\dagger = (A^\dagger)^H.$$

If U and V are unitary matrices for which $U^H A V$ is defined, then

$$(U^H A V)^\dagger = V^H A^\dagger U.$$

If B is the partitioned matrix

$$B = (A, 0),$$

then

$$B^\dagger = \begin{pmatrix} A^\dagger \\ 0 \end{pmatrix}.$$

If A and B are matrices for which AB and BA are both defined, then AB and BA have the same nonzero eigenvalues.

The Euclidean vector norm is defined by

$$\|x\|^2 = x^H x.$$

The subordinate spectral norm of a matrix is defined by

$$\|A\| = \sup_{x \neq 0} \frac{\|Ax\|}{\|x\|}.$$

Although it is customary to take A as square in defining this norm, the definition still holds for rectangular matrices. As such the norm has the following properties: $\|A\|^2$ is the largest eigenvalue of $A^H A$ or AA^H; it is consistent with the Euclidean vector norm in the sense that the inequality

$$\|Ax\| \leqq \|A\| \, \|x\|$$

holds for any vector x. This last fact implies that

$$\|AB\| \leqq \|A\| \, \|B\|$$

whenever AB is defined. If B is a square matrix with $\|B\| < 1$, then all the eigenvalues of B are less than unity in absolute value, $I + B$ is nonsingular, and

(2.1) $$\|(I + B)^{-1}\| \leqq (1 - \|B\|)^{-1}.$$

If U has orthonormal columns, then

$$\|U\| = 1.$$

If UA is defined, then

$$\|UA\| = \|A\|.$$

If P is a nonzero projector for which PA is defined, then

$$\|P\| = 1$$

and

$$\|PA\| \leqq \|A\|.$$

Additional material on the generalized inverse may be found in [2] and [4]. Projectors and norms are treated in [8].

3. The Case $P_AE = ER_A = E$. Ben-Israel's theorem is based on the following lemma.

LEMMA 3.1. *Let*

$$P_AE = ER_A = E \tag{3.1}$$

and

$$\|A^\dagger E\| < 1. \tag{3.2}$$

Then

$$A^\dagger = (I + A^\dagger E)(A + E)^\dagger.$$

Proof. Since

$$A + E = A + P_AE = A(I + A^\dagger E) \tag{3.3}$$

and $I + A^\dagger E$ is nonsingular by virtue of (3.2), the columns of A and those of $A + E$ span the same space. Also the eigenvalues of $A^\dagger E$ are less than unity in absolute value, and hence so are those of $EA^\dagger$. Thus $I + EA^\dagger$ is nonsingular. Since

$$A + E = A + ER_A = (I + EA^\dagger)A, \tag{3.4}$$

the rows of A and those of $A + E$ span the same space. Hence

$$(A + E)(A + E)^\dagger = P_A = AA^\dagger, \tag{3.5}$$

and

$$(A + E)^\dagger(A + E) = R_A = A^\dagger A. \tag{3.6}$$

The rest of the proof was suggested by one of the referees. By (1.4),

$$(A + E)^\dagger(A + E)(A + E)^\dagger = (A + E)^\dagger,$$

and hence, by (3.6),

$$A^\dagger A(A + E)^\dagger = (A + E)^\dagger. \tag{3.7}$$

Then

$$\begin{aligned} A^\dagger &= A^\dagger AA^\dagger && \text{(by (1.4))} \\ &= A^\dagger(A + E)(A + E)^\dagger && \text{(by (3.5))} \\ &= (A^\dagger A + A^\dagger E)(A + E)^\dagger \\ &= (I + A^\dagger E)(A + E)^\dagger - (I - A^\dagger A)(A + E)^\dagger \\ &= (I + A^\dagger E)(A + E)^\dagger && \text{(by (3.7))}, \end{aligned}$$

which is the desired result.

The main result of this section is the following theorem.

THEOREM 3.2 (Ben-Israel). *Let A and E satisfy* (3.1) *and* (3.2), *and let H be defined by*

$$H = (A + E)^{\dagger} - A^{\dagger}.$$

Then

$$\frac{\|H\|}{\|A^{\dagger}\|} \leqq \frac{\|A^{\dagger}E\|}{1 - \|A^{\dagger}E\|}. \tag{3.8}$$

Proof. From Lemma 3.1,

$$(A + E)^{\dagger} = (I + A^{\dagger}E)^{-1}A^{\dagger}.$$

Hence from (2.1),

$$\|(A + E)^{\dagger}\| \leqq \frac{\|A^{\dagger}\|}{1 - \|A^{\dagger}E\|}. \tag{3.9}$$

Also from Lemma 3.1,

$$H = (A + E)^{\dagger} - A^{\dagger} = -A^{\dagger}E(A + E)^{\dagger}.$$

Hence

$$\|H\| \leqq \|A^{\dagger}E\| \, \|(A + E)^{\dagger}\|.$$

If this inequality and (3.9) are combined, the result is (3.8).

COROLLARY 3.3. *If, in addition to the hypotheses of Theorem* 2, $\|A^{\dagger}\| \, \|E\| < 1$, *then*

$$\frac{\|H\|}{\|A^{\dagger}\|} \leqq \frac{\kappa\|E\|/\|A\|}{1 - \kappa\|E\|/\|A\|}, \tag{3.10}$$

where

$$\kappa = \|A\| \, \|A^{\dagger}\|. \tag{3.11}$$

The bound (3.10) is perfectly analogous to the bound (1.5) for the ordinary inverse. In (3.11), κ has been defined in analogy with (1.6) and continues to serve as a condition number, at least for the class of perturbations treated in the theorem.

4. Perturbation bounds for matrices of full rank. In this section the matrix A will be taken to have full rank. For this case it is not necessary to restrict E in order to obtain perturbation bounds for $(A + E)^{\dagger}$. However, the bounds are a good deal more complicated, and they already illustrate most of the features of the general case. The next two lemmas are devoted to getting bounds for the products of two projectors. They are also consequences of more general theorems in Afriat [1].

For definiteness assume that A is an $m \times n$ matrix of rank n. This is equivalent to saying that the columns of A are linearly independent or that $A^H A$ is nonsingular. Moreover,

$$A^\dagger A = R_A = I. \tag{4.1}$$

Let B be another $m \times n$ matrix of rank n, and let the columns of U and V form an orthonormal basis for the column spaces of A and B. Then

$$U^H U = I, \qquad V^H V = I.$$

The matrices S and T defined by

$$S = U^H A, \qquad T = V^H B$$

are the unique matrices satisfying

$$A = US, \qquad B = VT.$$

Moreover,

$$A^H A = S^H S, \qquad B^H B = T^H T,$$

which implies that S and T are nonsingular. Finally,

$$P_A = UU^H, \qquad P_B = VV^H.$$

LEMMA 4.1. *Let μ^2 be the smallest eigenvalue of $U^H V V^H U$. Then*

$$\|P_A(I - P_B)\|^2 = 1 - \mu^2.$$

Proof. $\|P_A(I - P_B)\|^2$ is equal to the largest eigenvalue of

$$P_A(I - P_B)P_A = UU^H(I - VV^H)UU^H.$$

But this is the same as the largest eigenvalue of $U^H(I - VV^H)U$, which is equal to $1 - \mu^2$, where μ^2 is the smallest eigenvalue of $U^H V V^H U$.

LEMMA 4.2. *If $B = A + E$ and $P_A E = 0$, then*

$$\mu^2 = (1 + \beta^2)^{-1},$$

where β^2 is the largest eigenvalue of $(A^H A)^{-1} E^H E$.

Proof. Since

$$B^H B = (A + E)^H (A + E) = A^H A + E^H E,$$

$B^H B$ is the sum of a positive definite matrix and a positive semidefinite matrix and is thereby nonsingular. Thus the columns of B are linearly independent. Now since $A^H B = S^H S$, it follows from the definition of S and T that

$$U^H V V^H U = (S^{-1})^H A^H B T^{-1} (T^{-1})^H B^H A S^{-1} = S T^{-1} (T^{-1})^H S^H.$$

Hence

$$S^{-1} U^H V V^H U S = (A^H A + E^H E)^{-1} (A^H A) = (I + (A^H A)^{-1} E^H E)^{-1}.$$

Hence the smallest eigenvalue of $U^H V V^H U$ is the reciprocal of the largest eigenvalue of $I + (A^H A)^{-1} E^H E$, which is $1 + \beta^2$.

Under the assumptions of this section, $A^\dagger$ is given by

$$A^\dagger = (A^H A)^{-1} A^H.$$

Hence $A^\dagger (A^\dagger)^H = (A^H A)^{-1} A^H A (A^H A)^{-1} = (A^H A)^{-1}$, and

$$\|A^\dagger\|^2 = \|(A^H A)^{-1}\|.$$

This gives the following bound for β:

$$\beta \leqq \|A^\dagger\| \, \|E\|. \tag{4.2}$$

The main result of this section is the following theorem.

THEOREM 4.3. *Let A be an $m \times n$ matrix of rank n, and let E be decomposed into its components lying in and lying orthogonal to the column space of A:*

$$E = E_1 + E_2 = P_A E + (I - P_A) E. \tag{4.3}$$

Let $\kappa = \|A\| \, \|A^\dagger\|$ and $H = (A + E)^\dagger - A^\dagger$. If

$$\|A^\dagger\| \, \|E_1\| < 1, \tag{4.4}$$

then

$$\frac{\|H\|}{\|A^\dagger\|} \leqq \beta_1 + \gamma \left(\frac{\beta_2^2}{1 + \beta_2^2} \right)^{1/2}, \tag{4.5}$$

where

$$\gamma = \left(1 - \frac{\kappa \|E_1\|}{\|A\|} \right)^{-1}$$

and

$$\beta_i = \frac{\gamma \kappa \|E_i\|}{\|A\|}, \qquad i = 1, 2. \tag{4.6}$$

Proof. Because of (4.1), (4.3) and (4.4), A and E_1 satisfy the hypothesis of Theorem 3.2. Moreover the columns of A and

$$A_1 = A + E_1$$

span the same spaces so that

$$P_A = P_{A_1}.$$

Now

$$P_{A+E} = (A + E)(A + E)^\dagger = (A + E)(A^\dagger + H).$$

Since $AA^\dagger = P_A$,

$$(A + E)H = P_{A+E} - P_A - EA^\dagger. \tag{4.7}$$

If (4.7) is multiplied on both sides by P_A, the result is

$$A_1 H = P_{A_1} P_{A_1 + E_2} - P_{A_1} - E_1 A^\dagger. \tag{4.8}$$

Since A_1 is of full rank, $A_1^\dagger A_1 = I$; hence multiplying (4.8) on both sides by $A_1^\dagger$ and taking norms gives

$$\|H\| \leqq \|A_1^\dagger\| \{ \|E_1\| \, \|A^\dagger\| + \|P_{A_1}(I - P_{A_1 + E_2})\| \}.$$

From (3.9) in the proof of Theorem 3.2,

$$\|H\| \leqq \frac{\|A^\dagger\|}{1 - \|A^\dagger E_1\|}\{\|E_1\|\;\|A^\dagger\| + \|P_{A_1}(I - P_{A_1+E_2})\|\},$$

or

$$\frac{\|H\|}{\|A^\dagger\|} \leqq \beta_1 + \gamma\|P_{A_1}(I - P_{A_1+E_2})\|.$$

Now from Lemma 4.1, Lemma 4.2 (with $A = A_1$ and $B = A_1 + E_2$), and (4.2),

$$\|P_{A_1}(I - P_{A_1+E_2})\|^2 \leqq \frac{\beta^2}{1+\beta^2}, \tag{4.9}$$

where

$$\beta \leqq \|A_1^\dagger\|\;\|E_2\|.$$

By (3.9), $\beta \leqq \beta_2$ and the result follows.

The first term on the right-hand side of (4.5) is analogous to the right-hand sides of (1.5) and (3.10). The second term, which vanishes when $E_2 = 0$, is due to the difference in the spaces spanned by the columns of A and those of $A + E$. Again κ, defined by (3.11), serves as a condition number for the problem. If A has orthonormal columns and E is small, then κ is unity, γ is near unity, and β_2 is small.

5. The case rank $(A + E)$ = rank (A). The following lemma gives an inequality for matrices of full rank which is analogous to the bound (3.9).

LEMMA 5.1. *Let A and E satisfy the conditions of Theorem* 4.3. *Then*

$$\|(A + E)^\dagger\| \leqq \frac{\|A^\dagger\|}{1 - \|A^\dagger\|\;\|E_1\|}.$$

Proof. Since

$$(A + E)(A + E)^\dagger = P_{A+E},$$

it follows that

$$(I + A^\dagger E_1)(A + E)^\dagger = A^\dagger P_{A+E}.$$

Since $\|A^\dagger E_1\| < 1$, $I + A^\dagger E_1$ is nonsingular and

$$\|(A + E)^\dagger\| \leqq \|(I + A^\dagger E_1)^{-1}\|\;\|A^\dagger\|\;\|P_{A+E}\|.$$

The result follows from the inequality (2.1) and the fact that $\|P_{A+E}\| = 1$.

THEOREM 5.2. *Let A be of rank r and let E have the same dimensions as A. Let*

$$E_{11} = P_A E R_A,$$
$$E_{12} = P_A E(I - R_A),$$
$$E_{21} = (I - P_A)ER_A,$$
$$E_{22} = (I - P_A)E(I - R_A),$$

and suppose that

$$\|A^\dagger\| \, \|E_{11}\| < 1.$$

Let $H = (A + E)^\dagger - A^\dagger$ *and* $\kappa = \|A\| \, \|A^\dagger\|$. *If*

$$\operatorname{rank}(A + E) = \operatorname{rank}(A), \tag{5.1}$$

then

$$\frac{\|H\|}{\|A^\dagger\|} \leqq \beta_{11} + \gamma \sum_{(i,j)\neq(1,1)} \left[\frac{\beta_{ij}^2}{1 + \beta_{ij}^2}\right]^{1/2}, \tag{5.2}$$

where

$$\gamma = \left(1 - \frac{\kappa\|E_{11}\|}{\|A\|}\right)^{-1}$$

and

$$\beta_{ij} = \frac{\gamma\kappa\|E_{ij}\|}{\|A\|}, \qquad i,j = 1, 2.$$

On the other hand, if

$$\operatorname{rank}(A + E) \neq \operatorname{rank}(A), \tag{5.3}$$

then

$$\|(A + E)^\dagger\| \geqq \|E\|^{-1}. \tag{5.4}$$

Proof. Suppose that (5.1) holds. Let U and V be unitary matrices such that the first r columns of U span the column space of A and the first r rows of V^H span the row space of A. Then U^HAV has the form

$$U^HAV = B = (B_1, 0) = \begin{pmatrix} B_{11} & 0 \\ 0 & 0 \end{pmatrix},$$

where B_{11} is a nonsingular matrix of order r. Let

$$U^H(A + E)V = B + F = (B_1 + F_1, F_2) = \begin{pmatrix} B_{11} + F_{11} & F_{12} \\ F_{21} & F_{22} \end{pmatrix}. \tag{5.5}$$

Then

$$\|F_{ij}\| = \|E_{ij}\|, \qquad i,j = 1, 2,$$

and

$$\|H\| = \|(B + F)^\dagger - B^\dagger\|. \tag{5.6}$$

Since

$$\|B_{11}^{-1}\| \, \|F_{11}\| = \|A^\dagger\| \, \|E_{11}\| < 1,$$

the matrix $B_1 + F_1$ has rank r. Let the first r columns of the unitary matrix W

span the column space of $B_1 + F_1$. Then since

$$\text{rank}(B + F) = \text{rank}(B),$$

the matrix $W^H(B + F)$ has the form

$$W^H(B + F) = \begin{pmatrix} C & G \\ 0 & 0 \end{pmatrix},$$

where

$$\begin{pmatrix} C \\ 0 \end{pmatrix} = W^H(B_1 + F_1)$$

and

$$\|G\| = \|F_2\|.$$

Since $B_1 + F_1$ has rank r, C is nonsingular and $(C, 0)$ has full rank.

Now, from Theorem 4.3,

$$\|(B + F)^\dagger - (B_1 + F_1, 0)^\dagger\| = \|(C, G)^\dagger - (C, 0)^\dagger\| \leqq \|C^{-1}\| \left[\frac{\beta^2}{1 + \beta^2}\right]^{1/2},$$

where

$$\beta = \|C^{-1}\| \, \|G\|.$$

By Lemma 5.1,

$$\|C^{-1}\| = \|(B_1 + F_1)^\dagger\| \leqq \frac{\|B_1^\dagger\|}{1 - \|B_1^\dagger\| \, \|F_{11}\|} = \gamma \|A^\dagger\|.$$

Hence

$$\beta \leqq \gamma \|A^\dagger\| \, \|F_2\| \leqq \gamma \|A^\dagger\| (\|F_{12}\| + \|F_{22}\|) = \beta_{12} + \beta_{22}.$$

Since the function $\nu(x) = [x^2/(1 + x^2)]^{1/2}$ satisfies the triangle inequality on the line,

$$\nu(\beta) \leqq \nu(\beta_{12}) + \nu(\beta_{22}).$$

All this gives

$$\|(B + F)^\dagger - (B_1 + F_1, 0)^\dagger\| \leqq \gamma \|A^\dagger\| [\nu(\beta_{12}) + \nu(\beta_{22})]. \tag{5.7}$$

Since $B_1 + F_1$ has full rank, by Theorem 4.3

$$\begin{aligned} \|(B_1 + F_1)^\dagger - B_1^\dagger\| &= \|(B_1 + F_1, 0)^\dagger - (B_1, 0)^\dagger\| \\ &\leqq \frac{\|B_1^\dagger\|}{1 - \|B_1^\dagger\| \, \|F_{11}\|} [\|F_{11}\| \, \|B_1^\dagger\| + \nu(\beta)], \end{aligned}$$

where

$$\beta = \frac{\|B_1^\dagger\| \, \|F_{21}\|}{1 - \|B_1^\dagger\| \, \|F_{11}\|} = \beta_{21}.$$

Hence

$$\|(B_1 + F_1, 0)^\dagger - (B_1, 0)^\dagger\| \leqq \|A^\dagger\|[\beta_{11} + \gamma\nu(\beta_{21})]. \tag{5.8}$$

If (5.6), (5.7) and (5.8) are combined, the result is (5.2).

Now suppose that (5.3) holds. Then it is evident from the decomposition (5.5) that

$$\operatorname{rank}(A + E) > \operatorname{rank}(A).$$

Hence there is a nonzero vector x such that

$$Ax = 0$$

and

$$R_{A+E}x = x.$$

But then

$$x = R_{A+E}x = (A + E)^\dagger(A + E)x = (A + E)^\dagger Ex.$$

Hence,

$$\|x\| \leqq \|(A + E)^\dagger\| \, \|E\| \, \|x\|,$$

and since $x \neq 0$, the lower bound (5.4) follows.

The assertion made in § 1 concerning necessary and sufficient conditions for (1.9) follows immediately from this theorem. The bound (5.2) in Theorem 5.2 differs from the bound (4.5) in Theorem 4.3 only by the additional terms corresponding to the more complicated decomposition of E. If either E_{21} or E_{12} is zero, then so is E_{22}, and the bound in Theorem 5.2 reduces to that of Theorem 4.3. If $P_A E = ER_A = E$, then $E_{21} = E_{12} = E_{22} = 0$, and the bound (5.2) reduces to Ben-Israel's bound (3.10). The lower bound (5.4) says in effect that if a matrix is near A and differs in rank from A then its generalized inverse must be large. However, the property of being "near" is not symmetric, since it is defined in terms of $A^\dagger$.

6. Application to the linear least squares problem. The linear least squares problem is that of determining a vector x for which

$$\|b - Ax\|^2 \text{ is minimum,}$$

where A is an $m \times n$ matrix and b is an n-vector. It is well known [11] that the linear least squares problem always has at least one solution given by

$$x = A^\dagger b. \tag{6.1}$$

If A is of rank n, as will be assumed throughout this section, then the solution is unique.

A new method for solving the linear least squares problem is based on Householder's reduction of a matrix to triangular form [7]. This method has been treated in detail by Golub [5], and in a recent paper Golub and Wilkinson [6] have shown that in the presence of rounding error it yields a solution $x + h$ satisfying

$$x + h = (A + E)(b + k),$$

where $\|E\|$ and $\|k\|$ have small bounds depending on the kind of arithmetic used. In order to complete the analysis it is necessary to obtain a bound for $\|h\|$. Golub and Wilkinson give a bound which is accurate to quantities of order $\|E\|^2$. In this section a strict bound will be given. The notation and definitions of § 4 will be freely used.

LEMMA 6.1.

$$\|P_A(I - P_B)P_A\| = \|P_A(I - P_B)\|^2.$$

Proof. The proof is computational:

$$\begin{aligned}\|P_A(I - P_B)P_A\|^2 &= \|UU^H(I - VV^H)UU^H(I - VV^H)UU^H\| \\ &= \|U^H(I - VV^H)UU^H(I - VV^H)U\| \\ &= \|U^H(I - VV^H)U\|^2 \\ &= \|UU^H(I - VV^H)UU^H\|^2 \\ &= \|P_A(I - P_B)\|^4.\end{aligned}$$

THEOREM 6.2. *Let A and E satisfy the conditions of Theorem* 4.3 *and let x satisfy* (6.1). *Let h be defined by*

$$x + h = (A + E)^\dagger b.$$

Then

$$\frac{\|h\|}{\|x\|} \leqq \beta_1 + \kappa\gamma\left\{\frac{\beta_2^2}{1 + \beta_2^2} + \left[\frac{\beta_2^2}{1 + \beta_2^2}\right]^{1/2}\frac{\|(I - P_A)b\|}{\|P_Ab\|}\right\}. \tag{6.2}$$

Proof. Since $h = Hb$, it follows from (4.8) that

$$\begin{aligned}h &= A_1^\dagger[P_A(P_{A+E} - I)b - E_1A^\dagger b] \\ &= A_1^\dagger[P_A(P_{A+E} - I)P_AP_Ab + P_A(P_{A+E} - I)(I - P_A)b - E_1x].\end{aligned}$$

Hence

$$\frac{\|h\|}{\|x\|} \leqq \|A_1^\dagger\|\left\{\|P_A(P_{A+E} - I)P_A\|\frac{\|P_Ab\|}{\|x\|} + \|P_A(P_{A+E} - I)\|\frac{\|(I - P_A)b\|}{\|x\|} + \|E_1\|\right\}. \tag{6.3}$$

Now

$$Ax = P_Ab;$$

whence

$$\|x\| \leqq \|P_Ab\|/\|A\|. \tag{6.4}$$

Also

$$\|A_1^\dagger\| \leqq \gamma\|A^\dagger\|. \tag{6.5}$$

Finally,

$$\|P_A(P_{A+E} - I)\|^2 \leqq \frac{\beta_2^2}{1 + \beta_2^2}, \tag{6.6}$$

and by Lemma 6.1,

$$\|P_A(P_{A+E} - I)P_A\| \leqq \frac{\beta_2^2}{1 + \beta_2^2}. \tag{6.7}$$

If the inequalities (6.3)–(6.7) are combined, the result is (6.2).

The bound (6.2) is in agreement with the bound given in Golub and Wilkinson. For small E_1 and E_2,

$$\beta_1 \sim \kappa\|E_1\|,$$

$$\frac{\kappa\gamma\beta_2^2}{1 + \beta_2^2} \sim \kappa^3\|E_2\|^2,$$

$$\kappa\gamma\left[\frac{\beta_2^2}{1 + \beta_2^2}\right]^{1/2} \sim \kappa^2\|E_2\|.$$

Thus the third term in the bound, which is of first order in $\|E_2\|$, depends on κ^2. Thus when $\|(I - P_A)b\|/\|P_Ab\|$ is not small, that is, when b has a significant component outside the column space of A, the number κ^2 rather than κ is the condition number for the problem.

7. Acknowledgments. I am happy to acknowledge the helpful comments of Mr. B. W. Rust and Professor A. S. Householder.

REFERENCES

[1] S. N. AFRIAT, *Orthogonal and oblique projectors and characteristics of pairs of vector spaces*, Proc. Cambridge Philos. Soc., 53 (1967), pp. 800–816.

[2] A. BEN-ISRAEL AND A. CHARNES, *Contributions to the theory of generalized inverses*, this Journal, 11 (1963), pp. 667–699.

[3] ———, *On error bounds for the generalized inverse*, SIAM J. Numer. Anal., 3 (1966), pp. 585–592.

[4] G. GOLUB AND W. KAHAN, *Calculating the singular values and pseudo-inverse of a matrix*, Ibid., 2 (1965), pp. 205–224.

[5] G. GOLUB, *Numerical methods for solving linear least squares problems*, Numer. Math., 7 (1965), pp. 206–216.

[6] G. GOLUB AND J. H. WILKINSON, *Note on iterative refinement of least squares solution*, Ibid., 9 (1966), pp. 139–148.

[7] A. S. HOUSEHOLDER, *Unitary triangularization of a nonsymmetric matrix*, J. Assoc. Comput. Mach., 5 (1958), pp. 339–342.

[8] ———, *The Theory of Matrices in Numerical Analysis*, Blaisdell, New York, 1964.

[9] E. H. MOORE, *On the reciprocal of the general algebraic matrix*, Abstract, Bull. Amer. Math. Soc., 26 (1919–20), pp. 394–395.

[10] R. PENROSE, *A generalized inverse for matrices*, Proc. Cambridge Philos. Soc., 51 (1955), pp. 406–413.

[11] ———, *On best approximate solution of linear matrix equations*, Ibid., 52 (1956), pp. 17–19.

[12] J. H. WILKINSON, *Rounding Errors in Algebraic Processes*, Prentice-Hall, Englewood Cliffs, New Jersey, 1963.

14.2. [GWS-J35] "On the Perturbation of Pseudo-inverses, Projections and Linear Least Squares Problems"

[GWS-J35] "On the Perturbation of Pseudo-inverses, Projections and Linear Least Squares Problems," *SIAM Review* **19** (1977) 634–662.
http://dx.doi.org/10.1137/1019104

SIAM REVIEW
Vol. 19, No. 4, October 1977

ON THE PERTURBATION OF PSEUDO-INVERSES, PROJECTIONS AND LINEAR LEAST SQUARES PROBLEMS*

G. W. STEWART†

Abstract. This paper surveys perturbation theory for the pseudo-inverse (Moore–Penrose generalized inverse), for the orthogonal projection onto the column space of a matrix, and for the linear least squares problem.

1. Introduction. The pseudo-inverse (or Moore–Penrose generalized inverse) of a matrix A may be defined as the unique matrix $A^\dagger$ satisfying the following conditions [due to Penrose (1955)]:

$$A^\dagger A A^\dagger = A^\dagger, \tag{1.1a}$$

$$A A^\dagger A = A, \tag{1.1b}$$

$$(AA^\dagger)^H = AA^\dagger, \tag{1.1c}$$

$$(A^\dagger A)^H = A^\dagger A. \tag{1.1d}$$

The pseudo-inverse and its generalizations have been extensively investigated and widely applied. One reason for this interest in the pseudo-inverse is that it permits the succinct expression of some important geometric constructions in n-dimensional space. This paper will be concerned with the pseudo-inverse and two related geometric constructions: the orthogonal projection onto a subspace and the linear least squares problem.

The orthogonal projection onto a subspace $\mathscr{X}$ is the unique Hermitian, idempotent matrix P whose column space [denoted by $\mathscr{R}(P)$] is $\mathscr{X}$. It follows from (1.1a) that the matrix

$$P_A = AA^\dagger$$

is Hermitian and from (1.1b) that P_A is idempotent and $\mathscr{R}(P_A) = \mathscr{R}(A)$. Hence P_A is the orthogonal projection onto $\mathscr{R}(A)$. A similar argument shows that

$$R_A = A^\dagger A \tag{1.2}$$

is the projection onto $\mathscr{R}(A^H)$, the row space of A.

The second construction is the solution of the linear least squares problem of choosing a vector x to minimize

$$\rho(x) = \|b - Ax\|_2, \tag{1.3}$$

where b is a fixed vector and $\|\cdot\|_2$ denotes the usual Euclidean norm. The solutions of this problem are given by

$$x = A^\dagger b + (I - R_A)z, \tag{1.4}$$

---

* Received by the editors August 18, 1975, and in revised form February 15, 1976.

† Computer Science Department, University of Maryland, College Park, Maryland 20742. This work was supported in part by the Office of Naval Research.

where z is arbitrary. When A has full column rank, $R_A = I$ and the solution $x = A^\dagger b$ is unique. Otherwise, it is easily verified from (1.1) and (1.2) that $A^\dagger b$ is orthogonal to $(I - R_A)z$, so that by the Pythagorean theorem

$$\|x\|_2^2 = \|A^\dagger b\|_2^2 + \|(I - R_A)z\|_2^2.$$

It follows that $x = A^\dagger b$ is the unique solution of (1.3) that has minimal norm.

The object of this paper is to describe the effects of perturbations in A on $A^\dagger$, on P_A, and on $A^\dagger b$; i.e., on the pseudo-inverse, on the projection onto $\mathcal{R}(A)$, and on the solution of the linear least squares problem. Such descriptions are important for three reasons. First, the results are useful mathematical tools. Second, in numerical applications the elements of A will seldom be known exactly, and it is necessary to have bounds on the effects of the uncertainties in A. Finally, many numerical processes for computing projections and least squares solutions behave as if exact computations had been performed on a perturbed matrix $A + E$, where E is a small matrix whose size depends on the algorithm and the arithmetic used in its execution.

We shall be concerned with three kinds of results: perturbation bounds, asymptotic expressions, and derivatives. The perturbation bounds are needed in the applications mentioned above. Asymptotic expressions and derivatives are useful computational tools when the perturbation is actually known. Moreover they can be used to check the sharpness of the perturbation bounds. Not surprisingly it is rather difficult to obtain a reasonably sharp perturbation bound that tells the complete story of the effects of the perturbations. Asymptotic forms and derivatives are easier to come by.

In order to make this survey reasonably self-contained, we begin in § 2 with a review of the necessary background. In § 3 we develop the perturbation theory for the pseudo-inverse, in § 4 for the projection P_A, and in § 5 for the least squares solution $A^\dagger b$.

Notes and references. For background on the generalized inverse see the books by Ben-Israel and Greville (1974), Boullion and Odell (1971), and Rao and Mitra (1971). The expression (1.1) is due to Penrose (1955), (1956), whose papers initiated the current interest in the pseudo-inverse.

Many articles on perturbation theory for pseudo-inverses and related problems have appeared in the literature. To date the most complete survey of the problem has been given by Wedin (1973). In addition to collecting and unifying earlier material, this paper will present some new results.

2. Preliminaries.

Notation. Throughout this paper we shall use the notational conventions of Householder (1964). Specifically, matrices are denoted by upper case italic and Greek letters, vectors by lower case italic letters, and scalars by lower case Greek letters. The symbol $\mathbb{C}$ denotes the set of complex numbers, $\mathbb{C}^n$ the set of complex n-vectors, and $\mathbb{C}^{m\times n}$ the set of complex $m \times n$ matrices. The matrix A^H is the conjugate transpose of A. The column space of A is denoted by $\mathcal{R}(A)$, and its orthogonal complement by $\mathcal{R}(A)^\perp$.

We shall be concerned with a fixed matrix $A \in \mathbb{C}^{m\times n}$ with

$$\operatorname{rank}(A) = r.$$

The matrix $E \in \mathbb{C}^{m\times n}$ will denote a perturbation of A and we shall set

$$B = A + E.$$

Since we are concerned with the geometry of $\mathbb{C}^n$, we shall be at some pains to cast our results in such a way that they are not affected by unitary transformations (cf. the section on unitarily invariant norms below). We may use this fact to transform our perturbation problems into a simpler form. Specifically, let $U = (U_1, U_2) \in \mathbb{C}^{m\times m}$ be a unitary matrix with $\mathcal{R}(U_1) = \mathcal{R}(A)$ and let $V = (V_1, V_2)$ be a unitary matrix with $\mathcal{R}(V_1) = \mathcal{R}(A^H)$. Then U^HAV has the form

$$U^HAV = \begin{pmatrix} A_{11} & 0 \\ 0 & 0 \end{pmatrix} \tag{2.1}$$

where $A_{11} \in \mathbb{C}^{r\times r}$ is nonsingular. We shall partition U^HEV and U^HBV conformally with U^HAV:

$$U^HEV = \begin{pmatrix} E_{11} & E_{12} \\ E_{21} & E_{22} \end{pmatrix}, \qquad U^HBV = \begin{pmatrix} B_{11} & B_{12} \\ B_{21} & B_{22} \end{pmatrix} = \begin{pmatrix} A_{11}+E_{11} & E_{12} \\ E_{21} & E_{22} \end{pmatrix}.$$

These forms will be called *reduced* forms of the matrices A, B, and E, and in the sequel we shall often assume that the matrices are in reduced form. In this case, the pseudo-inverse is given by

$$A^\dagger = \begin{pmatrix} A_{11}^{-1} & 0 \\ 0 & 0 \end{pmatrix}. \tag{2.2}$$

Singular values. It is a well-known result that in the reduced form (2.1) the matrices U_1 and V_1 may be chosen so that

$$A_{11} \equiv \Sigma = \operatorname{diag}(\sigma_1, \sigma_2, \cdots, \sigma_r),$$

where

$$\sigma_1 \geqq \sigma_2 \geqq \cdots \geqq \sigma_r > 0.$$

This reduced form is called the *singular value decomposition* of the matrix A, and the numbers σ_i are called the singular values of A. From the relation (2.2) and the fact that $(U^HAV)^\dagger = V^HA^\dagger U$, it follows that

$$A^\dagger = V\begin{pmatrix} \Sigma^{-1} & 0 \\ 0 & 0 \end{pmatrix} U^H.$$

The ith singular value of a matrix A, which will be denoted by $\sigma_i(A)$, can be written in the form

$$\sigma_i(A) = \sup_{\dim(\mathcal{X})=i} \inf_{\substack{x\in\mathcal{X} \\ \|x\|_2=1}} \|Ax\|_2 \qquad (i = 1, 2, \cdots, n), \tag{2.3}$$

where

$$\|y\|_2 = \sqrt{y^Hy} \tag{2.4}$$

is the usual Euclidean norm. This characterization provides a natural convention for numbering the singular values of a rectangular matrix: $A \in \mathbb{C}^{m\times n}$ has n

singular values of which $n-r$ are zero; A^H has m singular values of which $m-r$ are zero. The nonzero singular values of A and A^H are the same.

Two inequalities that we shall need in the sequel follow fairly directly from (2.3). They are

$$\sigma_i(A)-\sigma_1(E)\leqq\sigma_i(A+E)\leqq\sigma_i(A)+\sigma_i(E)$$

and

$$\sigma_i(AC)\leqq\sigma_i(A)\sigma_1(C), \qquad \sigma_1(A)\sigma_i(C). \tag{2.5}$$

Unitarily invariant matrix norms. A *norm* on $\mathbb{C}^{m\times n}$ is a function $\|\cdot\|: \mathbb{C}^{m\times n}\to\mathbb{R}$ that satisfies the conditions

$$\begin{aligned} &1.\quad A\neq 0 \Rightarrow \|A\|>0,\\ &2.\quad \|\alpha A\|=|\alpha|\,\|A\|,\\ &3.\quad \|A+B\|\leqq\|A\|+\|B\|. \end{aligned} \tag{2.6}$$

A norm $\|\cdot\|$ is *unitarily invariant* if

$$\|U^H A V\|=\|A\|$$

for all unitary matrices U and V. The perturbation bounds in this paper will be cast in terms of unitarily invariant norms, whose properties will now be described.

Let U and V be the unitary matrices realizing the singular value decomposition of the matrix $A\in\mathbb{C}^{m\times n}$. Then for any unitarily invariant norm $\|\cdot\|_{m,n}$

$$\|A\|_{m,n}=\|U^H A V\|_{m,n}=\left\|\begin{pmatrix}\Sigma & 0\\ 0 & 0\end{pmatrix}\right\|_{m,n}. \tag{2.7}$$

Thus $\|A\|_{m,n}$ is a function of the singular values of A, say

$$\|A\|_{m,n}=\varphi_{m,n}(\sigma_1,\sigma_2,\cdots,\sigma_n). \tag{2.8}$$

It follows from (2.6) that $\varphi_{m,n}$ regarded as a function on $\mathbb{R}^n$ is a norm. Since the interchange of two rows or two columns of a matrix is a unitary transformation of the matrix, the function $\varphi_{m,n}$ is symmetric in its arguments $\sigma_1,\sigma_2,\cdots,\sigma_n$. It can also be shown that $\varphi_{m,n}$ is nondecreasing in the sense that

$$0\leqq\sigma_i\leqq\sigma_i' \quad (i=1,2,\cdots,n) \;\Rightarrow\; \varphi_{m,n}(\sigma_1,\cdots,\sigma_n) \leqq\varphi_{m,n}(\sigma_1',\cdots,\sigma_n'). \tag{2.9}$$

We shall say that the norm $\|\cdot\|_{m,n}$ is *generated* by $\varphi_{m,n}$.

An important norm is the *spectral* norm $\|\cdot\|_2$ generated by the function φ defined by

$$\varphi_2(\sigma_1,\sigma_2,\cdots,\sigma_n)=\max\{|\sigma_1|,\cdots,|\sigma_n|\}.$$

This norm can also be defined by the equation

$$\|A\|_2=\sup_{\|x\|_2=1}\|Ax\|_2, \tag{2.10}$$

where $\|\cdot\|_2$ on the right denotes the Euclidean norm defined by (2.4).

The spectral norm satisfies an important consistency relation with other unitarily invariant norms. If $\|\cdot\|$ is a unitarily invariant norm generated by φ, then it follows from (2.5) and (2.9) that

$$\|CD\| \leqq \|C\|_2\|D\|, \|C\|\|D\|_2 \tag{2.11}$$

whenever $\|CD\| \in \mathbb{C}^{m\times n}$ and respectively $\|D\| \in \mathbb{C}^{m\times n}$ or $\|C\| \in \mathbb{C}^{m\times n}$.

A second example of a unitarily invariant norm is the Frobenius norm generated by the function

$$\varphi_F(\sigma_1, \sigma_2, \cdots, \sigma_n) = (\sigma_1^2 + \sigma_2^2 + \cdots + \sigma_n^2)^{1/2}.$$

For any matrix $A \in \mathbb{C}^{m\times n}$

$$\|A\|_F^2 = \sum_{i=1}^{m}\sum_{j=1}^{n} |\alpha_{ij}^2| = \text{trace}\,(A^H A).$$

The Frobenius norm satisfies the consistency relation

$$\|CD\|_F \leqq \|C\|_F\|D\|_F.$$

Since we shall be dealing with matrices of varying dimensions, we shall work with a family of unitarily invariant norms defined on $\bigcup_{m,n=1}^{\infty} \mathbb{C}^{m\times n}$. It is important that the individual norms so defined interact with one another properly. Accordingly, we make the following definition.

DEFINITION 2.1. Let $\|\cdot\|: \bigcup_{m,n=1}^{\infty} \mathbb{C}^{m\times n} \to \mathbb{R}$ be a family of unitarily invariant norms. Then $\|\cdot\|$ is *uniformly generated* if there is a symmetric function φ, defined for all infinite sequences with only a finite number of nonzero terms, such that

$$\|A\| = \varphi(\sigma_1(A), \sigma_2(A), \cdots, \sigma_n(A), 0, 0, \cdots)$$

for all $A \in \mathbb{C}^{m\times n}$. It is *normalized* if

$$\|x\| = \|x\|_2$$

for any vector x considered as a matrix.

The function φ in the above definition must satisfy the conditions (2.6). Any norm defined by such a function can be normalized. Indeed we have

$$\|x\| = \varphi(\sigma_1(x), 0, 0, \cdots) = \varphi(\|x\|_2, 0, 0, \cdots),$$

from which it follows that $\|x\| = \mu\|x\|_2$ for some constant μ that is independent of the dimension of x. The function $\mu^{-1}\varphi$ then generates the normalized family of norms.

A uniformly generated family of norms has some nice properties. First, since the nonzero singular values of a matrix and its conjugate transpose are the same, we have

$$\|A^H\| = \|A\|.$$

Second, if a matrix is bordered by zero matrices, its norm remains unchanged; i.e.,

$$A = \left\|\begin{pmatrix} 0 & 0 & 0 \\ 0 & A & 0 \\ 0 & 0 & 0 \end{pmatrix}\right\|. \tag{2.12}$$

In particular if A is in reduced form, then

$$\|A\| = \|A_{11}\| \quad \text{and} \quad \|A^{\dagger}\| = \|A_{11}^{-1}\|.$$

It is also a consequence of (2.12) that (2.11) holds for a uniformly generated family of norms whenever the product CD is defined, as may be seen by bordering C and D with zero matrices until they are both square.

A third property is that if $\|\cdot\|$ is normalized then

$$\|A\|_2 \leqq \|A\|. \tag{2.13}$$

In fact from (2.11) and the fact that $\|x\| = \|x\|_2$, we have

$$\|Ax\|_2 = \|Ax\| \leqq \|A\|\|x\|_2 \tag{2.14}$$

for all x. But by (2.10) $\|A\|_2$ is the smallest number for which (2.14) holds for all x, from which (2.13) follows. A trivial corollary of (2.11) and (2.13) is that $\|\cdot\|$ is consistent:

$$\|CD\| \leqq \|C\|\|D\|.$$

Finally we observe that

$$\forall x \|Cx\|_2 \leqq \|Dx\|_2 \Rightarrow \|C\| \leqq \|D\|. \tag{2.15}$$

To prove this implication note that by (2.3) the hypothesis implies that $\sigma_i(C) \leqq \sigma_i(D)$. Hence the inequality $\|C\| \leqq \|D\|$ follows from (2.9).

In the sequel $\|\cdot\|$ will always refer to a uniformly generated, normalized, unitarily invariant norm.

Perturbation of matrix inverses. We shall later need some results on the inverses of perturbations of nonsingular matrices. These are summarized in the following theorem.

THEOREM 2.2 *If A and $B = A + E$ are nonsingular, then*

$$\|B^{-1} - A^{-1}\|/\|A^{-1}\| \leqq \bar{\kappa}\|E\|/\|A\|, \tag{2.16}$$

where

$$\bar{\kappa} = \|A\|\|B^{-1}\|_2. \tag{2.17}$$

If A is nonsingular and

$$\|A^{-1}\|_2\|E\| < 1, \tag{2.18}$$

then B is a fortiori nonsingular. In this case

$$\|B^{-1}\| \leqq \|A^{-1}\|/\gamma, \tag{2.19}$$

and

$$\frac{\|B^{-1} - A^{-1}\|}{\|A^{-1}\|} \leqq \frac{\kappa}{\gamma}\frac{\|E\|}{\|A\|}, \tag{2.20}$$

where

$$\kappa = \|A\|\|A^{-1}\|_2 \tag{2.21}$$

and

$$\gamma = 1 - \kappa \|E\|/\|A\| > 0.$$

The bound (2.16) places no restrictions on the size of E; however, its use requires some estimate of the size of B^{-1}. When E satisfies (2.18) one such estimate is given by (2.19), from which the bound (2.20) follows. This bound has the advantage that it can be stated entirely in terms of the matrix A. Pairs of bounds analogous to (2.16) and (2.20) will repeat themselves through a number of subsequent theorems, as will the pairs $\bar{\kappa}$ and κ. The number κ measures the sensitivity of A^{-1} to perturbations in A and is usually called the *condition number* of A (with respect to inversion).

Projections. We have already observed that the orthogonal projections P_A and R_A onto the column space and the row space of A can be expressed in terms of the pseudo-inverse. The projection onto $\mathcal{R}(A)^\perp$ will be denoted by

$$P_A^\perp \equiv I - P_A.$$

Likewise

$$R_A^\perp \equiv I - R_A$$

will denote the projection onto $\mathcal{R}(A^H)^\perp$.

When A is in reduced form, its projections can be easily written out:

$$P_A = \begin{pmatrix} I_r & 0 \\ 0 & 0 \end{pmatrix} \in \mathbb{C}^{m \times m}, \qquad R_A = \begin{pmatrix} I_r & 0 \\ 0 & 0 \end{pmatrix} \in \mathbb{C}^{n \times n}.$$

It follows that

$$\|P_A A R_A\| = \|A_{11}\|$$

and

$$\|P_A E R_A\| = \|E_{11}\|, \qquad \|P_A E R_A^\perp\| = \|E_{12}\|,$$
$$\|P_A^\perp E R_A\| = \|E_{21}\|, \qquad \|P_A^\perp E R_A^\perp\| = \|E_{22}\|.$$

These identities enable us to pass from results for the reduced form to general results stated in terms of projections of A and E.

We shall need some properties of norms of projections later. These are summarized in the following theorem.

THEOREM 2.3. *For any A and B the following statements are true.*

1. *If* rank (A) = rank (B), *then the singular values of $P_A P_B^\perp$ and $P_B P_A^\perp$ are the same so that*

$$\|P_A P_B^\perp\| = \|P_B P_A^\perp\|.$$

Moreover the nonzero singular values σ of $P_A P_B^\perp$ correspond to pairs $\pm\sigma$ of eigenvalues of $P_B - P_A$, so that

$$\|P_B - P_A\|_2 = \|P_A P_B^\perp\|_2 = \|P_B P_A^\perp\|_2.$$

2. *If* $\|P_B - P_A\|_2 < 1$, *then* rank (A) = rank (B).
3. *If* rank $(B) \geqq$ rank (A), *then*

$$\|P_B P_A^\perp\| \geqq \|P_B^\perp P_A\|.$$

Proof. Proofs of parts 1 and 2 are readily found in the literature; however, a proof, based on a useful matrix decomposition, of part 1 is given in the Appendix to this paper. For part 3 write $P_B = P_1 + P_2$ where $\text{rank}(P_1) = \text{rank}(A)$ and $P_A P_2 = 0$ (i.e., $\mathcal{R}(P_2)$ is orthonormal to $\mathcal{R}(A)$). Then

$$\|P_A P_B^\perp\| = \|P_A(I - P_1 - P_2)\| = \|P_A(I - P_1)\| = \|P_1 P_A^\perp\|,$$

the last equality following from part 1. Now for any x

$$\|P_1 P_A^\perp x\| \leqq \|P_B P_A^\perp x\|,$$

and the result follows from (2.15). □

When $B = A + E$, we can estimate $\|P_B P_A^\perp\|$ in terms of E.

THEOREM 2.4. *The product* $P_B P_A^\perp$ *can be written in the form*

$$P_B P_A^\perp = (B^\dagger)^H R_B E^H P_A^\perp. \tag{2.22}$$

Hence

$$\|P_B P_A^\perp\| \leqq \|B^\dagger\|_2 \|E\|, \tag{2.23}$$

and if $\text{rank}(A) = \text{rank}(B)$, *then*

$$\|P_B P_A^\perp\| \leqq \min\{\|B^\dagger\|_2, \|A^\dagger\|_2\}\|E\|. \tag{2.24}$$

Proof. We have

$$\begin{aligned} P_B P_A^\perp = P_B^H P_A^\perp &= (B^\dagger)^H B^H P_A^\perp \\ &= (B^\dagger)^H (A + E)^H P_A^\perp = (B^\dagger)^H E^H P_A^\perp \\ &= (B^\dagger)^H B^H (B^\dagger)^H E^H P_A^\perp = (B^\dagger)^H R_B E^H P_A^\perp, \end{aligned}$$

which establishes (2.22). The inequality (2.23) follows upon taking norms in (2.22). Finally (2.24) follows from part 1 of Theorem 2.3. □

Theorems 2.3 and 2.4 have obvious analogues for other combinations of projectors (e.g. $R_B^\perp R_A = -A^\dagger E R_B^\perp$). In the sequel a reference to these theorems will also cover any trivial variants.

The case when $\|P_B - P_A\|_2 < 1$ will be particularly important later. We have seen in part 2 of Theorem 2.3 that in this case $\text{rank}(A) = \text{rank}(B)$. However more is true: no vector in $\mathcal{R}(A)$ can be orthogonal to $\mathcal{R}(B)$ and vice versa. For suppose that $x \neq 0$ satisfies $P_A x = x$ and $P_B x = 0$. Then $(P_B - P_A)x = -x$, which implies that $\|P_B - P_A\|_2 \geqq 1$. Conversely if $\|P_B - P_A\|_2 = 1$ then there is a vector in $\mathcal{R}(A)$ or $\mathcal{R}(B)$ that is orthogonal to $\mathcal{R}(B)$ or $\mathcal{R}(A)$. To see this, note that by Theorem 2.3, part 1, there is a vector x such that $(P_B - P_A)x = x$. If $P_A x = 0$ then $P_B x = x$, which shows that $x \in \mathcal{R}(B)$ and $x \in \mathcal{R}(A)^\perp$. If, on the other hand, $P_A x \neq 0$, then since $P_A x = -(I - P_B)x$ we have $P_B(P_A x) = 0$, which shows that $P_A x \in \mathcal{R}(A)$ and $P_A x \in \mathcal{R}(B)^\perp$.

Because of the above considerations we shall say that $\mathcal{R}(A)$ and $\mathcal{R}(B)$ are *acute* whenever $\|P_B - P_A\|_2 < 1$. We shall say that the matrices A and B are acute if $\mathcal{R}(A)$ and $\mathcal{R}(B)$ are acute and $\mathcal{R}(A^H)$ and $\mathcal{R}(B^H)$ are acute. In this case we shall also say that B is an *acute perturbation* of A. The following theorem gives necessary and sufficient conditions for A and B to be acute.

THEOREM 2.5. *The matrices A and B are acute if and only if*

$$\operatorname{rank}(A) = \operatorname{rank}(B) = \operatorname{rank}(P_A B R_A). \tag{2.25}$$

Proof. We shall use the reduced forms of A and B. First suppose (2.25) holds. Then $\operatorname{rank}(B_{11}) = \operatorname{rank}(A_{11})$, and B_{11} is nonsingular. Thus

$$\mathcal{R}(B) = \mathcal{R}\left[\begin{pmatrix} B_{11} \\ E_{21} \end{pmatrix}\right] = \mathcal{R}\left[\begin{pmatrix} I_r \\ E_{21}B_{11}^{-1} \end{pmatrix}\right].$$

But

$$\mathcal{R}(A) = \mathcal{R}\left[\begin{pmatrix} I_r \\ 0 \end{pmatrix}\right],$$

from which it is easily seen that no vector in $\mathcal{R}(A)$ can be orthogonal to $\mathcal{R}(B)$ and vice versa. A similar argument shows that $\mathcal{R}(A^H)$ and $\mathcal{R}(B^H)$ are also acute.

Now assume that A and B are acute. Then $\operatorname{rank}(A) = \operatorname{rank}(B) \geqq \operatorname{rank}(B_{11})$. Assume that $\operatorname{rank}(B_{11}) < \operatorname{rank}(A)$, so that B_{11} is singular. Let p and q be left and right null vectors of B of 2-norm unity, and let P and Q be unitary matrices whose first columns are p and q. Consider the reduced forms

$$\begin{pmatrix} P^H A_{11} Q & 0 \\ 0 & 0 \end{pmatrix}, \qquad \begin{pmatrix} P^H B_{11} Q & P^H E_{12} \\ E_{21} Q & E_{22} \end{pmatrix}.$$

The first row and column of $P^H B_{11} Q$ is zero. If $E_{21} q \neq 0$, then the nonzero vector

$$\begin{pmatrix} 0 \\ E_{21} q \end{pmatrix}$$

is in $\mathcal{R}(B)$ and is orthogonal to $\mathcal{R}(A)$. Similarly if $p^H E_k \neq 0$, then the nonzero vector

$$\begin{pmatrix} 0 \\ E_{12}^H p \end{pmatrix}$$

is in $\mathcal{R}(B^H)$ and is orthogonal to $\mathcal{R}(A^H)$. If $E_{21} q = 0$ and $p^H E_{12} = 0$, then the unit vector e_1 is in $\mathcal{R}(A)$ and $\mathcal{R}(A^H)$ and is orthogonal to $\mathcal{R}(B)$ and $\mathcal{R}(B^H)$. In all cases the contradiction establishes that B_{11} is nonsingular, or equivalently that $\operatorname{rank}(A) = \operatorname{rank}(B) = \operatorname{rank}(B_{11})$. □

Beyond the requirement that $\operatorname{rank}(B) = \operatorname{rank}(A)$, Theorem 2.5 shows that B_{11} must be nonsingular for A and B to be acute. By Theorem 2.2 this will be true whenever $\|A_{11}^{-1}\|_2 \|E_{11}\| < 1$ or equivalently whenever $\|A^\dagger\|_2 \|P_A E R_A\| < 1$. This condition is always satisfied when E_{11} is sufficiently small.

Notes and references. The properties of singular values are well known. See Stewart (1973) for an introduction and Gohberg and Krein (1969) for a more detailed treatment in an infinite dimensional setting.

Von Neumann (1937) was the first to prove that unitarily invariant norms can be written as a function of singular values (the function $\varphi_{m,n}$ in (2.7) is usually called a *symmetric gauge* function). Systematic treatments of unitarily invariant norms may be found in Mirsky (1960) and Gohberg and Krein (1969).

The treatment of unitarily invariant norms in finite dimensional spaces has often been a little sloppy. In infinite dimensional settings there is usually only one space and one generating function, and the same is true in a finite dimensional setting when one is concerned with square matrices. However, when one considers rectangular matrices with varying dimensions, different norms can be used for different dimensions, and there is no reason why these norms should interact nicely. How bad things can get is illustrated by the family of norms $\|\cdot\|$ defined for $A \in \mathbb{C}^{m\times n}$ by

$$\|A\| = \frac{m}{n}\|A\|_2.$$

This family is unitarily invariant and consistent, but $\|A^H\| \neq \|A\|$, unless A is square, and the relation (2.13) does not hold in general. Definition 2.1 represents a return to the simplicity of the infinite dimensional case.

Theorem 2.2 is classical and is usually proved by an appeal to the Neumann series representation $(I-A)^{-1} = I + A + A^2 + \cdots$. Wilkinson (1965) gives a proof that does not use series and discusses at some length the notion of condition number. The result is usually proved under the assumption that $\|I\| = 1$; however, the proofs can be extended to establish the result for any consistent norm.

The results in Theorem 2.3 are well known to people who work closely with orthogonal projectors (e.g., see Afriat (1957) or Wedin (1969)). The decomposition in Theorem 2.4 was established in a slightly weaker form by Wedin (1973). In some cases, when E is small, R_B will be near R_A and the approximation $\|P_A^\perp E R_B\| \cong \|E_{21}\|$ will be more realistic in (2.23).

The number $\|P_B - P_A\|_2$ is closely related to various measures of separation between subspaces. See Kato (1966) and especially Davis and Kahan (1970) where further references may be found. Theorem 2.4, with $\|P_A E R_A\|$ replaced by $\|E\|$, is proved by Wedin (1973). The term "acute" ordinarily refers to the angle subtended by two line segments, not to the segments themselves, and it is technically misapplied when subspaces are said to be acute. But this usage will cause no confusion and it is better than the ugly phrase "in the acute case." The term "acute perturbation" is new, but the notion is introduced in Wedin (1973).

3. The pseudo-inverse. In this section we shall consider the problem of bounding $\|B^\dagger - A^\dagger\|$ in terms of $\|E\|$. We shall obtain three basic theorems: one for when rank $(A) \neq$ rank (B), one for when rank $(A) =$ rank (B), and one for when B is an acute perturbation of A. All these theorems are based on expressions for $B^\dagger$, which also yield asymptotic expressions for $B^\dagger$ and expressions for the derivative of $A^\dagger$.

Lower bounds. Before proceeding to obtain bounds on $\|B^\dagger - A^\dagger\|$, we shall show how bad things can be by deriving lower bounds.

THEOREM 3.1. *If A and B are not acute, then*

$$\|B^\dagger - A^\dagger\|_2 \geqq 1/\|E\|_2. \tag{3.1}$$

If, further, rank $(B) \geqq$ rank (A), *then*

$$\|B^\dagger\|_2 \geqq 1/\|E\|_2. \tag{3.2}$$

Proof. Suppose for definiteness that rank $(B) \geqq$ rank (A). Then there is, say, a vector $y \in \mathcal{R}(B)$ with $\|y\|_2 = 1$ such that $y \in \mathcal{R}(A)^\perp$ (otherwise work with A^H and B^H). Thus

$$1 = y^H y = y^H P_B y = y^H B B^\dagger y = y^H (A+E) B^\dagger y$$
$$= y^H E B^\dagger y \leqq \|E\|_2 \|B^\dagger y\|_2,$$

which shows that $\|B^\dagger y\|_2$, and hence $\|B^\dagger\|_2$ is not less than $1/\|E\|_2$. From this and the fact that $A^\dagger y = A^\dagger P_A y = 0$ we have

$$\frac{1}{\|E\|_2} \leqq \|B^\dagger y\|_2 \leqq \|(B^\dagger - A^\dagger) y\|_2 \leqq \|B^\dagger - A^\dagger\|. \quad \square$$

Theorem 3.1 shows that the pseudo-inverse of a general matrix is not a continuous function of its elements, unless the class of perturbations is restricted. It also says that if two nearby matrices do not have acute column and row spaces, then one of them at least must have a large pseudo-inverse. Moreover if they are of the same rank, then both of them must have large pseudo-inverses.

A decomposition of $B^\dagger - A^\dagger$. In spite of the negative results in Theorem 3.1, it is possible to obtain bounds on $\|B^\dagger - A^\dagger\|$ in the general case, although these bounds need not remain finite as B approaches A. The basis for obtaining such bounds is contained in the following theorem.

THEOREM 3.2. *The following two decompositions of $B^\dagger - A^\dagger$ are valid:*

$$B^\dagger - A^\dagger = -B^\dagger P_B E R_A A^\dagger + B^\dagger P_B P_A^\perp - R_B^\perp R_A A^\dagger, \tag{3.3}$$

$$B^\dagger - A^\dagger = -B^\dagger P_B E R_A A^\dagger + (B^H B)^\dagger R_B E^H P_A^\perp - R_B^\perp E^H P_A (A A^H)^\dagger. \tag{3.4}$$

Proof. Both expressions can be verified directly by replacing E with $B - A$, replacing the projectors by their expressions in terms of pseudo-inverses, and simplifying. $\square$

It should be noted that (3.4) can be obtained directly from (3.3) by using Theorem 2.4 to express $P_B P_A^\perp$ and $R_B^\perp R_A$ in terms of E.

The general theorem. We are now in a position to prove the general theorem bounding $\|B^\dagger - A^\dagger\|$.

THEOREM 3.3. *For any A and B with $B = A + E$,*

$$\|B^\dagger - A^\dagger\| \leqq \mu \max\{\|A^\dagger\|_2^2, \|B^\dagger\|_2^2\} \|E\|,$$

where μ is given in the following table:

$\|\cdot\|$	*arbitrary*	*spectral*	*Frobenius*
μ	3	$\frac{1+\sqrt{5}}{2}$	$\sqrt{2}$

Proof. The proof is a slight modification of the proof given by Wedin (1973). We shall give only the proof for the Frobenius norm.

Suppose for definiteness rank $(B) \leqq$ rank (A). Let F_1, F_2, and F_3 denote the three terms on the right-hand side of (3.3). Then the column spaces of F_1 and F_2 are orthogonal to the column space of F_3. Hence

(3.5) $$\|B^\dagger - A^\dagger\|_F^2 = \|F_1 + F_2\|_F^2 + \|F_3\|_F^2.$$

Now since $F_1 + F_2 = B^\dagger(P_B D A^\dagger P_A + P_B P_A^\perp)$,

$$\|F_1 + F_2\|_F^2 \leqq \|B^\dagger\|_2^2 (\|P_B E A^\dagger P_A\|_F^2 + \|P_B P_A^\perp\|_F^2).$$

But from Theorems 2.4 and 2.5

$$\begin{aligned}\|P_B E A^\dagger P_A\|_F^2 + \|P_B P_A^\perp\|_F^2 &\leqq \|P_B E A^\dagger\|_F^2 + \|P_B^\perp P_A\|_F^2 \\ &= \|P_B E A^\dagger\|_F^2 + \|P_B^\perp E A^\dagger\|_F^2 = \|E A^\dagger\|_F^2 \leqq \|E\|_F^2 \|A^\dagger\|_2^2.\end{aligned}$$

Hence

(3.6) $$\|F_1 + F_2\|_F \leqq \|A^\dagger\|_2 \|B^\dagger\|_2 \|E\|_F.$$

Also from Theorem 2.5

(3.7) $$\begin{aligned}\|F_3\|_F = \|A^\dagger\|_2 \|R_B^\perp R_A\|_F &= \|A^\dagger\|_2 \|R_A R_B^\perp\|_F \\ &= \|A^\dagger\|_2 \|A^\dagger E R_B^\perp\|_F \leqq \|A^\dagger\|_2^2 \|E\|_F,\end{aligned}$$

and the result follows on combining (3.3), (3.6), and (3.7). Since the final bound is symmetric in A and B, it also holds when rank $(B) \geqq$ rank (A). □

It should be noted that these bounds do not imply that $\|B^\dagger - A^\dagger\|$ is small when $\|E\|$ is small, since $B^\dagger$ may grow unboundedly as E approaches zero.

The case rank (A) = rank (B). When A and B have the same rank, we can strengthen Theorem 3.3 in two ways. First, we can replace the term $\max\{\|A^\dagger\|_2^2, \|B^\dagger\|_2^2\}$ with the product $\|A\|_2 \|B\|_2$. Second we can distinguish more cases for the constant μ. In the following theorem recall that $A \in \mathbb{C}^{m \times n}$ with $m \geqq n$.

THEOREM 3.4. *If* rank (A) = rank (B), *then*

(3.8) $$\|B^\dagger - A^\dagger\| \leqq \mu \|A^\dagger\|_2 \|B^\dagger\|_2 \|E\|.$$

where μ is given in the following table.

rank \ $\|\cdot\|$	*Arbitrary*	*Spectral*	*Frobenius*
rank $(A) < \min(m, n)$	3	$(1+\sqrt{5})/2$	$\sqrt{2}$
rank $(A) = m \neq n = \min(m, n)$	2	$\sqrt{2}$	1
rank $(A) = m = n$	1	1	1

The proof of this theorem may be found in Wedin (1973). The bound (3.8) may be recast in the form

(3.9) $$\frac{\|B^\dagger - A^\dagger\|}{\|B^\dagger\|_2} \leqq \mu\kappa \frac{\|E\|}{\|A\|},$$

where

$$\kappa = \|A\|\|A^\dagger\|_2.$$

In this form the result is almost analogous to the bound (2.20) for the inverse in Theorem 2.2. The bound (3.9) also implies that as E approaches zero, the relative error in $B^\dagger$ approaches zero, which further implies that $B^\dagger$ approaches $A^\dagger$. Remembering, on the other hand, that if $\operatorname{rank}(B) \neq \operatorname{rank}(A)$ then A and B cannot be acute, we have from Theorem 3.1 the following corollary of Theorem 3.4.

COROLLARY 3.5. *A necessary and sufficient condition that*

$$\lim_{B\to A} B^\dagger = A^\dagger$$

is that $\operatorname{rank}(B) = \operatorname{rank}(A)$ *as* B *approaches* A.

Acute perturbations. It is evident from the proofs of Theorems 3.3 and 3.4 that we have given away much in deriving the bounds. In particular, if B is a small acute perturbation of A then P_A and P_B are nearly equal, and the same is true of R_A and R_B. Thus it follows from (3.4) that $B^\dagger - A^\dagger$ can be decomposed into three terms—one essentially depending on P_AER_A, one on $P_AER_A^\perp$, and one on $P_A^\perp ER_A$. However, this does not tell the whole story; for we shall show that the dependency of $B^\dagger - A^\dagger$ on $P_AER_A^\perp$ and $P_A^\perp ER_A$ is bounded, no matter how large these projections may be.

In order to state our theorems concisely, we must first introduce some additional notation. Let $\|\cdot\|$ be generated by φ and for any $F \in \mathbb{C}^{k\times r}$ $(k \geqq r)$ define

$$\psi_\varphi(F) = \varphi\left[\frac{\sigma_1(F)}{[1+\sigma_1^2(F)]^{1/2}}, \cdots, \frac{\sigma_r(F)}{[1+\sigma_r^2(F)]^{1/2}}\right]. \tag{3.10}$$

The function ψ_φ is not a norm; however, it has some useful properties. First, from (2.5) and the monotonicity of φ,

$$\psi(GF) \leqq \psi(\|G\|_2 F) \leqq \psi(\|G\|F).$$

Second, since for $\alpha \geqq 1$

$$\frac{\alpha\sigma}{(1+\alpha\sigma^2)^{1/2}} \leqq \frac{\alpha\sigma}{(1+\sigma^2)^{1/2}},$$

we have

$$\alpha \geqq 1 \Rightarrow \psi_\varphi(\alpha F) \leqq \alpha\psi_\varphi(F).$$

For small F, $\psi_\varphi(F)$ is asymptotic to $\|F\|$:

$$\psi_\varphi(F) = \|F\| + o(\|F\|).$$

For large F, $\psi_\varphi(F)$ is bounded:

$$\psi_\varphi(F) \leqq \|I_r\|.$$

Finally, for the spectral norm

$$\psi_2(F) = \|F\|_2/(1+\|F\|_2^2)^{1/2}.$$

Our first result concerns a rather special matrix

LEMMA 3.6. *The matrix*

$$\begin{pmatrix} I \\ F \end{pmatrix}$$

satisfies

$$\left\| \begin{pmatrix} I \\ F \end{pmatrix}^{\dagger} \right\|_2 \leqq 1 \tag{3.11}$$

and

$$\left\| \begin{pmatrix} I \\ F \end{pmatrix}^{\dagger} - (I \quad 0) \right\| = \psi_{\varphi}(F). \tag{3.12}$$

Proof. It is easily verified that

$$\begin{pmatrix} I \\ F \end{pmatrix}^{\dagger} = (I + F^H F)^{-1}(1 \quad F^H), \tag{3.13}$$

whose singular values are

$$\frac{1}{[1+\sigma_i^2(F)]^{\frac{1}{2}}} \leqq 1,$$

from which (3.11) follows. Also if

$$G = \begin{pmatrix} I \\ F \end{pmatrix}^{\dagger} - (I \quad 0),$$

then

$$GG^H = I - (I + F^H F)^{-1}.$$

It follows that the singular values of G are given by

$$\frac{\sigma_i(F)}{[1+\sigma_i(F)]^{1/2}},$$

which establishes (3.12). □

The main result is based on an explicit representation of $B^{\dagger}$. We shall work with the reduced forms of A and B.

THEOREM 3.7. *Let B be an acute perturbation of A. Then*

$$B^{\dagger} = (I \quad F_{12})^{\dagger} B_{11}^{-1} \begin{pmatrix} I \\ F_{21} \end{pmatrix}^{\dagger}, \tag{3.14}$$

where

$$F_{21} = E_{21} B_{11}^{-1}, \qquad F_{12} = B_{11}^{-1} E_{12}.$$

Proof. As in the proof of Theorem 3.4, we have

$$\mathscr{R}(B) = \mathscr{R}\left[\begin{pmatrix} B_{11} \\ E_{21} \end{pmatrix}\right].$$

Thus the columns of

$$\begin{pmatrix} E_{12} \\ E_{22} \end{pmatrix}$$

can be expressed as a linear combination of the columns of

$$\begin{pmatrix} B_{11} \\ E_{21} \end{pmatrix}.$$

Since $B_{11}(B_{11}^{-1}E_{12}) = E_{12}$, we must have

$$\begin{pmatrix} E_{12} \\ E_{22} \end{pmatrix} = \begin{pmatrix} B_{11} \\ E_{21} \end{pmatrix} B_{11}^{-1}E_{12},$$

from which it follows that

$$B = \begin{pmatrix} I \\ F_{21} \end{pmatrix} B_{11}^{-1}(I \quad F_{12}). \tag{3.15}$$

The result now follows from Penrose's conditions. □

It is interesting to observe that, from (3.15),

$$B_{22} = E_{22} = F_{21}B_{11}^{-1}F_{12},$$

which is of second order in $\|E\|$. In other words, if rank $(A+E) = $ rank (A), then $P_A^{\perp}ER_A^{\perp}$ must approach zero quadratically as E approaches zero.

We turn now to the perturbation theorem.

THEOREM 3.8. *Let B be an acute perturbation of A, and let*

$$\bar{\kappa} = \|A\|\|B_{11}^{-1}\|_2.$$

Then

$$\frac{\|B^{\dagger} - A^{\dagger}\|}{\|A^{\dagger}\|} \leqq \bar{\kappa}\frac{\|E_{11}\|}{\|A\|} + \psi_{\varphi}\left(\bar{\kappa}\frac{E_{12}}{\|A\|}\right) + \psi_{\varphi}\left(\bar{\kappa}\frac{E_{12}}{\|A\|}\right), \tag{3.16}$$

where ψ_{φ} is defined by (3.10).

Proof. Let F_{ij} be defined as in Theorem 3.7. Let

$$I_{21} = \begin{pmatrix} I_r \\ 0 \end{pmatrix}, \qquad I_{12} = (I_r \quad 0),$$

$$J_{21} = \begin{pmatrix} I_r \\ F_{21} \end{pmatrix}, \qquad J_{12} = (I_r \quad F_{12}).$$

From (3.14), $B^{\dagger} = J_{12}^{\dagger}B_{11}^{-1}J_{21}^{\dagger}$; hence

$$B^{\dagger} - A^{\dagger} = (J_{12}^{\dagger} - I_{12}^{\dagger})A_{11}^{-1}I_{21}^{\dagger} + J_{12}^{\dagger}A_{11}^{-1}(J_{21}^{\dagger} - I_{21}^{\dagger}) + J_{12}^{\dagger}(B_{11}^{-1} - A_{11}^{-1})J_{21}^{\dagger}. \tag{3.17}$$

From Theorem 2.2 we have the following bound:

$$\|J_{12}^{\dagger}(B_{11}^{-1} - A_{11}^{-1})\| \leqq \|A_{11}^{-1}\|\bar{\kappa}\frac{\|E_{11}\|}{\|A_{11}\|}. \tag{3.18}$$

By Lemma 3.6

$$\begin{aligned} \|(J_{12}^{\dagger}-I_{12}^{\dagger})A_{11}^{-1}I_{21}^{\dagger}\| &\leqq \|A_{11}^{-1}\|\,\|J_{12}^{\dagger}-I_{12}^{\dagger}\| \\ &= \|A_{11}^{-1}\|\psi_{\varphi}(F_{12}) = \|A_{11}^{-1}\|\psi_{\varphi}(B_{11}^{-1}E_{12}) \\ &\leqq \|A_{11}^{-1}\|\psi_{\varphi}\left(\bar{\kappa}\frac{E_{12}}{\|A\|}\right), \end{aligned} \tag{3.19}$$

and likewise

$$\|J_{12}^{\dagger}A_{11}^{-1}(J_{21}^{\dagger}-I_{21}^{\dagger})\| \leqq \|A_{11}^{-1}\|\psi_{\varphi}\left(\bar{\kappa}\frac{E_{21}}{\|A\|}\right). \tag{3.20}$$

The bound (3.16) follows on combining (3.17), (3.18), (3.19), and (3.20) and remembering that $\|A_{11}^{-1}\| = \|A^{\dagger}\|$. □

The bound (3.16) gives a rather nice dissection of $\|B^{\dagger}-A^{\dagger}\|$. Asymptotically, for E_{12} and E_{21} small, it reduces to the bound that would be obtained by taking norms in (3.4), i.e.,

$$\frac{\|B^{\dagger}-A^{\dagger}\|}{\|A^{\dagger}\|} \leqq \bar{\kappa}\frac{\|E_{11}\|+\|E_{12}\|+\|E_{21}\|}{\|A\|}.$$

However, the bound additionally shows that E_{12} and E_{21} can have at most a bounded effect on $\|B^{\dagger}-A^{\dagger}\|$.

When A is square and nonsingular, E_{12} and E_{21} are void, and the bound reduces to that of Theorem 2.2. Note that the number $\bar{\kappa}$, defined in analogy with (2.17), plays an analogous role here.

As in the second part of Theorem 3.2, if E_{11} is sufficiently small, we can estimate $\|B_{11}^{-1}\|_2$ in terms of $\|A_{11}^{-1}\|_2$ and $\|E\|$. This gives the following corollary.

COROLLARY 3.9. *In Theorem* 3.8, *let*

$$\kappa = \|A\|\|A^{\dagger}\|_2 \tag{3.21}$$

and suppose that

$$\|A^{\dagger}\|_2\|E_{11}\| < 1,$$

so that

$$\gamma \equiv 1-\kappa\|E_{11}\|/\|A\| > 0.$$

Then

$$\|B^{\dagger}\| \leqq \|A^{\dagger}\|/\gamma, \tag{3.22}$$

and

$$\frac{\|B^{\dagger}-A^{\dagger}\|}{\|A^{\dagger}\|} \leqq \frac{\kappa}{\gamma}\frac{\|E_{11}\|}{\|A\|} + \psi_{\varphi}\left(\frac{\kappa}{\gamma}\frac{E_{21}}{\|A\|}\right) + \psi_{\varphi}\left(\frac{\kappa}{\gamma}\frac{E_{12}}{\|A\|}\right). \tag{3.23}$$

Proof. From the equation $B^{\dagger} = J_{12}^{\dagger}B_{11}^{-1}J_{21}^{\dagger}$ we have

$$\|B^{\dagger}\| \leqq \|J_{12}^{\dagger}\|\|B_{11}^{-1}\|\|J_{21}^{\dagger}\|_2 \leqq \|B_{11}^{-1}\|.$$

By Theorem 2.2

$$\|B_{11}^{-1}\| \leqq \|A_{11}^{-1}\|/\gamma = \|A^{\dagger}\|/\gamma,$$

which establishes (3.21). Also $\bar{\kappa} \leqq \kappa/\gamma$, and (3.23) follows from (3.16).

The number κ is defined in analogy with (2.21). For E_{11} sufficiently small, $\bar{\kappa} \cong \kappa$, and (3.16) and (3.23) give essentially the same bound.

Asymptotic forms and derivatives. Asymptotic forms for B may be obtained from either (3.4) or (3.14). Of course for $B^{\dagger}$ to approach $A^{\dagger}$ we must have rank $(A) =$ rank (B); and since we are assuming that E is arbitrarily small, B may be assumed to be an acute perturbation of A. In this case

$$B^{\dagger} = A^{\dagger} + O(\|E\|),$$

and

$$P_B = BB^{\dagger} = (A+E)[A^{\dagger} + O(\|E\|)] = P_A + O(\|E\|)$$

with similar expressions for the other projections. Hence from (3.4)

$$B^{\dagger} = A^{\dagger} - A^{\dagger}P_A E R_A A^{\dagger} + (A^H A)^{\dagger} R_A E^H P_A^{\perp} - R_A^{\perp} E^H P_A (AA^H)^{\dagger} + O(\|E\|^2). \tag{3.24}$$

It follows immediately from (3.24) that if $A(\tau)$ is a differentiable function of τ with

$$\text{rank}\,[A(\tau)] = \text{rank}\,[A(\tau')]$$

for all τ, then $A(\tau)^{\dagger}$ is a differentiable function of τ and

$$\frac{dA^{\dagger}}{d\tau} = -A^{\dagger}P_A \frac{dA}{d\tau} R_A A^{\dagger} + (A^H A)^{\dagger} R_A \frac{dA^H}{d\tau} P_A^{\perp} - R_A^{\perp} \frac{dA^H}{d\tau} P_A (AA^H)^{\dagger}. \tag{3.25}$$

The asymptotic form obtained from (3.14) can be useful computationally when A has been put in reduced form as a preliminary to computing $A^{\dagger}$. We have from (3.24) that

$$B_{11}^{-1} = A_{11}^{-1} - A_{11}^{-1} E_{11} A_{11}^{-1} + O(\|E_{11}\|^2).$$

From (3.13) in the proof of Lemma 3.6 we have

$$\begin{pmatrix} I \\ F_{21} \end{pmatrix}^{\dagger} = (I \quad A_{11}^{-H} E_{21}^H) + O(\|E_{11}\| \|E_{21}\|)$$

and

$$(I \quad F_{12})^{\dagger} = \begin{pmatrix} I \\ E_{12}^H A_{11}^{-H} \end{pmatrix} + O(\|E_{11}\| \|E_{12}\|).$$

Hence from (3.14)

$$B^{\dagger} = \begin{pmatrix} A_{11}^{-1} - A_{11}^{-1} E_{11} A_{11}^{-1} + O(\|E_{11}\|^2) & (A_{11}^H A_{11})^{-1} E_{21}^H + O(\|E_{11}\| \|E_{21}\|) \\ E_{12}^H (A_{11} A_{11}^H)^{-1} + O(\|E_{12}\| \|E_{12}\|) & E_{12}^H (A_{11}^H A_{11} A_{11}^H)^{-1} E_{21}^H \\ & + O(\|E_{11}\|^2 \|E_{12}\| \|E_{21}\|) \end{pmatrix}.$$

This expression is in perfect agreement with (3.24) when the E_{ij} are interpreted appropriately as projections of E.

Notes and references. For expository reasons the results of this section have not been presented in the historical order of their development. Penrose (1955) established Corollary 3.4 using techniques that do not give explicit perturbation bounds. The subject was revived by Golub and Wilkinson (1966), whose interest in stable algorithms for solving least squares problems [cf. Golub (1965)] led them to derive first-order perturbation bounds for least squares solutions (more of this later). The first perturbation bounds for the pseudo-inverse itself were given by Ben-Israel (1966), who restricts his class of perturbations so that (in reduced form) only E_{11} is nonzero. More general theorems for acute perturbations were established by Hanson and Lawson (1969), Pereyra (1969), and Stewart (1969). Theorem 3.7 is a refinement and extension of Stewart's bound. An identity in terms of projections related to (3.14) is given by Wedin (1973), who uses it to derive bounds for acute perturbations.

The decompositions (3.3) and (3.4) and the consequent Theorem 3.4 are due to Wedin (1973). Theorem 3.3 is a slight extension of these results. Theorem 3.1 is also due to Wedin (1973), although a slightly restricted form of the result may be found in Stewart (1969). In an earlier report Wedin (1969) considers the sharpness of the constants μ in Theorem 3.4 and shows that for the spectral norm μ cannot be made smaller.

Early differentiability results have been given by Pavel-Parvu and Korganoff (1969) and Hearon and Evans (1968). Wedin (1969) derived the formula (3.25) as we did from the decomposition (3.4). The same result for functions of several variables was derived independently by Golub and Pereyra (1973) in connection with separable nonlinear least squares problems. For further references see Golub and Pereyra (1975).

4. Projections. In this section we shall consider how the projection P_A varies with A. Since $P_A = AA^\dagger$, it might be thought that the perturbation theory for P_A could be derived from the theory developed in the last section for $A^\dagger$. However this approach gives too much away, and sharper bounds may be obtained by working directly with one of the decompositions of $B^\dagger$. In particular we shall work with the decomposition (3.15) based on the reduced forms of A and B.

If $\mathcal{R}(A)$ and $\mathcal{R}(B)$ are not acute, then $\|P_B - P_A\|_2 = 1$. Consequently we can restrict ourselves to the case where $\mathcal{R}(A)$ and $\mathcal{R}(B)$ are acute. More particularly we shall only consider the case where B is an acute perturbation of A.

THEOREM 4.1. *Let B be an acute perturbation of A, and let $\bar{\kappa}$ be defined as in Theorem* 3.8. *Then*

$$\|P_B - P_A\|_2 \leq \frac{\bar{\kappa}\|E_{21}\|_2/\|A\|_2}{[1+(\bar{\kappa}\|E_{21}\|_2/\|A\|_2)^2]^{1/2}} < 1. \tag{4.1}$$

Proof. With F_{21} defined as in the last section we have [cf. (3.15)]

$$\mathcal{R}(B) = \mathcal{R}\begin{pmatrix} I \\ F_{21} \end{pmatrix}.$$

The matrix

$$\begin{pmatrix} I \\ F_{21} \end{pmatrix}(I + F_{21}^H F_{21})^{-1}(I \quad F_{21}^H)$$

is a Hermitian idempotent whose column space is $\mathcal{R}(B)$; hence it is P_B. It follows that

$$(4.2)\qquad P_B - P_A = \begin{pmatrix} (I+F_{21}^H F_{21})^{-1} - I & (I+F_{21}^H F_{21})^{-1} F_{21}^H \\ F_{21}(I+F_{21}^H F_{21})^{-1} & F_{21}(I+F_{21}^H F_{21})^{-1} F_{21}^H \end{pmatrix},$$

from which it is easily verified that

$$(4.3)\qquad (P_B - P_A)^2 = \begin{pmatrix} F_{21}^H F_{21}(I+F_{21}^H F_{21})^{-1} & 0 \\ 0 & F_{21}(I+F_{21}^H F_{21})^{-1} F_{21}^H \end{pmatrix}.$$

Now the nonzero singular values of the diagonal blocks in (4.3) are given by

$$\sigma_i^2(F_{21})/[1+\sigma_i^2(F_{21})]$$

where the $\sigma_i(F_{21})$ are the nonzero singular values of F_{21}. The result follows from the fact that the largest singular value σ_1 of F_{21} satisfies

$$\sigma_1(F_{21}) = \|F_{21}\|_2 \leqq \bar{\kappa}\|E_{21}\|_2/\|A\|_2. \quad \square$$

In terms of projections, the bound (4.1) can be written in the form

$$\|P_B - P_A\|_2 \leqq \frac{\bar{\kappa}\|P_A^\perp E R_A\|_2/\|A\|_2}{[1+(\bar{\kappa}\|P_A^\perp E R_A\|_2/\|A\|_2)^2]^{1/2}} < 1.$$

The bound is interesting in several ways. First it depends not at all on E_{12} and E_{22}. Second its dependence on E_{11} is only through the constant $\bar{\kappa}$. Third the bound is always less than unity. Finally, it goes to zero along with E_{21}. We may summarize this last observation in the following corollary.

COROLLARY 4.2. *Regarding B as variable, a sufficient condition for*

$$\lim_{B\to A} P_B = P_A$$

is that A and B are acute and

$$\lim_{B\to A} P_A^\perp B R_A = 0.$$

If the hypotheses of Corollary 3.9 are satisfied (i.e., if $\|A_{11}^{-1}\|_2\|E_{11}\| < 1$) then we may replace $\bar{\kappa}$ by κ/γ in (4.1).

Asymptotic forms and derivatives. Asymptotic forms may be obtained in the usual way from (4.2). Indeed

$$(4.4)\qquad P_B - P_A = \begin{pmatrix} O(\|E_{21}\|^2) & F_{21}^H + O(\|E_{21}\|^3) \\ F_{21} + O(\|E_{21}\|^3) & O(\|E_{21}\|^2) \end{pmatrix}.$$

In terms of projections

$$(4.5)\qquad P_B = P_A + P_A^\perp E R_A A^\dagger + A^{\dagger H} R_A E^H P_A^\perp + O(\|P_A^\perp E R_A\|^2).$$

It follows that if $A(\tau)$ is differentiable and varies without changing rank, then $P_{A(\tau)}$ is differentiable and

$$\frac{dP_A}{d\tau}=P_A^\perp\frac{dA}{d\tau}R_AA^\dagger+A^{\dagger H}R_A\frac{dA^H}{d\tau}P_A^\perp. \tag{4.6}$$

Notes and references. Theorem 4.1 and its corollary appear to be new. The expression (4.4) for the derivative of P_A was first given by Golub and Pereyra (1973). Per-Åke Wedin has pointed out to the author that the asymptotic form (4.5) and the expression (4.6) can be derived from the identity

$$P_B-P_A=P_B(I-P_A)+(I-P_B)P_A=B^{\dagger H}R_BE^HP_A^\perp+P_B^\perp ER_AA^\dagger.$$

5. The linear least squares problem. In this section we shall derive perturbation bounds for the least squares problem of minimizing $\|b-Ax\|_2$. Although the solution of minimum norm is given by $x=A^\dagger b$, the perturbation theory of § 3 again does not give the best possible results.

We shall assume throughout this section that B is an acute perturbation of A, and we shall work with the reduced form of the problem. In this form x is replaced by V^Hx and b is replaced by U^Hb (cf. § 2). If x and b are partitioned in the forms

$$x=\begin{pmatrix}x_1\\x_2\end{pmatrix},\qquad b=\begin{pmatrix}b_1\\b_2\end{pmatrix}$$

where $x_1, b_1\in\mathbb{C}^r$, then

$$x_1=A_{11}^{-1}b_1 \tag{5.1}$$

and

$$x_2=0.$$

Moreover the norm of the residual vector

$$r=b-Ax$$

is given by

$$\|r\|_2=\|b_2\|_2.$$

In the theorems to follow we shall freely use the definitions made in the previous sections (e.g., $\bar\kappa$, κ and γ). As in §§ 3 and 4 the number $\bar\kappa$ may be replaced by κ/γ whenever $\|A^\dagger\|_2\|E_{11}\|<1$. One additional piece of notation will be needed; namely, we shall define η as that nonnegative constant such that

$$\|b_1\|_2=\eta\|A\|_2\|x\|_2.$$

Since $b_1=A_{11}x_1$, we have $\eta\leqq 1$. Also $\|x\|\leqq\|A^\dagger\|\|b_1\|$, which shows that $\eta\geqq\kappa^{-1}$. When A is ill-conditioned, that is when $A^\dagger$ is large, the vector x may be either large or small. In the first case η is near zero, and we shall say that "x reflects the ill-condition of A."

We first consider perturbations in the vector b.

THEOREM 5.1. *Let $x=A^\dagger b$ and $x+h=A^\dagger(b+k)$. Then*

$$\|h\|_2/\|x\|_2\leqq\kappa\eta\|P_Ak\|_2/\|P_Ab\|_2. \tag{5.2}$$

Proof. With the obvious partitioning of k we have $h = A_{11}^{-1}k_1$, so that

$$\|h\|_2 \leqq \|A_{11}^{-1}\|\|k_1\|. \tag{5.3}$$

But $\|x\|_2 = \eta^{-1}\|b_1\|_2/\|A\|_2$, which combined with (5.3) yields (5.2). □

Theorem 5.1 shows that the perturbation in x is determined by the projection of k onto $\mathcal{R}(A)$. However, $P_A k$ is normalized by $\|P_A b\|_2$, and if this latter quantity is small, the perturbation may be large. Since

$$\|b\|_2^2 = \|P_A b\|_2^2 + \|r\|_2^2,$$

this observation may be summarized by saying that large residuals are troublesome, a statement which will be amply supported later.

Since η can be as small as κ^{-1}, the number κ cannot be taken as a condition number for perturbations in b without further qualification. If x does not reflect the ill-conditioning of A, then η is near unity and κ is a condition number. Otherwise the solution will be relatively insensitive to perturbations in b.

We next turn to assessing the effects on x of a perturbation in A.

THEOREM 5.2. *Let $x = A^\dagger b$ and $x + h = B^\dagger b$, where $B = A + E$ is an acute perturbation of A. Then*

$$\frac{\|h\|_2}{\|x\|_2} \leqq \bar{\kappa}\frac{\|E_{11}\|_2}{\|A\|_2} + \psi_2\left(\bar{\kappa}\frac{E_{12}}{\|A\|_2}\right) + \bar{\kappa}^2\frac{\|E_{12}\|_2}{\|A\|_2}\left(\eta\frac{\|b_2\|_2}{\|b_1\|_2} + \frac{\|E_{21}\|_2}{\|A\|_2}\right). \tag{5.4}$$

Proof. Write

$$h = J_{12}^\dagger(B_{11}^{-1} - A_{11}^{-1})b_1 + (J_{12}^\dagger - I_{12}^\dagger)A_{11}^{-1}b_1 + J_{12}^\dagger B_{11}^{-1}(J_{21}^\dagger - I_{21}^\dagger)b. \tag{5.5}$$

Then

$$\|J_{12}^\dagger(B_{11}^{-1} - A_{11}^{-1})b_1\|_2 \leqq \bar{\kappa}\frac{\|E_{11}\|_2}{\|A\|_2}\|x\|_2, \tag{5.6}$$

and

$$(J_{12}^\dagger - I_{12}^\dagger)A_{11}^{-1}b_{12} \leqq \psi_2\left(\bar{\kappa}\frac{E_{12}}{\|A\|_2}\right)\|x\|_2. \tag{5.7}$$

Now

$$\begin{aligned} J_{12}^\dagger B_{11}^{-1}(J_{21}^\dagger - I_{21}^\dagger)b = J_{12}^\dagger B_{11}^{-1}[(I + F_{21}^H F_{21})^{-1} - I]b_1 \\ + J_{12}^\dagger B_{11}^{-1}(I + F_{21}^H F_{21})^{-1}F_{21}^H b_2. \end{aligned} \tag{5.8}$$

To bound the first term in (5.8), note that

$$(I + F_{21}^H F_{21})^{-1} - I = -(I + F_{21}^H F_{21})^{-1}F_{21}^H F_{21}.$$

Hence

$$\begin{aligned} &\|J_{12}^\dagger B_{11}^{-1}[(I + F_{21}^H F_{21}) - I]b_1\|_2 \\ &\leqq \|B_{11}^{-1}\|_2\|(I + F_{21}^H F_{21})^{-1}\|_2\|F_{21}^H\|_2\|F_{21}b_1\|_2 \\ &\leqq \|B_{11}^{-1}\|_2^2\|E_{21}\|_2\|E_{21}B_{11}^{-1}b_1\|_2 \leqq \|B_{11}^{-1}\|_2^2\|E_{21}\|_2^2\|x\|_2 \leqq \left[\bar{\kappa}\frac{\|E_{21}\|_2}{\|A\|_2}\right]^2\|x\|_2. \end{aligned} \tag{5.9}$$

For the second term in (5.8) we have

$$
\begin{aligned}
\|J_{21}^{\dagger}B_{11}^{-1}(I+F_{21}^{H}F_{21})^{-1}F_{21}b_2\|_2 &\leqq \|B_{11}^{-1}\|^2\|E_{21}\|\|b_2\| \\
&= \|B_{11}^{-1}\|_2^2\|E_{21}\|_2\frac{\|b_2\|_2}{\|b_1\|_2}\eta\|x\|_2\|A\|_2 \qquad (5.10)\\
&\leqq \eta\bar{\kappa}^2\frac{\|E_{21}\|_2}{\|A\|_2}\frac{\|b_2\|}{\|b_1\|}\|x\|_2.
\end{aligned}
$$

The bound (5.4) follows on combining (5.5)–(5.10). □

The first two terms in (5.4) are unexceptionable. The first term corresponds to the classical result for linear systems and is the only nonzero term when A is square and nonsingular. The second term depends on $P_AER_A^{\perp}$ and vanishes when A is of full column rank, as it is in many applications.

The third term requires more explanation. If terms of second order in $\|E_{21}\|$ are ignored, this expression becomes essentially

$$
\bar{\kappa}^2\eta\frac{\|b_2\|_2}{\|b_1\|_2}\frac{\|E_{21}\|_2}{\|A\|_2}\equiv\bar{\kappa}^2\eta\tan\theta\frac{\|E_{21}\|_2}{\|A\|_2}, \qquad (5.11)
$$

where θ is the angle subtended by b and $\mathcal{R}(A)$. The number $\bar{\kappa}\eta\tan\theta$ can vary from 0 to ∞. It is small when θ is small (i.e. the residual vector is small). It is also reduced in size when $\|E_{11}\|_2$ is small and x reflects the ill-conditioning of A so that $\eta\cong\kappa^{-1}\cong\bar{\kappa}^{-1}$. When x does not reflect the ill-conditioning of A and θ is significant, it is of order $\bar{\kappa}^2$, thus making the third term in (5.4) the dominant one.

We have bounded the third term in the decomposition (5.5) in such a way as to reflect its behavior when E_{21} is small. In fact it is bounded for all values of E_{21}, and the third term in (5.4) may be replaced by

$$
\bar{\kappa}\eta\frac{\|b\|_2}{\|b_1\|_2}\psi_2\left(\bar{\kappa}\frac{E_{21}}{\|A\|_2}\right).
$$

The residual. Since the residual vector is given by $r=P_Ab$, the theory of § 4 may be applied to give perturbation bounds for the residual. Specifically, if

$$
\hat{x}=B^{\dagger}b
$$

and

$$
\hat{r}=b-B\hat{x}=P_Bb,
$$

then

$$
\|\hat{r}-r\|_2\leqq\|P_B-P_A\|_2\|b\|_2
$$

and $\|P_B-P_A\|_2$ can be bounded by (4.1) in Theorem 4.1.

In applications one may not be interested in $\hat{r}$; rather one is interested in the residual $\bar{r}$ of $\hat{x}$ with respect to the matrix A:

$$
\bar{r}=b-A\hat{x}.
$$

If we write

$$
\bar{r}-r=(P_B-P_A)b-E\hat{x},
$$

then

$$\|\tilde{r}-r\|_2 \geqq \|P_B - P_A\|_2\|b\|_2 + \|E\|_2\|\hat{x}\|_2.$$

Theorem 5.1 provides the necessary estimate of $\|\hat{x}\|_2$.

If we concern ourselves with only the change in $\|r\|_2$ we can derive a slightly stronger result. Since r is the minimizing residual, we have $\|r\|_2 \leqq \|\tilde{r}\|_2$. Likewise $\|b-(A+E)\hat{x}\|_2 \leqq \|bb-(A+E)x\|_2$, from which it follows that

$$\|r\|_2 \leqq \|\tilde{r}\|_2 \leqq \|r\|_2 + \|E\|_2(\|x\|_2 + \|\hat{x}\|_2).$$

Asymptotic forms and derivatives. An asymptotic form for the perturbed least squares solution $\hat{x}$ can be obtained from (3.4):

$$\hat{x} = x - A^\dagger P_A E R_A x - R_A^\perp E^H P_A (A^H)^\dagger x + (A^H A)^\dagger R_A E^H P_A^\perp b + O(\|E\|^2). \tag{5.12}$$

An equivalent asymptotic formula, which may be useful in computational work, can be derived from the reduced form (3.23). The derivative formula corresponding to (5.12) is

$$\frac{dx}{d\tau} = -A^\dagger P_A \frac{dA}{d\tau} R_A x - R_A^\perp \frac{dA^H}{d\tau} P_A (A^H)^\dagger x + (A^H A)^\dagger R_A \frac{dA^H}{d\tau} P_A b.$$

An inverse perturbation theorem. Theorem 5.2 shows how a perturbation in A can affect the least squares solution. Here we consider the question: given a vector $\hat{x}$, under what conditions is $\hat{x}$ the least squares solution of a slightly perturbed problem? One such condition is given in the following theorem.

THEOREM 5.3. *Let $\hat{x} \in \mathbb{C}^n$ be given. Let $x = A^\dagger b$, $r = b - Ax$, and $\hat{r} = b - A\hat{x}$. If*

$$\|\hat{r}\|_2^2 = \|r\|_2^2 + \varepsilon^2,$$

then there is a matrix E of rank unity with

$$\|E\|_2 = \varepsilon/\|\hat{x}\|_2 \tag{5.13}$$

such that $\|b-(A+E)\hat{x}\|_2$ is a minimum.

Proof. Let

$$e = \hat{r} - r = A(x - \hat{x}) \in \mathcal{R}(A).$$

Since $r \in \mathcal{R}(A)^\perp$,

$$\|\hat{r}\|_2^2 = \|r\|_2^2 + \|e\|_2^2,$$

which shows that $\|e\|_2 = \varepsilon$. Let

$$E = e\hat{x}^H/\|\hat{x}\|_2^2.$$

Then E satisfies (5.13) and $\mathcal{R}(E) \subset \mathcal{R}(A)$. Hence $\mathcal{R}(A+E) \subset \mathcal{R}(A)$. But

$$b - (A+E)\hat{x} = r \in \mathcal{R}(A)^\perp,$$

which shows that the residual $b-(A+E)\hat{x} \in \mathcal{R}(A+E)^\perp$, and $\hat{x}$ solves the required least squares problem. □

A consequence of this theorem is that there is little use hunting for the exact minimizing x. Provided the residual is nearly minimal, the approximate solution $\hat{x}$, however inaccurate, is the exact solution of a slightly perturbed problem.

It is sometimes desirable that the perturbation matrix E in Theorem 5.3 not alter some of the columns of A (e.g. a column may be dates in years). This can be done as follows. Let $\tilde{x}$ be the vector obtained from $\hat{x}$ by setting to zero the components corresponding to the columns that are not to be disturbed. Then

$$\tilde{E} = -e\tilde{x}^H/\|\tilde{x}\|_2^2$$

is the required matrix. Of course $\|\tilde{x}\|_2 \leqq \|x\|_2$ so that $\|\tilde{E}\|_2 \geqq \|E\|_2$; however $\|\tilde{E}\|_2$ may still be small enough for practical purposes.

Notes and references. Much of the perturbation theory for pseudo-inverses has been a byproduct of the search for bounds for the linear least squares problem. Golub and Wilkinson (1966) gave a first order analysis of the problem and were the first to note the dependence of the solution on κ^2. Rigorous upper bounds were derived by Hanson and Lawson (1969), Pereyra (1969), and Stewart (1969). Wedin (1969) also gives bounds. More recent treatments have been given by Lawson and Hanson (1974) and Abdelmalek (1974). Van der Sluis (1975) was the first to point out the mitigating effect of η in (5.11).

The inverse perturbation theorem is new.

Appendix. In this Appendix, we shall give a proof of part one of Theorem 2.3 that is based on a general decomposition of unitary matrices, a decomposition that is of independent interest. In establishing the decomposition, we shall use the notation ${}_m\overset{n}{A}$ to mean $A \in \mathbb{C}^{m\times n}$.

THEOREM A.1. *Let the unitary matrix* $W \in \mathbb{C}^{n\times n}$ *be partitioned in the form*

$$W=\begin{pmatrix} W_{11} & W_{12} \\ W_{21} & W_{22}\end{pmatrix},$$

where $W_{11} \in \mathbb{C}^{r\times r}$ *with* $r \leqq n/2$. *Then there are unitary matrices* $U = \operatorname{diag}(\overset{r}{U_1}, \overset{n-r}{U})$ *and* $V = \operatorname{diag}(\overset{r}{V_1}, \overset{n-r}{V_2})$ *such that*

$$\text{(A.1)} \qquad U^HWV = \begin{array}{c} \\ r \\ r \\ n-2r \end{array}\!\!\begin{array}{c} \begin{array}{ccc} r & r & n-2r \end{array} \\ \begin{pmatrix} \Gamma & -\Sigma & 0 \\ \Sigma & \Gamma & 0 \\ 0 & 0 & I \end{pmatrix}\end{array}$$

where

$$\Gamma = \operatorname{diag}(\gamma_1, \gamma_2, \cdots, \gamma_r) \geqq 0$$

and

$$\Sigma = \operatorname{diag}(\sigma_1, \sigma_2, \cdots, \sigma_r) \geqq 0.$$

Proof. Let

$$\Gamma = U_1^H W_{11} V_1$$

be the singular value decomposition of W_{11} with the diagonal elements of Γ ordered so that $\gamma_1 \leqq \gamma_2 \leqq \cdots \leqq \gamma_k < 1 = \gamma_{k+1} = \cdots = \gamma_r$; i.e.,

$$\Gamma = \operatorname{diag}(\Gamma', I_{r-k}),$$

where the diagonal elements of Γ' are less than unity. The matrix

$$\begin{pmatrix} W_{11} \\ W_{21} \end{pmatrix} V_1$$

has orthonormal columns. Hence

$$I = \left[\begin{pmatrix} W_{11} \\ W_{21} \end{pmatrix} V_1\right]^H \left[\begin{pmatrix} W_{11} \\ W_{21} \end{pmatrix} V_1\right] = \Gamma^2 + (W_{21}V_1)^H(W_{21}V_1).$$

Since I and Γ^2 are diagonal matrices, so is $(W_{21}V_1)^H(W_{21}V_1)$, which says that the columns of $W_{21}V_1$ are orthogonal. Since the ith diagonal entry of $I - \Gamma^2$ is the norm of the ith column of $W_{21}V_1$, only the first $k \leqq R \leqq n - r$ columns of $W_{21}V_1$ are nonzero. Let $\hat{U}_2 \in C^{(n-r)\times(n-r)}$ be any unitary matrix whose first k columns are the normalized columns of $W_{21}V_1$. Then

$$\hat{U}_2^H W_{21} V_1 = \begin{pmatrix} \Sigma \\ 0 \end{pmatrix}$$

where

$$\Sigma = \operatorname{diag}(\sigma_1, \sigma_2, \cdots, \sigma_k, 0, \cdots, 0) \equiv \operatorname{diag}\begin{pmatrix} \overset{k}{\Sigma'}, & \overset{r-k}{0} \end{pmatrix}.$$

Since

$$\operatorname{diag}(U_1, \hat{U}_2)^H \begin{pmatrix} W_{11} \\ W_{21} \end{pmatrix} V_1 = \begin{pmatrix} \Gamma \\ \Sigma \\ 0 \end{pmatrix}$$

has orthonormal columns, we must have

$$\text{(A.2)} \qquad \gamma_i^2 + \sigma_i^2 = 1 \qquad (i = 1, 2, \cdots, r).$$

In particular, Σ' is nonsingular.

In a like manner we may determine a unitary matrix $V_2 \in \mathbb{C}^{(n-r)\times(n-r)}$ such that

$$U_1^H W_{12} V_2 = (T, 0)$$

where $T = \operatorname{diag}(\tau_1, \tau_2, \cdots, \tau_r)$ and $\tau_i \leqq 0$ $(i = 1, 2, \cdots, r)$. Since, as above, $\gamma_i^2 + \tau_i^2 = 1$, it follows from (A.2) that $T = -\Sigma$.

Set $\hat{U} = \operatorname{diag}(U_1, \hat{U}_2)$ and $V = \operatorname{diag}(V_1, V_2)$. Then the foregoing shows that the matrix

$$X = \hat{U}^H W V$$

can be partitioned in the form

$$
\text{(A.3)} \qquad X = \begin{array}{c} \\ k \\ r-k \\ k \\ r-k \\ n-2r \end{array}
\begin{array}{c} \begin{array}{ccccc} k & r-k & k & r-k & n-2r \end{array} \\
\begin{pmatrix} \Gamma' & 0 & -\Sigma' & 0 & 0 \\ 0 & I & 0 & 0 & 0 \\ \Sigma' & 0 & X_{33} & X_{34} & X_{35} \\ 0 & 0 & X_{43} & X_{44} & X_{45} \\ 0 & 0 & X_{53} & X_{54} & X_{55} \end{pmatrix} \end{array}.
$$

Since columns 1 and 4 in the partition (A.3) are orthogonal, we have

$$\Sigma' X_{34} = 0,$$

and since Σ' is nonsingular, we have $X_{34} = 0$. Likewise X_{35}, X_{43}, and X_{53} are zero. From the orthogonality of columns 1 and 3 in (A.3) it follows that

$$-\Gamma'\Sigma' + \Sigma' X_{33} = 0,$$

from which it follows that $X_{33} = \Gamma'$.

The matrix X is thus seen to have the form

$$
X = \begin{array}{c} \\ k \\ r-k \\ k \\ r-k \\ n-2r \end{array}
\begin{array}{c} \begin{array}{ccccc} k & r-k & k & r-k & n-2r \end{array} \\
\begin{pmatrix} \Gamma' & 0 & -\Sigma' & 0 & 0 \\ 0 & I & 0 & 0 & 0 \\ \Sigma' & 0 & \Gamma' & 0 & 0 \\ 0 & 0 & 0 & X_{44} & X_{45} \\ 0 & 0 & 0 & X_{54} & X_{55} \end{pmatrix} \end{array}.
$$

The matrix

$$\hat{U}_3 = \begin{pmatrix} X_{44} & X_{45} \\ X_{54} & X_{55} \end{pmatrix}$$

is unitary. Set

$$U_2 = \operatorname{diag}(I_k, \hat{U}_3)\hat{U}_2$$

and $U = \operatorname{diag}(U_1, U_2)$. Then

$$U^H W V = \operatorname{diag}(I_{r+k}, \hat{U}_3) X,$$

$$
X = \begin{array}{c} \\ k \\ r-k \\ k \\ r-k \\ n-2r \end{array}
\begin{array}{c} \begin{array}{ccccc} k & r-k & k & r-k & n-2r \end{array} \\
\begin{pmatrix} \Gamma' & 0 & -\Sigma' & 0 & 0 \\ 0 & I & 0 & 0 & 0 \\ \Sigma' & 0 & \Gamma' & 0 & 0 \\ 0 & 0 & 0 & I & 0 \\ 0 & 0 & 0 & 0 & I \end{pmatrix} \end{array},
$$

which, considering the dimensions of the matrices involved is precisely (A.1). □

To establish part one of Theorem 2.3, let $\mathscr{X}=\mathscr{R}(A)$ and $\mathscr{Y}=\mathscr{R}(B)$ and let $\mathscr{X}^\perp$ and $\mathscr{Y}^\perp$ denote their orthogonal complements. Assume that

$$r=\dim(\mathscr{X})=\dim(\mathscr{Y})\leqq m/2$$

(there is no loss of generality in assuming the last inequality, since in the sequel we can also work with $\mathscr{X}^\perp$ and $\mathscr{Y}^\perp$). Let $X=(X_1, X_2)$ and $Y=(Y_1, Y_2)$ be unitary matrices with $\mathscr{R}(X_1)=\mathscr{X}$ and $\mathscr{R}(Y_1)=\mathscr{Y}$. Let

$$W=X^HY=\begin{pmatrix} W_{11} & W_{12}\\ W_{21} & W_{22}\end{pmatrix}$$

be partitioned conformally with X and Y. If $U=\operatorname{diag}(U_1, U_2)$ and $V=\operatorname{diag}(V_1, V_2)$ are the matrices whose existence is insured by Theorem A.1 and we set

$$\hat{X}_i=X_iU_i \qquad (i=1,2), \qquad \hat{X}=(\hat{X}_1,\hat{X}_2)$$

and

$$\hat{Y}_i=Y_iV_i \qquad (i=1,2), \qquad \hat{Y}=(\hat{Y}_1,\hat{Y}_2),$$

then

$$\hat{X}_1^H\hat{Y}_1=\Gamma\hat{X}_1^H\hat{Y}_2=(-\Sigma, 0),$$

$$\hat{X}_2^H\hat{Y}_1=\begin{pmatrix}\Sigma\\0\end{pmatrix}\hat{X}_2^H\hat{Y}_2=\begin{pmatrix}\Gamma & 0\\ 0 & I_{n-2r}\end{pmatrix}.$$

Note that $\mathscr{R}(\hat{X}_1)=\mathscr{X}$ and $\mathscr{R}(\hat{Y}_1)=\mathscr{Y}$.

If we now make the transformation $\mathbb{C}^n\to\hat{X}^H\mathbb{C}^n$, the bases $\hat{X}$ and $\hat{Y}$ become

$$\hat{X}_1=\begin{pmatrix}I_r\\0\\0\end{pmatrix},\qquad \hat{X}_2=\begin{pmatrix}0&0\\I_r&0\\0&I_{n-2r}\end{pmatrix},$$

(A.4)
$$\hat{Y}_1=\begin{pmatrix}\Gamma\\\Sigma\\0\end{pmatrix},\qquad \hat{Y}_2=\begin{pmatrix}-\Sigma&0\\\Gamma&0\\0&I_{n-2r}\end{pmatrix},$$

and it is with these bases that we shall prove the first part of Theorem 2.3. First note that

$$P_AP_B^\perp=(\hat{X}_1\hat{X}_1^H)(\hat{Y}_2\hat{Y}_2^H)=\begin{pmatrix}\Sigma^2&-\Sigma\Gamma&0\\0&0&0\\0&0&0\end{pmatrix}.$$

Likewise

$$P_BP_A^\perp=(\hat{Y}_1\hat{Y}_1^H)(\hat{X}_2\hat{X}_2^H)=\begin{pmatrix}0&\Gamma\Sigma&0\\0&\Sigma^2&0\\0&0&0\end{pmatrix},$$

and the nonzero singular values of both matrices are easily seen to be the numbers σ_i. Now consider

$$P_B - P_A = Y_1 Y_1^H - X_1 X_1^H = \begin{pmatrix} \Gamma^2 - I & \Gamma\Sigma & 0 \\ \Sigma\Gamma & \Gamma^2 & 0 \\ 0 & 0 & 0 \end{pmatrix} = \begin{pmatrix} -\Sigma^2 & \Gamma\Sigma & 0 \\ \Sigma\Gamma & \Sigma^2 & 0 \\ 0 & 0 & 0 \end{pmatrix}.$$

The nonzero eigenvalues of this matrix are the eigenvalues of the 2×2 matrices

$$\begin{pmatrix} -\sigma_i^2 & \gamma_i\sigma_i \\ \sigma_i\gamma_i & \sigma_i^2 \end{pmatrix},$$

which are easily seen to be $\pm\sigma_i$.

Notes and references. The matrix decomposition in Theorem A.1 has apparently not been explicitly stated before; however, it is implicit in the works of Davis and Kahan (1970) and Björk and Golub (1973). The diagonal elements of Γ are the cosines of the "canonical angles" between the subspaces $\mathcal{R}(A)$ and $\mathcal{R}(B)$ and the columns of $\hat{X}_1$ and $\hat{Y}_1$ form biorthogonal bases subtending these angles. The use of these canonical bases, particularly when they have been transformed into the forms (A.4), often enables one to obtain routine computational proofs of geometric theorems that would otherwise require considerable ingenuity to establish.

REFERENCES

N. N. ABDELMALEK (1974), *On the solution of the linear least squares problem and pseudo-inverses*, Computing, 13, pp. 215–228.

S. N. AFRIAT (1957), *Orthogonal and oblique projectors and the characteristics of pairs of vector spaces*, Proc. Cambridge Philos. Soc., 53, pp. 800–816.

A. BEN-ISRAEL (1966), *On error bounds for generalized inverses*, SIAM J. Numer. Anal., 3, pp. 585–592.

A. BEN-ISRAEL AND T. N. E. GREVILLE (1974), *Generalized Inverses: Theory and Applications*, John Wiley, New York.

Å. BJÖRK AND G. H. GOLUB (1973), *Numerical methods for computing angles between linear subspaces*, Math. Comp., 27, pp. 579–594.

T. L. BOULLION AND P. L. ODELL (1971), *Generalized Inverse Matrices*, John Wiley, New York.

CHANDLER DAVIS AND W. M. KAHAN (1970), *The rotation of eigenvectors by a perturbation. III*, SIAM J. Numer. Anal., 7, pp. 1–46.

I. C. GOHBERG AND M. G. KREIN (1969), *Introduction to the Theory of Nonself-adjoint Operators*, American Mathematical Society, Providence, R.I.

G. H. GOLUB (1965), *Numerical methods for solving linear least squares problems*, Numer. Math., 7, pp. 206–216.

G. H. GOLUB AND J. H. WILKINSON (1966), *Note on the iterative refinement of least squares solution*, Numer. Math., 9, pp. 139–148.

G. H. GOLUB AND V. PEREYRA (1973), *The differentiation of pseudoinverses and nonlinear least squares problems whose variables separate*, SIAM J. Numer. Anal., 10, pp. 413–432.

——— (1975), *Differentiation of pseudoinverses, separable nonlinear least squares problems, and other tales*, manuscript.

R. J. HANSON AND C. L. LAWSON (1969), *Extensions and applications of the Householder algorithm for solving linear least squares problems*, Math. Comp., 23, pp. 787–812.

J. Z. HEARON AND J. W. EVANS (1968), *Differentiable generalized inverses*, J. Res. Nat. Bur. Stand., Sect. B, 72B, pp. 109–113.

A. S. Householder (1964), *The Theory of Matrices in Numerical Analysis*, Dover, New York.
T. Kato (1966), *Perturbation Theory for Linear Operators*, Springer-Verlag, Berlin.
C. L. Lawson and R. J. Hanson (1974), *Solving Least Squares Problems*, Prentice-Hall, Englewood Cliffs, N.J.
L. Mirsky (1960), *Symmetric gauge functions and unitarily invariant norms*, Quart. J. Math. Oxford Ser., 11, no. 2, pp. 55–59.
J. von Neumann (1937), *Some matrix-inequalities and metrization of matric-space*, Tomsk. Univ. Rev., 1, pp. 286–300.
M. Pavel-Parvu and A. Korganoff (1969), *Iteration functions for solving polynomial equations*, Constructive Aspects of the Fundamental Theorem of Algebra, B. Dejon and P. Henrici, eds., John Wiley, New York.
R. Penrose (1955), *A generalized inverse for matrices*, Proc. Cambridge Philos. Soc., 51, pp. 506–513.
——— (1956), *On best approximate solution of linear matrix equations*, Ibid., 52, pp. 17–19.
V. Pereyra (1969), *Stability of general systems of linear equations*, Aequat. Math., 2, pp. 194–206.
C. R. Rao and S. K. Mitra (1971), *Generalized Inverse of Matrices and Its Applications*, John Wiley, New York.
A. van der Sluis (1975), *Stability of the solutions of linear least squares problems*, Numer. Math., 23, pp. 241–254.
G. W. Stewart (1969), *On the continuity of the generalized inverse*, SIAM J. Appl. Math., 17, pp. 33–45.
——— (1973), *Introduction to Matrix Computations*, Academic Press, New York.
P.-Å. Wedin (1969), *On pseudo-inverses of perturbed matrices*, Lund Univ. Comput. Sci. Tech. Rep., Lund, Sweden.
——— (1973), *Perturbation theory for pseudo-inverses*, BIT, 13, pp. 217–232.
J. H. Wilkinson (1965), *The Algebraic Eigenvalue Problem*, Oxford University Press, London.

14.3. [GWS-J65] "On Scaled Projections and Pseudoinverses"

[GWS-J65] "On Scaled Projections and Pseudoinverses," *Linear Algebra and its Applications* **112** (1989) 189–193.

On Scaled Projections and Pseudoinverses

G. W. Stewart*
Department of Computer Science
and Institute for Physical Science and Technology
University of Maryland
College Park, Maryland 20742

Submitted by Richard A. Brualdi

ABSTRACT

Let X be a matrix of full column rank, and let D be a diagonal matrix with positive diagonal elements. The weighted pseudoinverse defined by $X_D^{\dagger} = (X^{\mathrm{T}}DX)^{-1}X^{\mathrm{T}}D$ and the associated oblique projection $P_D = XX_D^{\dagger}$ arise in many applications. In this paper, we show that the norms of both matrices are bounded by numbers that are independent of D.

Let X be a matrix with linearly independent columns. In a number of applications [1–4] it is necessary to solve weighted least-squares problems of the form[1]

$$\|D^{1/2}(y - Xb)\|^2 = \min,$$

where D is a diagonal matrix with positive diagonal elements (we write $D \in \mathscr{D}_+$). The solution is given by $b = X_D^{\dagger}y$, where

$$X_D^{\dagger} = (X^{\mathrm{T}}DX)^{-1}X^{\mathrm{T}}D \tag{1}$$

is the weighted pseudoinverse of X.

---

*This work was supported in part by the Air Force Office of Sponsored Research under grant AFOSR-82-0078.

[1]Throughout this note, $\|\cdot\|$ will stand for the Euclidean vector norm or spectral matrix norm defined by $\|X\| = \max_{\|x\|=1}\|Xx\|$.

655 Avenue of the Americas, New York, NY 10010 0024-3795/89/$3.50

In some situations it is desirable to have a bound on the norm of $X_D^\dagger$. Unfortunately it is possible to choose D with $\|D\|=1$ to make the factor $(X^{\mathrm{T}}DX)^{-1}$ in (1) arbitrarily large. For example, suppose that the first row x_1^{T} of X is nonzero, and let $D=\mathrm{diag}(1,\epsilon,\ldots,\epsilon)$. Then as $\epsilon\to 0$, the matrix $X^{\mathrm{T}}DX$ approaches the rank-one matrix $x_1x_1^{\mathrm{T}}$. Since this matrix is singular, $(X^{\mathrm{T}}DX)^{-1}$ must grow unboundedly as ϵ approaches zero.

Nonetheless, $X_D^\dagger$ remains bounded as D varies over $\mathscr{D}_+$. The key to seeing this is to introduce the matrix

$$P_D = XX_D^\dagger. \tag{2}$$

It is easily verified that P_D projects a vector obliquely onto the column space $\mathscr{R}(X)$ of X along the orthogonal complement of $\mathscr{R}(DX)$. Now a little geometric reflection suggests that the more oblique a projector, the larger its norm. Thus we can show that P_D is bounded by showing that D cannot cause the orthogonal complement of $\mathscr{R}(DX)$ to lean too far toward $\mathscr{R}(X)$. This is the idea behind the proof of the following theorem.

THEOREM 1. *Let X be of full column rank, and for $D\in\mathscr{D}_+$ let $X_D^\dagger$ and P_D be defined by* (1) *and* (2). *Then there is a number $\rho>0$ such that*

$$\sup_{D\in\mathscr{D}_+}\|P_D\|\leqslant\rho^{-1} \tag{3}$$

and

$$\sup_{D\in\mathscr{D}_+}\|X_D^\dagger\|\leqslant\rho^{-1}\|X^\dagger\|. \tag{4}$$

Proof. Let

$$\mathscr{X}=\{x\in\mathscr{R}(X):\|x\|=1\}$$

and

$$\mathscr{Y}=\left\{y:\exists D\in\mathscr{D}_+ \text{ such that } X^{\mathrm{T}}Dy=0\right\}.$$

We claim that $\bar{\mathscr{Y}}\cap\mathscr{X}=\varnothing$. Suppose on the contrary that $y_k\in\mathscr{Y}$ $(k=1,2,\ldots)$

and $\lim y_k = x \in \mathscr{X}$. Then there are matrices $\mathscr{D}_k \in \mathscr{D}_+$ such that

$$0 = x^{\mathrm{T}} D_k y_k = \sum_{x_i \neq 0} x_i d_i^{(k)} y_i^{(k)}. \tag{5}$$

But if $x_i \neq 0$, then for k sufficiently large $x_i y_i^{(k)} > 0$. Hence the right-hand side of (5) must eventually become positive. The contradiction establishes the claim.

Since $\mathscr{X}$ is closed and bounded and does not intersect $\bar{\mathscr{Y}}$, it follows that

$$\rho \overset{\text{def}}{=} \inf_{\substack{y \in \mathscr{Y} \\ x \in \mathscr{X}}} \|y - x\| > 0.$$

In other words, the orthogonal complement of $\mathscr{R}(DX)$ remains uniformly bounded away from $\mathscr{R}(X)$.

Now for fixed $D \in \mathscr{D}_+$, let $\|z\| = 1$, and write $z = x + y$, where $x \in \mathscr{R}(X)$ and $X^{\mathrm{T}} D y = 0$. Assume without loss of generality that $x \neq 0$. Since $x/\|x\| \in \mathscr{X}$, it follows that

$$\|x\|^{-1} = \frac{\|z\|}{\|x\|} = \frac{\|x + y\|}{\|x\|} \geqslant \rho,$$

or $\|x\| \leqslant \rho^{-1}$. It now follows that $\|P_D z\| = \|x\| \leqslant \rho^{-1}$, which establishes (3). The inequality (4) follows from the fact that $X_D^{\dagger} = X^{\dagger} P_D$. ■

As the proof of Theorem 1 suggests, there is a combinatorial aspect to the result: it depends on the fact that D is diagonal. In fact we can make $\|X_D^{\dagger}\|$ arbitrarily large by means of a positive definite D that is arbitrarily close to being diagonal. To see this, let X be the vector $(0\ 1)^{\mathrm{T}}$ and let

$$D = \begin{pmatrix} 1 & -\epsilon \\ \epsilon & 1 \end{pmatrix} \begin{pmatrix} 1 & 0 \\ 0 & \delta \end{pmatrix} \begin{pmatrix} 1 & \epsilon \\ -\epsilon & 1 \end{pmatrix}.$$

Then it is easily verified that

$$\lim_{\delta \to 0} \left(X^{\mathrm{T}} D X\right)^{-1} X^{\mathrm{T}} D = \left(\epsilon^{-1}\ 1\right).$$

Thus by taking δ and ϵ small enough, we can make $\|(X^{\mathrm{T}} D X)^{-1} X^{\mathrm{T}} D\|$ as large as we please while making D arbitrarily close to diag(1,0).

We next derive a lower bound the number ρ.

THEOREM 2. *For any matrix* $A \neq 0$, *let* $\inf_+(A)$ *denote the smallest nonzero singular value of* A. *Let the columns of* U *form an orthonormal basis for* $\mathcal{R}(X)$. *Then*

$$\rho \leqslant \min \inf_+(U_I), \tag{6}$$

where U_I *denotes any submatrix formed from a set of rows of* U.

Proof. Since ρ is invariant under transformations of the form $U = XS$, where S is nonsingular, we may choose S so that the columns of U are orthonormal and work with U.

By rearranging the rows of U, we may write

$$U = \begin{pmatrix} U_1 \\ U_2 \end{pmatrix},$$

where U_1 is the submatrix for which the minimum in (6) is attained. Let V be the orthogonal matrix of right singular vectors of U_1, so that $U_1V = (u_1^{(1)} \cdots u_k^{(1)}\ 0 \cdots 0)$, where the $u_i^{(1)}$ are orthogonal and $\|u_k^{(1)}\|$ is the minimum in (6). By the observation in the last paragraph, we may replace U with UV. Write

$$U_2V = \begin{pmatrix} u_1^{(2)} & \cdots & u_k^{(2)} & u_{k+1}^{(2)} & \cdots & u_p^{(2)} \end{pmatrix}.$$

Then it is easily verified that the $u_i^{(2)}$ are also orthogonal.

Since 0 belongs to the space $\mathcal{Y}$ defined above, one is an upper bound on ρ. Consequently, we may assume without loss of generality that $\|u_k^{(1)}\| < 1$, or equivalently $\|u_k^{(2)}\| \neq 0$. For $\epsilon > 0$ write

$$y = \begin{pmatrix} -\epsilon u_k^{(1)} \\ u_k^{(2)} \end{pmatrix}.$$

If we set

$$D = \operatorname{diag}\left(I, \epsilon \frac{\|u_k^{(1)}\|}{\|u_k^{(2)}\|} I \right),$$

it is seen that $U^{\mathrm{T}}Dy = 0$; i.e., $y \in \mathscr{Y}$. But

$$\left\| y - \begin{pmatrix} u_k^{(1)} \\ u_k^{(2)} \end{pmatrix} \right\| = (1+\epsilon)\|u_k^{(1)}\|,$$

and the upper bound follows by letting ϵ approach zero. ∎

The bound on ρ exhibits the discontinuous behavior found in the above example. For example, it says that small rows are bad, but zero rows are harmless. Whether $\rho = \min \inf_{+}(U_1)$ is an open question.

My first proof of Theorem 1 was by a messy induction. Later Michael Powell came up with an elegant proof based on considerations from mathematical programming, which caused me to rethink my approach. I would like to thank him for the inspiration.

REFERENCES

1 S. J. Haberman, *The Analysis of Frequency Data*, Univ. of Chicago Press, Chicago, 1974.
2 P. J. Huber, *Robust Statistics*, Wiley, New York, 1981.
3 N. Marmarker, A new polynomial time algorithm for linear programming, *Combinatorica* 4:373–395 (1984).
4 G. W. Stewart, An Iterative Method for Solving Linear Inequalities, Univ. of Maryland Computer Science Technical Report TR-1833, 1987.

Received 2 May 1988; final manuscript accepted 15 May 1988

15

Papers on the Eigenproblem and Invariant Subspaces: Perturbation Theory

---

1. [GWS-J15] "Error Bounds for Approximate Invariant Subspaces of Closed Linear Operators," *SIAM Journal on Numerical Analysis* **8** (1971) 796–808.
2. [GWS-J19] "Error and Perturbation Bounds for Subspaces Associated with Certain Eigenvalue Problems'," *SIAM Review* **15** (1973) 727–764.
3. [GWS-J48] "Computable Error Bounds for Aggregated Markov Chains," *Journal of the ACM* **30** (1983) 271–285.
4. [GWS-J70] "Two Simple Residual Bounds for the Eigenvalues of a Hermitian Matrix", *SIAM Journal on Matrix Analysis and Applications* **12** (1991) 205–208.
5. [GWS-J71] (with G. Zhang) "Eigenvalues of Graded Matrices and the Condition Numbers of a Multiple Eigenvalue", *Numerische Mathematik* **58** (1991) 703–712.
6. [GWS-J108] "A Generalization of Saad's Theorem on Rayleigh-Ritz Approximations", *Linear Algebra and its Applications* **327** (2001) 115–119.
7. [GWS-J109] "On the Eigensystems of Graded Matrices", *Numerische Mathematik* **90** (2001) 349–370.
8. [GWS-J114] "On the Powers of a Matrix with Perturbations", *Numerische Mathematik* **96** (2003) 363–376.

15.1. [GWS-J15] "Error Bounds for Approximate Invariant Subspaces of Closed Linear Operators"

[GWS-J15] "Error Bounds for Approximate Invariant Subspaces of Closed Linear Operators," *SIAM Journal on Numerical Analysis* **8** (1971) 796–808.
http://dx.doi.org/10.1137/0708073

SIAM J. NUMER. ANAL.
Vol. 8, No. 4, December 1971

ERROR BOUNDS FOR APPROXIMATE INVARIANT SUBSPACES, OF CLOSED LINEAR OPERATORS*

G. W. STEWART†

Abstract. Let A be a closed linear operator on a separable Hilbert space $\mathscr{H}$ whose domain is dense in $\mathscr{H}$. Let $\mathscr{X}$ be a subspace of $\mathscr{H}$ contained in the domain of A and let $\mathscr{Y}$ be its orthogonal complement. Let B and C be the compressions of A to $\mathscr{X}$ and $\mathscr{Y}$ respectively, let $G = Y^*AX$, where X and Y are the injections of $\mathscr{X}$ and $\mathscr{Y}$ into $\mathscr{H}$. It is shown that if B and C have disjoint spectra and $\|G\|$ is sufficiently small, then there is an invariant subspace $\mathscr{X}'$ of A near $\mathscr{X}$. Bounds for the distance between $\mathscr{X}'$ and $\mathscr{X}$ are given, and the spectrum of A is related to the spectra of B and C. In the development a measure of the separation of the spectra of B and C which is insensitive to small perturbations in B and C is introduced and analyzed.

1. Introduction. It is well known to specialists in matrix computations that the eigenvectors of a matrix corresponding to a set of poorly separated eigenvalues are quite sensitive to perturbations in the elements of the matrix. It is also known that for Hermitian matrices the invariant subspace corresponding to a cluster of eigenvalues is insensitive to such perturbations. It is the object of this paper to extend such results to nonnormal matrices and, more generally, to closed operators in Hilbert space.

Davis and Kahan [1] give an extensive treatment of the problem for self-adjoint operators. The problem of calculating well-conditioned invariant subspaces of a matrix has inspired some work for general matrices, most notably by Varah [8] and Ruhe [6], [7].

Our approach is as follows. Let A be a closed linear operator on a separable Hilbert space $\mathscr{H}$. Let $\mathscr{X}$ be a subspace (closed linear manifold) of $\mathscr{H}$, and let $\mathscr{Y}$ be the orthogonal complement of $\mathscr{X}$. Let X and Y be the injections of $\mathscr{X}$ and $\mathscr{Y}$ into $\mathscr{H}$. Then $\mathscr{X}$ is an invariant subspace of A if and only if $\mathscr{X} \subset \mathscr{D}(A)$ (the domain of A) and

$$G = Y^*AX = 0.$$

It is natural to conjecture that if G is small, then $\mathscr{X}$ is near an invariant subspace of A. We shall show that under certain conditions it is possible to construct an isometry $X':\mathscr{X} \to \mathscr{H}$ such that $\mathscr{R}(X')$ (the range of X') is an invariant subspace of A. Moreover $\|X-X'\|$ approaches zero as $\|G\|$ approaches zero.

The results depend on a measure of the separation of the spectra of two operators. For Hermitian matrices, the distance between the spectra is an adequate measure and, in fact, is the one used by Davis and Kahan. In the general case, however, the spectra, and hence the distance between them, may vary violently with small perturbations in the operators. Accordingly, we begin in § 2 by defining

* Received by the editors March 31, 1971.

† Center for Numerical Analysis, University of Texas, Austin, Texas 78712. This work was supported in part by the U.S. Army Research Office (Durham) under Grant DA-ARO(D)-31-124-G1050 and by the National Science Foundation under Grant GP-8442.

a more stable measure of the separation and deriving its properties. In § 3 we establish the main results and in § 4 discuss some of their consequences.

In the sequel, $\|\cdot\|$ will denote the norm of a normed linear space. If $\mathscr{X}$ and $\mathscr{Y}$ are normed linear spaces, we denote by $B(\mathscr{X}, \mathscr{Y})$ the set of all linear operators $P:\mathscr{X} \to \mathscr{Y}$ with the norm

$$\|P\| = \sup_{\|x\|=1} \|Px\|.$$

We set $B(\mathscr{X}) = B(\mathscr{X}, \mathscr{X})$. The adjoints of the Banach spaces $\mathscr{X}$ and $\mathscr{Y}$ are written $\mathscr{X}^*$ and $\mathscr{Y}^*$. The adjoint of $P \in B(\mathscr{X}, \mathscr{Y})$ is written P^*.

If $\mathscr{H}$ is a Hilbert space with inner product $(\cdot,\cdot)$, we identify $\mathscr{H}^*$ with $\mathscr{H}$ in the usual way; i.e., y corresponds to y^* if and only if $(x, y) = y^*x$ for all $x \in \mathscr{H}$. Thus if $\mathscr{X}$ and $\mathscr{Y}$ are Hilbert spaces and $P \in B(\mathscr{X}, \mathscr{Y})$, then $P^* \in B(\mathscr{Y}, \mathscr{X})$.

If $\mathscr{Y}$ is a Banach space and $C:\mathscr{Y}_A \to \mathscr{Y}$ a closed operator whose domain is dense in $\mathscr{Y}$, then λ is a regular point of C ($\lambda \in \rho(C)$) if $R(\lambda; C) = (\lambda I - C)^{-1}$ is a bounded operator defined on all of $\mathscr{Y}$. The complement of $\rho(C)$, $\sigma(C)$, is a closed set called the spectrum of C. The set $\sigma(C)$ is subdivided as follows.

(a) The point spectrum,

$$\sigma_p(C) = \{\lambda \in \sigma(C) : \lambda I - C \text{ is not 1-1}\}.$$

(b) The continuous spectrum,

$$\sigma_c(C) = \{\lambda \in \sigma(C) : \lambda I - C \text{ is 1-1 and } \mathscr{R}(\lambda I - C) \text{ is dense in } \mathscr{Y}\}.$$

(c) The residual spectrum,

$$\sigma_r(C) = \{\lambda \in \sigma(C) : \lambda I - C \text{ is 1-1 and } \mathscr{R}(\lambda I - C) \text{ is not dense in } \mathscr{Y}\}.$$

If $\lambda \in \sigma_p(C) \cup \sigma_c(C)$, then there are vectors $y_i \in \mathscr{Y}_A$ such that $\|y_i\| = 1$ and $\|(\lambda I - C)y_i\| \to 0$. If $\lambda \in \sigma_r(C)$, then $\lambda \in \sigma_p(C^*)$.

Let $\mathscr{X}$ and $\mathscr{Y}$ be Hilbert spaces and let $\langle x_i \rangle$ and $\langle y_i \rangle$ be orthonormal bases for $\mathscr{X}$ and $\mathscr{Y}$. The linear operator $P:\mathscr{X} \to \mathscr{Y}$ is said to be a Hilbert–Schmidt operator ($P \in \mathrm{HS}(\mathscr{X}, \mathscr{Y})$) if

$$|P|^2 = \sum_{i,j} |(Px_i, y_j)|^2 < \infty.$$

The Hilbert–Schmidt norm $|P|$ does not depend on the choice of the orthonormal bases $\langle x_i \rangle$ and $\langle y_i \rangle$. If $P \in \mathrm{HS}(\mathscr{X}, \mathscr{Y})$, $B \in B(\mathscr{X})$, and $C \in B(\mathscr{Y})$, then $BPC \in \mathrm{HS}(\mathscr{X}, \mathscr{Y})$ and

$$|BPC| \leqq \|B\|\,|P|\,\|C\|.$$

The space $\mathrm{HS}(\mathscr{X}, \mathscr{Y})$ is a Hilbert space with the inner product

$$[P, Q] = \sum_{i,j} (Px_i, y_j)(y_j, Qx_i).$$

An orthonormal basis for $\mathrm{HS}(\mathscr{X}, \mathscr{Y})$ is given by the operators $E_{ij} = y_j x_i^*$ (for details on Hilbert–Schmidt operators see [3, Chap. XI]).

2. The separation of two operators. Let $\mathscr{X}$ and $\mathscr{Y}$ be Hilbert spaces. Let $B \in B(\mathscr{X})$ and $C \in B(\mathscr{Y})$. Then B and C define an operator $T \in B[B(\mathscr{X}, \mathscr{Y})]$ by

$$T(P) = PB - CP, \qquad P \in B(\mathscr{X}, \mathscr{Y}).$$

Likewise an operator $\tau \in B[\mathrm{HS}(\mathscr{X}, \mathscr{Y})]$ is defined by

$$\tau(P) = PB - CP, \qquad P \in \mathrm{HS}(\mathscr{X}, \mathscr{Y}).$$

Lumer and Rosenblum [2] have studied the operator T when B, C and P belong to a Banach algebra. Their techniques, which apply here, show that

$$\sigma(T) = \sigma(B) - \sigma(C) \equiv \{\beta - \gamma : \beta \in \sigma(B), \gamma \in \sigma(C)\}.$$

Moreover Rosenblum [5] has shown that if $\lambda \in \rho(T)$, then

$$(2.1) \qquad (T - \lambda I)^{-1}(Q) = \frac{1}{2\pi i} \int R(z; C)QR(\lambda + z; B)\, dz,$$

where the integral is taken over a suitable contour.

It is also true that $\sigma(\tau) = \sigma(B) - \sigma(C)$. First note that if $Q \in \mathrm{HS}(\mathscr{X}, \mathscr{Y})$, then the integrand of (2.1) is continuous in the Hilbert–Schmidt norm. Hence, if $\lambda \in \rho(T)$, the integral defines an operator in $B[\mathrm{HS}(\mathscr{X}, \mathscr{Y})]$ which is the inverse of $\tau - \lambda I$, so that $\lambda \in \rho(\tau)$. On the other hand if $\lambda \in \sigma(T)$, then the second half of the proof of Theorem 2.1 below implies that $\lambda \in \sigma(\tau)$.

It follows that if $\sigma(B) \cap \sigma(C) = \varnothing$, then

$$0 < \|T^{-1}\|^{-1} \leqq \inf |\sigma(B) - \sigma(C)| \equiv \{\beta - \gamma : \beta \in \sigma(B), \gamma \in \sigma(C)\}$$

with a similar inequality holding for τ. Hence it is natural to take $\|T^{-1}\|^{-1}$ or $\|\tau^{-1}\|^{-1}$ as a measure of the separation of the spectra of B and C. We shall do this; however, we first extend the above theory to cover the case where C is unbounded.

Let C be a closed operator on $\mathscr{Y}$ whose domain $\mathscr{Y}_A$ is dense in $\mathscr{Y}$. If $P \in B(\mathscr{X}, \mathscr{Y}_A)$, then CP is a closed operator defined on all of $\mathscr{X}$ and hence is bounded. Thus the mapping $P \to PB - CP$ defines a linear operator $T : B(\mathscr{X}, \mathscr{Y}_A) \to B(\mathscr{X}, \mathscr{Y})$. Likewise an operator τ mapping into $\mathrm{HS}(\mathscr{X}, \mathscr{Y})$ is defined on the set

$$\mathscr{D}(\tau) = \{P \in \mathrm{HS}(\mathscr{X}, \mathscr{Y}_A) : CP \in \mathrm{HS}(\mathscr{X}, \mathscr{Y})\}.$$

THEOREM 2.1.

$$\sigma(T) = \sigma(\tau) = \sigma(B) - \sigma(C).$$

Proof. We establish the result for $\sigma(T)$, the proof being the same for $\sigma(\tau)$. Clearly it suffices to show that $0 \in \sigma(T)$ if and only if $\sigma(B) \cap \sigma(C) \neq \varnothing$.

Suppose $\sigma(B) \cap \sigma(C) = \varnothing$. Then since $\sigma(B)$ and $\sigma(C)$ are closed, there is a point $\lambda \in \rho(B) \cap \rho(C)$. Let $Q \in B(\mathscr{X}, \mathscr{Y})$ and consider the equation

$$(2.2) \qquad T_\lambda(P) \equiv PR(\lambda; B) - R(\lambda; C)P = R(\lambda; C)QR(\lambda; B).$$

Since $R(\lambda; B)$ and $R(\lambda; C)$ are bounded operators with disjoint spectra, T_λ has a bounded inverse. Moreover if P satisfies (2.2), then P maps into $\mathscr{Y}_A$ and satisfies

$T(P) = Q$. Hence T has a bounded inverse satisfying

$$\|T^{-1}\| \leqq \|T_\lambda^{-1}\| \, \|R(\lambda; C)\| \, \|R(\lambda; B)\|,$$

and $0 \in \dot{\rho}(T)$.

The remainder of the proof is adapted from [2]. Let $\lambda \in \sigma(B) \cap \sigma(C)$. Since $T(P) = P(B - \lambda I) - (C - \lambda I)P$, we may assume without loss of generality that $\lambda = 0$. There are four cases.

(i) $0 \in \sigma_p(B^*) \cup \sigma_c(B^*)$, $0 \in \sigma_p(C) \cup \sigma_c(C)$. Then there are unit vectors x_i and y_i such that $x_i^*B \to 0$ and $Cy_i \to 0$. Let $P_i = y_i x_i^*$. Then $\|P_i\| = 1$ and $T(P_i) = y_i(x_i^*B) + (Cy_i)x_i^* \to 0$. Hence $0 \in \sigma(T)$.

(ii) $0 \in \sigma_r(B^*)$, $0 \in \sigma_p(C) \cup \sigma_c(C)$. Let x and y_i be unit vectors such that $Bx = 0$ and $Cy_i \to 0$. Suppose that $0 \notin \sigma(T)$. Then there are $P_i \in B(\mathscr{X}, \mathscr{Y}_A)$ such that $T(P_i) = y_i x^*$. Now $CP_i = P_iB - y_ix^*$. Hence $CP_i \in B(\mathscr{X}, \mathscr{Y}_A)$, and $C^2P_i = CP_iB - Cy_ix^*$ is bounded. Thus $CP_i \in \mathscr{D}(T)$. But $T(CP_i) = CT(P_i) = Cy_ix^* \to 0$. Hence $CP_i \to 0$. Then $1 = y_i^*T(P)x = y_i^*CP_ix \to 0$, a contradiction.

(iii) $0 \in \sigma_p(B^*) \cup \sigma_c(B^*)$, $0 \in \sigma_r(C)$. Let x_i and y be unit vectors such that $x_i^*B \to 0$ and $y^*C = 0$. If $0 \notin \sigma(T)$, there are $P_i \in B(\mathscr{X}, \mathscr{Y}_A)$ such that $T(P_i) = yx_i^*$. Then $T(P_iB) = T(P_i)B = yx_i^*B \to 0$. Hence $P_iB \to 0$. But then $1 = y^*T(P_i)x = y^*P_iBx \to 0$, a contradiction.

(iv) $0 \in \sigma_r(B^*)$, $0 \in \sigma_r(C)$. Let x and y be unit vectors such that $Bx = 0$ and $y^*C = 0$. If $0 \notin \sigma(T)$, then there is a $P \in B(\mathscr{X}, \mathscr{Y}_A)$ such that $T(P) = yx^*$. But then $1 = y^*T(P)x = y^*PBx - y^*CPx = 0$, a contradiction. This completes the proof.

We now define

$$\operatorname{sep}(B, C) = \begin{cases} \|T^{-1}\|^{-1} & \text{if } 0 \notin \sigma(T), \\ 0 & \text{if } 0 \in \sigma(T), \end{cases}$$

and

$$\operatorname{sep}_{\mathrm{HS}}(B, C) = \begin{cases} \|\tau^{-1}\|^{-1} & \text{if } 0 \notin \sigma(\tau), \\ 0 & \text{if } 0 \in \sigma(\tau). \end{cases}$$

The following theorem is a corollary of Theorem 2.1.

THEOREM 2.2. *The separation of B and C satisfies the inequality*

$$\operatorname{sep}(B, C) \leqq \inf|\sigma(B) - \sigma(C)|. \tag{2.3}$$

If $\operatorname{sep}(B, C) \neq 0$, *then*

$$\operatorname{sep}(B, C) = \inf_{\|P\| = 1} \|T(P)\|.$$

The Hilbert–Schmidt separation also satisfies (2.3), *and if* $\operatorname{sep}_{\mathrm{HS}}(B, C) \neq 0$, *then*

$$\operatorname{sep}_{\mathrm{HS}}(B, C) = \inf_{|P| = 1} |\tau(P)|.$$

As was mentioned in the introduction, $\operatorname{sep}(B, C)$ is insensitive to small perturbations in B and C. The following theorem is a precise statement of this fact.

THEOREM 2.3. *If* $E \in B(\mathscr{X})$ *and* $F \in B(\mathscr{Y})$, *then*

$$\operatorname{sep}(B - E, C - F) \geqq \operatorname{sep}(B, C) - \|E\| - \|F\|$$

and

$$\operatorname{sep}_{\mathrm{HS}}(B - E, C - F) \geqq \operatorname{sep}_{\mathrm{HS}}(B, C) - \|E\| - \|F\|.$$

Proof. Again the proofs are the same for sep and $\operatorname{sep}_{\mathrm{HS}}$. If $\operatorname{sep}(B, C) - \|E\| - \|F\| \leqq 0$, then the theorem is trivially true. Hence suppose that $\operatorname{sep}(B, C) > \|E\| + \|F\|$. Let $S \in B[B(\mathscr{X}, \mathscr{Y})]$ be defined by $S(P) = PE - FP$. Then if the inverse of $T - S$ exists and is bounded, $\operatorname{sep}(B - E, C - F) = \|(T - S)^{-1}\|^{-1}$. Now $\|S\| \leqq \|E\| + \|F\|$. Hence

$$\|ST^{-1}\| \leqq \frac{\|E\| + \|F\|}{\operatorname{sep}(B, C)} < 1.$$

It follows that $I - ST^{-1}$ has a bounded inverse and $\|(I - ST^{-1})^{-1}\| \leqq (1 - \|ST^{-1}\|)^{-1}$. Thus $(T - S)^{-1} = T^{-1}(I - ST^{-1})^{-1}$ is bounded. Moreover,

$$\begin{aligned}\operatorname{sep}(B - E, C - F) &= \|(T - S)^{-1}\|^{-1} \geqq \|T^{-1}\|^{-1}\|(I - ST^{-1})^{-1}\|^{-1}\\ &\geqq \operatorname{sep}(B, C)(1 - \|ST^{-1}\|)\\ &\geqq \operatorname{sep}(B, C)\left[1 - \frac{\|E\| + \|F\|}{\operatorname{sep}(B, C)}\right]\\ &= \operatorname{sep}(B, C) - \|E\| - \|F\|,\end{aligned}$$

thus completing the proof.

In § 3 we shall prove two versions of the principal theorem, one using sep and the other $\operatorname{sep}_{\mathrm{HS}}$. Because the range of τ is contained in the range of T, the theorem involving $\operatorname{sep}_{\mathrm{HS}}$ is weaker than the one involving sep. In compensation $\operatorname{sep}_{\mathrm{HS}}$ has a number of nice properties not enjoyed by sep. The remainder of this section will be devoted to establishing these properties.

Let $\mathscr{Y}_1, \mathscr{Y}_2, \cdots, \mathscr{Y}_n$ be orthogonal subspaces of $\mathscr{Y}$ with $\mathscr{Y} = \mathscr{Y}_1 \oplus \mathscr{Y}_2 \oplus \cdots \oplus \mathscr{Y}_n$. Let S_i be the projector onto $\mathscr{Y}_i$. The spaces $\mathscr{Y}_i$ reduce C if $S_i\mathscr{Y}_A \subset \mathscr{Y}_A$ and $AS_i\mathscr{Y}_A \subset \mathscr{Y}_i$. In this case we write $C = C_1 \oplus C_2 \oplus \cdots \oplus C_n$, where C_i is the restriction of C to $S_i\mathscr{Y}_A$.

THEOREM 2.4.

$$\begin{aligned}&\operatorname{sep}_{\mathrm{HS}}(B_1 \oplus B_2 \oplus \cdots \oplus B_m, C_1 \oplus C_2 \oplus \cdots \oplus C_n)\\ &\qquad = \min\{\operatorname{sep}_{\mathrm{HS}}(B_i, C_j) : i = 1, \cdots, m; j = 1, \cdots, n\}.\end{aligned}$$

Proof. We first show that

$$\operatorname{sep}_{\mathrm{HS}}(B, C_1 \oplus C_2) = \min\{\operatorname{sep}_{\mathrm{HS}}(B, C_1), \operatorname{sep}_{\mathrm{HS}}(B, C_2)\}. \tag{2.4}$$

Note that $\sigma(B) \cap \sigma(C) \neq \varnothing$ if and only if $\sigma(B) \cap \sigma(C_1) \neq \varnothing$ and $\sigma(B) \cap \sigma(C_2) \neq \varnothing$; i.e., both sides of (2.4) must be zero simultaneously. We may thus assume that $\operatorname{sep}_{\mathrm{HS}}(B, C)$, $\operatorname{sep}_{\mathrm{HS}}(B, C_1)$, $\operatorname{sep}_{\mathrm{HS}}(B, C_2) \neq 0$.

For any $P_i \in \mathrm{HS}(\mathscr{X}, S_i\mathscr{Y})$, $i = 1, 2$, we have $|P_1 + P_2|^2 = |P_1|^2 + |P_2|^2$. Given $P \in \mathrm{HS}(\mathscr{X}, \mathscr{Y}_A)$, let $P_i = S_iP$, $i = 1, 2$. Since $CP_i = C_iP_i$, $i = 1, 2$, it follows that

$$\begin{aligned}\mathrm{sep}^2_{\mathrm{HS}}(B, C) &= \inf_{\|P\|=1} \|PB - CP\|^2 \\ &= \inf_{\|P\|=1} \|(P_1B - C_1P_1) + (P_2B - C_2P_2)\|^2 \\ &= \inf_{\|P\|=1} \{\|P_1B - C_1P_1\|^2 + \|P_2B - C_2P_2\|^2\} \\ &\geqq \inf_{\|P_1\|^2+\|P_2\|^2=1} \{\mathrm{sep}^2_{\mathrm{HS}}(B, C_1)\|P_1\|^2 + \mathrm{sep}^2_{\mathrm{HS}}(B, C_2)\|P_2\|^2\} \\ &\geqq \min\{\mathrm{sep}^2_{\mathrm{HS}}(B, C_1), \mathrm{sep}^2_{\mathrm{HS}}(B, C_2)\}.\end{aligned}$$

On the other hand suppose, say, $\mathrm{sep}_{\mathrm{HS}}(B, C_1) \leqq \mathrm{sep}_{\mathrm{HS}}(B, C_2)$. Choose $P_1 \in \mathrm{HS}(\mathscr{X}, S_1\mathscr{Y}_A)$ so that $\|P_1\| = 1$ and $\|P_1B - C_1P_1\| \leqq \mathrm{sep}_{\mathrm{HS}}(B, C_1) + \varepsilon$. Then

$$\mathrm{sep}_{\mathrm{HS}}(B, C) \leqq \|P_1B - CP_1\| = \|P_1B - C_1P_1\| \leqq \mathrm{sep}_{\mathrm{HS}}(B, C_1) + \varepsilon.$$

Thus $\mathrm{sep}_{\mathrm{HS}}(B, C) \leqq \mathrm{sep}_{\mathrm{HS}}(B, C_1) = \min\{\mathrm{sep}_{\mathrm{HS}}(B, C_1), \mathrm{sep}_{\mathrm{HS}}(B, C_2)\}$, which establishes (2.4).

The equality $\mathrm{sep}_{\mathrm{HS}}(B_1 \oplus B_2, C) = \min\{\mathrm{sep}_{\mathrm{HS}}(B_1, C), \mathrm{sep}_{\mathrm{HS}}(B_2, C)\}$ is established similarly. An induction completes the proof of the theorem.

When C is unbounded, the definition of $\mathrm{sep}_{\mathrm{HS}}(B, C)$ is necessarily unsymmetric. When C is bounded, however, $\mathrm{sep}_{\mathrm{HS}}(B, C) = \mathrm{sep}_{\mathrm{HS}}(C, B)$. To see this note that the correspondence $P \to PB$ defines an operator $R_B \in B[\mathrm{HS}(\mathscr{X}, \mathscr{Y})]$. Similarly, if C is bounded, the correspondence $P \to CP$ defines an operator $L_C \in B[\mathrm{HS}(\mathscr{X}, \mathscr{Y})]$. Then $\tau = R_B - L_C$.

LEMMA 2.5. *If B and C are bounded, then $R_B^* = R_{B^*}$ and $L_C^* = L_{C^*}$.*

Proof. Let $\langle x_i \rangle$ and $\langle y_i \rangle$ be orthonormal bases for $\mathscr{X}$ and $\mathscr{Y}$, and let $E_{ij} = y_jx_i^*$. To show that $R_B^* = R_{B^*}$, we need only show that $[R_BE_{ij}, E_{kl}] = [E_{ij}, R_{B^*}E_{kl}]$ for all i, j, k, l. Now

$$\begin{aligned}[R_BE_{ij}, E_{kl}] &= \sum_{\alpha,\beta} (E_{ij}Bx_\alpha, y_\beta)(y_\beta, E_{kl}x_\alpha) \\ &= (E_{ij}Bx_k, y_l).\end{aligned}$$

Hence,

$$[R_BE_{ij}, E_{kl}] = \begin{cases} 0, & l \neq j, \\ (Bx_k, x_i), & l = j. \end{cases}$$

Similarly,

$$[E_{ij}, R_{B^*}E_{kl}] = \begin{cases} 0, & l \neq j, \\ (x_k, B^*x_i), & l = j, \end{cases}$$

which establishes the equality. The proof that $L_C^* = L_{C^*}$ is similar.

THEOREM 2.6. *If $B \in B(\mathscr{X})$ and $C \in B(\mathscr{Y})$, then*

$$\mathrm{sep}_{\mathrm{HS}}(B, C) = \mathrm{sep}_{\mathrm{HS}}(C, B).$$

Proof. If $\text{sep}_{\text{HS}}(B, C) = 0$, then $\sigma(B) \cap \sigma(C) \neq \emptyset$ and $\text{sep}_{\text{HS}}(C, B) = 0$. Otherwise $\tau = R_B - L_C$ has a bounded inverse and

$$\begin{aligned}\text{sep}_{\text{HS}}(B, C) &= \|\tau^{-1}\|^{-1} = \|\tau^{-1*}\|^{-1} \\ &= \inf\{|\tau^* P| : P \in \text{HS}(\mathscr{X}, \mathscr{Y}), |P| = 1\} \\ &= \inf\{|PB^* - C^*P| : P \in \text{HS}(\mathscr{X}, \mathscr{Y}), |P| = 1\} \\ &= \inf\{|BP^* - P^*C| : P \in \text{HS}(\mathscr{X}, \mathscr{Y}), |P| = 1\} \\ &= \inf\{|QC - BQ| : Q \in \text{HS}(\mathscr{Y}, \mathscr{X}), |Q| = 1\} \\ &= \text{sep}(C, B).\end{aligned}$$

We conclude this section with an investigation of $\text{sep}_{\text{HS}}(B, C)$ for B and C self-adjoint.

LEMMA 2.7. *If C is self-adjoint,* $\text{sep}_{\text{HS}}(\lambda I - C) = \inf|\sigma(C) - \{\lambda\}|$.

Proof. By Theorem 2.2, $\text{sep}_{\text{HS}}(\lambda I - C) \leqq |\sigma(C) - \{\lambda\}|$, with equality if $\lambda \in \sigma(C)$. On the other hand if $\lambda \notin \sigma(C)$, $P \in \mathscr{D}(\tau)$, and $|P| = 1$, then $Q = \tau(P) = (\lambda I - C)P$. Hence $|Q| \geqq \|R(\lambda; C)^{-1}\|^{-1} = \inf|\sigma(C) - \{\lambda\}|$ and $\text{sep}_{\text{HS}}(\lambda I - C) = \inf|Q| \geqq \inf|\sigma(C) - \{\lambda\}|$.

THEOREM 2.8. *Let B and C be self-adjoint. Then* $\text{sep}_{\text{HS}}(B, C) = \inf|\sigma(B) - \sigma(C)|$.

Proof. From Theorem 2.1, $\text{sep}_{\text{HS}}(B, C) = 0$ if and only if $\inf|\sigma(B) - \sigma(C)| = 0$. Hence assume that $\inf|\sigma(B) - \sigma(C)| = \delta > 0$. Let $B = \int \lambda\, dE_\lambda$, where E_λ is the resolution of the identity corresponding to B (e.g., see [4, § 107]). Let $0 < \varepsilon < \delta$. Since $\sigma(B)$ is closed and bounded, it can be covered by a finite number of intervals of length $\varepsilon/2$ each of which contains an element of $\sigma(B)$. The union of these intervals can be written as the union of a finite number of disjoint intervals, $U_1, U_2, \cdots, U_n$. For each $i = 1, 2, \cdots, n$ let $\mu_0^{(i)} < \mu_1^{(i)} < \cdots < \mu_{k_i}^{(i)}$ be a partition of U_i such that $|\mu_l^{(i)} - \mu_{l-1}^{(i)}| < \varepsilon/2$. Let $\mu_{l-1}^{(i)} < \lambda_l^{(i)} < \mu_l^{(i)}$, and let

$$D = \sum_{i=1}^{n} \sum_{l=1}^{k_i} \lambda_l^{(i)} (E_{\mu_l}^{(i)} - E_{\mu_{l-1}}^{(i)}).$$

Then $\|B - D\| < \varepsilon/2$. Now each $\lambda_l^{(i)}$ lies within $\varepsilon/2$ of $\sigma(B)$. Hence $\inf|\sigma(C) - \lambda_l^{(i)}| > \delta - \varepsilon/2$. It follows from Lemma 2.7 that if $I_l^{(i)}$ is the identity on the range of $E_{\mu_l}^{(i)} - E_{\mu_{l-1}}^{(i)}$, then

$$\text{sep}_{\text{HS}}(\lambda_l^{(i)} I_l^{(i)}, C) > \delta - \varepsilon/2.$$

But $D = \oplus_{i=1}^{n} \oplus_{l=1}^{k_i} \lambda_l^{(i)} I_l^{(i)}$. Hence by Theorem 2.4,

$$\text{sep}_{\text{HS}}(D, C) > \delta - \varepsilon/2.$$

Finally since $\|B - D\| < \varepsilon/2$, it follows from Theorem 2.3 that

$$\text{sep}_{\text{HS}}(B, C) \geqq \text{sep}_{\text{HS}}(D, C) - \varepsilon/2 > \delta - \varepsilon.$$

Hence $\text{sep}_{\text{HS}}(B, C) \geqq \delta$. But it is always true that $\text{sep}_{\text{HS}}(B, C) \leqq \delta$, which establishes the theorem.

3. The principal theorems. Let A be a closed operator defined on a separable Hilbert space $\mathscr{H}$ whose domain is dense in $\mathscr{H}$. Let $\mathscr{X} \subset \mathscr{D}(A)$ be a subspace, and let $\mathscr{Y}$ be its orthogonal complement. Let $\mathscr{Y}_A$ be the projection of $\mathscr{D}(A)$ onto $\mathscr{Y}$. Let X, Y, and Y_A be the insertions of $\mathscr{X}$, $\mathscr{Y}$, and $\mathscr{Y}_A$ into $\mathscr{H}$. The following two lemmas lay the basis for the central construction of the theorems.

LEMMA 3.1. *The linear manifold $\mathscr{Y}_A \subset \mathscr{D}(A)$ and is dense in $\mathscr{Y}$.*

Proof. If $y \in \mathscr{Y}_A$, then for some $z \in \mathscr{D}(A)$ and $x \in \mathscr{X} \subset \mathscr{D}(A)$, $z = x + y$. Hence $y = z - x \in \mathscr{D}(A)$. Since $\mathscr{D}(A)$ is dense in $\mathscr{H}$, for any $y \in \mathscr{Y}$ there is a sequence of vectors $z_i \in \mathscr{D}(A)$ such that $z_i \to y$. If y_i is the projection of z_i on $\mathscr{Y}$, then $y_i \in \mathscr{Y}_A$, and $y_i \to y$.

LEMMA 3.2. *Let $P \in B(\mathscr{X}, \mathscr{Y}_A)$. Let*

$$X' = (X + Y_A P)(I + P^*P)^{-1/2},$$

and

$$Y'_A = (Y_A - XP^*)(I + PP^*)^{-1/2}.$$

Let $\mathscr{X}' = \mathscr{R}(X')$ and $\mathscr{Y}'_A = \mathscr{R}(Y'_A)$. Then

(i) *X' and Y' are isometries,*

(ii) *$\mathscr{X}' \subset \mathscr{D}(A)$ is a subspace,*

(iii) *$\mathscr{Y}'_A$ is the projection of $\mathscr{D}(A)$ onto the orthogonal complement of $\mathscr{X}'$.*

Proof. From Lemma 3.1 it follows that Y_A, and hence both X' and Y'_A have ranges in $\mathscr{D}(A)$. In view of the identities $X^*X = I$, $Y_A^*Y_A = I_A$ (the identity on $\mathscr{Y}_A$), $X^*Y_A = 0$, and $Y_A^*X = 0$, it is easily verified that X' and Y'_A are isometries and that $\mathscr{X}'$ and $\mathscr{Y}'_A$ are orthogonal. Since X' is an isometry defined on a subspace, its range $\mathscr{X}'$ is a subspace.

It is obvious that $\mathscr{Y}'_A$ is contained in the projection of $\mathscr{D}(A)$ onto the orthogonal complement of $\mathscr{X}'$. Conversely, suppose y' belongs to the projection of $\mathscr{D}(A)$ onto the orthogonal complement of $\mathscr{X}'$. Then by Lemma 3.1 it follows that $y' \in \mathscr{D}(A)$ and hence $y' = x + y$, where $x \in \mathscr{X}$ and $y \in \mathscr{Y}_A$. Since $X'^*y' = 0$, it follows that $(I + P^*P)^{-1/2}(x + P^*y) = 0$, or $x = -P^*y$. Thus,

$$y' = y + x = (Y_A - XP^*)y = Y'_A(I + PP^*)^{1/2}y,$$

and $y' \in \mathscr{R}(Y'_A)$.

COROLLARY 3.3. *The subspace $\mathscr{X}'$ is an invariant subspace of A if and only if $Y_A'^*AX' = 0$.*

Proof. Since $\mathscr{Y}'_A$ is dense in the orthogonal complement of $\mathscr{X}'$, the condition $Y_A'^*AX' = 0$ merely says that $A\mathscr{X}' \subset \mathscr{X}'$.

LEMMA 3.4. *The operator $AY'_A : \mathscr{Y}_A \to \mathscr{H}$ is closed.*

Proof. Let $y_i \in \mathscr{Y}_A$, $y_i \to y$ and $AY'_A y_i \to h$. We must show that $y \in \mathscr{Y}_A$ and $AY'_A y = h$. Let $y'_i = Y'_A y_i$. Then $y'_i \in \mathscr{D}(A)$, and since Y'_A is an isometry, $y'_i \to y'$ for some y' in the closure of $\mathscr{Y}'_A$. Since A is closed $y' \in \mathscr{Y}'_A$ and $Ay' = h$. Thus $y' = Y'_A\bar{y}$ for some $\bar{y} \in \mathscr{Y}_A$. Since Y'_A is an isometry, $\bar{y} = y \in \mathscr{Y}_A$, and $AY'_A y = Ay' = h$.

The main result of this paper is contained in the following theorem, in which we recapitulate somewhat.

THEOREM 3.5. *Let $A : \mathscr{D}(A) \to \mathscr{H}$ be a closed linear operator with domain dense in $\mathscr{H}$. Let $\mathscr{X} \subset \mathscr{D}(A)$ be a subspace and $\mathscr{Y}_A$ the projection of $\mathscr{D}(A)$ onto the orthogonal* *complement of* $\mathscr{X}$. *Let* X *and* Y_A *be the injections of* $\mathscr{X}$ *and* $\mathscr{Y}_A$ *into* $\mathscr{H}$, *and let*

$$B = X^*AX, \qquad H = X^*AY_A,$$
$$G = Y_A^*AX, \qquad C = Y_A^*AY_A.$$

Set

$$\gamma = \|G\|, \quad \eta = \|H\|, \quad \delta = \mathrm{sep}(B, C).$$

Then if

$$\kappa_1 = \gamma\eta/\delta^2 < 1/4, \tag{3.1}$$

there is a $P \in B(\mathscr{X}, \mathscr{Y}_A)$ *satisfying*

$$\|P\| \leqq \frac{\gamma}{\delta}(1 + \kappa) \equiv \frac{\gamma}{\delta}\frac{1 + \sqrt{1 - 4\kappa_1}}{1 - 2\kappa_1 + \sqrt{1 - 4\kappa_1}} < 2\frac{\gamma}{\delta} \tag{3.2}$$

such that $\mathscr{R}(X + Y_AP)$ *is an invariant subspace of* A. *Moreover* $\sigma(A)$ *is the disjoint union*

$$\sigma(A) = \sigma(B + HP) \cup \sigma(C - PH). \tag{3.3}$$

Proof. First note that since A is closed and $\mathscr{X}$ is a subspace, B is bounded. Hence δ is well-defined.

Let X' and Y'_A be defined as in Lemma 3.2. Then from Corollary 3.3 we need only find P such that $G' = Y'^*_AAX' = 0$. Now

$$G' = (I + PP^*)^{-1/2}(Y_A^* - PX^*)A(X + Y_AP)(I + P^*P)^{-1/2}$$
$$= (I + PP^*)^{-1/2}(CP - PB + G - PHP)(I + P^*P)^{-1/2}.$$

Hence the requirement that $G' = 0$ is equivalent to the requirement that

$$T(P) \equiv PB - CP = G - PHP. \tag{3.4}$$

We shall solve (3.4) by successive substitutions. Since $\delta > 0$, T^{-1} exists and $\|T^{-1}\| = \delta^{-1}$. Set

$$P_0 = T^{-1}(G), \tag{3.5}$$

and given $P_i, i \geqq 0$, set

$$P_{i+1} = T^{-1}(G - P_iHP_i) = P_0 - T^{-1}(P_iHP_i). \tag{3.6}$$

To show this iteration converges we first obtain a bound on $\|P_i\|$. From (3.5),

$$\|P_0\| \leqq \delta^{-1}\gamma = \pi_0.$$

From (3.6), if $\|P_i\| \leqq \pi_i$, then

$$\|P_{i+1}\| \leqq \pi_0 + \delta^{-1}\eta\pi_i^2 = \pi_{i+1}.$$

Now π_i can be written in the form

$$\pi_i = \pi_0(1 + \kappa_i),$$

where κ_i is defined by the recursion

$$\kappa_1 = \eta\pi_0/\delta = \eta\gamma/\delta,$$
$$\kappa_{i+1} = \kappa_1(1 + \kappa_i)^2.$$

By inspecting the graphs of $y = \kappa_1(1 + x)^2$ and $y = x$, it is easy to see that if (3.1) is satisfied, then

$$\kappa = \lim \kappa_i = \frac{2\kappa_1}{1 - 2\kappa_1 + \sqrt{1 - 4\kappa_1}} < 1. \tag{3.7}$$

Hence the numbers $\|P_i\|$ remain bounded.

To show that the P_i converge, let $D_i = P_{i+1} - P_i$. Then

$$\begin{aligned}\|D_i\| &\leqq \delta^{-1}\|P_iHP_i - P_{i-1}HP_{i-1}\| \\ &= \delta^{-1}\|D_{i-1}HP_i + P_{i-1}HD_{i-1}\| \\ &\leqq 2\delta^{-1}\eta\|P_i\|\,\|D_{i-1}\| \\ &\leqq 2\delta^{-1}\eta\pi_i\|D_{i-1}\| \leqq 2\kappa_1(1 + \kappa_i)\|D_{i-1}\|.\end{aligned}$$

Hence $\sum \|D_i\| < \infty$, and the P_i converge, provided $2\kappa_1(1 + \kappa) < 1$. But from (3.7) this is true whenever (3.1) is satisfied. Since the limit P satisfies

$$P = T^{-1}(G - PHP),$$

$P \in B(\mathscr{X}, \mathscr{Y}_A)$.

To prove the assertions concerning $\sigma(A)$, let $\mathscr{Y}$ ($\mathscr{Y}'$) be the closure of $\mathscr{Y}_A$ ($\mathscr{Y}'_A$) and $Y(Y')$ the extension of $Y_A(Y'_A)$ to $\mathscr{Y}$. Then the transformation

$$U = X'X^* + Y'Y^*$$

is a unitary transformation that maps $\mathscr{D}(A)$ onto $\mathscr{D}(A)$. Hence if $A' = U^*AU$, $\sigma(A') = \sigma(A)$. Now with respect to the decomposition $\mathscr{H} = \mathscr{X} \oplus \mathscr{Y}$, the operator A' has the matrix representation

$$A' \cong \begin{pmatrix} B' & H' \\ 0 & C' \end{pmatrix},$$

where $B' = X'^*A'X'$, $H' = X'^*A'Y'_A$, and $C' = Y'^*_AA'Y'_A$. Thus if $\lambda \in \rho(B') \cap \rho(C')$, then $R(\lambda; A')$ has the representation

$$R(\lambda; A') \cong \begin{pmatrix} R(\lambda; B') & R(\lambda; B')H'R(\lambda; C') \\ 0 & R(\lambda; C') \end{pmatrix},$$

which is bounded if $R(\lambda; B')H'R(\lambda; C')$ is bounded. But by Lemma 3.4, AY'_A is closed. Since $R(\lambda; C') \in B(\mathscr{Y}; \mathscr{Y}_A)$, $AY'_AR(\lambda; C')$ is bounded. Hence

$$R(\lambda; B')X'^*AY'_AR(\lambda; C') = R(\lambda; B')H'R(\lambda; C')$$

is bounded. All this shows that $\sigma(A) = \sigma(B') \cup \sigma(C')$.

Now

$$\begin{aligned}B' &= (I + P^*P)^{-1/2}(X^* + P^*Y^*_A)A(X + Y_AP)(I + P^*P)^{-1/2} \\ &= (I + P^*P)^{-1/2}(B + P^*G + HP + P^*CP)(I + P^*P)^{-1/2}.\end{aligned}$$

Since G satisfies (3.4), $P^*G = P^*PB - P^*CP + P^*PHP$. Hence,

$$B' = (I + P^*P)^{1/2}(B + HP)(I + P^*P)^{-1/2}.$$

Thus $\sigma(B') = \sigma(B + HP)$. Likewise $\sigma(C') = \sigma(C - PH)$. This establishes (3.3).

To show that $\sigma(B + HP) \cap \sigma(C - PH) = \emptyset$, note that $\|HP\| \leqq 2\eta\gamma/\delta$ and $\|PH\| \leqq 2\eta\gamma/\delta$. Thus

$$\begin{aligned} \operatorname{sep}(B + HP, C - PH) &\geqq \operatorname{sep}(B, C) - \frac{4\eta\gamma}{\delta} \\ &= \delta - \frac{4\eta\gamma}{\delta} = \frac{\delta^2 - 4\eta\gamma}{\delta} > 0, \end{aligned}$$

the last inequality following from (3.1).

When $G \in \mathrm{HS}(\mathscr{X}, \mathscr{Y})$, the proof of Theorem 3.5 can be modified to give the following result.

THEOREM 3.6. *Let* $G \in \mathrm{HS}(\mathscr{X}, \mathscr{Y})$. *Then Theorem* 3.5 *holds with* $\gamma = |G|$ *and* $\delta = \operatorname{sep}_{\mathrm{HS}}(B, C)$. *In this case* $P \in \mathrm{HS}(\mathscr{X}, \mathscr{Y}_A)$ *and*

$$|P| \leqq (\gamma/\delta)(1 + \kappa).$$

4. Discussion. Among other things, the theorems of the last section state that P becomes small as G becomes small. This means that the invariant subspace $\mathscr{X}'$ is near $\mathscr{X}$; for if $x \in \mathscr{X}$, then the projection of x on $\mathscr{Y}'_A$ is given by

$$-(I + PP^*)^{-1/2}Px,$$

which is small. Davis and Kahan [1] describe the separation of subspaces in terms of the norm of a nonnegative Hermitian operator θ. Specifically, for the spaces $\mathscr{X}$ and $\mathscr{X}'$,

$$\|\sin\theta\| \leqq \|P\|.$$

The results concerning the spectrum of A can be interpreted as follows. Let

$$\mathscr{B} = \bigcup \left\{\sigma(B + E) : \|E\| \leqq \frac{\eta\gamma}{\delta}(1 + \kappa)\right\}$$

and

$$\mathscr{C} = \bigcup \left\{\sigma(C + F) : \|F\| \leqq \frac{\eta\gamma}{\delta}(1 + \kappa)\right\}.$$

Then under the conditions of the theorem, $\mathscr{B}$ and $\mathscr{C}$ are two disjoint sets that separate the spectrum of A. When A is a matrix of order two, $\mathscr{B}$ and $\mathscr{C}$ are disks whose radii are almost as small as the minimal Gershgorin disks obtained by applying a suitable diagonal similarity transformation to A (e.g., see [9, Chap. 2]).

Our theorems bound the separation of a subspace $\mathscr{X}$ from an invariant subspace of A. A related problem is the following: given an invariant subspace $\mathscr{X}$ of A and a perturbation E, how far is $\mathscr{X}$ from an invariant subspace of $A + E$? This question may be answered by applying Theorem 3.5 to $A + E$.

THEOREM 4.1. *Let* $\mathscr{X}$, A, B, C, *and* H *be as in Theorem* 3.5. *Let* $E \in B(\mathscr{H})$ *and*

$$\begin{aligned} E_{11} &= X^*EX, & E_{12} &= X^*EY, \\ E_{21} &= Y^*EX, & E_{22} &= Y^*EY. \end{aligned}$$

Let $\varepsilon_{ij} = \|E_{ij}\|$, $\eta = \|H\| + \varepsilon_{12}$, *and* $\delta = \text{sep}(B, C) - \varepsilon_{11} - \varepsilon_{22}$. *If* $\mathscr{X}$ *is an invariant subspace of* A *and* $\kappa_1 = \delta^{-2}\eta\varepsilon_{21} < 1/4$, *there is a* $P \in B(\mathscr{X}, \mathscr{Y}_A)$ *satisfying*

$$\|P\| \leqq \frac{\varepsilon_{21}}{\delta}(1 + \kappa) < 2\frac{\varepsilon_{21}}{\delta}$$

such that $\mathscr{R}(X + Y_A P)$ *is an invariant subspace of* $A + E$. *Moreover* $\sigma(A + E)$ *is the disjoint union*

$$\sigma(A + E) = \sigma(B + E_{11} + HP) \cup \sigma(C + E_{22} - HP).$$

When $E_{21} \in \text{HS}(\mathscr{X}, \mathscr{Y})$, an analogue of Theorem 3.6 also holds.

It is mildly surprising that the bounds on $\|P\|$ are proportional to ε_{21} while the bounds appearing in the definitions of $\mathscr{B}$ and $\mathscr{C}$ above are proportional to $\varepsilon_{21}\eta$. Since the latter may be significant when the former is insignificant, the results suggest the possibility of, say, matrices with ill-conditioned eigenvalues and well-conditioned eigenvectors. The following example illustrates this point.

Let

$$A = \begin{pmatrix} 1 & 10^4 & 0 \\ 0 & 0 & 0 \\ 0 & 0 & 2 \end{pmatrix}.$$

Then A has eigenvalues 1, 0 and 2 corresponding to the eigenvectors

$$\begin{pmatrix} 1 \\ 0 \\ 0 \end{pmatrix}, \quad \begin{pmatrix} 1 \\ -10^{-4} \\ 0 \end{pmatrix}, \quad \begin{pmatrix} 0 \\ 0 \\ 1 \end{pmatrix}.$$

On the other hand, the perturbed matrix

$$A + E = \begin{pmatrix} 1 & 10^4 & 0 \\ 1.1 \times 10^{-5} & 0 & 0 \\ 2 \times 10^{-5} & 0 & 2 \end{pmatrix}$$

has eigenvalues 1.1, $-.1$, and 2 corresponding to the eigenvectors

$$\begin{pmatrix} 1 \\ 10^{-5} \\ -\frac{2}{9} \times 10^{-4} \end{pmatrix}, \quad \begin{pmatrix} 1 \\ 1.1 \times 10^{-4} \\ -10^{-5} \end{pmatrix}, \quad \begin{pmatrix} 0 \\ 0 \\ 1 \end{pmatrix}.$$

It is seen that while the eigenvalues of A and $A + E$ differ significantly, the corresponding eigenvectors have hardly changed at all. The orders of magnitudes of the changes are predicted by our bounds.

However, our bounds also indicate that the invariant subspace corresponding to the first two eigenvectors can change significantly. In fact if

$$B = \begin{pmatrix} 1 & 10^4 \\ 0 & 0 \end{pmatrix}, \qquad C = (2),$$

then sep(B, C) is of order 10^{-4}. According to our bounds, a perturbation of order

10^{-4} in A should correspond to a change of order unity in the invariant subspace corresponding to B.

When A is self-adjoint $G = H^*$, and H and G must be large or small together. Moreover the bounds defining $\mathscr{B}$ and $\mathscr{C}$ are of order $\|G\|^2$. This corresponds to the well-known fact that if A is Hermitian and x is an approximate normalized eigenvector with an error of order ε, then x^*Ax is an approximate eigenvalue with an error of order ε^2.

Davis and Kahan [1] have investigated the self-adjoint case from a somewhat different point of view. They require that reducing subspaces X and X' be known for A and $A + E$ and that the spectra of $B = X^*AX$ and $C' = Y'^*(A + E)Y'$ lie in, e.g., disjoint intervals. Theorem 4.1 requires only that a reducing subspace for A be known and deduces the existence of one for $A + E$. Moreover, our results require only that $\sigma(B) \cap \sigma(C) = \emptyset$. However, to use the convenient measure $\inf|\sigma(B) - \sigma(C)|$ in Theorem 4.1, we must assume that G is a Hilbert–Schmidt operator, while Davis and Kahan use the same measure without such restrictions; and their theorems, when applicable, permit a more delicate examination of the invariant subspaces of A.

5. Acknowledgments. I would like to thank Professor A. S. Householder for his detailed comments on an earlier version of this paper, and Dr. Richard Tapia for his comments.

Correction added in proof. The domain of the operator Y'_A is $(I + PP^*)^{1/2}\mathscr{Y}_A \subset \mathscr{Y}$ rather than $\mathscr{Y}_A$. This necessitates some slight changes in the proof of Lemma 3.4 and the proof of (3.3).

REFERENCES

[1] C. Davis and W. Kahan, *The rotation of eigenvectors by a perturbation. III*, this Journal, 7 (1970), pp. 1–46.
[2] G. Lumer and M. Rosenblum, *Linear operator equations*, Proc. Amer. Math. Soc., 10 (1969), pp. 32–41.
[3] N. Dunford and J. Schwartz, *Linear Operators, Part II*, Interscience, New York, 1963.
[4] F. Riesz and B. Sz.-Nagy, *Functional Analysis*, Frederick Ungar, New York, 1955.
[5] M. Rosenblum, *On the operator equation $BX - XA = Q$*, Duke Math. J., 23 (1956), pp. 263–269.
[6] A. Ruhe, *An algorithm for numerical determination of the structure of a general matrix*, BIT, 10 (1970), pp. 196–216.
[7] ———, *Perturbation bounds for means of eigenvalues and invariant subspaces*, Ibid., 10 (1970), pp. 343–354.
[8] J. M. Varah, *The computation of bounds for the invariant subspaces of a general matrix operator*, Tech. Rep. CS66, Stanford Univ., Stanford, Calif., 1967.
[9] J. H. Wilkinson, *The Algebraic Eigenvalue Problem*, Clarendon, Oxford, 1965.

15.2. [GWS-J19] "Error and Perturbation Bounds for Subspaces Associated with Certain Eigenvalue Problems"

[GWS-J19] "Error and Perturbation Bounds for Subspaces Associated with Certain Eigenvalue Problems," *SIAM Review* **15** (1973) 727–764.
http://dx.doi.org/10.1137/1015095

SIAM REVIEW
Vol. 15, No. 4, October 1973

ERROR AND PERTURBATION BOUNDS FOR SUBSPACES ASSOCIATED WITH CERTAIN EIGENVALUE PROBLEMS*

G. W. STEWART†

Abstract. This paper describes a technique for obtaining error bounds for certain characteristic subspaces associated with the algebraic eigenvalue problem, the generalized eigenvalue problem, and the singular value decomposition. The method also gives perturbation bounds for isolated eigenvalues and useful information about clusters of eigenvalues. The bounds are obtained from an iterative process for generating the subspaces in question, and one or more steps of the iteration can be used to construct perturbation estimates whose error can be bounded.

1. Introduction. Let A be a Hermitian matrix having an eigenvalue λ of multiplicity two. If A is perturbed by the addition of a small Hermitian matrix E, the eigenvalue λ of A will generally split into two distinct eigenvalues λ_1 and λ_2 of $A + E$, each of which will have a unique normalized eigenvector. It is well known that the eigenvalues λ_1 and λ_2 must lie near λ; in fact $|\lambda - \lambda_i| \leqq \|E\|_2$ $(i = 1, 2)$, where $\|\cdot\|_2$ denotes the spectral norm. The eigenvectors, however, are not well determined and may vary drastically with a change in E. For example, the matrices

$$A' = \frac{1}{1 + 2\varepsilon^2}\begin{pmatrix} 1 + 3\varepsilon^2 + \varepsilon^3 & \varepsilon - \varepsilon^2 + \varepsilon^3 & -\varepsilon - \varepsilon^2 \\ \varepsilon - \varepsilon^2 + \varepsilon^3 & 1 + 3\varepsilon^2 + \varepsilon^3 & \varepsilon + \varepsilon^2 \\ -\varepsilon - \varepsilon^2 & \varepsilon + \varepsilon^2 & 2 + 2\varepsilon^2 - 2\varepsilon^3 \end{pmatrix}$$

and

$$A'' = \frac{1}{5(1 + 5\varepsilon^2)}$$
$$\begin{pmatrix} 5 - 3\varepsilon + 30\varepsilon^2 - 20\varepsilon^3 & 4\varepsilon + 10\varepsilon^2 + 10e^3 & 5\varepsilon(1 - \varepsilon) \\ 4\varepsilon + 10\varepsilon^2 + 10\varepsilon^3 & 5 + 3\varepsilon + 45\varepsilon^2 - 5\varepsilon^2 & 10\varepsilon(1 - \varepsilon) \\ 5\varepsilon(1 - \varepsilon) & 10\varepsilon(1 - \varepsilon) & 10 + 25\varepsilon^2 + 25\varepsilon^3 \end{pmatrix}$$

both differ from the matrix $A = \operatorname{diag}(1, 1, 2)$ by terms of order ε, and both have eigenvalues $1 + \varepsilon$, $1 - \varepsilon$, and 2. However, the matrix of eigenvectors of A' (normalized so that the largest element in each column is unity) is

$$\begin{pmatrix} 1 & 1 & -\varepsilon \\ 1 & -1 & \varepsilon \\ 0 & 2\varepsilon & 1 \end{pmatrix},$$

* Received by the editors July 11, 1972.

† Department of Computer Science, Carnegie–Mellon University, Schenley Park, Pittsburgh, Pennsylvania 15213. This research was supported in part by the Office of Naval Research under Contract 14-67-A-0126-0016 at the University of Texas at Austin.

and the matrix of eigenvectors of A'' is

$$\begin{pmatrix} \frac{1}{2} & 1 & \varepsilon \\ 1 & -\frac{1}{2} & 2\varepsilon \\ -\frac{5}{2}\varepsilon & 0 & 1 \end{pmatrix}.$$

No matter how small is ε, and hence the difference between A' and A'', the eigenvectors of A' and A'' corresponding to the eigenvalues $1 + \varepsilon$ and $1 - \varepsilon$ differ by quantities of order unity.

The above example suggests that in the absence of special information, one cannot expect the eigenvectors of nearby matrices to lie near one another when their corresponding eigenvalues belong to clusters of poorly separated eigenvalues (in the above example, however, the eigenvector corresponding to the isolated eigenvalue two does not change much). The reason for this is that both A' and A'' are near the matrix A, which has the double eigenvalue unity. Since any vector in the plane spanned by the unit vectors e_1 and e_2 is an eigenvector of A, it is not surprising that two different perturbations will cause this plane of eigenvectors to coalesce into two different sets of two distinct eigenvectors.

However, this coalescence is not entirely arbitrary. The eigenvectors of A' or A'' that correspond to the eigenvalues $1 + \varepsilon$ and $1 - \varepsilon$ lie near the plane spanned by e_1 and e_2. In other words, although the eigenvectors of A' and A'' may differ greatly, the spaces spanned by them are almost the same. This is generally true. Although the eigenvectors corresponding to a cluster of eigenvalues of a Hermitian matrix are sensitive to perturbations in the elements of the matrix, the subspace spanned by them is relatively insensitive.

The space $\mathscr{X}$ spanned by a set of eigenvectors of a Hermitian matrix A is, of course, an invariant subspace of A; i.e., it satisfies the relation

$$A\mathscr{X} \subset \mathscr{X}.$$

The import of the above discussion is that in perturbation problems it may make more sense to try to obtain bounds for invariant subspaces than for eigenvectors. In this connection there is no need to restrict oneself to Hermitian matrices. For a general matrix A, it will often be possible to obtain useful information about its invariant subspaces even though it is impossible to say much about its eigenvectors.

The algebraic eigenvalue problem is not the only one in which confluences cause difficulties. Another is the generalized eigenvalue problem of determining the nontrivial solutions of the equation

$$Ax = \lambda Bx, \tag{1.1}$$

where A and B are square. When B is the identity matrix, (1.1) reduces to the ordinary eigenvalue problem, and, in analogy with the ordinary problem, the scalar λ in (1.1) is called an eigenvalue of $A - \lambda B$ and the vector x an eigenvector. If the generalized eigenvalue problem (1.1) has a cluster of eigenvalues, then the corresponding eigenvectors will be ill determined; however, the corresponding subspace may be well determined. Of course this subspace cannot be regarded as an invariant subspace, but we shall show in § 5 how the notion of invariant subspace may be extended to cover the generalized eigenvalue problem.

Still another problem in which multiplicities cause difficulties is the singular value problem. Let A be an $m \times n$ matrix with, say, $m \geqq n$. It is well known (see, e.g., [11, p. 31], [9]) that there are unitary matrices U and V such that

$$V^H A U = \begin{pmatrix} \Sigma \\ 0 \end{pmatrix}, \tag{1.2}$$

where

$$\Sigma = \operatorname{diag}(\sigma_1, \sigma_2, \cdots, \sigma_n)$$

has nonnegative diagonal elements (here V^H denotes the conjugate transpose of V). The columns of U and V satisfy the relations

$$Au_i = \sigma_i v_i, \qquad i = 1, 2, \cdots, n,$$

and reciprocally

$$A^H v_i = \sigma_i u_i, \qquad i = 1, 2, \cdots, n.$$

The scalar σ_i is called a singular value of A with right singular vector u_i and left singular vector v_i. If the singular values of a matrix are distinct, its corresponding singular vectors are unique, and a singular vector corresponding to an isolated singular value is relatively insensitive to perturbations in the matrix. The singular vectors corresponding to a cluster of singular values are not well determined, but, as is true of the eigenvalue and generalized eigenvalue problems, the subspace corresponding to the cluster is well determined.

The object of this paper is to describe and apply a method for obtaining error and perturbation bounds for invariant subspaces associated with the three problems described above. The author has applied this technique to the eigenvalue problem and to the generalized eigenvalue problem in earlier papers [27], [28]; the results for the singular value decomposition are new. In their raw form, the bounds are seemingly unrelated to classical results, and one of the purposes of this paper is to relate the bounds, especially the bounds for the eigenvalue problem, to bounds already in the literature. Beyond the specific results presented here, it is hoped that the unified exposition of the technique and the examples of its application will encourage people to adapt it to their own problems. Also we shall point out some of the unsolved problems associated with the technique.

A brief sketch of the technique for the invariant subspace problem will motivate some of the material to be presented later. Let A be a matrix of order n with complex elements ($A \in \mathbb{C}^{n \times n}$), and let $\mathscr{X}$ be an invariant subspace of A of dimension l. Let X be a unitary matrix whose first l columns span $\mathscr{X}$, and partition X in the form

$$X = (X_1, X_2),$$

where $X_1 \in \mathbb{C}^{n \times l}$. Then $X^H A X$ may be partitioned in the form

$$X^H A X = \begin{pmatrix} A_{11} & A_{12} \\ A_{21} & A_{22} \end{pmatrix},$$

where

$$A_{ij} = X_i^H A X_j, \qquad i,j = 1, 2. \tag{1.3}$$

Since the column space of X_1 is an invariant subspace of A, it must contain the column space of AX_1:

$$\mathscr{R}(AX_1) \subset \mathscr{R}(X_1).$$

Since X is unitary, $\mathscr{R}(X_2)$ is orthogonal to $\mathscr{R}(X_1)$, and hence to $\mathscr{R}(AX_1)$. It follows that

$$A_{21} = X_2^H(AX_1) = 0.$$

Conversely, if $X = (X_1, X_2)$ is unitary and A_{21} is defined by (1.3), then $\mathscr{R}(X_1)$ will be an invariant subspace of A whenever $A_{21} = 0$.

Now, in the above discussion, suppose that A_{21}, instead of being zero, is merely small. Then it is reasonable to expect that $\mathscr{R}(X_1)$ will lie near an invariant subspace of A. To determine just how near, we shall attempt to find a unitary matrix U, differing only slightly from the identity matrix, such that the first l columns of

$$X' = XU \tag{1.4}$$

span an invariant subspace of A. To do this, we take U in the form

$$U = \begin{pmatrix} I_l & -P^H \\ P & I_{n-l} \end{pmatrix} \begin{pmatrix} (I + P^HP)^{-1/2} & 0 \\ 0 & (I + PP^H)^{-1/2} \end{pmatrix}, \tag{1.5}$$

where $P \in \mathbb{C}^{(n-l)\times l}$ and the square roots denote the unique positive definite square roots of the positive definite matrices $I + P^HP$ and $I + PP^H$. It is easily verified that U is unitary.

If X' is partitioned in the form $X' = (X'_1, X'_2)$ and the matrices $A'_{ij}\,(i,j = 1, 2)$ are formed in analogy with (1.3), then a necessary and sufficient condition for $\mathscr{R}(X'_1)$ to be an invariant subspace of A is that $A'_{21} = 0$. Now from (1.4) and (1.5), we can express X'_1 and X'_2, and hence A'_{21}, in terms of X_1, X_2 and P. If this is done, the equation $A'_{21} = 0$ becomes (after some simplification)

$$PA_{11} - A_{22}P = A_{21} - PA_{12}P,$$

or

$$\mathbf{T}P = A_{21} - \varphi(P), \tag{1.6}$$

where the mappings $\mathbf{T}, \varphi: \mathbb{C}^{(n-l)\times l} \to \mathbb{C}^{(n-l)\times l}$ are defined by

$$\mathbf{T}P = PA_{11} - A_{22}P \tag{1.7}$$

and

$$\varphi(P) = PA_{12}P. \tag{1.8}$$

Thus the problem of assessing the accuracy of $\mathscr{R}(X_1)$ as an invariant subspace of A reduces to the problem of showing the existence of a solution of (1.6) and determining a bound on its size.

A development similar to the foregoing is possible for subspaces associated with the generalized eigenvalue problem and the singular value decomposition. In both cases one ends up with an equation of the form (1.6), where the operators

$\mathbf{T}$ and $\boldsymbol{\varphi}$ share certain common properties. It is therefore possible to obtain error bounds for all three problems from a single analysis of (1.6).

It is evident from the above sketch that the technique described here will yield error bounds for approximate subspaces (e.g., in the case of invariant subspace the size of A_{21} is the measure of the accuracy of the approximation). Perturbation bounds can be obtained by regarding a subspace of the given problem as an approximate subspace of the perturbed problem and applying the approximation theorems. One complication is that the operators $\mathbf{T}$ and $\boldsymbol{\varphi}$ will be defined in terms of the perturbed problem rather than the original problem. It is therefore necessary to investigate how $\mathbf{T}$ and $\boldsymbol{\varphi}$ are affected by perturbations.

Although we shall be chiefly concerned with subspaces, the technique also gives useful information about eigenvalues and singular values. In the above sketch, if a matrix P satisfying (1.6) can be found, then $X'^H AX'$ has the form

$$X'^H AX' = \begin{pmatrix} A'_{11} & A'_{12} \\ 0 & A'_{22} \end{pmatrix}.$$

Thus the eigenvalues of A are just those of A'_{11} and A'_{22}. Knowing bounds for P, we may calculate bounds for the differences $A'_{11} - A_{11}$ and $A'_{22} - A_{22}$, and thence, from standard perturbation theory applied to A_{11} and A_{22}, bounds on the eigenvalues of A. In particular, if $l = \dim(\mathscr{X}) = 1$, we immediately obtain a bound for a single eigenvalue.

By replacing matrices with suitably defined operators, it is possible to extend much of what will be presented here to separable Hilbert spaces, with applications to differential and integral operators. Because the details of such extensions obscure the simplicity of the technique itself, this survey will be restricted to finite-dimensional spaces; however, the reader should bear in mind that such a restriction may not be essential in particular cases.

A final observation. The function $\boldsymbol{\varphi}$ in (1.8) is quadratic in P and will be negligible when P is small. Thus the expression $\mathbf{T}^{-1}A_{21}$ may be an accurate estimate of P. This suggests that some accurate approximation to $\mathbf{T}$ that is also easily invertible could be used to calculate approximate eigenvectors. We shall not pursue this line here; the reader is referred to [3], where the idea is applied to the computation of approximate eigenvectors of diagonally dominant matrices (also see Theorem 3.2 below).

The structure of this survey is as follows. In § 2 we shall discuss some of the notions of distance and angles between subspaces and convergence of subspaces, relating them to the matrix P that results from our approach. In § 3, the solution of equations of the form (1.6) will be analyzed in very general terms. In §§ 4, 5, and 6 we shall establish error and perturbation bounds for the eigenvalue problem, the generalized eigenvalue problem, and the singular value decomposition respectively. In connection with the eigenvalue problem we shall give a brief summary of the important theorems of Davis and Kahan [6].

We shall follow Householder's notational conventions [11], with the following exceptions. The space of all complex n-vectors will be denoted by $\mathbb{C}^n$, and the space of all complex $m \times n$ matrices by $\mathbb{C}^{m\times n}$. The set of eigenvalues of a matrix A

will be denoted by $\lambda(A)$, the set of singular values by sing (A). The set of eigenvalues of the generalized eigenvalue problem $Ax = \lambda Bx$ will be denoted by $\lambda(A, B)$, and, when B is singular, $\lambda(A, B)$ may include ∞.

We shall not often need to distinguish between a linear transformation and its matrix representation, but when it is necessary, we shall denote the former by boldface letters. Thus A is a matrix representation of the transformation $\mathbf{A}$.

We shall use the symbol $\|\cdot\|$ to denote a consistent family of norms on $\bigcup_{m,n=1}^{\infty} \mathbb{C}^{m\times n}$; that is, the restriction of $\|\cdot\|$ to the space $\mathbb{C}^{m\times n}$ is a norm and

$$\|AB\| \leqq \|A\|\,\|B\|$$

whenever the product AB is defined. Applied to operators such as $\mathbf{T}$ in (1.7), $\|\cdot\|$ will denote a norm satisfying

$$\|\mathbf{T}P\| \leqq \|\mathbf{T}\|\,\|P\|.$$

The most natural such norm is the subordinate operator norm defined by

$$\|\mathbf{T}\| = \sup_{\|P\|=1} \|\mathbf{T}P\|.$$

Two particular families of norms that will be used are the Frobenious norm defined by

$$\|A\|_F^2 = \operatorname{trace}(A^H A),$$

and the spectral norm (2-norm), defined by

$$\|A\|_2 = \sup_{\|x\|_F=1} \|Ax\|_F.$$

Although perturbation theory for eigenvalues and eigenvectors has a long history, relatively little attention has been paid to the problem of obtaining bounds for subspaces, and that has been confined to the eigenvalue problem. Kato [13], who takes an approach different from ours, gives series expansions for projectors corresponding to invariant subspaces when the operator depends analytically on a parameter (there is much more in this comprehensive work). Swanson [30] gives results which lead directly to error bounds for approximate invariant subspaces of operators having a complete orthonormal system of eigenvectors. These results were extended by Stewart [26] in connection with the orthogonal variant of Bauer's treppeniteration for Hermitian matrices (see also [4], [24], [25], [31]). Davis and Kahan [6] have given very general theorems relating the invariant subspaces of two self-adjoint operators. Ruhe [23] uses the singular value decomposition of $A - \lambda I$ to obtain bounds for invariant subspaces corresponding to the approximate eigenvalue λ.

In a series of papers, Varah [32], [33] has described techniques for computing invariant subspaces and assessing their accuracy a posteriori. Extending an algorithm of Kublanovskaya [14], Ruhe [22] has given another method for computing invariant subspaces. Householder [11, p. 182] describes a technique for refining approximate invariant subspaces that is closely related to the method of this paper.

Recently Wedin [35] has generalized part of the Davis–Kahan results to cover the singular value decomposition.

2. Notions of distance and convergence for subspaces. As was indicated in the Introduction, the error bounds obtained in this paper will exhibit an invariant subspace as the column space of the matrix

$$X'_1 = (X_1 + X_2P)(I + P^HP)^{-1/2}, \tag{2.1}$$

where $\mathscr{R}(X_1)$ is an approximate invariant subspace and (X_1, X_2) is unitary. The error bounds themselves will bound $\|P\|$. In some applications this may be sufficient; in others a bound on $\|P\|$ alone may be rather awkward to use. It turns out that the number $\|P\|$ is very closely connected with other measures of the distance between the subspaces $\mathscr{R}(X)$ and $\mathscr{R}(X')$. This section is therefore devoted to surveying briefly some of the notions of distance and convergence for subspaces. For definiteness we shall work with two fixed subspaces $\mathscr{X}$ and $\mathscr{Y}$ of $\mathbb{C}^n$.

2.1. The gap between $\mathscr{X}$ and $\mathscr{Y}$. A natural way of describing the nearness of $\mathscr{X}$ and $\mathscr{Y}$ is to define a distance function between any two subspaces. This may be done as follows (see [13] for proofs and references).

DEFINITION 2.1. Let $\mathscr{X}, \mathscr{Y} \subset \mathbb{C}^n$ be subspaces. The *gap* between $\mathscr{X}$ and $\mathscr{Y}$ is the number

$$\gamma(\mathscr{X}, \mathscr{Y}) = \max\{ \sup_{\substack{\|x\|=1\\ x\in\mathscr{X}}} \inf_{y\in\mathscr{Y}} \|x - y\|, \sup_{\substack{\|y\|=1\\ y\in\mathscr{Y}}} \inf_{x\in\mathscr{X}} \|y - x\|\}.$$

The gap function is not quite a metric (it may be turned into one by taking the infima over unit vectors), for it does not satisfy the triangle inequality. However, the neighborhoods $\mathscr{N}(\mathscr{X}; \delta)$ defined by

$$\mathscr{N}(\mathscr{X}; \delta) = \{\mathscr{Y} : \gamma(\mathscr{X}, \mathscr{Y}) < \delta\}$$

form a basis for a topology on the set of all subspaces of $\mathbb{C}^n$. The convergence of a sequence $\mathscr{X}_n$ to $\mathscr{X}$ in this topology is equivalent to the convergence of $\gamma(\mathscr{X}_n, \mathscr{X})$ to zero. The topology is complete in the sense that γ-Cauchy sequences have a limit. Convergence in the gap topology also preserves dimension. This is a consequence of the fact that

$$\gamma(\mathscr{X}, \mathscr{Y}) < 1 \Rightarrow \dim(\mathscr{X}) = \dim(\mathscr{Y}).$$

In the important case where the norm in Definition 2.1 is the 2-norm, the gap function is a metric. More important for our purposes is the following theorem that relates the gap between $\mathscr{X}$ and $\mathscr{Y}$ to the projectors onto $\mathscr{X}$ and $\mathscr{Y}$.

THEOREM 2.2. *Let $P_{\mathscr{X}}$ and $P_{\mathscr{Y}}$ be the (orthogonal) projectors onto the subspaces $\mathscr{X}$ and $\mathscr{Y}$ of $\mathbb{C}^n$. If the 2-norm is used to define the gap in Definition 2.1, then*

$$\gamma(\mathscr{X}, \mathscr{Y}) = \|P_{\mathscr{X}} - P_{\mathscr{Y}}\|_2.$$

Actually Theorem 2.2 holds for any norm defined by an inner product on $\mathbb{C}^n$; however, the projectors P and Q will then be orthogonal projectors with respect to the new inner product, i.e., they will be oblique projectors with respect to the "natural" inner product.

2.2. The canonical angles between $\mathscr{X}$ and $\mathscr{Y}$. The idea that the relation between subspaces can be characterized by a set of angles is quite old. A good set of references may be found in [6], where much of the material presented here is treated in an infinite-dimensional setting.

We begin with a theorem that distinguishes a special pair of bases for the subspaces $\mathscr{X}$ and $\mathscr{Y}$. For simplicity we assume that $\mathscr{X}$ and $\mathscr{Y}$ have the same dimension $l \leqq n/2$. If $l > n/2$ we may work with the orthogonal complements of $\mathscr{X}$ and $\mathscr{Y}$.

THEOREM 2.3. *Let $\mathscr{X}$, $\mathscr{Y} \subset \mathbb{C}^n$ be subspaces with $\dim(\mathscr{X}) = \dim(\mathscr{Y}) = l$. Let $2l \leqq n$. Then there are unitary matrices $X = (X_1, X_2)$ and $Y = (Y_1, Y_2)$ such that $\mathscr{R}(X_1) = \mathscr{X}$, $\mathscr{R}(Y_1) = \mathscr{Y}$, and*

$$Y^H X = \begin{pmatrix} \Gamma & \Sigma & 0 \\ -\Sigma & \Gamma & 0 \\ 0 & 0 & I_{n-2l} \end{pmatrix}, \tag{2.2}$$

where $\Gamma = \operatorname{diag}(\gamma_1, \gamma_2, \cdots, \gamma_l)$ and $\Sigma = \operatorname{diag}(\sigma_1, \sigma_2, \cdots, \sigma_l)$ have nonnegative diagonal elements.

A proof may be found in [2], where numerical methods for computing the bases are discussed. The significance of the theorem is the following. Since $Y^H X$ is unitary,

$$\Gamma^H \Gamma + \Sigma^H \Sigma = I_l,$$

and since $\Gamma^H\Gamma$ and $\Sigma^H\Sigma$ are real and diagonal,

$$\gamma_i^2 + \sigma_i^2 = 1, \qquad i = 1, 2, \cdots, l.$$

This suggests that we define the *canonical angles* between $\mathscr{X}$ and $\mathscr{Y}$ to be the numbers

$$\theta_i = \cos^{-1}\gamma_i, \qquad i = 1, 2, \cdots, l.$$

If we define

$$\Theta = \Theta(\mathscr{X}, \mathscr{Y}) \equiv \operatorname{diag}(\theta_1, \theta_2, \cdots, \theta_l),$$

then

$$\Gamma = \cos\Theta, \qquad \Sigma = \sin\Theta.$$

The columns of X_1 and Y_1 form a biorthogonal set of basis vectors for $\mathscr{X}$ and $\mathscr{Y}$. The angle between the ith column of X_1 and the ith column of Y_1 is the ith canonical angle θ_i.

The cosines γ_i of the canonical angles can be characterized as the singular values of an easily constructed matrix.

THEOREM 2.4. *In Theorem 2.3 let the orthonormal columns of X', $Y' \in \mathbb{C}^{n \times l}$ form bases for $\mathscr{X}$ and $\mathscr{Y}$. Then the numbers $\gamma_i (i = 1, 2, \cdots, l)$ are the singular values of $Y'^H X'$.*

Proof. Because $\mathscr{R}(X') = \mathscr{R}(X_1)$, we can write X_1 in the form $X_1 = X'U$, where U is nonsingular. Because X_1 and X' have orthonormal columns, U is unitary. Similarly $Y_1 = Y'V$, where V is unitary. Then

$$\Gamma = Y_1^H X_1 = V^H(Y'^H X')U,$$

and since Γ is diagonal, its diagonal elements must be singular values of $Y'^H X'$. ∎

For our purposes the importance of the canonical angles is that the gap between $\mathscr{X}$ and $\mathscr{Y}$ is just the sine of the largest canonical angle. This fact is a consequence of the following theorem.

THEOREM 2.5. *In Theorem 2.3 let $P_{\mathscr{X}}$ and $P_{\mathscr{Y}}$ be the projectors onto $\mathscr{X}$ and $\mathscr{Y}$. Then the nonzero eigenvalues of $P_{\mathscr{X}} - P_{\mathscr{Y}}$ are $\pm \sin \theta_i$ $(i = 1, 2, \cdots, l)$.*

Proof. By an obvious change of coordinates, we may assume that the columns of $(I_l, 0, 0)^T$ and $(\Gamma, -\Sigma, 0)^T$ form orthonormal bases for $\mathscr{X}$ and $\mathscr{Y}$, respectively. Then

$$P_{\mathscr{X}} - P_{\mathscr{Y}} = \begin{pmatrix} I_l \\ 0 \\ 0 \end{pmatrix}(I_l, 0, 0) - \begin{pmatrix} \Gamma \\ -\Sigma \\ 0 \end{pmatrix}(\Gamma, -\Sigma, 0)$$

$$= \begin{pmatrix} I_l - \Gamma^2 & \Gamma\Sigma & 0 \\ \Sigma\Gamma & -\Sigma^2 & 0 \\ 0 & 0 & 0 \end{pmatrix}$$

$$= \begin{pmatrix} \Sigma^2 & \Gamma\Sigma & 0 \\ \Sigma\Gamma & -\Sigma^2 & 0 \\ 0 & 0 & 0 \end{pmatrix}.$$

Since Γ and Σ are diagonal, the nonzero eigenvalues of $P_{\mathscr{X}} - P_{\mathscr{Y}}$ are the eigenvalues of the 2×2 matrices

$$\begin{pmatrix} \sigma_i^2 & \gamma_i\sigma_i \\ \gamma_i\sigma_i & -\sigma_i^2 \end{pmatrix},$$

which are easily seen to be $\pm\sigma_i = \pm\sin\theta_i$. ∎

Since the 2-norm of a matrix is equal to its largest singular value and the square of its Frobenius norm is the sum of squares of its singular values, we have the following corollary.

COROLLARY 2.6. *Let $\theta_1 \geqq \theta_2 \geqq \cdots \geqq \theta_l \geqq 0$. Then*

$$\gamma(\mathscr{X}, \mathscr{Y}) = \|P_{\mathscr{X}} - P_{\mathscr{Y}}\|_2 = \sin\theta_1$$

and

$$\sqrt{2}\gamma(\mathscr{X}, \mathscr{Y}) \leqq \|P_{\mathscr{X}} - P_{\mathscr{Y}}\|_F = \sqrt{2\sin^2\theta_1 + \cdots + 2\sin^2\theta_l}.$$

One final construction may be obtained from Theorem 2.3. From (2.2) it follows that the matrix

$$R = X\begin{pmatrix} \Gamma & -\Sigma & 0 \\ \Sigma & \Gamma & 0 \\ 0 & 0 & I \end{pmatrix}X^H \tag{2.3}$$

is a unitary matrix that transforms $\mathscr{X}$ onto $\mathscr{Y}$. It is called the *direct rotation* from $\mathscr{X}$ to $\mathscr{Y}$. It is uniquely defined by (2.3), provided that Γ is nonsingular. Davis and Kahan [6] have shown that of all unitary transformations taking $\mathscr{X}$ onto $\mathscr{Y}$, the direct rotation differs least from the identity in the Frobenius norm or the 2-norm; i.e., $\|I - R\|_p$ is minimal for $p = 2, F$. Other extremal properties of the direct rotation are also investigated in [6].

2.3. Bounds on $\tan \Theta$. To return to the considerations that motivated this section, let $X = (X_1, X_2)$ be unitary with $\mathscr{R}(X_1) = \mathscr{X}$ and suppose that the columns of

$$Y_1 = (X_1 + X_2 P)(I + P^H P)^{-1/2} \tag{2.4}$$

span $\mathscr{Y}$ (as above we suppose that $\dim(\mathscr{X}) = \dim(\mathscr{Y}) = l$). Then by Theorem 2.4, the cosines of the canonical angles between $\mathscr{X}$ and $\mathscr{Y}$ are the singular values of

$$Y_1^H X_1 = (I + P^H P)^{-1/2}.$$

Let τ_i ($i = 1, 2, \cdots, l$) be the singular values of P. Then

$$\cos \theta_i = (1 + \tau_i^2)^{-1/2},$$

whence

$$\tau_i = \tan \theta_i.$$

Thus the singular values of P are just the tangents of the canonical angles between $\mathscr{X}$ and $\mathscr{Y}$. In terms of norms, we may state the following theorem.

THEOREM 2.7. *Let $X = (X_1, X_2)$ be unitary and let Y_1 be defined by* (2.4). *Let $\mathscr{X} = \mathscr{R}(X_1)$ and $\mathscr{Y} = \mathscr{R}(Y_1)$, and let $\Theta = \Theta(\mathscr{X}, \mathscr{Y})$. Then*

$$\gamma(\mathscr{X}, \mathscr{Y}) = \|\sin \Theta\|_2 \leqq \|\tan \Theta\|_2 = \|P\|_2$$

and

$$\gamma(\mathscr{X}, \mathscr{Y}) \leqq \|\sin \Theta\|_F \leqq \|\tan \Theta\|_F = \|P\|_F.$$

2.4. Other notions of convergence. For completeness we conclude this section with a brief description of two other notions of convergence of subspaces. Although both notions lead to the topology generated by the gap function, they both have something of a qualitative, as opposed to a quantitative flavor.

One of the definitions of convergence is based on the natural expectation that converging subspaces should have converging bases. Specifically, let $\mathscr{X}_1, \mathscr{X}_2, \cdots$ be a sequence of subspaces. Then we say that $\lim \mathscr{X}_k = \mathscr{X}$, provided there are matrices $X_1, X_1^{(1)}, X_1^{(2)}, \cdots$ whose columns form bases for $\mathscr{X}, \mathscr{X}_1, \mathscr{X}_2, \cdots$ such that $\lim X_1^{(k)} = X_1$. The advantage of this approach is that a sequence of subspaces may be constructed by generating a sequence of bases for the subspaces. If these bases can be arranged so that they converge, then the subspaces converge. Parlett and Poole [19] have used this idea in an analysis of matrix power methods.

The final notion of convergence is obtained by placing a norm on the exterior product $\Lambda_l(\mathbb{C}^n)$ (see [10] for definitions). A vector of the form $x_1 \wedge x_2 \wedge \cdots \wedge x_l$ is called a decomposable l-vector and is nonzero if and only if $x_1, x_2, \cdots, x_l$ are linearly independent. If a nonzero decomposable l-vector $x_1 \wedge x_2 \wedge \cdots \wedge x_l$ is identified with the subspaces spanned by $x_1, x_2, \cdots, x_l$ then it can be shown that a sequence of subspaces converges if and only if a sequence of suitably scaled, corresponding l-vectors converges in $\Lambda_l(\mathbb{C}^n)$.

3. Iterative solution of $Tx = g - \varphi(x)$. The method for obtaining bounds for approximate invariant subspaces outlined in § 1 reduced the problem to one of solving the equation

$$(3.1) \qquad \mathbf{T}P = A_{21} - \varphi(P),$$

where $\mathbf{T}$ is the linear operator on $\mathbb{C}^{(n-l)\times l}$ defined by (1.7) and φ is the nonlinear operator defined by (1.8). In this section we shall show how (3.1) may be solved by successive substitutions and obtain bounds on the error in the approximate solutions. The hypothesis of finite-dimensionality is used nowhere in the analysis. Accordingly, we shall take $\mathbf{T}$ and φ to be operators on a Banach space $\mathscr{B}$ and work with the equation

$$(3.2) \qquad \mathbf{T}x = g - \varphi(x),$$

where $g \in \mathscr{B}$.

Obviously the existence and size of a solution of (3.2) will depend on the nature of the nonlinear function φ. An examination of (1.8) suggests that $\varphi(x)$ may be assumed to approach zero quadratically as x approaches zero. In this case, if x is small, the term $\varphi(x)$ in (3.2) will be negligible, and the solution (3.2) will be given approximately by

$$x \cong \mathbf{T}^{-1}g.$$

The quantity $\|x\|$ will then be bounded by some multiple of order unity of $\|\mathbf{T}^{-1}\|\,\|g\|$. Actually, the functions φ that we shall encounter in applications will satisfy even stronger conditions than quadratic behavior at zero (cf. condition (ii) in Theorem 3.1), and this permits the computation of rather sharp bounds on the solution x and its approximations.

The conditions under which (3.2) has a solution that can be approximated by successive substitutions are given in the following theorem.

THEOREM 3.1. *Let $\mathbf{T}$ be a bounded linear operator on the Banach space $\mathscr{B}$ that has a bounded inverse on $\mathscr{B}$. Let φ be a continuous function $\mathscr{B}$ that satisfies*

$$\text{(i)} \quad \|\varphi(x)\| \leqq \eta\|x\|^2,$$

$$\text{(ii)} \quad \|\varphi(x) - \varphi(y)\| \leqq 2\eta \max(\|x\|, \|y\|)\|x - y\|,$$

for some $\eta \geqq 0$. Let $g \in \mathscr{B}$, and let

$$\gamma = \|g\|$$

and

$$\delta = \|\mathbf{T}^{-1}\|^{-1}.$$

Let $x_0 = 0$, and define the sequence $x_1, x_2, x_3, \cdots$ by

$$(3.3) \qquad x_{i+1} = \mathbf{T}^{-1}(g - \varphi(x_i)) \equiv x_1 - \mathbf{T}^{-1}\varphi(x_i).$$

Then if

$$(3.4) \qquad \kappa_2 = \gamma\eta/\delta < 1/4,$$

the x_i converge to the unique solution x of (3.2) *that satisfies*

$$\|x\| \leqq \frac{\gamma}{\delta}(1 + \kappa) \equiv \frac{\gamma}{\delta}\frac{1 + \sqrt{1 - 4\kappa_2}}{1 - 2\kappa_2 + \sqrt{1 - 4\kappa_2}} < 2\frac{\gamma}{\delta}. \tag{3.5}$$

Moreover,

$$\|x - x_i\| \leqq \frac{\|x_{i+1} - x_i\|}{1 - \rho} \leqq \frac{\rho^{i-k}\|x_{k+1} - x_k\|}{1 - \rho}, \qquad k \leqq i, \tag{3.6}$$

where

$$\rho = 4(\eta\gamma)/\delta < 1.$$

Proof. As in [27] and [28], it can be shown that

$$\|x_i\| \leqq \frac{\gamma}{\delta}(1 + \kappa_i), \qquad i = 1, 2, \cdots, \tag{3.7}$$

where $\kappa_1 = 0$, κ_2 is given by (3.4), $\kappa_{i+1} = \kappa_2(1 + \kappa_i)^2$. If $\kappa_2 < 1/4$, the κ_i can be shown to converge to the number κ defined by (3.5). Thus all the iterates lie in the closed disk $\mathscr{D} = \{x : \|x\| \leqq \delta^{-1}\gamma(1 + \kappa)\}$.

Now the x_i satisfy the equation

$$x_{i+1} = \mathbf{T}^{-1}(g - \boldsymbol{\varphi}(x_i)) \equiv \boldsymbol{\theta}(x_i).$$

But property 2 of $\boldsymbol{\varphi}$ and the linearity of $\mathbf{T}^{-1}$ imply that $\boldsymbol{\varphi}$ is a contraction (with constant $4\kappa_2$) in $\mathscr{D}$. It follows from the contraction mapping theorem [18, p. 120] that the x_i converge to the unique fixed point of $\boldsymbol{\theta}$ in $\mathscr{D}$. The error estimates (3.6) follow in an elementary way from the proof of the contraction mapping theorem. ∎

Theorem 3.1 confirms the informal discussion preceding the theorem. The smaller g is, the more accurately x is approximated by $x_1 = \mathbf{T}^{-1}g$. In all cases where the theorem is applicable, $\|x\|$ is bounded by $2\|\mathbf{T}^{-1}\|\,\|g\|$.

It should be noted that the theorem is constructive in the sense that the sequence $x_1, x_2, x_3, \cdots$ can be constructed from (3.3), provided we are able to evaluate $\boldsymbol{\varphi}(x)$ and $\mathbf{T}^{-1}x$ for any $x \in \mathscr{B}$. In applications, $\boldsymbol{\varphi}$ will be relatively simple to evaluate (cf. (1.8)); however, $\mathbf{T}^{-1}$ will generally be difficult to evaluate. Now it may happen that for the problem at hand we can find an approximation $\mathbf{S}$ to $\mathbf{T}$ that is easily invertible. For example, if the matrices A_{11} and A_{22} in (1.8) are diagonally dominant, $\mathbf{S}$ might be taken to be the operator obtained by replacing the off-diagonal elements in A_{11} and A_{22} by zero. We may then ask how one may use this approximation $\mathbf{S}$ to compute a solution of (3.2).

It is not enough simply to replace $\mathbf{T}^{-1}$ by $\mathbf{S}^{-1}$ in (3.2); for then the x_i will converge, if at all, to a solution of $\mathbf{S}x = g - \boldsymbol{\varphi}(x)$. Instead we consider the following technique, described in algorithmic form, which is analogous to the method of iterative refinement for linear systems [34, p. 255].

1) $x_0 = 0$
2) For $i = 0, 1, 2, \cdots$,
    1) $r_i = g - \boldsymbol{\varphi}(x_i) - \mathbf{T}x_i$
    2) $d_i = \mathbf{S}^{-1}r_i$
    3) $x_{i+1} = x_i + d_i$

This iteration has been analyzed in [3], where it is used to compute eigenvectors and invariant subspaces of diagonally dominant matrices.

THEOREM 3.2. *Let* $\mathbf{T}$, $\boldsymbol{\varphi}$, g, η, *and* γ *be as in Theorem* 3.1 (*except* $\mathbf{T}$ *is not required to have a bounded inverse*). *Let* $\mathbf{S}:\mathscr{B} \to \mathscr{B}$ *be linear and have a bounded inverse on* $\mathscr{B}$. *Let*

$$\varepsilon = \|\mathbf{T} - \mathbf{S}\| \tag{3.8}$$

and

$$\delta_s = \|\mathbf{S}^{-1}\|^{-1}. \tag{3.9}$$

Then if

$$\delta_s - \varepsilon > 2\sqrt{\eta\gamma},$$

the sequence $x_0, x_1, x_2, \cdots$ *generated by* (3.7) *converges to the unique solution* x *of* (3.2) *that satisfies*

$$\|x\| \leqq \frac{\gamma}{\delta_s}(1 + \kappa) < 2\frac{\gamma}{\delta_s - \varepsilon},$$

where κ *is the smallest root of the equation*

$$\kappa = \frac{\varepsilon}{\delta_s}(1 + \kappa) + \frac{\eta\gamma}{\delta_s^2}(1 + \kappa)^2.$$

Moreover,

$$\|x - x_i\| \leqq \frac{\|x_{i+1} - x_i\|}{1 - \rho} \leqq \frac{\rho^{i-k}\|x_{k+1} - x_k\|}{1 - \rho}, \qquad k \leqq i,$$

where

$$\rho = \frac{\varepsilon}{\delta_s} + \frac{4\eta\gamma}{\delta_s(\delta_s - \varepsilon)} < 1.$$

The proof of this theorem is given in [3]. It should be noted that this reduces to Theorem 3.1 when $\mathbf{S} = \mathbf{T}$, so that $\varepsilon = 0$ and $\delta_s = \delta$.

4. The eigenvalue problem.

4.1. The approximation theorem. In § 1 we showed how the problem of bounding the error in an approximate invariant subspace can be reduced to one of solving a nonlinear equation, and in § 3 we gave bounds for the norm of the solution. These results are combined in the following theorem.

THEOREM 4.1. *Let* $A \in \mathbb{C}^{n\times n}$ *and let* $X = (X_1, X_2)$ *be unitary with* $X_1 \in \mathbb{C}^{n\times l}$. *Partition* $X^H A X$ *conformally with* X *in the form*

$$X^H A X = \begin{pmatrix} A_{11} & A_{12} \\ A_{21} & A_{22} \end{pmatrix}.$$

Define the operator $\mathbf{T}:\mathbb{C}^{(n-l)\times l} \to \mathbb{C}^{(n-l)\times l}$ *by*

$$\mathbf{T}P = PA_{11} - A_{22}P, \tag{4.1}$$

and suppose that **T** *is invertible. Let* $\delta = \|\mathbf{T}^{-1}\|^{-1}$, $\gamma = \|A_{21}\|$, *and* $\eta = \|A_{12}\|$. *Let* κ *be defined by* (3.5). *Then if*

$$\eta\gamma/\delta^2 < 1/4, \tag{4.2}$$

there is a matrix $P \in \mathbb{C}^{(n-l)\times l}$ *with*

$$\|P\| \leqq \frac{\gamma}{\delta}(1+\kappa) < 2\frac{\gamma}{\delta}$$

such that the columns of

$$X_1' = (X_1 + X_2P)(I + P^HP)^{-1/2}$$

span an invariant subspace of A.

Proof. In view of the developments in § 1 and § 3, it is necessary only to show that the function $\boldsymbol{\varphi}$ defined by (1.8) satisfies the conditions of Theorem 3.1 with $\eta = \|A_{12}\|$, which is easily done (see [27]). ∎

Although Theorem 4.1 gives mathematically valid conditions under which a subspace of $\mathbb{C}^n$ is near an invariant subspace of A, it is incomplete unless the quantities appearing in it, such as γ, δ, and $\|P\|$, can be meaningfully explicated. So far as $\|P\|$ is concerned, this has already been done in § 2. If $\|\cdot\|$ is taken to be either the spectral norm or the Frobenius norm, then by Theorem 2.7, the number $\|P\|_2$ bounds the tangent of the largest canonical angle between $\mathcal{R}(X_1)$ and the invariant subspace $\mathcal{R}(X_1')$, and hence $\|P\|_2$ bounds the gap between $\mathcal{R}(X_1)$ and $\mathcal{R}(X_1')$. We shall now consider the number γ, which is related to the notion of a generalized Rayleigh quotient, and then the number δ, which measures the separation of the eigenvalue of A_{11} and A_{22}.

4.2. Compressions, Rayleigh quotients, and residuals. If x is a nonzero vector, the Rayleigh quotient associated with x is the number

$$\rho = \frac{x^HAx}{x^Hx}.$$

If x is normalized so that $\|x\|_2 = 1$, ρ can be written simply as

$$\rho = x^HAx.$$

If x is an eigenvector of A corresponding to the eigenvalue λ, then $\rho = \lambda$. If x is an approximate eigenvector of A, then ρ is an approximate eigenvalue. In fact, it is well known that if A is Hermitian and the approximate eigenvector x is accurate to terms of order ε, then ρ will be an approximate eigenvalue accurate to terms of order ε^2 (see, e.g., [34, p. 173]).

This notion of a Rayleigh quotient can be generalized in two ways. The number ρ can be regarded as defining a linear operator that maps $\mathcal{R}(x)$ into $\mathcal{R}(x)$, or ρ can be regarded as a matrix representation of such an operator. Depending on which point of view is taken, the generalization yields either an operator, called a compression of A, or a matrix representation of that operator, called a Rayleigh quotient.

DEFINITION 4.2. Let $\mathcal{X} \subset \mathbb{C}^n$ be a subspace. The *compression of* **A** *to* $\mathcal{X}$ is the operator $\mathbf{A}_{11}:\mathcal{X} \to \mathcal{X}$ defined by

$$\mathbf{A}_{11}x = \mathbf{P}_{\mathcal{X}}\mathbf{A}x, \qquad x \in \mathcal{X},$$

where $\mathbf{P}_{\mathscr{X}}$ is the projection onto $\mathscr{X}$. If $X_1 \in \mathbb{C}^{n \times l}$ has orthonormal columns, the *Rayleigh quotient*, or *section* (see [11, p. 74]), corresponding to X_1 is the matrix

$$A_{11} = X_1^H A X_1.$$

If $\mathscr{R}(X_1) = \mathscr{X}$, the Rayleigh quotient A_{11} is simply the matrix representation of the compression A_{11} in the basis determined by the columns of X_1. We have seen in § 1 that if $\mathscr{X}$ is an invariant subspace of A the eigenvalues of A_{11} are eigenvalues of A. In this case if y is an eigenvector of A_{11}, then $X_1 y$ is an eigenvector of A. We shall see below that if $\mathscr{X}$ is an approximate invariant subspace of A, then the eigenvalues of A_{11} may approximate those of A.

The notation A_{11} in Definition 4.2 is not coincidental. The matrix A_{11} of Theorem 4.1 is the Rayleigh quotient corresponding to X_1.

The significance of the matrix $A_{21} = X_2^H A X_1$ in Theorem 4.1 may be explained as follows. If $\mathscr{R}(X_1)$ is an invariant subspace of A, then AX_1 can be written as a linear combination of the columns of X_1; that is there is a matrix B such that

$$AX_1 = X_1 B.$$

If $\mathscr{R}(x_1)$ is an approximate invariant subspace of A, then it should be possible to choose B so that the residual

$$R = AX_1 - X_1 B$$

is small in some norm. If either the spectral norm or the Frobenius norm is used, $\|R\|$ is minimized when $B = A_{11}$.

THEOREM 4.3. *Let A, X, and A_{ij} be as in Theorem 4.1. Then for $p = 2, F$,*

$$\|R\|_p = \|AX_1 - X_1 B\|_p$$

is minimized when $B = A_{11}$. In this case,

$$\|R\|_p = \|A_{21}\|_p.$$

Proof. Since $\|\cdot\|_p$ is unitarily invariant,

$$\|R\|_p = \|X^H R\|_p = \left\| \begin{pmatrix} X_1^H \\ X_2^H \end{pmatrix} AX_1 - \begin{pmatrix} X_1^H \\ X_2^H \end{pmatrix} X_1 B \right\|_p$$

$$= \left\| \begin{pmatrix} A_{11} - B \\ A_{21} \end{pmatrix} \right\|_p .$$

This last quantity is minimized when $A_{11} - B = 0$, and its value is then $\|A_{21}\|_p$. ∎

Theorem 4.3 is due to Kahan [12], who also points out that for $p = F$, the minimizing matrix A_{11} is unique. The significance of this theorem is that the number $\gamma = \|A_{21}\|$ of Theorem 4.1 is a legitimate measure of the deviation of $\mathscr{R}(X_1)$ from an approximate invariant subspace of A; $\gamma = \|A_{21}\|_p (p = 2, F)$ is the norm of the smallest possible residual for the matrix X_1.

4.3. The separation of two operators. The quantity δ of Theorem 4.1 is defined in terms of the operator T, and ultimately in terms of the matrices A_{11} and A_{22} (see (1.8)). We shall show that δ can be regarded as a measure of the separation between the eigenvalues of A_{11} and A_{22}.

The operator $\mathbf{T}$ arises in a number of connections and has been studied in both finite- and infinite-dimensional settings ([1], [7], [16], [21], [36]). For our purposes, the most important property of $\mathbf{T}$ is stated in the following theorem, whose proof may be found in [8, Chap. VI].

THEOREM 4.4. *The spectrum of* $\mathbf{T}$ *is the set*

$$\lambda(\mathbf{T}) = \lambda(A_{11}) - \lambda(A_{22}) \equiv \{\lambda - \lambda' : \lambda \in \lambda(A_{11}), \lambda' \in \lambda(A_{22})\}.$$

A consequence of Theorem 4.4 is that $\mathbf{T}$ is invertible if and only if A_{11} and A_{22} have no eigenvalues in common. In this case, $\|\mathbf{T}^{-1}\|$ will be greater than the modulus of the largest eigenvalue of $\mathbf{T}^{-1}$; in other words,

$$0 < \delta = \|T^{-1}\|^{-1} \leqq \min |\lambda(A_{11}) - \lambda(A_{22})|,$$

so that δ is a lower bound on the separation between the eigenvalues of A_{11} and those of A_{22}. This motivates the following definition.

DEFINITION 4.5. Let $B \in \mathbb{C}^{l \times l}$ and $C \in \mathbb{C}^{m \times m}$. Define the operator $\mathbf{T}: \mathbb{C}^{m \times l} \to \mathbb{C}^{m \times l}$ by

$$\mathbf{T}P = PB - CP, \qquad P \in \mathbb{C}^{m \times l}.$$

The *separation* of B and C is the number $\operatorname{sep}(B, C)$ defined by

$$\operatorname{sep}(B, C) = \begin{cases} \|\mathbf{T}^{-1}\|^{-1}, & 0 \notin \lambda(\mathbf{T}), \\ 0, & 0 \in \lambda(\mathbf{T}). \end{cases} \tag{4.3}$$

Of course sep (B, C) is defined for any norm on the space of linear operators on $\mathbb{C}^{m \times l}$, and it follows from the discussion above that sep (B, C) satisfies

$$\operatorname{sep}(B, C) \leqq \min |\lambda(B) - \lambda(C)|.$$

In practice, the norm in (4.3) will usually be taken to be the operator norm subordinate to some norm, $\|\cdot\|_p$, on $\mathbb{C}^{m \times l}$, and where it is required to display the choice of norm, we shall write $\operatorname{sep}_p(B, C)$. If a subordinate operator norm is used to define the separation, we have the alternative definition

$$\begin{aligned} \operatorname{sep}_p(B, C) &= \inf_{\|P\|_p = 1} \|\mathbf{T}P\|_p \\ &= \inf_{\|P\|_p = 1} \|PB - CP\|_p. \end{aligned} \tag{4.4}$$

In the sequel we shall assume that sep has been defined in terms of an operator norm subordinate to a consistent family of matrix norms.

As was mentioned in § 1, an approximation theorem such as Theorem 4.1 can be turned into a perturbation theorem by regarding an invariant subspace of A as an approximate invariant subspace of the perturbed matrix $A + E$. However, the number δ required by the theorem will be (with the obvious definitions of E_{11} and E_{22}) $\operatorname{sep}(A_{11} + E_{11}, A_{22} + E_{22})$, which cannot be computed unless E is known. Thus it is necessary to obtain a lower bound on $\operatorname{sep}(A_{11} + E_{11}, A_{22} + E_{22})$ in terms of $\operatorname{sep}(A_{11}, A_{22})$ and the norms of E_{11} and E_{22}. The proof of the following theorem may be found in [27]. It also follows easily from (4.4).

THEOREM 4.6. *Let* $B, E \in \mathbb{C}^{l \times l}$ *and* $C, F \in \mathbb{C}^{m \times m}$. *Then*

$$\operatorname{sep}(B, C) + \|E\| + \|F\| \geqq \operatorname{sep}(B + E, C + F) \geqq \operatorname{sep}(B, C) - \|E\| - \|F\|.$$

The stability of the function sep, as displayed by Theorem 4.6, is quite important. Theorem 4.1 suggests that an invariant subspace will become increasingly unstable as the eigenvalues of the Rayleigh quotients A_{11} and A_{22} approach one another. It is natural to ask if an analogue of Theorem 4.1 can be found in which the minimum distance between $\lambda(A_{11})$ and $\lambda(A_{22})$ is used rather than $\operatorname{sep}(A_{11}, A_{22})$, and indeed Varah [32] has given such a development. However since the eigenvalues of A_{11} and A_{22} may change greatly with slight perturbations, the number $\min|\lambda(A_{11}) - \lambda(A_{22})|$ can vary greatly. The function sep, on the other hand, behaves predictably well under perturbations of its arguments. The price that is paid for this stability is that $\operatorname{sep}(A_{11}, A_{22})$ may be much smaller than $\min|\lambda(A_{11}) - \lambda(A_{22})|$. And sep is harder to calculate.

The function sep_F defined by the Frobenius norm has several nice properties. Proofs of the following assertions may be found in [27].

THEOREM 4.7. *Let* $B \in \mathbb{C}^{l \times l}$ *and* $C \in \mathbb{C}^{m \times m}$. *Then*

$$\operatorname{sep}_F(B, C) = \operatorname{sep}_F(C, B).$$

If $B = \operatorname{diag}(B_1, B_2, \cdots, B_p)$ *and* $C = \operatorname{diag}(C_1, C_2, \cdots, C_q)$, *then*

$$\operatorname{sep}_F(B, C) = \min\{\operatorname{sep}_F(B_i, C_j) : i = 1, 2, \cdots, p; j = 1, 2, \cdots, q\}.$$

If B and C are Hermitian, then

$$\operatorname{sep}_F(B, C) = \min|\lambda(B) - \lambda(C)|.$$

The problem of estimating sep is a difficult one. Even when B and C are Hermitian, the calculation of $\operatorname{sep}_F(B, C)$ requires the calculation of the eigenvalues of B and C. When B and C are not Hermitian, there is no simple general expression for $\operatorname{sep}(B, C)$ in terms $\lambda(B) - \lambda(C)$. However, if B and C are diagonalizable, a bound for $\operatorname{sep}(B, C)$ can be obtained in terms of $|\lambda(B) - \lambda(C)|$ and the matrices of eigenvectors of B and C. We begin with the following theorem, which shows what happens when B and C are transformed by similarity transformations.

THEOREM 4.8. *Let* $B, X \in \mathbb{C}^{l \times l}$ *and* $C, Y \in \mathbb{C}^{m \times m}$ *with* X *and* Y *nonsingular. Then*

$$\operatorname{sep}(XBX^{-1}, YCY^{-1}) \geqq \frac{\operatorname{sep}(B, C)}{\kappa(X)\kappa(Y)}, \tag{4.5}$$

where

$$\kappa(X) = \|X\|\,\|X^{-1}\|, \quad \kappa(Y) = \|Y\|\,\|Y^{-1}\|.$$

If X *and* Y *are unitary, then*

$$\operatorname{sep}_p(XBX^H, YCY^H) = \operatorname{sep}_p(B, C), \qquad p = 2, F. \tag{4.6}$$

Proof. We prove the theorem for the case in which only B is transformed, the general case being similar. Let P be a matrix with $\|P\| = 1$ such that if we set

$$Q = PXBX^{-1} - CP$$

then $\operatorname{sep}(XBX^{-1}, C) = \|Q\|$. Then

$$(PX)B - C(PX) = QX.$$

Hence

$$\begin{aligned}\|X\| \operatorname{sep}(XBX^{-1}, C) &\geqq \|QX\| \\ &= \|(PX)B - C(PX)\| \\ &\geqq \operatorname{sep}(B, C)\|PX\| \geqq \operatorname{sep}(B, C)\frac{\|P\|}{\|X^{-1}\|} \\ &= \frac{\operatorname{sep}(B, C)}{\|X^{-1}\|},\end{aligned}$$

which establishes (4.5) with $Y = I$. Equation (4.6) follows from the fact that $\|\cdot\|_2$ and $\|\cdot\|_F$ are unitarily invariant. ∎

Theorems 4.6 and 4.8 can be combined to give the following corollary.

COROLLARY 4.9. *Let the columns of X and Y form a complete system of eigenvectors for B and C respectively. Then*

$$\operatorname{sep}_F(B, C) \geqq \frac{\min|\lambda(B) - \lambda(C)|}{\kappa(X)\kappa(Y)}.$$

Theorem 4.7 gives a simple expression for $\operatorname{sep}_F(B, C)$ when B and C are Hermitian. For some applications, it may be more convenient to have a bound for $\operatorname{sep}_2(B, C)$. Such a bound may be obtained from the following theorem, which bounds the change in sep when the defining norm is changed.

THEOREM 4.10. *Let $B \in \mathbb{C}^{l\times l}$ and $C \in \mathbb{C}^{m\times m}$. Let the norms $\|\cdot\|_r$ and $\|\cdot\|_s$ satisfy*

$$\sigma\|P\|_r \leqq \|P\|_s \leqq \tau\|P\|_r, \qquad P \in \mathbb{C}^{m\times l}.$$

Then

$$\operatorname{sep}_r(B, C) \geqq \frac{\sigma}{\tau}\operatorname{sep}_s(B, C).$$

Proof. We have the inequalities

$$\begin{aligned}\tau\|PB - CP\|_r &\geqq \|PB - CP\|_s \\ &\geqq \operatorname{sep}_s(B, C)\|P\|_s \geqq \sigma \operatorname{sep}_s(B, C)\|P\|_r,\end{aligned}$$

or, if $P \neq 0$,

$$\frac{\|PB - CP\|_r}{\|P\|_r} \geqq \frac{\sigma}{\tau}\operatorname{sep}_s(B, C). \tag{4.7}$$

Taking the infimum of $\|P\|_r = 1$ in (4.7) gives the result. ∎

Since

$$\|P\|_2 \leqq \|P\|_F \leqq \sqrt{\min\{l, m\}}\,\|P\|_2,$$

whenever $P \in \mathbb{C}^{m \times l}$, we have, for matrices $B \in \mathbb{C}^{l \times l}$ and $C \in \mathbb{C}^{m \times m}$, the relation

$$\operatorname{sep}_2(B, C) \geqq \frac{\operatorname{sep}_F(B, C)}{\sqrt{\min\{l, m\}}},$$

and the same inequality holds with the subscripts 2 and F reversed.

4.4. The perturbation theorem. Theorems 4.1 and 4.6 may be combined to yield a theorem describing the behavior of an invariant subspace of the matrix A under a perturbation E.

THEOREM 4.11. *Let $A, E \in \mathbb{C}^{n \times n}$. Let $X = (X_1, X_2)$ be unitary with $X_1 \in \mathbb{C}^{n \times l}$, and suppose $\mathcal{R}(X_1)$ is an invariant subspace of A. Let $X^H AX$ and $X^H EX$ be partitioned conformally with X in the forms*

$$X^H AX = \begin{pmatrix} A_{11} & A_{12} \\ 0 & A_{22} \end{pmatrix}$$

and

$$X^H EX = \begin{pmatrix} E_{11} & E_{12} \\ E_{21} & E_{22} \end{pmatrix}.$$

Let

$$\delta = \operatorname{sep}(A_{11}, A_{22}) - \|E_{11}\| - \|E_{22}\|.$$

If

$$\frac{\|E_{21}\|(\|A_{12}\| + \|E_{12}\|)}{\delta^2} \leqq \frac{1}{4},$$

there is a matrix P satisfying

$$\|P\| \leqq 2\frac{\|E_{21}\|}{\delta}$$

such that the columns of $X_1' = (X_1 + X_2P)(I + P^H P)^{-1/2}$ span an invariant subspace of $A + E$.

4.5. Bounds for the eigenvalues of A. In Theorem 4.1 let

$$X' = \begin{pmatrix} I & -P^H \\ P & I \end{pmatrix} \begin{pmatrix} (I + P^H P)^{-1/2} & 0 \\ 0 & (I + PP^H)^{-1/2} \end{pmatrix}. \tag{4.8}$$

Then, as was observed in § 1, X' is unitary, and the matrix

$$A' = X'^H AX' = \begin{pmatrix} A'_{11} & A'_{12} \\ 0 & A'_{22} \end{pmatrix}$$

is block upper triangular. Hence $\lambda(A) = \lambda(A') = \lambda(A'_{11}) \cup \lambda(A'_{22})$. Explicit expressions for A'_{11} and A'_{22} in terms of A_{11} and A_{22} may be obtained by using the definitions of X'_1 and X'_2. These expressions may be further simplified by using the relation

$$PA_{11} - A_{22}P = A_{21} - PA_{12}P.$$

A similar development can be carried out for Theorem 4.11 (for details see [27]). The result is the following theorem.

THEOREM 4.12. *Under the hypotheses of Theorem* 4.1, *the matrices* A'_{11} *and* A'_{22} *are given by*

$$A'_{11} = (I + P^H P)^{1/2}(A_{11} + A_{12}P)(I + P^H P)^{-1/2}$$

and

$$A'_{22} = (I + PP^H)^{-1/2}(A_{22} - PA_{12})(I + PP^H)^{1/2}.$$

The set $\lambda(A)$ *is the disjoint union of* $\lambda(A'_{11})$ *and* $\lambda(A'_{22})$.

Under the hypotheses of Theorem 4.11, *the matrices* A'_{11} *and* A'_{22} *are given by*

$$A'_{11} = (I + P^H P)^{1/2}[A_{11} + E_{11} + (A_{12} + E_{12})P](I + P^H P)^{-1/2}$$

and

$$A'_{22} = (I + PP^H)^{-1/2}[A_{22} + E_{22} - P(A_{12} + E_{12})](I + PP^H)^{1/2}.$$

The set $\lambda(A + E)$ *is the disjoint union of* $\lambda(A'_{11})$ *and* $\lambda(A'_{22})$.

In the first part of Theorem 4.12, the matrix A'_{11} is similar to $A_{11} + A_{12}P$. From Theorem 4.1,

$$\|A_{12}P\| < 2\frac{\gamma\eta}{\delta}.$$

Hence bounds on the eigenvalues of A can be established by applying standard perturbation theory to the matrix A_{11}. In particular, if $l = 1$, the matrix A_{11} is a scalar, the Rayleigh quotient $\alpha_{11} = X_1^H A X_1$. The result then assures us that there is an eigenvalue λ of A satisfying

$$|\lambda - \alpha_{11}| < 2\frac{\eta\gamma}{\delta}.$$

Similar observations hold for the second part of Theorem 4.12. It should be remembered that if the eigenvalues of A_{11} are ill conditioned, they may be changed greatly by the perturbation $A_{12}P$, even when $\gamma\eta/\delta$ is small.

An unsavory aspect of the bounds obtained from Theorem 4.12 is their dependence on δ^{-1}. This means that they cannot be used to obtain bounds for a single eigenvalue that is poorly separated from its neighbors, even though the eigenvalue may be quite insensitive to perturbations in A. In part, this dependence is an essential feature of our approach. Theorem 4.12 is an incidental consequence of the proofs of Theorems 4.1 and 4.11, which give bounds for invariant subspaces. Since invariant subspaces are truly ill determined when δ is small, it is not surprising that Theorem 4.12 should reflect this ill-determination.

Although we cannot completely remove the dependence on δ^{-1} in Theorem 4.12, we can reduce it to a second order effort by introducing the notion of the left-invariant subspace associated with the invariant subspace $\mathscr{X}$. A left-invariant subspace $\mathscr{Y}$ of A is a nontrivial subspace satisfying

$$A^H\mathscr{Y} \subset \mathscr{Y}.$$

When $\dim(\mathscr{Y}) = 1$, a nonzero vector in $\mathscr{Y}$ is a left eigenvector of A. If λ is a simple eigenvalue of A, it is possible to find right and left eigenvectors x and y corresponding to λ such that $y^Hx = 1$. The following theorem generalizes this construction.

THEOREM 4.13. *Let X, A, and A_{ij} be as in Theorem 4.11. If $\operatorname{sep}(A_{11}, A_{22}) \neq 0$, then there are matrices $X_1 \in \mathbb{C}^{n\times l}$ and $Y_2 \in \mathbb{C}^{n\times(n-1)}$ such that*

$$(X_1, Y_2)^{-1} = (Y_1, X_2)^H \tag{4.9}$$

and

$$(Y_1, X_2)^H A(X_1, Y_2) = \begin{pmatrix} A_{11} & 0 \\ 0 & A_{22} \end{pmatrix}. \tag{4.10}$$

Moreover if $\Theta = \Theta[\mathscr{R}(X_1), \mathscr{R}(Y_1)]$, we have

$$\|Y_1\|_2 = \|Y_2\|_2 = \|\sec\Theta\|_2 \leqq \|\sec\Theta\|_F = \|Y_1\|_F = \|Y_2\|_F. \tag{4.11}$$

Proof. We seek the matrix (X_1, Y_2) in the form

$$(X_1, Y_2) = (X_1, X_2)\begin{pmatrix} I_l & Q \\ 0 & I_{n-l} \end{pmatrix},$$

the matrix $(Y_1, X_2)^H$ then being determined by (4.9) in the form

$$(Y_1, X_2)^H = \begin{pmatrix} I & -Q \\ 0 & I \end{pmatrix}\begin{pmatrix} X_1^H \\ X_2^H \end{pmatrix}.$$

From (4.10) and the definition of the matrices A_{ij}, we get the relation

$$\begin{pmatrix} I & -Q \\ 0 & I \end{pmatrix}\begin{pmatrix} A_{11} & A_{12} \\ 0 & A_{22} \end{pmatrix}\begin{pmatrix} I & Q \\ 0 & I \end{pmatrix} = \begin{pmatrix} A_{11} & 0 \\ 0 & A_{22} \end{pmatrix}. \tag{4.12}$$

Upon computing the (1, 2)-element of the partition (4.12), we get

$$A_{11}Q - QA_{22} = -A_{12}. \tag{4.13}$$

Now the operator $Q \mapsto A_{11}Q - QA_{22}$ is not the operator $\mathbf{T}$ of (1.7); however, because $\operatorname{sep}(A_{11}, A_{22}) \neq 0$, the sets $\lambda(A_{11})$ and $\lambda(A_{22})$ are disjoint, and hence $Q \mapsto A_{11}Q - QA_{22}$ is nonsingular. Thus the matrix Q is uniquely defined by (4.13).

The inequalities (4.11) follow from Theorem 2.7 and the fact that $Y_1 = X_1 + X_2Q^H$. ∎

The following terminology is suggested by Theorem 4.13. The subspace $\mathscr{X} = \mathscr{R}(X_1)$ is called a *simple invariant subspace* of A. To any simple invariant subspace there corresponds a unique left-invariant subspace $\mathscr{Y} = \mathscr{R}(Y_1)$. If the orthonormal columns of X_1 form a basis for $\mathscr{X}$, there is a unique matrix Y_1, whose columns span $\mathscr{Y}$, such that $Y_1^HX_1 = I$. The spectrum of the *generalized Rayleigh quotient* $A_{11} = Y_1^H A X_1$ is a subset of $\lambda(A)$. The invariant subspaces $\mathscr{X}$ and $\mathscr{Y}$ have unique invariant complements spanned respectively by the columns of Y_2 and X_2. The spectrum of the complementary generalized Rayleigh quotient $A_{22} = Y_2^H A X_2$ is disjoint from $\lambda(A)$. We shall define

$$\mathrm{sep}_p(\mathscr{X}) = \mathrm{sep}_p(A_{11}, A_{22}).$$

In this notation we may state the following generalization of a well-known result [34, Chap. II].

THEOREM 4.14. *Let $A, E \in \mathbb{C}^{n\times n}$. Let $\mathscr{X}$ be a simple invariant subspace of A and let $\mathscr{Y}$ be its corresponding left-invariant subspace. Let $\Theta = \Theta(x, y)$. Let*

$$\delta = \mathrm{sep}(\mathscr{X}) - 2\|\sec\Theta\|_p\|E\|_p,$$

and suppose that

$$\frac{\|\sec\Theta\|_p\|E_p\|}{\delta} < \frac{1}{2}. \tag{4.14}$$

Then to any generalized Rayleigh quotient A_{11} of A corresponding to $\mathscr{X}$ and $\mathscr{Y}$, there is a generalized Rayleigh quotient A'_{11} of $A + E$ corresponding to right- and left-invariant subspaces $\mathscr{X}'$ and $\mathscr{Y}'$ such that

$$\begin{aligned}\|A_{11} - A'_{11}\|_p &\leqq \|\sec\Theta\|_p\|E\|_p\left[1 + \frac{2\|\sec\Theta\|_p\|E\|_p}{\delta}\right] \\ &< 2\|\sec\Theta\|_p\|E\|_p.\end{aligned} \tag{4.15}$$

Proof. In the notation of Theorem 4.13,

$$(X_1, Y_2)^{-1}(A + E)(X_1, Y_2) = \begin{pmatrix} A_{11} + F_{11} & F_{12} \\ F_{21} & A_{22} + F_{22} \end{pmatrix},$$

where

$$\begin{pmatrix} F_{11} & F_{12} \\ F_{21} & F_{22} \end{pmatrix} = \begin{pmatrix} Y_1^H E X_1 & Y_1^H E Y_2 \\ X_2^H E X_1 & X_2^H E Y_2 \end{pmatrix}.$$

Because of (4.14), Theorem 4.11 applies, with F_{ij} replacing E_{ij}, to give a matrix P satisfying

$$\|P\|_p \leqq 2\frac{\|F_{21}\|_p}{\delta} \leqq \frac{2\|E\|_p}{\delta}$$

such that

$$A'_{11} = A_{11} + F_{11} + F_{12}P \tag{4.16}$$

is a generalized Rayleigh quotient for $A + E$. The bound (4.15) follows upon taking norms in (4.16). ∎

Theorem 4.14 generalizes the result that the condition number of a simple eigenvalue is the secant of the angle between its left and right eigenvectors.

Actually the proof of the theorem yields a little more. For it follows from (4.16) that

$$\begin{aligned} A'_{11} &= A_{11} + F_{11} + O(\|E\|_p^2) \\ &= Y_1^H(A + E)X_1 + O(\|E\|_p^2). \end{aligned}$$

In other words, the generalized Rayleigh quotient on $A + E$ is accurate to terms of the second order in E.

4.6. Hermitian matrices. The hypothesis that A is Hermitian does not much affect the conclusions concerning invariant subspaces of Theorems 4.1 and 4.11. However, because $A_{21} = A_{12}^H$, the constants η and γ must be small together, and the applicability of the theorem may be determined by examining η or γ alone. If $\|\cdot\|_F$ is used, the number δ can be taken to be the distance between $\lambda(A_{11})$ and $\lambda(A_{22})$.

The hypothesis that A is Hermitian does make an essential difference in bounding perturbations of the spectrum of A. In the first part of Theorem 4.12, we have the following bounds:

$$\|A_{12}P\| \leqq 2\frac{\gamma^2}{\delta}$$

and

$$\|P^H P\| \leqq 4\frac{\gamma^2}{\delta^2}.$$

It follows that $\|A'_{11} - A_{11}\|$ is of order γ^2 as γ approaches zero. Moreover since A_{11} and A'_{11} are Hermitian, the eigenvalues $\mu_1, \mu_2, \cdots, \mu_l$ of A_{11} and the eigenvalues $\lambda_1, \lambda_2, \cdots, \lambda_l$ of A' (which are also eigenvalues of A) can be ordered in such a way that

$$|\lambda_i - \mu_i| \leqq \|A'_{11} - A_{11}\|_2, \qquad i = 1, 2, \cdots, l, \tag{4.17}$$

and

$$\sum_{i=1}^{l} (\lambda_i - \mu_i)^2 \leqq \|A'_{11} - A_{11}\|_F^2 \tag{4.18}$$

(e.g., see [34, Chap. 2]).

The use of a direct estimate of $\|A'_{11} - A_{11}\|$ in (4.17) or (4.18) is somewhat unsatisfactory, since, when δ is small, terms in $P^H P$ will dominate those in $A_{12}P$, even though the matrix $P^H P$ appears only as part of a similarity transformation that does not change the eigenvalues of A'. However, by applying an extension [29] of the theorems that lead to (4.17) and (4.18), the dependence on $P^H P$ can be reduced.

THEOREM 4.15. *Let A be Hermitian. In the first part of Theorem* 4.12 *let the eigenvalues $\mu_1, \mu_2, \cdots, \mu_l$ of A_{11} and $\lambda_1, \lambda_2, \cdots, \lambda_l$ of A'_{11} be ordered so that*

$$\mu_1 \leqq \mu_2 \leqq \cdots \leqq \mu_l$$

and

$$\lambda_1 \leqq \lambda_2 \leqq \cdots \leqq \lambda_l.$$

If $\|\cdot\|_2$ *is used, then*

$$|\lambda_i - \mu_i| \leqq (1 + \kappa)\left(1 + \frac{\gamma^2}{\delta^2}\right)\frac{\gamma^2}{\delta} + 3\|A_{11}\|_2\frac{\gamma^4}{\delta^4},$$

where $\kappa < 1$ *is defined as in Theorem* 4.1. *If* $\|\cdot\|_F$ *is used, the same bound holds for* $\sum_{i=1}^{l}(\lambda_i - \mu_i)^2$. *If, in the second part of Theorem* 4.12, *the* μ_i *denote eigenvalues of* $A_{11} + E_{11}$, *then the same bounds hold with*

$$\gamma = \|E_{21}\|$$

and

$$\delta = \operatorname{sep}(A_{11}, A_{22}) - \|E_{11}\| - \|E_{22}\|.$$

4.7. Other theorems on Hermitian matrices. For completeness a brief summary of results of Kahan [12] on the spectrum of Hermitian matrices and results of Davis and Kahan [6] on invariant subspaces of Hermitian matrices will be given here.

In Theorem 4.15 the price one pays for the $O(\gamma^2)$ bound is the dependence of the bound on δ^{-1}. The following theorem, due to Kahan, shows that this dependence can be removed.

THEOREM 4.16. *Let* $A \in \mathbb{C}^{n\times n}$ *and* $B \in \mathbb{C}^{l\times l}$ *be Hermitian. Let* $X_1 \in \mathbb{C}^{n\times l}$ *have orthonormal columns and set*

$$R = AX_1 - X_1B.$$

Then to the eigenvalues $\mu_1, \mu_2, \cdots, \mu_l$ *of* B *there correspond* l *eigenvalues* $\lambda_1, \lambda_2, \cdots, \lambda_l$ *of* A *such that*

$$|\lambda_i - \mu_i| \leqq \|R\|_2, \qquad i = 1, 2, \cdots, l.$$

There are also eigenvalues $\lambda_1, \lambda_2, \cdots, \lambda_l$ *of* A *such that*

$$\sum_{i=1}^{l}(\lambda_i - \mu_i)^2 \leqq \|R\|_F^2.$$

If, in Theorem 4.16, $B = X_1^HAX_1$ then B is the A_{11} of Theorem 4.15, and, by Theorem 4.3, $\|R\|_p = \gamma, (p = 2, F)$. The bounds are independent of the separation of the spectrum of A_{11} from the spectrum of its complementary Rayleigh quotient A_{22}; however, as γ approaches zero the bounds are no longer $O(\gamma^2)$.

The perturbation bounds for invariant subspaces of Davis and Kahan differ from ours in that they do not establish the existence of an invariant subspace of $A + E$ lying near that of A. Rather they assume that invariant subspaces of A and $A + E$ are known and establish conditions under which the canonical angles between them are small. Specifically, in our notation, suppose that $X = (X_1, X_2)$ and $X' = (X'_1, X'_2)$ are unitary matrices such that

$$X^HAX = \begin{pmatrix} A_{11} & 0 \\ 0 & A_{22} \end{pmatrix}$$

and

$$X'^H A' X' = \begin{pmatrix} A'_{11} & 0 \\ 0 & A'_{22} \end{pmatrix}.$$

Let

$$X^H E X = \begin{pmatrix} E_{11} & E_{12} \\ E_{21} & E_{22} \end{pmatrix},$$

and let

$$R = (A + E)X_1 - X_1 A_{11}.$$

Let $\Theta = \Theta[\mathscr{R}(X_1), \mathscr{R}(X'_1)]$ and let $\|\cdot\|$ denote a family of unitarily invariant norms. The four basic theorems of Davis and Kahan bound the norms of $\sin\Theta$, $\sin 2\Theta$, $\tan\Theta$, and $\tan 2\Theta$.

<u>Sin Θ</u>: Let the spectra of A_{11} and A'_{22} lie in intervals thus:

—[——]——[——]——[——]—.
$\lambda(A'_{22})$ $\lambda(A_{11})$ $\lambda(A'_{22})$

Let $\delta = \min|\lambda(A_{11}) - \lambda(A'_{22})|$. Then

$$\|\sin\Theta\| \leqq \|R\|/\delta.$$

<u>Tan Θ</u>: Let the spectra of A_{11} and A'_{22} lie in intervals thus:

—[——]——[——]—.
$\lambda(A'_{22})$ $\lambda(A_{11})$

Let $\delta = \min|\lambda(A_{11}) - \lambda(A'_{22})|$. Then if $E_{22} = 0$,

$$\|\tan\Theta\| \leqq \frac{\|R\|}{\delta}, \frac{\|E\|}{\delta}.$$

<u>Sin 2Θ</u>: Let the spectra of A_{11} and A_{22} lie in intervals thus:

—[——]——[——]——[——]—.
$\lambda(A_{22})$ $\lambda(A_{11})$ $\lambda(A_{22})$

Let $\delta = \min|\lambda(A_{11}) - \lambda(A_{22})|$. Then

$$\|\sin 2\Theta\| \leqq 2\frac{\|R\|}{\delta}, 2\frac{\|E\|}{\delta}.$$

<u>Tan 2Θ</u>: Let the spectra of A_{11} and A_{22} lie in intervals thus:

—[——]——[——]—.
$\lambda(A_{22})$ $\lambda(A_{11})$

Let $\delta = \min|\lambda(A_{11}) - \lambda(A_{22})|$. Then if $E_{11} = 0$ and $E_{22} = 0$,

$$\|\tan 2\Theta\| \leqq 2\frac{\|R\|}{\delta}, 2\frac{\|E\|}{\delta}.$$

The above summary hardly does justice to the paper of Davis and Kahan, which contains much more.

5. The generalized eigenvalue problem. In this section we shall consider the generalized eigenvalue problem of finding the nontrivial solutions of the equation

$$Ax = \lambda Bx, \tag{5.1}$$

where $A, B \in \mathbb{C}^{n \times m}$. A solution $x \neq 0$ of (5.1) is called an eigenvector of the problem $A - \lambda B$ corresponding to the eigenvalue λ. The algebraic theory underlying this problem is well developed (see [8]), and there exist satisfactory algorithms for its numerical solution (see [17] and [20]). Perturbation theory for the problem is less well developed. Crawford [5] has derived bounds for the case where A is Hermitian and B is positive definite. Kato [13, § VII.6] has considered the problem in an infinite-dimensional setting. The results of this section are embellishments of earlier work by the author [28].

The algebraic theory reveals two unusual features of the problem. In the first place, the set of eigenvalues of $A - \lambda B$ (denoted by $\lambda(A, B)$) may be the whole complex plane. This occurs, for example, when A and B have a common null vector, although this is by no means the only way it can happen. We shall call such a problem, and any problem near it, an *ill-disposed* problem. Ill-disposed problems have pathological features, some of which are discussed below, and our bounds will not apply to them. The general nature of ill-disposed problems has not been satisfactorily determined.

The second unusual feature is that when B is singular, the problem $A - \lambda B$ will have fewer than n eigenvalues. It is customary to regard the missing eigenvalues as infinite, for the reciprocal problem $B - \lambda A$ has zero eigenvalues. The presence of infinite or nearly infinite eigenvalues does not in itself represent a pathological situation; the remaining eigenvalues may be quite well conditioned, that is, they may be quite insensitive to perturbations in A and B. However, infinite eigenvalues must be handled with some care. For example, the natural approach of working with the related eigenvalue problem for the matrix $B^{-1}A$ breaks down, both theoretically and computationally, when B is nearly singular. One of the advantages of the approach taken here is that it deals effectively with singular B.

5.1. Deflating pairs of subspaces. The notion of invariant subspace does not carry over to the generalized eigenvalue problem $A - \lambda B$. For even if $Ax = \lambda Bx$, the direction of Ax and Bx may be different from that of x. What is true, however, is that Ax and Bx lie in the same direction. Thus what is required of a subspace $\mathscr{X}$ is that both $A\mathscr{X}$ and $B\mathscr{X}$ should lie in a subspace $\mathscr{Y}$ of the same dimension as $\mathscr{X}$. These considerations motivate the following definition.

DEFINITION 5.1. The subspaces $\mathscr{X}$ and $\mathscr{Y}$ form a *deflating pair* for the generalized eigenvalue problem $A - \lambda B$ if

$$\dim(\mathscr{X}) = \dim(\mathscr{Y})$$

and

$$A\mathscr{X}, B\mathscr{X} \subset \mathscr{Y}.$$

The term deflating pair originates as follows. For definiteness let $\dim(\mathscr{X}) = l$. Let $X = (X_1, X_2)$ and $Y = (Y_1, Y_2)$ be unitary matrices with $\mathscr{R}(X_1) = \mathscr{X}$ and

$\mathscr{R}(Y_1) = \mathscr{Y}$. Since $A\mathscr{X} \subset \mathscr{Y}$, it follows that $\mathscr{R}(AX_1) \subset \mathscr{R}(Y_1)$. Since $Y_2^H Y_1 = 0$, we have $Y_2^H AX_1 = 0$. Similarly, $Y_2^H BX_1 = 0$. Thus if $Y^H AX$ and $Y^H BX$ are partitioned conformally with X and Y in the forms

$$Y^H AX = \begin{pmatrix} A_{11} & A_{12} \\ A_{21} & A_{22} \end{pmatrix} \tag{5.2}$$

and

$$Y^H BX = \begin{pmatrix} B_{11} & B_{12} \\ B_{21} & B_{22} \end{pmatrix}, \tag{5.3}$$

then a necessary condition that $\mathscr{X}$ and $\mathscr{Y}$ form a deflating pair is that

$$A_{21} = B_{21} = 0.$$

This condition is obviously also sufficient.

If $\mathscr{X}$ and $\mathscr{Y}$ do not form a deflating pair, A_{21} and B_{21} will be nonzero. In line with the technique used for approximate invariant subspaces, we shall attempt to rotate X and Y so that their first l columns span a deflating pair. Specifically, let

$$X' = (X'_1, X'_2) = (X_1, X_2)\begin{pmatrix} I & -P^H \\ P & I \end{pmatrix}\begin{pmatrix} (I + P^H P)^{-1/2} & 0 \\ 0 & (I + PP^H)^{-1/2} \end{pmatrix}$$

and

$$Y' = (Y'_1, Y'_2) = (Y_1, Y_2)\begin{pmatrix} I & -Q^H \\ Q & I \end{pmatrix}\begin{pmatrix} (I + Q^H Q)^{1/2} & 0 \\ 0 & (I + QQ^H) \end{pmatrix}.$$

Let $A' = Y'^H AX'$ and $B' = Y'^H BX'$ be partitioned as in (5.2) and (5.3) conformally with X' and Y'. Then for $\mathscr{X}' = \mathscr{R}(X'_1)$ and $\mathscr{Y}' = \mathscr{R}(Y'_1)$ to form a deflating pair, it is necessary and sufficient that

$$A'_{21} = B'_{21} = 0. \tag{5.4}$$

Thus we must determine P and Q so that (5.4) is satisfied. This leads to the pair of equations

$$\begin{aligned} QA_{11} - A_{22}P &= A_{21} - QA_{12}P, \\ QB_{11} - B_{22}P &= B_{21} - QB_{12}P. \end{aligned} \tag{5.5}$$

If we define the linear operator $\mathbf{T}: \mathbb{C}^{(n-l)\times 2l} \to \mathbb{C}^{(n-l)\times 2l}$ by

$$\mathbf{T}(P, Q) = (QA_{11} - A_{22}P, QB_{11} - B_{22}P), \tag{5.6}$$

where $P, Q \in \mathbb{C}^{(n-l)\times l}$, and define the operator φ by

$$\varphi[(P, Q)] = (QA_{12}P, QB_{12}P),$$

then the system (5.5) can be rewritten in the form

$$\mathbf{T}(P, Q) = (A_{21}, B_{21}) - \varphi[(P, Q)].$$

Moreover, if the Frobenius norm is used, as it will be throughout the rest of this section, conditions (i) and (ii) of Theorem 3.1 are satisfied with

$$\eta = \|(A_{12}, B_{12})\|_F.$$

Thus Theorem 3.1 applies directly to give an analogue for deflating pairs of Theorem 4.1. However, reversing the order of presentation in § 4, we shall first describe the properties of the operator $\mathbf{T}$ defined by (5.6).

5.2. The function dif. The invertibility of the operator $\mathbf{T}$ is a necessary condition for the applicability of Theorem 3.1. The following theorem, whose proof is given in [28], is an analogue of Theorem 4.4.

THEOREM 5.2. *The operator* $\mathbf{T}$ *defined by* (5.6) *is invertible if and only if*

$$\lambda(A_{11}, B_{11}) \cap \lambda(A_{22}, B_{22}) = \emptyset.$$

Thus $\mathbf{T}$ is invertible if and only if the generalized eigenvalue problems $A_{11} - \lambda B_{11}$ and $A_{22} - \lambda B_{22}$ have disjoint sets of eigenvalues. In analogy with the characterization (4.4) of sep, we define the function dif by the equation

$$\operatorname{dif}(A_{11}, B_{11}; A_{22}, B_{22}) = \inf_{\|(P,Q)\|_F = 1} \|\mathbf{T}(P, Q)\|_F.$$

If $\mathbf{T}$ is invertible, then

$$\operatorname{dif}(A_{11}, B_{11}; A_{22}, B_{22}) = \|\mathbf{T}^{-1}\|^{-1}.$$

This definition is slightly different from the one given in [28], where

$$\|(P, Q)\|_F' = \max\{\|P\|_F, \|Q\|_F\}$$

is used; however, this difference does not change the results in any essential way.

Unlike Theorem 4.4, Theorem 5.2 does not state what the spectrum of $\mathbf{T}$ is. In particular, $\operatorname{dif}(A_{11}, B_{11}; A_{22}, B_{22})$ is not a lower bound for $|\lambda(A_{11}, B_{11}) - \lambda(A_{22}, B_{22})|$. In fact, for any scalar $\sigma \neq 0$, we have $\lambda(\sigma A_{11}, \sigma B_{11}) = \lambda(A_{11}, B_{11})$. But

$$\operatorname{dif}(\sigma A_{11}, \sigma B_{11}; \sigma A_{22}, \sigma B_{22}) = \sigma \operatorname{dif}(A_{11}, B_{11}; A_{22}, B_{22}),$$

and by varying σ we can make dif as small or as large as we like without changing the eigenvalues of the two problems that are being compared. Even more is true. From evaluating first $\mathbf{T}(P, 0)$ and then $\mathbf{T}(0, Q)$ with $\|P\|_F = \|Q\|_F = 1$, it follows that

$$\operatorname{dif}(A_{11}, B_{11}; A_{22}, B_{22}) \leqq \min\{\|(A_{11}, B_{11})\|_F, \|(A_{22}, B_{22})\|_F\}. \tag{5.7}$$

Thus if, say, A_{11} and B_{11} are both small, the number dif will be small.

The inequality (5.7) is closely related to the notion of ill-disposed problems. For if A_{11} and B_{11} are both small, then the matrices

$$\begin{pmatrix} A_{11} & A_{12} \\ 0 & A_{22} \end{pmatrix}, \begin{pmatrix} B_{11} & B_{12} \\ 0 & B_{22} \end{pmatrix}$$

almost have a common null space. A perturbation that is small compared to A_{22} and B_{22} may completely change the eigenvalues of the problem $A_{11} - \lambda B_{11}$, causing them to approach eigenvalues of the problem $A_{22} - \lambda B_{22}$. This can have disastrous consequences, although they are unlikely to occur for a random perturbation. For an example, see [28].

The function dif does have some nice properties, however. It is insensitive to perturbations, it transforms nicely, and it can be computed when its arguments are diagonal matrices. The proofs of the following theorems are similar to the proofs of the corresponding theorems in § 4. In their statements, it is assumed that the dimensions of all matrices conform.

THEOREM 5.3.

$$\operatorname{dif}(A_1 - E_1, B_1 - F_1; A_2 - E_2, B_2 - F_2) \geqq \operatorname{dif}(A_1, B_1; A_2, B_2) - \|(E_1, F_1)\|_F - \|(E_2, F_2)\|_F.$$

THEOREM 5.4. *Let U_1, V_1, U_2, and V_2 be nonsingular. Then*

$$\operatorname{dif}(V_1^{-1}A_1U_1, V_1^{-1}B_1U_1; V_2^{-1}A_2U_2, V_2^{-1}B_2U_2) \geqq \frac{\operatorname{dif}(A_1, B_1; A_2, B_2)}{\|U_1^{-1}\|_F \max\{\|U_1\|_F \|V_1\|_F\} \|U_2^{-1}\|_F \max\{\|U_2\|_F \|V_2\|_F\}}.$$

THEOREM 5.5. *If $A_1 = \operatorname{diag}(A_1^{(1)}, A_2^{(1)}, \cdots, A_k^{(1)})$ and $B_1 = \operatorname{diag}(B_1^{(1)}, B_2^{(1)}, \cdots, B_k^{(1)})$, then*

$$\operatorname{dif}(A_1, B_1; A_2, B_2) = \min\{\operatorname{dif}(A_i^{(1)}, B_i^{(1)}; A_2, B_2) : i = 1, 2, \cdots, k\},$$

with a similar identity holding for the pair A_2, B_2.

Since many generalized eigenvalue problems can be reduced to diagonal form by equivalences, Theorems 5.4 and 5.5 provide a means of obtaining a lower bound on dif for such problems. In particular, Theorem 5.5 reduces the problem of estimating dif for diagonal matrices to that of calculating dif for a number of one-dimensional problems. This is easily done; for the operator **T** corresponding to $\operatorname{dif}(\alpha_1, \beta_1; \alpha_2, \beta_2)$ has a matrix representation

$$\begin{pmatrix} \alpha_1 & -\alpha_2 \\ \beta_1 & -\beta_2 \end{pmatrix}$$

whose inverse is given by

$$(\alpha_1\beta_2 - \alpha_2\beta_1)^{-1}\begin{pmatrix} \beta_2 & -\alpha_2 \\ \beta_1 & -\alpha_1 \end{pmatrix},$$

from which estimates of $\operatorname{dif}(\alpha_1, \beta_1; \alpha_2, \beta_2)$ may easily be obtained. Note that $\operatorname{dif}(\alpha_1, \beta_1; \alpha_2, 0)$ is finite; i.e., infinite eigenvalues will not necessarily affect the bounds to follow.

5.3. Theorems for deflating pairs. We are now in a position to state the following analogues of Theorems 4.1, 4.11, and 4.12 [28].

THEOREM 5.6. *Let $A, B \in \mathbb{C}^{n \times n}$ and let $X = (X_1, X_2)$ and $Y = (Y_1, Y_2)$ be unitary with $X_1, Y_1 \in \mathbb{C}^{n \times l}$. Let*

$$A_{ij} = Y_i^H A X_j, \quad B_{ij} = Y_i^H B X_j, \quad i, j = 1, 2.$$

Let

$$\gamma = \|(A_{21}, B_{21})\|_F, \qquad \eta = \|(A_{12}, B_{12})\|_F$$

and

$$\delta = \operatorname{dif}(A_{11}, B_{11}; A_{22}, B_{22}).$$

Then if

$$\gamma\eta/\delta^2 < 1/4,$$

there are matrices $P, Q \in \mathbb{C}^{(n-l)\times l}$ *satisfying*

$$\|(P, Q)\|_F \leqq (1 + \kappa)\frac{\gamma}{\delta} < 2\frac{\gamma}{\delta}$$

such that

$$\mathscr{R}(X_1 + X_2P), \mathscr{R}(Y_1 + Y_2Q)$$

form a deflating pair of subspaces for the generalized eigenvalue problem $A - \lambda B$. *Moreover,* $\lambda(A, B)$ *is the disjoint union*

$$\lambda(A, B) = \lambda(A_{11} + A_{12}P, B_{11} + B_{12}P) \cup \lambda(A_{22} + QA_{12}, B_{22} + QB_{12}).$$

THEOREM 5.7. *Let* A, B, E, $F \in \mathbb{C}^{n\times n}$ *and let* $X = (X_1, X_2)$ *and* $Y = (Y_1, Y_2)$ *be unitary matrices such that* $\mathscr{R}(X_1)$ *and* $\mathscr{R}(Y_1)$ *form a deflating pair of dimension* l *for the problem* $A - \lambda B$. *For* $i, j = 1, 2$, *let*

$$A_{ij} = X_i^H A X_j, \qquad B_{ij} = Y_i^H B X_j,$$

$$E_{ij} = Y_i^H E X_j, \qquad F_{ij} = Y_i^H F X_j.$$

Let

$$\gamma = \|(E_{21}, F_{21})\|_F, \qquad \eta = \|(E_{12}, F_{12})\|_F,$$

$$\delta = \operatorname{dif}(A_{11}, B_{11}; A_{22}, B_{22}) - \|(E_{11}, F_{11})\|_F - \|(E_{22}, F_{22})\|_F.$$

Then if

$$\eta\gamma/\delta^2 < 1/4,$$

there are matrices $P, Q \in \mathbb{C}^{(n-l)\times l}$ *satisfying*

$$\|(P, Q)\|_F \leqq (1 + \kappa)\frac{\gamma}{\delta} < \frac{2\gamma}{\delta}$$

such that

$$\mathscr{R}(X_1 + X_2P), \mathscr{R}(Y_1 + Y_2Q)$$

form a deflating pair for the problem $(A + E) - \lambda(B + F)$. *Moreover,* $\lambda(A + E, B + F)$ *is the disjoint union*

$$\lambda(A + E, B + F) = \lambda(A_{11} + E_{11} + E_{12}P, B_{11} + F_{11} + F_{12}P)$$
$$\cup\, \lambda(A_{22} + E_{22} + QE_{12}, B_{22} + F_{22} + QF_{12}).$$

So far as deflating pairs of subspaces are concerned, Theorems 5.6 and 5.7 are straightforward. When $l = 1$, they provide explicit bounds for the accuracy of an approximate eigenvector or for the amount an eigenvector will be changed by perturbations in A and B. The results on $\lambda(A, B)$, however, require further explanation and elaboration.

5.4. Rayleigh components and the spectrum. The most important feature of Theorems 5.6 and 5.7 is that they do not provide explicit bounds for the spectrum. For example, if $l = 1$, then Theorem 5.7 asserts that an eigenvalue λ' of $(A + E) - \lambda(B + F)$ can be written in the form $(\alpha_{11} + \varepsilon_{11})/(\beta_{11} + \varphi_{11})$, where ε_{11} and φ_{11} are small quantities with known bounds. This in no way guarantees that λ' will be near $\lambda = \alpha_{11}/\beta_{11}$. Indeed, if $\beta_{11} + \varphi_{11} = 0$, then λ' will be infinite. It is characteristic of our theorems that they express bounds for the spectrum in terms of a subproblem whose components A_{11} and B_{11} are treated separately. We shall call the matrices A_{11} and B_{11} Rayleigh components of the problem $A - \lambda B$.

DEFINITION 5.8. Let X_1, $Y_1 \in \mathbb{C}^{n \times l}$ be of full rank. For any $A, B \in \mathbb{C}^{n \times n}$ the matrices $A_{11} = Y_1^H A X_1$ and $B_{11} = Y_1^H B X_1$ are called *Rayleigh components* of the generalized eigenvalue problem $A - \lambda B$.

It follows from the discussion at the beginning of this section that if $\mathscr{R}(X_1)$ and $\mathscr{R}(Y_1)$ form a deflating pair of subspaces for $A - \lambda B$, then the Rayleigh components satisfy

$$\lambda(A_{11}, B_{11}) \subset \lambda(A, B).$$

In particular, if $l = 1$, α_{11}/β_{11} is an eigenvalue of $A - \lambda B$.

As with the ordinary eigenvalue problem, the perturbation bounds of Theorem 5.7 depend on δ^{-1}. In a development parallel to that of § 4, we can make this dependence a second order effect by introducing the notion of a left deflating pair of subspaces. A pair of subspaces $\mathscr{V}$ and $\mathscr{U}$ will be called a left deflating pair for the problem $A - \lambda B$ if they are a deflating pair for the problem $A^H - \lambda B^H$. The following theorem shows how a left deflating pair corresponding to a given right deflating pair may be constructed.

THEOREM 5.9. *Let $X = (X_1, X_2)$ and $Y = (Y_1, Y_2)$ be unitary matrices such that $\mathscr{R}(X_1)$ and $\mathscr{R}(Y_1)$ form a deflating pair of dimension l for $A - \lambda B$. Let $A_{ij} = Y_i^H A X_j$ and $B_{ij} = Y_i^H B X_j (i, j = 1, 2)$. Then if $\operatorname{dif}(A_{11}, B_{11}; A_{22}, B_{22}) \neq 0$, there are unique matrices V_1 and U_2 such that (X_1, U_2) and (V_1, Y_2) are nonsingular and*

$$(V_1, Y_2)^H A(X_1, U_2) = \begin{pmatrix} A_{11} & 0 \\ 0 & A_{22} \end{pmatrix},$$

$$(V_1, Y_2)^H B(X_1, U_2) = \begin{pmatrix} B_{11} & 0 \\ 0 & B_{22} \end{pmatrix}.$$

Moreover,

$$\|V_1\|_F = \|\sec \Theta[\mathscr{R}(V_1), \mathscr{R}(Y_1)]\|_F. \tag{5.8}$$

Proof. The proof parallels that of Theorem 4.13. We seek (X_1, U_2) and (V_1, Y_2) in the forms

$$(X_1, U_2) = (X_1, X_2)\begin{pmatrix} I & -P \\ 0 & I \end{pmatrix}$$

and

$$(V_1, Y_2) = (Y_1, Y_2)\begin{pmatrix} I & 0 \\ Q^H & I \end{pmatrix}.$$

This leads to the equations

$$A_{11}P - QA_{22} = A_{12},$$

$$B_{11}P - QB_{22} = A_{12},$$

which have a unique solution, since $\operatorname{dif}(A_{11}, B_{11}; A_{22}, B_{22}) \neq 0$. The equality (5.8) follows from the equation $V_1 = Y_1 + Y_2Q^H$. ∎

In Theorem 5.9, the spaces $\mathscr{R}(V_1)$ and the orthogonal complement of $\mathscr{R}(U_2)$ form a left deflating pair for $A - \lambda B$. We shall call this deflating pair the left deflating pair corresponding to the right deflating pair $\mathscr{R}(X_1), \mathscr{R}(Y_1)$. The theorem further shows that it is possible to choose the columns of V_1 so that the Rayleigh components $V_1^H A X_1$ and $V_1^H B X_1$ are the same as the natural Rayleigh components A_{11} and B_{11} associated with the deflating pair $\mathscr{R}(X_1), \mathscr{R}(Y_1)$. With this particular "scaling" of the left deflating pair, one can prove an analogue of Theorem 4.14.

THEOREM 5.10. *In addition to the hypotheses of Theorem 5.9, let E and F be given. Suppose that*

$$\delta \equiv \operatorname{dif}(A_{11}, B_{11}; A_{22}, B_{22}) - (\|V_1\|_2 + \|U_2\|_2)(\|E\|_2 + \|F\|_2) > 0.$$

Further, suppose that

$$\frac{\sqrt{\|V_1\|_F\|U_2\|_F}\|(E, F)\|_F}{\delta} < \frac{1}{4}.$$

Then there exists a deflating pair for $(A + E) - \lambda(B + F)$ and corresponding Rayleigh components A'_{11} and B'_{11} such that

$$\|A'_{11} - A_{11}\|_F \leqq \|V_1\|_F\|E\|_F\left[1 + \frac{2\|U_1\|_F\|E\|_F}{\delta}\right] \tag{5.9}$$

and

$$\|B'_{11} - B_{11}\|_F \leqq \|V_1\|_F\|F\|_F\left[1 + \frac{2\|U_1\|_F\|F\|_F}{\delta}\right]. \tag{5.10}$$

The proof of Theorem 5.10 is perfectly analogous to the proof of Theorem 4.14. It shows, in addition, that

$$\|A'_{11} - V_1^H(A + E)X_1\|_F = O(\|E\|_F^2)$$

and

$$\|B'_{11} - V_1^H(B + F)X_1\|_F = O(\|F\|_F^2).$$

In other words, the Rayleigh components $V_1^H(A + E)X_1$ and $V_1^H(B + E)X_1$ are accurate to terms of order $\|E\|_F^2$.

When second order terms are ignored in (5.9) and (5.10), the relation (5.8) implies that the secants of the canonical angles between $\mathscr{R}(V_1)$ and $\mathscr{R}(Y_1)$ determine the sensitivity of the Rayleigh components to perturbations in A and B. For $l = 1$ and $\lambda = \alpha_{11}/\beta_{11}$, a simple eigenvalue with right and left eigenvectors x_1 and v_1, the condition number for α_{11} and β_{11} is the secant of the angle between v_1 and Ax_1 (or Bx_1 if $Ax_1 = 0$). However, it should be remembered that one must use the "natural" Rayleigh components associated with the deflating pair of which $\mathscr{R}(x_1)$ is a part; i.e., $y_1^H Ax_1$ and $y_1^H Bx_1$. It should also be stressed that the secant is a condition number for α_{11} and β_{11}, not for λ. If β_{11} is small, λ may vary greatly with small perturbations in A and B.

5.5. The case of Hermitian A and positive definite B. When A is Hermitian and B is positive definite, it is well known that the problem $A - \lambda B$ has a set of B-orthogonal eigenvectors; that is, there are linearly independent vectors $x_1, x_2, \cdots, x_n$ such that if $X = (x_1, x_2, \cdots, x_n)$ then

$$X^H AX = \operatorname{diag}(\alpha_1, \alpha_2, \cdots, \alpha_n) \tag{5.11}$$

and

$$X^H BX = \operatorname{diag}(\beta_1, \beta_2, \cdots, \beta_n), \qquad \beta_i > 0. \tag{5.12}$$

If a set of eigenvectors is known, Theorem 5.5 allows us to obtain lower bounds for the values of dif that appear in the second order terms of Theorem 5.10. Specifically, in [28], it is shown that if for $i = 1, 2, \cdots, n$ the vector x_i is scaled so that $\max\{|\alpha_i|, |\beta_i|\} = 1$, then

$$\operatorname{dif}^{-1}(\alpha_1, \beta_1; \beta_i, \beta_i) \leqq \sqrt{2}(1 + |\lambda_1|)\frac{1 + |\lambda_1|}{|\lambda_1 - \lambda_i|}.$$

If Theorem 5.7 is applied to the problem $X^H AX - \lambda X^H BX$ with perturbations $X^H EX$ and $X^H FH$, there results the following theorem.

THEOREM 5.11. *Let $A, B \in \mathbb{C}^{n \times n}$ with A Hermitian and B positive definite. Let the columns of $X = (x_1, x_2, \cdots, x_n)$ form a complete set of eigenvectors for $A - \lambda B$ so that* (5.11) *and* (5.12) *are satisfied with*

$$\max\{|\alpha_i|, |\beta_i|\} = 1, \qquad i = 1, 2, \cdots, n.$$

Let

$$\rho_i = \|x_i\|_2, \qquad \sigma_i = \max\{\rho_j : j \neq i\},$$

and let

$$\delta_i^{-1} = \sqrt{2}(1 + |\lambda_i|)\max\left\{\frac{1 + |\lambda_j|}{|\lambda_i - \lambda_j|} : j \neq i\right\}.$$

For any $E, F \in \mathbb{C}^{n \times n}$ let $\varepsilon = \|(E, F)\|_F$. Then if

$$2\sqrt{n-1}\rho_i\sigma_i\varepsilon < \delta_i/2,$$

there is an eigenvector x_i' of the problem $(A + E) - \lambda(B + F)$ with Rayleigh components α_i' and β_i' such that

$$\tan \angle(x_i, x_i') < 2\sqrt{n-1}\rho_i\sigma_i\varepsilon$$

and

$$|\alpha_i' - \alpha_i|, |\beta_i' - \beta_i| < \rho_i^2\varepsilon\left[1 + \frac{4(n-1)\sigma_i^2\varepsilon}{\delta_i}\right].$$

In the proof of Theorem 5.11, the upper bounds

$$\eta, \gamma < \sqrt{n-1}\rho_i\sigma_i\varepsilon$$

are used. The numbers $\rho_i^2\varepsilon$ and $(n-1)\sigma_i^2\varepsilon$ bound the diagonal part of the perturbations. It should be noted that $B + F$ need not be positive definite; indeed, E and F need not be Hermitian. For some examples of the application of this theorem, especially in connection with an ill-disposed problem, see [28].

6. The singular value decomposition. In this section we shall apply the theory of § 3 to the singular value decomposition. The results for the singular value decomposition are somewhat simpler than the corresponding results for the eigenvalue and generalized eigenvalue problems. This is because the number $\|\mathbf{T}^{-1}\|$ can be written simply in terms of the singular values of the matrices defining $\mathbf{T}$. As in § 5, we shall work exclusively with the Frobenius norm $\|\cdot\|_F$ and its subordinate operator norm. For definiteness we shall work with a fixed matrix $A \in \mathbb{C}^{m\times n}$ with $m \geqq n$.

The first step is to find an appropriate notion of a subspace associated with the singular value decomposition. As with the generalized eigenvalue problem, we shall actually be concerned with a pair of subspaces.

DEFINITION 6.1. Let $A \in \mathbb{C}^{m\times n}$ and let $\mathscr{X} \subset \mathbb{C}^n$ and $\mathscr{Y} \subset \mathbb{C}^m$ be subspaces of dimension l. Then $\mathscr{X}$ and $\mathscr{Y}$ form a pair of *singular subspaces for* A if

(i) $A\mathscr{X} \subset \mathscr{Y}$,
(ii) $A^H\mathscr{Y} \subset \mathscr{X}$.

If in Definition 6.1 we let $X = (X_1, X_2)$ and $Y = (Y_1, Y_2)$ be unitary matrices such that $\mathscr{R}(X_1) = \mathscr{X}$ and $\mathscr{R}(Y_1) = \mathscr{Y}$ and set

$$Y^HAX = \begin{pmatrix} A_{11} & A_{12} \\ A_{21} & A_{22} \end{pmatrix},$$

then

$$A_{21} = 0, \qquad A_{12} = 0. \tag{6.1}$$

Conversely, if (6.1) is satisfied, then $\mathscr{R}(X_1)$ and $\mathscr{R}(Y_1)$ form a singular pair of subspaces for A.

Now suppose that A_{21} and A_{12} are not zero. We seek new unitary matrices X' and Y' in the forms

$$X' = (X_1', X_2') = (X_1, X_2)\begin{pmatrix} I & -P^H \\ P & I \end{pmatrix}\begin{pmatrix} (I + P^HP)^{-1/2} & 0 \\ 0 & (I + PP^H)^{-1/2} \end{pmatrix}$$

and

$$Y' = (Y_1', Y_2') = (Y_1, Y_2)\begin{pmatrix} I & Q^H \\ -Q & I \end{pmatrix}\begin{pmatrix} (I + Q^HQ)^{-1/2} & 0 \\ 0 & (I + QQ^H)^{-1/2} \end{pmatrix}$$

such that

$$A'_{21} \equiv Y'^H_2 A X'_1 = 0$$

and

$$A'_{12} \equiv Y'^H_1 A X'_2 = 0.$$

This leads to the equations

$$\begin{aligned} QA_{11} - A_{22}P &= A_{21} - QA_{12}P, \\ PA^H_{11} - A^H_{22}Q &= A^H_{12} - PA^H_{21}Q. \end{aligned} \tag{6.2}$$

If we define the operator $\mathbf{T}: \mathbb{C}^{(m-l)\times l} \oplus \mathbb{C}^{(n-l)\times l} \to \mathbb{C}^{(m-l)\times l} \oplus \mathbb{C}^{(n-l)\times l}$ by

$$\mathbf{T}(Q, P) = (QA_{11} - A_{22}P, PA^H_{11} - A^H_{22}Q)$$

and the function $\boldsymbol{\varphi}$ by

$$\boldsymbol{\varphi}[(Q, P)] = (QA_{12}P, PA^H_{21}Q),$$

then the system (6.2) can be written in the form

$$\mathbf{T}(Q, P) = (A_{21}, A^H_{12}) - \boldsymbol{\varphi}[(Q, P)].$$

Moreover, if we equip the space $\mathbb{C}^{(m-l)\times l} \oplus \mathbb{C}^{(n-l)\times l}$ with the product Frobenius norm defined by

$$\|(Q, P)\|^2_F = \|Q\|^2_F + \|P\|^2_F,$$

then $\boldsymbol{\varphi}$ satisfies the condition of Theorem 3.1 with

$$\eta = \|(A_{21}, A^H_{12})\|_F.$$

The application of Theorem 3.1 requires that $\|\mathbf{T}^{-1}\|^{-1}$ be known. For the singular value problem, we have the following simple expression.

THEOREM 6.2.

$$\|\mathbf{T}^{-1}\|^{-1} = \min |\operatorname{sing}(A_{11}) - \operatorname{sing}(A_{22})|,$$

where, if $m > n$, A_{22} is understood to have $m - n$ zero singular values.

Proof. For simplicity, we assume that $m = n$, so that A_{11} and A_{22} are both square. The modifications for the general case are straightforward.

Let U_1, V_1, U_2, and V_2 be unitary. Then

$$\|(V^H_2 Q V_1, U^H_2 P U_1)\|_F = \|(Q, P)\|_F$$

and

$$\begin{aligned} \|\mathbf{T}(Q, P)\|_F = \|&((V^H_2 Q V_1)(V^H_1 A_{11} U_1) - (V^H_2 A_{22} U_2)(U^H_2 P U_1), \\ &(U^H_2 P U_1)(V^H_1 A_{11} U_1)^H - (V^H_2 A_{22} U_2)^H (V^H_2 Q V_1))\|_F. \end{aligned}$$

It follows that we may replace A_{11} by $V^H_1 A_{11} U_1$ and A_{22} by $V^H_2 A_{22} U_2$ in the definition of T. In particular, we may take

$$A_{11} = \operatorname{diag}(\sigma_1, \sigma_2, \cdots, \sigma_l)$$

and

$$A_{22} = \operatorname{diag}(\tau_1, \tau_2, \cdots, \tau_{n-l}),$$

where $\sigma_1, \sigma_2, \cdots, \sigma_l$ and $\tau_1, \tau_2, \cdots, \tau_{n-l}$ are the singular values of A_{11} and A_{22}.

Now if $\mathbf{T}(Q, P) = (S, R)$, then the (i, j)-elements of P, Q, R, and S satisfy

$$q_{ij}\sigma_j - \tau_i P_{ij} = s_{ij},$$
$$p_{ij}\sigma_j - \tau_i q_{ij} = r_{ij}.$$

It follows that

$$\|\mathbf{T}^{-1}\| = \max_{i,j} \left\| \begin{pmatrix} \sigma_j - \tau_i \\ \tau_i - \sigma_j \end{pmatrix}^{-1} \right\|_2$$
$$= \max_{i,j} |\sigma_j - \tau_i|^{-1}. \quad \blacksquare$$

We may now state the theorem for approximate pairs of singular subspaces of a matrix.

THEOREM 6.3. *Let $A \in \mathbb{C}^{m \times n}$, assume $m \geqq n$, and let $X = (X_1, X_2) \in \mathbb{C}^{n \times n}$ and $Y = (Y_1, Y_2) \in \mathbb{C}^{m \times n}$ be unitary with X_1 and X_2 having l columns. Let $Y^H AX$ be partitioned conformally with X and Y in the form*

$$Y^H AX = \begin{pmatrix} A_{11} & A_{12} \\ A_{21} & A_{22} \end{pmatrix}.$$

Let

$$\eta = \gamma = \|(A_{21}, A_{12}^H)\|_F$$

and

$$\delta = \min |\operatorname{sing}(A_{11}) - \operatorname{sing}(A_{22})|.$$

Then if

$$\gamma/\delta < 1/2,$$

there are matrices $P \in \mathbb{C}^{(n-l) \times l}$ and $Q \in \mathbb{C}^{(m-l) \times l}$ satisfying

$$\|(Q, P)\|_F \leqq (1 + \kappa)\frac{\gamma}{\delta} < \frac{2\gamma}{\delta}$$

such that $\mathscr{R}(X_1 + X_2 P)$ and $\mathscr{R}(Y_1 - Y_2 Q)$ form a pair of singular subspaces for A. Moreover, sing (A) *is the disjoint union of*

$$\operatorname{sing}[(I + Q^H Q)^{1/2}(A_{11} + A_{12}P)(I + P^H P)^{-1/2}]$$
$$= \operatorname{sing}[(I + Q^H Q)^{-1/2}(A_{11} + Q^H A_{21})(I + P^H P)^{1/2}]$$

and

$$\operatorname{sing}[(I + QQ^H)^{1/2}(A_{22} - A_{21}P^H)(I + PP^H)^{-1/2}]$$
$$= \operatorname{sing}[(I + QQ^H)^{-1/2}(A_{22} - QA_{12})(I + PP^H)^{1/2}].$$

In order to convert Theorem 6.3 into a perturbation theorem, we must know how the number δ is effected when A_{11} and A_{22} are perturbed; or equivalently how the singular values of A_{11} and A_{22} change under perturbations. It is well known [9] that if A_{11} has singular values $\sigma_1, \sigma_2, \cdots, \sigma_l$ then the singular values of $\sigma'_1, \sigma'_2, \cdots, \sigma'_l$ of $A_{11} + E_{11}$ can be ordered so that

$$|\sigma'_i - \sigma_i| \leqq \|E_{11}\|_2, \qquad i = 1, 2, \cdots, l.$$

With this result we can prove the following theorem.

THEOREM 6.4. *In Theorem* 6.3, *let* $\mathscr{R}(X_1)$ *and* $\mathscr{R}(Y_1)$ *form a pair of singular subspaces for* A. *Let* $E \in \mathbb{C}^{m \times n}$ *be given and partition* $Y^H E X$ *conformally with* X *and* Y *in the form*

$$Y^H E X = \begin{pmatrix} E_{11} & E_{12} \\ E_{21} & E_{22} \end{pmatrix}.$$

Let

$$\eta = \gamma = \|(A_{21}, A_{12}^H)\|_F + \|(E_{21}, E_{12}^H)\|_F$$

and let

$$\delta = \min |\operatorname{sing}(A_{11}) - \operatorname{sing}(A_{22})| - \|E_{11}\|_2 - \|E_{22}\|_2.$$

If

$$\gamma/\delta < 1/2,$$

then there are matrices $P \in \mathbb{C}^{(n-l) \times l}$ *and* $Q \in \mathbb{C}^{(m-l) \times l}$ *satisfying*

$$\|(Q, P)\|_F \leqq (1 + \kappa)\frac{\gamma}{\delta} < 2\frac{\gamma}{\delta}$$

such that $\mathscr{R}(X_1 + X_2 P)$ *and* $\mathscr{R}(Y_1 + Y_2 Q)$ *form a pair of singular subspaces for* $A + E$. *Moreover, the set* sing $(A + E)$ *is the disjoint union of*

$$\operatorname{sing}[(I + Q^H Q)^{1/2}(A_{11} + E_{11} + E_{12}P)(I + P^H P)^{-1/2}]$$
$$= \operatorname{sing}[(I + Q^H Q)^{-1/2}(A_{11} + E_{11} + Q^H E_{21})(I + P^H P)^{1/2}]$$

and

$$\operatorname{sing}[(I + QQ^H)^{1/2}(A_{22} + E_{22} - E_{21}P^H)(I + PP^H)^{-1/2}$$
$$= \operatorname{sing}[(I + QQ^H)^{-1/2}(A_{22} + E_{22} - QE_{12})(I + PP^H)^{1/2}].$$

In an analogy with the results for the Hermitian eigenvalue problem, we find that up to terms in $\|E\|^2$ the singular values of $A + E$ are those of $A_{11} + E_{11}$ and $A_{22} + E_{22}$. However, the error depends now on δ^{-2} rather than δ^{-1}. Whether an analogue of Theorem 4.15 holds is an open question.

REFERENCES

[1] R. H. BARTELS AND G. W. STEWART, *Algorithm 432, solution of the matrix equation* $AX + XB = C$, Comm. ACM, 15 (1972), pp. 820–826.

[2] Å. BJORK AND G. H. GOLUB, *Numerical methods for computing angles between linear subspaces*. Department of Applied Mathematics, Linköplug Institute of Technology, 1971, to appear in Math. Comp.

[3] M. M. BLEVINS AND G. W. STEWART, *Calculating eigenvectors of diagonally dominant matrices*, Report CNA-47, Center for Numerical Analysis, University of Texas, Austin, 1972, in Comm. ACM.

[4] M. CLINT AND A. JENNINGS, *The evaluation of eigenvalues and eigenvectors of real symmetric matrices by simultaneous iteration*, Comput. J., 13 (1970), pp. 76–80.

[5] C. R. Crawford, *The numerical solution of the generalized eigenvalue problem*, Ph.D. dissertation, The University of Michigan, Ann Arbor, 1970.

[6] C. Davis and W. Kahan, *The rotation of eigenvectors by a perturbation III*, SIAM J. Numer. Anal., 7 (1970), pp. 1–46.

[7] Fred W. Dorr, *The direct solution of the discrete Poisson equation on a rectangle*, this Review, 12 (1970), pp. 248–263.

[8] F. R. Gantmacher, *The Theory of Matrices I, II*, Chelsea, New York, 1959.

[9] I. C. Gohberg and M. G. Krein, *Introduction to the Theory of Linear Non-selfadjoint Operators*, American Mathematical Society, Providence, 1969.

[10] W. H. Grueb, *Multilinear Algebra*, Springer, New York, 1967.

[11] A. S. Householder, *The Theory of Matrices in Numerical Analysis*, Blaisdell, New York, 1964.

[12] W. Kahan, *Inclusion theorems for clusters of eigenvalues of Hermitian matrices*, Department of Computer Sciences, University of Toronto, 1967.

[13] T. Kato, *Perturbation Theory for Linear Operators*, Springer, New York, 1966.

[14] V. N. Kublanovskaya, *On a method of solving the complete eigenvalue problem for a degenerate matrix*, U.S.S.R. Comput. Math. and Math. Phys., 6 (1968), pp. 1–14.

[15] Peter Lancaster, *Lambda-matrices and Vibration Systems*, Pergamon Press, Oxford, 1966.

[16] G. Lumer and M. Rosenblum, *Linear operator equations*, Proc. Amer. Math. Soc., 10 (1969), pp. 32–41.

[17] C. B. Moler and G. W. Stewart, *An algorithm for the generalized matrix eigenvalue problem* $Ax = \lambda Bx$, SIAM J. Numer. Anal., 10 (1973), pp. 241–256.

[18] J. M. Ortega and W. C. Rheinboldt, *Iterative Solution of Nonlinear Equations in Several Variables*, Academic Press, New York, 1970.

[19] B. N. Parlett and W. G. Poole, Jr., *A geometric theory for the QR, LU and power iterations*, SIAM J. Numer. Anal., 10 (1973), pp. 389–412.

[20] G. Peters and J. H. Wilkinson, $Ax = \lambda Bx$ *and the generalized eigenproblem*, Ibid., 7 (1970), pp. 479–492.

[21] M. Rosenblum, *On the operator equation* $BX - XA = Q$, Duke Math. J., 23 (1956), pp. 263–269.

[22] A. Ruhe, *An algorithm for numerical determination of the structure of a general matrix*, BIT, 10 (1970), pp. 196–216.

[23] ———, *Perturbation bounds for means of eigenvalues and invariant subspaces*, Ibid., 10 (1970), pp. 343–354.

[24] H. Rutishauser, *Computational aspects of F. L. Bauer's simultaneous iteration method*, Numer. Math., 13 (1969), pp. 4–13.

[25] ———, *Simultaneous iteration method for symmetric matrices*, Ibid., 16 (1970), pp. 205–223.

[26] G. W. Stewart, *Accelerating the orthogonal iteration for the eigenvectors of a Hermitian matrix*, Ibid., 13 (1969), pp. 362–376.

[27] ———, *Error bounds for invariant subspaces of closed operators*, SIAM J. Numer. Anal., 8 (1971), pp. 796–808.

[28] ———, *On the sensitivity of the eigenvalue problem* $Ax = \lambda Bx$, Ibid., 9 (1972), pp. 669–686.

[29] ———, *A note on non-Hermitian perturbations of Hermitian operators*, Rep. CNA-41, Center for Numerical Analysis, The University of Texas, Austin, 1972.

[30] C. A. Swanson, *An inequality for linear transformations with eigenvalues*, Bull. Amer. Math. Soc., 67 (1961), pp. 607–608.

[31] J. S. Vandergraft, *Generalized Rayleigh methods with applications to finding eigenvalues of large matrices*, Linear Algebra and Appl., 4 (1971), pp. 353–368.

[32] James M. Varah, *The computation of bounds for the invariant subspaces of a general matrix operator*, Tech. Rep. CS66, Computer Science Department, Stanford University, Stanford, Calif., 1967.

[33] ———, *Computing invariant subspaces of a general matrix when the eigensystem is poorly conditioned*, Math. Comp., 24 (1970), pp. 137–149.

[34] J. H. Wilkinson, *The Algebraic Eigenvalue Problem*, Clarendon Press, Oxford, 1965.

[35] Per-Åke Wedin, *Perturbation bounds in connection with singular value decomposition*, BIT, 12 (1972), pp. 99–111.

[36] Peter Lancaster, *Explicit solutions of linear matrix equations*, this Review, 12 (1970), pp. 544–566.

15.3. [GWS-J48] "Computable Error Bounds for Aggregated Markov Chains"

[GWS-J48] "Computable Error Bounds for Aggregated Markov Chains," *Journal of the ACM* **30** (1983) 271–285.
http://doi.acm.org/10.1145/322374.322377

Computable Error Bounds for Aggregated Markov Chains

G. W. STEWART

The University of Maryland, College Park, Maryland

Abstract A method is described for computing the steady-state probability vector of a nearly completely decomposable Markov chain The method is closely related to one proposed by Simon and Ando and developed by Courtois However, the method described here does not require the determination of a completely decomposable stochastic approximation to the transition matrix, and hence it is applicable to nonstochastic matrices An error analysis of the procedure which results in effectively computable error bounds is given

Categories and Subject Descriptors D.4 8 [**Operating Systems**] Performance—*queuing theory*; *stochastic analysis*, G 1 3 [**Numerical Analysis**]: Numerical Linear Algebra—*eigenvalues*

General Terms Algorithms

Additional Key Words and Phrases: Markov chains, decomposability, stationary probability, eigenvalues

1. *Introduction*

This paper is concerned with techniques for treating a discrete, finite Markov chain whose matrix of transition probabilities can, after a suitable renumbering of the states, be written in the form

$$A = \begin{bmatrix} A_{11} & E_{12} & \cdots & E_{1l} \\ E_{21} & A_{22} & \cdots & E_{2l} \\ \vdots & \vdots & & \vdots \\ E_{l1} & E_{l2} & \cdots & A_{ll} \end{bmatrix}, \qquad (1.1)$$

where the matrices E_{ij} are small. The matrix A is nonnegative and stochastic; that is,

$$A\mathbf{1} = \mathbf{1},$$

so that the vector $\mathbf{1}$ consisting of all 1's is a right eigenvector of A corresponding to the eigenvalue 1. If, in addition, A is irreducible,[1] the eigenvalue 1 is simple and there is a unique, normalized, positive left eigenvector y corresponding to the eigenvalue 1 (in the irreducible case we call the eigenvalue the Perron root). If A is acyclic and y is normalized so that $\mathbf{1}^{\mathrm{T}}y = 1$, then y is the vector of steady-state probabilities for the chain. One of the chief computational problems associated with Markov chains is the determination of the vector y.

[1] For terminology see [14]

This work was supported in part by the Office of Naval Research under Contract No. N00014 76 C 0391

Author's address: Department of Computer Science and Institute for Physical Science and Technology, The University of Maryland, College Park, MD 20742

Journal of the Association for Computing Machinery, Vol 30, No 2, April 1983, pp 271–285

Chains with transition matrices of the form (1.1) are said to be nearly completely decomposable. They arise naturally as models of systems whose states can be clustered into aggregates that are loosely connected to one another. They were first studied by Simon and Ando [7], who had applications to economic systems in mind. A recent monograph by Courtois [3] contains a history of the subject and extensive applications in the computer sciences.

The usual computational procedure goes as follows. The off-diagonal blocks E_{ij} are amalgamated into the diagonal blocks A_{ii} to produce a block diagonal approximation A^* to A that has the form

$$A^* = \operatorname{diag}(A_{11}^*, A_{22}^*, \ldots, A_{ll}^*).$$

This decomposition is done in such a way that each block A_{ii}^* is stochastic and irreducible. The steady-state vectors y_i of the A_{ii}^* are then computed and the steady-state vector of the original system approximated in the form

$$y \cong \begin{bmatrix} \nu_1 y_1^* \\ \nu_2 y_2^* \\ \vdots \\ \nu_l y_l^* \end{bmatrix}. \tag{1.2}$$

The quantities ν_i are calculated as the components of an eigenvector of a matrix of order l whose elements may be easily calculated from the vectors y_i^* and the original matrix A. The computational advantages of this method are obvious, since it reduces the solution of a large eigenvalue problem to that of several potentially much smaller ones.

The purpose of this paper is to resolve two difficulties with the method as it is currently practiced. The first concerns the determination of the approximating decomposed matrix A^*, a process sometimes referred to by the unfortunate term "aggregation."[2] There are infinitely many ways to incorporate the off-diagonal blocks of A into the diagonal blocks in order to get an approximation A^*. In some instances this flexibility may be useful. For example [13, 15], in certain highly structured systems it is possible to determine the diagonal blocks A_{ii}^* so that the eigenvectors y_i^* are exactly proportional to the corresponding pieces of y (cf. (1.2)). In general, however, the indeterminacy of A^* is a nuisance; some choices of A^* may be better than others, but without further information there is no way of knowing. In particular, the derivation of any general error bound for the approximation (1.2) must necessarily entail the assumption that the worst choice has been made. In this paper a new method is proposed that does not require intermediate approximations but works directly with the original matrix A.

The second problem treated here is that of showing how reasonably sharp error bounds may be computed. Courtois [2, 3] has given an error analysis for the procedure sketched above, which shows in part how it behaves as the off-diagonal blocks become small. However, the analysis is not suitable for computing error bounds for two reasons. First, the analysis is based on asymptotic approximations, and it is not shown how small the off-diagonal blocks must be in norm for the approximations to be accurate. Second, the analysis assumes that all the matrices involved have complete systems of eigenvectors. Although it is unlikely that any given problem will fail to have this property, it is not at all unlikely that it will be near a problem that does, in which case an analysis based on eigenvector expansions

[2] This usage is disparaged because it conflicts with the original use of the term in economics to refer to the clustering together of states.

will give unrealistic results owing to the ill condition of the matrix of normalized eigenvectors.

The techniques developed in this paper are not restricted to stochastic matrices; rather, they can be applied to find the dominant eigenvalue of almost any matrix of the form (1.1). What is required is that the dominant eigenvalues of the A_{ii} be simple and have sufficiently well-conditioned eigenvectors and that the E_{ij} be sufficiently small. If A is stochastic, these conditions are likely to be satisfied; but as will be seen, the computational techniques test the conditions directly, without reference to the properties of A. In particular, if one of the A_{ii} has the form (1.1), its dominant eigenvectors can be found independently of A by the method described in this paper. This observation has important consequences for the process of multilevel aggregation described by Courtois [3].

This paper is organized as follows. Sections 2 and 3 lay the theoretical foundations for the techniques to follow; Section 2 describes the deflation of a simple eigenvalue, and Section 3 reviews perturbation theory for invariant subspaces. In Section 4 the technique is sketched broadly, and in Section 5 it is justified in detail by the derivation of effectively computable error bounds. In Section 6 the practical techniques from numerical analysis required to implement the method are discussed. The paper concludes with a numerical example.

Many of the results of this paper are cast in terms of vector and matrix norms. The symbol $\|\cdot\|$ denotes either the Euclidean vector norm defined by

$$\|x\|^2 = x^{\mathrm{T}}x$$

or the spectral matrix norm defined by

$$\|A\| = \max_{\|x\|=1} \|Ax\|.$$

The symbol $\|\cdot\|_{\mathrm{F}}$ denotes the Frobenius matrix norm defined by

$$\|A\|_{\mathrm{F}}^2 = \sum_{i,j} a_{ij}^2.$$

Note that for any vector x,

$$\|x\| = \|x\|_{\mathrm{F}}.$$

For more on these norms see [9].

It is important not to expect too much of error bounds cast in terms of norms. In the first place, repeated use of inequalities such as the triangle inequality tends to make them pessimistic. In the second place, such a bound can be difficult to interpret in terms of the components of the vector thus bounded. For example, if $\|e\| \leq \epsilon$, then any component of e can be as large as ϵ. But other things being equal, it is more likely that each component is of order $\epsilon/\sqrt{n}$.

In cases where an error bound is unsatisfactory, it may be necessary to calculate an error estimate, in which an attempt is made to approximate the error vector itself. For many problems this can be done, although frequently a heavy computational price will be paid, as it must be for the problem treated here. Moreover, once an error estimate has been calculated, it is hard to resist the temptation to use it to improve the putative solution, which will set off another round of error bounding. The error estimates in [9, p. 187] are an example of how compelling this temptation can be.

2. *The Constructive Theory of a Simple Eigenvalue*

In this section are collected a number of results about a simple eigenvalue and its eigenvectors which can be found in one form or another scattered throughout the literature. The results follow from a constructive reduction of A to block diagonal form by means of rather simple similarity transformations.

Let A be a matrix of order n with a real simple eigenvalue β corresponding to the eigenvector x. Since x is nonzero, it may be normalized so that $\|x\| = 1$. Let the columns of $n \times (n-1)$ matrix Y form an orthonormal basis for the subspace orthogonal to x; that is,

$$Y^{\mathrm{T}}Y = I \tag{2.1}$$

and

$$Y^{\mathrm{T}}x = 0. \tag{2.2}$$

This implies that the matrix $(x\ \ Y)$ is orthogonal.

Consider the similarity transformation

$$\begin{aligned}(x\ \ Y)^{\mathrm{T}}A(x\ \ Y) &= \begin{bmatrix} x^{\mathrm{T}}Ax & x^{\mathrm{T}}AY \\ Y^{\mathrm{T}}Ax & Y^{\mathrm{T}}AY \end{bmatrix} \\ &= \begin{bmatrix} \beta x^{\mathrm{T}}x & x^{\mathrm{T}}AY \\ \beta Y^{\mathrm{T}}x & Y^{\mathrm{T}}AY \end{bmatrix}. \end{aligned} \tag{2.3}$$

It follows from (2.2) and the fact that $x^{\mathrm{T}}x = 1$ that

$$(x\ \ Y)^{\mathrm{T}}A(x\ \ Y) = \begin{bmatrix} \beta & g^{\mathrm{T}} \\ 0 & C \end{bmatrix},$$

where

$$g^{\mathrm{T}} = x^{\mathrm{T}}AY, \qquad C = Y^{\mathrm{T}}AY.$$

The matrix has for its eigenvalues the eigenvalues of A other than β, hence $C - \beta I$ is nonsingular.

Consider now the further similarity transformation,

$$\begin{bmatrix} 1 & -q^{\mathrm{T}} \\ 0 & I \end{bmatrix}\begin{bmatrix} \beta & g^{\mathrm{T}} \\ 0 & C \end{bmatrix}\begin{bmatrix} 1 & q^{\mathrm{T}} \\ 0 & I \end{bmatrix} = \begin{bmatrix} \beta & \beta q^{\mathrm{T}} + g^{\mathrm{T}} - q^{\mathrm{T}}C \\ 0 & C \end{bmatrix}. \tag{2.4}$$

Since $C - \beta I$ is nonsingular, q may be chosen to satisfy

$$(C - \beta I)^{\mathrm{T}}q = g,$$

from which it follows that the row vector in the upper right of (2.4) is zero. Thus the two similarity transformations (2.3) and (2.4) reduce A to the block diagonal form $\mathrm{diag}(\beta, C)$.

The composite similarity transformation that reduces A can be found by multiplying the two transformations (2.3) and (2.4). Specifically, set

$$(x\ \ X) = (x\ \ Y)\begin{bmatrix} 1 & q^{\mathrm{T}} \\ 0 & I \end{bmatrix} = (x\ \ Y + xq^{\mathrm{T}})$$

and

$$(y\ \ Y) = (x\ \ Y)\begin{bmatrix} 1 & 0 \\ -q & I \end{bmatrix} = (x - Yq\ \ Y). \tag{2.5}$$

Then

$$(x\ \ X)^{-1} = \begin{bmatrix} y^{\mathrm{T}} \\ Y^{\mathrm{T}} \end{bmatrix} \tag{2.6}$$

and

$$\begin{bmatrix} y^{\mathrm{T}} \\ Y^{\mathrm{T}} \end{bmatrix} A(x \quad X) = \begin{bmatrix} y^{\mathrm{T}}Ax & 0 \\ 0 & Y^{\mathrm{T}}AX \end{bmatrix} = \begin{bmatrix} \beta & 0 \\ 0 & C \end{bmatrix}. \tag{2.7}$$

A number of important facts can be read from this reduction. In the first place, y is the left eigenvector of A corresponding to β. Since, from (2.6), $y^{\mathrm{T}}x = 1$, it follows that y is not orthogonal to x. Since, from (2.5),

$$y = x - Yq,$$

an alternate expression for q follows from (2.1) and (2.2):

$$q = -Y^{\mathrm{T}}y.$$

Moreover,

$$\|y\|^2 = 1 + \|q\|^2.$$

Similarly,

$$\|X\|^2 = 1 + \|q\|^2.$$

All these results may be summarized in the following theorem.

THEOREM 2.1. *Let β be a simple eigenvalue of a matrix A of order n, and let the corresponding eigenvector x be normalized so that*

$$\|x\| = 1. \tag{a}$$

Let y be the left eigenvector corresponding to β. Then y is not orthogonal to x and may be normalized so that

$$y^T x = 1. \tag{b}$$

Moreover there are $n \times (n-1)$ matrices X and Y such that

$$Y^T Y = Y^T X = I, \tag{c}$$

$$y^T X = x^T Y = 0, \tag{d}$$

$$(x \quad X)^{-1} = (y \quad Y)^T, \tag{e}$$

$$(y \quad Y)^T A(x \quad X) = \begin{bmatrix} \beta & 0 \\ 0 & C \end{bmatrix}, \tag{f}$$

where

$$C = Y^T AX = Y^T AY. \tag{g}$$

The eigenvalues of the matrix C are the eigenvalues of A other than β. If q is defined by either of the expressions

$$q = (C - \beta I)^{-T} Y^T A^T x, \tag{h}$$

$$q = -Y^T y, \tag{i}$$

then

$$X = Y + xq^T, \tag{j}$$

$$y = x - Yq, \tag{k}$$

$$\|X\|^2 = \|y\|^2 = 1 + \|q\|^2. \tag{l}$$

3. *Perturbation Theory*

In this section the following problem is addressed: Given a matrix A partitioned in the form

$$A = \begin{bmatrix} B & G \\ H & C \end{bmatrix},$$

find a matrix U as near as possible to the identity such that the transformed matrix $\bar{A} = U^{-1}AU$ has the form

$$\bar{A} = \begin{bmatrix} \bar{B} & 0 \\ \bar{H} & \bar{C} \end{bmatrix}.$$

The importance of this problem lies in the following observation. If v is a left eigenvector of $\bar{B}$, then $(v^{\mathrm{T}}\ 0)$ will be a left eigenvector of $\bar{A}$, and $(v^{\mathrm{T}}\ 0)U$ will be a left eigenvector of A. Since

$$\frac{\|(v^{\mathrm{T}}\quad 0)U - (v^{\mathrm{T}}\quad 0)\|}{\|(v^{\mathrm{T}}\quad 0)\|} \leq \|I - U\|,$$

the vector $(v^{\mathrm{T}}\ 0)$ will be a good approximate left eigenvector of A in proportion as U is near the identity matrix.

This problem has been treated in [8], and the following is a summary of the results required in this paper. The reader is referred to the reference for proofs.

The problem will have a solution only if the eigenvalues of B and C are separated and G is sufficiently small. Unfortunately, the minimum of the distances between the eigenvalues of B and C is too crude a measure of separation to give satisfactory bounds. Instead the measure

$$\delta(B, C) = \inf_{\|P\|_F = 1} \|BP - PC\|_F$$

will be used. The properties of $\delta(B, C)$ are summarized in the following theorem.

THEOREM 3.1. *The number $\delta(B, C)$ is zero if and only if B and C have an eigenvalue in common. Moreover,*

$$\delta(B, C) \leq \min\{|\beta - \gamma| : \beta \text{ an eigenvalue of } B,\ \gamma \text{ an eigenvalue of } C\}, \qquad (a)$$

$$\delta(B + E, C + F) \geq \delta(B, C) - \|E\| - \|F\|, \qquad (b)$$

$$\delta[\mathrm{diag}(B_1, \ldots, B_p), \mathrm{diag}(C_1, \ldots, C_q)] = \min\{\delta(B_i, C_j) : i = 1, \ldots, p;\ j = 1, \ldots, q\}, \qquad (c)$$

$$\delta(\beta, C) = \|(\beta I - C)^{-1}\|^{-1}. \qquad (d)$$

Expressions (b)–(d) in the theorem are particularly important in computational practice. Expression (b) says that small changes in B and C make equally small changes in $\delta(B, C)$, a property not shared by the minimum distance between the eigenvalues of B and C. Expression (c) shows how δ for a block diagonal matrix can be found from the δ's between the blocks. Finally, expression (d) gives an explicit expression for δ when one of the matrices is a scalar. These properties will be used extensively in the derivation of the error bounds in Section 5.

The solution to the problem posed at the beginning of this section requires that U be chosen in a specific form. Specifically, U will be written

$$U = \begin{bmatrix} I & -P \\ P^{\mathrm{T}} & I \end{bmatrix} \begin{bmatrix} (I + PP^{\mathrm{T}})^{-1/2} & 0 \\ 0 & (I + P^{\mathrm{T}}P)^{-1/2} \end{bmatrix}. \tag{3.1}$$

Here $(I + PP^{\mathrm{T}})^{-1/2}$ is the inverse of the unique positive definite square root of the positive definite matrix $I + PP^{\mathrm{T}}$—similarly for $(I + P^{\mathrm{T}}P)^{-1/2}$. It is easily verified that U is orthogonal; that is, $U^{\mathrm{T}}U = I$ or $U^{-1} = U^{\mathrm{T}}$. Thus the problem becomes that of determining P so that

$$U^{\mathrm{T}} \begin{bmatrix} B & G \\ H & C \end{bmatrix} U = \begin{bmatrix} \bar{B} & 0 \\ \bar{H} & \bar{C} \end{bmatrix}. \tag{3.2}$$

Conditions under which this can be done are contained in the following theorem.

THEOREM 3.2. *In the notation introduced above, let*

$$\gamma \geq \|G\|_F, \qquad \eta \geq \|H\|_F, \qquad \delta \leq \delta(B, C). \tag{3.3}$$

If

$$\tau \equiv \frac{\eta\gamma}{\delta^2} < \frac{1}{4}, \tag{3.4}$$

then there exists a unique matrix P satisfying

$$\|P\|_F \leq \frac{\gamma}{\delta} \frac{1 + \sqrt{1 - 4\tau}}{1 - 2\tau + \sqrt{1 - 4\tau}} < 2\frac{\gamma}{\delta} \tag{3.5}$$

such that U defined by (3.1) *satisfies* (3.2). *Moreover,*

$$\bar{B} = (I + PP^T)^{-1/2}(B + PH)(I + PP^T)^{1/2}. \tag{3.6}$$

When P is small, the distance of U from the identity matrix is roughly $\sqrt{2}\|P\|_F$. The bound (3.5) shows that this distance depends linearly on $\|G\|_F$ and inversely on $\delta(B, C)$. In other words, the bound becomes smaller as G becomes small and larger as the spectra of B and C approach one another; however, the bounds can be considerably worse than a naive inspection of the spectra would indicate (cf. Theorem 3.1(a)).

The expression for $\bar{B}$ is particularly interesting. When $\|H\| = O(\|G\|)$, as it will in Section 5, both PP^{T} and PH are $O(\|G\|^2)$, and as G approaches zero, the eigenvalues of B approximate those at the original matrix up to terms of $O(\|G\|^2)$. This quadratic behavior will prove critical in deriving workable error bounds.

4. *The Approximation Algorithm*

In this section an algorithm for approximating the dominant eigenvector of a matrix of the form (1.1) is described. It is assumed that the diagonal blocks A_{ii} are all irreducible. In order to keep the exposition simple, the algorithm is described for a 3×3 partitioning, that is, $l = 3$. The general case is a trivial extension.

For each i, let $\beta_i > 0$ be the Perron root of A_{ii}, and let $x_i > 0$ be its corresponding right eigenvector. Since β_i is simple, A_{ii} has a decomposition of the form described in Theorem 2.1, namely,

$$(y_i \quad Y_i)^{\mathrm{T}} A_{ii} (x_i \quad X_i) = \begin{bmatrix} \beta_i & 0 \\ 0 & C_i \end{bmatrix}.$$

It then follows from Theorem 2.1 that the inverse of the matrix

$$X = \begin{bmatrix} x_1 & 0 & 0 & X_1 & 0 & 0 \\ 0 & x_2 & 0 & 0 & X_2 & 0 \\ 0 & 0 & x_3 & 0 & 0 & X_3 \end{bmatrix}$$

is

$$Y^{\mathrm{T}} = \begin{bmatrix} y_1^{\mathrm{T}} & 0 & 0 \\ 0 & y_2^{\mathrm{T}} & 0 \\ 0 & 0 & y_3^{\mathrm{T}} \\ Y_1^{\mathrm{T}} & 0 & 0 \\ 0 & Y_2^{\mathrm{T}} & 0 \\ 0 & 0 & Y_3^{\mathrm{T}} \end{bmatrix}$$

Consider the matrix

$$Y^{\mathrm{T}}AX = \begin{bmatrix} \beta_1 & \phi_{12} & \phi_{13} & 0 & g_{12}^{\mathrm{T}} & g_{13}^{\mathrm{T}} \\ \phi_{21} & \beta_2 & \phi_{23} & g_{21}^{\mathrm{T}} & 0 & g_{23}^{\mathrm{T}} \\ \phi_{31} & \phi_{32} & \beta_3 & g_{31}^{\mathrm{T}} & g_{32}^{\mathrm{T}} & 0 \\ 0 & h_{12} & h_{13} & C_1 & F_{12} & F_{13} \\ h_{21} & 0 & h_{23} & F_{21} & C_2 & F_{23} \\ h_{31} & h_{32} & 0 & F_{31} & F_{32} & C_3 \end{bmatrix} = \begin{bmatrix} B & G \\ H & C \end{bmatrix}, \tag{4.1}$$

where

$$\phi_{ij} = y_i^{\mathrm{T}} E_{ij} x_j, \qquad g_{ij}^{\mathrm{T}} = y_i^{\mathrm{T}} E_{ij} X_j,$$
$$h_{ij} = Y_i^{\mathrm{T}} E_{ij} x_j, \qquad F_{ij} = Y_i^{\mathrm{T}} E_{ij} X_j.$$

Because $y_i > 0$ and $x_j > 0$, it follows that $\phi_{ij} = 0$ if and only if $E_{ij} = 0$. Hence, B is irreducible, since A is. Let ρ be the Perron root of B with right eigenvector u and left eigenvector

$$\nu^{\mathrm{T}} = (\nu_1, \nu_2, \nu_3).$$

The approximation to the left eigenvector y of A is then given by

$$\tilde{y} = \begin{bmatrix} \nu_1 y_1 \\ \nu_2 y_2 \\ \nu_3 y_3 \end{bmatrix}. \tag{4.2}$$

This algorithm is extremely simple. All that it requires is the calculation of the left and right Perone vectors of the diagonal blocks of A, the formation of the matrix B from the E_{ij}, and the calculation of the left Perone vector of B. Except for the initial grouping of states to get the partition (1.1), the process is entirely deterministic, requiring no assimilation of the matrices E_{ij} into the diagonal blocks A_{ii}.

5. *Error Bounds*

In this section error bounds for the approximation (4.2) are derived. The bounds provide a formal proof of convergence of the algorithm, as well as considerable insight into its behavior. The practical computation of the bounds is discussed in the next section.

The approach is to use Theorem 3.2 to obtain an exact expression for y in terms of a vector $\bar{v}$ that is the left eigenvector of a matrix $\bar{B}$ lying very near to B. A second application of Theorem 3.2 bounds $\|v - \bar{v}\|$ and hence the error in the approximation (4.2) to y.

For the first step, the notation of Theorem 3.2 coincides exactly with the notation of the approximation algorithm. Consequently, if (3.4) is satisfied, there is a matrix U of the form (3.1) that reduces $Y^{\mathrm{T}}AX$ to the form (3.2). Now the eigenvalues of $\bar{B}$ are eigenvalues of A. Let $\bar{v}$ be the left eigenvector of $\bar{B}$ corresponding to the Perron root of A. Then

$$y^{\mathrm{T}} = (\bar{v}^{\mathrm{T}} \quad 0)U^{\mathrm{T}}Y^{\mathrm{T}} = \bar{v}^{\mathrm{T}}(I + PP^{\mathrm{T}})^{-1/2}(I \quad P)Y^{\mathrm{T}}. \tag{5.1}$$

The relation between $\bar{v}$ and v must now be considered. Let

$$\|P\|_{\mathrm{F}} \leq \pi \tag{5.2}$$

be any bound, presumably obtained by an application of Theorem 3.2. As in the first part of the development of Section 2, extend v to an orthogonal matrix $(v \ \ V)$ such that

$$(v \quad V)^{\mathrm{T}}B(v \quad V) = \begin{bmatrix} \rho & 0 \\ r & R \end{bmatrix}, \tag{5.3}$$

and let

$$\delta_\rho \leq \delta(\rho, R).$$

Let

$$(v \quad V)^{\mathrm{T}}(B + PH)(v \quad V) = \begin{bmatrix} \tilde{\rho} & s \\ \tilde{r} & \tilde{R} \end{bmatrix}.$$

By Theorem 3.1,

$$\delta(\tilde{\rho}, \tilde{R}) \geq \delta_\rho - 2\pi\eta.$$

Moreover,

$$\|\tilde{r}\| \leq \|r\| + \pi\eta, \tag{5.4}$$

and

$$\|s\| \leq \pi\eta. \tag{5.5}$$

Hence, by Theorem 3.2, if

$$\frac{\pi\eta(\|r\| + \pi\eta)}{(\delta_\rho - 2\pi\eta)^2} < \frac{1}{4},$$

there is an eigenvector

$$\tilde{v} = v + e$$

of $B + PH$ satisfying

$$\|v - \tilde{v}\| = \|e\| \leq \frac{2\pi\eta}{\delta_\rho - 2\pi\eta}. \tag{5.6}$$

From (3.6) we have

$$\bar{v} = \tilde{v}(I + PP^{\mathrm{T}})^{1/2} = (v + e)(I + PP^{\mathrm{T}})^{1/2}.$$

Hence, from (5.1),

$$\begin{aligned} y^{\mathrm{T}} &= (v+e)^{\mathrm{T}}(I \quad P)Y^{\mathrm{T}} \\ &= v^{\mathrm{T}} \begin{bmatrix} y_1^{\mathrm{T}} & 0 & 0 \\ 0 & y_2^{\mathrm{T}} & 0 \\ 0 & 0 & y_3^{\mathrm{T}} \end{bmatrix} + v^{\mathrm{T}}P \begin{bmatrix} Y_1^{\mathrm{T}} & 0 & 0 \\ 0 & Y_2^{\mathrm{T}} & 0 \\ 0 & 0 & Y_3^{\mathrm{T}} \end{bmatrix} \\ &\quad + e^{\mathrm{T}} \begin{bmatrix} y_1^{\mathrm{T}} & 0 & 0 \\ 0 & y_2^{\mathrm{T}} & 0 \\ 0 & 0 & y_3^{\mathrm{T}} \end{bmatrix} + e^{\mathrm{T}}P \begin{bmatrix} Y_1^{\mathrm{T}} & 0 & 0 \\ 0 & Y_2^{\mathrm{T}} & 0 \\ 0 & 0 & Y_3^{\mathrm{T}} \end{bmatrix}. \end{aligned}$$

But the first term in this sum is the approximation (4.2) to y. Hence, since $\|Y_i\| = 1$, the final bound becomes

$$\left\| y - \begin{bmatrix} \nu_1 y_1 \\ \nu_2 y_2 \\ \nu_3 y_3 \end{bmatrix} \right\| \le \pi + \frac{2\pi\eta}{\delta_\rho - 2\pi\eta} \max\{\|y_i\|\} + \frac{2\pi^2\eta}{\delta_\rho - 2\pi\eta}. \tag{5.7}$$

It is important to note that in deriving the bound (5.7) it has been implicitly assumed that the Perron root of A was to be found in $\bar{B}$ and that this root corresponded to the Perron root of B. This cannot be ensured a priori without making further assumptions. Essentially, what is required is that the eigenvalues of the C_{ii} be sufficiently removed from the β_i so that subsequent perturbations cannot turn one of them into a Perron root. Although a formal analysis is possible, it will not be given here, since in practice it is easy to see whether the largest eigenvalue of $\bar{B}$ is the Perron root.

When A is stochastic, the $O(\|E\|^2)$ perturbation in passing from B to $B + PH$ is critical to the analysis. This is because all the β_i are within $O(\|E\|)$ of unity, so that a perturbation of $O(\|E\|)$ will completely scramble the eigenvectors.

For computational purposes we have scaled the x_i and y_i so that $\|x_i\| = 1$ and

$$y_i^{\mathrm{T}} x_i = 1. \tag{5.8}$$

In fact, the approximation algorithm will yield the same results for any scaling, provided only that (5.8) is satisfied. To see this, suppose that x_i is replaced by $\hat{x}_i = \delta_i x_i$, where δ_i is a nonzero scaling factor. Then (5.8) requires that y_i be replaced by $\hat{y}_i = \delta_i^{-1} y_i$. It is easily seen that this results in B being replaced by $D^{-1}BD$, where $D = \operatorname{diag}(\delta_1, \delta_2, \delta_3)$. Consequently, the left eigenvector of $D^{-1}BD$ is $\nu^{\mathrm{T}}D$, and the approximation to the left eigenvector of the entire system is $(\delta_1\nu_1\hat{y}_1^{\mathrm{T}}\ \delta_2\nu_2\hat{y}_2^{\mathrm{T}}\ \delta_3\nu_3\hat{y}_3^{\mathrm{T}}) = (\nu_1 y_1^{\mathrm{T}}\ \nu_2 y_2^{\mathrm{T}}\ \nu_3 y_3^{\mathrm{T}})$, which is the same as (4.2).

For stochastic matrices there is a natural scaling of the y_i that leads to a beautiful interpretation of the approximation process. Specifically, let y_i be scaled so that

$$\mathbf{1}^{\mathrm{T}} y_i = 1, \tag{5.9}$$

that is, so that y_i can be interpreted as a vector of probabilities. Write

$$x_i = \mathbf{1} + p_i.$$

If x_i is given the scaling implied by (5.8), then

$$y_i^{\mathrm{T}} p_i = y_i^{\mathrm{T}} x_i - y_i^{\mathrm{T}} \mathbf{1} = 1 - 1 = 0,$$

and

$$y_i^{\mathrm{T}} A_{ii} p_i = \beta_i y_i^{\mathrm{T}} p_i = 0. \tag{5.10}$$

Moreover, since A_{ii} is within $O(\|E\|)$ of a stochastic matrix, it follows from Theorems 3.1 and 3.2 that if $\delta(\beta_i, C_i)$ is large enough the vector p_i will satisfy

$$p_i = O(\|E\|). \tag{5.11}$$

It will now be shown that up to terms of $O(\|E\|^2)$, the matrix B in (4.1) is stochastic. Consider the first row sum,

$$\begin{aligned}\beta_1 + \phi_{12} + \phi_{13} &= y_1^{\mathrm{T}} A_{11} x_1 + y_1^{\mathrm{T}} E_{12} x_2 + y_1^{\mathrm{T}} E_{13} x_3 \\ &= y_1^{\mathrm{T}}(A_{11}\mathbf{1} + E_{12}\mathbf{1} + E_{13}\mathbf{1}) + y_1^{\mathrm{T}} A_{11} p_1 + y_1^{\mathrm{T}}(E_{12}p_2 + E_{13}p_3) \\ &= y_1^{\mathrm{T}}(A_{11}\mathbf{1} + E_{12}\mathbf{1} + E_{13}\mathbf{1}) + O(\|E\|^2),\end{aligned}$$

by (5.10) and (5.11). Because A is stochastic, $A_{11}\mathbf{1} + E_{12}\mathbf{1} + E_{13}\mathbf{1} = \mathbf{1}$. Hence,

$$\beta_1 + \phi_{12} + \phi_{13} = y_1^{\mathrm{T}}\mathbf{1} + O(\|E\|^2) = 1 + O(\|E\|^2),$$

by (5.9). Thus the first row sum of B is within $O(\|E\|^2)$ of 1, and so are the other row sums.

The nearly stochastic matrix B—or rather, $\bar{B}$ which differs from B by $O(\|E\|^2)$—controls the long-term behavior of the Markov chain. To see this, note that by an application of Theorem 2.1 the matrix (4.1) can be reduced to the form $\operatorname{diag}(\bar{B}, \bar{C})$. Now the behavior of the Markov chain is controlled by the behavior of the powers A^k ($k = 1, 2, 3, \ldots$), and this behavior can be determined by examining the behavior of the powers of $\operatorname{diag}(\bar{B}, \bar{C})$.

Specifically, $\operatorname{diag}(\bar{B}, \bar{C})^k = \operatorname{diag}(\bar{B}^k, \bar{C}^k)$. Since the eigenvalues of $\bar{C}$ are less than those of $\bar{B}$, the powers will approach $\operatorname{diag}(\bar{B}^k, 0)$. Since $\bar{B}$ has a dominant eigenvalue of 1, $\bar{B}^k$ will tend more slowly to the matrix $\bar{v}\bar{w}^{\mathrm{T}}$, where $\bar{v}$ and $\bar{w}$ are the left and right eigenvectors corresponding to 1. In terms of the original Markov chain, if the state vector $y^{(k)}$ is written in the form

$$y^{(k)} = \begin{bmatrix} \nu_1^{(k)} y_1^{(k)} \\ \nu_2^{(k)} y_2^{(k)} \\ \nu_3^{(k)} y_3^{(k)} \end{bmatrix},$$

where $\mathbf{1}^{\mathrm{T}} y_i^{(k)} = 1$, then the $y_i^{(k)}$ will converge swiftly as $\bar{C}^k \to 0$ and the $\nu_i^{(k)}$ will converge more leisurely as $\bar{B}^k$ approaches its limit. This justifies calling B the long-term transition matrix of the chain.

Of course, this double-limit behavior of nearly completely decomposable chains has been remarked by numerous researchers, beginning with Simon and Ando [7]; the approach taken here merely makes explicit the factors that control the rates of convergence. Although the matter will not be pursued in this paper, it should be possible to obtain numerical convergence rates from an analysis of the behavior of the powers of the matrices B and C_i.

6. *Practical Details*

The details of the implementation of the approximation algorithm and the computation of the bounds will depend on the sizes of the matrices involved. Three classes of matrices may be distinguished.

(1) Matrices that can be stored as an array in the high-speed memory of a computer. Typically, an upper bound for the order of such matrices ranges from 50 to 500, depending on the computer.

(2) Matrices that cannot be stored in array form but whose structure permits the efficient solution of a system of linear equations with the matrix elements as coefficients. Examples of such matrices are band matrices and "sparse" matrices [4]. Their orders can be very large.
(3) Matrices that are so large that the only thing one can do with them is to form their product with a vector.

Each of these classes is discussed in turn.

If the matrices A_{ii} lie in the first class, the appropriate procedure is to use the QR algorithm to reduce A_{ii} to quasitriangular form. Specifically, there exists software [11] for computing an orthogonal matrix $(x_i \ \ Y_i)$, such that

$$(x_i \quad Y_i)^{\mathrm{T}} A_{ii} (x_i \quad Y_i) = \begin{bmatrix} \beta_i & g_i^{\mathrm{T}} \\ 0 & C_i \end{bmatrix},$$

where C_i is quasitriangular, that is, block upper triangular with 1×1 and 2×2 blocks on its diagonal. Because C_i is quasitriangular, it is extremely cheap to solve linear systems involving $C_i - \beta_i I$, which means that it is practical to compute q in Theorem 2.1(h), from which y_i can be calculated.

The next step is to compute B. Since this requires a pass over the matrices E_{ij}, this is also a logical time to compute the bounds η and γ in (3.3). In very large problems it is unlikely that there will be storage to contain the matrices X_i and Y_i, so that H and G cannot be computed explicitly. However, bounds may be obtained by the following procedure. For each E_{ij}, compute $E_{ij}^{\mathrm{T}} x_j$ and $y_i^{\mathrm{T}} E_{ij}$. Then set

$$\eta^2 = \sum_{i,j} \| Y_i^{\mathrm{T}} \|^2 \| E_{ij} x_j \|^2 = \sum_{i,j} \| E_{ij} x_j \|^2, \tag{6.1}$$

$$\gamma^2 = \sum_{i,j} \| y_i^{\mathrm{T}} E_{ij} \|^2 \| X_i \|^2 = \sum_{i,j} \| y_i^{\mathrm{T}} E_{ij} \|^2 \| y_i \|^2. \tag{6.2}$$

A bound ϕ_c on the off-diagonal elements of C will be needed later. It can be calculated at this point in the form

$$\phi_c = \sum_{i,j} \| Y_i^{\mathrm{T}} \| \| E_{ij} \| \| X_j \| = \sum_{i,j} \| E_{ij} \| \| X_j \|. \tag{6.3}$$

The application of the perturbation theorem requires a lower bound δ on $\delta(B, C)$. Because C_i is quasitriangular, $\|(\beta - C)^{-1}\|$ can be cheaply estimated by a variant of the inverse power method [1, 6]. Set $\beta^+ = \max\{\beta_i\}$ and $\beta^- = \min\{\beta_i\}$. Then set

$$\delta = \min \|(\beta^+ I - C_i)^{-1}\|^{-1} - (\beta^+ - \beta^-) - \phi_b - \phi_c, \tag{6.4}$$

where ϕ_b is any upper bound on the norm of the off-diagonal part of B. It can easily be shown by repeated appeals to Theorem 3.1 that δ is indeed a lower bound for $\delta(B, C)$.

Having computed η, γ, and δ, the bound π in (5.2) can be computed from (3.5). The matrix B may now be analyzed in the same manner as the A_{ii} (cf. (5.3)) to get a bound on $\delta(\rho, R)$. At this point the bounds (5.4) and (5.5) may also be computed. Finally, the bound (5.7) for the accuracy of the approximation may be computed.

For systems of large sparse matrices, it is possible to compute right and left eigenvectors x_i and y_i by means of the inverse power method [9]. However, it will not generally be possible to maintain the matrices X_i, Y_i, and C_i in the high-speed memory of the computer in question. Fortunately, the matrix $(x_i, \ Y_i)$ can be written as a Householder transformation in the form

$$(x_i \quad Y_i) = I - 2 w_i w_i^{\mathrm{T}},$$

$$
\begin{pmatrix}
0.85000 & 0.00000 & 0.14900 & 0.00090 & 0.00000 & 0.00005 & 0.00000 & 0.00005 \\
0.10000 & 0.65000 & 0.24900 & 0.00000 & 0.00090 & 0.00005 & 0.00000 & 0.00005 \\
0.10000 & 0.80000 & 0.09960 & 0.00030 & 0.00000 & 0.00000 & 0.00010 & 0.00000 \\
\hline
0.00000 & 0.00040 & 0.00000 & 0.70000 & 0.29950 & 0.00000 & 0.00010 & 0.00000 \\
0.00050 & 0.00000 & 0.00040 & 0.39900 & 0.60000 & 0.00010 & 0.00000 & 0.00000 \\
\hline
0.00000 & 0.00050 & 0.00000 & 0.00000 & 0.00005 & 0.60000 & 0.24990 & 0.15000 \\
0.00003 & 0.00000 & 0.00003 & 0.00004 & 0.00000 & 0.10000 & 0.80000 & 0.09990 \\
0.00000 & 0.00005 & 0.00000 & 0.00000 & 0.00005 & 0.19990 & 0.25000 & 0.55000
\end{pmatrix}
$$

FIG. 1 The matrix A and its partitioning [2, 3].

where w_i is a vector of norm unity that can be determined from x_i alone (for details see [9]). If $x_i^{\mathrm{T}}y_i = 1$, then it follows from Theorem 2.1 that $\|X_i\| = \|y_i\|$. Hence η, γ, and ϕ_c may be estimated as in (6.1)–(6.3). Finally, since from Theorem 2.1(g) it follows that $C_i = Y_i^{\mathrm{T}}A_{ii}Y_i$, the separation $\delta(B, C)$ may be bounded as described above, but this time using the technique of implicit deflation [12] to solve systems involving $Y_i^{\mathrm{T}}AY_i - \beta I$. The rest of the bounding process proceeds as above.

For matrices of the third class, that is, matrices for which only matrix–vector multiplication is possible, the eigenvectors x_i and y_i must be computed by the power method or its variants [5, 10]. The numbers η and γ may be computed as described above. Unfortunately, there is no way of computing a lower bound on $\delta(B, C)$, since there is no way of solving linear systems involving $C_i = Y_i^{\mathrm{T}}AY_i$. The best that can be done is to compute an approximate upper bound that may be a reasonable estimate. From Theorem 2.1(h) we have

$$\|q\| \le \|(C_i - \beta_i I)^{-\mathrm{T}}\| \|Y_i^{\mathrm{T}}A_{ii}^{\mathrm{T}}x_i\|,$$

and from Theorem 2.1(l),

$$\|(C_i - \beta_i I)^{-1}\|^{-1} \le \frac{\|Y_i^{\mathrm{T}}A_{ii}^{\mathrm{T}}x_i\|}{(\|y_i\|^2 - 1)^{1/2}}.$$

Hence, in analogy with (6.4), it may be hoped that the number

$$\delta = \min \frac{\|Y_i^{\mathrm{T}}A_{ii}^{\mathrm{T}}x_i\|}{(\|y_i\|^2 - 1)^{1/2}} - (\beta^+ - \beta^-) - \phi_b - \phi_c$$

will not differ too much from $\delta(B, C)$.

7. *A Numerical Example*

In this section the techniques of this paper are applied to a matrix analyzed previously by Courtois. The matrix with its partitioning is given in Figure 1.[3]

Figure 2 gives the details of the calculation of the approximation $\tilde{y}$ in (4.2). The vector is compared with the true vector y, which has been scaled so that $\|y - \tilde{y}\|$ is a minimum. Figure 3 gives the details of the computation of the error bound (5.7).

The error bound is good enough for practical purposes, even though it is an order of magnitude bigger than the actual error. This overestimate is in part due to the repeated use of norm inequalities in the derivation of the bound. It is also due to the fact that Theorem 3.2 bounds $\|P\|_F$, whereas it is clear that the smaller spectral norm could be used in the derivation of (5.7). As was pointed out in the introduction, a

[3] The matrix, as reported in [2, 3], is not stochastic; the sixth row does not sum to 1. Since the property of being stochastic is not necessary to the procedure described in this paper, the matrix is left as reported.

β_i	x_i	y_i
0.9991	0.5772	0.6954
	0.5773	0.7218
	0.5776	0.3150
0.9993	0.7074	0.8080
	0.7069	0.6061
0.9993	0.5774	0.4170
	0.5774	0.9624
	0.5774	0.3527

$$B = \begin{pmatrix} 0.9991 & 0.0010 & 0.0001 \\ 0.0005 & 0.9993 & 0.0001 \\ 0.0002 & 0.0001 & 0.9999 \end{pmatrix}$$

$$v^T = (0.4433 \quad 0.6130 \quad 0.6539)$$

$$\tilde{y} = \begin{pmatrix} 0.308281 \\ 0.320022 \\ 0.139653 \\ 0.495361 \\ 0.371565 \\ 0.272669 \\ 0.629330 \\ 0.230652 \end{pmatrix} \qquad y = \begin{pmatrix} 0.307971 \\ 0.320321 \\ 0.139726 \\ 0.495323 \\ 0.371616 \\ 0.272734 \\ 0.629282 \\ 0.230659 \end{pmatrix}$$

$$\frac{\|\tilde{y} - y\|}{4.5 \cdot 10^{-4}}$$

FIG. 2. Computation of $\tilde{y}$ (4.2).

Norms of off-diagonal matrices:
$\eta = \|H\|_F = 4.54 \cdot 10^{-4}$,
$\gamma = \|G\|_F = 2.14 \cdot 10^{-4}$,
$\phi_b = 9.76 \cdot 10^{-4}$,
$\phi_c = 9.29 \cdot 10^{-4}$.

Computation of δ (6.4) and π (3.5):
$\beta_+ = 0.9999$, $\beta_- = 0.9991$,
$\|(\beta_+ - C_{ii})^{-1}\|^{-1} = 0.2352, 0.6992, 0.4487$,
$\delta = 0.2325$,
$\tau = 0.0013$, (from (3.4))
$\pi = 9.22 \cdot 10^{-4}$.

Final error bound:
$\delta_p = 2.81 \cdot 10^{-4}$,
$\|r\| = 3.095$, (see (5.3))
$\tau = 0.0017$,
$\|e\| \leq 2.99 \cdot 10^{-3}$, (see (5.6))
$\|\tilde{y} - y\| \leq 4.23 \cdot 10^{-3}$.

FIG. 3. Computation of the error bound.

pessimistic view of the error is inevitable when one attempts to bound it rather than estimate it.

ACKNOWLEDGMENT. The author is grateful to Jon Agre for performing the numerical calculations reported in Section 7 and especially for his interest and encouragement.

REFERENCES

1. CLINE, A.K., MOLER, C.B., STEWART, G.W., AND WILKINSON, J.H. An estimate for the condition number. *SIAM J. Numer. Anal. 16* (1979), 368–375.
2. COURTOIS, P.J. Error analysis in nearly-completely decomposable stochastic systems. *Econ. 43* (1975), 691–709.
3. COURTOIS, P J *Decomposability: Queueing and Computer System Applications*. Academic Press, New York, 1977.
4. DUFF, I. AND STEWART, G.W , EDS. *Sparse Matrix Proc. 1978*. SIAM, Philadelphia, Pa. 1979.
5. JENNINGS, A. AND STEWART, W.J. Simultaneous iteration for partial eigensolution of real matrices. *J. Inst. Math. Appl. 15* (1975), 351–362.
6. O'LEARY, D.P., STEWART, G W., VANDERGRAFT, J.S. Estimating the largest eigenvalue of a positive definite matrix. *Math. Comput. 33* (1979), 1289–1292.
7. SIMON, H.A. AND ANDO, A. Aggregation of variables in dynamic systems. *Econ. 29* (1961), 111–138.
8. STEWART, G.W. Error and perturbation bounds for subspaces associated with certain eigenvalue problems. *SIAM Rev. 15* (1973), 727–764.
9. STEWART, G.W. *Introduction to Matrix Computations*. Academic Press, New York, 1974.
10. STEWART, G.W. Simultaneous iteration for computing invariant subspaces of non-Hermitian matrices. *Numer. Math. 25* (1976), 123–136.
11. STEWART, G.W. HQR3 and EXCHNG: Fortran subroutines for calculating and ordering the eigenvalues of a real upper Hessenberg matrix. *ACM Trans. Math. Softw. 2*, 3 (Sept. 1976), 275–280.

12. STEWART, G.W. On the implicit deflation of nearly singular systems of linear equations. *SIAM J. Scient Stat Comput* 2 (1981), 136–140
13. VANTILBORGH, H Exact aggregation in exponential queueing networks. *J. ACM. 25*, 4 (Oct. 1978), 620–629.
14 VARGA, R S *Matrix Iterative Analysis* Prentice-Hall, Englewood Cliffs, N.J., 1962.
15 ZARLING, R Numerical solution of nearly decomposable queueing networks. Ph.D. Dissertation, Univ of North Carolina, 1976. Available from University Microfilms, Ann Arbor, Mich.

RECEIVED AUGUST 1980, REVISED MAY 1982; ACCEPTED MAY 1982

15.4. [GWS-J70] "Two Simple Residual Bounds for the Eigenvalues of a Hermitian Matrix"

[GWS-J70] "Two Simple Residual Bounds for the Eigenvalues of a Hermitian Matrix", *SIAM Journal on Matrix Analysis and Applications* **12** (1991) 205–208.
http://dx.doi.org/10.1137/0612016

SIAM J. MATRIX ANAL. APPL.
Vol. 12, No. 2, pp. 205–208, April 1991

002

TWO SIMPLE RESIDUAL BOUNDS FOR THE EIGENVALUES OF A HERMITIAN MATRIX*

G. W. STEWART†

Abstract. Let A be Hermitian and let the orthonormal columns of X span an approximate invariant subspace of X. Then the residual $R = AX - XM$ ($M = X^{\mathrm{H}}AX$) will be small. The theorems of this paper bound the distance of the spectrum of M from the spectrum of A in terms of appropriate norms of R.

Key words. eigenvalue, invariant subspace, perturbation theory, residual bounds

AMS(MOS) subject classifications. 15A18, 15A42

Let A be a Hermitian matrix with eigenvalues $\lambda_1 \geq \cdots \geq \lambda_n$. If X is a matrix with orthonormal columns that spans an invariant subspace of A and

$$M = X^{\mathrm{H}}AX, \tag{1}$$

then $AX - XM = 0$.

Now suppose that the columns of X span an *approximate* invariant subspace of A. Then the matrix

$$R = AX - XM$$

will be small, say in the spectral norm $\|\cdot\|$ defined by $\|R\| = \max_{\|x\|=1} \|Rx\|$, where $\|x\|$ is the Euclidean norm of x.[1] If the eigenvalues of M are $\mu_1 \geq \cdots \geq \mu_k$, then we should expect the μ_i to be near k of the λ_i. The problem treated in this note is to derive a bound in terms of the matrix R.

An important result, due to Kahan [3] (see also [6, p. 219]), states that there are eigenvalues $\lambda_{j_1}, \cdots, \lambda_{j_k}$ of A such that

$$|\mu_i - \lambda_{j_i}| \leq \|R\|, \qquad i = 1, \cdots, k. \tag{2}$$

If nothing further is known about the spectrum of A, this bound is generally satisfactory, although it can be improved somewhat [5]. However, it frequently happens (e.g., in the Lanczos algorithm or simultaneous iteration [6, Chaps. 13,14]) that we know that $n - k$ of the eigenvalues of A are well separated from the eigenvalues of M: specifically, if we know that

(3) there is a number $\delta > 0$ such that exactly $n - k$ of the eigenvalues of A lie outside the interval $[\mu_k - \delta, \mu_1 + \delta]$,

then the bound in (2) can be replaced by a bound of order $\|R\|^2$. Bounds of this kind have been given by Temple, Kato, and Lehman (see [6, Chap. 10] and [1, §6.5]). Early bounds of this kind dealt only with a single eigenvalue and eigenvector. Lehman's bounds are in some sense optimal, but they are quite complicated.

---

* Received by the editors January 25, 1990; accepted for publication (in revised form) June 13, 1990.

† Department of Computer Science and Institute for Advanced Computer Studies, University of Maryland, College Park, Maryland 20742. This work was supported in part by Air Force Office of Scientific Research contract AFOSR-87-0188.

[1] In fact, the choice (1) of M minimizes $\|R\|$, although we will not make use of this fact here.

The purpose of this note is to give two other bounds derived from bounds on the accuracy of the column space of X as an invariant subspace of A. They are very simple to state and yet are asymptotically sharp. In addition, they can be established by appealing to results readily available in the literature.

THEOREM 1. *With the above definitions, assume that A and M satisfy* (3). *If*

$$\rho \equiv \frac{\|R\|}{\delta} < 1,$$

then there is an index j such that $\lambda_j, \cdots, \lambda_{j+k-1} \in (\mu_k - \delta, \mu_1 + \delta)$ and

$$|\mu_i - \lambda_{j+i-1}| \le \frac{1}{1-\rho^2}\frac{\|R\|^2}{\delta}, \qquad i = 1, \cdots, k.$$

Proof. Let $(X\ Y)$ be unitary. Then

$$\begin{pmatrix} X^{\mathrm{H}} \\ Y^{\mathrm{H}} \end{pmatrix} A(X\ Y) = \begin{pmatrix} M & S^{\mathrm{H}} \\ S & N \end{pmatrix}$$

where $\|S\| = \|R\|$. By the "$\sin\Theta$" theorem of Davis and Kahan [2], there is a matrix P satisfying

$$\|P(I + P^{\mathrm{H}}P)^{1/2}\| \le \rho \tag{4}$$

such that the columns of

$$\hat{X} = (X + YP)(I + P^{\mathrm{H}}P)^{-1/2}$$

(which are orthonormal) span an invariant subspace of A. From (4) it follows that

$$\frac{\|P\|}{\sqrt{1 + \|P\|^2}} \le \rho,$$

and since $\rho < 1$

$$\|P\| \le \frac{\rho}{\sqrt{1-\rho^2}}. \tag{5}$$

Let $\hat{Y} = (Y - XP^{\mathrm{H}})(I + PP^{\mathrm{H}})^{-1/2}$. Then $(\hat{X}\ \hat{Y})$ is unitary. Since the columns of $\hat{X}$ span an invariant subspace of A, we have $\hat{Y}^{\mathrm{H}}A\hat{X} = 0$. Hence

$$\begin{pmatrix} \hat{X}^{\mathrm{H}} \\ \hat{Y}^{\mathrm{H}} \end{pmatrix} A(\hat{X}\ \hat{Y}) = \begin{pmatrix} \hat{M} & 0 \\ 0 & \hat{N} \end{pmatrix}.$$

In [7] it is shown that

$$\hat{M} = (I + P^{\mathrm{H}}P)^{1/2}(M + S^{\mathrm{H}}P)(I + P^{\mathrm{H}}P)^{-1/2}.$$

The eigenvalues of $\hat{M}$ are eigenvalues of A. Since $\rho < 1$, it follows from (2) that they lie in the interval $(\mu_k - \delta, \mu_1 + \delta)$, and hence are $\lambda_j, \cdots, \lambda_{j+k-1}$ for some index j. By a result of Kahan [4] on non-Hermitian perturbations of Hermitian matrices,

$$|\mu_i - \lambda_{j+i-1}| \le \|(I + P^{\mathrm{H}}P)^{1/2}\|\|(I + P^{\mathrm{H}}P)^{-1/2}\|\|S\|\|P\|, \qquad i = 1, \cdots, k.$$

The theorem now follows on noting that $\|(I+P^{\mathrm{H}}P)^{-1/2}\| \le 1$ and inserting the bound (5) for $\|P\|$. □

There are two remarks to be made about this theorem. First, it extends to operators in Hilbert space, provided X (now itself an operator) has a finite-dimensional domain. Second, the bound is asymptotically sharp, as may be seen by letting $X = (1\ 0)^{\mathrm{T}}$ and

$$A = \begin{pmatrix} 0 & \epsilon \\ \epsilon & 1 \end{pmatrix}$$

(the eigenvalues of A are asymptotic to ϵ^2 and $1 - \epsilon^2$).

The requirement (3) unfortunately does not allow the eigenvalues of M to be scattered through the spectrum of A. If we pass to the Frobenius norm defined by $\|X\|_{\mathrm{F}}^2 = \mathrm{trace}(X^{\mathrm{H}}X)$, then we can obtain a Hoffman–Wielandt type residual bound. Specifically, if

$$\delta = \min\{|\lambda_i - \mu_j| : \lambda_i \in \lambda(A), \mu_j \in \lambda(M)\} > 0, \tag{6}$$

then a variant of the $\sin\Theta$ theorem shows that there is a matrix P satisfying

$$\|P(I + P^{\mathrm{H}}P)^{1/2}\| \le \|P(I + P^{\mathrm{H}}P)^{1/2}\|_{\mathrm{F}} \le \frac{\|R\|_{\mathrm{F}}}{\delta}$$

such that the columns of

$$\hat{X} = (X + YP)(I + P^{\mathrm{H}}P)^{-1/2}$$

span an invariant subspace of A. By a variant of Kahan's theorem due to Sun [9], [8], the eigenvalues $\lambda_{j_1}, \cdots, \lambda_{j_k}$ of $\hat{M}$ may be ordered so that

$$\sqrt{\sum_{i=1}^{k} (\mu_i - \lambda_{j_i})^2} \le \|(I + P^{\mathrm{H}}P)^{1/2}\| \|(I + P^{\mathrm{H}}P)^{-1/2}\| \|S\|_{\mathrm{F}} \|P\|.$$

Hence we have the following theorem.

THEOREM 2. *With the above definitions, assume that A and M satisfy* (6). *If*

$$\rho_{\mathrm{F}} \equiv \frac{\|R\|_{\mathrm{F}}}{\delta} < 1,$$

then there are eigenvalues $\lambda_{j_1}, \cdots, \lambda_{j_k}$ of A such that

$$\sqrt{\sum_{i=1}^{k} (\mu_i - \lambda_{j_i})^2} \le \frac{1}{1 - \rho_{\mathrm{F}}^2} \frac{\|R\|_{\mathrm{F}}^2}{\delta}.$$

REFERENCES

[1] F. CHATELIN, *Spectral Approximation of Linear Operators*, Academic Press, New York, 1983.

[2] C. DAVIS AND W. M. KAHAN, *The rotation of eigenvectors by a perturbation. III*, SIAM J. Numer. Anal., 7 (1970), pp. 1–46.

[3] W. KAHAN, *Inclusion theorems for clusters of eigenvalues of Hermitian matrices*, Tech. Report, Computer Science Department, University of Toronto, Toronto, Ontario, Canada, 1967.

[4] W. KAHAN, *Spectra of nearly Hermitian matrices*, Proc. Amer. Math. Soc., 48 (1975), pp. 11–17.
[5] N. J. LEHMANN, *Optimale Eigenwerteinschiessungen*, Numer. Math., 5 (1963), pp. 246–272.
[6] B. N. PARLETT, *The Symmetric Eigenvalue Problem*, Prentice-Hall, Englewood Cliffs, NJ, 1980.
[7] G. W. STEWART, *Error bounds for approximate invariant subspaces of closed linear operators*, SIAM J. Numer. Anal., 8 (1971), pp. 796–808.
[8] G. W. STEWART AND G.-J. SUN, *Matrix Perturbation Theory*, Academic Press, Boston, 1990.
[9] J.-G. SUN, *On the perturbation of the eigenvalues of a normal matrix*, Math. Numer. Sinica, 6 (1984), pp. 334–336.

15.5. [GWS-J71] (with G. Zhang) "Eigenvalues of Graded Matrices and the Condition Numbers of a Multiple Eigenvalue"

[GWS-J71] (with G. Zhang) "Eigenvalues of Graded Matrices and the Condition Numbers of a Multiple Eigenvalue", *Numerische Mathematik* **58** (1991) 703–712.
http://dx.doi.org/10.1007/BF01385650

Numer. Math. 58, 703–712 (1991)

Numerische Mathematik

Eigenvalues of graded matrices and the condition numbers of a multiple eigenvalue *

G.W. Stewart [1,2] and G. Zhang [2]

[1] Department of Computer Science and [2] Institute for Advanced Computer Studies, University of Maryland, College Park, MD 20742, USA

Received April 3, 1990 / September 20, 1990

Summary. This paper concerns two closely related topics: the behavior of the eigenvalues of graded matrices and the perturbation of a nondefective multiple eigenvalue. We will show that the eigenvalues of a graded matrix tend to share the graded structure of the matrix and give precise conditions insuring that this tendency is realized. These results are then applied to show that the secants of the canonical angles between the left and right invariant of a multiple eigenvalue tend to characterize its behavior when its matrix is slightly perturbed.

Subject classifications: AMS (MOS): 65F15; CR: G1.3.

1 Introduction

In this paper we will be concerned with the distribution of the eigenvalues of a graded matrix. The specific problem that gave rise to this investigation is that of explaining the behavior of a nondefective multiple eigenvalue of a general matrix when the matrix is slightly perturbed[1]. Under such circumstances, an eigenvalue of multiplicity m will typically spawn m simple eigenvalues, as might be expected. What requires explanation is that the new eigenvalues will be found at varying distances from the original eigenvalue, and these distances are more a characteristic of the matrix than of the perturbation. Thus, a multiple eigenvalue can have several condition numbers that reflect the different sensitivities of its progeny.

As an illustration, consider the variation of the eigenvalues of the matrix

$$A = \frac{1}{\varepsilon}\begin{pmatrix}1\\1\\\varepsilon\end{pmatrix}(\varepsilon\ \varepsilon\ 1),$$

* This work was supported in part by the Air Force Office of Sponsored Research under Contract AFOSR-87-0188

[1] There is a body of literature on the perturbation of multiple eigenvalues of Hermitian matrices or Hermitian pencils when the perturbation is an analytic function of one or more variables. For an entry into this literature, see [9]

where ε is small. This matrix has a simple eigenvalue 3 and a double eigenvalue 0. Let $\varepsilon = 10^{-5}$ and let

$$E = \begin{pmatrix} -6.7248\text{e}-13 & -3.2031\text{e}-11 & 1.6070\text{e}-10 \\ -5.5392\text{e}-11 & -1.0694\text{e}-10 & 6.2824\text{e}-12 \\ -4.0564\text{e}-11 & -5.0153\text{e}-11 & -1.6116\text{e}-10 \end{pmatrix}.$$

Then the eigenvalues of $A+E$ are

$$\begin{aligned} \lambda_1 &= -3.244714710216396\text{e}-12, \\ \lambda_2 &= 3.023765334447618\text{e}-06, \\ \lambda_3 &= 2.999996975969131\text{e}+00. \end{aligned} \tag{1.1}$$

Both λ_2 and λ_3 come from the unperturbed eigenvalue 0, however, λ_2 is six orders of magnitude greater than λ_1. This difference is not an artifice of the perturbation E: almost any randomly chosen perturbation would cause the same behavior.

Some insight into this phenomenon may be obtained by choosing a suitable set of eigenvectors for A. Specifically let

$$X = \begin{pmatrix} 1.0000\text{e}+00 & 1.0000\text{e}+02 & 1.0000\text{e}+02 \\ -1.0000\text{e}+00 & 1.0000\text{e}+02 & 1.0000\text{e}+02 \\ 0 & -2.0000\text{e}-03 & 1.0000\text{e}-03 \end{pmatrix}$$

be a matrix of right eigenvectors of A. Then

$$X^{-1}(A+E)X = \begin{pmatrix} -1.0096\text{e}-11 & 6.4813\text{e}-09 & 6.4816\text{e}-09 \\ -3.1963\text{e}-09 & 3.0238\text{e}-06 & 3.0239\text{e}-06 \\ 3.1967\text{e}-09 & -3.0239\text{e}-06 & 3.0000\text{e}+00 \end{pmatrix}.$$

Now from the theory of the perturbation of invariant subspaces, we know that up to terms of order 10^{-12} the eigenvalues of the leading principle matrix

$$\begin{pmatrix} -1.0096\text{e}-11 & 6.4813\text{e}-09 \\ -3.1963\text{e}-09 & 3.0238\text{e}-06 \end{pmatrix} \tag{1.2}$$

are eigenvalues of $A+E$. But this matrix is graded, and the theory to be developed in the next two sections will show that it must have a small eigenvalue of order 10^{-11} and a larger eigenvalue of order 10^{-6}. Hence, $A+E$ must have two eigenvalues of similar orders of magnitude. Since

$$X^{-1} = \begin{pmatrix} 5.0000\text{e}-01 & -5.0000\text{e}-01 & 0 \\ 1.6667\text{e}-03 & 1.6667\text{e}-03 & -3.3333\text{e}+02 \\ 3.3333\text{e}-03 & 3.3333\text{e}-03 & 3.3333\text{e}+02 \end{pmatrix},$$

it is easily seen that $X^{-1}(A+E)X$ has the same graded structure for almost any balanced perturbation E. Thus, the problem of assessing the effects of perturbations on the zero eigenvalues of A is reduced to the problem of characterizing the eigenvalues of graded matrices such as (1.2).

To investigate the distribution of the eigenvalues of a graded matrix, we need a characterization of graded matrices. At this point it is useful not to be too precise. We will call matrix A of order n a *graded matrix* if

$$A = DBD, \tag{1.3}$$

where

$$D = \operatorname{diag}(\delta_1, \delta_2, \ldots, \delta_n)$$

with

$$\delta_1 \geqq \delta_2 \geqq \ldots \geqq \delta_n > 0$$

and $B = (\beta_{ij})$ is "well behaved". The imprecision in this definition lies in the term "well behaved", which will be given specific meaning through hypotheses in the theorems of the next two sections. Note that this definition does not preclude some of the δ_i being equal, in which case the matrix is said to be *block graded.*

Throughout this paper, norm $\|\cdot\|$ denotes the vector 2-norm and the subordinate matrix operator norm. The magnitude of the largest element of B in (1.3) will be written

$$\beta_{\max} \stackrel{\text{def}}{=} \max_{i,j} \{|\beta_{ij}|\}.$$

We will also denote the ratio of δ_{i+1} to δ_i by

$$\rho_i \stackrel{\text{def}}{=} \frac{\delta_{i+1}}{\delta_i}.$$

In Sect. 2 of this paper, we give a lower bound on the largest eigenvalue of a graded matrix. In Sect. 3, we explore the relation between eigenvalues of a graded matrix and those of its Schur complements. These results are closely related to results obtained by one of the authors [7][2], and more distantly to results by Barlow and Demmel [1] and Demmel and Veselić [3] for graded symmetric matrices. Here it should be stressed that our goal is not so much to derive tight bounds on the eigenvalues as to make statements about their magnitudes – as befits our intended application. Finally, in Sect. 4, we analyze the perturbation of a multiple eigenvalue and show that the secants of the canonical angles between its left and right invariant subspaces form a set of condition numbers for the eigenvalue.

2 The largest eigenvalue of a graded matrix

It is well known that the elements of a matrix can be arbitrarily larger than its spectral radius. In this section we will show that under appropriate conditions

[2] Mention should also be made of two papers on low rank approximation [4, 2], whose results can be regarded as limiting cases of block scaling

this is not true of graded matrices. Specifically, if β_{11} is not too small compared to $\beta_{\max}$, the graded matrix A has an eigenvalue that approximates $\alpha_{11}=\delta_1^2\beta_{11}$. The basic tool used to establish this result is Gerschgorin's theorem (see, e.g., [5, p. 341]). Related results may be found in [1].

The center of the Gerschgorin disk from the first row of A is $\delta_1^2\beta_{11}$, and its radius is bounded by $\beta_{\max}\delta_1(\delta_2+\ldots+\delta_n)$. For each row other than the first, the sum of the absolute values of its elements is bounded by $\beta_{\max}\delta_2(\delta_1+\ldots+\delta_n)$. From these facts and the Gerschgorin theorem we have the following result.

Theorem 2.1. *If*

$$\frac{|\beta_{11}|}{\beta_{\max}} \geqq \frac{\delta_1(\delta_2+\ldots+\delta_n)+\delta_2(\delta_1+\ldots+\delta_n)}{\delta_1^2}, \tag{2.1}$$

then the largest eigenvalue $\lambda_{\max}$ of A is simple and satisfies

$$|\lambda_{\max}-\alpha_{11}| \leqq \beta_{\max}\delta_1(\delta_2+\ldots+\delta_n).$$

The other eigenvalues of A satisfy

$$|\lambda| \leqq \beta_{\max}\delta_2(\delta_1+\ldots+\delta_n) \leqq |\alpha_{11}|-\beta_{\max}\delta_1(\delta_2+\ldots+\delta_n) \leqq |\lambda_{\max}|.$$

To gain more insight into the condition (2.1), suppose that A is *uniformly graded* in the sense that the ratios ρ_i are constant – say they are equal to ρ. Then the condition (2.1) is certainly satisfied if

$$\frac{|\beta_{11}|}{\beta_{\max}} \geqq \frac{2\rho}{1-\rho}.$$

When $|\beta_{11}|/\beta_{\max}=1$, this condition is satisfied for $\rho \leqq \frac{1}{3}$. As $|\beta_{11}|/\beta_{\max}$ decreases, we must have

$$\rho \lesssim \frac{1}{2}\frac{|\beta_{11}|}{\beta_{\max}}.$$

Thus for the purpose of this section the "good behavior" of B means that the ratio $|\beta_{11}|/\beta_{\max}$ is near one. As this ratio grows smaller, the grading ratio ρ must decrease to compensate.

Theorem 2.1 is sufficient for assessing the magnitude of the largest eigenvalue. However, when the grading is strong, the bounds can be improved by the well-known technique of diagonal similarities. For example, ordinary Gerschgorin theory shows that the (2, 2)-element of the matrix (1.2) is at least a three digit approximation to an eigenvalue. However, if we multiply its second row by 10^{-2} and its second column by 10^2, we obtain the matrix

$$\begin{pmatrix} -1.0096\text{e}-11 & 6.4813\text{e}-07 \\ -3.1963\text{e}-11 & 3.0238\text{e}-06 \end{pmatrix},$$

from which it is seen that the (2, 2)-element approximates an eigenvalue to at least five digits. Note the agreement of this element with the second eigenvalue in the list (1.1).

Unfortunately, Gerschgorin theory can tell us little about the smaller eigenvalues. As an extreme example, the matrix

$$\begin{pmatrix} 1 & 10^{-2} & 10^{-4} \\ 10^{-2} & 10^{-4} & 10^{-6} \\ 10^{-4} & 10^{-6} & 10^{-8} \end{pmatrix}$$

has rank one, and hence its two smallest eigenvalues are zero. Nonetheless, it often happens that the eigenvalues of a graded matrix share its graded structure. In the next section we will show how this comes about.

3 Eigenvalues and Schur complements

The principal result of this section is that under appropriate hypotheses the $n-k$ smallest eigenvalues of a graded matrix A are approximated by the eigenvalues of the Schur complement of the $k\times k$ leading principal submatrix of A. Specifically, partition A in the form

$$A=\begin{pmatrix} A_{11} & A_{12} \\ A_{21} & A_{22} \end{pmatrix}=\begin{pmatrix} D_1 B_{11} D_1 & D_1 B_{12} D_2 \\ D_2 B_{21} D_1 & D_2 B_{22} D_2 \end{pmatrix},$$

where $D_1=\operatorname{diag}(\delta_1, \delta_2, \ldots, \delta_k)$ and $D_2=\operatorname{diag}(\delta_{k+1}, \ldots, \delta_n)$. Then the *Schur complement* of A_{11} is the matrix

$$A_{22}-A_{21} A_{11}^{-1} A_{12}=D_2(B_{22}-B_{21} B_{11}^{-1} B_{12}) D_2.$$

Note that the Schur complement of A_{11} is the graded Schur complement of B_{11}. Consequently, if the Schur complement of B_{11} is well behaved, by the results of the last section it will have an eigenvalue of magnitude $\tilde{\beta}_{11}\delta_{k+1}^2$, which, under conditions given below, will approximate an eigenvalue of A.

The approach taken here is to determine an $(n-k)\times k$ matrix P such that

$$\begin{pmatrix} I \\ P \end{pmatrix}$$

spans an invariant subspace corresponding to the k largest eigenvalues of A. It then follows that the eigenvalues of the matrix

$$\tilde{A}_{22}=A_{22}-PA_{12} \tag{3.1}$$

are the eigenvalues associated with the complementary invariant subspace; i.e., the $n-k$ smallest eigenvalues of A. For details see [6] or [8, Ch. V].

It can be shown that the matrix P must satisfy the equation

$$PA_{11}-A_{22} P=A_{21}-PA_{12} P,$$

or in terms of B and D

$$PD_1 B_{11} D_1-D_2 B_{22} D_2 P=D_2 B_{21} D_1-PD_1 B_{12} D_2 P.$$

In this form, the equation is badly scaled. A better scaling is obtained by replacing P with

$$\hat{P} = D_2^{-1} P D_1, \tag{3.2}$$

which satisfies the equation

$$\hat{P} B_{11} D_1 - B_{22} D_2^2 \hat{P} D_1^{-1} = B_{21} D_1 - \hat{P} B_{12} D_2^2 \hat{P} D_1^{-1},$$

or

$$\hat{P} = B_{21} B_{11}^{-1} + B_{22} D_2^2 \hat{P} D_1^{-2} B_{11}^{-1} - \hat{P} B_{12} D_2^2 \hat{P} D_1^{-2} B_{11}^{-1}. \tag{3.3}$$

The following theorem gives a sufficient condition for the existence of a solution of (3.3) and a bound on $\hat{P}$.

Theorem 3.1. *If*

$$\gamma \stackrel{\text{def}}{=} (\|B_{12}\| \, \|B_{21}\| \, \|B_{11}^{-1}\| + \tfrac{1}{2}\|B_{22}\|) \, \|B_{11}^{-1}\| \, \rho_k^2 < \tfrac{1}{4} \tag{3.4}$$

then equation (3.3) has a solution satisfying

$$\|\hat{P}\| \leqq 2 \|B_{21}\| \, \|B_{11}^{-1}\|. \tag{3.5}$$

Proof. Let $\hat{P}_0 = 0$, and for $k = 1, 2, \ldots$ let

$$\hat{P}_k = B_{21} B_{11}^{-1} + B_{22} D_2^2 \hat{P}_{k-1} D_1^{-2} B_{11}^{-1} - \hat{P}_{k-1} B_{12} D_2^2 \hat{P}_{k-1} D_1^{-2} B_{11}^{-1}.$$

We will show that if (3.4) is satisfied $\hat{P}_k$ converges to a solution of (3.3)[3]. For brevity set

$$\eta_1 = \|B_{21}\| \, \|B_{11}^{-1}\|, \quad \eta_2 = \rho_k^2 \|B_{22}\| \, \|B_{11}^{-1}\|, \quad \eta_3 = \rho_k^2 \|B_{12}\| \, \|B_{11}^{-1}\|.$$

By a simple induction

$$\|\hat{P}_k\| \leqq \eta_1 (1 + \eta_2 (1 + s_{k-1}) + \eta_1 \eta_3 (1 + s_{k-1})^2) = \eta_1 (1 + s_k), \tag{3.6}$$

where s_k satisfies the recursion

$$s_0 = 0,$$
$$s_k = \eta_2 (1 + s_{k-1}) + \eta_1 \eta_3 (1 + s_{k-1})^2.$$

Using (3.4), we can show by induction that the sequence $\{s_k\}$ is monotonically increasing and bounded by one. Let $s \leqq 1$ be the limit of the sequence $\{s_k\}$. Then from (3.6),

$$\|\hat{P}_k\| \leqq \eta_1 (1 + s). \tag{3.7}$$

---

[3] An alternative is to follow [6] and define $\hat{P}_k$ as the solution of

$$\hat{P}_k - B_{22} D_2^2 \hat{P}_k D_1^{-2} B_{11}^{-1} = B_{21} B_{11}^{-1} - \hat{P}_{k-1} B_{12} D_2^2 \hat{P}_{k-1} D_1^{-2} B_{11}^{-1}.$$

This approach gives a slightly less stringent condition for convergence but a slightly larger bound

Now

$$\begin{aligned}\|\hat{P}_{k+1}-\hat{P}_k\| &\leqq \eta_2 \|\hat{P}_k-\hat{P}_{k-1}\|+\eta_3 \|\hat{P}_k-\hat{P}_{k-1}\| (\|\hat{P}_k\|+\|\hat{P}_{k-1}\|)\\ &\leqq \{\eta_2+2\eta_1\eta_3(1+s)\} \|\hat{P}_k-\hat{P}_{k-1}\|\\ &=\eta \|\hat{P}_k-\hat{P}_{k-1}\| \leqq \eta^k \eta_1.\end{aligned}$$

Since $\eta=\eta_2+2\eta_1\eta_3(1+s)<(\eta_2+2\eta_1\eta_3)(1+s)<1$, the sequence $\{\hat{P}_k\}$ is a Cauchy sequence and has a limit $\hat{P}$, which by continuity must satisfy (3.3). The bound (3.5) follows from (3.7). □

For purposes of discussion, let

$$\kappa_k=\|B\| \, \|B_{11}^{-1}\|.$$

Then the condition (3.4) will be satisfied if

$$(\kappa_k^2+\tfrac{1}{2}\kappa_k)\,\rho_k^2<\tfrac{1}{4}.$$

Moreover, it follows from (3.2) and (3.5) that

$$\|P\| \leqq 2\kappa_k\rho_k.$$

Thus, for the purposes of our theorem, B is "well behaved" if κ_k is near one. As κ_k grows, it must be compensated for by a larger grading ratio ρ_k.

If ρ_k is sufficiently small, then all but the first term on the right hand side of (3.3) are insignificant and

$$\hat{P} \cong B_{21}B_{11}^{-1}.$$

It follows from (3.1) that

$$\tilde{A}_{22} \cong D_2(B_{22}-B_{21}B_{11}^{-1}B_{12})D_2,$$

which is just the Schur complement of A_{11}, or equivalently the graded Schur complement of B_{11}. If the Schur complement is well behaved in the sense of the last section, then A must have an eigenvalue the size of the leading diagonal element of the Schur complement. The chief way in which ill behavior manifests itself is through cancellation producing a small leading diagonal element of the Schur complement of B_{11}.[4]

Since grading enters the condition (3.4) only through the grading ratio ρ_k, Theorem 3.1 applies to block graded matrices. More important for our purposes is the fact that the condition of the theorem can fail for one value of k and hold for another, larger value of k. Thus it is possible for the descending sequence of eigenvalues to stutter, with occasional groups of eigenvalues having the wrong magnitudes while the sequence itself remains generally correct. The following example exhibits this behavior.

[4] We note in passing that the possibility of cancellation destroying the graded structure of a matrix is the bane of algorithmists who work with graded matrices

Consider the matrix A whose elements are given in the following array.

−7.1532e−01	4.1271e−02	−2.0433e−03	1.7447e−03	−1.4459e−04	4.0821e−06
4.1745e−02	3.7412e−03	−1.4124e−03	3.1508e−04	−8.6632e−06	3.2593e−07
−3.2573e−03	−3.5565e−04	8.7329e−05	1.0717e−05	7.6451e−07	−2.7899e−08
−1.5509e−03	−1.2599e−04	9.1642e−06	6.8861e−07	−1.1000e−08	−1.7092e−09
−2.5920e−05	−2.3092e−06	5.8399e−07	−3.0490e−08	−2.5573e−09	5.2496e−10
−4.2303e−06	−1.0778e−07	−6.2901e−08	4.3068e−09	−6.5392e−10	1.2152e−11

The matrix was generated by uniformly grading a matrix of normal random numbers with mean zero and standard deviation one. The grading ratio is $\rho = 0.1$.

The eigenvalues of A are:

$$\lambda_1 = -7.1771\mathrm{e}{-01}$$
$$\lambda_2 = 6.2493\mathrm{e}{-03}$$
$$\lambda_3 = -1.3472\mathrm{e}{-05} + 3.0630\mathrm{e}{-05}i$$
$$\lambda_4 = -1.3472\mathrm{e}{-05} - 3.0630\mathrm{e}{-05}i$$
$$\lambda_5 = -6.3603\mathrm{e}{-09}$$
$$\lambda_6 = 2.6849\mathrm{e}{-11}.$$

The table below exhibits the value of γ from (3.4) and the first diagonal element of the Schur complement.

k	γ	$\tilde{\alpha}_{11}$
1	0.10	6.1497e−03
2	0.35	−3.8747e−05
3	7.38	−2.9463e−05
4	0.05	−6.3180e−09
5	0.12	2.7030e−11

For $k=2$, 3, the distribution stutters. The reasons for the failure of the above theory to predict the eigenvalues are different for the two values of k. For $k=2$, the number γ can be made smaller than one-fourth by choosing a slightly different scaling matrix D. Thus, the failure is due to the cancellation in forming the leading diagonal element of the Schur complement of B_{11}. For $k=3$, the number γ is always greater than one-fourth, and Theorem 3.1 fails. Even so, the leading diagonal element of the Schur complement gives a ball-park estimate of the size of the complex pair – something not predicted by our theory. For the other values of k the leading diagonal element predicts the size of the corresponding eigenvalue very well. In fact, when γ is small it approximates the eigenvalue itself.

4 The condition numbers of a multiple eigenvalue

Let us now return to the problem that initiated the investigations of the last two sections: the perturbation of a multiple eigenvalue. Let A have a nondefective eigenvalue λ of multiplicity m. Since we may replace A with $\lambda I - A$, we may assume without loss of generality that $\lambda = 0$.

Since zero is a nondefective eigenvalue of multiplicity m, there are m-dimensional subspaces $\mathscr{X}$ and $\mathscr{Y}$ such that $A\mathscr{X} = 0$ and $A^H \mathscr{Y} = 0$: namely, the spaces of left and right eigenvectors corresponding to the eigenvalue zero. In Sect. 1 we saw that a judicious choice of eigenvectors led to a graded eigenvalue problem. The existence of a suitable choice for the general case is stated in the following theorem [6].

Theorem 4.1. *There are $n \times m$ matrices X and Y whose columns spaces are $\mathscr{X}$ and $\mathscr{Y}$ and which satisfy*

$$X^H X = Y^H Y = \operatorname{diag}(\sigma_1, \ldots, \sigma_m), \qquad \sigma_1 \geqq \ldots \geqq \sigma_m > 0,$$

and

$$Y^H X = I.$$

The number σ_i may be defined sequentially as follows. First,

$$\sigma_1 = \max_{x \in \mathscr{X}} \min_{y \in \mathscr{Y}} \sec \angle(x, y). \tag{4.1}$$

If x_1 and y_1 are vectors for which the extrema are attained in (4.1), then

$$\sigma_2 = \max_{\substack{x \in \mathscr{X} \\ x \perp x_1}} \min_{\substack{y \in \mathscr{Y} \\ y \perp y_1}} \sec \angle(x, y). \tag{4.2}$$

If x_2 and y_2 are vectors for which the extrema are attained in (4.2), then

$$\sigma_3 = \max_{\substack{x \in \mathscr{X} \\ x \perp x_1, x_2}} \min_{\substack{y \in \mathscr{Y} \\ y \perp y_1, y_2}} \sec \angle(x, y), \tag{4.3}$$

and so on. The maximizing angles are called the *canonical angles* between $\mathscr{X}$ and $\mathscr{Y}$. For more details see [8, Sect. I.5].

We must now relate this choice of basis to the eigenvalues of a perturbation $A + E$ of A. This is done in the following theorem [6].

Theorem 4.2. *Let*

$$C = Y^H E X.$$

Then there is a matrix $\hat{C} = C + O(\|E\|^2)$ whose eigenvalues are eigenvalues of $A + E$.

Since the eigenvalues of C approach zero, they must approximate the m eigenvalues spawned by the zero eigenvalue of A. Now the (i, j)-element of C has the form $y_i^H E x_j$. Hence,

$$|c_{ij}| \leqq \sqrt{\sigma_i \sigma_j}\, \|E\|.$$

Thus, unless E has special structure, C will tend to be a graded matrix with grading constants $\delta_i = \sqrt{\sigma_i}$, and by the characterizations (4.1)–(4.3) the grading will tend to be maximal. From the results of the last two sections, we know that the magnitudes of the eigenvalues of C will tend to be around $\sigma_i \|E\|$. For the example of Sect. 1, we calculated that $\|E\|$ is of order 10^{-10}, and σ_1 and σ_2 are of order one and 10^{+4}, respectively. This explains the behavior of the double eigenvalue 0.

It is unfortunate that we cannot say with complete rigor that the σ_i are condition numbers for λ. In the first place, without precise knowledge of E we cannot assert that C is graded. And even when C is graded, the example of the last section shows that the eigenvalues need not behave as we would like them to. But the phenomenon is no less real for having exceptions; and if we recognize that we are speaking about what is likely to be instead of what has to be, there can be no objection to calling the numbers σ_i, the condition numbers of the multiple eigenvalue λ.

References

1. Barlow, J., Demmel, J.: Computing accurate eigensystems of scaled diagonally dominant matrices. Technical report 421, Computer Science Department, Courant Institute (1988). SIAM J. Numer. Anal. (to appear)
2. J. Demmel, J.: The smallest perturbation of a submatrix which lowers the rank and constrained total least squares problems. SIAM J. Numer. Anal. **24**, 199–206 (1987)
3. Demmel, J., Veselić, K.: Jacobi's method is more accurate than QR. Technical report 468, Computer Science Department, New York University, 1989
4. Golub, G., Hoffman, A., Stewart, G.W.: A Generalization of the Echart-Young Matrix Approximation Theorem. Linear Algebra Appl. **88/89**, 317–327 (1987)
5. Golub, G.H., Van Loan, C.F.: Matrix computations, 2nd ed. Baltimore, Maryland: Johns Hopkins University Press 1989
6. Stewart, G.W.: Error and perturbation bounds for subspaces associated with certain eigenvalue problems. SIAM Review **15**, 727–764 (1973)
7. Stewart, G.W.: On the Asymptotic Behavior of Scaled Singular Value and QR Decompositions. Math. Comput. **43**, 483–489 (1984)
8. Stewart, G.W., Sun, J.-G.: Matrix perturbation theory. Boston: Academic Press 1990
9. Sun, J.-G.: A note on local behavior of multiple eigenvalues. SIAM J. Matrix Anal. Appl. **10**, 533–541 (1989)

15.6. [GWS-J108] "A Generalization of Saad's Theorem on Rayleigh-Ritz Approximations"

[GWS-J108] "A Generalization of Saad's Theorem on Rayleigh-Ritz Approximations", *Linear Algebra and its Applications* **327** (2001) 115–119.

ELSEVIER Linear Algebra and its Applications 327 (2001) 115–119

LINEAR ALGEBRA AND ITS APPLICATIONS

www.elsevier.com/locate/laa

A generalization of Saad's theorem on Rayleigh–Ritz approximations ☆

G.W. Stewart[1]

Department of Computer Science and Institute for Advanced Computer Studies, University of Maryland, College Park, MD 20742, USA

Received 10 February 2000; accepted 4 October 2000

Submitted by R.A. Brualdi

Abstract

Let (λ, x) be an eigenpair of the Hermitian matrix A of order n and let (μ, u) be a Ritz pair from a subspace $\mathscr{K}$ of $\mathbb{C}^2$. Y. Saad (Numerical Methods for Large Eigenvalue Problems: Theory and Algorithm, Wiley, New York, 1992) has given a simple inequality bounding $\sin \angle(x, u)$ in terms of $\sin \angle(x, \mathscr{K})$. In this paper, we show that this inequality can be extended to an equally simple inequality for eigenspaces of non-Hermitian matrices. © 2001 Elsevier Science Inc. All rights reserved.

Keywords: Large eigen problem; Rayleigh–Ritz approximation; Non-Hermitian matrix; Saad's theorem

Let A be a Hermitian matrix of order n. If n is very large, we will not be able to compute the eigensystem of A and must be content with computing a few of its eigenpairs. The usual methods for doing this—e.g., the Lanzcos and Jacobi–Davidson methods—produce a sequence of subspaces $\mathscr{K}_k$ that contain increasingly accurate approximations to the eigenvectors in question. Approximations to the corresponding eigenpairs of A are extracted from $\mathscr{K}$ by computing *Ritz pairs* (μ, u) defined by the conditions

$$\begin{aligned} &1.\ u \in \mathscr{K}_k, \\ &2.\ Au - \mu u \perp \mathscr{K}_k. \end{aligned} \tag{1}$$

☆ The report is available by anonymous ftp from thales.cs.umd.edu in the directory pub/reports or on the web at http:/www.cs.umd.edu/~stewart/.

[1] This work was supported in part by the National Science Foundation under Grant No. 970909-8426.

E-mail address: stewart@cs.umd.edu (G.W. Stewart).

PII: S 0 0 2 4 - 3 7 9 5 (0 0) 0 0 3 2 4 - 4

It is easy to see that if K_k is an orthonormal basis for $\mathscr{K}_k$, then every Ritz pair (μ, u) has the form $K_k y$, where $K_k^{\mathrm{H}} A K_k y = \mu y$.[2]

Saad [2, Theorem 4.6] has proven the following theorem relating eigenpairs of A to Ritz pairs. For now we drop the subscript on $\mathscr{K}_k$.

Theorem 1 (Saad [2]). *Let (λ, x) be an eigenpair of A and (μ, u) be a Ritz pair with respect to the subspace $\mathscr{K}$. Let $P_{\mathscr{K}}$ be the orthogonal projection on $\mathscr{K}$, and for non-zero v define*

$$\angle(v, \mathscr{K}) = \min_{\substack{w \in \mathscr{K} \\ w \neq 0}} \angle(v, w).$$

Then

$$\sin \angle(x, u) \leqslant \sin \angle(x, \mathscr{K}) \sqrt{1 + \frac{\gamma^2}{\delta^2}},$$

where

$$\gamma = \| P_{\mathscr{K}} A (I - P_{\mathscr{K}}) \|_2$$

and δ is the distance between λ and the Ritz values other than μ.

The purpose of this paper is to extend this theorem to eigenspaces (invariant subpaces) of non-Hermitian matrices. Specifically, $\mathscr{X}$ is an eigenspace of A if $A\mathscr{X} \subset \mathscr{X}$. If X is an orthonormal basis for $\mathscr{X}$, then

$$AX = XL, \quad \text{where } L = X.$$

We say that (L, X) is an eigenpair of A with eigenbasis X and eigenblock L. In what follows we will assume that all eigenbases are orthonormal.

Definition (1) of Ritz pairs extends to eigenspaces. Specifically, the pair (M, U), where U is orthonormal, is a Ritz pair if

1. $\mathscr{U} \subset \mathscr{K}$,
2. $AU - UM \perp \mathscr{K}$.

Again it is easy to see that (M, U) is a Ritz pair if and only if $U = KY$, where (M, Y) is an eigenpair of $K^{\mathrm{H}} A K$.[3]

In what follows we will use canonical angles to measure the distance between subspaces. Specifically, let (L, X) be an eigenpair of A and let W be an orthonormal basis for the orthogonal complement of $\mathscr{K}$. Then the singular values of $W^{\mathrm{H}} X$ are

---

[2] Throughout this paper a script letter will stand for a subspace and the corresponding capital Roman letter will stand for a matrix containing an orthonormal basis for the subspace.

[3] This method is variously associated with the names Rayleigh–Ritz and Galerkin. We have chosen to use the former because the terminology and computational processes from the Rayleigh–Ritz method for the symmetric eigenproblem generalize naturally. For example, the matrix $K^{\mathrm{H}} A K$ can be regarded as a generalized Rayleigh quotient, and it is natural to call (M, KY) a Ritz pair.

the sines of the canonical angles between $\mathscr{K}$ and $\mathscr{X}$. For more on canonical angles, see [3].

It will be easier to establish our generalization of Saad's theorem if we change our coordinate system to one in which the matrices bearing the canonical angles appear explicitly. As above, let (M, U) be a Ritz pair lying in $\mathscr{K}$. Let $(U \;\; V \;\; W)$ be a unitary matrix in which the columns of V span the orthogonal complement of $\mathscr{U}$ in $\mathscr{K}$. Then we may write

$$\begin{pmatrix} U^{\mathrm{H}} \\ V^{\mathrm{H}} \\ W^{\mathrm{H}} \end{pmatrix} A(U \;\; V \;\; W) = \begin{pmatrix} M & B_{12} & B_{13} \\ B_{21} & N & B_{23} \\ B_{31} & B_{32} & B_{33} \end{pmatrix}.$$

It is readily verified that, since (M, U) is a Ritz pair, B_{21} must be 0.

Now let (L, X) be an eigenpair of A and transform X into the UVW coordinate system

$$\begin{pmatrix} X^{\mathrm{H}} \\ V^{\mathrm{H}} \\ W^{\mathrm{H}} \end{pmatrix} U = \begin{pmatrix} P \\ Q \\ R \end{pmatrix}.$$

Then the singular values of R are the sines of the angles between $\mathscr{X}$ and $\mathscr{K}$, while the singular values of $(Q^{\mathrm{H}} \;\; R^{\mathrm{H}})^{\mathrm{H}}$ are the sines of the angles between $\mathscr{X}$ and $\mathscr{U}$. Thus, our problem is to bound

$$\left\| \begin{pmatrix} Q \\ R \end{pmatrix} \right\|$$

in terms of $\|R\|$. In what follows, $\|\cdot\|$ will denote either the 2-norm or the Frobenius norm. Note that for either norm

$$\left\| \begin{pmatrix} Q \\ R \end{pmatrix} \right\|^2 \leqslant \|Q\|^2 + \|R\|^2, \tag{2}$$

with equality holding for the Frobenius norm.

In the UVW coordinate system, the relation $AX = XL$ becomes

$$\begin{pmatrix} M & B_{12} & B_{13} \\ 0 & N & B_{23} \\ B_{31} & B_{32} & B_{33} \end{pmatrix} \begin{pmatrix} P \\ Q \\ R \end{pmatrix} = \begin{pmatrix} P \\ Q \\ R \end{pmatrix} L.$$

Hence,

$$\begin{pmatrix} M & B_{12} \\ 0 & N \end{pmatrix} \begin{pmatrix} P \\ Q \end{pmatrix} - \begin{pmatrix} P \\ Q \end{pmatrix} L = - \begin{pmatrix} B_{13} \\ B_{32} \end{pmatrix} R. \tag{3}$$

The right-hand side of (3) can be bounded above as follows:

$$\left\| \begin{pmatrix} B_{13} \\ B_{23} \end{pmatrix} R \right\| \leqslant \left\| \begin{pmatrix} B_{13} \\ B_{23} \end{pmatrix} \right\| \|R\| \equiv \eta \|R\|. \tag{4}$$

The left-hand side can be bounded below as follows:

$$\begin{aligned}\left\|\begin{pmatrix} M & B_{12} \\ 0 & N \end{pmatrix}\begin{pmatrix} P \\ Q \end{pmatrix} - \begin{pmatrix} P \\ Q \end{pmatrix} L\right\| &= \left\|\begin{pmatrix} MP - PL + B_{12}Q \\ NQ - QL \end{pmatrix}\right\| \\ &\geqslant \|NQ - QL\| \\ &\geqslant \mathrm{sep}(N, L)\|Q\|, \end{aligned} \tag{5}$$

where

$$\mathrm{sep}(N, L) = \inf_{\|Z\|=1} \|NZ - ZL\|.$$

Combining (5) and (4), we have

$$\|Q\| \leqslant \eta \frac{\|R\|}{\mathrm{sep}(N, L)}.$$

It follows from (2) that

$$\left\|\begin{pmatrix} Q \\ R \end{pmatrix}\right\| \leqslant \|R\| \sqrt{1 + \frac{\eta^2}{\mathrm{sep}(N, L)^2}}.$$

Written in terms of angles this inequality becomes

$$\sin \angle(\mathscr{X}, \mathscr{U}) \leqslant \sin \angle(\mathscr{X}, \mathscr{K}) \sqrt{1 + \frac{\eta^2}{\mathrm{sep}(N, L)^2}}.$$

We now will transform this inequality back into the original coordinate system. Clearly, $\eta = \|P_{\mathscr{K}} A(I - P_{\mathscr{K}})\|$. The quantity sep appears to depend on the choices of the bases X and V. However, because $\|\cdot\|$ is unitarily invariant, it is easy to show that in fact it depends only on the subspaces spanned by X and V. Hence, we will define

$$\mathrm{sep}(\mathscr{V}, \mathscr{X}) = \inf_{\|Z\|=1} \left\|(V^{\mathrm{H}} A V)Z - Z(X^{\mathrm{H}} A X)\right\|, \tag{6}$$

where V and X are arbitrary orthonormal bases for $\mathscr{V}$ and $\mathscr{X}$.

We are now in a position to state our final result.

Theorem 2. *Let $\mathscr{X}$ be an eigenspace of A. Let $\mathscr{U}$ be a Ritz subspace in $\mathscr{K}$ of the same dimension as $\mathscr{X}$ and let $\mathscr{V}$ be the orthogonal complement of $\mathscr{U}$ in $\mathscr{K}$. Then*

$$\sin \angle(\mathscr{X}, \mathscr{U}) \leqslant \sin \angle(\mathscr{X}, \mathscr{K}) \sqrt{1 + \frac{\eta^2}{\mathrm{sep}(\mathscr{V}, \mathscr{X})^2}},$$

where $\eta = \|P_{\mathscr{K}} A(I - P_{\mathscr{K}})\|$ and sep *is defined by* (6).

Theorem 2 is an exact analogue of Theorem 1. The major difference is the replacement of δ by $\mathrm{sep}(\mathscr{V}, \mathscr{X})$. This latter quantity is bounded above by the physical separation of the spectra of M and L, but it can be much smaller. If A is Hermitian and we use the Frobenius norm, then it is the physical separation, so that our generalization reduces to Theorem 1. This is not generally true of the 2-norm unless one assumes that the eigenvalues of $X^{\mathrm{H}}AX$ lie inside an interval and the eigenvalues of $V^{\mathrm{T}}AV$ lie outside that interval or vice versa (see [3, Lemma V.3.5]).

We now return to the case where we have a sequence $\mathscr{K}_k$ of subspaces having increasingly accurate approximations to the eigenspace $\mathscr{X}$, i.e., $\sin \angle(\mathscr{X}, \mathscr{K}_k) \to 0$. Theorem 2 is not strong enough to formally prove the convergence of Ritz pairs from the $\mathscr{K}_k$. There are two reasons. First, the theorem does not single out any one of the many possible Ritz pairs, so that one must be chosen by ad hoc methods. Second, the quantity $\operatorname{sep}(\mathscr{V}_k, \mathscr{X})$ may converge to 0, something that we cannot check, since we do not know $\mathscr{X}$.

A treatment of these problems is given in [1]. We should stress, however, that they are not of great concern in practice. Typically the person doing the computation will have predetermined the eigenpairs he or she wants—e.g., the eigenpairs with largest real parts—and this will direct the choice of Ritz pairs. Moreover, as the Ritz pairs converge, they can be used to approximate $\operatorname{sep}(\mathscr{V}_k, \mathscr{X})$, which is Lipschitz continuous with constant one.

References

[1] Z. Jia, G.W. Stewart, An analysis of the Rayleigh–Ritz method for approximating eigenspaces. Technical Report TR-4015, Department of Computer Science, University of Maryland, College Park, 1999, Math. Comp., to appear.

[2] Y. Saad, Numerical Methods for Large Eigenvalue Problems: Theory and Algorithm, Wiley, New York, 1992.

[3] G.W. Stewart, J.-G. Sun, Matrix Perturbation Theory, Academic Press, New York, 1990.

15.7. [GWS-J109] "On the Eigensystems of Graded Matrices"

[GWS-J109] "On the Eigensystems of Graded Matrices", *Numerische Mathematik* **90** (2001) 349–370.
http://dx.doi.org/10.1007/s002110100290

Numer. Math. (2001) 90: 349–370
Digital Object Identifier (DOI) 10.1007/s002110100290

Numerische Mathematik

On the eigensystems of graded matrices*

G.W. Stewart

University of Maryland, Department of Computer Science, College Park, MD 20742, USA;
e-mail: stewart@cs.umd.edu

Received January 18, 2000 / Revised version received September 29, 2000 /
Published online June 7, 2001 –

Dedicated to F. L. Bauer

What, then, is time?
If no one asks of me, I know;
If I wish to explain it to him who asks,
I know not.
St. Augustine

I shall not attempt further to define it . . .
but I know it when I see it.
Justice Potter Stewart (on pornography)

1. Introduction

Informally a graded matrix is one whose elements show a systematic decrease or increase as one passes across the matrix. Thus we would recognize a matrix whose elements have the magnitudes

$$\begin{pmatrix} 10^0 & 10^{-4} & 10^{-8} & 10^{-12} \\ 10^0 & 10^{-4} & 10^{-8} & 10^{-12} \\ 10^0 & 10^{-4} & 10^{-8} & 10^{-12} \\ 10^0 & 10^{-4} & 10^{-8} & 10^{-12} \end{pmatrix}$$

as being graded by columns. Similarly, the matrices

$$\begin{pmatrix} 10^0 & 10^0 & 10^0 & 10^0 \\ 10^{-4} & 10^{-4} & 10^{-4} & 10^{-4} \\ 10^{-8} & 10^{-8} & 10^{-8} & 10^{-8} \\ 10^{-12} & 10^{-12} & 10^{-12} & 10^{-12} \end{pmatrix} \quad \text{and} \quad \begin{pmatrix} 10^0 & 10^{-2} & 10^{-4} & 10^{-6} \\ 10^{-2} & 10^{-4} & 10^{-6} & 10^{-8} \\ 10^{-4} & 10^{-6} & 10^{-8} & 10^{-10} \\ 10^{-6} & 10^{-8} & 10^{-10} & 10^{-12} \end{pmatrix}$$

are respectively row graded and diagonally graded.

* This work was supported in part by the National Science Foundation under Grant No. 970909-8426

It has long been recognized that the eigenvalues and eigenvectors of graded matrices have special properties. For example, Martin, Reinsch, and Wilkinson writing in 1968 about Householder tridiagonalization [8] assert that a diagonally graded matrix will have small eigenvalues that are insensitive to small relative perturbations in the elements of the matrix. They go on to assert that if the direction of the grading is consonant with their implementation of the algorithm, then the small eigenvalues will be computed accurately.

Informally, then, we know what a graded matrix is — just as St. Augustine knows what time is and Potter Stewart knows pornography when he sees it. Unfortunately, when it comes to a formal definition we encounter difficulties.

1. If we attempt to define grading as a systematic change in the magnitude of the elements, we have to take into account exceptions to the pattern. Does an occasional zero element destroy the grading? What is a graded tridiagonal matrix?
2. Alternatively we can try to base a definition on the properties we observe in graded matrices — e.g., the possession of small, well determined eigenvalues. Unfortunately, matrices that are nicely graded in the informal sense can fail to have these properties.

In this paper we will take an intermediate course. We will define grading as a scaling of a base matrix B. Then we will determine the properties of B that insure that the graded matrix has the properties we want. In particular, we will be concerned with the structure and perturbation theory of the eigensystem of graded matrices.

The paper is organized as follows. We begin with a numerical example that illustrates some of the typical properties of a graded matrix. In the same section we will also give an example in which these properties fail. The heart of the paper is §3, where we establish the structure of the eigensystem of a graded matrix. In §4 we will derive condition numbers for the eigenvectors and eigenvectors of graded matrices. In §5 we will treat positive definite matrices and the singular value decomposition of general matrices. The paper concludes with bibliographical notes surveying previous and related work.

Throughout this paper, $\|\cdot\|$ will denote the Euclidean vector norm and its subordinate matrix norm. The conjugate transpose of a matrix A is denoted A^{H}. The reader is assumed to be familiar with the basic matrix decompositions, like Cholesky and the QR decompositions (see, e.g., [7, 13]). In partitioned matrices we will index each block by the indices of the element in the southeast corner. Thus if A is of order n, a partition of A in the form

$$A = \begin{pmatrix} A_{kk} & A_{kn} \\ A_{nk} & A_{nn} \end{pmatrix}$$

implies that A_{kk} is of order k.

2. Examples

Since the structure of the eigensystem of a graded matrix is not widely known, it is appropriate to set the stage with some examples. We begin with a matrix whose eigenvalues and eigenvectors exhibit the properties of a typical graded matrix. We will than present an example in which the properties fail. The computations were performed in MATLAB with rounding unit about 10^{-16}.

2.1. A typical graded matrix

The matrix

$$A = \begin{pmatrix} -6.5\text{e}{-01} & -5.0\text{e}{-05} & 4.4\text{e}{-09} & 4.1\text{e}{-14} & -9.8\text{e}{-17} \\ -1.1\text{e}{+00} & -3.6\text{e}{-06} & 1.3\text{e}{-08} & -7.6\text{e}{-13} & -6.9\text{e}{-17} \\ -4.8\text{e}{-02} & -1.7\text{e}{-05} & -5.0\text{e}{-09} & -8.9\text{e}{-14} & 1.3\text{e}{-16} \\ 3.8\text{e}{-01} & -9.6\text{e}{-05} & -1.1\text{e}{-08} & -2.0\text{e}{-12} & -9.1\text{e}{-17} \\ -3.3\text{e}{-01} & 1.3\text{e}{-04} & 8.1\text{e}{-09} & 1.1\text{e}{-12} & -4.1\text{e}{-17} \end{pmatrix} \tag{2.1}$$

was formed from a matrix B of standard normal deviates by postmultiplying it by

$$D = \text{diag}(1, 10^{-4}, 10^{-8}, 10^{-12}, 10^{-16}).$$

Thus A is column graded with grading ratio of 10^{-4} from column to column. (Here we only display two digits of the double precision representation of A.)

The eigenvalues of A are given by

$$-6.5\text{e}{-01}\ 7.9\text{e}{-05}\ -4.3\text{e}{-09}\ -3.3\text{e}{-12}\ -3.5\text{e}{-16}$$

It is seen that they share the grading of A, which is typical for such matrices.

The following is the matrix of eigenvectors of A, scaled so that the diagonal elements are one:

$$\begin{pmatrix} 1.0\text{e}{+00} & -7.6\text{e}{-05} & 1.2\text{e}{-08} & -1.4\text{e}{-12} & 3.8\text{e}{-16} \\ 1.7\text{e}{+00} & 1.0\text{e}{+00} & -7.0\text{e}{-05} & 1.4\text{e}{-08} & -3.7\text{e}{-12} \\ 7.3\text{e}{-02} & -1.8\text{e}{-01} & 1.0\text{e}{+00} & -5.5\text{e}{-05} & 3.6\text{e}{-08} \\ -5.8\text{e}{-01} & -1.6\text{e}{+00} & -2.5\text{e}{-02} & 1.0\text{e}{+00} & 9.9\text{e}{-07} \\ 5.0\text{e}{-01} & 2.0\text{e}{+00} & 1.1\text{e}{+00} & -9.0\text{e}{-01} & 1.0\text{e}{+00} \end{pmatrix}. \tag{2.2}$$

The behavior of these vectors is more complicated than the behavior of the eigenvalues. The subdiagonal elements are all of order one in magnitude. As we go upward from the diagonal, the components of the eigenvectors scale downward with ratios of about 10^{-4}. Once again this is typical behavior.

In the next section we will see that there is an intimate relation between the structure of the eigensystem of a graded matrix and the Schur complements of its leading principal submatrices. The following numbers illustrate this connection.

$$\begin{array}{ccccc} -6.5479\text{e}{-}01 & 7.8905\text{e}{-}05 & -4.3292\text{e}{-}09 & -3.2932\text{e}{-}12 & -3.5208\text{e}{-}16 \\ -6.5471\text{e}{-}01 & 7.8915\text{e}{-}05 & -4.3292\text{e}{-}09 & -3.2932\text{e}{-}12 & -3.5208\text{e}{-}16 \end{array}$$

The first row contains the eigenvalues of A, this time displayed to five figures. Below it are the diagonals from the U-factor of an unpivoted LU decomposition of A. The latter approximate the former to four or five figures. Since the U-factor is the triangular matrix computed by Gaussian elimination, the eigenvalues of a graded matrix can typically be approximated by performing Gaussian elimination on the matrix.

2.2. An atypical matrix

The matrix

$$\tilde{A} = \begin{pmatrix} -6.5\text{e}{-}01 & -5.0\text{e}{-}05 & 4.4\text{e}{-}09 & 4.1\text{e}{-}14 & -9.8\text{e}{-}17 \\ -1.1\text{e}{+}00 & -8.2\text{e}{-}05 & 1.3\text{e}{-}08 & -7.6\text{e}{-}13 & -6.9\text{e}{-}17 \\ -4.8\text{e}{-}02 & -1.7\text{e}{-}05 & -5.0\text{e}{-}09 & -8.9\text{e}{-}14 & 1.3\text{e}{-}16 \\ 3.8\text{e}{-}01 & -9.6\text{e}{-}05 & -1.1\text{e}{-}08 & -2.0\text{e}{-}12 & -9.1\text{e}{-}17 \\ -3.3\text{e}{-}01 & 1.3\text{e}{-}04 & 8.1\text{e}{-}09 & 1.1\text{e}{-}12 & -4.1\text{e}{-}17 \end{pmatrix}$$

was obtained from A by altering its $(2,2)$-element. The eigenvalues of this matrix are

$$-6.5\text{e}{-}01 \; 9.2\text{e}{-}07 \; 7.8\text{e}{-}08 \; 5.2\text{e}{-}12 \; -1.0\text{e}{-}15$$

It is seen that the second and third eigenvalues of A no longer track the original grading.

The matrix of eigenvectors for $\tilde{A}$ is

$$\begin{pmatrix} 1.0\text{e}{+}00 & -7.6\text{e}{-}05 & 4.6\text{e}{-}07 & 6.4\text{e}{-}12 & 1.1\text{e}{-}15 \\ 1.7\text{e}{+}00 & 1.0\text{e}{+}00 & -6.0\text{e}{-}03 & -6.9\text{e}{-}08 & -1.1\text{e}{-}11 \\ 7.3\text{e}{-}02 & -1.5\text{e}{+}01 & 1.0\text{e}{+}00 & 1.6\text{e}{-}04 & 4.9\text{e}{-}08 \\ -5.8\text{e}{-}01 & -1.4\text{e}{+}02 & 9.5\text{e}{+}00 & 1.0\text{e}{+}00 & 4.3\text{e}{-}04 \\ 5.0\text{e}{-}01 & 1.7\text{e}{+}02 & -1.2\text{e}{+}01 & -1.7\text{e}{+}00 & 1.0\text{e}{+}00 \end{pmatrix}.$$

The grading in the first, fourth and fifth columns is as above. However, the subdiagonal elements in the second column are considerably larger than one, and the grading of the superdiagonal elements in the third column is more gentle than above.

Finally, when we compare the eigenvalues of $\tilde{A}$ with the diagonals of its U-factors we get the following table.

$$-6.5479e{-}01\ 9.1704e{-}07\ 7.7527e{-}08\ 5.1728e{-}12\ -1.0073e{-}15$$
$$-6.5471e{-}01\ 1.0000e{-}06\ 7.1214e{-}08\ 5.1612e{-}12\ -1.0080e{-}15$$

The second and third eigenvalues are not well approximated by the diagonals of U.

This last set of numbers has two features well worth noting. First, only the approximations to the second and third eigenvalues are affected. The diagonals of U provide good approximations to the first, fourth and fifth eigenvalues. Somehow the atypical behavior is localized.

Second, the number $1.0000e{-}06$ (already suspect because of the string of zeros) is smaller than one would expect from performing Gaussian elimination on a random matrix scaled by D. This suggests that unusually small elements on the diagonal of U are associated with atypical behavior. We will make the connection clear in the next section.

3. The eigenstructure of a graded matrix

In this section we will describe the structure of the eigenvalues and eigenvectors of a graded matrix. The key idea is that when the grading is sufficiently strong, the matrix can be reduced by a similarity transformation to a block diagonal matrix. Moreover, as the grading increases, the reducing transformation approaches a fixed limit that is independent of the grading. By calculating the eigenvectors of the diagonal blocks of the block tridiagonal matrix we can compute approximations to the eigenvectors of the original matrix that amount to scaling certain essentially constant vectors.

We will begin this section with some definitions and observations. We will then show how to block-triangularize a graded matrix. We will then use the block triangular matrix to compute approximations to the eigensystem of the matrix. We conclude with an example of a gently graded matrix.

3.1. Definitions and observations

As we mentioned in the introduction, our approach to graded matrices amounts to grading a base matrix B and then determining what properties of B yield a tractable graded matrix. This approach leads to the following definition.

Definition 3.1. *Let B be a given base matrix of order n and let*

$$D = \operatorname{diag}(\delta_1, \delta_2, \ldots, \delta_n), \qquad \delta_1 \geq \delta_2 \geq \cdots \delta_n > 0.$$

Then

1. $A = BD$ *is column graded with respect to* B *and* D*,*
2. $A = DB$ *is row graded with respect to* B *and* D*,*
3. $A = D^{\frac{1}{2}}BD^{\frac{1}{2}}$ *is diagonally graded with respect to* B *and* D*,*

The numbers δ_k *are called the grading factors. The numbers*

$$\rho_k = \frac{\delta_{k+1}}{\delta_k}$$

are called the grading ratios.

There are four comments to make about this definition.

1. Because the grading factors decrease, we say the grading is downward. It is also possible to grade upward. Our results, derived here for downward grading, also apply with obvious modifications to upward grading.
2. Although we formally regard our graded matrices as coming from a base matrix B and a diagonal grading matrix D, in practice it will be the other way around. For example, given a column graded matrix A one might define B by normalizing the columns of A and define D to consist of the reciprocals of the normalizing factors.
3. No particular assumption is made about the structure of B, although it is natural to think of it as being in some sense balanced. In particular, the results we are going to establish hold if B is a band matrix or a Hessenberg matrix.
4. The grading ratios are never greater than one, but they are allowed to be equal to one. Thus our theory applies to block-graded matrices, as well as the more conventional grading appearing in the first two sections.

An important observation is that the three types of grading in Definition 3.1 can be obtained from one another by diagonal similarities. For example if $A = BD$ is column graded, then $DAD^{-1} = DB$ is row graded. This means that we are free to chose a style of grading and stick to it through our analysis, after which the results can be transfered to the other styles. It turns out that column grading gives the cleanest derivations.

Our main result will be cast in terms of partitioned matrices and certain numbers obtained from these partitions. In particular, we introduce the following notation and terminology.

Definition 3.2. *Let* $A = BD$ *be partitioned in the form*

$$\begin{pmatrix} A_{kk} & A_{kn} \\ A_{nk} & A_{nn} \end{pmatrix} = \begin{pmatrix} B_{kk}D_k & B_{kn}D_n \\ B_{nk}D_k & B_{nn}D_n \end{pmatrix},$$

where A_{kk} *is of order* k*. Then the number*

$$\kappa_k \stackrel{\text{def}}{=} \|B_{kk}^{-1}\|\|B\|$$

is called the kth grading impediment. *The number*

$$\gamma_k \stackrel{\text{def}}{=} \kappa_k \rho_k$$

is called the kth grading coefficient.

The grading impediments get their name as follows. It will turn out that the behavior of graded matrices is controled by the size of the grading coefficients γ_k—the smaller the better. These coefficients can be made arbitrarily small by making the grading ratios sufficiently small. But if κ_k (which is never less than one) is large, the grading ratios will have to be correspondingly small for the the grading coefficients to be small. For this reason we call the numbers κ_k grading impediments.

3.2. Block triangularization

In this subsection we will be concerned with reducing A to block triangular form by a triangular similarity transformation. Specifically, we will try to find a matrix P_{nk} such that

$$\begin{pmatrix} I & 0 \\ -P_{nk} & I \end{pmatrix} \begin{pmatrix} A_{kk} & A_{kn} \\ A_{nk} & A_{nn} \end{pmatrix} \begin{pmatrix} I & 0 \\ P_{nk} & I \end{pmatrix} = \begin{pmatrix} A_{kk} + A_{kn}P_{nk} & A_{kn} \\ 0 & A_{nn} - P_{nk}A_{kn} \end{pmatrix}. \tag{3.1}$$

From elementary linear algebra, we know that the eigenvalues of A are, counting multiplicities, the union of the eigenvalues of $A_{kk} + A_{kn}P_{nk}$ and those of $A_{nn} - P_{nk}A_{kn}$. Moreover, it is easily verified that if y is an eigenvector of $A_{kk} + A_{kn}P_{nk}$ then

$$\begin{pmatrix} y \\ P_{nk}y \end{pmatrix} \tag{3.2}$$

is an eigenvector of A and conversely.

More generally, it follows from (3.1) that

$$A\begin{pmatrix} I \\ P_{nk} \end{pmatrix} = \begin{pmatrix} I \\ P_{nk} \end{pmatrix}(A_{kk} + A_{kn}P_{nk}). \tag{3.3}$$

We say that the columns of $(I \; P_{nk}^{\mathrm{H}})^{\mathrm{H}}$ span an eigenspace (or invariant subspace) of A and that $A_{kk} + A_{kn}P_{nk}$ is the representation of A on that subspace. In particular, if $k = 1$, then $(1 \; p_{n1}^{\mathrm{H}})^{\mathrm{H}}$ is an eigenvector of A corresponding to the eigenvalue $a_{11} + a_{1n}^{\mathrm{H}}p_{n1}$.

Turning now to the existence of P_{nk}, if we write out the $(2,1)$-block of the right-hand side of (3.1) and set the result to zero, we get the equation

$$P_{nk}A_{kk} - A_{nn}P_{nk} = A_{nk} - P_{nk}A_{kn}P_{nk}.$$

Assuming that A_{kk} is nonsingular, we can write this equation in the form

$$P_{nk} - A_{nn}P_{nk}A_{kk}^{-1} = A_{nk}A_{kk}^{-1} - P_{nk}A_{kn}P_{nk}A_{kk}^{-1},$$

or in terms of B and D

$$P_{nk} - B_{nn}D_nP_{nk}D_k^{-1}B_{kk}^{-1} = B_{nk}B_{kk}^{-1} - P_{nk}B_{kn}D_nP_{nk}D_k^{-1}B_{kk}^{-1}. \tag{3.4}$$

This equation already exhibits the asymptotic form of P_{nk} as the grading ratio ρ_k approaches zero. Specifically, we have

$$\begin{aligned}\|B_{nn}D_nP_{nk}D_k^{-1}B_{kk}^{-1}\| &\le \|B_{kk}^{-1}\|\|B_{nn}\|\|D_k^{-1}\|\|D_n\|\|P_{nk}\| \\ &\le \kappa_k\rho_k\|P_{nk}\| = \gamma_k\|P_{nk}\|.\end{aligned}$$

Thus the second term on the left-hand side of (3.4) vanishes as $\rho_k \to 0$. Similarly,

$$\|P_{nk}B_{kn}D_nP_{nk}D_k^{-1}B_{kk}^{-1}\| \le \gamma_k\|P_{nk}\|^2,$$

so that the second term on the right-hand side vanishes as $\rho_k \to 0$. We are left with the equation

$$P_{nk} \cong B_{nk}B_{kk}^{-1}.$$

This remarkable approximation says that as the kth grading ratio approaches zero the block diagonalizing similarity transformation in (3.1) effectively depends only on the base matrix and not the grading. It also says that the norm of P_{nk} is asymptotically bounded by the kth grading impediment.

Regarding the existence of P_{nk}, the equation (3.4) is nonlinear and cannot be solved in closed form. Fortunately, similar equations appear in the perturbation theory of eigenspaces, and the analyses contained in that literature can be adapted to prove the following theorem[1].

Theorem 3.3. *If*

$$\frac{\gamma_k\|B_{nk}B_{kk}^{-1}\|}{(1-\gamma_k)^2} < \frac{1}{4},$$

then (3.4) *has a unique solution satisfying*

$$\|P_{nk}\| \le \frac{2\|B_{nk}B_{kk}^{-1}\|}{1-\gamma_k} \le \frac{2\kappa_k}{1-\gamma_k}.$$

Moreover,

$$\frac{\|B_{nk}B_{kk}^{-1} - P\|}{\|P\|} \le \gamma_k(1 + \|P_{nk}\|). \tag{3.5}$$

[1] Specifically, in Theorem V.2.1 in [14] take $T = P_{nk} \mapsto P_{nk} - B_{nn}D_nP_{nk}D_k^{-1}B_{kk}^{-1}$, $g = B_{nk}B_{kk}^{-1}$, and $\varphi(P_{nk}) = P_{nk}B_{kn}D_nP_{nk}D_k^{-1}B_{kk}^{-1}$

Here are some observations on this theorem.

1. Although the purpose of this paper is to investigate the behavior of graded matrices as the grading ratios become small, this theorem does presuppose any such dynamic behavior. In fact, given any matrix A we can factor $A = BD$ into a base matrix and a grading matrix in any way we like. As long as a grading coefficient from this factorization satisfies the conditions of the theorem, the matrix can be block triangularized. In the next subsection, we will encounter a situation in which the "base matrix" actually depends on the grading ratios.
2. The theorem is local, depending only on the grading from k to $k+1$— i.e., γ_k.
3. The bound (3.5) quantifies the fact that P_{nk} is approximated by $B_{nk}B_{kk}^{-1}$. Specifically, the bound on the normwise relative error in P_{nk} is proportional to the grading coefficient γ_k.
4. The matrix

$$A_{kk} + A_{kn}P_{nk} = B_{kk}D_k + B_{kn}D_nP_{nk}$$

contains the eigenvalues of A corresponding to the eigenspace spanned by $(I\ P_{nk}^{\mathrm{H}})^{\mathrm{H}}$. As γ_k decreases, the second term on the right becomes insignificant compared to the first. In other words, the eigenvalues of A_{kk} provide approximations to the largest k eigenvalues of A.
5. The matrix

$$A_{nn} - P_{nk}A_{kn} = (B_{nn} - P_{nk}B_{kn})D_n.$$

contains the remaining eigenvalues of A. As γ_k decreases, this matrix approaches

$$(B_{nn} - B_{nk}B_{kk}^{-1}B_{kn})D_n = A_{nn} - A_{nk}A_{kk}^{-1}A_{kn}.$$

The right hand side is the Schur complement of A_{kk} in A, which therefore contains approximations to the $n-k$ smallest eigenvalues of A.
6. Since the subspace spanned by the columns of a matrix does not change when the matrix is postmultiplied by a nonsingular matrix, the matrix

$$\begin{pmatrix} I \\ P_{nk} \end{pmatrix} B_{kk}D_k$$

spans an eigenspace of A. As γ_k decreases, this matrix approaches

$$\begin{pmatrix} A_{kk} \\ A_{nk} \end{pmatrix}.$$

Thus the span of the first k columns of A approximates an eigenspace of A.

In the special case where $k = 1$, and γ_1 is small, it follows from the above results that a_{11} is an approximate eigenvalue of A whose eigenvector is approximately the first column of A. To say something about the other eigenvalues and eigenvectors we must perform a further reduction, to which we now turn.

3.3. Eigenvalues and eigenvectors

If γ_k is sufficiently small, we can compute the first k eigenvalues of A from the matrix

$$C_{kk} = A_{kk} + A_{kn}P_{nk}$$

Moreover, any eigenvector y of C_{kk} can be converted into an eigenvector of A via the formula (3.2). We will be interested in the eigenvector corresponding to the kth eigenvalue λ_k. Our approach is to use Theorem 3.3 to pick off that eigenpair, under the hypothesis that γ_{k-1} is sufficiently small.

We will assume that $B_{k-1,k-1}$ is nonsingular. Write C_{kk} in the form

$$C_{kk} = (B_{kk} + B_{kn}D_nP_{nk}D_k^{-1})D_k. \tag{3.6}$$

Since $B_{kn}D_nP_{nk}D_k^{-1} \to 0$ as $\gamma_k \to 0$, if γ_k is sufficiently small, $C_{k-1,k-1}$ will also be nonsingular. In particular, the grading coefficient $\tilde{\gamma}_{k-1}$ for $C_{k-1,k-1}$ factored as in (3.6) is well defined. Moreover, this grading coefficient goes to zero at the same rate as γ_{k-1}, so we can write $O(\gamma_{k-1})$ in place of $O(\tilde{\gamma}_{k-1})$.

Partition C_{kk} in the form

$$C_{kk} = \begin{pmatrix} C_{k-1,k-1} & c_{k-1,k} \\ c_{k,k-1}^{\mathrm{H}} & c_{kk} \end{pmatrix}.$$

By Theorem 3.3, if $\tilde{\gamma}_{k-1}$ is sufficiently small, there is a vector $q_{k,k-1}^{\mathrm{H}}$ such that

$$\begin{pmatrix} I & 0 \\ -q_{k,k-1}^{\mathrm{H}} & 1 \end{pmatrix} \begin{pmatrix} C_{k-1,k-1} & c_{k-1,k} \\ c_{k,k-1}^{\mathrm{H}} & c_{kk} \end{pmatrix} \begin{pmatrix} I & 0 \\ q_{k,k-1}^{\mathrm{H}} & 1 \end{pmatrix}$$
$$= \begin{pmatrix} C_{k-1,k-1} + c_{k-1,k}q_{k,k-1}^{\mathrm{H}} & c_{k-1,k} \\ 0 & c_{kk} - q_{k,k-1}^{\mathrm{H}}c_{k-1,k} \end{pmatrix}.$$

The quantity

$$\lambda_k = c_{kk} - q_{k,k-1}^{\mathrm{H}}c_{k-1,k}$$

is the kth eigenvalue of A. The corresponding eigenvector of C_{kk} is

$$y = \begin{pmatrix} (\lambda_k I - C_{k-1,k-1} - c_{k-1,k}q_{k,k-1}^{\mathrm{H}})^{-1}c_{k-1,k} \\ 1 + q_{k,k-1}^{\mathrm{H}}(\lambda_k I - C_{k-1,k-1} - c_{k-1,k}q_{k,k-1}^{\mathrm{H}})^{-1}c_{k-1,k} \end{pmatrix}. \tag{3.7}$$

Now

$$q_{k,k-1}^{\mathrm{H}} = c_{k,k-1}^{\mathrm{H}} C_{k-1,k-1}^{-1} + O(\gamma_{k-1})$$

$$\lambda_k = c_{kk} - c_{k,k-1}^{\mathrm{H}} C_{k-1,k-1}^{-1} c_{k-1,k} + \delta_k O(\gamma_{k-1}).$$

These two equations, along with (3.6) imply that $C_{kk}D_k^{-1}$, $q_{k,k-1}$, δ_k^{-1} $c_{k-1,k}$, and $\delta_k^{-1}\lambda_k$ all approach constants as $\gamma_{k-1} \to 0$.

We now wish to simplify the the expression (3.7) for y. Write

$$w = (\lambda_k I - C_{k-1,k-1} - c_{k-1,k} q_{k,k-1}^{\mathrm{H}})^{-1} c_{k-1,k}$$

Then

$$w = (Q - I)^{-1} C_{k-1,k-1}^{-1} c_{k-1,k}$$

where

$$Q = \lambda_k C_{k-1,k-1}^{-1} - C_{k-1,k-1}^{-1} c_{k-1,k} q_{k,k-1}^{\mathrm{H}}.$$

But $(Q - I)^{-1} = -I + Q(Q - I)^{-1}$. Hence

$$w = -C_{k-1,k-1}^{-1} c_{k-1,k} + Q(Q - I) C_{k-1,k-1}^{-1} c_{k-1,k},$$

or

$$w = -C_{k-1,k-1}^{-1} c_{k-1,k} + \delta_k D_{k-1}^{-1} (\delta_k^{-1} D_{k-1} Q)(I + Q) C_{k-1,k-1}^{-1} c_{k-1,k}$$

Now as $\gamma_k \to 0$, $Q \to 0$ and $\delta_k^{-1} D_{k-1} Q$ approaches a constant. Moreover, the vector $C_{k-1,k-1}^{-1} c_{k-1,k}$ is $O(\gamma_{k-1})$. It follows that

$$w = -C_{k-1,k-1}^{-1} c_{k-1,k} + \delta_k D_{k-1}^{-1} O(\gamma_{k-1}),$$

and

$$y = \begin{pmatrix} -C_{k-1,k-1}^{-1} c_{k-1,k} + \delta_k D_{k-1}^{-1} O(\gamma_{k-1}) \\ 1 + O(\gamma_{k-1}) \end{pmatrix}.$$

From (3.2), the kth eigenvector of A is given by

$$x_k = \begin{pmatrix} -C_{k-1,k-1}^{-1} c_{k-1,k} + \delta_k D_{k-1}^{-1} O(\gamma_{k-1}) \\ 1 + O(\gamma_{k-1}) \\ -P_{n,k-1} C_{k-1,k-1}^{-1} c_{k-1,k} + p_{n,k} + O(\gamma_{k-1}) \end{pmatrix}.$$

We now observe that as $\gamma_k \to 0$, $C_{k-1,k-1} D_{k-1}^{-1} \to B_{k-1,k-1}$, and similarly for the other components in the partition of C_{kk}. Hence we have the following theorem.

Theorem 3.4. *As γ_{k-1} and γ_k approach zero, we have*

$$\lambda_k = \delta_k(b_{kk} - b_{k,k-1}^{\mathrm{H}} B_{k-1,k-1}^{-1} b_{k-1,k}) + \delta_k O(\max\{\gamma_{k-1}, \gamma_k\}) \tag{3.8}$$

and

(3.9)

$$x_k = \begin{pmatrix} -\delta_k D_{k-1}^{-1} B_{k-1,k-1}^{-1} b_{k-1,k} \\ 1 \\ B_{nk} B_{kk}^{-1} \mathrm{e}_k \end{pmatrix} + \begin{pmatrix} \delta_k D_{k-1}^{-1} O(\max\{\gamma_{k-1}, \gamma_k\}) \\ 0 \\ O(\max\{\gamma_{k-1}, \gamma_k\}). \end{pmatrix},$$

where e_k is the last column of the $k \times k$ identity matrix.

The expressions in Theorem 3.4 represent a nice division of labor. The matrix B determines the unscaled structure of the eigenvalue and eigenvector; the matrix D determines their scale. We will exploit this division of labor in the next section, where we determine condition numbers of eigenvalues and eigenvectors.

The expressions confirm the observations made in §2.1 and §2.2. Their validity depends only on the sizes of the local grading coefficients γ_{k-1} and γ_k. The approximation (3.8) to λ_k is δ_k times the Schur complement of $B_{k-1,k-1}$ in B_{kk}—precisely the kth diagonal element of the U-factor in the LU decomposition of A. The approximation (3.9) to the eigenvector has the scaling shown in (2.2). In fact the approximate eigenvector can be quite good. For example, here is x_3 from the matrix (2.1) compared with its approximation.

x_3	approximation
1.2086e−08	1.2087e−08
−7.0094e−05	−7.0097e−05
1.0000e+00	1.0000e+00
−2.4855e−02	−2.4947e−02
1.1495e+00	1.1496e+00

With the exception of the unusually small fourth component, the vectors agree to four figures, which is consonant with the grading ratios $\gamma_2 = 1.1 \cdot 10^{-3}$ and $\gamma_3 = 1.3 \cdot 10^{-3}$ for this example.

If A is real, then all the quantities in (3.8) are real, and consequently λ_k is real. Moreover, if we allow nonpositive scaling factors, we can change the sign of λ_k, or even make it complex with any argument we wish. In particular, a general complex matrix whose leading principal submatrices are nonsingular can be graded so that all its eigenvalues are real (Fisher and Fuller [6]).

The grading coefficients for A^{H} are the same as for A. Consequently, the left eigenvectors of A are as well (or ill) behaved as the right eigenvectors. However, the approximation (3.9) is no longer valid, since A^{H} is graded by

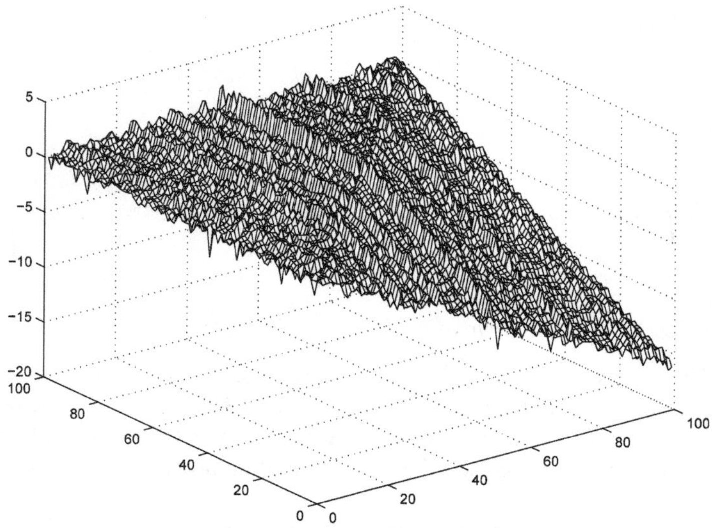

Fig. 1. Eigenvectors of a gently graded matrix

rows. However, the correct approximation can be recovered by transforming (3.9) into an approximation suitable for a matrix graded by rows; i.e., by multiplying it by D.

3.4. Gentle grading

In the foregoing we have assumed that the local grading coefficients were sufficiently small to allow the matrix to split as in Theorem 3.3. But even when the grading is gentle, the structure we have described persists, albeit in a rough way. Figure 1 contains a mesh plot of the common logarithms of the absolute values of the components of the eigenvectors of a matrix obtained by column grading a random matrix of order 100 with grading ratios of 0.69 (i.e., grading factors running from 1 to 10^{-16}). As above the first k components tend to be constant and then the components show a decrease at a rate determined by the grading ratio. The behavior is not uniform: note the ridges formed groups of the eigenvectors. But the plot never deviates far from the normative behavior. Why this should be so is an open research question.

4. Condition numbers

In this section we will derive approximate perturbation bounds for the eigenvalues and eigenvectors of a column graded matrix. The bounds themselves are certainly overestimates. But they give us reason to believe that, baring untoward circumstances, the small eigenvalues and the small components of their eigenvectors are determined to high relative accuracy.

An interesting fact that will emerge from our analysis is that the condition of the eigenvalues and eigenvectors of a graded matrix depends on the grading impediments, not the grading coefficients. Of course, the grading coefficients must be small enough for our approximations to be valid. But once they are, further reducing the grading coefficients by reducing the grading ratios has little effect on the condition of the eigenvalues and eigenvectors.

4.1. Generalities

In order to derive condition numbers for graded matrices, we must first decide what it means to perturb a graded matrix. It seems natural that such a perturbation should itself be graded, we will adopt the following definition.

Definition 4.1. *Let $A = BD$ be a graded matrix and let $B + E$ be a perturbation of B. Then $BD + ED$ is a graded perturbation of A.*

Just as a graded matrix is generated by grading a base matrix, a graded perturbation is generated by grading a base perturbation of the base matrix. Our analysis will be entirely in terms of this base perturbation. Note that a graded perturbation need not represent a small componentwise perturbation in the elements of A since we do not exclude large relative perturbations in small elements of B.

As we shall see, it is easy to plug E into the expressions in Theorem 3.4 to get first order perturbation expansions. Taking norms in these expansions gives first order bounds from which we can derive putative condition numbers. Unfortunately, we are not working with eigenvalues and eigenvectors of A but with approximations to them, and these approximations may be in greater error than the first order error bounds we derive. It is therefore uncertain what such bounds actually mean.

To see what is going on, let us suppose we have a function $\varphi(\gamma, \epsilon)$, where $\varphi(0,0)$ represents our eigenvalue or eigenvector and $\varphi(\gamma, 0)$ represents its approximation. The quantities $\varphi(0, \epsilon)$ and $\varphi(\gamma, \epsilon)$ represents the perturbations of the original quantity and its approximation. Now a first order perturbation expansion gives

$$\varphi(0, \epsilon) \cong \varphi(0,0) + \varphi_\epsilon(0,0)\epsilon,$$

where φ_ϵ is the derivative of φ with respect to ϵ. What we actually compute is

$$\varphi(\gamma, \epsilon) \cong \varphi(\gamma, 0) + \varphi_\epsilon(\gamma, 0)\epsilon.$$

Depending on the size of ϵ the perturbations $\varphi_\epsilon(0,0)\epsilon$ and $\varphi_\epsilon(\gamma,0)\epsilon$ may be far smaller than $\varphi(\gamma,0)-\varphi(0,0)$. Nonetheless, if φ_ϵ is differentiable with respect to γ, $\varphi_\epsilon(\gamma,0)-\varphi_\epsilon(0,0)=O(\gamma)$. Hence

$$\varphi_\epsilon(\gamma,0)\epsilon-\varphi_\epsilon(0,0)\epsilon=O(\gamma\epsilon).$$

It follows that if γ is small enough, whether we base our perturbation theory on $\varphi(0,0)$ or $\varphi(\gamma,0)$ they give essentially the same correction.

4.2. Eigenvalues

Since the approximation (3.8) for the eigenvalue is derived by multiplying the Schur complement

$$\mu_k=b_{kk}-b_{k,k-1}^{\mathrm{H}}B_{k-1,k-1}^{-1}b_{k-1,k}$$

by δ_k, it is sufficient to derive a perturbation expansion for the Schur complement. If E is partitioned conformally with B, the perturbed Schur complement becomes

$$\begin{aligned}\tilde{\mu}_k=b_{kk}+e_{kk}-(b_{k,k-1}^{\mathrm{H}}+e_{k,k-1}^{\mathrm{H}})(B_{k-1,k-1}+E_{k-1,k-1})^{-1}\\ \times(b_{k-1,k}+e_{k-1,k}).\end{aligned}$$

Replacing $(B_{k-1,k-1}+E_{k-1,k-1})^{-1}$ by the first order expansion

$$(B_{k-1,k-1}+E_{k-1,k-1})^{-1}=B_{k-1,k-1}^{-1}-B_{k-1,k-1}^{-1}E_{k-1,k-1}B_{k-1,k-1}^{-1}$$

and dropping second order terms, we get

$$\begin{aligned}\tilde{\mu}_k\cong\mu_k+e_{kk}+b_{k,k-1}^{\mathrm{H}}B_{k-1,k-1}^{-1}e_{k-1,k}+e_{k,k-1}^{\mathrm{H}}B_{k-1,k-1}\\ -b_{k,k-1}^{\mathrm{H}}B_{k-1,k-1}^{-1}E_{k-1,k-1}B_{k-1,k-1}^{-1}b_{k-1,k}.\end{aligned}$$

Taking norms we get and bounding terms like $B_{k-1,k-1}^{-1}b_{k-1,k}$ by κ_{k-1}, we get

$$|\tilde{\mu}_k-\mu_k|\lesssim(1+\kappa_{k-1})^2\|E\|.$$

Hence

$$\frac{|\tilde{\mu}_k-\mu_k|}{|\mu_k|}\lesssim(1+\kappa_{k-1})^2\frac{\|B\|}{|\mu_k|}\frac{\|E\|}{\|B\|}.$$

Multiplying the numerator and denominator of the left-hand side of this relation by δ_k, we get

$$\frac{|\tilde{\lambda}_k-\lambda_k|}{|\lambda_k|}\lesssim(1+\kappa_{k-1})^2\frac{\|B\|}{|\mu_k|}\frac{\|E\|}{\|B\|}.$$

This bound shows that the relative condition of λ_k is governed by two factors. The first is essentially the square of the grading impediment κ_{k-1}.

The second is the ratio of $\|B\|$ to the kth diagonal element of the U-factor of B. Since μ_k^{-1} is the (k,k)-element of B_{kk}^{-1}, this ratio is bounded by κ_k.

The second factor has an interesting interpretation. In an ordinary ungraded eigenvalue problem, a small eigenvalue, even if it is well conditioned in an absolute sense, will be ill-conditioned in a relative sense. An analogous phenomenon holds for eigenvalues of graded matrices, but it is not the size of the eigenvalue that determines the ill-conditioning but the size of the Schur complement μ_k with respect to the base matrix.

4.3. Eigenvectors

Bounds for eigenvectors are complicated by the fact that the expression (3.9) has two distinct formulas. We begin by writing

$$x_k = \begin{pmatrix} -\delta_k D_{k-1}^{-1} y \\ 1 \\ z \end{pmatrix},$$

where $y = B_{k-1,k-1}^{-1} b_{k-1,k}$ and $z = B_{nk} B_{kk}^{-1} \mathbf{e}_k$.

The perturbation of y is the same as the perturbation of the system $B_{k-1,k-1} y = b_{k-1,k}$. We can therefore use standard perturbation theory [13, Sect. 3.3.1]to get

$$\frac{\|\tilde{y} - y\|}{\|y\|} \lesssim \kappa_{k-1} \left(\frac{\|E_{k-1,k-1}\|}{\|B_{k-1,k-1}\|} + \frac{\|e_{k-1,k}\|}{\|b_{k-1,k}\|} \right).$$

Since the ith component of x_i is approximated by $x_i^{(k)} \cong -\delta_k \delta_i^{-1} y_i$, we have

$$\frac{|\tilde{x}_i^{(k)} - x_i^{(k)}|}{\delta_k^{-1} \delta_i \|y\|} \lesssim \kappa_{k-1} \left(\frac{\|E_{k-1,k-1}\|}{\|B_{k-1,k-1}\|} + \frac{\|e_{k-1,k}\|}{\|b_{k-1,k}\|} \right).$$

This bound has the following interpretation. The number κ_k is the relative condition number for all components of the upper half of the eigenvector for which y_i is not much smaller as $\|y\|$. However, as y_i becomes smaller, its relative accuracy deteriorates.

The perturbation expansion for z does not simplify as nicely as that of y. A straightforward analysis yields the following bound:

$$\frac{\|\tilde{z} - z\|}{\|z\|} \lesssim (1 + \kappa_k) \frac{\|b_{kk}^{(-1)}\| \|B\|}{\|z\|} \frac{\|E\|}{\|B\|},$$

where $b_{kk}^{(-1)}$ denotes the kth column of B_{kk}^{-1}. There is no need to break this bound into components, since z is not graded. The first factor in this bound is essentially the kth grading impediment. The second factor, which is bounded by $\kappa_k/\|z\|$, has the following interpretation. Since $z = B_{nk}b_{kk}^{(-1)}$, we have $\|z\| \le \|B_{nk}\|\|b_{kk}^{(-1)}\|$, so that the factor is always greater than one. It is much greater than one when z is atypically small; i.e., when it does not reflect the size of $b_{kk}^{(-1)}$. As with the bound for y, only the larger components of z are determined with high relative accuracy.

Most bounds for eigenvectors, whether normwise or componentwise, invoke a gap hypothesis that says that the eigenvalue in question is sufficiently separated from its neighbors. No explicit gap appears in our expressions. The reason is that we have assumed that γ_{k-1} and γ_k are small. This forces the eigenvalue λ_k to be well enough separated from its neighbors for our bounds to hold.

5. Positive definite matrices and the singular value decomposition

In this section we will show how our theory applies to positive definite matrices. We will then use these results to describe the behavior of the singular value decomposition of a graded matrix.

5.1. Positive definite matrices

In treating positive definite matrices, it is natural to pass on the symmetry (and positive-definiteness) of B to A by grading B by diagonals so that $A = D^{\frac{1}{2}}BD^{\frac{1}{2}}$. The expression (3.9) for the eigenvectors must then be multiplied by $D^{\frac{1}{2}}$. When the kth component of x_k is normalized to one, the result is

(5.1)

$$x_k = \begin{pmatrix} -\delta_k^{\frac{1}{2}} D_{k-1}^{-\frac{1}{2}} B_{k-1,k-1}^{-1} b_{k-1,k} \\ 1 \\ \delta_k^{-\frac{1}{2}} D_n^{\frac{1}{2}} B_{nk} B_{kk}^{-1} e_k \end{pmatrix} + \begin{pmatrix} \delta_k^{\frac{1}{2}} D_{k-1}^{-\frac{1}{2}} O(\max\{\gamma_{k-1}, \gamma_k\}) \\ 0 \\ \delta_k^{-\frac{1}{2}} D_n^{\frac{1}{2}} O(\max\{\gamma_{k-1}, \gamma_k\}) \end{pmatrix}.$$

Thus, when the grading ratios are constant, each eigenvector x_k exhibits constant grading downward above and below its kth component.

The grading impediments of a graded positive definite matrix are better behaved that those of a general graded matrix. Because of the interlacing properties of the eigenvalues of symmetric matrices, we have

$$\|B_{11}^{-1}\| \le \|B_{22}^{-1}\| \le \cdots \le \|B_{nn}^{-1}\|.$$

Hence the grading impediments κ_k are nondecreasing and are bounded by κ_n. In particular, graded positive definite matrices cannot exhibit the intermediate ill behavior found in §2.2.

There is also a computational difference between graded positive definite matrices and graded general matrices. We have seen that for a general graded matrix A the eigenvalues corresponding to sufficiently small grading coefficients are approximated by the diagonal elements of the U-factor of A. Unfortunately, this U-factor must be computed by Gaussian elimination without pivoting, which will be unstable if any of the grading impediments are large. With positive definite matrices pivoting is unnecessary for a stable reduction.

5.2. The singular value decomposition

In this subsection we will derive the structure of the singular value decomposition of graded matrices. For definiteness we will consider an $m \times n$ $(m \geq n)$ base matrix Y and a column graded matrix $X = YD^{\frac{1}{2}}$. (Results for row graded matrices can be obtained by considering the transpose matrix.) We will write the singular value decomposition of X as

$$X = U\Sigma V^{\mathrm{H}},$$

where $\Sigma = \mathrm{diag}(\sigma_1, \ldots, \sigma_n)$ $(\sigma_1 \geq \cdots \geq \sigma_n \geq 0)$ and U and V are orthonormal. The columns u_i of U are called the right singular vectors of X and the columns v_i of V are called the left singular vectors of X.

We will be chiefly concerned with a qualitative description of the structure of the graded singular value decomposition; however, formulas and bounds can easily be obtained from our previous results. The key observation is that the squares of the singular values of X are the eigenvalues of $A = X^{\mathrm{H}}X$ and the right singular vectors of X are the eigenvectors of A. Moreover if v_k is a singular vector of X corresponding to the singular value σ_k, then $u_k = \sigma_k^{-1}Xv_k$.

Now let $B = Y^{\mathrm{H}}Y$, so that $A = D^{\frac{1}{2}}BD^{\frac{1}{2}}$, and let $B = S^{\mathrm{H}}S$ be the Cholesky factorization of B. Assuming that the grading coefficients γ_{k-1} and γ_k of A are sufficiently small, we have the following results.

1. The square of the kth singular value of X is approximately

$$\delta_k s_{kk}^2 + \delta_k O(\max\{\gamma_{k-1}, \gamma_k\}).$$

   It follows that

$$\sigma_k = \delta_k^{\frac{1}{2}}(s_{kk} + O(\max\{\gamma_{k-1}, \gamma_k\})).$$

2. The left singular vector v_k has the structure given in (5.1).

To determine the behavior of the right singular vectors, we will exploit a connection between the singular value decomposition of a graded matrix and its QR decomposition. Specifically, suppose that $\gamma_k \to 0$. Then the columns of $(I \ \ P_{nk}^{\mathrm{H}})^{\mathrm{H}}$ [see (3.3)] span the eigenspace of A corresponding to the first k right singular vectors. It follows that the columns of

$$X_{nk} + X_{nn}P_{nk} = Y_{nk}D_k + Y_{nk}D_n B_{nk} B_{kk}^{-1} + O(\gamma_k)$$

span the space $\mathcal{U}_k$ spanned by the first k right singular vectors of X. Post-multiplying by D_k^{-1}, we find that the columns of

$$Y_{nk} + O(\gamma_k)$$

span the same subspace. Thus, in the limit $\mathcal{U}_k$ is the column space of Y_{nk} or X_{nk}.

Now suppose that γ_{k-1} also approaches zero. Then $\mathcal{U}_{k-1}$ is well approximated by the column space of $Y_{n,k-1}$. Since $\mathcal{U}_k$ obtained by appending u_k to $\mathcal{U}_{k-1}$, up to terms of order $\max\{\gamma_{k-1}, \gamma_k\}$ the vector u_k must be the result of orthogonalizing the kth column of Y against $Y_{n,k-1}$. This is just the kth vector in the orthogonal part of the QR factorization of Y or X.

Recalling that the R-factor R in the QR decomposition of X is the Cholesky factor of $X^{\mathrm{T}}X$ and that $R = SD^{\frac{1}{2}}$, we have the following theorem.

Theorem 5.1. *Let $X = QR$ be the QR factorization of X. If γ_{k-1} and γ_k are sufficiently small then*

$$\sigma_k = r_{kk} + \delta_k^{\frac{1}{2}} O(\max\{\gamma_{k-1}, \gamma_k\}),$$

and

$$u_k = q_k + O(\max\{\gamma_{k-1}, \gamma_k\}).$$

We have scaled X by $D^{\frac{1}{2}}$ to retain consistency with our earlier results. However, it is the γ_i, which are proportional to the δ_i, that control the convergence of our approximations in Theorem 5.1. Thus with respect to actual gradings, approximations for the singular value decomposition converge faster than approximations for the eigenvalue problem.

To illustrate this phenomenon consider the following MATLAB code.

```
err = [];
for l = 1:5
   D = diag(logspace(0,-1,2));
   X = Y*D;
   [Q, R] = qr(X);
   [U,S,V] = svd(X);
   U(:,2) = U(:,2)/sign(U(1,2));
```

```
    Q(:,2) = Q(:,2)/sign(Q(1,2));
    err = [err;
           [norm(U(:,2)-Q(:,2)),
            abs(abs(R(2,2))-S(2,2))/S(2,2)]];
end
```

It generates a random base matrix, successively scales the second column by 10^{-1} through 10^{-5}, and computes the error in the QR approximation to the second right singular vector and the corresponding singular value. The array `err` is

5.9262e−03	2.9763e−03
5.9264e−05	2.9633e−05
5.9264e−07	2.9632e−07
5.9264e−09	2.9632e−09
5.9264e−11	2.9631e−11

We see that the approximations are converging as 100^{-i}. The fact that the ratios of these errors quickly become constant suggests that the second order terms in γ are also converging.

6. Bibliographical notes

Problems involving scaling matrices by diagonal matricesf have a long history in modern matrix computations. Early work was directed to the effects of scaling on the condition of the matrix in question and the effects of rounding error in Gaussian elimination. Although this work does not concern us directly, it is appropriate to mention the seminal papers by Bauer [3, 1966], van der Sluis [16, 1969], and Skeel [12, 1979].

In 1958 Fisher and Fuller [6, 1958] showed that if the leading principle submatrices of a matrix B are nonsingular, there is a diagonal matrix D such that the eigenvalues of DB are positive. Although they do not mention grading explicitly, their construction amounts to chosing grading ratios so small that the eigenvalues of the resulting matrix are real. By allowing D to have negative elements, the eigenvalues can be made positive. Later Ballantine [1, 1970] gave a simple proof of the theorem.[2]

The first reference I can find to graded matrices as such is by Martin, Reinsch, and Wilkinson [8, 1968], who warned that their version of Householder

[2] Fisher and Fuller had in mind the solution of the linear system $Bx = c$ by an iterative method of the form $x_{k+1} = (I - DB)x_k + Dc$. If w choose D so that the eigenvalues of the iteration matrix lie in $[0, 2)$, the iteration matrix has spectral radius less than one. However, if the grading of B is strong, $I - DB$ will have eigenvalues very near one, and the iteration will converge very slowly

tridiagonalization would destroy the accuracy of the small eigenvalues of a downward graded matrix. However, the analysis of the eigensystems of graded matrices began with Dalquist [4, 1985], whose application was to stiff ordinary equations. He introduces block grading in terms of a base matrix, and shows that under certain conditions a graded matrix A can be written in the form $A = LRL^{-1}$, where L and R are close to the (block) L- and U-factors of A. He establishes this fact using a block LR-algorithm ([17, Ch. 8]) to triangularize A. By accumulating the transformations he gets error bounds on his approximations.

The idea of grading a base matrix was rediscovered by Barlow and Demmel [2, 1990] and Stewart and Zhang [15, 1991]. The latter paper, which like Dalquist's dealt only with eigenvalues, established its result by a direct block triangularization of the kind described here. This paper also introduced (though not by name) the grading impediments κ_k. Mathias [9, 1996] gave eigenvalue and eigenvector bounds for positive definite matrices and at at the thirteenth Householder Symposium in Pontresina (1996) and observed, without proof, that similar results hold for general graded matrices.

When the base matrix has special structure — e.g., when it is symmetric diagonally dominant or positive definite — different kinds of bounds can be obtained. This line of investigation was initiated by Barlow and Demmel [2, 1990] and continued by Demmel and Veselić [5, 1992], Mathias and Stewart [11, 1993], and Mathias [9, 1996] [10, 1997]. A typical result for eigenvalues is the following [5, Theorem 2.3].

Theorem 6.1. *Let B be positive definite with unit diagonal elements, and let $A = D^{\frac{1}{2}}BD^{\frac{1}{2}}$. Let $\|F\| \leq \lambda_{\min}(B)$. Let λ_i be the ith eigenvalue of A (in descending order), and let $\tilde{\lambda}_i$ be the ith eigenvalue of $A + E$, where $E = D^{\frac{1}{2}}FD^{\frac{1}{2}}$. Then if $\|F\| < \lambda_{\min}(B)$,*

$$\frac{|\tilde{\lambda}_i - \lambda_i|}{\lambda_i} \leq \frac{\|E\|}{\lambda_{\min}(B)}. \tag{6.1}$$

In addition to perturbation bounds for eigenvalues, this sequence of papers develops componentwise perturbation theory for eigenvectors and for the singular value decomposition.

It is evident that Theorem 6.1 has a different flavor than the expansions and bounds derived in this paper. It is global and simple in the sense that one bound serves all eigenvalues, whereas our condition numbers vary with the eigenvalue. The price to be paid for this simplicity is that the results can be quite pessimistic. Our analysis makes it clear that the sensitivity of an eigenvalue depends only on the local grading impediments; whereas the reciprocal of $\lambda_{\min}(B)$ in the bound (6.1) represents the largest grading impediment. Thus an open problem in the perturbation theory of graded positive definite matrices is to derive bounds in the spirit of (6.1) that are local in nature.

References

1. C. S. Ballantine: Stabilization by a diagonal matrix. Proc. Ame. Math. Soc. **25**, 728–734 (1970)
2. J. Barlow, J. Demmel: Computing accurate eigensystems of scaled diagonally dominant matrices. SIAM J. Numerical Anal. **27**, 762–791 (1988)
3. F. L. Bauer: Genauigkeitsfragen bei der Lösung linear Gleichungssysteme. Zeitschrift für angewandte Mathematik und Mechanik **46**, 409–421 (1966)
4. G. Dahlquist: On transformations of graded matrices, with applications to stiff ODE's. Numerische Mathematik **47**, 363–385 (1995)
5. J. Demmel, K. Veselić: Jacobi's method is more accurate than QR. SIAM J. Matrix Anal. Appl. **13**, 1204–1245 (1992)
6. M. E. Fisher, A. T. Fuller: On the convergence of matrices and the convergence of linear iterative processes. Proc. Cambri. Phil. Soc. **45**, 417–425 (1958)
7. G. H. Golub, C. F. Van Loan: Matrix Computations. Johns Hopkins University Press, Baltimore, MD, second edition, 1989
8. R. S. Martin, C. Reinsch, J. H. Wilkinson: Householder tridiagonalization of a real symmetric matrix. Numerische Mathematik **11**, 181–195 (1968). Also in [18, pp. 212–226]
9. R. Mathias: Fast accurate eigenvalue methods for graded positive definite matrices. Numerische Mathematik **74**, 85–104 (1996)
10. R. Mathias: Spectral perturbation bounds for positive definite matrices. SIAM J. Matrix Anal. Appl. **18**, 959–980 (1997)
11. R. Mathias, G. W. Stewart: A block QR algorithm and the singular value decomposition. Linear Algebra Its Appl. **182**, 91–100 (1993)
12. R. D. Skeel: Scaling for numerical stability in Gaussian elimination. J. ACM **26**, 494–526 (1979)
13. G. W. Stewart: Matrix Algorithms I: Basic Decompositions. SIAM, Philadelphia, 1998
14. G. W. Stewart, J.-G. Sun: Matrix Perturbation Theory. New York: Academic Press, 1990
15. G. W. Stewart, G. Zhang: Eigenvalues of graded matrices and the condition of numbers of a multiple eigenvalue. Numerische Mathematik **58**, 703–712 (1991)
16. A. van der Sluis: Condition numbers and equilibration of matrices. Numerische Mathematik **14**, 14–23 (1969)
17. J. H. Wilkinson: The Algebraic Eigenvalue Problem. Oxford: Clarendon Press 1965
18. J. H. Wilkinson, C. Reinsch: Handbook for Automatic Computation. Vol. II Linear Algebra. New York: Springer 1971

15.8. [GWS-J114] "On the Powers of a Matrix with Perturbations"

[GWS-J114] "On the Powers of a Matrix with Perturbations", *Numerische Mathematik* **96** (2003) 363–376.
http://dx.doi.org/10.1007/s00211-003-0470-0

Numer. Math. (2003) 96: 363–376
Digital Object Identifier (DOI) 10.1007/s00211-003-0470-0

Numerische Mathematik

On the powers of a matrix with perturbations

G.W. Stewart

Department of Computer Science and Institute for Advanced Computer Studies, University of Maryland, College Park, MD 20742; e-mail: stewart@cs.umd.edu

Received February 25, 2002 / Revised version received February 7, 2003 /
Published online June 6, 2003 –

Summary. Let A be a matrix of order n. The properties of the powers A^k of A have been extensively studied in the literature. This paper concerns the perturbed powers

$$P_k = (A + E_k)(A + E_{k-1}) \cdots (A + E_1),$$

where the E_k are perturbation matrices. We will treat three problems concerning the asymptotic behavior of the perturbed powers. First, determine conditions under which $P_k \to 0$. Second, determine the limiting structure of P_k. Third, investigate the convergence of the power method with error: that is, given u_1, determine the behavior of $u_k = \nu_k P_k u_1$, where ν_k is a suitable scaling factor.

Mathematics Subject Classification (2000): 15A60, 65F15

1 Introduction

Let A be a matrix of order n with eigenvalues $\lambda_1, \ldots, \lambda_n$ ordered so that

$$|\lambda_1| \geq |\lambda_2| \geq \cdots \geq |\lambda_n|$$

and let $\rho(A) = |\lambda_1|$ denote the spectral radius of A. We will be concerned with extending the following three results about the behavior of the powers A^k of A. (These results are easily proved by exploiting the relations between norms and spectral radii; see [12, Sections I.2, II.1]. For earlier work on powers of a matrix see [3,4,6,10].)

- The first result is classic. If $\rho(A) < 1$, then $\lim_{k\to\infty} A^k = 0$. Moreover, in any norm $\|A^k\|^{1/k} \to \rho(A)$, or equivalently the convergence of A^k to zero is faster than than that of $[\rho(A) + \eta]^k$ for any $\eta > 0$. We say that the *root convergence index* of A^k is $\rho(A)$.

• The second result concerns the asymptotic form of A^k. Let $|\lambda_1| > |\lambda_2|$, and let the right and left eigenvectors corresponding to λ_1 be x and y, normalized so that $y^{\mathrm{H}}x = 1$. Then

$$\lambda_1^{-k} A^k \to xy^{\mathrm{H}}.$$

Moreover, the root convergence index is $|\lambda_2/\lambda_1|$.

• The third result concerns the convergence of the power method. Specifically, let u_1 be given and define

$$u_{k+1} = \nu_k A u_k, \qquad k = 1, 2, \dots,$$

where ν_k is a normalizing factor (e.g., $\|Au_k\|^{-1}$). In the above notation, if $|\lambda_1| > |\lambda_2|$ and $y^{\mathrm{H}}u_1 \neq 0$, then u_k, suitably scaled, converges to a multiple of x. The root index of convergence is $|\lambda_2/\lambda_1|$.

Now let $E_1, E_2, \dots$ be a sequence of perturbation matrices and let

$$P_k = (A + E_k)(A + E_{k-1}) \cdots (A + E_1).$$

The purpose of this paper is to extend the three results above to the *perturbed powers* P_k.[1] Regarding the first result, we will show that if $\rho(A) < 1$ then for sufficiently small E_k, the P_k approach zero. Moreover, by making the E_k small enough we can bring the convergence ratio arbitrarily near $\rho(A)$. Regarding the second result, we will show that if

$$\sum_k \|E_k\| < \infty \tag{1.1}$$

in any norm then the P_k converge to xz^{H} for some z (which may be zero), and we will investigate the convergence rate. Finally, a finite-precision implementation of the power method results in perturbations E_k that are of the order of the rounding unit and therefore do not satisfy (1.1). Thus, we cannot use the second result to analyze the convergence (actually nonconvergence) of the power method in the presence of rounding error. However, using other techniques we can show that in the presence of rounding error the power method will converge up to a point and then stagnate.

There is not a large literature on perturbed matrix powers. Ostrowski [10, Chapter 20] gives bounds on the size of the perturbed powers — results that imply that if $\rho(A) < 1$ and the perturbations are sufficiently small, then $P_k \to 0$. Higham and Knight [8] give a very detailed investigation of when

[1] The phrase "perturbed powers" is, strictly speaking, a misnomer, since it is the factors, not the powers that are perturbed.

$P_k \to 0$, along with some useful references. These results relate most naturally to our Theorem 3.1. However, the results of Sections 4–5 do more than treat the convergence of P_k to zero — they reveal the structure of the matrices P_k, at the cost of additional hypotheses on the nature of the spectrum of A.

This paper is organized as follows. In establishing our extensions it will prove convenient to transform our matrices by certain similarity transformations, and any conditions placed on the transformed perturbations must be translated back to the original problem. Since the transformations can be ill conditioned, it is important to understand the source of the ill-conditioning. Accordingly, the next section is devoted to describing the two transformations we will use. In Section 3 we will establish our extension of the first result, and in Section 4 the extension of the second. In Section 5 we will give an analysis of the power method. It is worth noting that the last two sections end with a little hook: each provides a new result about the problem treated in the preceding section.

Throughout this paper the jth column of the identity matrix will be denoted by $\mathbf{e}_j$. In addition, $\|\cdot\|$ will denote a consistent family of norms such that that $\|A\|$ bounds the norm of any submatrix of A and $\|\mathrm{diag}(d_1, \ldots, d_n)\| = \max_i |d_i|$. This class includes the 1-, 2-, ∞- norms but excludes the Frobenious norm [5,11].

As above, we will use the *root convergence index* to measure speed of convergence. Specifically, if a_k is a sequence converging to zero, and $\rho = \limsup_k |a|^{1/k} < 1$ we say that a_k converges with root index ρ.

2 Two transformations

In deriving our results we will have to transform A into $\hat{A} = X^{-1}AX$, for some X appropriate to the problem at hand. In this case, we must also transform the perturbation matrices: $\hat{E}_k = X^{-1}E_kX$. Now $\|\hat{E}_k\| \le \kappa(X)\|E_k\|$, where $\kappa(X) = \|X\|\|X^{-1}\|$ is the condition number of X. Hence in order to insure that a bound like $\|\hat{E}_k\| \le \gamma$ holds we have to require that $\|E_k\| \le \gamma/\kappa(X)$; that is, the bound on $\|E_k\|$ must be stronger by a factor of $\kappa^{-1}(X)$. Thus it is appropriate to examine the conditions under which $\kappa(X)$ is large — that is, under which the transformations are ill conditioned. There are two classes of transformations.

The first transformation is described in the following classic theorem (see, e.g., [12, Theorem I.2.8]).

Theorem 2.1 *For any $\eta > 0$ there is a matrix X such that*

$$\|X^{-1}AX\| \le \rho(A) + \eta. \tag{2.1}$$

The theorem is proved by first transforming A to Schur form; i.e.,

$$U^{\mathrm{H}}AU = T,$$

where U is unitary and $T = (\tau_{ij})$ is upper triangular. If we then set

$$D_\alpha = \mathrm{diag}(1, \alpha, \ldots, \alpha^{n-1}),$$

the superdiagonal elements of $D^{-1}TD$ are $\alpha^{j-i}\tau_{ij}$. If we define $X_\alpha = UD_\alpha$, then as $\alpha \to 0$ we have $\|X_\alpha^{-1}AX_\alpha\| \to \|\mathrm{diag}(\tau_{11}, \ldots, \tau_{nn})\| = \rho(A)$. Consequently, by the continuity of norms we may set $X = X_\alpha$, where α is chosen so that (2.1) is satisfied.

As α decreases, X_α^{-1} becomes large. If the off-diagonal elements of T are nonzero, α becomes small along with ϵ, the more so in proportion as the off diagonal elements of T are large. Large off-diagonal elements in T are associated with Henrici's measure of nonnormality [6]. Hence, nonnormality in A may weaken our theorems. However, it should be stressed, that the above construction of X is designed to accommodate the worst possible case, and in particular situations there may be a better way to construct T. For example, if A has a complete, well-conditioned system of eigenvectors, then the matrix X of eigenvectors reduces A to diagonal form, so that (2.1) is satisfied for $\eta = 0$.

The following useful theorem describes the second of our transformations.

Theorem 2.2 *Let λ_1 be a simple eigenvalue of A with right and left eigenvectors x_1 and y_1 normalized so that $\|x_1\|_2 = \|y_1\|_2 = 1$ and $\gamma = y_1^{\mathrm{H}}x_1$ is positive. Let $\sigma = \sqrt{1-\gamma^2}$. Then there is a matrix U with*

$$\kappa(U) = \frac{1+\sigma}{\gamma} \tag{2.2}$$

in the 2-norm such that

$$U^{-1}AU = \begin{pmatrix} \lambda_1 & 0 \\ 0 & B \end{pmatrix}. \tag{2.3}$$

Proof. By appealing to the CS decomposition [5, 11], we can find orthonormal matrices $(x_1\ x_2\ X_3)$ and $(y_1\ y_2\ Y_3)$ such that

$$(x_1\ x_2\ X_3)^{\mathrm{H}}(y_1\ y_2\ Y_3) = \begin{pmatrix} \gamma & \sigma & 0 \\ \sigma & \gamma & 0 \\ 0 & 0 & I \end{pmatrix}.$$

Let $U = (\gamma^{-\frac{1}{2}}x_1\ \gamma^{-\frac{1}{2}}y_2\ X_3)$ and $V = (\gamma^{-\frac{1}{2}}y_1\ \gamma^{-\frac{1}{2}}x_2\ Y_3)$. Then it is easily verified that $V^{\mathrm{H}} = U^{-1}$. Moreover,

$$U^{\mathrm{H}}U = \begin{pmatrix} 1/\gamma & \sigma/\gamma & 0 \\ \sigma/\gamma & 1/\gamma & 0 \\ 0 & 0 & I \end{pmatrix}$$

Thus $\|U\|_2^2 = \|U^{\mathrm{H}}U\|_2 = (1+\sigma)/\gamma$. Similarly $\|V\|_2^2 = (1+\sigma)/\gamma$, which establishes (2.2). The fact that (2.3) is satisfied follows immediately from the fact that the first column of U is the right eigenvector x_1 and the first column of V is the left eigenvector y_1. □

The matrix U will be ill conditioned when γ^{-1} is small. But γ^{-1} — the secant of the angle between x_1 and y_1 — is a condition number for the eigenvalue λ_1 [5, Section 7.2.2], [12, Section I.3.2]. Thus U_1 will be ill conditioned precisely when λ_1 is.

3 Convergence to zero

We are now in the position to state and prove our extension of the first result. In fact, thanks to results of the last section, it is trivial to establish the following theorem. As we indicated in the introduction, this theorem is essentially due to Ostrowski [10, Chapter 20] and has been extended by Higham and Knight [8].

Theorem 3.1 *Let $\rho(A) < 1$ and consider the perturbed products*

$$P_k = (A + E_k)(A + E_{k-1}) \cdots (A + E_1). \tag{3.1}$$

For every $\eta > 0$ there is an $\epsilon > 0$ such that if $\|E_k\| \le \epsilon$ then

$$\limsup_k \|P_k\|^{1/k} \le \rho(A) + \eta. \tag{3.2}$$

Hence if $\rho(A) + \eta < 1$ then P_k converges to zero with root convergence index at greatest $\rho(A) + \eta$.

Proof. By Theorem 2.1 there is a nonsingular matrix X such that if $\hat{A} = X^{-1}AX$ then $\|\hat{A}\| \le \rho(A) + \eta/2$. Let $\hat{E}_k = X^{-1}E_kX$, $\hat{P}_k = X^{-1}P_kX$, and $\hat{\epsilon} = \eta/2$. Then if $\|\hat{E}_k\| \le \hat{\epsilon}$ we have $\|\hat{A} + \hat{E}_k\| \le \rho(A) + \eta$, and $\|\hat{P}_k\| \le [\rho(A) + \eta]^k$.

Transforming back to the original problem, we see that if $\|E_k\| \le \epsilon \equiv \hat{\epsilon}/\kappa(X)$ then $\|P_k\| \le \kappa(X)[\rho(A) + \eta]^k$. The inequality (3.2) now follows on taking kth roots and observing that $\limsup \|P\|^{1/k} \le \lim \kappa(X)^{1/k}[\rho(A) + \eta] = \rho(A) + \eta$. □

There is little to add to this theorem. The price we pay for the perturbations is that to make the root convergence index approach $\rho(A)$ we must increasingly restrict the size of the perturbations. This is unavoidable. For if we fix the size of the error at ϵ we can always find E such that the largest eigenvalue of $A + E$ has magnitude $\rho(A) + \epsilon$. If we set $E_k = E$, then $P_k = (A + E)^k$, and the best root convergence index we can hope for is $\rho(A) + \epsilon$. A referee has pointed out that results in [9] imply that the worst case occurs when E is constant, in which case E should be taken to be the matrix of norm not greater than ϵ that maximizes $\rho(A + E)$.

4 Convergence with a simple dominant eigenvalue

In this section we will treat the behavior of the perturbed powers when A has a single, simple dominant eigenvalue λ; i.e., when $|\lambda_1| > |\lambda_2|$. By dividing by A by λ_1, we may assume that $\lambda_1 = 1$. The basic result is given in the following theorem.

Theorem 4.1 *Let* $1 = \lambda_1 > |\lambda_2|$ *and let the right eigenvector corresponding to* λ_1 *be* x. *Let* P_k *be defined as in* (3.1). *If*

$$\sum_{i=1}^{\infty} \|E_k\| < \infty, \tag{4.1}$$

then for some (possibly zero) vector z *we have*

$$\lim_{k\to\infty} P_k = xz^{\mathrm{H}}.$$

The root convergence index is not greater than

$$\max\{\rho, \sigma\},$$

where ρ *is the largest of the magnitudes of the subdominant eigenvalues of* A *and*

$$\sigma = \limsup \|E_k\|^{1/k}. \tag{4.2}$$

Proof. By Theorem 2.2, we may transform A so that it has the form $\operatorname{diag}(1, B)$, where $\rho(B) < 1$. By Theorem 2.1, we may assume that $\|B\| \le \beta < 1$. Note that the right eigenvector of the transformed matrix is $\mathbf{e}_1$.

The theorem is best established in a vectorized form. First write

$$P_{k+1} = (A + E_k)P_k = AP_k + E_k P_k$$

Let $u \neq 0$ be given and let $p_k = P_k u$. Then

$$p_{k+1} = Ap_k + E_k p_k. \tag{4.3}$$

We will use this recurrence to show that the p_k approach a multiple of $\mathbf{e}_1$.

Our first job is to find a condition that insures that the p_k remain bounded. To this end, partition (4.3) in the form

$$\begin{pmatrix} p_1^{(k+1)} \\ p_2^{(k+1)} \end{pmatrix} = \begin{pmatrix} 1 + e_{11}^{(k)} & e_{12}^{(k)\mathrm{H}} \\ e_{21}^{(k)} & B + E_{22}^{(k)} \end{pmatrix} \begin{pmatrix} p_1^{(k)} \\ p_2^{(k)} \end{pmatrix}. \tag{4.4}$$

Now let $\epsilon_k = \|E_k\|$ and let $\pi_1^{(k)}$ and $\pi_2^{(k)}$ be upper bounds on $\|p_1^{(k)}\|$ and $\|p_2^{(k)}\|$. Then by taking norms in (4.4) we see that the components of

$$\begin{pmatrix} \pi_1^{(k+1)} \\ \pi_2^{(k+1)} \end{pmatrix} = \begin{pmatrix} 1+\epsilon_k & \epsilon_k \\ \epsilon_k & \beta+\epsilon_k \end{pmatrix} \begin{pmatrix} \pi_1^{(k)} \\ \pi_2^{(k)} \end{pmatrix} \tag{4.5}$$

are upper bounds on $\|p_1^{(k+1)}\|$ and $\|p_2^{(k+1)}\|$. Thus if we write

$$\begin{pmatrix} 1+\epsilon_k & \epsilon_k \\ \epsilon_k & \beta+\epsilon_k \end{pmatrix} = \operatorname{diag}(1,\beta) + \begin{pmatrix} \epsilon_k & \epsilon_k \\ \epsilon_k & \epsilon_k \end{pmatrix} \equiv \operatorname{diag}(1,\beta) + H_k,$$

then the p_k will be bounded provided the product

$$\prod_{k=1}^{\infty} \|\operatorname{diag}(1,\beta) + H_k\| < \infty.$$

Now

$$\prod_{k=1}^{\infty} \|\operatorname{diag}(1,\beta) + H_k\| \le \prod_{k=1}^{\infty} (\|\operatorname{diag}(1,\beta)\| + \|H_k\|) \le \prod_{k=1}^{\infty} (1+4\epsilon_k).$$

It is well known that the product on the right is finite if and only if the series $\sum_k \epsilon_k$ converges [1, Section 5.2.2]. Hence a sufficient condition for the p_k to remain bounded is for (4.1) to be satisfied.

The next step is to show that $p_2^{(k)}$ converges to zero. Let π be a uniform upper bound on $\|p_1^{(k)}\|$ and $\|p_2^{(k)}\|$. From (4.5) we have

$$\pi_2^{(k+1)} \le (2\epsilon_k \pi + \beta \pi_2^{(k)}).$$

Hence if we set $\pi = \hat{\pi}_1$ and define $\hat{\pi}_k$ by the recurrence

$$\hat{\pi}_{k+1} = \beta \hat{\pi}_k + 2\epsilon_k \pi,$$

we have that $\|p_2^{(k)}\| \le \hat{\pi}_k$. But it is easy to see that if we define $\epsilon_0 = \frac{1}{2}$ then

$$\hat{\pi}_{k+1} = 2\pi(\beta^k \epsilon_0 + \beta^{k-1}\epsilon_1 + \cdots + \beta^1 \epsilon_{k-1} + \epsilon_k). \tag{4.6}$$

It follows that

$$\hat{\pi}_1 + \hat{\pi}_2 + \cdots = 2\pi(1+\beta+\beta^2+\cdots)(\epsilon_0+\epsilon_1+\epsilon_2+\cdots). \tag{4.7}$$

But the geometric series in β on the right is absolutely convergent, and by (4.1) the series in ϵ_k is also. Thus the series on the left is absolutely convergent, and its terms $\hat{\pi}_k$ must converge to zero.

We must next show that $p_1^{(k)}$ converges. From the first row of (4.4) we have

$$p_1^{(k+1)} - p_1^{(k)} = e_{11}^{(k)} p_1^{(k)} + e_{12}^{(k)} p_2^{(k)}, \tag{4.8}$$

whence

$$|p_1^{(k+1)} - p_1^{(k)}| \le 2\epsilon_k \pi. \tag{4.9}$$

Since $\sum_k \epsilon_k$ converges, if we set $p_1^{(0)} = 0$, the telescoping series $\sum_{j=0}^{k} (p_1^{(j+1)} - p_1^{(j)}) = p_1^{(k+1)}$ converges.

By taking $u = \mathbf{e}_i$, we find that the ith column of P_k converges to $\bar{w}_i \mathbf{e}_1$ for some w_i. Consequently, if we set $w^{\mathrm{H}} = (\bar{w}_1 \ \cdots \ \bar{w}_n)$, then P_k converges to $\mathbf{e}_1 w^{\mathrm{H}}$.

Finally, in assuming that $A = \operatorname{diag}(1, B)$ we transformed the original matrix A by a similarity transformation X whose first column was the dominant eigenvector of A. It follows that the original P_k converge to

$$X\mathbf{e}_1 w^{\mathrm{H}} X^{-1} = x z^{\mathrm{H}},$$

where $z^{\mathrm{H}} = X^{-1} w^{\mathrm{H}}$.

We now turn to the rates of convergence. The inequality (4.9) shows that $p_1^{(k)}$ converges as fast as $\|E_k\|$ approaches zero; i.e., its root convergence index is not greater than σ defined by (4.2).

To analyze the convergence of $p_2^{(k)}$ we make the observation that the reciprocal of the radius of convergence of any function $f(z)$ that is analytic at the origin is $\alpha = \limsup_k |a_k|^{1/k}$, where a_k is the kth coefficient in the power series of f [1, Secton II.2.4]. We also note that in the expression (4.6) for $\hat{\pi}_{k+1}$ we can replace β by $\|B^k\|$ and still have an upper bound on $\|p_2^{(k+1)}\|$. Now let $r(\zeta) = \sum_k \|B^k\| \zeta^k$ and $s(\zeta) = \sum_k \|E_k\| \zeta^k$. Since $\|B^k\|^{1/k} \to \rho(B) = \rho$, we know that the radius of convergence of r is ρ^{-1}. By definition the radius of convergence of s is σ^{-1}. But by (4.7), $\limsup \hat{\pi}_k^{1/k}$ is the reciprocal of the radius of convergence of the function $p(\zeta) = r(\zeta)s(\zeta)$. Since the radius of convergence of p is at least as great as the smaller of ρ^{-1} and σ^{-1}, the root index of convergence of $p_2^{(k)}$ is not greater than $\max\{\rho, \sigma\}$. □

There are four comments to be made about this theorem.

- By the equivalence of norms, if the condition (4.1) on the E_k holds for one norm, it holds for any norm. Thus, the condition on the errors does not depend on the similarity transformation we used to bring A into the form $\operatorname{diag}(1, B)$. But this happy state of affairs obtains only because (4.1) is an asymptotic statement. In practice, the sizes of the initial errors, which do depend on the transformation, may be important.
- Since P_k converges to xz^{H}, if $z \ne 0$, at least one column of P_k contains an increasingly accurate approximation to x. In the error free case, z is equal to the left eigenvector of A, which is by definition nonzero. In general, however, we cannot guarantee that $z \ne 0$, and indeed it is easy to contrive examples for which z is zero. For example, in the transformed problem take

$$E_1 = \begin{pmatrix} -1 & 0 \\ 0 & 0 \end{pmatrix} \quad \text{and} \quad E_k = \begin{pmatrix} 0 & 0 \\ e_{21}^{(k)} & E_{22}^{(k)} \end{pmatrix} \quad (k > 1).$$

However, it follows from (4.8) that

$$|p_1^{(k+1)}| \geq |p_1^{(k)}| - 2\pi\epsilon_k \geq |p_k^{(1)}| - 2\pi(\epsilon_k + \cdots + \epsilon_1).$$

Hence if $2\pi \sum_k \epsilon_k < \|p_1^{(1)}\|$, then $\lim_k p_1^k \neq 0$, and hence $\lim_k P_k \neq 0$.

• The proof can be extended to the case where A has more than one dominant eigenvalue, provided they are all simple. The key is to use a generalization of Theorem 2.2 that uses bases for the left and right dominant eigenspaces of A, to reduce A to the form $\mathrm{diag}(D, B)$, where $|D| = I$. The quantities $p_1^{(k)}$ and $p_1^{(k+1)}$ in (4.4) are no longer scalars, but the recursion (4.5) for upper bounds remains the same, as does the subsequent analysis.

• We have been interested in the case where A has a simple dominant eigenvalue of one. However, the proof of the theorem can easily be adapted to the case where $\rho(A) < 1$ with no hypothesis of simplicity (it is essentially the analysis of $p_2^{(k)}$ without the contributions from $p_1^{(k)}$). The result is the following corollary.

Corollary 4.2 *Let* $\rho(A) < 1$ *and let* E_k *satisfy* (4.1). *Then* $P_k \to 0$ *and the root convergence index is not greater than* $\max\{\rho, \sigma\}$.

5 The power method

The power method starts with a vector u_1 and generates a sequence of vectors according to the formula

$$u_{k+1} = \nu_k A u_k,$$

where ν_k is a normalizing factor. If A has a simple dominant eigenvalue (which we may assume to be one), under mild restrictions on u_1, the u_k converge to the dominant eigenvector of A.

A backward rounding-error analysis shows that in the presence of rounding error we actually compute

$$u_{k+1} = \nu_k(A + E_k)u_k = (\nu_k \cdots \nu_1)P_k u_1.$$

where $\|E_k\|/\|A_k\|$ is of the order of the rounding unit [7, 11]. Theorem 4.1 is not well suited to analyzing this method for two reasons. First the E_k will all be roughly the same size, so that the condition (4.1) is not satisfied. But even if it were, it is possible for the P_k to approach zero while at the same time the normalized vectors u_k converge to a nonzero limit, in which case Theorem 4.1 says nothing useful. Accordingly, in this section we give a different convergence analysis for the power method.

As in the last section we will assume that $A = \mathrm{diag}(1, B)$, where $\|B\| = \beta < 1$. Let $\epsilon_k = \|E_k\|$. We will normalize the u_k so that the first component is one and write

$$u_k = \begin{pmatrix} 1 \\ h_k \end{pmatrix}.$$

It is important to have some appreciation of the magnitudes of the quantities involved. If the computations are being done in IEEE double precision, ϵ will be around $\sqrt{n}\cdot 10^{-16}$; e.g., 10^{-14} if $n = 10{,}000$. If u_1 is a random vector, we can expect $\|h_1\|$ to be of order $\sqrt{n}$; e.g., 100, if $n = 10{,}000$. Finally, since the ratio of convergence of the power method is approximately β, β must not be too near one; e.g., 0.99 gives unacceptably slow convergence. Thus, in interpreting our results, we may assume that $\epsilon\|h_1\|$ and $\epsilon/(1-\beta)$ are small.

Let η_k be an upper bound for $\|h_k\|$. We will derive an upper bound η_{k+1} for $\|h_{k+1}\|$, in the form of the quotient of a lower bound on the first component of $(A + E_k)u_k$ and and upper bound on the rest of the vector. We have

$$(A + E_k)u_k = \begin{pmatrix} 1 + e_{11}^{(k)} & e_{12}^{(k)\mathrm{H}} \\ e_{21}^{(k)} & B + E_{22}^{(k)} \end{pmatrix} \begin{pmatrix} 1 \\ h_k \end{pmatrix}.$$

The first component γ_1 of this vector is $1 + e_{11}^{(k)} + e_{12}^{(k)\mathrm{H}} h_k$. Hence

$$|\gamma_1| \ge 1 - |e_{11}^{(k)}| - \|e_{12}^{(k)}\| \|h_k\| \ge 1 - (1 + \eta_k)\epsilon_k.$$

Similarly an upper bound on the remaining part is

$$\beta\eta_k + \epsilon_k(1 + \eta_k).$$

Hence

$$\|h_{k+1}\| \le \eta_{k+1} \equiv \frac{\beta\eta_k + \epsilon_k(1 + \eta_k)}{1 - (1 + \eta_k)\epsilon_k}$$

Let

$$\varphi_\epsilon(\eta) = \frac{\beta\eta + \epsilon(1 + \eta)}{1 - (1 + \eta)\epsilon}.$$

so that

(5.1) $$\eta_{k+1} = \varphi_{\epsilon_k}(\eta_k).$$

If we multiply the equation $\eta = \varphi_\epsilon(\eta)$ by $1 - (1 + \eta)\epsilon$, we obtain a quadratic equation for the fixed points of η. From the quadratic formula it is easily verified that if

(5.2) $$c = \frac{1 - \beta - 2\epsilon}{\epsilon} \ge 2$$

then φ_ϵ has a minimal fixed point

$$\eta_* = \frac{2}{c + \sqrt{c^2 - 4}}. \tag{5.3}$$

Moreover,

$$\varphi'_\epsilon(\eta) = \frac{\beta}{1 - (1+\eta)\epsilon} + \frac{\beta\eta + \epsilon(1+\eta)}{[1 - (1+\eta)\epsilon]^2}\epsilon. \tag{5.4}$$

The following theorem gives conditions under which we can iterate for the fixed point η_*.

Theorem 5.1 *If*

$$\epsilon < \frac{1}{4}(1 - \beta), \tag{5.5}$$

then φ_ϵ has a unique smallest fixed point η_* given by (5.3) *with* $0 < \varphi'_\epsilon(\eta_*) < 1$. *Moreover, if*

$$\eta_* < \eta_1 < \min\left\{\frac{1}{2\epsilon} - 1,\ \frac{1}{\epsilon}\cdot\frac{1 - \beta - \epsilon - 2\epsilon^2}{1 + 2\beta + 2\epsilon}\right\}, \tag{5.6}$$

then the iteration

$$\eta_{k+1} = \varphi_\epsilon(\eta_k), \qquad k = 1, 2, \dots, \tag{5.7}$$

converges from above to η_*.

Proof. The condition (5.5) insures that c defined by (5.2) is less than two. Hence η_* given by (5.3) is the required fixed point. We have $\varphi_\epsilon(0) = \epsilon/(1-\epsilon) > 0$, and

$$\varphi'_\epsilon(0) = \frac{\beta + \epsilon}{1 - \epsilon} + \frac{\epsilon^2}{(1-\epsilon)^2} > 0.$$

Moreover $\varphi'(\eta)$ is strictly increasing. Hence the curve $\varphi(\eta)$ cannot cross the line $y = x$ unless its derivative at the crossing point is positive and less than one.

From the theory of fixed point iteration [2, Theorem 6.5.1] we know that the iteration (5.7) converges provided we start with η_1 in an interval $[\eta_*, \tau)$ for which φ'_ϵ is less than one. To compute such an interval, restrict η so that

$$\eta < \frac{1}{2\epsilon} - 1,$$

whence

$$\varphi'_\epsilon(\eta_*) \le \varphi'_\epsilon(\eta) < \frac{\beta}{1 - (1+\eta)\epsilon} + 2\frac{\beta\eta + \epsilon(1+\eta)}{1 - (1+\eta)\epsilon}\epsilon.$$

Setting this bound equal to one and solving for η, we get a solution τ satisfying

$$\tau = \frac{1}{\epsilon} \cdot \frac{1 - \beta - \epsilon - 2\epsilon^2}{1 + 2\beta + 2\epsilon}. \tag{5.8}$$

Note that since $1 - \beta > 4\epsilon$ and $\epsilon < \frac{1}{4}$, the numerator in (5.8) is positive. It follows that for η in $[\eta_*, \tau)$, we have $\varphi_\epsilon'(\eta) < 1$. □

This theorem does not apply directly to the power method, since the bounding iteration (5.1) involves the varying errors ϵ_k. But owing to the monotonicity of φ_ϵ, if ϵ is an upper bound on the ϵ_k, then the iteration (5.7) provides a upper bounds on the quantities $\|h_k\|$. To the extent that these upper bounds reflect reality, the theorem has important things to say about the power method.

We are chiefly interested in the behavior of the iteration for small ϵ. In this case, the nearness of β to one influences the iteration in four ways.

- **Tolerable error** The condition (5.5)—$\epsilon < \frac{1}{4}(1 - \beta)$—suggests that for the power method to be effective the size of the error in A must decrease with $1 - \beta$. For our typical values, this is no problem, since $\epsilon = 10^{-14} < .01 = 1 - \beta$.

- **Size of the fixed point** When ϵ is small, $\eta_* \cong \epsilon/(1 - \beta)$. Thus as β approaches 1, the limiting accuracy of the power method will be degraded.

- **Rate of convergence** For small ϵ, φ_ϵ essentially a straight line with slope β over a wide range above η_*. Thus a β near one implies slow convergence.

- **Size of the convergence region** For small ϵ the two expressions for the upper end τ of the convergence region in (5.6) become essentially

$$\frac{1}{2\epsilon} \quad \text{and} \quad \frac{1}{\epsilon} \cdot \frac{1 - \beta}{1 + 2\beta}$$

It is seen that for $\beta < \frac{1}{4}$, the first expression determines τ, while otherwise the second expression determines τ. In particular, as β approaches one, the size of the convergence region decreases. For our typical parameters this fact is unimportant, since $\tau \cong 3 \cdot 10^{11}$, which is far greater than our estimate of 100 for $\|h_1\|$ when $n = 10{,}000$.

—

It is important to keep in mind that we have analyzed the diagonalized problem whose matrix is diag$(1, B)$. As we pointed out in Section 2, the norms of the errors in A must be multiplied by the condition number of the diagonalizing transformation. In particular, ill-conditioning in λ_1 will limit the accuracy of the final solution.

Although we have naturally focused on errors whose norms are bounded away from zero, we can use our analysis to show that if $\epsilon_k = \|E_k\|$ converges monotonically to zero and ϵ_1 is suitably small, then the power method converges. Specifically, we have the following theorem.

Theorem 5.2 *In the above notation, let* $0 < \beta < 1$. *For any* η_1, *there is an* ϵ_1 *such that if the sequence* $\epsilon_1, \epsilon_2, \ldots$ *approaches zero monotonically then the sequence defined by*

$$\eta_{k+1} = \varphi_{\epsilon_k}(\eta_k), \qquad k = 1, 2, \ldots,$$

converges monotonically to zero.

Proof. From (5.4) it is clear that if ϵ_1 is sufficiently small then $\varphi'_\epsilon(\eta) \le \alpha < 1$ for any $\epsilon < \epsilon_1$ and $\eta < \eta_1$. It then follows from the theory of fixed point iterations that the sequence $\eta_1, \eta_2, \ldots$ is monotonic decreasing. Let its limit be $\hat{\eta}$.

We must show that $\hat{\eta} = 0$. Let $\delta > 0$ be given. Now $\lim_{\epsilon \to 0} \varphi_\epsilon(\eta) = \beta\eta$ uniformly on $[0, \eta_1]$. Hence there is an integer $K > 0$ such that

$$k \ge K \implies |\varphi_{\epsilon_k}(\eta_k) - \beta\eta_k| < \frac{\delta}{2}.$$

We may also assume that K is so large that

$$k \ge K \implies |\beta\eta_k - \beta\hat{\eta}| < \frac{\delta}{2}.$$

Then for $k \ge K$

$$|\eta_{k+1} - \beta\hat{\eta}| = |\varphi_{\epsilon_k}(\eta_k) - \beta\hat{\eta}| \le |\varphi_{\epsilon_k}(\eta_k) - \beta\eta_k| + |\beta\eta_k - \beta\hat{\eta}_k| < \delta.$$

It follows that $\eta_k \to \beta\hat{\eta}$. But since $\eta_k \to \hat{\eta}$ and $\beta \ne 0$, we must have $\hat{\eta} = 0$. □

This theorem has an important implication for the behavior of the perturbed powers P_k, which was treated in the previous section. The jth column of P_k, suitably scaled, is just the result of applying the unscaled power method with error to $\mathbf{e}_j$. Now suppose that $y^{\mathrm{H}}\mathbf{e}_j \ne 0$, where y is the dominant left eigenvector. Then if $\epsilon_1 \ge \epsilon_2 \ge \cdots$ and ϵ_1 is sufficiently small, the jth column of P_k, suitably scaled, approximates the dominant eigenvector of A, even if P_k converges to zero. Thus if we are interested only in the behavior of the columns of P_k, we can relax the condition that $\sum_k \epsilon_k < \infty$. However, the price we pay is a less clean estimate of the asymptotic convergence rate.

Acknowledgements. I would like to thank Donald Estep and Sean Eastman for their comments on this paper, and especially Sean Eastman for the elegant proof of Theorem 5.2. I am indebted to the Mathematical and Computational Sciences Division of the National Institute of Standards and Technology for the use of their research facilities.

References

[1] Ahlfors, L.V.: Complex Analysis. McGraw–Hill, New York, 1966

[2] Dahlquist, G., Björck, Å.: Numerical Methods. Prentice–Hall, Englewood Cliffs, New Jersey, 1974

[3] Gautschi, W.: The asymptotic behavior of powers of a matrix. Duke Mathematical Journal **20**, 127–140 (1953)

[4] Gautschi, W.: The asymptotic behavior of powers of a matrix. II. Duke Mathematical Journal **20**, 275–279 (1953)

[5] Golub, G.H., Van Loan, C.F.: Matrix Computations. Johns Hopkins University Press, Baltimore, MD, third edition, 1996

[6] Henrici, P.: Bounds for iterates, inverses, spectral variation and fields of values of nonnormal matrices. Numerische Mathematik **4**, 24–39 (1962)

[7] Higham, N.J.: Accuracy and Stability of Numerical Algorithms. SIAM, Philadelphia, 1996

[8] Higham, N.J., Knight, P.A.: Matrix powers in finite precision arithmetic. SIAM Journal on Matrix Analysis and Appliciations **16**, 343–358 (1995)

[9] Oarǒ, C., Van Dooren, P.: Stability of discrete time-varying systems of descriptor form. In: Proceedings of the 36th IEEE Conference on Decision and Control, pages 4541–4542. IEEE, Washington, DC, 1997

[10] Ostrowski, A.M.: Solution of Equations and Systems of Equations. Academic Press, New York, second edition, 1966

[11] Stewart, G.W.: Matrix Algorithms I: Basic Decompositions. SIAM, Philadelphia, 1998

[12] Stewart, G.W.: Matrix Algorithms II: Eigensystems. SIAM, Philadelphia, 2001

16

Papers on the SVD, Eigenproblem and Invariant Subspaces: Algorithms

1. [GWS-J5] "Accelerating the Orthogonal Iteration for the Eigenvectors of a Hermitian Matrix," *Numerische Mathematik* **13** (1969) 362–376.
2. [GWS-J30] "Simultaneous Iteration for Computing Invariant Subspaces of Non-Hermitian Matrices," *Numerische Mathematik* **25** (1976) 123–136.
3. [GWS-J33] "Algorithm 506: HQR3 and EXCHNG: FORTRAN Subroutines for Calculating and Ordering the Eigenvalues of a Real Upper Hessenberg Matrix," *ACM Transactions on Mathematical Software* **2** (1976) 275–280.
4. [GWS-J37] (with C. A. Bavely) "An Algorithm for Computing Reducing Subspaces by Block Diagonalization," *SIAM Journal on Numerical Analysis* **16** (1979) 359–367.
5. [GWS-J75] (with R. Mathias) "A Block QR Algorithm and the Singular Value Decomposition," *Linear Algebra and its Applications* **182** (1993) 91–100.
6. [GWS-J102] "The QLP Approximation to the Singular Value Decomposition," *SIAM Journal on Scientific Computation* **20** (1999) 1336–1348.
7. [GWS-J107] (with Z. Jia) "An Analysis of the Rayleigh-Ritz Method for Approximating Eigenspaces," *Mathematics of Computation* **70** (2001) 637–647.

16.1. [GWS-J5] "Accelerating the Orthogonal Iteration for the Eigenvectors of a Hermitian Matrix"

[GWS-J5] "Accelerating the Orthogonal Iteration for the Eigenvectors of a Hermitian Matrix," *Numerische Mathematik* **13** (1969) 362–376.
http://dx.doi.org/10.1007/BF02165413

Numer. Math. 13, 362—376 (1969)

Accelerating the Orthogonal Iteration for the Eigenvectors of a Hermitian Matrix

G. W. STEWART*

Received August 22, 1968

I. Introduction

Let A be a nonsingular, normalizable matrix of order n. Then to the n eigenvalues, $\lambda_1, \ldots, \lambda_n$, of A there corresponds a set of linearly independent eigenvectors $x_1, \ldots, x_n$. Assume that the eigenvalues of A have been ordered so that

$$|\lambda_1| \geqq |\lambda_2| \geqq \cdots \geqq |\lambda_n| > 0 \tag{1.1}$$

and that the eigenvectors have been scaled so that

$$\|x_1\| = \|x_2\| = \cdots = \|x_n\| = 1.$$

BAUER's treppen-iteration [1] and its orthogonal variant [5, p. 607] to be considered here are based on the following fact. *Let Q be an $n \times r$ matrix $(r < n)$ and suppose that $|\lambda_r| > |\lambda_{r+1}|$. Then, under mild restrictions on Q, as k increases the column space of $A^k Q$ approaches the invariant subspace spanned by $x_1, x_2, \ldots, x_r$.* Both methods generate a sequence of iterates $Q^{(k)}$ as follows. First $Q^{(k)}$ is multiplied by the matrix A. Then the product $AQ^{(k)}$ is reduced by column operations to a normal form to give $Q^{(k+1)}$. The normal form is chosen so that the columns of $Q^{(k)}$ remain strongly independent. For the treppen-iteration $Q^{(k)}$ is required to be unit lower trapezoidal; for the orthogonal iteration the columns of $Q^{(k)}$ are required to be orthonormal.

Both iterations, with their two steps of multiplication followed by normalization, are generalizations of the power method [5, p. 571]. Like the power method they are most useful when only a few of the dominant eigenvalues and eigenvectors of A are required. However, for very large sparse matrices they may be the only feasible methods.

The orthogonal iteration starts with a matrix $Q^{(0)}$ having orthonormal columns and generates a sequence of iterates by the formula

$$Q^{(k)} R^{(k)} = A Q^{(k-1)}, \tag{1.2}$$

where $R^{(k)}$ is upper triangular with positive diagonal elements and $Q^{(k)}$ has orthonormal columns. Since A is nonsingular and $Q^{(k-1)}$ has full rank, this decomposition

* Oak Ridge Graduate Fellow from the University of Tennessee under appointment from the Oak Ridge Associated Universities.

Operated by Union Carbide Corporation for the U.S. Atomic Energy Commission.

of $AQ^{(k-1)}$ is always possible, and moreover it is unique. The columns of the matrix $Q^{(k)}$ are the result of applying the Gram-Schmidt orthogonalization to the columns of $AQ^{(k-1)}$ [2, pp. 134—137].

By applying the iteration formula (1.2) repeatedly, it is easy to show that

$$Q^{(k)}\tilde{R}^{(k)} = A^k Q^{(0)}, \tag{1.3}$$

where

$$\tilde{R}^{(k)} = R^{(k)} R^{(k-1)} \dots R^{(1)}.$$

Since each of the matrices $R^{(1)}, \dots, R^{(k)}$ is upper triangular with positive diagonal, so is $\tilde{R}^{(k)}$. Hence $Q^{(k)}$ is the matrix obtained by orthogonalizing the columns of $A^k Q^{(0)}$.

If A is Hermitian, $i \leqq j \leqq r$, and

$$|\lambda_{i-1}| > |\lambda_i| = \dots = |\lambda_j| > |\lambda_{j+1}|,$$

then the space spanned by the vectors $q_i^{(k)}, \dots, q_j^{(k)}$ of $Q^{(k)}$ converges to the space spanned by $x_i, \dots, x_j$. If $i=j$, then $q_i^{(k)}$ approaches an eigenvector of A. Thus for Hermitian matrices the orthogonal iteration produces vectors which converge to eigenvectors of A or at least, in the limit, span invariant subspaces corresponding to eigenvalues of equal modulus. However, for eigenvalues of nearly equal modulus, the convergence to the individual eigenvectors is slow, and it is the object of this paper to examine a device for accelerating the convergence.

In the next section the accelerating procedure will be described. It produces a set of refined eigenvalues and eigenvectors, and the remainder of the paper is devoted to determining their accuracy. In order to do this, it is necessary to examine the convergence of the orthogonal iteration in detail.

Throughout this section the notational conventions of [2] will be followed. The symbol $\|\cdot\|$ will always denote the Euclidean vector norm,

$$\|x\|^2 = x^H x,$$

or the spectral matrix norm,

$$\|A\| = \sup_{\|x\|=1} \|Ax\|.$$

The space spanned by the columns of a matrix will be called the space of the matrix.

II. A Refinement Procedure for Approximate Eigenvectors

Let

$$\Lambda = \operatorname{diag}(\lambda_1, \dots, \lambda_n), \quad \Lambda_1 = \operatorname{diag}(\lambda_1, \dots, \lambda_r), \quad \Lambda_2 = \operatorname{diag}(\lambda_{r+1}, \dots, \lambda_n),$$

and

$$X = (x_1, \dots, x_n), \quad X_1 = (x_1, \dots, x_r), \quad X_2 = (x_{r+1}, \dots, x_n).$$

Because A is Hermitian, the λ_i are real and the x_i may be chosen so that

$$X^H X = I.$$

Moreover

$$AX = X\Lambda$$

with similar relations holding for X_1, Λ_1 and X_2, Λ_2.

The following refinement procedure may be applied to any matrix Q with orthonormal columns whose space approximates the space of X_1. Let

$$P = X^H Q = \begin{pmatrix} X_1^H Q \\ X_2^H Q \end{pmatrix} = \begin{pmatrix} P_1 \\ P_2 \end{pmatrix}, \tag{2.1}$$

and consider the matrix

$$B = Q^H A Q = P^H \Lambda P = P_1^H \Lambda_1 P_1 + P_2^H \Lambda_2 P_2. \tag{2.2}$$

Let

$$Y^H B Y = M = \operatorname{diag}(\mu_1, \ldots, \mu_r), \tag{2.3}$$

where Y is the unitary matrix whose columns are the eigenvectors of B. If P_2 is zero and the eigenvalues $\lambda_1, \ldots, \lambda_r$ are distinct, then the μ_i may be ordered so that $M = \Lambda_1$, $Y = P_1^H$, and

$$QY = XPP_1^H = X_1.$$

Hence if P_2 is small, which means the space of Q is a good approximation to the space of X_1, the matrices M and QY should be good approximations to Λ_1 and X_1.

It is proposed that for some suitable k this refinement process be applied to the matrix $Q^{(k)}$ generated in the course of the orthogonal iteration. To evaluate the amount of work required to perform this acceleration step, note that three distinct calculations are involved:

1) the calculation of $Q^{(k)H} A Q^{(k)}$,
2) the calculation of Y and M,
3) the calculation of $Q^{(k)} Y$.

If $n \gg r$, then the first calculation will dominate the other two. But the bulk of this calculation lies in computing $AQ^{(k)}$, which must be done anyway to find $Q^{(k+1)}$. Hence the amount of work involved in one acceleration step is about equal to the amount of work required to perform one step of the orthogonal iteration.

WILKINSON [5, p. 609] has proposed a related technique for finding complex conjugate eigenvalues of a real nonsymmetric matrix. In his method the first two columns of $Q^{(k)}$ are used to determine a 2×2 nonsymmetric matrix from which the eigenvalues are calculated.

III. Convergence of the Orthogonal Iteration

The convergence proof in this section is adapted from WILKINSON's proof of the convergence of the QR algorithm [4]. The idea of the proof is to exhibit $A^k Q^{(0)}$ as the product of a matrix with orthonormal columns and an upper triangular matrix with positive diagonal elements. Since such a decomposition is unique, it follows from (1.3) that the factor with orthonormal columns must coincide with $Q^{(k)}$. The properties of $Q^{(k)}$ may then be read off from the factorization.

Let

$$P^{(0)} = X^H Q^{(0)}.$$

Suppose that $P^{(0)}$ can be written in the form

$$P^{(0)} = L^{(0)} U^{(0)}, \tag{3.1}$$

where $L^{(0)}$ is lower trapezoidal with diagonal elements equal to unity in absolute value and $U^{(0)}$ is upper triangular with positive diagonal elements. Then

$$\begin{aligned} A^k Q^{(0)} &= X \Lambda^k P^{(0)} = X \Lambda^k L^{(0)} U^{(0)} \\ &= X(\Lambda^k L_k^{(0)} |\Lambda_1^{-k}|) (|\Lambda_1^k| U^{(0)}) \\ &= X L^{(k)} U^{(k)}. \end{aligned}$$

Now $L^{(k)}$ may be decomposed into the product of a matrix with orthonormal columns and an upper triangular matrix with positive diagonal:

$$L^{(k)} = P^{(k)} \tilde{U}^{(k)}. \tag{3.2}$$

Hence

$$A^k Q^{(0)} = (X P^{(k)}) (\tilde{U}^{(k)} U^{(k)}).$$

But $X P^{(k)}$ has orthonormal columns and $\tilde{U}^{(k)} U^{(k)}$ is upper triangular with positive diagonal. Hence by the foregoing comments

$$Q^{(k)} = X P^{(k)}.$$

Now the (j, i) element of $L^{(k)}$ is

$$\begin{aligned} l_{ji}^{(k)} &= l_{ji}^{(0)} (\lambda_j / |\lambda_i|)^k, & j &\geqq i, \\ l_{ji}^{(k)} &= 0, & j &< i. \end{aligned}$$

Since the λ_i are real and $|\lambda_j| \leqq |\lambda_i|$ for $j \geqq i$, it follows that the elements of $L^{(2k)}$ must approach zero or remain constant with increasing k. In particular suppose that for some $i \leqq r$

$$|\lambda_{i-1}| > |\lambda_i| \geqq \cdots \geqq |\lambda_j| > |\lambda_{j+1}| \tag{3.3}$$

and let

$$j' = \min\{j, r\}.$$

Then the elements of $L^{(2k)}$ in rows i through j and columns i through j' tend toward zero with the exception of the elements in their intersection, which remain constant. In the limit $P^{(2k)}$ must have the same block structure, and the nonvanishing block has a limit. If $P^{(2k)}$ is premultiplied by X and the block structure of $P^{(2k)}$ taken into account, the result is

Theorem 3.1. *If (3.3) is satisfied then the columns $q_i^{(2k)}, \ldots, q_{j'}^{(2k)}$ of $Q^{(2k)}$ each approach a limit which is a linear combination of $x_i, \ldots, x_j$.*

The same result holds for the columns of $Q^{(2k+1)}$. From the proof it is evident that the rate at which the limit is attained depends on the larger of the two ratios $|\lambda_i/\lambda_{i-1}|$ and $|\lambda_{j+1}/\lambda_j|$.

Theorem 3.1 is true only under the assumption that $P^{(0)}$ has a decomposition of the form (3.1). When this fails, the iteration is said to be disordered. WILKINSON handles the problem of disorder by permuting the rows of $P^{(0)}$ in such a way that the resulting matrix has a decomposition of the form (3.1) *and* the above convergence proof goes through. For the orthogonal iteration there are two types of disorder. Let

$$P^{(k)} = \begin{pmatrix} P_1^{(k)} \\ P_2^{(k)} \end{pmatrix} \tag{3.4}$$

where $P_1^{(k)}$ is square. If $P_1^{(0)}$ is nonsingular, then in event of disorder the space of $Q^{(k)}$ converges to the space of X_1, but the eigenvectors are found in a different order. When $P_1^{(0)}$ is singular, the space of $Q^{(k)}$ will converge to a different invariant subspace. The case of disorder will not be treated here; however, all the results of this paper remain essentially unaltered for the first kind of disorder.

Some auxiliary quantities will be needed later. Let $L^{(k)}$ be partitioned in the form

$$L^{(k)} = \begin{pmatrix} L_1^{(k)} \\ L_2^{(k)} \end{pmatrix},$$

where $L_1^{(k)}$ is square. Define

$$K^{(k)} = L_2^{(k)} (L_1^{(k)})^{-1}.$$

Then

$$\begin{aligned} K^{(k)} &= \Lambda_2^k L_2^{(0)} |\Lambda_1^{-k}| \, (\Lambda_1^k L_1^{(0)} |\Lambda_1^{-k}|)^{-1} \\ &= \Lambda_2^k L_2^{(0)} (L_1^{(0)})^{-1} \Lambda_1^{-k} = \Lambda_2^k K^{(0)} \Lambda_1^{-k}. \end{aligned} \tag{3.5}$$

From (3.2)

$$L_i^{(k)} = P_i^{(k)} \tilde{U}^{(k)}, \qquad (i = 1, 2),$$

where $P_i^{(k)}$ $(i = 1, 2)$ is defined by (3.4). Hence

$$K^{(k)} = P_2^{(k)} (P_1^{(k)})^{-1}. \tag{3.6}$$

Moreover since

$$\begin{pmatrix} I \\ K^{(k)} \end{pmatrix} = P (P_1^{(k)})^{-1},$$

$$I + K^{(k)H} K^{(k)} = (P_1^{(k)} P_1^{(k)H})^{-1}. \tag{3.7}$$

IV. Some Miscellaneous Theorems

Some theorems that will be needed later will be developed in this section.

It is a well known fact that, for $0 \leqq \theta \leqq 1$, $\cos\theta$ is nearer to unity than $\sin\theta$ is to zero. Namely

$$1 - \cos\theta = 1 - \sqrt{1 - \sin^2\theta} \leqq \sin^2\theta.$$

The following theorem generalizes this fact.

Theorem 4.1. *Let the matrix*

$$Q = \begin{pmatrix} Q_1 \\ Q_2 \end{pmatrix}$$

have orthonormal columns, and suppose that Q_1 has at least as many rows as columns. Then

$$Q_1 = N + F, \tag{4.1}$$

where N has orthonormal columns and

$$\|F\| \leqq \|Q_2\|^2.$$

Proof. The matrix Q_1 has the singular value decomposition

$$Q_1 = U \Gamma V^H,$$

where V^H is unitary, Γ is nonnegative diagonal, and U has orthonormal columns [2, p. 31, Ex. 19]. Let

$$N = U V^H$$

and

$$F = U(\Gamma - I) V^H.$$

Then $N^H N = I$ and (4.1) is satisfied. Now

$$\begin{aligned} I = (Q V^H)^H (Q V^H) &= \Gamma^2 + V Q_2^H Q_2 V^H \\ &= \Gamma^2 + \Sigma^2. \end{aligned}$$

Hence Σ^2 is diagonal and

$$I - \Gamma = I - (I - \Sigma^2)^{\frac{1}{2}}.$$

Since $1 - \sqrt{1 - x^2}$ is an increasing function of x,

$$\|I - \Gamma\| = 1 - \sqrt{1 - \|\Sigma^2\|}.$$

But

$$\|I - \Gamma\| = \|F\|$$

and

$$\|\Sigma^2\| = \|Q_2\|^2.$$

Hence

$$\|F\| = 1 - \sqrt{1 - \|Q_2\|^2} \leqq \|Q_2\|^2.$$

Consider the eigenvalue problem

$$B Y - Y M = 0, \tag{4.2}$$

where B is Hermitian, Y is unitary, and M is diagonal, all of order r. Suppose that the eigenvalues of B have been ordered so that

$$\mu_1 \geqq \mu \geqq \cdots \geqq \mu_r.$$

The following theorem is well known.

Theorem 4.2. *Let F be Hermitian. Then the eigenvalues of $B+F$ satisfy*

$$\mu_r - \|F\| \leqq \lambda(B+F) \leqq \mu_1 + \|F\|.$$

The hard part of the famous minimax theorem is the following

Theorem 4.3. *Let the $r\times s$ matrix Q have orthonormal columns and let*

$$\nu_1 \geqq \nu_2 \geqq \cdots \geqq \nu_s$$

be the eigenvalues of $Q^H BQ$. Then

$$\nu_1 \geqq \mu_{r-s+1}$$

and

$$\nu_s \leqq \mu_s.$$

This theorem may be applied repeatedly to give

Corollary 4.4. *Under the hypothesis of Theorem* 4.3

$$\nu_i \leqq \mu_i, \qquad (i=1,2,\ldots,s),$$

and

$$\nu_{s-i+1} \geqq \mu_{r-i+1}, \qquad (i=1,2,\ldots,s).$$

Let G be a nonsingular Hermitian matrix. With the substitutions

$$Y = GZ,$$

and

$$C = GBG$$

the eigenvalue problem (4.2) becomes

$$CZ - G^2 ZM = 0. \tag{4.3}$$

Obviously

$$Z^H G^2 Z = I. \tag{4.4}$$

Any matrix (possibly rectangular) satisfying (4.4) will be said to have columns that are orthonormal with respect to G, or for short G-orthonormal columns. If V has G-orthonormal columns, then GV has orthonormal columns.

Given any $r\times s$ matrix V of rank s, there is an $s\times s$ matrix T such that VT has G-orthonormal columns. In fact let the unitary matrix H diagonalize the matrix $V^H G^2 V$:

$$H^H V^H G^2 V H = \Delta^2. \tag{4.5}$$

Since G is nonsingular and V is of full rank, Δ is nonsingular. Then

$$T = H\Delta^{-1} \tag{4.6}$$

is the required matrix. If V has orthonormal columns then Corollary 4.4 shows that the smallest eigenvalue of Δ^2 is not less than the smallest eigenvalue of G^2. Hence for this case

$$\|T\| \leqq \|G^{-1}\|.$$

The following notation will be used. Let

$$\mathcal{J} = \{i_1, \ldots, i_s\}$$

be a set of distinct integers taken from $\{1, 2, \ldots, r\}$ and let $\bar{\mathcal{J}}$ be its complementary subset. If V is any matrix having at least r columns, define

$$V_{\mathcal{J}} = (v_{i_1}, v_{i_2}, \ldots, v_{i_s}).$$

The matrix

$$Z_{\mathcal{J}} Z_{\mathcal{J}}^H G^2$$

is the oblique projector onto the space of $Z_{\mathcal{J}}$ along the space of $Z_{\bar{\mathcal{J}}}$. Since

$$(Z_{\mathcal{J}}, Z_{\bar{\mathcal{J}}}) \begin{pmatrix} Z_{\mathcal{J}}^H \\ Z_{\bar{\mathcal{J}}}^H \end{pmatrix} G^2$$

is an $r \times r$ matrix projecting onto r space, it must be the identity. If V has G-orthonormal columns, then

$$\begin{aligned} \left[\begin{pmatrix} Z_{\mathcal{J}}^H \\ Z_{\bar{\mathcal{J}}}^H \end{pmatrix} G^2 V\right]^H \left[\begin{pmatrix} Z_{\mathcal{J}}^H \\ Z_{\bar{\mathcal{J}}}^H \end{pmatrix} G^2 V\right] &= V^H G^2 (Z_{\mathcal{J}}, Z_{\bar{\mathcal{J}}}) \begin{pmatrix} Z_{\mathcal{J}}^H \\ Z_{\bar{\mathcal{J}}}^H \end{pmatrix} G^2 V \\ &= V^H G^2 V = I. \end{aligned}$$

Hence

$$\begin{pmatrix} Z_{\mathcal{J}}^H \\ Z_{\bar{\mathcal{J}}}^H \end{pmatrix} G^2 V$$

has orthonormal columns.

The following analogue of Theorem 4 holds for the eigenvalue problem (4.3).

Theorem 4.5. *Let the $r \times s$ matrix V have G-orthonormal columns and let*

$$\nu_1 \geqq \nu_2 \geqq \cdots \geqq \nu_s$$

be the eigenvalues of $V^H C V$. Then

$$\nu_1 \geqq \mu_{r-s+1}$$

and

$$\nu_s \leqq \mu_s.$$

Proof. Let

$$Q = G V.$$

Then $Q^H Q = I$ and

$$V^H C V = Q^H B Q.$$

Hence Theorem 4.3 applies to give the result.

Although the individual eigenvectors corresponding to a set of clustered eigenvalues are poorly determined by the elements of the matrix, the invariant subspace spanned by them is well determined. The following generalization of a theorem of Swanson [3] gives this assertion a quantitative form.

Theorem 4.6. *Let V be an $r\times s$ matrix and $D=\operatorname{diag}(\delta_{i_1}, \delta_{i_2}, \ldots, \delta_{i_s})$. Let $\mathscr{I}=\{i_1, i_2, \ldots, i_s\}$ and $\mathscr{J}=\{1, 2, \ldots, r\}\backslash\mathscr{I}$. Suppose*

$$|\delta_i-\mu_j|\geqq\alpha>0, \quad (i\in\mathscr{I},\ j\in\mathscr{J}).$$

If

$$\|CV-G^2VD\|\leqq\eta,$$

then

$$\|Z_{\mathscr{J}}^H G^2 V\|\leqq\frac{\sqrt{s}\,\eta}{\alpha}\|G^{-1}\|. \tag{4.7}$$

Proof. Let $M'=\operatorname{diag}(\mu_j)$ $(j\in\mathscr{J})$. Then

$$Z_{\mathscr{J}}^H(CV-G^2VD)=M'Z_{\mathscr{J}}^HG^2V-Z_{\mathscr{J}}^HG^2VD.$$

Hence

$$\|(M'Z_{\mathscr{J}}^HG^2V-Z_{\mathscr{J}}^HG^2VD)\,e_i\|\leqq\eta\,\|Z_{\mathscr{J}}\|,$$

where e_i is the i-th column of the $s\times s$ identity matrix.

But

$$\begin{aligned}\|(M'Z_{\mathscr{J}}^HG^2V-Z_{\mathscr{J}}^HG^2VD)\,e_i\|&=\|(M'-\delta_iI)\,Z_{\mathscr{J}}^HG^2Ve_i\|\\&\geqq\alpha\|Z_{\mathscr{J}}^HG^2Ve_i\|.\end{aligned}$$

Thus the norm of the i-th column of $Z_{\mathscr{J}}^HG^2V$ is not greater than $\eta\|Z_{\mathscr{J}}\|/\alpha$. Since $Z_{\mathscr{J}}^HG^2V$ has s columns,

$$\|Z_{\mathscr{J}}^HG^2V\|\leqq\frac{\sqrt{s}\,\eta}{\alpha}\|Z_{\mathscr{J}}\|. \tag{4.8}$$

But

$$\|Z_{\mathscr{J}}\|\leqq\|G^{-1}\|\|GZ_{\mathscr{J}}\|=\|G^{-1}\|, \tag{4.9}$$

since GZ has orthonormal columns. The inequality (4.7) follows from (4.8) and (4.9).

Corollary 4.7. *If VT has G-orthonormal columns, then there is a unitary matrix N such that*

$$\|VTN-Z_{\mathscr{I}}\|\leqq\frac{2\sqrt{s}\,\eta}{\alpha}\|G^{-1}\|^2\|T\|.$$

Proof. Since VT has G-orthonormal columns, the matrix

$$\begin{pmatrix}Z_{\mathscr{I}}^H G^2VT\\ Z_{\mathscr{J}}^HG^2VT\end{pmatrix}=\begin{pmatrix}Z_{\mathscr{I}}^H\\ Z_{\mathscr{J}}^H\end{pmatrix}G^2VT$$

has orthonormal columns. Moreover by Theorem 4.6

$$\|Z_{\mathscr{J}}^HG^2VT\|\leqq\frac{\sqrt{s}\,\eta}{\alpha}\|G^{-1}\|\|T\|.$$

Hence by Theorem 4.1

$$\begin{pmatrix}Z_{\mathscr{I}}^H\\ Z_{\mathscr{J}}^H\end{pmatrix}G^2VT=\begin{pmatrix}N^H+F\\ Z_{\mathscr{J}}^HG^2VB\end{pmatrix} \tag{4.10}$$

where N^H is unitary and

$$\|F\| \leq \|Z_{\mathscr{J}}^H G^2 V T\|^2 \leq \frac{\sqrt{s}\,\eta}{\alpha} \|G^{-1}\| \|T\|.$$

Premultiply (4.10) by $(Z_{\mathscr{I}}, Z_{\mathscr{J}})$ and postmultiply by N to get

$$V T N = Z_{\mathscr{I}}(I + F N) + Z_{\mathscr{J}}(Z_{\mathscr{J}}^H G^2 V T N).$$

Hence upon taking norms

$$\|V T N - Z_{\mathscr{I}}\| \leq (\|Z_{\mathscr{I}}\| + \|Z_{\mathscr{J}}\|) \frac{\sqrt{s}\,\eta}{\alpha} \|G^{-1}\| \|T\|.$$

Since

$$\|Z_{\mathscr{I}}\| \leq \|G^{-1}\|, \quad \|Z_{\mathscr{J}}\| \leq \|G^{-1}\|,$$

the result follows.

V. Accuracy of the Refined Eigenvalues

Suppose now that the acceleration step is applied at the k-th step of the orthogonal iteration so that matrices B, Y, and M are determined from A and $Q^{(k)}$ by Eqs. (2.2) and (2.3). Note that the auxiliary matrix P defined by (2.1) is identical to the matrix $P^{(k)}$ of Eq. (3.2). For brevity the iteration superscripts will be dropped in the next two sections.

The first step in assessing the accuracy of the refined eigenvalues and eigenvectors is to reduce the eigenvalue problem to a more tractable form. Let

$$Z = P_1 Y.$$

Then the eigenvalue problem

$$P^H \Lambda P Y = Y M$$

may be written in the form

$$(I, K^H) \Lambda \begin{pmatrix} I \\ K \end{pmatrix} Z = (P_1 P_1^H)^{-1} Z M$$

where K is defined by (3.6). By Eq. (3.7)

$$(I, K^H) \Lambda \begin{pmatrix} I \\ K \end{pmatrix} Z = (I, K^H) \begin{pmatrix} I \\ K \end{pmatrix} Z M. \tag{5.1}$$

If

$$C = (I, K^H) \Lambda \begin{pmatrix} I \\ K \end{pmatrix} = \Lambda_1 + K^H \Lambda_2 K$$

and a Hermitian matrix G is determined [2] so that

$$G^2 = I + K^H K,$$

then Eq. (5.1) takes the form of the eigenvalue problem (4.3) of the last section. Moreover

$$P Y = \begin{pmatrix} I \\ K \end{pmatrix} Z. \tag{5.2}$$

As in the last section let $\mathcal{J}$ denote a set of s integers taken from $\{1, 2, \ldots, r\}$. Let the columns of $E_{\mathcal{J}}$ be taken from the $r\times r$ identity matrix. Then there is a matrix T, defined in the last section, such that $E_{\mathcal{J}}T$ has G-orthonormal columns.

Lemma 5.1. *For $V=E$ let T be defined by Eqs.* (4.5) *and* (4.6). *Then*

$$T=H(I-\Gamma) \tag{5.3}$$

where Γ is diagonal and

$$\|\Gamma\|\leq\frac{\|K_{\mathcal{J}}\|^2}{1+\|K_{\mathcal{J}}\|^2}. \tag{5.4}$$

Moreover

$$\|K_{\mathcal{J}}T\|^2=\frac{\|K_{\mathcal{J}}\|^2}{1+\|K_{\mathcal{J}}\|^2}. \tag{5.5}$$

Proof. By the definitions of H, G, and Δ^2,

$$\Delta^2=I+H^H E_{\mathcal{J}}^H K^H K E_{\mathcal{J}} H=I+H^H K_{\mathcal{J}}^H K_{\mathcal{J}} H=I+\Theta^2.$$

Since Δ^2 is diagonal, so is Θ^2, and

$$\|\Theta\|=\|K_{\mathcal{J}}\|.$$

Also if

$$\Gamma=I-(I+\Theta^2)^{-\frac{1}{2}}=I-\Delta^{-1}$$

then T is given by (5.3). Since $1-(1+x^2)^{-\frac{1}{2}}$ is an increasing function of x

$$\|\Gamma\|=1-(1+\|K_{\mathcal{J}}\|^2)^{-\frac{1}{2}}\leq\frac{\|K_{\mathcal{J}}\|^2}{1+\|K_{\mathcal{J}}\|^2}.$$

Finally

$$\begin{aligned} T^H K_{\mathcal{J}}^H K_{\mathcal{J}} T &= \Delta^{-1} H^H K_{\mathcal{J}}^H K_{\mathcal{J}} H \Delta^{-1} \\ &= \Delta^{-2}\Theta^2=(I+\Theta^2)^{-1}\Theta^2, \end{aligned}$$

and since $x^2/(1+x^2)$ is an increasing function of x, (5.5) holds.

In order to compare the elements μ_i of M with the λ_i, it is important that the μ_i be ordered properly. Let the λ_i be ordered as in (1.1) and let σ be a permutation of the integers $1, 2, \ldots, r$ such that

$$\lambda_{\sigma(1)}\geqq\lambda_{\sigma(2)}\geqq\cdots\geqq\lambda_{\sigma(r)}.$$

Then the μ_i are to be ordered so that

$$\mu_{\sigma(1)}\geqq\mu_{\sigma(2)}\geqq\cdots\geqq\mu_{\sigma(r)},$$

and μ_i will be compared with λ_i.

Let $\tau=\sigma^{-1}$. The μ_i are the eigenvalues of the section $Q^H A Q$ of the matrix A. Suppose that $\lambda_m>0$. Then λ_m is the $\tau(m)$-th largest eigenvalue of A and μ_m is the $\tau(m)$-th largest eigenvalue of B. Hence by Corollary 4.4

$$\lambda_m\geqq\mu_m.$$

Similarly if λ_m is negative then it is the $(r-\tau(m)+1)$-th smallest eigenvalue of A while μ_m is the $(r-\tau(m)+1)$-th smallest eigenvalue of B. Hence

$$\lambda_m\leqq\mu_m.$$

Thus to determine the accuracy of μ_m it is only necessary to determine a sharp lower bound for μ_m when λ_m is positive or a sharp upper bound when λ_m is negative.

The case $\lambda_m < 0$ is typical. Let

$$\mathscr{I} = \{i\colon \lambda_i \leqq \lambda_m\},$$

and let the columns of $E_{\mathscr{I}}$ be taken from the $r \times r$ identity matrix. Let T be defined as in Lemma 5.1 so that $E_{\mathscr{I}} T$ has G-orthonormal columns. Then the matrix

$$S = (E_{\mathscr{I}} T)^H C (E_{\mathscr{I}} T)$$

is of order $(r - \tau(m) + 1)$. Hence by Theorem 4.5 its largest eigenvalue is greater than the $(r - \tau(m) + 1)$-th smallest μ_i; that is the largest eigenvalue of S is greater than μ_m.

Let

$$\Lambda' = \operatorname{diag}(\lambda_i) \quad (i \in \mathscr{I}).$$

Then

$$S = T^H \Lambda' T + T^H K_{\mathscr{I}}^H \Lambda_2 K_{\mathscr{I}} T.$$

From the preceding lemma

$$\begin{aligned} S &= H^H \Lambda' H + \Gamma H^H \Lambda' H + H^H \Lambda' H \Gamma + T^H K_{\mathscr{I}}^H \Lambda_2 K^H T \\ &= H^H \Lambda' H + F. \end{aligned}$$

Now $H^H \Lambda' H$ is a Hermitian matrix whose largest eigenvalue is λ_m, and F is a Hermitian matrix. Thus by Theorem 4.2 the largest eigenvalue ν of S satisfies

$$\mu_m \leqq \nu \leqq \lambda_m + \|F\|.$$

But by Lemma 5.1

$$\begin{aligned} \|F\| &\leqq 2\|\Gamma H^H \Lambda' H\| + \|T^H K_{\mathscr{I}}^H \Lambda_2' K_{\mathscr{I}} T\| \\ &\leqq \frac{3\|A\| \, \|K_{\mathscr{I}}\|^2}{1 - \|K_{\mathscr{I}}\|^2}. \end{aligned}$$

If a similar argument is carried out for $\lambda_m \geqq 0$ with

$$\mathscr{I} = \{i\colon \lambda_i \geqq \lambda_m\},$$

then the result is

Theorem 5.2. *Let*

$$\mathscr{I} = \{i\colon \operatorname{sign}(\lambda_i) = \operatorname{sign}(\lambda_m) \text{ and } |\lambda_i| \geqq |\lambda_m|\},$$

and

$$\varepsilon = \frac{3\|A\| \, \|K_{\mathscr{I}}\|^2}{1 + \|K_{\mathscr{I}}\|^2}.$$

Then if $\lambda_m > 0$,

$$\lambda_m - \varepsilon \leqq \mu_m \leqq \lambda_m.$$

If $\lambda_m < 0$

$$\lambda_m \leqq \mu_m \leqq \lambda_m + \varepsilon.$$

Thus the error in μ_m is proportional to the square of $\|K_{\mathscr{I}}\|$ when $\|K_{\mathscr{I}}\|$ is small. At the k-th iteration $\|K_{\mathscr{I}}\|$ may be estimated from Eq. (3.5):

$$\|K_{\mathscr{I}}^{(k)}\| \leqq |\lambda_{r+1}/\lambda_m|^k \|K^{(0)}\|.$$

Since $|\lambda_m|$ decreases with increasing m, the μ_m may be expected to show a progressive loss in accuracy from μ_1 to μ_r, with μ_r being least accurate. In fact if $\lambda_{r+1}=-\lambda_r$, the value of μ_r will be entirely spurious. However, if $|\lambda_s|$ is significantly less than $|\lambda_r|$, then the columns of $Q^{(k)}$ will tend to lie in the space spanned by $x_1, \ldots, x_{s-1}$. Hence if $\lambda_1, \ldots, \lambda_{s-1}$ all have the same sign, as when A is positive definite, μ_r will tend to lie between λ_r and λ_{s-1} and may not be too inaccurate.

VI. Accuracy of the Refined Eigenvectors

In assessing the accuracy of the refined eigenvectors, some care must be taken to treat clusters of eigenvalues together, for it is only the subspace corresponding to a cluster of poorly separated eigenvalue that is really well determined. Specifically, let $\mathscr{I}$ be the index set of such a cluster. Then the question to be answered in this section is how well do the spaces of $QY_{\mathscr{I}}$ and $X_{\mathscr{I}}$ compare.

As in the last section, it is convenient to phrase the question in terms of the transformed problem (5.1). Let the columns of $I_{\mathscr{I}}$ be taken from the $n\times n$ identity matrix. Then the above question becomes one of comparing the spaces of $PY_{\mathscr{I}}=X^H Q Y_{\mathscr{I}}$ and $I_{\mathscr{I}}=X^H X_{\mathscr{I}}$. The question will be answered by showing that under suitable restrictions, there is a unitary matrix S such that $\|I_{\mathscr{I}}S-PY_{\mathscr{I}}\|$ is small. By virtue of Eq. (5.2) this is equivalent to showing that

$$\left\| I_{\mathscr{I}} S - \begin{pmatrix} I \\ K \end{pmatrix} Z_{\mathscr{I}} \right\|$$

is small.

Let the s columns of $E_{\mathscr{I}}$ be taken from the $r\times r$ identity matrix. If T is the matrix of Lemma 5.1, then the columns of $E_{\mathscr{I}}T$ are G-orthonormal. Let

$$\mathscr{J}=\{1, 2, \ldots, r\}\setminus\mathscr{I}$$

be the index set complementary to $\mathscr{I}$. Now because $\{\lambda_i\colon i\in\mathscr{I}\}$ is a cluster of eigenvalues, they are well separated from the other eigenvalues λ_j $(j\in\mathscr{J})$. Suppose that the orthogonal iteration has proceeded so far that the λ_i $(i\in\mathscr{I})$ are also separated from the μ_j $(j\in\mathscr{J})$, say

$$|\lambda_i-\mu_j|\geqq\alpha, \quad (i\in\mathscr{I},\ j\in\mathscr{J}). \tag{6.1}$$

Let

$$\Lambda'=\operatorname{diag}(\lambda_i), \quad (i\in\mathscr{I}).$$

Then

$$(I, K^H)\Lambda\begin{pmatrix} I \\ K \end{pmatrix} E_{\mathscr{I}} - (I, K^H)\begin{pmatrix} I \\ K \end{pmatrix} E_{\mathscr{I}}\Lambda'$$
$$= K^H \Lambda_2 K_{\mathscr{I}} - K^H K_{\mathscr{I}}\Lambda'.$$

In the notation of the last section (and Theorem 4.6)

$$\begin{aligned}\|C E_{\mathcal{J}} - G^2 E_{\mathcal{J}} \Lambda'\| &= \|K^H \Lambda_2 K_{\mathcal{J}} - K^H K_{\mathcal{J}} \Lambda'\| \\ &\leqq 2\lambda \|K\| \, \|K_{\mathcal{J}}\|,\end{aligned}$$

where

$$\lambda = \max |\lambda_i|, \quad (i \in \mathcal{J}). \tag{6.2}$$

Hence by Corollary 4.7 there is a unitary matrix N such that

$$\|E_{\mathcal{J}} T N - Z_{\mathcal{J}}\| \leqq \frac{4\sqrt{s}\,\lambda}{\alpha} \|K\| \, \|K_{\mathcal{J}}\| \, \|G^{-1}\|^2 \, \|T\|.$$

But since E has orthonormal columns $\|T\| \leqq \|G^{-1}\|$. Hence

$$\|E_{\mathcal{J}} T N - Z_{\mathcal{J}}\| \leqq \frac{4\sqrt{s}\,\lambda}{\alpha} \|G^{-1}\|^3 \, \|K\| \, \|K_{\mathcal{J}}\|.$$

Let

$$S = H N$$

where H is the matrix of Eq. (5.3). Since H and N are unitary, so is S. Moreover

$$\begin{aligned}\left\| I_{\mathcal{J}} S - \begin{pmatrix} I \\ K \end{pmatrix} Z_{\mathcal{J}} \right\| &\leqq \|I_{\mathcal{J}} H N - I_{\mathcal{J}} T N\| \\ &\quad + \left\| I_{\mathcal{J}} T N - \begin{pmatrix} I \\ K \end{pmatrix} E_{\mathcal{J}} T N \right\| + \left\| \begin{pmatrix} I \\ K \end{pmatrix} (E_{\mathcal{J}} T N - Z_{\mathcal{J}}) \right\| \\ &= \varepsilon_1 + \varepsilon_2 + \varepsilon_s.\end{aligned}$$

Thus the problem is to find bounds for ε_1, ε_2, and ε_3.

Now

$$\|I_{\mathcal{J}} H N - I_{\mathcal{J}} T N\| = \|I_{\mathcal{J}} H \Gamma N\|$$

where Γ satisfies (5.4). Hence

$$\varepsilon_1 \leqq \frac{\|K_{\mathcal{J}}\|^2}{1 + \|K_{\mathcal{J}}\|^2} \leqq \|K_{\mathcal{J}}\|^2.$$

Also

$$I_{\mathcal{J}} T N - \begin{pmatrix} I \\ K \end{pmatrix} E_{\mathcal{J}} T N = \begin{pmatrix} 0 \\ K_{\mathcal{J}} T N \end{pmatrix}.$$

Hence

$$\varepsilon_2 \leqq \|G^{-1}\| \, \|K_{\mathcal{J}}\|.$$

Finally

$$\varepsilon_3 \leqq \|G\| \, \|E_{\mathcal{J}} T N - Z_{\mathcal{J}}\|.$$

In terms of the original eigenvectors, the result of all this is

Theorem 6.1. *Let the index set $\mathcal{J}$ be chosen so that* (6.1) *is satisfied. Then there is a unitary matrix S such that*

$$\|X S - Q Y\| \leqq \left[\|K_{\mathcal{J}}\| + \|G^{-1}\| + \frac{4\sqrt{s}\,\lambda}{\alpha} \|G\| \, \|G^{-1}\|^3 \, \|K\| \right] \|K_{\mathcal{J}}\|,$$

where λ is defined by (6.2).

26*

Thus the accuracy of the space of QY is approximately proportional to $\|K_{\mathscr{J}}\|$ when $\|K_{\mathscr{J}}\|$ is small. The quantity λ/α is large when there is poor relative separation between the cluster of eigenvalues indexed by $\mathscr{J}$ and its neighbors.

Acknowledgement. I wish to thank Professor A. S. HOUSEHOLDER for his guidance and Professor G. GOLUB for a stimulating conversation.

Bibliography

1. BAUER, F. L.: Das Verfahren der Treppeniteration und verwandte Verfahren zur Lösung algebraischer Eigenwertprobleme. Z. Angew. Math. Phys. **8**, 214—235 (1957).
2. HOUSEHOLDER, A. S.: The theory of matrices in numerical analysis. New York: Blaisdell Publishing Co. 1964.
3. SWANSON, C. A.: An inequality for linear transformations with eigenvalues. Bull. Amer. Math. Soc. **67**, 607—608 (1961).
4. WILKINSON, J. H.: Convergence of the LR, QR, and related algorithms. Comp. J. **8**, 77—84 (1965).
5. — The algebraic eigenvalue problem. Oxford: Clarendon Press 1965.

G. W. STEWART
University of Texas at Austin
Department of Mathematics
Austin, Texas 78712, USA

16.2. [GWS-J30] "Simultaneous Iteration for Computing Invariant Subspaces of Non-Hermitian Matrices"

[GWS-J30] "Simultaneous Iteration for Computing Invariant Subspaces of Non-Hermitian Matrices," *Numerische Mathematik* **25** (1976) 123–136.
http://dx.doi.org/10.1007/BF01462265

Numer. Math. 25, 123—136 (1976)

Simultaneous Iteration for Computing Invariant Subspaces of Non-Hermitian Matrices

G. W. Stewart *

Received November 22, 1974

Summary. This paper describes a simultaneous iteration technique for computing a nested sequence of orthonormal bases for the dominant invariant subspaces of a non-Hermitian matrix. The method is particularly suited to large sparse eigenvalue problems, since it requires only that one be able to form the product of the matrix in question with a vector. A convergence theory for the method is developed and practical details are discussed.

1. Introduction

Let A be a matrix of order n with eigenvalues $\lambda_1, \lambda_2, \ldots, \lambda_n$ ordered so that

$$|\lambda_1| \geqq |\lambda_2| \geqq \cdots \geqq |\lambda_n|. \tag{1.1}$$

In this paper we shall be concerned with the computational consequences of the following observation. *Let $\mathcal{Q}$ be a subspace of dimension r. If $|\lambda_r| > |\lambda_{r+1}|$, then under mild restrictions on $\mathcal{Q}$ the subspaces $A^\nu \mathcal{Q}$ tend toward the invariant subspace of A corresponding to $\lambda_1, \lambda_2, \ldots, \lambda_r$.* The proof of this observation [12] shows that the convergence will not be uniform; the subspace $A^\nu \mathcal{Q}$ will tend to contain more accurate approximations to the eigenvector corresponding to λ_1 than to the eigenvector corresponding to λ_r (see also [5], where the convergence is investigated in the case $|\lambda_r| = |\lambda_{r+1}|$).

When $r=1$, the above observation leads directly to the power method in which the dominant eigenvector of A is approximated by a sequence of vectors $A^\nu q$. The method is particularly attractive when A is large and sparse, since it requires only that one be able to multiply a vector by A. It has the drawback that it may converge slowly when $|\lambda_1|$ is nearly equal to $|\lambda_2|$ and not at all when $|\lambda_1| = |\lambda_2|$. Moreover the method finds only the dominant eigenvector.

When $r > 1$, the observation leads to a class of methods, now generally known as simultaneous iteration methods, whose prototype is Bauer's *Treppeniteration* [1]. In their most general form these methods start with an $n \times r$ matrix Q_0 whose columns form a basis for $\mathcal{Q}$ and generate a sequence of matrices Q_ν according to the formula

$$A Q_\nu = Q_{\nu+1} R_{\nu+1}, \tag{1.2}$$

* This work was supported in part by the Office of Naval Research under Contract No. N00014-67-A-0128-0018.

where $R_{\nu+1}$ is a nonsingular matrix chosen variously by the different methods (since the column spaces of Q_ν and $Q_\nu R_\nu$ are the same, R_ν can be regarded as a scaling factor). It is easily verified that

$$A^\nu Q_0 = Q_\nu R_\nu R_{\nu-1} \cdots R_1$$

so that the columns of Q_ν form a basis of $A^\nu \mathfrak{Q}$, which is approaching the dominant invariant subspace of A.

When A is Hermitian and hence has a set of orthonormal eigenvectors, it is natural to choose R_ν so that Q_ν has orthonormal columns. The matrix R_ν is usually taken to be upper triangular, which corresponds to applying the Gram-Schmidt orthogonalization to the columns of AQ_ν. When this is done, the columns of Q_ν usually approach the eigenvectors of A in the order prescribed by (1.1). However, the convergence may be slow, far slower than the oblique convergence of $\mathfrak{Q}_\nu$ would suggest is possible. This difficulty may be circumvented by employing a device that is essentially the Rayleigh-Ritz method for approximating eigenvectors. Namely the Hermitian matrix

$$B_\nu = Q_\nu^H A Q_\nu \tag{1.3}$$

is formed and diagonalized by a unitary matrix Y_ν:

$$T_\nu = Y_\nu^H B_\nu Y_\nu, \tag{1.4}$$

$$|\tau_{11}| \geqq |\tau_{22}| \geqq \cdots \geqq |\tau_{rr}|. \tag{1.5}$$

The columns of the matrix $Q_\nu Y_\nu$ are in general better approximate eigenvectors than those of Q_ν. In particular, under mild restrictions on Q_0 the first column of $Q_\nu Y_\nu$ usually approaches a dominant eigenvector of A linearly with ratio $|\lambda_{r+1}/\lambda_1|$, as opposed to a ratio of $|\lambda_2/\lambda_1|$ for the first column of Q_ν. The theoretical and practical aspects of this method have been treated in several places [2, 4, 6, 9, 14], and Rutishauser [6, 7] has published a program.

The method can of course be applied to non-Hermitian matrices; however the columns of Q_ν will not in general approach eigenvectors of A since the latter need not be orthogonal. To obtain eigenvectors from the columns of Q_ν some generalization of the Rayleigh-Ritz method is needed. The natural generalization is to attempt to diagonalize B_ν by a similarity transformation: $Y_\nu^{-1} B_\nu Y_\nu = \mathrm{diag}(\mu_1, \mu_2, \ldots, \mu_r)$. The columns of $Q_\nu Y_\nu$ may then be taken as approximations to eigenvectors of A. A more elaborate variant has been proposed by Bauer under the name bi-iteration [1]. Here a second sequence defined by $P_{\nu+1} S_{\nu+1} = A^H P_\nu$ is generated, and P_ν and Q_ν are bi-orthogonalized so that $P_\nu^H Q_\nu = I$. The columns of P_ν and Q_ν generally converge to left and right eigenvectors of A; however, the convergence can be accelerated by diagonalizing $B_\nu = P_\nu^H A Q_\nu$ and proceeding as above. Some of the practical details of this variant have been discussed by Clint and Jennings [3].

The principal drawback to the above schemes is that they attempt to compute the eigenvectors of A directly. If A is defective, then it may not have enough eigenvectors to span the dominant subspace corresponding to $\lambda_1, \lambda_2, \ldots, \lambda_r$. Even when A is not defective, its eigenvectors may be very nearly dependent so that they form a poor numerical basis for this subspace. The object of this paper is to examine a variant of simultaneous iteration for computing a nested sequence of orthonormal bases for the dominant invariant subspaces of A. Such a basis is furnished by the following theorem due to Schur (for a proof see [11]).

Theorem 1.1. Let $A \in \mathbb{C}^{n\times n}$. Then there is a unitary $X \in \mathbb{C}^{n\times n}$ such that

$$S = X^H A X$$

is upper triangular. The matrix X may be chosen so that the diagonal elements of S, which are eigenvalues of A, appear in descending order of absolute value.

With the ordering specified in Theorem 1.1, the columns x_i of X enjoy many of the properties of the eigenvectors of a Hermitian matrix. If $|\lambda_{i-1}| > |\lambda_i| > |\lambda_{i+1}|$, then x_i is uniquely determined up to a scalar factor of absolute value unity. If $|\lambda_{i-1}| > |\lambda_i| = |\lambda_{i+1}| = \dots = |\lambda_j| > |\lambda_{j+1}|$, the vectors $x_i, x_{i+1}, \dots, x_j$ are not in general uniquely determined; however the subspace spanned by them is. Because of the uniqueness properties of the vectors x_i, we shall call them the *Schur vectors* of A.

If $|\lambda_r| > |\lambda_{r+1}|$ then the vectors $x_1, x_2, \dots, x_r$ form an orthonormal basis for the invariant subspace corresponding to $\lambda_1, \lambda_2, \dots, \lambda_r$. To see this let $X^{|r} = (x_1, x_2, \dots, x_r)$ and let $S^{\lceil r}$ denote the $r\times r$ leading principal submatrix of S. Then from the equation

$$AX = XS$$

and the upper triangularity of S, we have

$$A X^{|r} = X^{|r} S^{\lceil r}. \tag{1.6}$$

This implies that the columns of $X^{|r}$ span an invariant subspace of A whose eigenvalues are those of $S^{\lceil r}$, i.e., $\lambda_1, \lambda_2, \dots, \lambda_r$.

Moreover a knowledge of $S^{\lceil r}$ and $X^{|r}$ permits the computation of the eigenvectors of A corresponding to $\lambda_1, \lambda_2, \dots, \lambda_r$. For if $S^{\lceil r} z_i = \lambda_i z_i$, then from (1.6)

$$A(X^{|r} z_i) = X^{|r} S^{\lceil r} z_i = \lambda_i X^{|r} z_i,$$

and $X^{|r} z_i$ is an eigenvector of A corresponding to λ_i.

In this paper we shall be concerned with analyzing an algorithm for non-Hermitian matrices based on the above considerations. Specifically, starting with an $n\times r$ matrix Q_0, we iterate according to the formula (1.2), where the upper triangular matrix $R_{\nu+1}$ is chosen so that $Q_{\nu+1}$ has orthonormal columns. It can be shown that the columns of Q_ν will usually tend toward the Schur vectors $x_1, x_2, \dots, x_r$; however, as in the Hermitian case, the convergence may be slow. To accelerate the process we shall occasionally perform an analogue of the Rayleigh-Ritz step in which B_ν is formed according to (1.3) and reduced by the unitary matrix Y_ν to Schur form with its diagonal elements in descending order of absolute value (cf. (1.4) and (1.5)). The matrix Q_ν is then replaced by the matrix $Q_\nu Y_\nu$. We shall call this process a Schur-Rayleigh-Ritz step and abbreviate it to SRR step.

The principal theoretical difficulties of the algorithm concern the behavior of the matrices generated by the SRR step. This behavior is analyzed in the next two sections. Some of the practical considerations that must enter any implementation of the method are discussed in Section 4. Throughout this paper we shall use the notational conventions of [11]. The symbol $\|\cdot\|$ will denote the Frobenius

10*

norm [11, p. 173] defined by

$$\|A\|^2 = \text{trace}\,(A^H A).$$

We shall use the notation $A^{|l}$ for the matrix consisting of the first l columns of A and $A^{l|}$ for the matrix consisting of the last l columns.

2. Some Preliminaries

In this section we shall collect some results that will be needed in Section 3, where the behavior of the SRR iterates will be analyzed. The first result concerns the behavior of powers of matrices, the second the stability of the QR factorization, and the third the stability of invariant subspaces. The reason for stating the results in detail is to give the reader some basis for evaluating the order constants that are implicit in the results of Section 3.

The first result is well known.

Theorem 2.1. Let $S_1 \in \mathbb{C}^{m\times m}$, $K \in \mathbb{C}^{m\times n}$, and $S_2 \in \mathbb{C}^{n\times n}$. Let λ denote an eigenvalue of S_1 that is largest in magnitude and μ an eigenvalue of S_2 that is smallest in magnitude. If

$$|\mu| > |\lambda|,$$

then for any $\varepsilon > 0$

$$\|S_1^\nu K S_2^{-\nu}\| = O[(|\lambda/\mu| + \varepsilon)^\nu]. \tag{2.1}$$

Proof. It is well known (see, e.g., [11, p. 284]) that there is a norm, $\|\cdot\|_{(1)}$, such that $\|S_1\|_{(1)} \leq |\lambda| + \varepsilon$. Since μ^{-1} is a largest eigenvalue of S_2^{-1}, there is also a norm, $\|\cdot\|_{(2)}$, such that $\|S_2^{-1}\|_{(2)} \leq |\mu^{-1}| + \varepsilon$. By the equivalence of norms there are constants τ_1 and τ_2 such that

$$\|A\| \leq \tau_i \|A\|_{(i)}, \quad (i = 1, 2). \tag{2.2}$$

Now

$$\begin{aligned}\|S_1^\nu K S_2^{-\nu}\| &\leq \|S_1\|\,\|K\|\,\|S_2^{-\nu}\| \\ &\leq \tau_1\tau_2 \|S_1\|_{(1)} \|K\|\,\|S_2^{-\nu}\|_{(2)} \\ &\leq \tau_1\tau_2 \|K\| [(|\lambda| + \varepsilon)\,(|\mu|^{-1} + \varepsilon)]^\nu,\end{aligned}$$

which implies (2.1). □

According to (2.1) the convergence in Theorem 2.1 is almost linear with a ratio $|\lambda/\mu|$. We shall signal this kind of convergence by writing

$$\|S_1^\nu K S_2^{-\nu}\| = O_\varepsilon(|\lambda/\mu|^\nu).$$

The order constants depend on τ_1 and τ_2 in (2.2), and they can be large. Large values of τ_1 and τ_2 are associated in an obscure way with ill conditioning of the eigenvalue problems for the matrices S_1 and S_2. This is a point that can stand further investigation.

The next result concerns the QR factorization of a perturbation of a matrix whose columns are taken from the columns of the identity matrix. The following theorem follows easily from the results given in [13].

Theorem 2.2. Let $E \in \mathbb{C}^{m\times n}$ have the form

$$E = \begin{pmatrix} I_n \\ 0 \end{pmatrix}.$$

Let $K \in \mathbb{C}^{m \times n}$ satisfy $\|K\| < 1/8$. Then there is an upper triangular matrix $R \in \mathbb{C}^{n \times n}$ and a matrix $W \in \mathbb{C}^{m \times n}$ such that $E + W$ has orthonormal columns, $E + K = (E + W) R$, and

$$\|W\| \leq \frac{3\|K\|}{1 - 2\|K\|}.$$

Finally we shall need a perturbation theorem for invariant subspaces. We state a special case of a theorem in [10].

Theorem 2.3. Let $B \in \mathbb{C}^{r \times r}$ be partitioned in the form

$$B = \begin{pmatrix} B_{11} & B_{12} \\ B_{21} & B_{22} \end{pmatrix}$$

where $B_{11} \in \mathbb{C}^{l \times l}$. Define δ by

$$\delta = \inf\{\|P B_{11} - B_{22} P\|: P \in \mathbb{C}^{(r-l) \times l},\ \|P\| = 1\}.$$

If

$$\frac{\|B_{21}\|\,\|B_{12}\|}{\delta^2} < \frac{1}{4}, \tag{2.3}$$

then there is a unique $P \in \mathbb{C}^{(r-l) \times r}$ satisfying

$$\|P\| < 2\frac{\|B_{21}\|}{\delta} \tag{2.4}$$

such that

$$\mathscr{R}\begin{pmatrix} I_l \\ P \end{pmatrix} \tag{2.5}$$

is an invariant subspace of B. Moreover $\lambda(B)$ is the disjoint union

$$\lambda(B) = \lambda(B_{11} + B_{12} P) \cup \lambda(B_{22} - P B_{12}), \tag{2.6}$$

where $\lambda(B_{11} + B_{12} P)$ corresponds to the invariant subspace (2.5).

If B_{21} were zero the choice $P = 0$ would make (2.5) an invariant subspace of B. The inequality (2.4) then shows how much the invariant subspace is perturbed by the presence of B_{21}. The number δ appearing in the bound, which in [10] is written sep (B_{11}, B_{22}), measures the separation between the spectra of B_{11} and B_{22}. It is zero if and only if B_{11} and B_{22} have common eigenvalues. It also has the easily verified property that

$$\operatorname{sep}(B_{11} + G_{11}, B_{22} + G_{22}) \geq \operatorname{sep}(B_{11}, B_{22}) - \|G_{11}\| - \|G_{22}\|. \tag{2.7}$$

The presence of B_{12} in (2.3) and (2.6) suggests that a large value of $\|B_{12}\|$ will be associated with ill conditioning of the eigensystem of B, even though $\|B_{12}\|$ does not appear in the bound (2.4).

3. Convergence of the SRR Iterates

In this section we shall be concerned with describing the behavior of the SRR iterates defined in Section 1. It is important to realize that these matrices depend only on the subspaces $A^\nu \mathfrak{Q}$ and not on a particular choice of basis. For let Q and $\hat{Q}$ be matrices with orthonormal columns such that $\mathscr{R}(Q) = \mathscr{R}(\hat{Q}) = A^\nu \mathfrak{Q}$. Then there is a unitary matrix U such that $\hat{Q} = QU$. Let Y be the unitary matrix that reduces $Q^H A Q$ to its Schur form T. Then

$$\begin{aligned}(U^H Y)^H (\hat{Q}^H A \hat{Q})(U^H Y) &= Y^H (U\hat{Q}^H) A (\hat{Q} U^H) Y \\ &= Y^H (Q^H A Q) Y = T\end{aligned}$$

so that $U^H Y$ is the unitary matrix that reduces $\hat{Q}^H A \hat{Q}$ to Schur form. The SRR iterate for $\hat{Q}$ is then

$$\hat{Q}(U^H Y) = (\hat{Q} U^H) Y = QY,$$

which is also the SRR iterate for Q.

We shall show that for $1 \leqq l \leqq r$ if $|\lambda_l| > |\lambda_{l+1}|$ and $|\lambda_r| > |\lambda_{r+1}|$ then $R(Q_\nu^{|l}) \to R(X^{|l})$ and the convergence is $O_s(|\lambda_{r+1}/\lambda_l|^\nu)$. Here we have let Q_ν denote the SRR approximation associated with $A^\nu \mathfrak{Q}$, and as in Section 1 the matrix X is the matrix of Schur vectors of A. There are two steps in the proof. First we shall exhibit a basis for $A^\nu \mathfrak{Q}$ which contains very good approximations to the columns of $X^{|r}$. Then we shall use Theorem 2.3 to show that the SRR step retrieves these approximations.

Our notational problems will be considerably simplified if we work in the coordinate system defined by X. This means that we can assume that $A = S$ is already upper triangular so that the Schur vectors we are seeking are the columns of the identity matrix. The first part of our program can then be summarized in the following lemma (n.b. there are easily rectified abuses in the statement and proof when $l = r$ or $r = n$).

Lemma 3.1. Let $S \in \mathbb{C}^{n\times n}$ be an upper triangular matrix with its diagonal elements $\sigma_{ii} = \lambda_i$ ordered in decreasing order of magnitude. For $1 \leqq l \leqq r \leqq n$ let E_1 denote the matrix consisting of columns 1 through l of the identity matrix and E_2 denote the matrix consisting of columns $l+1$ through r. Let $\dim(\mathfrak{Q}) = r$. Under mild restrictions on $\mathfrak{Q}$ (see (3.2) below) if $|\lambda_l| > |\lambda_{l+1}|$ and $|\lambda_r| > |\lambda_{r+1}|$, there are matrices $W_\nu \in \mathbb{C}^{n\times r}$ with orthonormal columns such that

$$\begin{aligned}&1.\ R(W_\nu) = S^\nu \mathfrak{Q},\\ &2.\ W_\nu^{|l} = E_1 + O_s(|\lambda_{r+1}/\lambda_l|^\nu),\\ &3.\ W_\nu^{r-l|} = E_2 + O_s(|\lambda_{r+1}/\lambda_r|^\nu).\end{aligned} \tag{3.1}$$

Proof. Partition S in the form

$$S = \begin{pmatrix} S_{11} & S_{12} & S_{13} \\ 0 & S_{22} & S_{23} \\ 0 & 0 & S_{33} \end{pmatrix},$$

where $S_{11} \in \mathbb{C}^{l\times l}$ and $S_{22} \in \mathbb{C}^{(r-l)\times(r-l)}$. By hypothesis the spectra of S_{11}, S_{22}, and S_{33} are disjoint. Hence there is an upper triangular matrix

$$Z = \begin{pmatrix} I_l & Z_{12} & Z_{13} \\ 0 & I_{r-l} & Z_{23} \\ 0 & 0 & I_{n-r} \end{pmatrix}$$

such that

$$Z^{-1}SZ = \operatorname{diag}(S_{11}, S_{22}, S_{33}).$$

Let the columns of Q form an orthonormal basis for $\mathcal{Q}$ and set

$$P = Z^{-1}Q = \begin{pmatrix} P_1 \\ P_2 \end{pmatrix},$$

where $P_1 \in \mathbb{C}^{r \times r}$. We now assume

$$P_1 \text{ is nonsingular} \tag{3.2}$$

and set

$$K = P P_1^{-1} = \begin{pmatrix} I_l & 0 \\ 0 & I_{r-l} \\ K_{31} & K_{32} \end{pmatrix}.$$

Then

$$\begin{aligned}(S^\nu Q)\,P_1^{-1} &= Z \operatorname{diag}(S_{11}^\nu, S_{22}^\nu, S_{33}^\nu)\,K \\ &= \begin{pmatrix} S_{11}^\nu + Z_{13}\,S_{33}^\nu\,K_{31} & Z_{12}\,S_{22}^\nu + Z_{13}\,S_{33}^\nu\,K_{32} \\ Z_{23}\,S_{33}^\nu\,K_{31} & S_{22}^\nu + Z_{23}\,S_{23}^\nu\,K_{32} \\ S_{33}^\nu\,K_{31} & S_{33}^\nu\,K_{32} \end{pmatrix}.\end{aligned}$$

Hence

$$(S^\nu Q)\,P_1^{-1} \operatorname{diag}(S_{11}^{-\nu}, S_{22}^{-\nu}) = \begin{pmatrix} I_l & Z_{12} \\ 0 & I_{r-l} \\ 0 & 0 \end{pmatrix} + \begin{pmatrix} Z_{13}\,K_{31}^{(\nu)} & Z_{13}\,K_{32}^{(\nu)} \\ Z_{23}\,K_{31}^{(\nu)} & Z_{23}\,K_{32}^{(\nu)} \\ K_{31}^{(\nu)} & K_{32}^{(\nu)} \end{pmatrix},$$

where

$$K_{3i}^{(\nu)} = S_{33}^\nu\,K_{3i}\,S_{ii}^{-\nu} \qquad (i = 1, 2). \tag{3.3}$$

From Theorem 2.1

$$\begin{aligned} \|K_{31}^{(\nu)}\| &= O_\varepsilon(|\lambda_{r+1}/\lambda_l|^\nu), \\ \|K_{32}^{(\nu)}\| &= O_\varepsilon(|\lambda_{r+1}/\lambda_r|^\nu). \end{aligned} \tag{3.4}$$

Set

$$\begin{aligned} V &= (S^\nu Q)\,P_1^{-1} \operatorname{diag}(S_{11}^{-\nu}, S_{22}^{-\nu}) \begin{pmatrix} I_l & -Z_{12} \\ 0 & I_{r-l} \end{pmatrix} \\ &= (E_1, E_2) + \begin{pmatrix} Z_{11}\,K_{31}^{(\nu)} & -Z_{13}\,K_{31}^{(\nu)}\,Z_{12} + Z_{13}\,K_{32}^{(\nu)} \\ Z_{23}\,K_{31}^{(\nu)} & -Z_{23}\,K_{31}^{(\nu)}\,Z_{12} + Z_{23}\,K_{32}^{(\nu)} \\ K_{31}^{(\nu)} & -K_{31}^{(\nu)}\,Z_{12} \quad + K_{32}^{(\nu)} \end{pmatrix}. \end{aligned} \tag{3.5}$$

Now $R(V) = S^\nu \mathcal{Q}$. From (3.4), $V^{l|} = E_1 + O_\varepsilon(|\lambda_{r+1}/\lambda_l|^\nu)$ and $V^{r-l|} = E_2 + O_\varepsilon(|\lambda_{r+1}/\lambda_r|^\nu)$. Theorem 2.2 then assures us that if we set $W = VR$ where R is the upper triangular matrix that orthogonalizes the columns of V, the conditions (3.1) will be satisfied. □

Lemma 3.1 states precisely the assertions about oblique convergence made in Section 1. However, it is important to assess what can make this convergence slow. There are two factors, as may be seen from (3.3) and (3.5): first the size of $K_3 = (K_{31}, K_{32})$ and second the speed with which the asymptotic behavior in (3.3) is attained. As we indicated in the discussion following Theorem 2.1, the second factor is imperfectly understood, although a failure to achieve asymptotic behavior early is in some way associated with ill conditioning of the eigensystems of S_{11} and S_{22}.

The explication of the first factor requires further analysis. Partition Z and Q in the forms

$$Z=\begin{pmatrix}\hat{Z}_{11} & \hat{Z}_{12}\\ 0 & I_{n-r}\end{pmatrix}$$

and

$$Q=\begin{pmatrix}Q_1\\ Q_2\end{pmatrix}$$

where $\hat{Z}_{11}, Q_1 \in \mathbb{C}^{r\times r}$. Then

$$K_3=Q_2(Q_1-\hat{Z}_{12}Q_2)^{-1}\hat{Z}_{11}$$

and

$$\|K_3\| \leqq \|(Q_1-\hat{Z}_{12}Q_2)^{-1}\| \, \|\hat{Z}_{11}\|. \tag{3.6}$$

If $\hat{Z}_{11}=(I, Z_{12})$ is large then Z_{12} is large, which means that the eigenvalue problem for S is ill conditioned.

To see the meaning of the factor $\|(Q_1-\hat{Z}_{12}Q_2)^{-1}\|$ in (3.6), suppose it is large. Then there is a vector p with $\|p\|=1$ such that

$$r=(Q_1-\hat{Z}_{12}Q_2)\,p$$

is small. But

$$r=(I, \, -\hat{Z}_{12})\begin{pmatrix}Q_1\\ Q_2\end{pmatrix}p=(I, \, -\hat{Z}_{12})\,q,$$

where the $q\in \mathcal{Q}$ so defined also has norm unity. Now $(I+\hat{Z}_{12}\,\hat{Z}_{12}^H)^{-1}\,(I, \, -\hat{Z}_{12})$ has orthonormal rows that span the orthogonal complement of

$$\mathscr{R}\begin{pmatrix}\hat{Z}_{12}\\ I_{n-n}\end{pmatrix},$$

which is the subspace complementary to the invariant subspace $\mathscr{R}(I_n^{|r})$ corresponding to S_{11} and S_{22}. Since $\|(I+\hat{Z}_{12}\,Z_{12})^{-1}\|_2<1$, to say that $\|(Q_1-\hat{Z}_{12}Q)^{-1}\|$ is large is to say that there is a vector $q\in\mathcal{Q}$ that almost lies in the invariant subspace corresponding to S_{33}, which agrees with our naive intuition about how the iteration should behave.

Incidentally, note that the hypotheses $|\lambda_l|>|\lambda_{l+1}|$ and $|\lambda_r|>|\lambda_{r+1}|$ were used only to insure the existence of the uncoupling matrix Z. If S is not defective, then such a Z always exists and the proof goes through without these hypotheses.

We are now in a position to state the main result of this section.

Theorem 3.2. With the hypotheses and notation of Lemma 3.1, let Q_ν denote the SRR approximation corresponding to $S^\nu\mathcal{Q}$. Then

$$\mathscr{R}(Q_\nu^{|l}) = \mathscr{R}(E_1)+O_s(|\lambda_{r+1}/\lambda_l|^\nu). \tag{3.7}$$

Proof. Let W_ν be the matrix of Lemma 3.1. Then

$$\begin{aligned}B_\nu=W_\nu^H\,SW_\nu&=\begin{pmatrix}S_{11}+O_s(|\lambda_{r+1}/\lambda_l|^\nu) & S_{12}+O_s(|\lambda_{r+1}/\lambda_r|^\nu)\\ O_s(|\lambda_{r+1}/\lambda_l|^\nu) & S_{22}+O_s(|\lambda_{r+1}/\lambda_r|^\nu)\end{pmatrix}\\ &=\begin{pmatrix}B_{11}^{(\nu)} & B_{12}^{(\nu)}\\ B_{21}^{(\nu)} & B_{22}^{(\nu)}\end{pmatrix}.\end{aligned}$$

Since from (2.7),

$$\operatorname{sep}(B_{11}^{(\nu)}, B_{22}^{(\nu)}) \geqq \operatorname{sep}(S_{11}, S_{22}) - O_\varepsilon(|\lambda_{r+1}/\lambda_r|^\nu),$$

Theorem 3.3 eventually applies to give a matrix $P_\nu = O_\varepsilon(|\lambda_{r+1}/|\lambda_l|^\nu)$ such that $\mathscr{R}[(I_l, P_\nu^H)^H]$ is an invariant subspace of B_ν. Moreover from (2.6) the spectrum of B_ν corresponding to this subspace is the spectrum of $S_{11} + O_\varepsilon(|\lambda_{r+1}/\lambda_l|^\nu)$ while the remaining spectrum is the spectrum of $S_{22} + O_\varepsilon(|\lambda_{r+1}/\lambda_r|^\nu)$. Since the eigenvalues of S_{11} are all greater in absolute value than those of S_{22}, the space $R[(I_l, P_\nu^H)^H]$ must ultimately be the space $\mathscr{R}(Y_\nu^{|l})$, where Y_ν is the matrix of Schur vectors of B_ν. Since $Q_\nu^{|l} = W_\nu Y_\nu^{|l}$, the result (3.7) follows. □

The recasting of these results in terms of the X coordinate system is trivial. Two applications of the theorem, one to $Q_\nu^{|l-1}$ and one to $Q_\nu^{|l}$, show that if $|\lambda_{l-1}| > |\lambda_l| > |\lambda_{l+1}|$, then the l-th column of Q_ν converges to x_l at a rate that is $O_\varepsilon(|\lambda_{r+1}/\lambda_l|)$. The proof of the theorem also shows that the diagonal elements of the Schur form of B_ν estimate the eigenvalues of A.

An unsavory aspect of Theorem 3.2 is the requirement $|\lambda_r| > |\lambda_{r+1}|$. It is needed to insure that ultimately all the eigenvalues of B_{11} are greater in absolute value than the eigenvalues of B_{22} so that the Schur vectors are computed in the right order. Unfortunately this hypothesis is necessary as the following example shows.

Example 3.3. Let $l = 1$ and $r = 2$. Let

$$A = \begin{pmatrix} 2 & 0 & 0 \\ 0 & 1 & -6 \\ 0 & 0 & -1 \end{pmatrix}$$

and

$$Q = \begin{pmatrix} 1 & 0 \\ 1 & 1 \\ 1 & -1 \end{pmatrix}.$$

Then

$$A^{2\nu} Q = \begin{pmatrix} 2^{2\nu} & 0 \\ 1 & 1 \\ 1 & -1 \end{pmatrix}, \quad A^{2\nu+1} Q = \begin{pmatrix} 2^{2\nu+1} & 0 \\ -5 & 7 \\ -1 & 1 \end{pmatrix}.$$

The columns of $A^{2\nu} Q$ are already orthogonal. They may be normalized by post-multiplying by the inverse of $\operatorname{diag}(\sqrt{2^{4\nu}+2}, \sqrt{2})$. We then find that

$$B_{2\nu} = \begin{pmatrix} \dfrac{2 - 6\varepsilon_\nu^2}{1 + 2\varepsilon_\nu^2} & \dfrac{8\varepsilon_\nu}{\sqrt{2(1+2\varepsilon_\nu^2)}} \\ \dfrac{-4\varepsilon_\nu}{\sqrt{2(1+2\varepsilon_\nu^2)}} & 3 \end{pmatrix},$$

where $\varepsilon_\nu = 2^{-2\nu}$. The dominant eigenvalue of $B_{2\nu}$ approaches 3 and its eigenvector, suitably scaled, approaches $(0, 1)^T$. Thus the first column of the even SRR approximations approach $(0, 1/\sqrt{2}, -1/\sqrt{2})^T$, which is not a good approximation to the dominant eigenvector $(1, 0, 0)^T$ of A.

There is nothing pathological about the above example. The eigenvalues and eigenvectors of A are all well conditioned. The columns space of Q is not deficient in the eigenvector $(1, 0, 0)^T$, (3.2) holds, and successive iterates contain increasingly accurate approximations. Moreover the example is stable; changing Q slightly will not affect the results.

However a number of features of the above example suggest that the phenomenon poses no serious practical difficulties; i.e. we are in no real danger of accepting a spurious Schur vector of A. In the first place, the matrix A is rather contrived. The eigenvalues of A of absolute value unity are distinct and the block of the Schur decomposition corresponding to them is involutory. Second, the odd iterates $B_{2\nu+1}$ behave quite differently from the even ones, and one would have to inspect only the even iterates to be fooled into thinking convergence has occurred. Finally, the approximate eigenvalue 3 and eigenvector (0, 1, −1) clearly do not belong to A; they have a large residual.

4. Practicalities

In this section we shall discuss some of the more important points that must enter into any practical implementation of the algorithm sketched in Section 1.

Real matrices. In most applications it can be expected that the matrix A will be real. Since the complex eigenvalues of a real matrix occur in conjugate pairs, the necessity of using complex arithmetic can be circumvented by working with a quasi-Schur form in which A is reduced by an orthogonal transformation to a block triangular matrix. The largest diagonal blocks are 2×2 and contain the complex eigenvalues of A. The theory of the last two sections is directly applicable to this modified form.

The SRR step requires that one be able to calculate the quasi-Schur decomposition of the matrix $B_\nu = Q_\nu^H A Q_\nu$. The QR algorithm with double implicit shifts can be used to reduce B_ν to the required block triangular form; however, the eigenvalues may not appear in correct order on the diagonal. Fortunately it is not difficult to modify the standard QR programs [8, 15] to interchange disordered blocks as they emerge (this is done by using one or two QR iterations with a fixed shift). Programs for calculating eigenvectors of the Schur form are also given in [8, 15].

General structure of the algorithm. From the description of the algorithm in Section 1 it is evident that the bulk of the work will be contained in three steps:

1. Calculate AQ_ν,
2. Orthogonalize AQ_ν,
3. Compute the SRR approximation.

The first step represents a fixed overhead for the algorithm; it must be done at every iteration. The purpose of the orthogonalization is to maintain linear independence among the columns of Q_ν, and it should not be done until there is reason to believe that the columns have degenerated. Similarly there is no reason to perform an SRR step unless it is expected that some of the columns of Q_ν have converged, after which they may be held fixed with considerable savings in work. These considerations lead to the following general structure for the algorithm.

```
1. While some columns of Q have not converged
   1. While convergence is not expected
      1. While the columns of Q are sufficiently
         independent
         1. Q ← AQ
      2. Orthogonalize
   2. SRR step
   3. Test convergence                                  (4.1)
```

The rest of this section is devoted to exploring the problems that are implicit in (4.1). For convenience we shall drop the iteration subscripts.

Orthogonalization. The k-fold iteration of statement 1.1.1.1 in (4.1) results in the matrix A^kQ, whose columns will tend toward linear dependence. An estimate of the degree of dependence may be obtained from quantities computed in the SRR step. Specifically, as the process begins to converge, the SRR approximation will satisfy the approximate equality

$$AQ \cong QT,$$

where T is the triangular matrix of (1.4). It follows that $A^kQ \cong QT^k$ and hence approximately

$$\|A^kQ\| \lesssim \|T\|^k.$$

On the other hand A^kQT^{-k} has approximately orthonormal columns, so that if R is the triangular matrix such that A^kQR has orthogonal columns, then

$$\|R\| \lesssim \|T^{-1}\|^k.$$

It follows that in passing from A^kQ to A^kQR we may lose as many as $k \cdot \log_{10}(\|T\|\|T^{-1}\|)$ decimal digits. Thus if we set a maximum number t of significant digits that we shall allow to be lost, we may choose k so that

$$k \leqq \frac{t}{\log_{10}(\|T\|\|T^{-1}\|)}. \tag{4.2}$$

It should be noted that the occasional computation of $\|T^{-1}\|$ is not inordinately expensive.

There are two ways of orthogonalizing the columns of Q: the Gram-Schmidt method and reduction by Householder transformations [11, Ch. 5]. Although the latter method is unconditionally stable, the former is simpler, cheaper, and, in view of the conservative criterion (4.2), stable enough for our purposes, especially when it is supplemented by judicious reorthogonalizations (see, e.g., the program in [7]).

Convergence. The natural place to test for convergence is right after an SRR step. The theory of Section 3 suggests that the first columns of Q will tend to converge before the later ones. However, the part of the Schur decomposition associated with a multiple eigenvalue is not uniquely determined, and the corresponding columns of Q may fail to converge, even though the subspace they span does converge. To circumvent this difficulty, let $Q=(Q_1, Q_2)$, where Q_1 contains the columns that are thought to have converged. Let

$$T=Q^H A Q=\begin{pmatrix} T_{11} & T_{12} \\ 0 & T_{22} \end{pmatrix}$$

be partitioned conformally with Q. We shall say that Q_1 has converged if the residual

$$R_1=A Q_1-Q_1 T_{11} \tag{4.3}$$

is smaller than a prespecified tolerance. Although this criterion cannot guarantee the accuracy of the columns of Q, it does guarantee that there is a matrix

$$E=R_1 Q_1^H \tag{4.4}$$

such that $\|E\|=\|R_1\|$ and

$$(A-E)\,Q_1=Q_1 T_{11}.$$

Thus if the dominant eigenvalues of $A+E$ and the eigenvalues of T_{11} are the same, $\mathcal{R}(Q_i)$ will span the dominant subspace of $A+E$. We suggest that in order to avoid premature acceptance of a disordered set of vectors, the test be applied only after the moduli of the eigenvalues of T_{11} have settled down somewhat. But this criterion should not be too stringent, since individual eigenvalues of T may be ill-conditioned.

As was mentioned in Section 1, the eigenvectors of T_{11} can be used to calculate eigenvectors of A. We cannot in general expect to calculate accurate eigenvectors, only eigenvectors with small residuals. Specifically, let z be an approximate eigenvector of T_{11} with $\|z\|=1$ corresponding to the approximate eigenvalue λ, and let

$$r=T_{11}z-\lambda z. \tag{4.5}$$

The corresponding approximate eigenvector of A is given by Qz, and from (4.3) and (4.5) we have

$$\begin{aligned} R_1 z &= A Q_1 z - Q_1 T_{11} z \\ &= A(Q_1 z)-\lambda(Q_1 z)-Q_1 r. \end{aligned}$$

Hence

$$\|A(Q_1 z)-\lambda(Q_1 z)\| \leqq \|R_1\|+\|r\|, \tag{4.6}$$

which shows that the residual for the approximate eigenvector of A is also small.

There still remains the problem of when to perform an SRR step. Theorem 3.2 implies that the columns of Q will have residuals that decrease at an essentially linear rate. If the norms of the residuals are computed after two SRR steps, the convergence ratio can be estimated and used to predict when the residual will fall below a predetermined level. This technique must of course be supplemented by *ad hoc* devices to reject unreasonable predictions. In numerical experiments the technique has been found to be quite effective.

Deflation. Once Q_1 has been accepted, it may be implicily deflated from the problem. Specifically, during the power steps one forms the matrices $(Q_1, A^{\nu} Q_2)$, with a corresponding savings in matrix multiplications. At the orthogonalization step further savings may be effected by observing that the columns of Q_1 are already orthogonal. In the SRR step, the matrix B of (1.3) has the form

$$B=\begin{pmatrix} T_{11} & Q_1^H A Q_2 \\ Q_2^H A Q_1 & Q_2^H A Q_2 \end{pmatrix}.$$

At first glance, there would appear to be no savings in the SSR step, since $Q_2^H A Q_1$ is in general not zero. However, if E is the matrix defined by (4.4), then

$$B'=Q^H(A-E)\,Q=\begin{pmatrix} T_{11} & Q_1^H A Q_2 \\ 0 & Q_2^H A Q_2 \end{pmatrix}. \tag{4.7}$$

Thus we may work with the partly triangularized matrix B'.

Strictly speaking the convergence theory of § 3 applies to the deflated iteration only when an orthogonalization is performed at each step. In this case we are effectively computing $(PA\,P)^\nu Q_1$, where

$$P=I-Q_1Q_1^H$$

is the projection onto the orthogonal complement of the column space of Q_1. Since

$$PA\,P=P(A-E)\,P,$$

we are in effect iterating with the matrix $P(A-E)\,P$ whose eigenvalues associated with Q_1 are zero. Since by (4.7) the SRR step is also performed with $A+E$, the convergence theory of § 3 applies.

When orthogonalization is not performed at each step, the iteration is equivalent to multiplying by the sequence of matrices $PA^{k_1}P$, $PA^{k_2}P, \ldots$, where the integers k_i are the number of steps between orthogonalizations. Since in general $PA^{k_i}P \neq P(A+E)^{k_i}P$, we cannot claim that we are iterating with a slightly perturbed operator. However, note that

$$A^{k_i}=(A+E)^{k_i}+O(\|A\|^{k_i-1}\|E\|),$$

so that if E is fairly small and k_i is not too large, the spaces spanned by $A^{k_i}Q_2$ and $(A+E)^{k_i}Q_2$ will not differ by much. This lends some support to the use of the deflation process without successive orthogonalizations, and numerical experience confirms this conclusion.

Acknowledgments. I have benefited from discussions with Professor Gene Golub and Mr. David Eklund. I am indebted to Dr. John Reid for suggesting that it was necessary to establish (4.6). Finally, the practical details in § 4 draw so heavily on Rutishauser's program [7, 15] that his contribution must be specially acknowledged.

References

1. Bauer, F. L.: Das Verfahren der Treppeniteration und verwandte Verfahren zur Lösung algebraischer Eigenwertprobleme. Z. Angew. Math. Phys. **8**, 214–235 (1957)
2. Clint, M., Jennings, A.: The evaluation of eigenvalues and eigenvectors of real symmetric matrices by simultaneous iteration. Comp. J. **13**, 76–80 (1970)
3. Clint, M., Jennings, A.: A simultaneous iteration method for the unsymmetric eigenvalue problem, J. Inst. Math. Appl. **8**, 111–121 (1971)

4. Jennings, A.: A direct iteration method of obtaining latent roots and vectors of a symmetric matrix, Proc. Cambridge Philos. Soc. **63**, 755–765 (1967)
5. Parlett, B. N., Poole, W. G.: A geometric theory for the QR, LU, and power iterations, SIAM J. Numer. Anal. **10**, 389–412 (1973)
6. Rutishauser, H.: Computational aspects of F. L. Bauer's simultaneous iteration method, Numer. Math. **13**, 4–13 (1969)
7. Rutishauser, H.: Simultaneous iteration method for symmetric matrices. Numer. Math. **16**, 205–223 (1970)
8. Smith, B. T., Boyle, J. M., Garbow, B. S., Ikebe, Y., Klema, V. C., Moler, C. B.: Lecture Notes in Computer Science V6: Matrix Eigensystem Routines—EISPACK Guide. New York: Springer 1974
9. Stewart, G. W.: Accelerating the orthogonal iteration for the eigenvectors of a Hermitian matrix. Numer. Math. **13**, 362–376 (1969)
10. Stewart, G. W.: Error bounds for approximate invariant subspaces of closed linear operators. SIAM J. Numer. Anal. **8**, 796–808 (1971)
11. Stewart, G. W.: Introduction to Matrix Computations. New York: Academic Press 1973
12. Stewart, G. W.: Methods of simultaneous iteration for calculating eigenvectors of matrices. In: Topics in Numerical Analysis II (John J. H. Miller, ed.). New York: Academic Press 1975, pp. 185–169
13. Stewart, G. W.: Perturbation bounds for the QR factorization of a matrix. University of Maryland Computer Science Department Technical Report TR-323 (1974). To appear in SIAM F. Numer. Anali
14. Vandergraft, J. S.: Generalized Rayleigh methods with applications to finding eigenvalues of large matrices. Lin. Alg. and Appl. (1971) p. 353–368
15. Wilkinson, J. H., Reinsch, C. (eds.): Handbook for Automatic Computation VII Linear Algebra. New York: Springer 1971

Prof. Dr. G. W. Stewart
University of Maryland
Dept. of Computer Science
College Park, Ma. 20742
USA

16.3. [GWS-J33] "Algorithm 506: HQR3 and EXCHNG: FORTRAN Subroutines for Calculating and Ordering the Eigenvalues of a Real Upper Hessenberg Matrix"

[GWS-J33] "Algorithm 506: HQR3 and EXCHNG: FORTRAN Subroutines for Calculating and Ordering the Eigenvalues of a Real Upper Hessenberg Matrix," *ACM Transactions on Mathematical Software* **2** (1976) 275–280.
http://doi.acm.org/10.1145/355694.355700

ALGORITHM 506
HQR3 and EXCHNG: Fortran Subroutines for Calculating and Ordering the Eigenvalues of a Real Upper Hessenberg Matrix [F2]

G. W. STEWART
University of Maryland

Key Words and Phrases: eigenvalue, QR-algorithm
CR Categories: 5.14
Language: Fortran

DESCRIPTION

1. Usage

HQR3 is a Fortran subroutine to reduce a real upper Hessenberg matrix A to quasi-triangular form B by a unitary similarity transformation U:

$$B = U^{T}AU.$$

The diagonal of B consists of 1×1 and 2×2 blocks as illustrated below:

$$\begin{bmatrix} x & x & x & x & x & x \\ 0 & x & x & x & x & x \\ 0 & x & x & x & x & x \\ 0 & 0 & 0 & x & x & x \\ 0 & 0 & 0 & 0 & x & x \\ 0 & 0 & 0 & 0 & x & x \end{bmatrix}$$

The 1×1 blocks contain the real eigenvalues of A, and the 2×2 blocks contain the complex eigenvalues, a conjugate pair to each block. The blocks are ordered so that the eigenvalues appear in descending order of absolute value along the diagonal. The transformation U is postmultiplied into an array V, which presumably contains earlier transformations performed on A.

The decomposition produced by *HQR3* differs from the one produced by the *EISPACK* subroutine *HQR2* [2] in that the eigenvalues of the final quasi-triangular

---

Received 22 August 1975 and 14 October 1975.

This work was supported in part by the Office of Naval Research under Contract N0014-67-A-0218-0018.
Author's address: Department of Computer Science, University of Maryland, College Park, MD 20742.

matrix are ordered. This ordering makes the decomposition essentially unique, which is important in some applications (e.g. see [4]). It should also be noted that when the eigenvalues $\lambda_1, \lambda_2, \ldots, \lambda_n$ of A are ordered so that $|\lambda_1| \geq |\lambda_2| \geq \cdots \geq |\lambda_n|$, and if $|\lambda_i| > |\lambda_{i+1}|$, then the first i columns of U form an orthonormal basis for the invariant subspace corresponding to $\lambda_1, \lambda_2, \ldots, \lambda_i$. In applications where it is desired to work with the matrix A in such a dominant invariant subspace, *HQR*3 provides a convenient means for calculating a basis. The corresponding leading principal submatrix of B is a representation of A in that subspace. When an ordered quasi-triangular form is not required, the *EISPACK* program *HQR*2 will be slightly more efficient than *HQR*3.

The calling sequence for *HQR*3 is:

CALL HQR3 (A,V,N,NLOW,NUP,EPS,ER,EI,TYPE,NA,NV)

with (parameters preceded by an asterisk are altered by the subroutine):

*A	A doubly subscripted real array containing the matrix to be reduced On return, A contains the final quasi-triangular matrix. The elements of the array below the third subdiagonal are unaltered by the subroutine.
*V	A doubly subscripted real array into which the reducing transformation is postmultiplied.
N	An integer containing the order of A and V.
NLOW, NUP	Integers prescribing what part of A is to be reduced. Specifically A(NLOW,NLOW−1) and A(NUP+1,NUP) are assumed zero and only the block from NLOW through NUP is reduced. However, the transformation is performed on all of the matrix A so that the final result is similar to A.
EPS	Convergence criterion. Maximal accuracy will be attained if EPS is set to β^{-t}, where β is the base of the floating-point word and t is the length of its fraction. Smaller values of EPS will increase the amount of work without significantly improving the accuracy.
*ER	A singly subscripted real array containing the real parts of the eigenvalues.
*EI	A singly subscripted real array containing the imaginary parts of the eigenvalues.
*TYPE	A singly subscripted integer array whose ith entry is 0 if the ith eigenvalue is real; 1 if the ith eigenvalue is complex with positive imaginary part; 2 if the ith eigenvalue is complex with negative imaginary part; −1 if the ith eigenvalue was not successfully calculated. The entry 1 is always followed by a 2. Only elements NLOW through NUP of ER, EI, and TYPE are set by *HQR*3.
NA	The first dimension of the array A.
NV	The first dimension of the array V.

*HQR*3 can be used together with the *EISPACK* programs *ORTHES* and *ORTRAN* [2] to reduce a full matrix A to quasi-triangular form with the eigenvalues appearing in descending order of absolute value along the diagonal,

```
CALL ORTHES(NA,N,NLOW,NUP,A,P)
CALL ORTRAN(NA,N,NLOW,NUP,A,P,V)
CALL HQR3(A,V,N,NLOW,NUP,EPS,ER,EI,TYPE,NA,NA)
```

where P is a singly subscripted scratch array of order N.

*HQR*3 requires the subroutines *EXCHNG*, *SPLIT*, and *QRSTEP*.

EXCHNG is a Fortran subroutine to interchange consecutive 1×1 and 2×2 blocks of an upper Hessenberg matrix. Specifically it is supposed that the upper Hessenberg matrix A has a block of order $b1$ starting at the lth diagonal element and a block of order $b2$ starting at the $(l + b1)$-th diagonal element (illustrated below for $n = 5, l = 2, b1 = 2, b2 = 1$):

$$\begin{array}{cccccc} & & l & & & \\ & x & x & x & x & x \\ l & 0 & x & x & x & x \\ & 0 & x & x & x & x \\ & 0 & 0 & 0 & x & x \\ & 0 & 0 & 0 & 0 & x \end{array}$$

EXCHNG produces an orthogonal similarity transformation W such that W^TAW has consecutive blocks of order $b2$ and $b1$ starting at the lth diagonal element (illustrated from the example above):

$$\begin{array}{cccccc} & & l & & & \\ & x & x & x & x & x \\ l & 0 & x & x & x & x \\ & 0 & 0 & x & x & x \\ & 0 & 0 & x & x & x \\ & 0 & 0 & 0 & 0 & x \end{array}$$

The eigenvalues associated with each block are interchanged along with the blocks. The transformation W is postmultiplied into the matrix V.

EXCHNG can be used to rearrange the blocks of the quasi-triangular matrix produced by *HQR3*. For example, one might wish to cluster a group of nearly equal eigenvalues at the top of the matrix before applying a deflation technique to uncouple them from the rest of the problem.

The calling sequence for *EXCHNG* is:

CALL EXCHNG(A,V,N,L,B1,B2,EPS,FAIL,NB,NV)

with (parameters preceded by an asterisk are altered by the subroutine):

*A	A doubly subscripted real array containing the matrix whose blocks are to be interchanged. Only rows and columns L through L + B1 + B2 − 1 are transformed. Elements of the array A below the third subdiagonal are not altered.
*V	A doubly subscripted array containing the matrix V into which the reducing transformation is to be accumulated. Only columns L through L + B1 + B2 − 1 are altered.
N	An integer containing the order of A and V.
L	An integer containing the leading diagonal position of the blocks.
B1	An integer containing the size of the first block.
B2	An integer containing the size of the second block.
EPS	A convergence criterion (cf. EPS in the calling sequence for *HQR3*).
*FAIL	A logical variable that on normal return is false. If the iteration to interchange the blocks failed, it is set to true.
NA	The first dimension of the array A.
NV	The first dimension of the array V.

By repeated applications of *EXCHNG* the eigenvalues of a quasi-triangular matrix can be arranged in any desired order.

EXCHNG requires the subroutine *QRSTEP*.

2. Method and Programming Details

HQR3 uses the implicit double-shift *QR* algorithm to reduce A to quasi-triangular form (for the theoretical background see [3, 5]). The program is essentially a

Fortran variant of part of the Algol program *hqr*2 in the *Numerische Mathematik* handbook series [1] with these differences:

(1) The calling sequence is somewhat different.

(2) Eigenvectors are not computed.

(3) The parameters NLOW and NUP are not comparable to the parameters *low* and *upp* of *hqr*2. Specifically, in *HQR*3 rows 1 through N of V are transformed; whereas in *hqr*2 only rows *low* through *upp* are transformed.

(4) The code that performs the *QR* iterations in *hqr*2 is replaced by a call to the subroutine *QRSTEP*.

(5) After a 1×1 or 2×2 block has been isolated, the subroutine *EXCHNG* is used to position it correctly among the previously isolated blocks. It should be realized that *EXCHNG* is a numerical algorithm and may change an ill-conditioned eigenvalue in a block significantly. This means that after interchanging two blocks with ill-conditioned eigenvalues that are very nearly equal in absolute value, the eigenvalues may still not be in descending order of absolute value. Since numerically the absolute values of these eigenvalues cannot be told apart, this phenomenon is not of much importance.

The convergence criterion is the same for both programs. A subdiagonal element A(I + 1, I) is regarded as negligible it it is less than or equal to EPS * (ABS(A(I, I)) + ABS(A(I + 1, I + 1))). If 10 or 20 iterations are performed without convergence, an ad hoc shift is introduced to break up any cycling. If 30 iterations are performed without convergence, the subroutine gives up. Although when this happens the matrix returned is not quasi-triangular, it is still almost exactly similar to the original matrix, and the similarity transformation has been accumulated in V.

EXCHNG works as follows. The first block is used to determine an implicit *QR* shift. An arbitrary *QR* step is performed on both blocks to eliminate the uncoupling between them. Then a sequence of *QR* steps using the previously determined shift is performed on both blocks. Except in ill-conditioned cases, a block of size *B*1 having the eigenvalues of the first block will emerge in the lower part of the array occupied by both blocks, usually in one or two steps. If 30 iterations pass without convergence (the criterion is the same as in *HQR*3), the subroutine gives an error return.

Both *HQR*3 and *EXCHNG* use the subroutine *QRSTEP* to perform the *QR* iterations. In addition, *HQR*3 uses the subroutine *SPLIT* to separate real eigenvalues of a 2×2 block.

REFERENCES

1. Peters, G., and Wilkinson, J.H. Eigenvectors of real and complex matrices by LR and QR triangularizations. *Numer. Math. 16* (1970), 181–204.
2. Smith, B.T., Boyle, J.M., Garbow, B.S., Ikebe, Y., Klema, V.C., and Moler, C.B. *Matrix Eigensystem Routines—EISPACK Guide.* Lecture Notes in Computer Science, Vol. 6, Springer, New York, 1974.
3. Stewart, G.W. *Introduction to Matrix Computations.* Academic Press, New York, 1974.
4. Stewart, G.W. Simultaneous iteration for computing invariant subspaces of non-Hermitian matrices. *Numer. Math. 25* (1976), 123–136.
5. Wilkinson, J.H. *The Algebraic Eigenvalue Problem.* Clarendon, New York, 1965.

ALGORITHMS

[Only that portion of each listing which gives the introductory comments explaining the algorithm is printed here. The complete listing is available from the ACM Algorithms Distribution Service (see inside back cover for order form), or may be found in "Collected Algorithms from ACM."]

```
      SUBROUTINE HQR3(A, V, N, NLOW, NUP, EPS, ER, EI, TYPE, NA, NV)
      INTEGER N, NA, NLOW, NUP, NV, TYPE(N)
      REAL A(NA,N), EI(N), ER(N), EPS, V(NV,N)
C HQR3 REDUCES THE UPPER HESSENBERG MATRIX A TO QUASI-
C TRIANGULAR FORM BY UNITARY SIMILARITY TRANSFORMATIONS.
C THE EIGENVALUES OF A, WHICH ARE CONTAINED IN THE 1X1
C AND 2X2 DIAGONAL BLOCKS OF THE REDUCED MATRIX, ARE
C ORDERED IN DESCENDING ORDER OF MAGNITUDE ALONG THE
C DIAGONAL.  THE TRANSFORMATIONS ARE ACCUMULATED IN THE
C ARRAY V.  HQR3 REQUIRES THE SUBROUTINES EXCHNG,
C QRSTEP, AND SPLIT.  THE PARAMETERS IN THE CALLING
C SEQUENCE ARE (STARRED PARAMETERS ARE ALTERED BY THE
C SUBROUTINE)
C    *A         AN ARRAY THAT INITIALLY CONTAINS THE N X N
C               UPPER HESSENBERG MATRIX TO BE REDUCED.  ON
C               RETURN A CONTAINS THE REDUCED, QUASI-
C               TRIANGULAR MATRIX.
C    *V         AN ARRAY THAT CONTAINS A MATRIX INTO WHICH
C               THE REDUCING TRANSFORMATIONS ARE TO BE
C               MULTIPLIED.
C     N         THE ORDER OF THE MATRICES A AND V.
C     NLOW      A(NLOW,NLOW-1) AND A(NUP,+1,NUP) ARE
C     NUP       ASSUMED TO BE ZERO, AND ONLY ROWS NLOW
C               THROUGH NUP AND COLUMNS NLOW THROUGH
C               NUP ARE TRANSFORMED, RESULTING IN THE
C               CALCULATION OF EIGENVALUES NLOW
C               THROUGH NUP.
C     EPS       A CONVERGENCE CRITERION.
C    *ER        AN ARRAY THAT ON RETURN CONTAINS THE REAL
C               PARTS OF THE EIGENVALUES.
C    *EI        AN ARRAY THAT ON RETURN CONTAINS THE
C               IMAGINARY PARTS OF THE EIGENVALUES.
C    *TYPE      AN INTEGER ARRAY WHOSE I-TH ENTRY IS
C                  Ø   IF THE I-TH EIGENVALUE IS REAL,
C                  1   IF THE I-TH EIGENVALUE IS COMPLEX
C                      WITH POSITIVE IMAGINARY PART.
C                  2   IF THE I-TH EIGENVALUE IS COMPLEX
C                      WITH NEGATIVE IMAGINARY PART,
C                 -1   IF THE I-TH EIGENVALUE WAS NOT
C                      CALCULATED SUCCESSFULLY.
C     NA        THE FIRST DIMENSION OF THE ARRAY A.
C     NV        THE FIRST DIMENSION OF THE ARRAY V.
C THE CONVERGENCE CRITERION EPS IS USED TO DETERMINE
C WHEN A SUBDIAGONAL ELEMENT OF A IS NEGLIGIBLE.
C SPECIFICALLY A(I+1,I) IS REGARDED AS NEGLIGIBLE
C IF
C         ABS(A(I+1),I)) .LE. EPS*(ABS(A(I,I))+ABS(A(I+1,I+1))).
C THIS MEANS THAT THE FINAL MATRIX RETURNED BY THE
C PROGRAM WILL BE EXACTLY SIMILAR TO A + E WHERE E IS
C OF ORDER EPS*NORM(A), FOR ANY REASONABLY BALANCED NORM
C SUCH AS THE ROW-SUM NORM.
```

```
      SUBROUTINE EXCHNG(A, V, N, L, B1, B2, EPS, FAIL, NA, NV)
      INTEGER B1, B2, L, NA, NV
      REAL A(NA,N), EPS, V(NV,N)
      LOGICAL FAIL
C GIVEN THE UPPER HESSENBERG MATRIX A WITH CONSECUTIVE
C B1XB1 AND B2XB2 DIAGONAL BLOCKS (B1,B2 .LE. 2)
C STARTING AT A(L,L), EXCHNG PRODUCES A UNITARY
C SIMILARITY TRANSFORMATION THAT EXCHANGES THE BLOCKS
C ALONG WITH THEIR EIGENVALUES.  THE TRANSFORMATION
C IS ACCUMULATED IN V.  EXCHNG REQUIRES THE SUBROUTINE
C QRSTEP.  THE PARAMETERS IN THE CALLING SEQUENCE ARE
C (STARRED PARAMETERS ARE ALTERED BY THE SUBROUTINE)
C    *A        THE MATRIX WHOSE BLOCKS ARE TO BE
C              INTERCHANGED.
C    *V        THE ARRAY INTO WHICH THE TRANSFORMATIONS
C              ARE TO BE ACCUMULATED.
C     N        THE ORDER OF THE MATRIX A.
C     L        THE POSITION OF THE BLOCKS.
C     B1       AN INTEGER CONTAINING THE SIZE OF THE
C              FIRST BLOCK.
C     B2       AN INTEGER CONTAINING THE SIZE OF THE
C              SECOND BLOCK.
C     EPS      A CONVERGENCE CRITERION (CF. HQR3).
C    *FAIL     A LOGICAL VARIABLE WHICH IS FALSE ON A
C              NORMAL RETURN.  IF THIRTY ITERATIONS WERE
C              PERFORMED WITHOUT CONVERGENCE, FAIL IS SET
C              TO TRUE AND THE ELEMENT
C              A(L+B2,L+B2-1) CANNOT BE ASSUMED ZERO.
C     NA       THE FIRST DIMENSION OF THE ARRAY A.
C     NV       THE FIRST DIMENSION OF THE ARRAY V.
```

16.4. [GWS-J37] (with C. A. Bavely) "An Algorithm for Computing Reducing Subspaces by Block Diagonalization"

[GWS-J37] (with C. A. Bavely) "An Algorithm for Computing Reducing Subspaces by Block Diagonalization," *SIAM Journal on Numerical Analysis* **16** (1979) 359–367.
http://dx.doi.org/10.1137/0716028

AN ALGORITHM FOR COMPUTING REDUCING SUBSPACES BY BLOCK DIAGONALIZATION*

CONNICE A. BAVELY† AND G. W. STEWART†

Abstract. This paper describes an algorithm for reducing a real matrix A to block diagonal form by a real similarity transformation. The columns of the transformation corresponding to a block span a reducing subspace of A, and the block is the representation of A in that subspace with respect to the basis. The algorithm attempts to control the condition of the transformation matrices, so that the reducing subspaces are well conditioned and the basis vectors are numerically independent.

1. Introduction. The purpose of this report is to describe an algorithm for reducing a real matrix A of order n to a block diagonal form by a real similarity transformation. Specifically, the algorithm attempts to compute a real nonsingular matrix X such that $X^{-1}AX$ has the form

$$X^{-1}AX = B = \operatorname{diag}(B_1, B_2, \cdots, B_s), \tag{1.1}$$

where each matrix B_i is square of order n_i.

A decomposition such as (1.1) has many applications. When the blocks B_i are small, powers of A can be economically calculated in the form

$$A^k = X \operatorname{diag}(B_1^k, B_2^k, \cdots, B_s^k)X^{-1},$$

and this fact can be used to simplify the computation of functions of A defined by power series (e.g. see [7]). If X is partitioned in the form

$$X = (X_1, X_2, \cdots, X_s),$$

where each X_i has n_i columns, then the columns of X_i form a basis for a reducing subspace of A, and B_i is the representation of A with respect to that basis. The associated spectral projector is given by $X_iX_i^{(-1)}$, where $X_i^{(-1)}$ is formed from the corresponding rows of X^{-1} (for definitions and applications see [4]).

There are theoretical and practical limitations on how small the blocks in (1.1) can be. Theoretically, they can be no smaller than the blocks in the Jordan canonical form of A. Practically, they may have to be larger. The numerical problems associated with decompositions such as (1.1) have been examined in detail in [3]. Here we give only a brief summary.

The principal difficulty is that the Jordan form of a matrix need not be numerically well determined; very small perturbations in the matrix may cause blocks to split or coalesce. Any attempt to separate two such "nearby" blocks will result in a transformation matrix X whose columns are nearly linearly dependent, or equivalently X will be ill-conditioned in the sense that the product $\|X\|\,\|X^{-1}\|$ is large (here $\|\cdot\|$ denotes a suitable matrix norm). In this case, it will be impossible to form X^{-1} or solve linear systems involving X accurately [10], [14]. The phenomenon is loosely associated with close eigenvalues; but there are matrices with equal eigenvalues, e.g. symmetric matrices, that can be split completely into 1×1 blocks, and there are matrices with well separated eigenvalues that cannot be split except by very ill-conditioned transformations.

* Received by the editors October 28, 1976. This work was supported in part by the Office of Naval Research under Contract N00014-76-C-0391.

† Department of Computer Science, University of Maryland, College Park, Maryland 20742.

Our algorithm attempts to avoid these difficulties by working only with well-conditioned transformations. If a group of eigenvalues cannot be split off into a block by a transformation whose condition observes a tolerance provided by the user, the block is enlarged until a well-conditioned reducing transformation can be found. In principle this does not insure that the final transformation will be well-conditioned, since it is formed as the product of a number of reducing transformations; however, we have found that when a matrix possesses a well-conditioned decomposition of the form (1.1), our algorithm generally finds it. And the exceptions have not so much to do with the ill-conditioning of X as with the failure of the algorithm to split the matrix completely owing to the comingling of degenerate eigenvalues with well-conditioned ones.

A good deal of work has been done on the numerically stable simplification of matrices by similarity transformations [5], [6], [8], [12], [13], most of which has been summarized in [3]. For the most part, these algorithms attempt to go farther than ours in reducing the matrix, however at considerable cost in complexity and computation. The virtues of the algorithm proposed here are its simplicity and economy. When it is required to reduce a matrix beyond what is done by our algorithm, the other techniques can be applied to the blocks produced by our algorithm. The algorithm also has the advantage that it works entirely with real matrices by the device of grouping pairs of complex conjugate eigenvalues in the same block.

In the next section of this paper the algorithm is described. In § 3 some numerical examples are given. Programming details and a listing are given in an appendix. (See the microfiche section in the back of this volume.)

2. The algorithm. The first part of the algorithm uses orthogonal transformations to reduce the matrix A to quasi-triangular form, that is to a block upper-triangular form in which the diagonal blocks are of order at most two. The blocks of order one contain real eigenvalues of A and the blocks of order two contain complex conjugate pairs of eigenvalues. The ordering of the blocks is arbitrary, and the order can be changed by applying appropriate orthogonal transformations. Since this reduction of A can be effected by standard techniques [9], [11], we may assume that A is already in quasi-triangular form.

The subsequent block diagonalization is accomplished as follows. The matrix A is partitioned in the form

$$A=\begin{pmatrix} A_{11} & A_{12} \\ 0 & A_{22} \end{pmatrix},$$

where initially A_{11} is 1×1 or 2×2 depending on the dimension of the leading diagonal block of A. An attempt is then made to find a similarity transformation X such that

$$X^{-1}AX=\begin{pmatrix} A_{11} & 0 \\ 0 & A_{22} \end{pmatrix}.$$

If such a transformation can be found and if it is not too ill-conditioned, the reduction proceeds with the submatrix A_{22}. If not, a suitable 1×1 or 2×2 diagonal block from A_{22} is located and moved by means of orthogonal transformations to the leading position of A_{22}. The block is then adjoined to A_{11} by increasing the order of A_{11} by one or two, as is appropriate, and another attempt is made to find a reducing matrix X.

The implementation of such an algorithm requires the answers to two questions.

1. How may the transformation X be computed?
2. In the event of failure, which block of A_{22} is to be incorporated into A_{11}?

We shall now answer these questions.

We seek the transformation X in the form

$$X=\begin{pmatrix} I & P \\ 0 & I \end{pmatrix}, \tag{2.1}$$

where the identity matrices are of the same orders as A_{11} and A_{22}. The inverse of X is easily seen to be

$$X^{-1}=\begin{pmatrix} I & -P \\ 0 & I \end{pmatrix}. \tag{2.2}$$

Hence

$$X^{-1}AX=\begin{pmatrix} A_{11} & A_{11}P-PA_{22}+A_{12} \\ 0 & A_{22} \end{pmatrix},$$

and the problem of determining X becomes that of solving the equation

$$A_{11}P-PA_{22}=-A_{12}. \tag{2.3}$$

Because A_{11} and A_{22} are quasi-triangular, this equation can be solved by a back-substitution algorithm of Bartels and Stewart [1], provided the eigenvalues of A_{11} and A_{22} are disjoint.

From (2.1) and (2.2) it follows that X will be ill-conditioned whenever P is large. As each element of P is generated, it is tested to see if its magnitude exceeds a bound provided by the user. If it does, the attempt to compute X is abandoned and a new, larger block A_{11} is formed. If no element of P exceeds the bound, the matrix X is accepted and the matrix A is deflated as described above. The transformation X is postmultiplied into a matrix that accumulates all the transformations made on the matrix A.

The process for selecting a 1×1 or 2×2 block of A_{22} to incorporate into A_{11} goes as follows. We compute the mean of those eigenvalues of A_{11} having nonnegative imaginary part. A block is chosen from A_{22} whose eigenvalue with nonnegative imaginary part is nearest this mean. This block is moved, as described above, by orthogonal transformations to the leading position in A_{22}, where it is incorporated into A_{11}. The program HQR3 [11], which can be used to obtain the initial quasi-triangular form, has a subroutine which will compute these orthogonal transformation. The transformations are of course postmultiplied into the accumulating matrix.

We summarize our algorithm in the following informal code. Further details can be found in the appendix to this paper, where a FORTRAN subroutine implementing this code is given. The code takes as input an array A of order N containing the matrix A and an array X in which the transformations are accumulated. In addition the user must provide a tolerance to bound the size of the elements of the deflating transformations. The integers $L11$ and $L22$ point to the beginnings of the current blocks A_{11} and A_{22} in the array A. The informal code should be self-explanatory. Comments are delineated by the symbol #.

1 : reduce A to quasitriangular form, accumulating the transformations in X [10];
2 : $L11=1$;
3 : **loop** # until the matrix is diagonalized #
3.1 :   **if** $L11>N$ **then leave** 3 **fi**;
3.2 : $L22=L11$;

3.3 : **loop** # until a block has been deflated #
3.3.1 : **if** $L22 = L11$ **then** # use the first 1×1 or 2×2 block #
3.3.1t.1 : M = order of the block at $L11$;
3.3.1t.2 : $L22 = L22 + M$;
3.3.1 : **else** # augment A_{11} with a 1×1 or 2×2 block from A_{22} #
3.3.1e.1: compute the mean of the eigenvalues of A_{11} with nonnegative imaginary parts;
3.3.1e.2: find the $M \times M$ # $M = 1$ or 2 # block of A_{22} whose eigenvalue with nonnegative imaginary part is nearest the mean;
3.3.1e.3: move the block to the leading position of A_{22} accumulating the transformations in X;
3.3.1e.4: $L22 = L22 + M$ # which incorporates the block in A_{11} #;
3.3.1 : **fi**;
3.3.2 : **if** $L22 > N$ **then leave** 3.3 **fi**;
3.3.3 : attempt to split off A_{11} [1];
3.3.4 : **if** the attempt was successful **then leave** 3.3 **fi**;
3.3.5 : restore A_{12};
3.3 : **end loop**;
3.4 : **if** $L22 \leqq N$ **then** accumulate the deflating transformation in X **fi**;
3.5 : scale columns $L11$ through $L22-1$ of X so that they have 2-norm unity, adjusting A accordingly;
3.6 : $L11 = L22$;
3 : **end loop**;

Several comments should be made about the algorithm. First, it uses only real arithmetic, even when A has complex eigenvalues. Second, the algorithm cannot guarantee that the final transformation is well conditioned, since the bound on the elements of P restricts the condition of only the individual transformations comprising the final one. Nonetheless, we have found little tendency toward excessive growth in the condition of the transformation. Third, no attempt is made to segregate nearly equal eigenvalues initially into clusters; whether an eigenvalue splits off or not depends entirely on its determining a suitably small matrix P. This is important, since it means that the algorithm can compute well-conditioned eigenvectors for multiple eigenvalues (a little help is required from rounding error; see § 3).

The strategy for determining what eigenvalues to add to the current group has the defect that it can mix well-conditioned and ill-conditioned eigenvalues that are nearly equal, thus missing the possibility of a more complete reduction. This is not a serious problem when the blocks are small. However, if a finer resolution of the structure of the matrix is required, the techniques referenced in § 1 may be applied to the blocks produced by our algorithm. In fact our algorithm can be regarded as a preprocessing step to reduce the problem to a size where it can be attacked by more sophisticated, but more expensive methods.

We note that the output of our algorithm may depend on the user supplied tolerance for the elements of P. In general the larger the tolerance, the smaller the blocks but the more ill conditioned the transformation. This tradeoff is an inevitable consequence of the poor determination of the structure of a matrix in the presence of errors, and we know of no algorithm that relieves the user of the necessity of making a decision of this kind.

So far as storage is concerned, the algorithm requires $2n^2$ locations to contain the matrices A and X and a number of arrays of order n. This is the same as the storages required by algorithms that compute the eigenvectors of a general matrix.

Excluding the initial reduction of A to quasitriangular form, the bulk of the computations in the algorithm occur at statement 3.3.3, where an attempt is made to solve the equation (2.3), and at statement 3.4, where the transformation is accumulated. The multiplication count for the algorithm for solving (2.3) is of order $(l^2m+m^2l)/2$, where l is the size of A_{11} and m is the size of A_{22}. The cost of accumulating this transformation in X is of order $l \cdot m \cdot n$. Thus, at one extreme, if all the eigenvalues of A are real and they all can be deflated, the cost in multiplications will be of order $n^3/2$, which compares favorably with algorithms for computing the eigenvectors of A from its quasitriangular form. On the other hand, if the eigenvalues of A are real and none of them can be deflated, the algorithm will require on the order of $n^4/12$ multiplications.

Although an n^4 operation count is disturbing, there are several mitigating factors. First, we do not expect the algorithm to find many applications to matrices that cannot be substantially reduced by it, since the object of using it is to save on subsequent calculations with the matrix. Second, the count assumes that the algorithm for solving (2.3) must be executed to completion before it is found that P is unacceptably large. This is not likely; in fact because the algorithm [11] for reducing A to quasitriangular form arranges the eigenvalues in decreasing order of magnitude, it is rather unlikely. Finally the order constant $\frac{1}{12}$ is modest; for n less than 60 the bound is less than $5n^3$.

3. Numerical results. In this section we summarize the results of some numerical experiments. Two of the tests were performed with a class of test matrices generated as follows. Let J be a given matrix whose structure is known (e.g. J could be in Jordan canonical form). Let

$$H_1 = I - \frac{2}{2} ee^T,$$

where $e = (1, 1, \cdots, 1)^T$, and

$$H_2 = I - \frac{2}{n} ff^T,$$

where $f = (1, -1, \cdots, (-1)^{n-1})^T$. Note that H_1 and H_2 are symmetric and orthogonal. Let

$$S = \operatorname{diag}(1, \sigma, \sigma^2, \cdots, \sigma^{n-1})$$

where σ is a given parameter. Then we take A in the form

$$A = (H_2SH_1)J(H_2SH_1)^{-1} = (H_2SH_1)J(H_1S^{-1}H_2). \tag{3.1}$$

The matrix A, which is cheap to compute, has the same structure as J. The transformation H_2SH_1 can be made as ill-conditioned as desired by varying σ.

In describing the results of the tests we report two numbers. The first is the maximum ρ of the scaled residuals

$$\rho_i = \frac{\|AX_i - XB_i\|_\infty}{\|A\|_\infty \|X_i\|_\infty},$$

where X_i is composed of the columns of X corresponding to the block B_i. Second we report $\|X^{-1}\|_\infty$. Together these numbers give an idea of how stably the reduction proceeded; for if

$$R = AX - XB$$

then, with $E = RX^{-1}$, we have that

$$(A+E)X - XB = 0;$$

that is B is exactly similar to $A+E$. The relative error that this perturbation represents in A is

$$\frac{\|E\|}{\|A\|} \leqq \frac{\|R\| \, \|X^{-1}\|}{\|A\|}.$$

Since $\|X\| \cong 1$, the relative error will be of the order $\rho \|X^{-1}\|$.

The first test case illustrates the ability of the algorithm to split apart nearly equal eigenvalues with independent eigenvectors. We took

$$J = \text{diag}\left[1, 1-\varepsilon, 1+\varepsilon, \begin{pmatrix} 1 & -1 \\ 1 & 1 \end{pmatrix}, .3, .4, .5, .6, .7\right].$$

The algorithm was applied for various values of ε, σ, and RMAX, the bound on the size of the deflating transformations. The results are summarized in Table 1, in which the eigenvalues of A are numbered as follows:

1	2	3	4	5	6	7	8	9	10
$1+i$	$1-i$	$1+\varepsilon$	1	$1-\varepsilon$	.7	.6	.5	.5	.3

Complex eigenvalues are indicated by a circumflex.

TABLE 1

σ	ε	RMAX	Block structure	ρ	$\|X^{-1}\|_\infty$
1.0	10^{-1}	10.	$(\hat{1}, \hat{2})$, 3, 4, 5, 6, 7, 8, 9, 10	2.1×10^{-7}	7.7
	10^{-5}	10.	$(\hat{1}, \hat{2})$, 3, 4, 5, 6, 7, 8, 9, 10	1.4×10^{-7}	6.7
	0.0	10.	$(\hat{1}, \hat{2})$, 3, 4, 5, 6, 7, 8, 9, 10	1.1×10^{-7}	8.0
10.0	10^{-1}	10.	$(\hat{1}, \hat{2})$, 3, 4, 5, 6, 7, 8, 9, 10	1.5×10^{-7}	25.6
	10^{-5}	10.	$(\hat{1}, \hat{2})$, 3, 4, 5, 6, 7, 8, 9, 10	1.8×10^{-7}	25.4
	0.0	10.	$(\hat{1}, \hat{2})$, 3, 4, 5, 6, 7, 8, 9, 10	1.6×10^{-7}	24.2
100.0	10^{-1}	10.	$(\hat{1}, \hat{2})$, 3, 4, 5, 6, 7, 8, 9, 10	1.1×10^{-7}	187.9
	10^{-5}	10.	$(\hat{1}, \hat{2})$, 3, 4, 5, 6, 7, 8, 9, 10	1.3×10^{-7}	196.6
	0.0	10.	$(\hat{1}, \hat{2})$, 3, 4, 5, 6, 7, 8, 9, 10	1.2×10^{-7}	157.9
1000.0	10^{-1}	10.	$(\hat{1}, \hat{2}$, 4, 3, 5), (6, 7, 8, 9), 10	1.1×10^{-7}	326.0
		100.	$(\hat{1}, \hat{2})$, 3, 4, 5, 6, 7, 8, 9, 10	1.0×10^{-8}	1537.9
	10^{-5}	10.	$(\hat{1}, \hat{2}$, 4, 3, 5), (6, 7, 8, 9), 10	7.8×10^{-8}	333.8
		100.	$(\hat{1}, \hat{2})$, 3, 4, 5, 6, 7, 8, 9, 10	7.8×10^{-8}	1559.1
	0.0	10.	$(\hat{1}, \hat{2}$, 3), 4, 5, (6, 7, 8, 9), 10	8.2×10^{-8}	394.1
		100.	$(\hat{1}, \hat{2})$, 3, 4, 5, 6, 7, 8, 9, 10	8.2×10^{-8}	1355.7

Provided RMAX was large enough to allow a sufficiently ill-conditioned transformation, all the cases were split completely. The condition of the transformation was in no case much greater than the condition of the scrambling transformation in (3.1). It is of interest to note that the algorithm was successful when $\varepsilon = 0$, that is when A has three equal eigenvalues. Mathematically, the algorithm for solving (2.3) breaks down when A_{11} and A_{22} have common eigenvalues; however, the experiments indicate that if the equal eigenvalues have independent vectors, rounding error will perturb them enough for the algorithm to work.

The second example shows the failure of the algorithm's strategy for selecting the next eigenvalue to be adjoined to A_{11}. Here J has the form

$$J = \operatorname{diag}\left[1, \begin{pmatrix} 1 & 1 & 0 \\ 0 & 1 & 1 \\ 0 & 0 & 1 \end{pmatrix}, .3, .4, .5, .6, .7, .8\right]$$

and σ was taken to be one. Of the four eigenvalues at unity, one is perfectly conditioned, and the other three, which belong to a single Jordan block, are very ill-conditioned. To six figures the computed eigenvalues were

$$\begin{array}{l} 1.00073+0.001303i \\ 1.00073-0.001303i \\ 1.00000 \\ 0.99853 \\ \;\,.80000 \\ \quad\vdots \end{array}$$

The pair of complex eigenvalues could not be deflated, since they were coupled to the third member of the block. But this member would not be adjoined without first adjoining the well-conditioned eigenvalue at unity. Consequently, the algorithm produced a single block of order four, rather than two blocks of orders one and three. This block of order four could be reduced further by the more sophisticated techniques described in the references.

The third test case is the Frank matrix of order 12 which has appeared frequently in the literature [2], [3]. The smaller eigenvalues of this matrix are extremely ill conditioned. The results of test runs on this matrix are summarized below.

RMAX	Block structure	ρ	$\|X^{-1}\|_\infty$
10	1, 2, 3, 4, 5, (6, 7, 8, 9, 10, 11, 12)	2.9×10^{-7}	93.6
50	1, 2, 3, 4, 5, 6, (7, 8, 9, 10, 11, 12)	2.9×10^{-7}	2920.2
100	1, 2, 3, 4, 5, 6, (7, 8, 9, 10, 11, 12)	2.9×10^{-7}	2920.2
1000	1, 2, 3, 4, 5, 6, (7, 8, 9, 10, 11, 12)	2.9×10^{-7}	2920.2

The algorithm performed much as expected, separating the larger eigenvalues and grouping the smaller eigenvalues together. This grouping is consistent with the precision of the computation.

The final test case is included because the failure of our algorithm to decompose it reveals the shaky foundations of a fairly common numerical practice. Specifically we generated the companion matrix of the polynomial given in [14, p. 74]. Since the zeros of this polynomial are not very ill-conditioned, we were surprised when the algorithm failed to split off so much as one. Some further computations revealed that the matrix of left eigenvectors (the inverse Vandermode of the zeros) had rows of order 10^6–10^8. This explains the failure to reduce the matrix. More important, though, it shows that the eigenvalues of the companion matrix are much more ill-conditioned than the zeros of the polynomial and suggests that the practice of using eigenvalue routines to find zeros of polynomials can result in an unnecessary loss in accuracy.

Appendix. Programming details and program listing.

A1. Usage. The calling sequence for BDIAG is

CALL BDIAG(A, LDA, N, EPSHQR, RMAX, ER, EI, TYPE, BS, X, LDX, FAIL).

The parameters in the calling sequence are (starred parameters are altered by the subroutine)

*A(LDA, N)	an array that initially contains the N×N matrix to be reduced. On return A contains the reduced block diagonal matrix.
LDA	the leading dimension of A.
N	the order of the matrices A and X.
EPSHQR	a real variable containing a convergence criterion for the subroutines HQR3 and EXCHNG [11].
RMAX	a real variable containing a bound on the absolute values of the elements in the reducing matrices.
*ER(N)	a real array containing the real parts of the eigenvalues of A.
*EI(N)	a real array containing the imaginary parts of the eigenvalues.
*TYPE(N)	an integer array whose *i*th entry is 0 if the *i*th eigenvalue is real 1 if the *i*th eigenvalue is complex with positive imaginary part 2 if the *i*th eigenvalue is complex with negative imaginary part −1 if the *i*th eigenvalue could not be computed
*BS(N)	a singly subscripted array that contains information on the block structure of the matrix returned by the program. If there is a block of order K at A(L, L), then BS(L), BS(L+1), ··· BS(L+K−1) contain the integers K, −(K−1), ···, −1. Thus a positive entry of K indicates the start of a block of order K.
*X(LDX, N)	an array into which the reducing transformations are accumulated.
LDX	the leading dimension of X.
FAIL	a logical variable which is false on a normal return and is true on return if an error has occurred.

The program requires the programs ORTHES [9], ORTRAN [9], HQR3 [11], EXCHNG [11], SHRSLV [1], DAD and SPLIT [11].

A suitable choice for the parameter EPSHQR is the rounding unit of the computer on which the program is being run; i.e. if one is working in a precision of about t decimal figures EPSHQR should be of order 10^{-t}.

A.2. Programming details. In this section we shall describe some of the details of the implementation of the algorithm. Throughout this section we refer to the outline of the algorithm in § 2.

Statement 1. This is accomplished by the EISPACK routines ORTHES and ORTRAN [9] and the QR routine HQR3 [11].

Statement 2. L11 points to the leading position of the current block A11, which is of order DA11.

Statement 3. This is the main loop of the program. It ends when L11>N, indicating that there are no further blocks to deflate.

Statement 3.2. L22 points to the leading position of the current block A22, which is of order DA22.

Statement 3.3. In this loop A_{11} is repeatedly augmented until it can be deflated or until A_{22} is void.

Statements 3.3.1t. A_{11} is initially void. Here it is taken to be the 1×1 or 2×2 block starting at L11.

Statements 3.3.1e. This is the search for a 1×1 or 2×2 block described in § 2.

Statement 3.3.1e.3. The subroutine EXCHNG [11] is used repeately to move the block just located to the beginning of A_{22}. After each exchange of a complex block SPLIT [11] is called to recompute its eigenvalues and to see if, owing to rounding error, it can be split into a pair of real eigenvalues.

Statement 3.3.2. Because A_{22} is void, A_{11} is effectively deflated.

Statement 3.3.3. The matrix A_{12} is saved below the lower subdiagonal of A, in case the attempt to deflate A_{11} is unsuccessful. Since the routine SHRSLV [1], which computes the deflating transformation, requires A_{11} to be lower Hessenberg, the subroutine DAD is called to transform A_{11} and A_{12}. SHRSLV has been modified to return with a signal if some element of the deflating transformation exceeds the bound RMAX. Otherwise the matrix P that determines the transformation overwrites A_{12}. DAD is once again called to restore A_{11} to its original form.

Statement 3.3.5. The submatrix A_{22}, which was overwritten by SHRSLV, must be restored before another attempt to deflate is made.

Statement 3.4. Only if A_{22} is not void was a deflating transformation generated.

Statement 3.6. Set L11 to point to the next submatrix before continuing.

A.3. Program listing. See the microfiche section in the back of this volume.

REFERENCES

[1] R. H. BARTELS AND G. W. STEWART, *Algorithm 432, the solution of the matrix equation $AX - XB = C$*, Comm. A.C.M. (1972), pp. 820–826.

[2] R. T. GREGORY AND D. L. KARNEY, *A Collection of Matrices for Testing Computational Algorithms*, Wiley–Interscience, New York, 1969.

[3] G. H. GOLUB AND J. H. WILKINSON, *Ill-conditioned eigensystems and the computation of the Jordan canonical form*, Stanford University Report STAN-CS-75-478, Stanford, CA, 1975.

[4] T. KATO, *Perturbation Theory for Linear Operators*, Springer-Verlag, New York, 1966.

[5] V. N. KUBLANOVSKAYA, *On a method of solving the complete eigenvalue problem for a degenerate matrix*, Ž. Vyčisl. Mat. I Mat. Fiz., 6 (1966), pp. 611–620; translation in USSR Computational Math. and Math. Phys., 6 (1968), pp. 1–14.

[6] B. KÅGSTRÖM AND AXEL RUHE, *An algorithm for numerical computation of the Jordan normal form of a complex matrix*, University of Umeå, UMINF-51.74, Sweden, 1977.

[7] C. B. MOLER AND C. F. VAN LOAN, *Nineteen ways to compute the exponential of a matrix*, Cornell University Computer Science Technical Report TR 76-283, Ithaca, NY, 1976.

[8] A. RUHE, *An algorithm for numerical determination of the structure of a general matrix*, BIT, 10 (1970), pp. 196–216.

[9] B. T. SMITH, J. M. BOYLE, J. J. DONGARRA, B. S. GARBOW, Y. IKEBE, V. C. KLEMA AND C. B. MOLER, *Matrix Eigensystem Routines, EISPACK Guide*, Second Ed., Springer-Verlag, New York, 1976.

[10] G. W. STEWART, *Introduction to Matrix Computations*, Academic Press, New York, 1973.

[11] ———, *Algorithm* 506: *HQR*3 *and EXCHNG*: *Fortran subroutines for calculating and ordering the eigenvalues of a real upper Hessenberg matrix*, ACM Trans. Math. Software, 3 (1976), pp. 275–280.

[12] J. M. VARAH, *Rigorous machine bounds for the eigensystem of a general complex matrix*, Math. Comput., 22 (1968), pp. 293–801.

[13] ———, *Computing invariant subspaces of a general matrix when the eigensystem is poorly conditioned*, Math. Comp., 24 (1970), pp. 137–149.

[14] J. H. WILKINSON, *Rounding Errors in Algebraic processes*, Prentice-Hall, Englewood Cliffs, NJ, 1963.

Appendix A3. Program Listing*

Connice A. Bavely and G. W. Stewart, *An Algorithm for Computing Reducing Subspaces by Block Diagonalization*, SIAM J. Numer. Anal., 16 (1979), pp. 359-367.

*No warranties, express or implied, are made by the publisher that this program is free of error. It should not be relied on as the sole basis to solve a problem whose incorrect solution could result in injury to person or property. If the program is employed in such a manner, it is at the user's own risk and the publisher disclaims all liability for such misuse.

```
      SUBROUTINE BDIAG (A,LDA,N,EPSHQR,RMAX,ER,EI,TYPE,BS,X,LDX,FAIL)

      INTEGER LDA, LDX, N, TYPE(N), BS(N)
      REAL A(LDA,N), ER(N), EI(N), X(LDX,N), EPSHQR, RMAX
      LOGICAL FAIL

C     BDIAG REDUCES A MATRIX A TO BLOCK DIAGONAL FORM BY FIRST
C     REDUCING IT TO QUASI-TRIANGULAR FORM BY HQR3 AND THEN BY
C     SOLVING THE MATRIX EQUATION  -A11*P+P*A22=A12 TO INTRODUCE ZEROS
C     ABOVE THE DIAGONAL.  THE PARAMETERS IN THE CALLING SEQUENCE
C     ARE (STARRED PARAMETERS ARE ALTERED BY THE SUBROUTINE):

C     *A         AN ARRAY THAT INITIALLY CONTAINS THE N X N MATRIX
C                TO BE REDUCED. ON RETURN, A CONTAINS THE REDUCED
C                BLOCK DIAGONAL MATRIX.

C     LDA        THE LEADING DIMENSION OF ARRAY A.

C     N          THE ORDER OF THE MATRICES A AND X.

C     EPSHQR     THE CONVERGENCE CRITERION FOR SUBROUTINE HQR3.

C     RMAX       THE MAXIMUM SIZE ALLOWED FOR ANY ELEMENT OF THE
C                TRANSFORMATIONS.

C     *ER        A SINGLY SUBSCRIPTED REAL ARRAY CONTAINING THE REAL
C                PARTS OF THE EIGENVALUES.

C     *EI        A SINGLY SUBSCRIPTED REAL ARRAY CONTAING THE IMAGINARY
C                PARTS OF THE EIGENVALUES.

C     *TYPE      A SINGLY SUBSCRIPTED INTEGER ARRAY WHOSE I-TH ENTRY IS
C                0   IF THE I-TH EIGENVALUE IS REAL
C                1   IF THE I-TH EIGENVALUE IS COMPLEX WITH POSITIVE
C                    IMAGINARY PART
C                2   IF THE I-TH EIGENVALUE IS COMPLEX WITH NEGATIVE
C                    IMAGINARY PART
C               -1   IF THE I-TH EIGENVALUE COULD NOT BE COMPUTED

C     *BS        A SINGLY SUBSCRIPTED INTEGER ARRAY THAT CONTAINS BLOCK
C                STRUCTURE INFORMATION.  IF THERE IS A BLOCK OF ORDER
C                K STARTING AT A(L,L) IN THE OUTPUT MATRIX A, THEN
C                BS(L) CONTAINS THE POSITIVE INTEGER K, BS(L+1) CONTAINS
C                -(K-1), BS(LA+2) = -(K-2), ..., BS(L+K-1) = -1.
C                THUS A POSITIVE INTEGER IN THE L-TH ENTRY OF BS
C                INDICATES A NEW BLOCK OF ORDER BS(L) STARTING AT A(L,L).

C     *X         AN ARRAY INTO WHICH THE REDUCING TRANSFORMATIONS
C                ARE TO BE MULTIPLIED.

C     LDX        THE LEADING DIMENSION OF ARRAY X.

C     *FAIL      A LOGICAL VARIABLE WHICH IS FALSE ON NORMAL RETURN AND
C                TRUE IF THERE IS ANY ERROR IN BDIAG.


C     BDIAG USES SUBROUTINES ORTHES, ORTRAN, HQR3, EXCHNG, SHRSLV, DAD, AND SPLIT.

      INTEGER DA11, DA22, I, J, K, KM1, KM2, L, LEAVE, LOOP,
     1 L11, L22, L22M1, NK
      REAL C, CAV, D, E1, E2, RAV,  SC, TEMP

      FAIL = .TRUE.

C     CONVERT A TO UPPER HESSENBERG FORM.
```

```
      CALL ORTHES (LDA,N,1,N,A,ER)
      CALL ORTRAN (LDA,N,1,N,A,ER,X)

      CONVERT A TO QUASI-UPPER TRIANGULAR FORM BY QR METHOD.

      CALL HQR3 (A,X,N,1,N,EPSHQR,ER,EI,TYPE,LDA,LDX)
      CALL ARYOUT ('BDIAG4',A,N,N,LDA,10,'E12.7')

      CHECK TO SEE IF HQR3 FAILED IN COMPUTING ANY EIGENVALUE


      DO 5 I = 1,N
        IF (TYPE(I).EQ.-1) GO TO 900
 5    CONTINUE

      REDUCE A TO BLOCK DIAGONAL FORM

      SEGMENT A INTO 4 MATRICES: A11, A DA11 X DA11 BLOCK
      WHOSE (1,1)-ELEMENT IS AT A(L11,L11); A22, A DA22 X DA22
      BLOCK WHOSE (1,1)-ELEMENT IS AT A(L22,L22); A12,
      A DA11 X DA22 BLOCK WHOSE (1,1)-ELEMENT IS AT A(L11,L22);
      AND A21, A DA22 X DA11 BLOCK = 0 WHOSE (1,1)-
      ELEMENT IS AT A(L22,L11).


      THIS LOOP USES L11 AS LOOP INDEX AND SPLITS OFF A BLOCK
      STARTING AT A(L11,L11).


      L11 = 1
10    ASSIGN 550 TO LOOP
        ASSIGN 600 TO LEAVE
        IF (L11 .GT. N) GO TO LEAVE
        L22 = L11

        THIS LOOP USES DA11 AS LOOP VARIABLE AND ATTEMPTS TO SPLIT
        OFF A BLOCK OF SIZE DA11 STARTING AT A(L11,L11)

100     ASSIGN 350 TO LOOP
          IF (L22 .NE. L11) GO TO 110
            DA11 = TYPE(L11) + 1
            L22 = L11 + DA11
            L22M1 = L22 - 1
          GO TO 290
110       CONTINUE

            COMPUTE THE AVERAGE OF THE EIGENVALUES IN A11

            RAV = 0.
            CAV = 0.
            DO 120 I = L11,L22M1
              RAV = RAV + ER(I)
              CAV = CAV + ABS(EI(I))
120         CONTINUE
            RAV = RAV / FLOAT(DA11)
            CAV = CAV / FLOAT(DA11)

            LOOP ON EIGENVALUES OF A22 TO FIND THE ONE CLOSEST TO THE AVERAGE.

            D = (RAV-ER(L22))**2 + (CAV-EI(L22))**2
            K = L22
            L = L22 + TYPE(L22) + 1
130         ASSIGN 145 TO LOOP
              IF (L.GT.N) GO TO 150
              C = (RAV-ER(L))**2 + (CAV-EI(L))**2
              IF (C.GE.D) GO TO 140
                K = L
                D = C
140           CONTINUE
```

```
          L = L + TYPE(L) + 1
145       GO TO 130
150       CONTINUE

          LOOP TO MOVE THE EIGENVALUE JUST LOCATED
          INTO FIRST POSITION OF BLOCK A22.

          ASSIGN 280 TO LOOP
             IF (TYPE(K).NE.0) GO TO 200

             THE BLOCK WE RE MOVING TO ADD TO A11 IS A 1 X 1

             NK = 1
160          CONTINUE
             IF (K .EQ. L22) GO TO 280
             KM1 = K - 1
             IF (TYPE(KM1).EQ.0) GO TO 190

             WE RE SWAPPING THE CLOSEST BLOCK WITH A 2 X 2

             KM2 = K - 2
             CALL EXCHNG (A,X,N,KM2,2,1,EPSHQR,FAIL,LDA,LDX)
             IF (FAIL) GO TO 900

             TRY TO SPLIT THIS BLOCK INTO 2 REAL EIGENVALUES.

             CALL SPLIT (A,X,N,KM1,E1,E2,LDA,LDX)
             IF (A(K,KM1).EQ.0.) GO TO 170

             BLOCK IS STILL COMPLEX.

             TYPE(KM2) = 0
             TYPE(KM1) = 1
             TYPE(K) = 2
             ER(KM2) = ER(K)
             EI(KM2) = 0.
             ER(K) = E1
             ER(KM1) = E1
             EI(KM1) = E2
             EI(K) = -E2
             GO TO 180

             COMPLEX BLOCK SPLIT INTO TWO REAL EIGENVALUES.

170          CONTINUE
             TYPE(KM1) = 0
             TYPE(KM2) = 0
             ER(KM2) = ER(K)
             ER(KM1) = E1
             ER(K) = E2
             EI(KM2) = 0.
             EI(KM1) = 0.
180          K = KM2
             IF (K .LE. L22) GO TO 280
             GO TO 160

             WE RE SWAPPING THE CLOSEST BLOCK WITH A 1 X 1.

190          CONTINUE
             CALL EXCHNG (A,X,N,KM1,1,1,EPSHQR,FAIL,LDA,LDX)
             IF (FAIL) GO TO 900
             TEMP = ER(K)
             ER(K ) = ER(KM1)
             ER(KM1) = TEMP
             K = KM1
             IF (K .LE. L22) GO TO 280
             GO TO 160

             THE BLOCK WE'RE MOVING TO ADD TO A11 IS A 2 X 2.
```

```
200       CONTINUE
          NK = 2
210       CONTINUE
          IF (K .EQ. L22) GO TO 280
          KM1 = K - 1
          IF (TYPE(KM1) .EQ. 0) GO TO 240

          WE'RE SWAPPING THE CLOSEST BLOCK WITH A 2 X 2 BLOCK.

          KM2 = K - 2
          CALL EXCHNG (A,X,N,KM2,2,2,EPSHQR,FAIL,LDA,LDX)
          IF (FAIL) GO TO 900

          TRY TO SPLIT SWAPPED BLOCK INTO T O REALS.

          CALL SPLIT (A,X,N,K,E1,E2,LDA,LDX)
          ER(KM2) = ER(K)
          ER(KM1) = ER(K+1)
          EI(KM2) = EI(K)
          EI(KM1) = EI(K+1)
          IF (A(K+1,K) .EQ. 0.) GO TO 220

          STILL COMPLEX BLOCK.

          ER(K) = E1
          ER(K+1) = E1
          EI(K) = E2
          EI(K+1) = -E2
          GO TO 230

          TWO REAL ROOTS.

220       CONTINUE
          TYPE(K) = 0
          TYPE(K+1) = 0
          ER(K) = E1
          ER(K+1) = E2
          EI(K) = 0.
          EI(K+1) = 0.
230       CONTINUE
          K = KM2
          IF (K.EQ.L22) GO TO 260
          GO TO 210

          WE'RE SWAPPING THE CLOSEST BLOCK WITH A 1 X 1.

240       CONTINUE
          CALL EXCHNG (A,X,N,KM1,1,2,EPSHQR,FAIL,LDA,LDX)
          IF (FAIL) GO TO 900
          TYPE (KM1) = 1
          TYPE(K) = 2
          TYPE(K+1) = 0
          ER(K+1) = ER(KM1)
          ER(KM1) = ER(K)
          EI(KM1) = EI(K)
          EI(K) = EI(K+1)
          EI(K+1) = 0.
          GO TO 250

250       CONTINUE
          K = KM1
          IF (K.EQ.L22) GO TO 260
          GO TO 210

          TRY TO SPLIT RELOCATED COMPLEX BLOCK.

260       CONTINUE
          CALL SPLIT (A,X,N,K,E1,E2,LDA,LDX)
          IF (A(K+1,K).EQ.0.) GO TO 270

          STILL COMPLEX.
```

```
          ER(K) = E1
          ER(K+1) = E1
          EI(K) = E2
          EI(K+1) = -E2
          GO TO 280

          SPLIT INTO TWO REAL EIGENVALUES.

270       CONTINUE
          TYPE(K) = 0
          TYPE(K+1) = 0
          ER(K) = E1
          ER(K+1) = E2
          EI(K) = 0.
          EI(K+1) = 0.

280     CONTINUE
        DA11 = DA11 + NK
        L22 = L11 + DA11
        L22M1 = L22 - 1
290   CONTINUE
      ASSIGN 400 TO LEAVE
      IF (L22 .GT. N) GO TO LEAVE

      ATTEMPT TO SPLIT OFF A BLOCK OF SIZE DA11.

      DA22 = N - L22 + 1

      SAVE A12 IN ITS TRANSPOSE FORM IN BLOCK A21.

      DO 300 J = L11,L22M1
        DO 300 I = L22,N
          A(I,J) = A(J,I)
300   CONTINUE


      CONVERT A11 TO LOWER QUASI-TRIANGULAR AND MULTIPLY IT BY -1 AND CHANGE
      A12 APPROPRIATELY (FOR SOLVING -A11*P+P*A22=A12).

      CALL DAD (A,LDA,L11,L22M1,L11,N,1.,0)
      CALL DAD (A,LDA,L11,L22M1,L11,L22M1,-1.,1)
      CALL ARYOUT('A11*D',A(L11,L11),DA11,DA11,LDA,10,'E13.7')

      SOLVE -A11*P + P*A22 = A12.

      CALL SHRSLV (A(L11,L11),A(L22,L22),A(L11,L22),
        DA11,DA22,LDA,LDA,LDA,RMAX,FAIL)
      ASSIGN 400 TO LEAVE
      IF (.NOT. FAIL) GO TO LEAVE

      CHANGE A11 BACK TO UPPER QUASI-TRIANGULAR.

      CALL DAD (A,LDA,L11,L22M1,L11,L22M1,1.,1)
      CALL DAD (A,LDA,L11,L22M1,L11,L22M1,-1.,0)

      WAS UNABLE TO SOLVE FOR P - TRY AGAIN


      MOVE SAVED A12 BACK INTO ITS CORRECT POSITION.

      DO 310 J=L11,L22M1
        DO 310 I = L22,N
          A(J,I) = A(I,J)
          A(I,J) = 0.
310   CONTINUE

350   GO TO 100
400   CONTINUE

      CHANGE SOLUTION TO P TO PROPER FORM.
```

```
      IF (L22 .GT. N) GO TO 440
        CALL DAD (A,LDA,L11,L22M1,L11,N,1.,0)
        CALL DAD (A,LDA,L11,L22M1,L11,L22M1,-1.,1)

        MULTIPLY TRANSFORMATION INTO X.

        ONLY COLUMNS L22 THRU N ARE AFFECTED.

        DO 410 J = L22,N
          DO 410 I = 1,N
            DO 410 K = L11,L22M1
              X(I,J) = X(I,J) + X(I,K) * A(K,J)
410     CONTINUE

        ZERO OUT A12 FOR EASE IN HANDLING.

        DO 420 J = L22,N
          DO 420 I = L11,L22M1
            A(I,J) = 0.
420     CONTINUE

        ZERO OUT TRIANGULAR BLOCK BELOW DIAGONAL.

        DO 430 J = L11,L22M1
          DO 430 I = L22,N
            A(I,J) = 0.
430     CONTINUE

      SCALE THOSE COLUMNS OF X THAT WON'T BE ALTERED AGAIN TO UNITY.
      CHANGE A11 APPROPRIATELY.

440   CONTINUE
      DO 500 J = L11,L22M1
        SC = 0.
        DO 450 I = 1,N
          SC = SC + (X(I,J))**2
450     CONTINUE
        SC = SQRT(SC)
        DO 460 I = 1,N
          X(I,J) = X(I,J) / SC
460     CONTINUE
         DO 470 I = L11,L22M1
           A(I,J) = A(I,J) / SC
470     CONTINUE
        DO 480 I = L11,L22M1
          A(J,I) = A(J,I) * SC
480     CONTINUE
500   CONTINUE

      STORE BLOCK SIZE IN ARRAY BS.

      BS(L11) = DA11
      J = DA11 - 1
      IF (J .EQ. 0) GO TO 520
        DO 510 I = 1,J
          BS(L11+I) = -(DA11-I)
510     CONTINUE
520   CONTINUE
      L11 = L22
550 GO TO 10
600 CONTINUE
    FAIL = .FALSE.
    RETURN

    ERROR RETURN.

900 CONTINUE
    FAIL = .TRUE.
    RETURN
    END
```

```
      SUBROUTINE DAD (A, NA, I1, I2, J1, J2, R,ISW)

          IF ISW = 0, SUBROUTINE DAD
          COMPUTES D*A WHERE D IS THE MATRIX WITH ONES DOWN
          THE MINOR DIAGONAL AND A IS THE INPUT MATRIX.
          IF ISW = 1, IT COMPUTES THE PRODUCT A*D.
          IT COMPUTES THIS PRODUCT FOR ROWS I1 THRU I2 AND COL-
          UMNS J1 THRU J2. IT ALSO MULTIPLIES EACH ELEMENT OF
          THE PRODUCT WITH THE CONSTANT R. NA IS THE FIRST
          DIMENSION OF THE MATRIX A.  THE PRODUCT
          OVERWRITES THE SPECIFIED LOCATIONS OF MATRIX A.

      REAL A(NA,1), R
      IF (ISW .EQ. 1) GO TO 200
      IF (I1 .EQ. I2) GO TO 150

      NRD2 = IFIX((I2 - I1 + 1) / 2
      DO 100 J = J1,J2
        DO 50 IP1 = 1,NRD2
          I = IP1 - 1
          TEMP = A(I1+I,J)
          A(I1+I,J) = A(I2-I,J) * R
          A(I2-I,J) = TEMP * R
50      CONTINUE
100   CONTINUE
      IF (MOD(I2-I1,2) .EQ. 1) RETURN
      I = I1 + NRD2
      DO 110 J=J1,J2
        A(I,J) = A(I,J) * R
110   CONTINUE
      RETURN
150   CONTINUE
      DO 160 J = J1,J2
        A(I1,J) = A(I1,J) * R
160   CONTINUE
      RETURN

          COMPUTES THE PRODUCT AD.WHERE D IS AS ABOVE.


200   CONTINUE
      IF (J1 .EQ. J2) GO TO 350
      NCD2 = IFIX((J2 - J1 + 1) / 2    )
      DO 300 JP1 = 1,NCD2
        J = JP1 - 1
        DO 250 I = I1,I2
          TEMP = A(I,J1+J)
          A(I,J1+J) = A(I,J2-J) * R
          A(I,J2-J) = TEMP * R
250     CONTINUE
300   CONTINUE
      IF ( MOD(J2-J1,2) .EQ. 1) RETURN
      J = J1 + NCD2
      DO 310 I=I1,I2
        A(I,J) = A(I,J) * R
310   CONTINUE
      RETURN
350   CONTINUE
      DO 360 I = I1,I2
        A(I,J1) = A(I,J1) * R
360   CONTINUE
      RETURN
      END
```

```
      SUBROUTINE SHRSLV (A,B,C,M,N,NA,NB,NC,RMAX,FAIL)

    SHRSLV IS A FORTRAN IV SUBROUTINE TO SOLVE THE REAL MATRIX
    EQUATION AX + XB = C, WH RE A IS IN LOWER REAL SCHUR FORM
    AND B IS IN UPPER REAL SCHUR FORM.  SHRSLV USES THE AUX-
    ILLIARY SUBROUTINE SYSSLV, WHICH IT COMMUNICATES WITH THROUGH
    THE COMMON BLOCK SLVBLK.  THE PARAMETERS IN THE CALLING
    SEQUENCE ARE

    A         A DOUBLY SUBSCRIPTED ARRAY CONTAING THE MATRIX A IN
              LOWER SCHUR FORM

    B         A DOUBLY SUBSCRIPTED ARRAY CONTAINING THE MATRIX B
              IN UPPER REAL SCHUR FORM

    C         A DOUBLY SUBSCRIPTED ARRAY CONTAINING THE MATRIX C.

    M         THE ORDER OF THE MATRIX A

    N         THE ORDER OF THE MATRIX B

    NA        THE FIRST DIMENSION OF THE ARRAY A

    NB        THE FIRST DIMENSION OF THE ARRAY B

    NC        THE FIRST DIMENSION OF THE ARRAY C

    RMAX     MAXIMUM ALLOWED SIZE OF ANY ELEMENT OF THE TRANSFORMATICN

    FAIL   INDICATES IF SHRSLV FAILED


      INTEGER M,N,NA,NB,NC,K,KM1,DK,KK,L,LM1,DL,LL,I,IB,J,JA,NSYS
      REAL A(NA,1), B(NB,1), C(NC,1), T, P
      LOGICAL FAIL, SING
      COMMON/SLVBLK/ T(5,5),P(5),NSYS,SING
      FAIL =  TRUE.
      L = 1
 10   LM1 = L - 1
      DL = 1
      IF (L .EQ. N) GO TO 15
      IF (B(L+1,L) .NE. 0.) DL = 2
 15   LL = L+DL-1
      IF (L .EQ. 1) GO TO 30
      DO 20 J = L,LL
        DO 20 I = 1,M
          DO 20 IB = 1,LM1
            C(I,J) = C(I,J) - C(I,IB)*B(IB,J)
 20   CONTINUE
 30   K = 1
 40   KM1 = K - 1
      DK = 1
      IF (K .EQ. M) GO TO 45
      IF (A(K,K+1) .NE. 0.) DK = 2
 45   KK = K+DK-1
      IF (K .EQ. 1) GO TO 60
      DO 50 I=K,KK
        DO 50 J=L,LL
          DO 50 JA=1,KM1
            C(I,J) = C(I,J) - A(I,JA)*C(JA,J)
 50   CONTINUE
 60   IF (DL .EQ. 2) GO TO 80
      IF (DK .EQ. 2) GO TO 70
      T(1,1) = A(K,K) + B(L,L)
      IF (T(1,1) .EQ. 0.) RETURN
      C(K,L) = C(K,L)   T(1,1)
      IF (ABS(C(K,L)) .GE. RMAX) RETURN
      GO TO 100
 70   T(1,1) = A(K,K) + B(L,L)
```

```
      T(1,2) = A(K,KK)
      T(2,1) = A(KK,K)
      T(2,2) = A(KK,KK) + B(L,L)
      P(1) = C(K,L)
      P(2) = C(KK,L)
      NSYS = 2
      CALL SYSSLV
      IF (SING) RETURN
      C(K,L) = P(1)
      IF (ABS(C(K,L)) .GE. RMAX) RETURN
      C(KK,L) = P(2)
      IF (ABS(C(KK,L)) .GE. RMAX) RETURN
      GO TO 100
 80   IF (DK .EQ. 2) GO TO 90
      T(1,1) = A(K,K) + B(L,L)
      T(1,2) = B(LL,L)
      T(2,1) = B(L,LL)
      T(2,2) = A(K,K) + B(LL,LL)
      P(1) = C(K,L)
      P(2) = C(K,LL)
      NSYS = 2
      CALL SYSSLV
      IF (SING) RETURN
      C(K,L) = P(1)
      IF (ABS(C(K,L)) .GE. RMAX) RETURN
      C(K,LL) = P(2)
      IF (ABS(C(K,LL)) .GE. RMAX) RETURN
      GO TO 100
 90   T(1,1) = A(K,K) + B(L,L)
      T(1,2) = A(K,KK)
      T(1,3) = B(LL,L)
      T(1,4) = 0.
      T(2,1) = A(KK,K)
      T(2,2) = A(KK,KK) + B(L,L)
      T(2,3) = 0.
      T(2,4) = T(1,3)
      T(3,1) = B(L,LL)
      T(3,2) = 0.
      T(3,3) = A(K,K) +  (LL,LL)
      T(3,4) = T(1,2)
      T(4,1) = 0.
      T(4,2) = T(3,1)
      T(4,3) = T(2,1)
      T(4,4) = A(KK,KK) + B(LL,LL)
      P(1) = C(K,L)
      P(2) = C(KK,L)
      P(3) = C(K,LL)
      P(4) = C(KK,LL)
      NSYS = 4
      CALL SYSSLV
      IF (SING) RETURN
      C(K,L) = P(1)
      IF (ABS(C(K,L)) .GE. RMAX) RETURN
      C(KK,L) = P(2)
      IF (ABS(C(KK,L)) .GE. RMAX) RETURN
      C(K,LL) = P(3)
      IF (ABS(C(K,LL)) .GE. RMAX) RETURN
      C(KK,LL) = P(4)
      IF (ABS(C(KK,LL)) .GE. RMAX) RETURN
 100  K = K + DK
      IF (K .LE. M) GO TO 40
      L = L + DL
      IF (L .LE. N) GO TO 10
      FAIL = .FALSE.
      RETURN
      END
```

```
      SUBROUTINE SYSSLV
C
C     SYSSLV IS A FORTRAN IV SUBROUTINE THAT SOLVES THE LINEAR
C     SYSTEM AX = B OF ORDER N LESS THAN 5 BY CROUT REDUCTION
C     FOLLOWED BY BLOCK SUBSTITUTION.  THE MATRIX A, THE VECTOR B,
C     AND THE ORDER N ARE CONTAINED IN THE ARRAYS A,B, AND THE
C     VARIABLE N OF THE COMMON BLOCK SLVBLK.  THE SOLUTION IS RETURNED
C     IN THE ARRAY B.
C
      COMMON /SLVBLK/ A(5,5), B(5), N, SING
      REAL MAX
      LOGICAL SING
      SING = .TRUE.
    1 NM1 = N - 1
      N1 = N + 1
C
C
C     COMPUTE THE LU FACTORIZATION OF A
C
      DO 80 K=1,N
         KM1 = K-1
         IF (K .EQ. 1) GO TO 20
         DO 10 I=K,N
            DO 10 J=1,KM1
               A(I,K) = A(I,K) - A(I,J)*A(J,K)
   10    CONTINUE
   20    IF (K .EQ. N) GO TO 100
         KP1 = K+1
         MAX = ABS(A(K,K))
         INTR = K
         DO 30 I=KP1,N
            AA = ABS(A(I,K))
            IF (AA .LE. MAX) GO TO 30
            MAX = AA
            INTR = I
   30    CONTINUE
         IF (MAX .EQ. 0.) RETURN
         A(N1,K) = INTR
         IF (INTR .EQ. K) GO TO 50
         DO 40 J=1,N
            TEMP = A(K,J)
            A(K,J) = A(INTR,J)
            A(INTR,J) = TEMP
   40    CONTINUE
   50    DO 80 J=KP1,N
            IF (K .EQ. 1) GO TO 70
            DO 60 I=1,KM1
               A(K,J) = A(K,J) - A(K,I)*A(I,J)
   60       CONTINUE
   70       A(K,J) = A(K,J) / A(K,K)
   80 CONTINUE
C
C     INTERCHANGE THE COMPONENTS OF B
C
  100 DO 110 J=1,NM1
         INTR = A(N1,J)
         IF (INTR .EQ. J) GO TO 110
         TEMP = B(J)
         B(J) = B(INTR)
         B(INTR) = TEMP
  110 CONTINUE
C
C     SOLVE SY = B
C
  200 B(1) = B(1) / A(1,1)
      DO 220 I=2,N
         IM1 = I-1
         DO 210 J=1,IM1
            B(I) = B(I) - A(I,J)*B(J)
  210       CONTINUE
```

```
      B(I) = B(I) / A(I,I)
220 CONTINUE

C   SOLVE UX=Y

300 DO 310 II=1,NM1
      I = NM1-II+1
      I1 = I+1
      DO 310 J=I1,N
        B(I) = B(I) - A(I,J)*B(J)
310 CONTINUE
    SING = .FALSE.
    RETURN
    END
```

```
      SUBROUTINE HQR3(A,V,N,NLOW,NUP,EPS,ER,EI,TYPE,NA,NV)

      INTEGER N,NA,NLOW,NUP,NV,TYPE(N)
      REAL A(NA,N),EI(N),ER(N),EPS,V(NV,N)

      HQR3 REDUCES THE UPPER HESSENBERG MATRIX A TO QUASI-
      TRIANGULAR FORM BY UNITARY SIMILARITY TRANSFORMATIONS.
      THE EIGENVALUES OF A, WHICH ARE CONTAINED IN THE 1X1
      AND 2X2 DIAGONAL BLOCKS OF THE REDUCED MATRIX, ARE
      ORDERED IN DESCENDING ORDER OF MAGNITUDE ALONG THE
      DIAGONAL.  THE TRANSFORMATIONS ARE ACCUMULATED IN THE
      ARRAY V.  HQR3 REQUIRES THE SUBROUTINES EXCHNG,
      QRSTEP, AND SPLIT  THE PARAMETERS IN THE CALLING
      SEQUENCE ARE (STARRED PARAMETERS ARE ALTERED BY THE
      SUBROUTINE)

         *A          AN ARRAY THAT INITIALLY CONTAINS THE N X N
                     UPPER HESSENBERG MATRIX TO BE REDUCED.  ON
                     RETURN A CONTAINS THE REDUCED, QUASI-
                     TRIANGULAR MATRIX.
         *V          AN ARRAY THAT CONTAINS A MATRIX INTO WHICH
                     THE REDUCING TRANSFORMATIONS ARE TO BE
                     MULTIPLIED.
          N          THE ORDER OF THE MATRICES A AND V.
          NLOW       A(NLOW-1,NLOW) AND A(NUP,NUP+U) ARE
          NUP        ASSUMED TO BE ZERO, AND ONLY ROWS NLOW
                     THROUGH NUP AND COLUMNS NLOW THROUGH
                     NUP ARE TRANSFORMED, RESULTING IN THE
                     CALCULATION OF EIGENVALUES NLOW
                     THROUGH NUP.
          EPS        A CONVERGENCE CRITERION.
         *ER         AN ARRAY THAT ON RETURN CONTAINS THE REAL
                     PARTS OF THE EIGENVALUES.
         *EI         AN ARRAY THAT ON RETURN CONTAINS THE
                     IMAGINARY PARTS OF THE EIGENVALUES.
         *TYPE       AND INTEGER ARRAY WHOSE I-TH ENTRY IS
                        0    IF THE I-TH EIGENVALUE IS REAL,
                        1    IF THE I-TH EIGENVALUE IS COMPLEX
                             WITH POSITIVE IMAGINARY PART.
                        2    IF THE I-TH EIGENVALUE IS COMPLEX
                             WITH NEGATIVE IMAGINARY PART,
                       -1    IF THE I-TH EIGENVALUE WAS NOT
                             CALCULATED SUCCESSFULLY.
          NA         THE FIRST DIMENSION OF THE ARRAY A.
          NV         THE FIRST DIMENSION OF THE ARRAY V.

      INTERNAL VARIABLES

      INTEGER I,IT,L,MU,NL,NU
      REAL E1,E2,P,Q,R,S,T,W,X,Y,Z
      LOGICAL FAIL

      INITIALIZE.

      DO 10 I=NLOW,NUP
         TYPE(I) = -1
   10 CONTINUE
      T = 0.

      MAIN LOOP. FIND AND ORDER EIGENVALUES.

      NU = NUP
  100 IF(NU .LT. NLOW) GO TO 500
         IT = 0

         QR LOOP.  FIND NEGLIGABLE ELEMENTS AND PERFORM
         QR STEPS.

  110    CONTINUE
```

```
C     SEARCH BACK FOR NEGLIGABLE ELEMENTS.
C
      L = NU
  120 CONTINUE
        IF(L .EQ. NLOW) GO TO 130
        IF(ABS(A(L,L-1)) .LT. EPS*(ABS(A(L-1,L-1))+ABS(A(L,L))))
     1    GO TO 130
        L = L-1
      GO TO 120
  130 CONTINUE
C
C     TEST TO SEE IF AN EIGENVALUE OR A 2X2 BLOCK
C     HAS BEEN FOUND.
C
      X = A(NU,NU)
      IF(L .EQ. NU) GO TO 300
      Y = A(NU-1,NU-1)
      W = A(NU,NU-1)*A(NU-1,NU)
      IF(L   .EQ. NU-1) GO TO 200
C
C     TEST ITERATION COUNT. IF IT IS 30 QUIT.  IF
C     IT IS 10 OR 20 SET UP AN AD-HOC SHIFT.
C
      IF(IT .EQ. 60) GO TO 500
      IF(IT.NE.40 .AND. IT.NE.50) GO TO 150
C
C       AD-HOC SHIFT.
C
        T = T + X
        DO 140 I=NLOW,NU
          A(I,I) = A(I,I) - X
  140   CONTINUE
        S = ABS(A(NU,NU-1)) + ABS(A(NU-1,NU-2))
        X = 0.75*S
        Y = X
        W = -0.4375*S**2
  150 CONTINUE
      IT = IT + 1
C
C     LOOK FOR TWO CONSECUTIVE SMALL SUB-DIAGONAL
C     ELEMENTS.
C
      NL= NU-2
  160 CONTINUE
        Z =  A(NL,NL)
        R = X - Z
        S = Y - Z
        P = (R*S-W)/A(NL+1,NL) + A(NL,NL+1)
        Q = A(NL+1,NL+1) - Z - R - S
        R = A(NL+2,NL+1)
        S = ABS(P) + ABS(Q) + ABS(R)
        P = P/S
        Q = Q/S
        R = R/S
        IF(NL .EQ. L) GO TO 170
        IF(ABS(A(NL,NL-1))*(ABS(Q)+ABS(R)) .LE.
     1      EPS*ABS(P)*(ABS(A(NL-1,NL-1))+ABS(Z)+ABS(A(NL+1,NL+1))))
     2      GO TO 170
        NL = NL-1
      GO TO 160
  170 CONTINUE
C
C     PERFORM A QR STEP BETWEEN NL AND NU.
C
      CALL QRSTEP(A,V,P,Q,R,NL,NU,N,NA,NV)
      GO TO 110
C
C     2X2 BLOCK FOUND.
C
  200 IF(NU .NE. NLOW+1) A(NU-1,NU-2) = 0.
```

```
      A(NU,NU) = A(NU,NU) + T
      A(NU-1,NU-1) = A(NU-1,NU-1) + T
      TYPE(NU) = 0
      TYPE(NU-1) = 0
      MU = NU

      LOOP TO POSITION  2X2 BLOCK.

210   CONTINUE
        NL = MU-1

        ATTEMPT  TO SPLIT THE BLOCK INTO TWO REAL
        EIGENVALUES.

        CALL SPLIT(A,V,N,NL,E1,E2,NA,NV)

        IF THE SPLIT WAS SUCCESSFUL, GO AND ORDER THE
        REAL EIGENVALUES.

        IF(A(MU,MU-1) .EQ. 0.) GO TO 310

        TEST TO SEE IF THE BLOCK IS PROPERLY POSITIONED,
        AND IF NOT EXCHANGE IT

        IF(MU .EQ. NUP) GO TO 400
        IF(MU .EQ. NUP-1) GO TO 220
        IF(A(MU+2,MU+1) .EQ. 0.) GO TO 220

          THE NEXT BLOCK IS 2X2.

          IF(A(MU-1,MU-1)*A(MU,MU)-A(MU-1,MU)*A(MU,MU-1)
     1       .GE. A(MU+1,MU+1)*A(MU+2,MU+2)-A(MU+1,MU+2)*
     2            A(MU+2,MU+1))
     3       GO TO 400
          CALL EXCHNG(A,V,N,NL,2,2,EPS,FAIL,NA,NV)
          IF(.NOT. FAIL) GO TO 215
            TYPE(NL) = -1
            TYPE(NL+1) = -1
            TYPE(NL+2) = -1
            TYPE(NL+3) = -1
            GO TO 500
215       CONTINUE
          MU = MU+2
          GO TO 230
220     CONTINUE

          THE NEXT BLOCK IS 1X1.

          IF(A(MU-1,MU-1)*A(MU,MU)-A(MU-1,MU)*A(MU,MU-1)
     1       .GE. A(MU+1,MU+1)**2)
     2       GO TO 400
          CALL EXCHNG(A,V,N,NL,2,1,EPS,FAIL,NA,NV)
          IF(.NOT. FAIL) GO TO 225
            TYPE(NL) = -1
            TYPE(NL+1) = -1
            TYPE(NL+2) = -1
            GO TO 500
225       CONTINUE
          MU = MU+1
230     CONTINUE
      GO TO 210

      SINGLE EIGENVALUE FOUND.

300   NL = 0
      A(NU,NU) = A(NU,NU) + T
      IF(NU .NE. NLOW) A(NU,NU-1) = 0.
      TYPE(NU) = 0
      MU = NU
```

```
C     LOOP TO POSITION ONE OR TWO REAL EIGENVALUES.
C
  310 CONTINUE
C
C       POSITION THE EIGENVALUE LOCATED AT A(NL,NL).
C
  320   CONTINUE
          IF(MU .EQ. NUP) GO TO 350
          IF(MU .EQ. NUP-1) GO TO 330
          IF(A(MU+2,MU+1) .EQ. 0.) GO TO 330
C
C           THE NEXT BLOCK IS 2X2.
C
            IF(A(MU,MU)**2 .GE.
     1         A(MU+1,MU+1)*A(MU+2,MU+2)-A(MU+1,MU+2)*A(MU+2,MU+1))
     2        GO TO 400
            CALL EXCHNG(A,V,N,MU,1,2,EPS,FAIL,NA,NV)
          IF(.NOT. FAIL) GO TO 325
            TYPE(MU) = -1
            TYPE(MU+1) = -1
            TYPE(MU+2) = -1
            GO TO 500
  325     CONTINUE
            MU = MU+2
            GO TO 340
  330     CONTINUE
C
C           THE NEXT BLOCK IS 1X1.
C
            IF(ABS(A(MU,MU)) .GE. ABS(A(MU+1,MU+1)))
     1        GO TO 350
            CALL EXCHNG(A,V,N,MU,1,1,EPS,FAIL,NA,NV)
            MU = MU+1
  340     CONTINUE
        GO TO 320
  350   CONTINUE
        MU = NL
        NL = 0
      IF(MU .NE. 0) GO TO 310
C
C       GO BACK AND GET THE NEXT EIGENVALUE.
C
  400   CONTINUE
       NU = L-1
     GO TO 100
C
C     ALL THE EIGNVALUES HAVE BEEN FOUND AND ORDERED.
C     COMPUTE THEIR VALUES AND TYPE.
C
  500 IF(NU .LT. NLOW) GO TO 507
        DO 503 I=1,NU
          A(I,I) = A(I,I) + T
  503   CONTINUE
  507 CONTINUE
      NU = NUP
  510 CONTINUE
        IF(TYPE(NU) .NE. -1) GO TO 515
          NU = NU-1
          GO TO 540
  515   CONTINUE
        IF(NU .EQ. NLOW) GO TO 520
        IF(A(NU,NU-1) .EQ. 0.) GO TO 520
C
C         2X2 BLOCK.
C
          CALL SPLIT(A,V,N,NU-1,E1,E2,NA,NV)
          IF(A(NU,NU-1) .EQ. 0.) GO TO 520
          ER(NU) = E1
          EI(NU-1) = E2
          ER(NU-1) = ER(NU)
```

```
         EI(NU) = -EI(NU-1)
         TYPE(NU-1) = 1
         TYPE(NU) = 2
         NU = NU-2
         GO TO 530
520   CONTINUE

         SINGLE ROOT.

         ER(NU) = A(NU,NU)
         EI(NU) = 0.
         NU = NU-1
530   CONTINUE
540 CONTINUE
    IF(NU .GE. NLOW) GO TO 510
    RETURN
    END
```

```
      SUBROUTINE EXCHNG(A,V,N,L,B1,B2,EPS,FAIL,NA,NV)

      INTEGER B1,B2,L,NA,NV
      REAL A(NA,N),EPS,V(NV,N)
      LOGICAL FAIL

      GIVEN THE UPPER HESSENBERG MATRIX A WITH CONSECUTIVE
      B1XB1 AND B2XB2 DIAGONAL BLOCKS (B1,B2 .LE. 2)
      STARTING-AT A(L,L), EXCHNG PRODUCES A UNITARY
      SIMILARITY TRANSFORMATION THAT EXCHANGES THE BLOCKS
      ALONG WITH THEIR EIGENVALUES.  THE TRANSFORMATION
      IS ACCUMULATED IN V.  EXCHNG REQUIRES THE SUBROUTINE
      QRSTEP.  THE PARAMETERS IN THE CALLING SEQUENCE ARE
      (STARRED PARAMETERS ARE ALTERED BY THE SUBROUTINE)

         *A        THE MATRIX WHOSE BLOCKS ARE TO BE
                   INTERCHANGED.
         *V        THE ARRAY INTO WHICH THE TRANSFORMATIONS
                   ARE TO BE ACCUMULATED.
          N        THE ORDER OF THE MATRIX A.
          L        THE POSITION OF THE BLOCKS.
          B1       THE SIZE OF THE FIRST BLOCK.
          B2       THE SIZE OF THE SECOND BLOCK.
          EPS      A CONVERGENCE CRITERION.
         *FAIL     A LOGICAL VARIABLE WHICH IS FALSE ON A
                   NORMAL RETURN.  IF THIRTY ITERATIONS WERE
                   PERFORMED WITHOUT CONVERGENCE, FAIL IS SET
                   TO TRUE AND THE ELEMENT
                   A(L+B2,L+B2-1) CANNOT BE ASSUMED ZERO.
          NA       THE FIRST DIMENSION OF THE ARRAY A.
          NV       THE FIRST DIMENSION OF THE ARRAY V.

      INTERNAL VARIABLES.

      INTEGER I,IT,J,L1,M
      REAL P,Q,R,S,W,X,Y,Z

      FAIL = .FALSE.
      IF(B1 .EQ. 2) GO TO 40
        IF(B2 .EQ. 2) GO TO 10

          INTERCHANGE 1X1 AND 1X1 BLOCKS.

          L1 = L+1
          Q = A(L+1,L+1) - A(L,L)
          P = A(L,L+1)
          R = AMAX1(P,Q)
          IF(R .EQ. 0.) RETURN
          P = P/R
          Q = Q/R
          R = SQRT(P**2 + Q**2)
          P = P/R
          Q = Q/R
          DO 3 J=L,N
            S = P*A(L,J) + Q*A(L+1,J)
            A(L+1,J) = P*A(L+1,J) - Q*A(L,J)
            A(L,J) = S
 3        CONTINUE
          DO 5 I=1,L1
            S = P*A(I,L) + Q*A(I,L+1)
            A(I,L+1) = P*A(I,L+1) - Q*A(I,L)
            A(I,L) = S
 5        CONTINUE
          DO 7 I=1,N
            S = P*V(I,L) + Q*V(I,L+1)
            V(I,L+1) = P*V(I,L+1) - Q*V(I,L)
            V(I,L) = S
 7        CONTINUE
          A(L+1,L) = 0.
          RETURN
```

```
10    CONTINUE

         INTERCHANGE 1X1 AND 2X2 BLOCKS.

         X = A(L,L)
         P = 1.
         Q = 1.
         R = 1.
         CALL QRSTEP(A,V,P,Q,R,L,L+2,N,NA,NV)
         IT = 0
20       IT = IT+1
           IF(IT .LE. 60) GO TO 30
             FAIL = .TRUE.
             RETURN
30         CONTINUE
           P = A(L,L) - X
           Q = A(L+1,L)
           R = 0.
           CALL QRSTEP(A,V,P,Q,R,L,L+2,N,NA,NV)
           IF(ABS(A(L+2,L+1)) .GT.
  1                             EPS*(ABS(A(L+1,L+1))+ABS(A(L+2,L+2))))
  1          GO TO 20
         A(L+2,L+1) = 0.
         RETURN
40 CONTINUE

       INTERCHANGE 2X2 AND B2XB2 BLOCKS.

       M = L+2
       IF(B2 .EQ. 2) M = M+1
       X = A(L+1,L+1)
       Y = A(L,L)
       W = A(L+1,L)*A(L,L+1)
       P = 1.
       Q = 1.
       R = 1.
       CALL QRSTEP(A,V,P,Q,R,L,M,N,NA,NV)
       IT = 0
50     IT = IT+1
         IF(IT .LE. 60) GO TO 60
           FAIL = .TRUE.
           RETURN
60       CONTINUE
         Z = A(L,L)
         R = X - Z
         S = Y - Z
         P = (R*S-W)/A(L+1,L) + A(L,L+1)
         Q = A(L+1,L+1) - Z - R - S
         R = A(L+2,L+1)
         S = ABS(P) + ABS(Q) + ABS(R)
         P = P/S
         Q = Q/S
         R = R/S
         CALL QRSTEP(A,V,P,Q,R,L,M,N,NA,NV)
       IF(ABS(A(M-1,M-2)) .GT. EPS*(ABS(A(M-1,M-1))+ABS(A(M-2,M-2))))
  1      GO TO 50
       A(M-1,M-2) = 0.
       RETURN
     CONTINUE
     END
```

```
      SUBROUTINE SPLIT(A,V,N,L,E1,E2,NA,NV)

      INTEGER L,N,NA,NV
      REAL A(NA,N),V(NV,N)

      GIVEN THE UPPER HESSENBERG MATRIX A WITH A 2X2 BLOCK
      STARTING AT A(L,L), SPLIT DETERMINES IF THE
      CORRESPONDING EIGENVALUES ARE REAL OR COMPLEX. IF THEY
      ARE REAL, A ROTATION IS DETERMINED THAT REDUCES THE
      BLOCK TO UPPER TRIANGULAR FORM WITH THE EIGENVALUE
      OF LARGEST ABSOLUTE VALUE APPEARING FIRST.  THE
      ROTATION IS ACCUMULATED IN V.  THE EIGENVALUES (REAL
      OR COMPLEX) ARE RETURNED IN E1 AND E2.  THE PARAMETERS
      IN THE CALLING SEQUENCE ARE (STARRED PARAMETERS ARE
      ALTERED BY THE SUBROUTINE)

         *A         THE UPPER HESSENVERG MATRIX WHOSE 2X2
                    BLOCK IS TO BE SPLIT.
         *V         THE ARRAY IN WHICH THE SPLITTING TRANS-
                    FORMATION IS TO BE ACCUMULATED.
          N         THE ORDER OF THE MATRIX A.
          L         THE POSITION OF THE 2X2 BLOCK.
         *E1        ON RETURN IF THE EIGENVALUES ARE COMPLEX
         *E2        E1 CONTAINS THEIR COMMON REAL PART AND
                    E2 CONTAINS THE POSITIVE IMAGINARY PART.
                    IF THE EIGENVALUES ARE REAL, E1 CONTAINS
                    THE ONE LARGEST IN ABSOLUTE VALUE AND E2
                    CONTAINS THE OTHER ONE.
          NA        THE FIRST DIMENSION OF THE ARRAY A.
          NV        THE FIRST DIMENSION OF THE ARRAY V.

      INTERNAL VARIABLES

      INTEGER I,J,L1
      REAL P,Q,R,T,U,W,X,Y,Z

      X = A(L+1,L+1)
      Y = A(L,L)
      W = A(L,L+1)*A(L+1,L)
      P = (Y-X)/2.
      Q = P**2 + W
      IF(Q .GE. 0.) GO TO 5

         COMPLEX EIGENVALUE.

         E1 = P + X
         E2 = SQRT(-Q)
         RETURN
    5 CONTINUE

      TWO REAL EIGENVALUES.  SET UP TRANSFORMATION.

      Z = SQRT(Q)
      IF(P .LT. 0.) GO TO 10
         Z = P + Z
         GO TO 20
   10 CONTINUE
         Z = P - Z
   20 CONTINUE
      IF(Z .EQ. 0.) GO TO 30
         R = -W/Z
         GO TO 40
   30 CONTINUE
         R = 0.
   40 CONTINUE
      IF(ABS(X+Z) .GE. ABS(X+R)) Z = R
      Y = Y - X - Z
      X = -Z
      T = A(L,L+1)
      U = A(L+1,L)
```

```
      IF(ABS(Y)+ABS(U) .LE. ABS(T)+ABS(X)) GO TO 60
         Q = U
         P = Y
         GO TO 70
   60 CONTINUE
         Q = X
         P = T
   70 CONTINUE
      R = SQRT(P**2 + Q**2)
      IF(R .GT. 0.) GO TO 80
      E1 = A(L,L)
      E2 = A(L+1,L+1)
         A(L+1,L) = 0.
         RETURN
   80 CONTINUE
      P = P/R
      Q = Q/R

      PREMULTIPLY.

      DO 90 J=L,N
         Z = A(L,J)
         A(L,J) = P*Z + Q*A(L+1,J)
         A(L+1,J) = P*A(L+1,J) - Q*Z
   90 CONTINUE

      POSTMULTIPLY.

      L1 = L+1
      DO 100 I=1,L1
         Z = A(I,L)
         A(I,L) = P*Z + Q*A(I,L+1)
         A(I,L+1) = P*A(I,L+1) - Q*Z
  100 CONTINUE

      ACCUMULATE THE TRANSFORMATION IN V.

      DO 110 I=1,N
         Z = V(I,L)
         V(I,L) = P*Z + Q*V(I,L+1)
         V(I,L+1) = P*V(I,L+1) - Q*Z
  110 CONTINUE
      A(L+1,L) = 0.
      E1 = A(L,L)
      E2 = A(L+1,L+1)
      RETURN
      END
```

```
      SUBROUTINE QRSTEP(A,V,P,Q,R,NL,NU,N,NA,NV)

      INTEGER N,NA,NL,NU,NV
      REAL A(NA,N),P,Q,R,V(NV,N)

      QRSTEP PERFORMS ONE IMPLICIT QR STEP ON THE
      UPPER HESSENBERG MATRIX A.  THE SHIFT IS DETERMINED
      BY THE NUMBERS P,Q, AND R, AND THE STEP IS APPLIED TO
      ROWS AND COLUMNS NL THROUGH NU.  THE TRANSFORMATIONS
      ARE ACCUMULATED IN V.  THE PARAMETERS IN THE CALLING
      SEQUENCE ARE (STARRED APRAMETERS ARE ALTERED BY THE
      SUBROUTINE)

         *A         THE UPPER HESSENBERG MATRIX ON WHICH THE
                    QR STEP IS TO BE PERFORMED.
         *V         THE ARRAY IN WHICH THE TRANSFORMATIONS
                    ARE TO BE ACCUMULATED
         *P         PARAMETERS THAT DETERMINE THE SHIFT.
         *Q
         *R
          NL        THE LOWER LIMIT OF THE STEP.
          NU        THE UPPER LIMIT OF THE STEP.
          N         THE ORDER OF THE MATRIX A.
          NA        THE FIRST DIMENSION OF THE ARRAY A.
          NV        THE FIRST DIMENSION OF THE ARRAY V.

      INTERNAL VARIABLES.

      INTEGER I,J,K,NL2,NL3,NUM1
      REAL S,X,Y,Z
      LOGICAL LAST

      NL2 = NL+2
      DO 10 I=NL2,NU
        A(I,I-2) = 0.
   10 CONTINUE
      IF(NL2 .EQ. NU) GO TO 30
        NL3 = NL+3
        DO 20 I=NL3,NU
          A(I,I-3) = 0.
   20   CONTINUE
   30 CONTINUE
      NUM1 = NU-1
      DO 130 K=NL,NUM1

        DETERMINE THE TRANSFORMATION.

        LAST = K .EQ. NUM1
        IF(K .EQ. NL) GO TO 40
          P = A(K,K-1)
          Q = A(K+1,K-1)
          R = 0.
          IF(.NOT.LAST) R = A(K+2,K-1)
          X = ABS(P) + ABS(Q) + ABS(R)
          IF(X .EQ. 0.) GO TO 130
          P = P/X
          Q = Q/X
          R = R/X
   40   CONTINUE
        S = SQRT(P**2 + Q**2 + R**2)
        IF(P .LT. 0.) S = -S
        IF(K .EQ. NL) GO TO 50
          A(K,K-1) = -S*X
          GO TO 60
   50   CONTINUE
          IF(NL .NE. 1) A(K,K-1) = -A(K,K-1)
   60   CONTINUE
        P = P + S
        X = P/S
        Y = Q/S
```

```
      Z = R/S
      Q = Q/P
      R = R/P

      PREMULTIPLY.

      DO 80 J=K,N
        P = A(K,J) + Q*A(K+1,J)
        IF(LAST) GO TO 70
          P = P + R*A(K+2,J)
          A(K+2,J) = A(K+2,J) - P*Z
   70   CONTINUE
        A(K+1,J) = A(K+1,J) - P*Y
        A(K,J) = A(K,J) - P*X
   80 CONTINUE

      POSTMULTIPLY.

      J = MINO(K+3,NU)
      DO 100 I=1,J
        P = X*A(I,K) + Y*A(I,K+1)
        IF(LAST) GO TO 90
          P = P + Z*A(I,K+2)
          A(I,K+2) = A(I,K+2) - P*R
   90   CONTINUE
        A(I,K+1) = A(I,K+1) - P*Q
        A(I,K) = A(I,K) - P
  100 CONTINUE

      ACCUMULATE THE TRANSFORMATION IN V.

      DO 120 I=1,N
        P = X*V(I,K) + Y*V(I,K+1)
        IF(LAST) GO TO 110
          P = P + Z*V(I,K+2)
          V(I,K+2) = V(I,K+2) - P*R
  110   CONTINUE
        V(I,K+1) = V(I,K+1) - P*Q
        V(I,K) = V(I,K) - P.
  120 CONTINUE
  130 CONTINUE
      RETURN
      END
```

```
      SUBROUTINE ORTHES(NM,N,LOW,IGH,A,ORT)
      INTEGER I,J,M,N,II,JJ,LA,MP,NM,IGH,KP1,LOW
      REAL A(NM,N),ORT(IGH)
      REAL F,G,H,SCALE
      LA = IGH - 1
      KP1 = LOW + 1
      IF(LA .LT. KP1) GO TO 200
      DO 180 M=KP1,LA
         H = 0.
         ORT(M) = 0.
         SCALE = 0.
         DO 90 I=1,IGH
 90      SCALE = SCALE + ABS(A(I,M-1))
         IF(SCALE .EQ. 0.) GO TO 180
         MP = M + IGH
         DO 100 II=M,IGH
            I = MP - II
            ORT(I) = A(I,M-1)/SCALE
            H = H + ORT(I)*ORT(I)
100      CONTINUE
         G = -SIGN(SQRT(H),ORT(M))
         H = H- ORT(M)*G
         ORT(M) = ORT(M) - G
         DO 130 J=M,N
            F = 0.
            DO 110 II=M,IGH
               I = MP - II
               F = F + ORT(I)*A(I,J)
110         CONTINUE
       IF (H .EQ.0.) GO TO 162
            F = F/H
            DO 120 I=M,IGH
               A(I,J) = A(I,J) - F*ORT(I)
120         CONTINUE
130      CONTINUE
         DO 160 I=1,IGH
            F = 0.
            DO 140 JJ=M,IGH
               J = MP - JJ
               F = F + ORT(J)*A(I,J)
140         CONTINUE
            F = F/H
            DO 150 J=M,IGH
               A(I,J) = A(I,J) - F*ORT(J)
150         CONTINUE
160      CONTINUE
162   CONTINUE
         ORT(M) = SCALE*ORT(M)
         A(M,M-1) = SCALE*G
180   CONTINUE
200   RETURN
      END
```

```
      SUBROUTINE ORTRAN(NM,N,LOW,IGH,A,ORT,Z)
      INTEGER I,J,N,KL,MM,MP,NM,IGH,LOW,MP1
      REAL A(NM,IGH),ORT(IGH),Z(NM,N)
      REAL G,H
      DO 80 I=1,N
        DO 60 J=1,N
          Z(I,J) = 0.
 60     CONTINUE
        Z(I,I) = 1.
 80   CONTINUE
      KL = IGH - LOW - 1
      IF(KL .LT. 1) GO TO 200
      DO 140 MM=1,KL
        MP = IGH - MM
        H = A(MP,MP-1)*ORT(MP)
        IF(H .EQ. 0.) GO TO 140
        MP1 = MP+1
        DO 100 I=MP1,IGH
          ORT(I) = A(I,MP-1)
100     CONTINUE
        DO 130 J=MP,IGH
          G = 0.
          DO 110 I=MP,IGH
            G = G + ORT(I)*Z(I,J)
110       CONTINUE
          G = G/H
          DO 120 I=MP,IGH
            Z(I,J) = Z(I,J) + G*ORT(I)
120       CONTINUE
130     CONTINUE
140   CONTINUE
200   RETURN
      END
```

16.5. [GWS-J75] (with R. Mathias) "A Block QR Algorithm and the Singular Value Decomposition"

[GWS-J75] (with R. Mathias) "A Block QR Algorithm and the Singular Value Decomposition," *Linear Algebra and its Applications* **182** (1993) 91–100.

A Block QR Algorithm and the Singular Value Decomposition

R. Mathias

Institute for Mathematics and its Applications
University of Minnesota
Minneapolis, Minnesota 55455

G. W. Stewart*

Department of Computer Science and
Institute for Advanced Computer Studies
University of Maryland
College Park, Maryland 20742

Submitted by Richard A. Brualdi

ABSTRACT

In this note we consider an iterative algorithm for moving a triangular matrix toward diagonality. The algorithm is related to algorithms for refining rank-revealing triangular decompositions, and in a variant form, to the QR algorithm. It is shown to converge if there is a sufficient gap in the singular values of the matrix, and the analysis provides a new approximation theorem for singular values and singular subspaces.

1. INTRODUCTION

Let R_0 be an $n \times n$ block-triangular matrix of the form

$$R_0 = \begin{pmatrix} S_0 & H_0 \\ 0 & E_0 \end{pmatrix}, \tag{1.1}$$

*This work was supported in part by the Air Force Office of Scientific Research under Contract AFOSR-87-0188.

655 Avenue of the Americas, New York, NY 10010 0024-3795/93/$6.00

where H_0 and E_0 are small compared to the smallest singular value of S_0. In this paper we will be concerned with the following two-stage iteration. For the first step, let Q_0 be a unitary matrix such that

$$R_1 \equiv R_0 Q_0 = \begin{pmatrix} S_1 & 0 \\ H_1 & E_1 \end{pmatrix} \tag{1.2}$$

is block lower triangular. Then let Q_1 be a unitary matrix such that

$$R_2 \equiv Q_1^{\mathrm{H}} R_1 = \begin{pmatrix} S_2 & H_2 \\ 0 & E_2 \end{pmatrix},$$

is block upper triangular, like R_1. The iteration is continued in the obvious way. Note that the matrices Q_0 and Q_1 are not unique; for example, Q_0 can be any unitary matrix of the form $Q = (Q_1\ Q_2)$, where the columns of Q_2 are orthogonal to the rows of $(S_0\ H_0)$.

This iteration arises in two connections. The one that motivated this paper is a refinement step in updating rank-revealing URV and ULV decompositions [3, 2]. Here H_0 is a vector and E_0 is a scalar and the purpose of the iteration is to make H small, so that R_2 is nearer a diagonal matrix.

The second connection is with a variant of the (unshifted) QR algorithm for Hermitian matrices. Specifically, suppose that in addition to the above requirements, we demand that $R_0, R_2, \ldots$ be upper triangular and that $R_1, R_3, \ldots$ be lower triangular. Then

$$A_0 \equiv R_0^{\mathrm{H}} R_0 = (R_0^{\mathrm{H}} R_1) Q^{\mathrm{H}}$$

is a factorization of the Hermitian matrix A_0 into the product of a lower triangular matrix and a unitary matrix — the first step of the LQ variant of the QR algorithm. If we perform the second step of the LQ algorithm by multiplying the factors in the reverse order, we get

$$\begin{aligned} Q^{\mathrm{H}}(R_0^{\mathrm{H}} R_1) &= R_1^{\mathrm{H}} R_1 \\ &= (R_1^{\mathrm{H}} P)(P^{\mathrm{H}} R_1) \\ &= R_2^{\mathrm{H}} R_2 \\ &\equiv A_2. \end{aligned}$$

Thus R_0 is the Cholesky factor of the Hermitian matrix A_0, and R_2 is the Cholesky factor of the matrix A_2 obtained by applying a step of the LQ algorithm to A_0. Since, under mild restrictions on A_0, the LQ algorithm converges to a diagonal matrix whose diagonal elements are the eigenvalues of A in descending order, the matrices $R_0, R_2, \ldots$ will converge to diagonal matrices whose diagonal elements are the singular values of R in descending order.

In this paper we will chiefly be concerned with the block variant of the algorithm, although our results will say something about the triangular LQ variant. In the next section we will analyze the convergence of the matrices H_i, an analysis which answers our concerns with the algorithm for refining rank-revealing decompositions. However, in the following section we will go on to show how our analysis can be applied to give a new approximation theorem for singular values and their associated subspaces.

Throughout the paper $\sigma_i(R)$ will denote the ith singular value of R in descending order. The quantity $\|R\|_2 = \sigma_1(R)$ is the spectral norm of R, $\|R\|$ denotes any unitarily invariant norm of R, and $\inf(R)$ is the smallest singular value of R.

We will later use the following lemma to obtain good relative bounds on all the singular values of R. It can be proved from the min-max characterization of singular values [1, Theorem 7.3.10], and a proof is outlined in [1, Problem 7.3.18].

LEMMA 1.1. *Let A and B be n by n matrices. Then*

$$\inf(A)\sigma_i(B) \leq \sigma_i(AB) \leq \|A\|_2\sigma_i(B).$$

This result can be used to prove that for any unitarily invariant norm

$$\|AB\| \leq \|A\|_2\|B\|. \tag{1.3}$$

See, for example, [1, Example 7.4.54] for a proof.

2. CONVERGENCE OF THE ITERATION

It turns out that the analysis of the passage from R_{2i} to R_{2i+1} of the refinement algorithm is *mutatis mutandis* the same as the analysis of the passage from R_{2i+1} to R_{2i+2}. We will therefore confine ourselves to the former, and in particular to the passage from R_0 to R_1. For notational convenience we will drop the subscripts and attach a prime to quantities associated with R_1.

Let

$$\begin{aligned} \epsilon &= \|E\|_2, \\ \eta &= \|H\|_2, \\ \gamma &= \inf(S), \end{aligned} \tag{2.1}$$

and assume that

$$\rho \equiv \frac{\epsilon}{\gamma} < 1. \tag{2.2}$$

Partition Q conformally with R and write

$$\begin{pmatrix} S & H \\ 0 & E \end{pmatrix} \begin{pmatrix} Q_{11} & Q_{12} \\ Q_{21} & Q_{22} \end{pmatrix} = \begin{pmatrix} S' & 0 \\ H' & E' \end{pmatrix}. \tag{2.3}$$

Now it is easily verified from the orthogonality of Q that

$$\|Q_{22}\|_2 \le 1$$

and that

$$Q_{21}^{\mathrm{H}} Q_{21} = I - Q_{11}^{\mathrm{H}} Q_{11} \quad \text{and} \quad Q_{12} Q_{12}^{\mathrm{H}} = I - Q_{11} Q_{11}^{\mathrm{H}}.$$

Since $I - Q_{11}^{\mathrm{H}} Q_{11}$ and $I - Q_{11} Q_{11}^{\mathrm{H}}$ have the same eigenvalues so do $Q_{21}^{\mathrm{H}} Q_{21}$ and $Q_{12} Q_{12}^{\mathrm{H}}$. Hence Q_{12} and Q_{21} have the same singular values, and so $\|Q_{12}\| = \|Q_{21}\|$ for any unitarily invariant norm. Consequently, from the equation

$$SQ_{12} + HQ_{22} = 0,$$

we obtain by two applications of (1.3)

$$\|Q_{21}\| = \|Q_{12}\| \le \|S^{-1}\|_2 \|H\| \|Q_{22}\|_2 \le \frac{\|H\|}{\gamma}.$$

It follows from the equation

$$H' = EQ_{21}$$

that

$$\|H'\| \le \frac{\epsilon \|H\|}{\gamma} = \rho \|H\|.$$

Since by assumption $\rho < 1$, the norm of H' is less than the norm of H by a factor of at least ρ.

But more is true. Let the quantities ϵ', η', and ρ' be defined in analogy with (2.1) and (2.2). We have already shown that $\eta' < \eta$. From (2.3) it follows that

$$E' = EQ_{22}, \tag{2.4}$$

and hence $\sigma_i(E') \le \sigma_i(E)$ by Lemma 1.1. In particular, $\epsilon' \le \epsilon$. Similarly, from $R' = RQ$ we have $R = R'Q^{\mathrm{H}}$, which implies

$$S' Q_{11}^{\mathrm{H}} = S. \tag{2.5}$$

Thus, $\sigma_i(S') \ge \sigma_i(S)$. From this it follows that $\rho' \le \rho < 1$. Since $\rho' < 1$, we may repeat the above argument to show that the passage from R_1 to R_2

will produce a matrix $H'' = H_2$ satisfying $\|H''\| \le \rho'\|H'\| \le \rho\|H'\|$; i.e., the left iteration reduces the norm of the off-diagonal block by at least ρ. The same is obviously true of subsequent iterations. Hence we have proved the following theorem. Here we drop the primes in favor of subscripts, with the convention that unadorned quantities refer to R_0.

THEOREM 2.1. *Let the matrices R_i $(i = 0, 1, \ldots)$ be partitioned in analogy with* (1.1) *or* (1.2) *according as i is even or odd. Assume that*

$$\rho \equiv \frac{\|E\|_2}{\inf(S)} < 1. \tag{2.6}$$

Then

$$\|H_i\| \le \rho^i\|H\|, \tag{2.7}$$
$$\sigma_j(S_{i+1}) \ge \sigma_j(S_i), \quad j = 1, \ldots, k, \tag{2.8}$$
$$\sigma_j(E_{i+1}) \le \sigma_j(E_i), \quad j = 1, \ldots, n-k. \tag{2.9}$$

The condition (2.6) is necessary; for if we start with the matrix

$$\begin{pmatrix} 1 & \eta \\ 0 & 1 \end{pmatrix},$$

then the first iteration produces the matrix

$$\begin{pmatrix} 1 & 0 \\ \eta & 1 \end{pmatrix},$$

and the next iteration restores the original matrix.

In practice one may not know $\inf(S)$ but may know $\sigma_k(R)$. In this case, one can still apply Theorem 2.1 since the theorem is true with ρ replaced by

$$\hat{\rho} = \frac{\|E\|_2}{\sigma_k(R) - \|H\|_2},$$

as we will now show. Suppose that $\hat{\rho} < 1$, then

$$\sigma_k(R) - \sigma_i(E) \ge \sigma_k(R) - \|E\|_2 > \|H\|_2. \tag{2.10}$$

We know that the singular values of R can be paired with those of S and E in such a way that the difference between the pairs is at most $\|H\|_2$. In view of (2.10), the k largest singular values of R must be paired with the singular values of S, and in particular, $|\sigma_i(R) - \sigma_k(S)| \le \|H\|_2$ for some $i \le k$. Thus, $\sigma_k(S) \ge \sigma_i(R) - \|H\|_2 \ge \sigma_k(R) - \|H\|_2$. Thus if $\hat{\rho} < 1$, then $\rho \le \hat{\rho} < 1$, and the theorem holds with ρ replaced by $\hat{\rho}$.

3. APPROXIMATION RESULTS

We now turn to the problem of assessing the accuracy of the singular values of

$$\hat{R} = \begin{pmatrix} S & 0 \\ 0 & E \end{pmatrix}$$

as approximations to singular values of R. We know from standard perturbation theory that they differ from singular values of R by quantities no greater than $\|H\|_2$. We will now show that under the condition (2.6) $\sigma_i(R)/\sigma_i(\hat{R}) = 1 + O(\|H\|_2^2)$.

The basic idea is to follow the iterates R_i of the iteration as the H_i approach zero. However, the approach is complicated by the fact that the R_i need not converge. Nonetheless, from the fact that $\|H_i\|_2 \to 0$ and from (2.8) and (2.9), we know that the singular values of S_i and E_i converge to those of R. Because

$$\sigma_k(S_i) \geq \sigma_k(S) > \sigma_1(E) \geq \sigma_1(E_i)$$

it follows that $\lim_{i\to\infty} \sigma_k(S_i) > \lim_{i\to\infty} \sigma_1(E_i)$, and hence

$$\begin{aligned} \sigma_j(R) &= \lim_{i\to\infty} \sigma_j(S_i), \qquad j = 1, \ldots, k, \\ \sigma_{k+j}(R) &= \lim_{i\to\infty} \sigma_j(E_i), \qquad j = 1, \ldots, n-k. \end{aligned}$$

We have shown in (2.4) that after one step of refinement $E_1 = EQ_{22}$. Since Q is unitary, $Q_{22}Q_{22}^{\mathrm{H}} = I - Q_{21}Q_{21}^{\mathrm{H}}$, and

$$\inf(Q_{22}) = (1 - \|Q_{21}\|_2^2)^{1/2} \geq (1 - (\eta/\gamma)^2)^{1/2}.$$

Hence, by Lemma 1.1,

$$\sigma_j(E) \geq \sigma_j(E_1) \geq \sigma_j(E)(1 - (\eta/\gamma)^2)^{1/2}.$$

Iterating this argument and recalling that $\|H_i\|_2 \leq \rho^i\eta$, we obtain the following lower bound on the smallest singular values of R:

$$\begin{aligned} \sigma_{k+j}(R) &= \lim_{i\to\infty} \sigma_j(E_i) \\ &\geq \lim_{i\to\infty} \left[\prod_{l=0}^{i} (1 - (\rho^l\eta/\gamma)^2)^{1/2}\right] \sigma_j(E) \\ &\geq \left(1 - (\eta/\gamma)^2 \sum_{l=0}^{\infty} \rho^{2l}\right)^{1/2} \sigma_j(E) \\ &= \left(1 - \frac{\eta^2}{\gamma^2(1-\rho^2)}\right)^{1/2} \sigma_j(E). \end{aligned}$$

One can show that

$$\left(1-\frac{\eta}{\gamma^2(1-\rho^2)}\right)^{-1/2}\sigma_j(S)\geq\sigma_j(R)\geq\sigma_j(S)$$

in the same way.

We can also bound the perturbation of the singular subspace associated with the smallest $n-k$ singular values. Here it is convenient to choose a specific unitary $Q=Q_0$ of the form

$$Q=\begin{pmatrix} I & -P \\ P^{\mathrm{H}} & I \end{pmatrix}\begin{pmatrix} (I+PP^{\mathrm{H}})^{-1/2} & 0 \\ 0 & (I+P^{\mathrm{H}}P)^{-1/2} \end{pmatrix} \tag{3.1}$$

where $P=S^{-1}H$. It is easy to check that

$$\begin{aligned}\|Q-I\|_2 &\leq \left\|\begin{pmatrix} 0 & -P \\ P^{\mathrm{H}} & 0 \end{pmatrix}\right\|_2 \\ &\quad+\left\|\begin{pmatrix} (I+PP^{\mathrm{H}})^{-1/2}-I & 0 \\ 0 & (I+P^{\mathrm{H}}P)^{-1/2}-I \end{pmatrix}\right\|_2 \\ &\leq 2\frac{\eta}{\gamma}. \end{aligned} \tag{3.2}$$

Let $Q_1,Q_2,Q_3,\ldots$ be defined analogously, and let $V_k=Q_0Q_2\cdots Q_{2k}$. Then by (2.7) and the analogue of (3.2),

$$\|V_{k+1}-V_k\|_2=\|I-Q_{2(k+1)}\|_2\leq 2\rho^{2(k+1)}\eta/\gamma.$$

Thus, $V_0,V_1,\ldots$ has a limit V, which is also unitary. Similarly, the limit $U=Q_1Q_3\cdots$ exists and is unitary.

Let

$$U^{\mathrm{H}}RV=R_\infty=\begin{pmatrix} S_\infty & 0 \\ 0 & E_\infty \end{pmatrix},$$

and let

$$V=\begin{pmatrix} V_{11} & V_{12} \\ V_{21} & V_{22} \end{pmatrix}.$$

What we require is a bound on the canonical angles between the spaces spanned by

$$\begin{pmatrix} 0 \\ I \end{pmatrix} \quad\text{and}\quad \begin{pmatrix} V_{21}^{\mathrm{H}} \\ V_{22}^{\mathrm{H}} \end{pmatrix}.$$

If Θ_R denotes the matrix of canonical angles, it is known that $\|\sin\Theta_R\| = \|V_{21}\|$. To get a bound on $\|V_{21}\|$, recall that $\|(Q_i)_{21}\| \le \rho^i\|H\|/\gamma$. Hence by an easy induction

$$\|\sin\Theta_R\| = \|V_{21}\| \le \frac{1}{(1-\rho^2)}\frac{\|H\|}{\gamma}.$$

In much the same way, one can obtain a bound for the matrix Θ_L of canonical angles for the left singular subspace corresponding to E_∞:

$$\|\sin\Theta_L\| \le \frac{\rho}{(1-\rho^2)}\frac{\|H\|}{\gamma}.$$

The extra factor ρ arises because the first iteration, which does not affect U, reduces $\|H\|$ by a factor of ρ.

We now summarize what we have proved.

THEOREM 3.1. *Let Θ_L and Θ_R be the matrices of canonical angles between the left and right singular subspace corresponding to the smallest $n-k$ singular values of R and $\hat{R}$. Under the hypotheses of Theorem 2.1,*

$$1 \ge \frac{\sigma_{k+i}(R)}{\sigma_i(E)} \ge \left(1 - \frac{\|H\|_2^2}{(1-\rho^2)\gamma^2}\right)^{1/2}, \quad i = 1,\ldots,n-k \tag{3.3}$$

$$1 \ge \frac{\sigma_i(S)}{\sigma_i(R)} \ge \left(1 - \frac{\|H\|_2^2}{(1-\rho^2)\gamma^2}\right)^{1/2}, \quad i = 1,\ldots,k \tag{3.4}$$

and for any unitarily invariant norm

$$\|\sin\Theta_R\| \le \frac{\|H\|}{(1-\rho^2)\gamma} \tag{3.5}$$

$$\|\sin\Theta_L\| \le \frac{\rho\|H\|}{(1-\rho^2)\gamma}. \tag{3.6}$$

There are a number of comments to be made about this theorem. First, the bounds (3.3) and (3.4) are remarkable in that they show that the *relative* error in the singular values is $O(\|H\|_2^2)$. Ordinarily, an off-diagonal perturbation of size H would obliterate singular values smaller than $\|H\|_2^2$. The fact that even the smallest singular values retain their accuracy is a consequence of the block triangularity of the starting matrix, as we shall see in a moment.

Second, Wedin ([5], [4, Theorem V.4.1]) has given a bound on the perturbation of singular subspaces. His bound does not assume that the matrix is in block triangular form, merely that the off diagonal blocks are small; but this bound, when specialized to our situation, gives the weaker inequality

$$\max\{\|\sin\Theta_R\|, \|\sin\Theta_L\|\} \leq \frac{\|H\|}{(1-\rho)\gamma}.$$

Third, the result can be cast as more conventional residual bounds for approximate null spaces. Specifically, suppose that for a given matrix B we have a unitary matrix $(V_1\ V_2)$ such that the residual

$$\|BV_2\|$$

is small. If $(U_1\ U_2)$ is a unitary matrix such that the column space of U_1 is the same as the column space of AV_1, then

$$U^{\mathrm{T}}AV = \begin{pmatrix} S & H \\ 0 & E \end{pmatrix},$$

where

$$\left\| \begin{pmatrix} H \\ E \end{pmatrix} \right\| = \|BV_2\|.$$

Thus if the singular values of S are not too small, the above theorem applies to bound the singular values and subspaces of B in terms of the residual norm $\|BV_2\|$.

Finally, we note that the approach yields approximation bounds for matrices of the form

$$\begin{pmatrix} \hat{S} & \hat{H} \\ \hat{G} & \hat{E} \end{pmatrix}, \qquad \|\hat{G}\| = O(\|\hat{H}\|)$$

by the expedient of first premultiplying to reduce the matrix to the form (1.1): i.e.,

$$\begin{pmatrix} P_{11} & P_{12} \\ P_{21} & P_{22} \end{pmatrix} \begin{pmatrix} \hat{S} & \hat{H} \\ \hat{G} & \hat{E} \end{pmatrix} = \begin{pmatrix} S & H \\ 0 & E \end{pmatrix}.$$

However, in this case the matrix

$$E = P_{21}\hat{H} + P_{22}\hat{E} = \hat{E} + O(\|\hat{H}\|_2^2)$$

is an $O(\|H\|_2^2)$ *additive* perturbation of of $\hat{E}$, so that the small singular values of $\hat{E}$ are no longer give $O(\|H\|_2^2)$ relative approximations to the small singular values of R. The block triangularity of R is really necessary.

The authors are grateful to W. Kahan for pointing out the relation of the iteration to the QR algorithm and to Marc Moonen for illuminating it.

REFERENCES

1 R. A. Horn and C. R. Johnson, *Matrix Analysis*, Cambridge U.P., Cambridge, 1985.
2 G. W. Stewart, *Updating a Rank-Revealing ULV Decomposition*, Technical Report CS-TR 2627, Dept. of Computer Science, Univ. of Maryland, 1991.
3 G. W. Stewart, *An Updating Algorithm for Subspace Tracking*, Technical Report CS-TR 2494, Dept. of Computer Science, Univ. of Maryland, 1990; IEEE Trans. Signal Pro., to appear.
4 G. W. Stewart and J.-G. Sun, *Matrix Perturbation Theory*, Academic, Boston, 1990.
5 P.-Å. Wedin, Perturbation bounds in connection with singular value decomposition, *BIT*, 12:99–111 (1972).

Received 17 April 1991; final manuscript accepted 9 July 1992

16.6. [GWS-J102] "The QLP Approximation to the Singular Value Decomposition"

[GWS-J102] "The QLP Approximation to the Singular Value Decomposition," *SIAM Journal on Scientific Computation* **20** (1999) 1336–1348.
`http://dx.doi.org/10.1137/S1064827597319519`

SIAM J. SCI. COMPUT.
Vol. 20, No. 4, pp. 1336–1348

THE QLP APPROXIMATION TO THE SINGULAR VALUE DECOMPOSITION*

G. W. STEWART†

Abstract. In this paper we introduce a new decomposition called the pivoted QLP decomposition. It is computed by applying pivoted orthogonal triangularization to the columns of the matrix X in question to get an upper triangular factor R and then applying the same procedure to the rows of R to get a lower triangular matrix L. The diagonal elements of R are called the R-values of X; those of L are called the L-values. Numerical examples show that the L-values track the singular values of X with considerable fidelity—far better than the R-values. At a gap in the L-values the decomposition provides orthonormal bases of analogues of row, column, and null spaces provided of X. The decomposition requires no more than twice the work required for a pivoted QR decomposition. The computation of R and L can be interleaved, so that the computation can be terminated at any suitable point, which makes the decomposition especially suitable for low-rank determination problems. The interleaved algorithm also suggests a new, efficient 2-norm estimator.

Key words. singular value decomposition, QLP decomposition, pivoted QR decomposition, rank determination

AMS subject classifications. 15A18, 15A23, 65F99

PII. S1064827597319519

1. Introduction. This paper concerns the problem of locating gaps in the singular values of a matrix. Specifically, suppose that X is an $n \times p$ matrix with $n \geq p$. Then there are orthogonal matrices U and V such that

$$U^{\mathrm{T}} X V = \begin{pmatrix} \Sigma \\ 0 \end{pmatrix}, \tag{1.1}$$

where

$$\Sigma = \operatorname{diag}(\sigma_1, \ldots, \sigma_p), \qquad \sigma_1 \geq \cdots \geq \sigma_p.$$

The decomposition (1.1) is called the singular value decomposition of X, and the scalars σ_i are called singular values. We say that X has a (relative) gap at m if σ_{m+1}/σ_m is small.

A gap in the singular values represents a natural point to reduce the dimensionality of a problem by setting the singular values below the gap to zero. In this case one is usually interested in certain subspaces associated with the decomposition. Specifically, partition the decomposition in the form

$$\begin{pmatrix} U_1^{\mathrm{T}} \\ U_2^{\mathrm{T}} \end{pmatrix} X (V_1 \ V_2) = \begin{pmatrix} \Sigma_1 & 0 \\ 0 & \Sigma_2 \\ 0 & 0 \end{pmatrix},$$

*Received by the editors April 4, 1997; accepted for publication November 17, 1997; published electronically March 17, 1999. This report is an extensive revision and expansion of TR-97-31, Department of Computer Science, University of Maryland. Both are available by anonymous ftp from thales.cs.umd.edu in the directory pub/reports or on the web at http://www.cs.umd/edu/~stewart/. This work was supported by National Science Foundation grant CCR-95-03126.

http://www.siam.org/journals/sisc/20-4/31951.html

†Department of Computer Science and Institute for Advanced Computer Studies, University of Maryland, College Park, MD 20742 (stewart@cs.umd.edu).

where Σ_1 is of order m. If $\Sigma_2 = 0$, then the column spaces of U_1 and V_1 are the column spaces of X and X^{T}, while the column spaces of U_1 and V_1 are the null spaces of X and X^{T}. When Σ_2 is merely small, the columns of U and V provide useful analogues of these spaces. To fix our nomenclature, we will call

$\mathcal{R}(U_1)$ the right superior singular subspace,
$\mathcal{R}(U_2)$ the right inferior singular subspace,
$\mathcal{R}(V_1)$ the left superior singular subspace,
$\mathcal{R}(V_2)$ the left inferior singular subspace

and collectively call the spaces *fundamental spaces.*[1]

Since gaps and singular subspaces are defined in terms of the singular value decomposition, the natural way to find them is to compute the decomposition. Unfortunately, this computation is expensive. Consequently, researchers have proposed many alternatives. Of these the pivoted QR decomposition is widely recommended because of its simplicity. As we shall see, however, the pivoted QR decomposition has certain drawbacks: it gives only fuzzy approximations to the singular values, and it fails to provide orthonormal bases for some of the fundamental subspaces. The purpose of this paper is to investigate the consequences of the empirical observation that if we reduce the R-factor of the pivoted QR decomposition to lower triangular form, then the diagonals track the singular values with considerable fidelity. We call this decomposition the pivoted QLP decomposition.

The paper is organized as follows. In section 2 we introduce the pivoted QR decomposition and discuss its properties. In section 3 we introduce the pivoted QLP decomposition and present some numerical experiments that show its abilities to track singular values. Section 4 is devoted to the approximations to the fundamental subspaces provided by the new decomposition. In section 5 we discuss implementation issues. It turns out that the computation of the R-factor and the L-factor can be interleaved, making the decomposition efficient for low-rank problems. The interleaving can also be used to derive an efficient 2-norm estimator, which is the subject of section 6. The paper concludes with a summary and a discussion of open problems associated with the decomposition.

Throughout this paper, $\|\cdot\|$ will denote the Euclidean vector norm and the subordinate spectral matrix norm. The smallest singular value of X will be denoted by $\inf(X)$.

2. The pivoted QR decomposition. As above let X be an $n{\times}p$ matrix with $n \geq p$. Then for any permutation matrix Π_{R} there is an orthogonal matrix Q such that

$$Q^{\mathrm{T}} X \Pi_{\mathrm{R}} = \begin{pmatrix} R \\ 0 \end{pmatrix}, \tag{2.1}$$

where R is upper triangular. The matrix Π_{R} can be chosen so that the elements of R satisfy

$$r_{kk}^2 \geq \sum_{i=k}^{p} r_{ij}^2, \qquad j = k{+}1, \ldots, p.$$

[1]Gil Strang [11] introduced the modifier "fundamental" to describe the column, row, and null spaces of a matrix. Per Christian Hansen extended the usage to their analogues. The use of "superior" and "inferior" in this connection is new.

In other words if R_{kk} denotes the trailing submatrix of R of order $p-k+1$, then the norm of the first column of R_{kk} dominates the norms of the other columns. We will call this decomposition the *pivoted QR decomposition.*

The pivoted QR decomposition was introduced by Golub [6], who computed it by a variant of Householder's orthogonal triangularization. Specifically, at the kth stage of the reduction, we will have computed Householder transformations $H_1, \ldots, H_{k-1}$ and permutations $\Pi_1, \ldots, \Pi_{k-1}$ such that

$$H_{k-1} \cdots H_1 X \Pi_1 \cdots \Pi_{k-1} = \begin{pmatrix} R_{11} & R_{12} \\ 0 & X_k \end{pmatrix}, \tag{2.2}$$

where R_{11} is upper triangular. Before the kth Householder transformation H_k is computed, the column of X_k of largest norm (along with the corresponding column of R_{12}) is swapped with the kth column. Since the norm of this column is $|r_{kk}|$, the pivoting strategy can be regarded as a greedy algorithm to make the leading principal submatrices of R as well conditioned as possible by making their trailing diagonals as large as possible.

We will call the diagonals of R the *R-values* of X. The folklore has it that the R-values track the singular values well enough to expose gaps in the latter, as illustrated in the following example. A matrix X of order 100 was generated in the form

$$X = U\Sigma V^{\mathrm{T}} + 0.1\sigma_{50}E,$$

where

1. Σ is formed by creating a diagonal matrix with diagonals decreasing geometrically from one to 10^{-3} and setting the last 50 diagonal elements to zero,
2. U and V are random orthogonal matrices,
3. E is a matrix of standard normal deviates.

Thus X represents a matrix of rank 50 perturbed by an error whose elements are one-tenth the size of the last nonzero singular value. Figure 2.1 plots the common logarithms of the singular values of X (continuous line) and R-values of X (dashed line) against their indices. The +'s indicate the values of $r_{50,50}$ and $r_{51,51}$. It is seen that there is a well-marked gap in the R-values, although not as marked as the gap in the singular values.

If we detect a gap at m and partition

$$\hat{X} \equiv X\Pi_{\mathrm{R}} = (Q_1 \quad Q_2 \quad Q_\perp) \begin{pmatrix} R_{11} & R_{12} \\ 0 & R_{22} \\ 0 & 0 \end{pmatrix}, \tag{2.3}$$

where R_{11} is $m\times n$, then Q_1 and $(Q_2\ Q_\perp)$ approximate the right fundamental subspaces of X. Perhaps more important, if we partition

$$\hat{X} = (\hat{X}_1\ \hat{X}_2), \tag{2.4}$$

it follows that $\hat{X}_1 = Q_1 R_{11}$ and hence that $\hat{X}_1$ is a basis for the left superior subspace of X. Thus the pivoted QR decomposition extracts a natural basis from among the columns of X. In applications where these columns represent variables in a model, this basis is a conceptual as well as a computational economization.

The decomposition also provides bases

$$\Pi_{\mathrm{R}} \begin{pmatrix} R_{11}^{\mathrm{T}} \\ R_{12}^{\mathrm{T}} \end{pmatrix} \quad \text{and} \quad \Pi_{\mathrm{R}} \begin{pmatrix} I \\ -R_{11}^{-1} R_{12} \end{pmatrix}$$

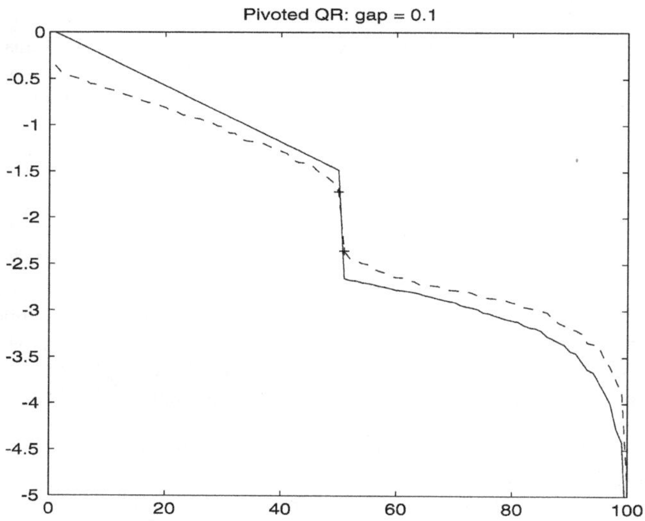

FIG. 2.1. *Gap revelation in pivoted QR.*

that approximate the right fundamental subspaces. Unfortunately, these bases are not orthonormal, and the basis for the right inferior subspace requires additional computation.

Although the R-values tend to reveal gaps in the singular values, they can fail spectacularly, as the following example due to Kahan shows. Let K_n be the upper triangular matrix illustrated below for $n = 6$:

$$K_6 = \begin{pmatrix} 1 & 0 & 0 & 0 & 0 & 0 \\ 0 & s & 0 & 0 & 0 & 0 \\ 0 & 0 & s^2 & 0 & 0 & 0 \\ 0 & 0 & 0 & s^3 & 0 & 0 \\ 0 & 0 & 0 & 0 & s^4 & 0 \\ 0 & 0 & 0 & 0 & 0 & s^5 \end{pmatrix} \begin{pmatrix} 1 & -c & -c & -c & -c & -c \\ 0 & 1 & -c & -c & -c & -c \\ 0 & 0 & 1 & -c & -c & -c \\ 0 & 0 & 0 & 1 & -c & -c \\ 0 & 0 & 0 & 0 & 1 & -c \\ 0 & 0 & 0 & 0 & 0 & 1 \end{pmatrix}.$$

Here $c^2 + s^2 = 1$. All the columns of the matrix have the same 2-norm—namely, 1—so that if ties in pivoting process are broken by choosing the first candidate, the first step of pivoted orthogonal triangularization leaves the matrix unchanged. This is similar for the remaining steps. Thus pivoted orthogonal triangularization leaves K_n unchanged, and the smallest R-value is s^{n-1}.

However, the matrix can have singular values far smaller than s^{n-1}. The table

(2.5)

c	σ_{99}	σ_{100}	$r_{99,99}$	$r_{100,100}$
0.0	1.0e+00	1.0e+00	1.0e+00	1.0e+00
0.1	6.4e−01	9.5e−05	6.1e−01	6.1e−01
0.2	1.5e−01	3.7e−09	1.4e−01	1.3e−01
0.3	1.1e−02	9.3e−14	9.8e−03	9.4e−03
0.4	2.3e−04	1.1e−18	1.9e−04	1.8e−04

presents the 99th and 100th singular and R-values of K_{100} for various values of c. When $c = 0$, $K_n = I$ and the R-values and singular values coincide. As c departs from zero, however, there is an increasingly great gap between the next-to-last and last singular values, while the ratio of the corresponding R-values remains near one.

This example has inspired researchers to look for other pivoting strategies under the rubric of rank revealing QR decompositions [2, 3, 4, 5, 7, 8]. There are, however, certain limitations to *any* pivoted QR decomposition. For example, the first R-value is the norm of the first column of $A\Pi_{\mathrm{R}}$. We hope this number will approximate σ_1, which, however, is the spectral norm of the entire matrix A. Thus r_{11} will in general underestimate σ_1. Similarly, $|r_{pp}|$ will generally overestimate σ_p.

For example, consider the rank-1 matrix $X = \mathbf{e}\mathbf{e}^{\mathrm{T}}$, where $\mathbf{e}$ is the vector whose components are all one. The norm of this matrix is $\sqrt{np}$. On the other hand, the columns of X all have norm $\sqrt{n}$. Thus if $n = p = 100$, the first R-value underestimates the corresponding singular value by a factor of 10—regardless of the pivoting strategy. Figure 2.1 also illustrates this tendency of the pivoted QR decomposition to underestimate the largest singular value. In fact, although the R-values reveal the gap, they do not give a very precise representation of the distribution of the singular values. We now turn to a postprocessing step that seems to clean up these values.

3. The pivoted QLP decomposition. To motivate our new decomposition consider the partitioned R-factor

$$R = \begin{pmatrix} r_{11} & r_{12}^{\mathrm{T}} \\ 0 & R_{22} \end{pmatrix}$$

of the pivoted QR decomposition. We have observed that r_{11} is an underestimate of $\|X\|_2$. A better estimate is the norm $\ell_{11} = \sqrt{r_{11}^2 + r_{12}^{\mathrm{T}} r_{12}}$ of the first row of R. We can calculate that norm by postmultiplying R by a Householder transformation H_1 that reduces the first row of R to a multiple of $\mathbf{e}_1$:

$$RH_1 = \begin{pmatrix} \ell_{11} & 0 \\ \ell_{12} & \hat{R}_{22} \end{pmatrix}.$$

We can obtain an even better value if we interchange the largest row of R with the first:

$$\Pi_1 R H_1 = \begin{pmatrix} \ell_{11} & 0 \\ \ell_{12} & \hat{R}_{22} \end{pmatrix}. \tag{3.1}$$

Now if we transpose (3.1), we see that it is the first step of pivoted Householder triangularization applied to R^{T} (cf. (2.2)). If we continue this reduction and transpose the result, we obtain a triangular decomposition of the form

$$\Pi_{\mathrm{L}}^{\mathrm{T}} Q^{\mathrm{T}} X \Pi_{\mathrm{R}} P = \begin{pmatrix} L \\ 0 \end{pmatrix}.$$

We will call this the *pivoted QLP decomposition* of X and will call the diagonal elements of L the *L-values* of X.[2]

[2] The decomposition has been introduced independently by Hosoda [9], who uses it to regularize ill-posed problems.

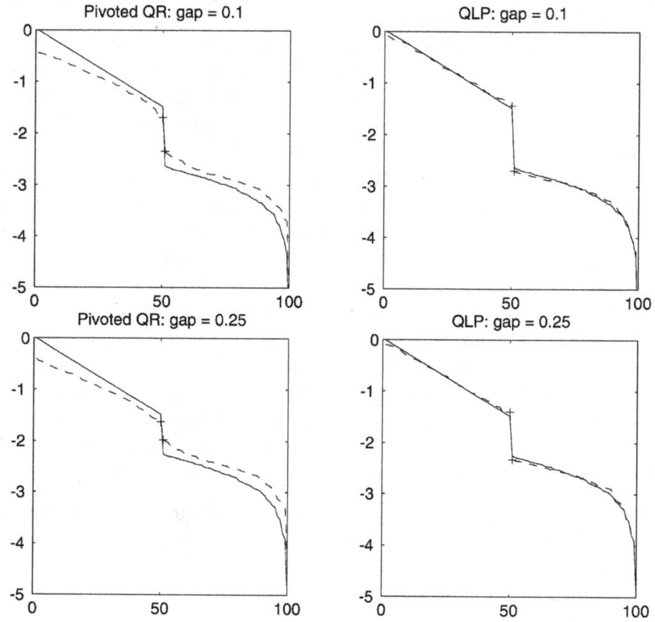

FIG. 3.1. *Pivoted QR and QLP decompositions compared.*

The way we motivated the pivoted QLP decomposition suggests that it might provide better approximations to the singular values of the original matrix X than does the pivoted QR decomposition. The top two graphs in Figure 3.1 compare performance of the two decompositions on the matrix of Figure 2.1. The solid lines, as above, indicate singular values and the dashed lines represent R-values on the left and L-values on the right. It is seen that in comparison with the R-values, the L-values track the singular values with remarkable fidelity.

The lower pair of graphs shows the behavior of the decomposition when the gap is reduced from a ratio of 0.1 to 0.25. Here the L-values perform essentially as well as the singular values. The gap in the R-values is reduced to the point where an automatic gap-detecting algorithm might fail to see it.

The superiority becomes even more marked in the low-rank case illustrated in Figure 3.2. The R-values only suggest the gap of 0.25 and reveal the gap of 0.1 weakly. The L-values are as good as one could hope for.

Figure 3.3 presents a more taxing example—called the devil's stairs—in which the singular values have multiple gaps. When the gaps are small, as in the top two graphs, neither decomposition does well at exhibiting their presence, although the L-values track the general trend of the singular values far better than the R-values. In the pair of graphs at the bottom, the gaps are fewer and bigger. Here the L-values clearly reveal the gaps, while the R-values do not.

The pivoted QLP decomposition takes Kahan's example in stride. The following table—the QLP analogue of (2.5)—shows that the last two L-values are good

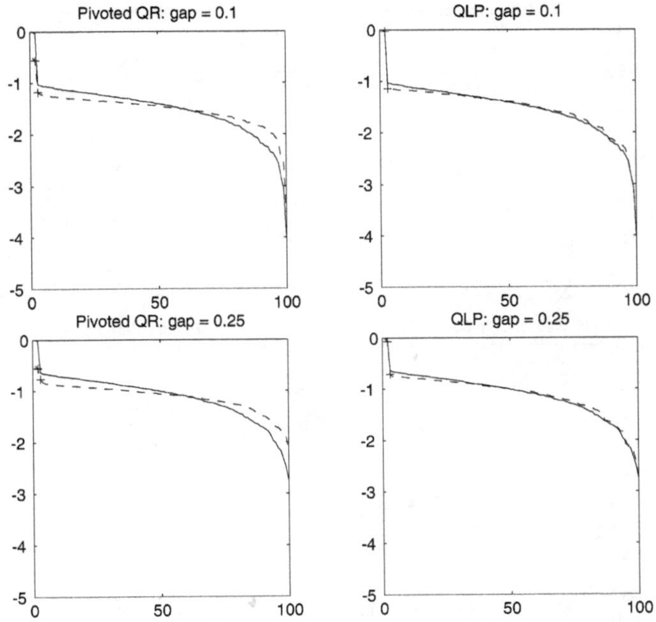

FIG. 3.2. *Rank determination: low rank.*

approximations of the corresponding singular values:

c	σ_{99}	σ_{100}	$\ell_{99,99}$	$\ell_{100,100}$
0.0	1.0e+00	1.0e+00	−1.0e+00	1.0e+00
0.1	6.4e−01	9.5e−05	−4.8e−01	2.2e−04
0.2	1.5e−01	3.7e−09	−1.1e−01	6.4e−09
0.3	1.1e−02	9.3e−14	−9.0e−03	1.4e−13
0.4	2.3e−04	1.1e−18	−1.9e−04	1.5e−18

The above examples suggest that not only does the pivoted QLP decomposition reveal gaps in the singular values better than the pivoted QR decomposition but that it is also superior in tracking the singular values. Pivoting plays a key role in this. Pivoting in the QR stage of the algorithm is absolutely necessary. For the above examples, the pivoting in the reduction to lower triangular form is largely cosmetic—it enforces monotonicity of the L-values. Without pivoting the L-values tend to behave like their pivoted counterparts. Nonetheless, pivoting at this stage may be necessary. For example, consider the following matrix of order n:

$$\begin{pmatrix} 1 & 0 \\ 0 & \frac{1}{\sqrt{n}}\mathbf{e}\mathbf{e}^{\mathrm{T}} \end{pmatrix}.$$

Without pivoting the first L-value is one, which underestimates the norm $(n-1)/\sqrt{n}$. On the other hand, the pivoted L-value reproduces the norm exactly.

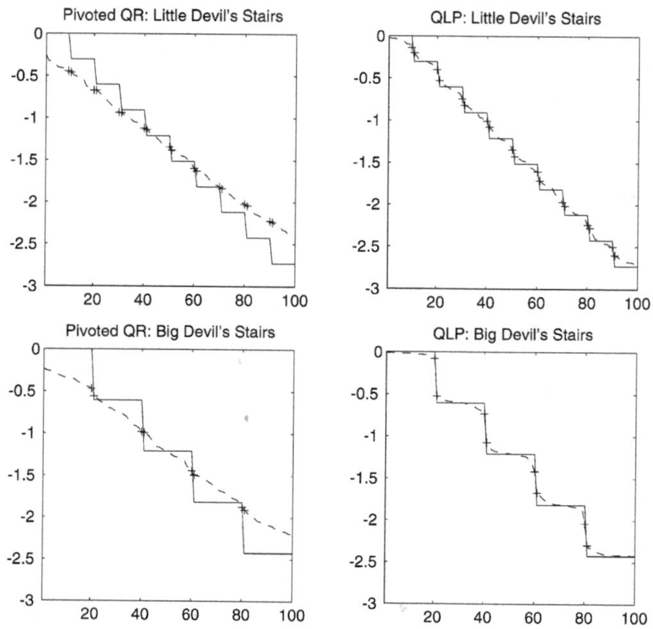

FIG. 3.3. *The devil's stairs.*

4. Fundamental subspaces. If we incorporate the pivots in the QLP decomposition into the orthogonal transformations by defining

$$\hat{Q} = Q\Pi_{\mathrm{L}} \quad \text{and} \quad \hat{P} = \Pi_{\mathrm{R}}P,$$

then the decomposition can be written in the partition form

$$X = (\hat{Q}_1 \quad \hat{Q}_2 \quad Q_\perp) \begin{pmatrix} L_{11} & 0 \\ L_{21} & L_{22} \\ 0 & 0 \end{pmatrix} \begin{pmatrix} \hat{P}_1^{\mathrm{T}} \\ \hat{P}_2^{\mathrm{T}} \end{pmatrix}, \tag{4.1}$$

where L_{11} is of order m. Since the L-values tend to track the singular values, if there is a gap in the latter at m, the partition of $\hat{P}$ and $\hat{Q}$ provide orthonormal bases approximating the four fundamental subspaces of X at m. Specifically,

1. $\mathcal{R}(\hat{Q}_1)$ approximates the left superior subspace of X,
2. $\mathcal{R}[(\hat{Q}_2 \ Q_\perp)]$ approximates the left inferior subspace of X,
3. $\mathcal{R}(\hat{P}_1)$ approximates the right superior subspace of X,
4. $\mathcal{R}(\hat{P}_2)$ approximates the right inferior subspace of X.

Thus the pivoted QLP decomposition, like the pivoted QR decomposition, furnishes orthonormal approximations to the left fundamental subspaces, but unlike the latter, it also furnishes orthonormal approximations to the right fundamental subspaces.

A nice feature of the decomposition is that the accuracy of the approximate subspaces can be estimated from the decomposition itself. We will measure the concordance of two subspaces $\mathcal{X}$ and $\mathcal{Y}$ by the sine of their largest canonical angle between them. This number is the largest singular value of $X^{\mathrm{T}}Y_{\perp}$, where X is an orthonormal basis for $\mathcal{X}$ and $Y_{\perp}$ is an orthonormal basis for the orthogonal complement of $\mathcal{Y}$. The following theorem follows from results in [10].

THEOREM 4.1. *Let s_{u} be the sine of the largest canonical angle between the left superior subspace of X and $\mathcal{R}(\hat{Q}_1)$, and let s_{v} be the sine of the largest canonical angle between the right superior subspace of X and $\mathcal{R}(\hat{P}_1)$. If*

$$\rho = \frac{\|L_{22}\|_2}{\inf(L_{11})} < 1,$$

then

$$s_{\mathrm{u}} \le \frac{1}{1-\rho^2}\,\frac{\|L_{21}\|}{\inf(L_{11})} \quad \textit{and} \quad s_{\mathrm{v}} \le \frac{\rho}{1-\rho^2}\,\frac{\|L_{21}\|}{\inf(L_{11})}. \tag{4.2}$$

The bounds for the left and right inferior subspaces are the same.

Although this theorem is cast in terms of quantities that are difficult to compute, $\inf(L_{11})$ can be estimated by the L-value ℓ_{mm}, while $\|L_{22}\|$ can be estimated by $\ell_{m+1,m+1}$. The quantity $\|L_{21}\|$ can be bounded by the Frobenius norm or can be estimated by the 2-norm estimator described in section 6.

In section 2 we stressed the desirability of having the left superior subspace at a gap associated with a specific set of columns of X. For the pivoted QR decomposition we showed that the columns of $\hat{X}_1$ in (2.4) spanned the same space as those of Q_1 in (2.3), and hence $\hat{X}_1$ provides such a basis. In the pivoted QRP decomposition we must replace Q by

$$\hat{Q} \equiv Q\Pi_{\mathrm{L}} = (\hat{Q}_1 \;\; \hat{Q}_2 \;\; Q_{\perp}).$$

In general, the pivoting in the reduction to lower triangular form mixes up the columns of Q so that $\hat{Q}_1$ cannot be associated with a set of columns of X. However, if the partition (4.1) corresponds to a substantial gap in the R-values, it is unlikely that the pivoting process will interchange columns across Q_1 and Q_2. In this case the column spaces of Q_1 and $\hat{Q}_1$ are the same and are spanned by the columns of $\hat{X}_1$. Thus in the applications we are most interested in, the left superior subspace will be associated with a specific set of columns of X.

5. Implementation issues. An advantage of the QLP decomposition is that it can be computed with off-the-shelf software. The standard matrix packages have programs for computing a pivoted QR decomposition. To compute the QLP decomposition one merely applies the program twice, once to X and once to R^{T}.

Householder triangularization requires about $np^2 - \frac{1}{3}p^3$ floating point additions and multiplications. The additional work in the reduction to lower triangular form is $\frac{2}{3}p^3$ additions and multiplications. Thus when $n = p$ the pivoted QLP decomposition requires twice as much work as the pivoted QR decomposition. As n increases, the additional work becomes negligible.

It should be pointed out that when n is much greater than p the work required for calculation of the singular value decomposition of R (and hence that of X) is also negligible (e.g., see [1]). However, an important aspect of the QLP decomposition—one not shared with the singular value decomposition—is that by interleaving the computation of R and L we can terminate the algorithm as soon as it has proceeded far enough for the purposes at hand. Specifically, at the kth step of the reduction to upper triangular form, we have

$$H_{k-1}\cdots H_1 X \Pi_1 \cdots \Pi_{k-1} = \begin{pmatrix} R_{11} & R_{1k} \\ 0 & X_{kk} \end{pmatrix}.$$

Since the first $k-1$ rows of R are present in this decomposition, we can reduce them to get the first $k-1$ rows of L. We can then compute more rows of R, and reduce them to get the corresponding rows of L. And so on. This process restricts the pivoting in forming L to the set of rows currently being reduced. However, as we have noted, except for contrived examples, pivoting in the formation of L does not much improve the L-values.

This interleaved pivoting is especially valuable in low-rank problems, since one can stop computing the decomposition after a gap has been found, with a potentially great savings in work. In fact, since the gap-revealing abilities of the R-values are themselves not negligible, one can treat the computation of R as a probe for potential gaps, which can then be confirmed by computing the corresponding rows of L. The interleaved pivoting can also be used to improve the efficiency of Hosoda's regularization algorithm [9].

6. A 2-norm estimator. We have seen that the first L-value in the pivoted QLP decomposition of a matrix X is generally a good estimate of $\|X\|_2$. Since we can interleave the computation of R and L, we can estimate $\|X\|_2$ by computing a few rows of the pivoted R-factor. Usually (but not always) the norm of the largest row will be the $(1,1)$-element of the pivoted L-factor—the first L-value. In this section we will describe how to implement this estimation scheme.

The basic idea is that the pivoted R-factor of X is the pivoted Cholesky factor of the cross-product matrix $A = X^{\mathrm{T}}X$. We will now show how to compute the factor row by row. We begin with the unpivoted version.

Suppose that we have computed the first $k-1$ rows of R and partition them in the form

$$(R_{11} \quad R_{12}),$$

where R_{11} is triangular of order $k-1$. Then

$$A - \begin{pmatrix} R_{11}^{\mathrm{T}} \\ R_{12}^{\mathrm{T}} \end{pmatrix} (R_{11} \quad R_{12}) = \begin{pmatrix} 0 & 0 \\ 0 & S \end{pmatrix},$$

where S is the Schur complement of $A_{11} = R_{11}^{\mathrm{T}} R_{11}$ in A. Since the kth row of R is the first row of S divided by the square root of the $(1,1)$-element of S, we can compute the elements of the kth row in a Crout-like algorithm as follows.

for $j = k$ **to** p
 $r_{kj} = x_k^{\mathrm{T}} x_j$
 $r_{kj} = r_{kj} - r_{1k}r_{1j} - \cdots - r_{k-1,k}r_{k-1,j}$
 $r_{kj} = r_{kj}/\sqrt{r_{kk}}$
end for k

The first statement in the loop generates an element in the kth row of A. The second statement computes the Schur complement of that element. The third statement scales the element so that it becomes an element of the R-factor.

Turning now to the pivoted algorithm, we must answer two questions.

1. How do we keep track of the norms to determine the pivots?

Given an $n \times p$ matrix X, this algorithm returns an estimate *norm2est* of $\|X\|_2$.

```
normx2[j] = ||X[:,j]||_2^2   (j = 1,...,p)
P = ∅
norm2est = 0
for k = 1 to kmax
   Choose pvt ∉ P so that nrmx2[pvt] ≥ nrmx2[j]  (j ∉ P)
   P = P ∪ {pvt}
   nr2 = nrmx2[pvt]
   rkk = sqrt(nr2)
   for j = 1 to p, j ∉ P
      R[k,j] = X[:,pvt]^T * X[:,j]
      for i = 1 to k-1
         R[k,j] = R[k,j] - R[i,pvt] * R[i,j]
      end for i
      R[k,j] = R[k,j]/rkk
      nr2 = nr2 + R[k,j]^2
      normx2[j] = normx2[j] - R[k,j]^2
   end for j
   norm2est = max{norm2est, nr2}
end for k
norm2est = sqrt(norm2est)
```

FIG. 6.1. *A 2-norm estimator.*

2. How do we organize the interchanges?

The answer to the first question is that the squares of the norms that determine the pivots are the diagonals of the Schur complement. These quantities can be formed initially and downdated as we add rows. The answer to the second question is that we perform no interchanges. Instead we keep track of indices of the columns we have selected as pivots and skip operations involving them as we add the kth row. For example, with a pivot sequence of $3, 2, 4, 1$ in a 4×4 matrix we would obtain an "R-factor" of the form

$$\begin{pmatrix} r_{14} & r_{12} & r_{11} & r_{13} \\ r_{24} & r_{22} & 0 & r_{23} \\ r_{34} & 0 & 0 & r_{33} \\ r_{44} & 0 & 0 & 0 \end{pmatrix}.$$

Figure 6.1 contains an implementation of this scheme. The heart of it is the loop on j, in which $R[k, j]$ is initialized, transformed into the corresponding element of the Schur complement, and normalized. The norms of the reduced columns of X, which are contained in the array *normx2*, are also updated at this point. The square of the norm of the row is accumulated in *nr2* and then compared with *norm2est*. Here are some comments.

1. The variable *kmax* is an upper bound on the number of rows of R to compute. Two or perhaps three is a reasonable number. Another possibility is to compute rows of R until a decrease in norm occurs.

2. The bulk of the work is in computing the norms of the columns of X and initializing R. Specifically if X is $n \times p$ and *kmax* is not too large, the algorithm requires about $(kmax+1)np$ additions and multiplications.

3. The algorithm can produce an underestimate. For example, consider the $n \times n$ matrix

$$X = \begin{pmatrix} I_{kmax} & 0 \\ 0 & \frac{1}{\sqrt{n}}\mathbf{e}\mathbf{e}^{\mathrm{T}} \end{pmatrix}.$$

Because the first *kmax* columns of X dominate the rest, the estimator will return a value of *norm2est* of one. But the norm of the matrix is $(n-kmax)/\sqrt{n}$.

7. Discussion. The pivoted QLP decomposition of an $n \times p$ matrix X is computed by applying pivoted orthogonal triangularization to the columns of X to get an upper triangular factor R and then applying the same procedure to the rows of R to get a lower triangular matrix L. The diagonal elements of R are called the R-values of X; those of L are called the L-values. This decomposition has several attractive properties.

1. The L-values track the singular values of X with considerable fidelity—far better than the R-values. This makes the decomposition particularly appropriate for rank determination.

2. The decomposition provides orthonormal bases for approximations to the four fundamental subspaces from the singular value decomposition. The accuracy of these approximations can be estimated from the decomposition itself. When the decomposition is divided by at a gap, it provides a well-conditioned subset of the columns of X that span the left superior subspace.

3. The decomposition requires no more than twice the work required for a pivoted QR decomposition. The additional work decreases as n increases.

4. The computation of R and L can be interleaved. This makes the decomposition suitable for low-rank problems. It can also be used to regularize systems resulting from ill-posed problems.

5. The decomposition suggests an efficient method for estimating the 2-norm of a matrix.

Why the L-values track the singular values so well is an open question. Without pivoting, the decomposition represents the first two steps in an iterative algorithm for computing the singular value decomposition by reducing a matrix successively to upper then lower triangular form. However, the convergence of this algorithm depends on the ratios of neighboring singular values, which in our examples are too near one to account for the behavior we have observed. The effectiveness of the decomposition is intimately tied to the pivoting, the pivoting in the formation of R being essential and the pivoting in the formation of L being necessary to avoid certain contrived counterexamples. I conjecture that the analysis of this decomposition will not be simple.

REFERENCES

[1] T. F. Chan, *An improved algorithm for computing the singular value decomposition*, ACM Trans. Math. Software, 8 (1982), pp. 72–83.

[2] T. F. Chan, *Rank revealing QR factorizations*, Linear Algebra Appl., 88/89 (1987), pp. 67–82.

[3] T. F. Chan and P. C. Hansen, *Computing truncated singular value decomposition least squares solutions by rank revealing QR-factorizations*, SIAM J. Sci. Statist. Comput., 11 (1990), pp. 519–530.

[4] T. F. Chan and P. C. Hansen, *Some applications of the rank revealing QR factorization*, SIAM J. Sci. Statist. Comput., 13 (1992), pp. 727–741.

[5] S. Chandrasekaran and I. C. F. Ipsen, *On rank-revealing factorisations*, SIAM J. Matrix Anal. Appl., 15 (1994), pp. 592–622.

[6] G. H. Golub, *Numerical methods for solving least squares problems*, Numer. Math., 7 (1965), pp. 206–216.

[7] M. Gu and S. C. Eisenstat, *Efficient algorithms for computing a strong rank-revealing QR-factorization*, SIAM J. Sci. Comput., 17 (1996), pp. 848–869.

[8] Y. P. Hong and C.-T. Pan, *Rank-revealing QR factorizations and the singular value decomposition*, Math. Comp., 58 (1992), pp. 213–232.

[9] Y. Hosoda, *A new method for linear ill-posed problems with double use of the QR-decomposition*, paper presented at the Kalamazoo Symposium on Matrix Analysis and Applications, Kalamazoo, MI, 1997.

[10] R. Mathias and G. W. Stewart, *A block QR algorithm and the singular value decomposition*, Linear Algebra Appl., 182 (1993), pp. 91–100.

[11] G. Strang, *Linear Algebra and Its Applications*, 3rd ed., Academic Press, New York, 1988.

16.7. [GWS-J107] (with Z. Jia) "An Analysis of the Rayleigh-Ritz Method for Approximating Eigenspaces"

[GWS-J107] (with Z. Jia) "An Analysis of the Rayleigh-Ritz Method for Approximating Eigenspaces," *Mathematics of Computation* **70** (2001) 637–647.
`http://dx.doi.org/10.1090/S0025-5718-00-01208-4`

MATHEMATICS OF COMPUTATION
Volume 70, Number 234, Pages 637–647
S 0025-5718(00)01208-4
Article electronically published on February 18, 2000

AN ANALYSIS OF THE RAYLEIGH–RITZ METHOD FOR APPROXIMATING EIGENSPACES

ZHONGXIAO JIA AND G. W. STEWART

ABSTRACT. This paper concerns the Rayleigh–Ritz method for computing an approximation to an eigenspace $\mathcal{X}$ of a general matrix A from a subspace $\mathcal{W}$ that contains an approximation to $\mathcal{X}$. The method produces a pair $(N, \tilde{X})$ that purports to approximate a pair (L, X), where X is a basis for $\mathcal{X}$ and $AX = XL$. In this paper we consider the convergence of $(N, \tilde{X})$ as the sine ϵ of the angle between $\mathcal{X}$ and $\mathcal{W}$ approaches zero. It is shown that under a natural hypothesis — called the uniform separation condition — the Ritz pairs $(N, \tilde{X})$ converge to the eigenpair (L, X). When one is concerned with eigenvalues and eigenvectors, one can compute certain refined Ritz vectors whose convergence is guaranteed, even when the uniform separation condition is not satisfied. An attractive feature of the analysis is that it does not assume that A has distinct eigenvalues or is diagonalizable.

1. INTRODUCTION

Many methods for finding eigenvalues and eigenvectors of a large matrix A proceed by generating a sequence of subspaces $\mathcal{W}_k$ containing increasingly accurate approximations to the desired eigenvectors. There are a number of methods for accomplishing this — e.g., the Arnoldi method, the nonsymmetric Lanczos method, subspace iteration, and the Jacobi Davidson method (for more on these methods see [12, 13]).

A central problem in all these methods is how to extract approximations to the desired eigenvalues and eigenvectors from the subspace $\mathcal{W}_k$. A widely used technique for accomplishing this is called the Rayleigh–Ritz procedure (it is also an example of the more general Galerkin technique). In its simplest form the technique retrieves an approximation to a simple eigenpair (λ, x) as follows (here we drop the iteration subscript).

(1.1)
1. Compute an orthonormal basis W for $\mathcal{W}$.
2. Compute $B = W^{\mathrm{H}}AW$.
3. Let (ν, z) be an eigenpair of B, where $\nu \cong \lambda$.
4. Take $(\nu, \tilde{x}) = (\nu, Wz)$ as the approximate eigenpair.

Received by the editor April 9, 1998 and, in revised form, May 5, 1999.

2000 *Mathematics Subject Classification.* Primary 15A18, 65F15, 65F50.

The first author's work was supported by the China State Major Key Project for Basic Researches, the National Natural Science Foundation of China, the Foundation for Excellent Young Scholars of the Ministry of Education and the Doctoral Point Program of the Ministry of Education, China.

The second author's work was supported by the National Science Foundation under Grant No. 970909-8562.

Of course, steps 3 and 4 can be repeated to extract approximations to other eigenpairs.

The matrix B is called a Rayleigh quotient. The number ν is called a Ritz value and the vector $\tilde{x} = Wz$ is called a Ritz vector. The informal justification for the method is that if $x \in \mathcal{W}$ then there is an eigenpair (λ, z) of B with $x = Wz$. Continuity suggests that if x is nearly in $\mathcal{W}$ then there should be an eigenpair (ν, z) of B with ν near λ and Wz near x.

When A is non-Hermitian, one of the authors (Jia [6, 7, 8, 9]) has established a priori error bounds for Ritz values and Ritz vectors in terms of the deviation of x from $\mathcal{W}$. The results show that Ritz values converge. The Ritz vectors, on the other hand, behave more erratically and may even fail to converge. This led the first author to introduce certain refined Ritz vectors for which the continuity argument is valid [5, 7, 8]. The refined Ritz vectors have been used in some other cases [10, 11]. Unfortunately, the results just cited were proved under the restrictive hypothesis that the eigenvalues of A are distinct or A is diagonalizable. One of the contributions of this paper is to remove this restriction.

When A has a cluster of very close or multiple eigenvalues, the corresponding eigenvectors are ill-determined, and it makes better sense to approximate the eigenspace $\mathcal{X}$ spanned by the vectors instead of the vectors themselves. The Rayleigh–Ritz procedure can be extended to do this [see (3.1) below]. A second contribution of this paper is to provide an analysis of this extended procedure. In fact, essentially the same results hold for eigenspaces as for eigenvectors, and we will present our results at this level of generality.

Our approach is to derive bounds in terms of the sine ϵ of the angle between $\mathcal{X}$ and the subspace $\mathcal{W}$. (In practice, of course, the size of ϵ must be established from the properties of the underlying algorithm that determines $\mathcal{W}$.) Sections 2 and 3 are devoted to background material and setting the stage for our analysis. In Section 4 we will consider the convergence of the Ritz values of the Rayleigh quotient. In Section 5 we will establish the convergence of Ritz pairs under a hypothesis that is computationally verifiable. In Section 6 we treat the relation of residual vectors to the accuracy of an approximate eigenspace, and in Section 7, we consider the convergence of the refined Ritz vectors mentioned above. The paper concludes with a brief summary.

2. Preliminaries

Background material for this paper can be found in [3, 14]. The norm $\|\cdot\|$ will denote both the Euclidean vector norm and the subordinate spectral matrix norm. We will denote the spectrum of a matrix A by the multiset $\Lambda(A)$.

A subspace $\mathcal{X}$ is an eigenspace (or invariant subspace) of A if $A\mathcal{X} \subset \mathcal{X}$. If X is a basis for $\mathcal{X}$, then there is a unique matrix L such that

$$AX = XL, \tag{2.1}$$

and conversely if X has linearly independent columns and satisfies (2.1), then the column space of X is an eigenspace of A. In this case, we say that (L, X) is an eigenpair of A, that the matrix X is an eigenbasis of A, and that L is its corresponding eigenblock. The eigenpair is simple if its eigenvalues are distinct from the other eigenvalues of A. It is orthonormal if $X^{\mathrm{H}}X = I$, in which case its eigenblock is given by the Rayleigh quotient $L = X^{\mathrm{H}}AX$. Throughout this paper we will assume that eigenpairs are orthonormal.

We will measure the deviation of a subspace $\mathcal{X}$ from a subspace $\mathcal{W}$ as follows. Let X be an orthonormal basis for $\mathcal{X}$, W an orthonormal basis for $\mathcal{W}$, and $W_\perp$ an orthonormal basis for the orthogonal complement of $\mathcal{W}$. Then we define

$$\sin\angle(\mathcal{X},\mathcal{W}) = \sin(X,W) = \|W_\perp^{\mathrm{H}} X\|. \tag{2.2}$$

This measure is a metric on any space of subspaces of fixed dimension. If the dimension of $\mathcal{W}$ is greater than that of $\mathcal{X}$, the measure is not symmetric in its arguments; in fact, $\sin\angle(\mathcal{W},\mathcal{X}) = 1$, although it may happen that $\sin\angle(\mathcal{X},\mathcal{W}) < 1$.

We will cast our results in terms of a function sep that in some sense measures the distance between the spectra of two matrices. Specifically, let L and M be matrices of order ℓ and m, and define

$$\operatorname{sep}(L,M) = \min_{\|P\|=1} \|PL - MP\|. \tag{2.3}$$

Alternatively, let $\mathbf{S}$ be the Sylvester operator defined by

$$\mathbf{S}P = PL - MP.$$

Then $\mathbf{S}$ is a linear operator whose eigenvalues are

$$\Lambda(\mathbf{S}) = \Lambda(L) - \Lambda(M);$$

i.e., its eigenvalues are the pairwise differences of the eigenvalues of L and M. If

$$\delta \equiv \min|\Lambda(\mathbf{S})| > 0,$$

then $\mathbf{S}$ is nonsingular, and it follows from (2.3) that $\operatorname{sep}^{-1}(L,M) = \|\mathbf{S}^{-1}\| \geq \delta$. Hence

$$\operatorname{sep}(L,M) = \|\mathbf{S}^{-1}\|^{-1} \leq \delta;$$

i.e., $\operatorname{sep}(L,M)$ is not greater than the physical separation δ of the spectra of L and M. Unfortunately, $\operatorname{sep}(L,M)$ can be much smaller than δ.

The function sep has an important advantage over δ — it is Lipschitz continuous. Specifically,

$$\operatorname{sep}(L,M) - \|E\| - \|F\| \leq \operatorname{sep}(L+E, M+F) \leq \operatorname{sep}(L,M) + \|E\| + \|F\|. \tag{2.4}$$

Finally, we note that, because $\|\cdot\|$ is unitarily invariant, if U and V are unitary, then

$$\operatorname{sep}(U^{\mathrm{H}}LU, V^{\mathrm{H}}MV) = \operatorname{sep}(L,M). \tag{2.5}$$

We conclude this section with a theorem of Elsner [2] (as improved in [1]), which will be used to establish the convergence of Ritz values.

Theorem 2.1. *Let the eigenvalues of A be $\lambda_1,\ldots,\lambda_n$ and let the eigenvalues of $\tilde{A} = A + E$ be $\tilde{\lambda}_1,\ldots,\tilde{\lambda}_n$. Then there is a permutation $j_1,\ldots,j_n$ of the integers $1,\ldots,n$ such that*

$$|\lambda_i - \tilde{\lambda}_{j_i}| \leq 4(\|A\| + \|\tilde{A}\|)^{1-\frac{1}{n}}\|E\|^{\frac{1}{n}}, \qquad i = 1,\ldots,n.$$

3. The setting

As we indicated in the introduction, we are concerned with the approximation of a simple eigenpair (L, X) by the Rayleigh–Ritz method applied to a subspace $\mathcal{W}$. We will consider the following generalization of the method (1.1) in the introduction:

(3.1)
1. Compute an orthonormal basis W for $\mathcal{W}$.
2. Compute $B = W^{\mathrm{H}}AW$.
3. Let (N, Z) be an eigenpair of B, where $\Lambda(N) \cong \Lambda(L)$.
4. Take $(N, \tilde{X}) = (N, WZ)$ as the approximate eigenpair.

For definiteness we will suppose that the dimensions of $\mathcal{X}$ and $\mathcal{W}$ are ℓ and p, so that the eigenblock L is of order ℓ and the Rayleigh quotient B is of order p. Denoting the column space of X by $\mathcal{X}$, we will set

$$\epsilon = \sin \angle(\mathcal{X}, \mathcal{W}) \tag{3.2}$$

and examine the behavior of the method as $\epsilon \to 0$.

In the sequel the representation of X in the coordinate system specified by W will play a central role. As above, let the columns of $W_\perp$ form an orthonormal basis for $\mathcal{W}_\perp$. Then X can be written in the form

$$X = WY + W_\perp Y_\perp,$$

where

$$Y = W^{\mathrm{H}}X \quad \text{and} \quad Y_\perp = W_\perp^{\mathrm{H}}X.$$

By definition, $\|Y_\perp\| = \epsilon$.

The columns of Y can be orthonormalized by setting

$$\hat{Y} = YQ, \quad \text{where} \quad Q = (Y^{\mathrm{H}}Y)^{-\frac{1}{2}}. \tag{3.3}$$

Since $Y^{\mathrm{H}}Y + Y_\perp^{\mathrm{H}}Y_\perp = I$, it follows that

$$\|I - Q\| = 1 - \frac{1}{\sqrt{1-\epsilon^2}} \cong \frac{1}{2}\epsilon^2 \quad \text{and} \quad \|Q^{-1}\| = \frac{1}{\sqrt{1-\epsilon^2}} \cong 1 + \frac{1}{2}\epsilon^2.$$

Note that, since $X - W\hat{Y} = WY_\perp + WY(I - Q)$,

$$\|X - W\hat{Y}\| \le \epsilon + O(\epsilon^2). \tag{3.4}$$

4. The convergence of Ritz values

Although we will be primarily concerned with Ritz pairs, the only effective way of choosing a pair from a Rayleigh quotient B is to examine the eigenvalues of B. We therefore need to know when the eigenvalues of a Rayleigh quotient converge.

It is a surprising fact that the hypothesis $\epsilon \to 0$ is by itself sufficient to insure that B in the algorithm (3.1) contains Ritz values that converge to the eigenvalues of L. We will establish this result in two stages. First we will show that if ϵ is small then $\Lambda(L)$ is a subset of the spectrum of a matrix $\tilde{B}$ that is near B. We will then use Elsner's theorem to show that B must have eigenvalues that are near those of L.

Theorem 4.1. *Let B be the Rayleigh quotient in* (3.1), *and let $\hat{Y}$ and Q be defined by* (3.3). *Then there is a matrix E satisfying*

$$\|E\| \le \frac{\epsilon}{\sqrt{1-\epsilon^2}}\|A\| \tag{4.1}$$

such that $(Q^{-1}LQ, \hat{Y})$ *is an eigenblock of* $B + E$.

Proof. From the relation $AX - XL = 0$ we have

$$W^{\mathrm{H}}A\ (W\quad W_{\perp})\begin{pmatrix} W^{\mathrm{H}} \\ W_{\perp}^{\mathrm{H}} \end{pmatrix} X - W^{\mathrm{H}}XL = 0.$$

Equivalently,

$$BY + W^{\mathrm{H}}AW_{\perp}Y_{\perp} - YL = 0.$$

Postmultiplying by Q, we have

$$B\hat{Y} + W^{\mathrm{H}}AW_{\perp}Y_{\perp}Q - \hat{Y}Q^{-1}LQ = 0. \tag{4.2}$$

Set

$$R = B\hat{Y} - \hat{Y}Q^{-1}LQ. \tag{4.3}$$

Then it follows from (4.2) that

$$\|R\| \le \frac{\epsilon}{\sqrt{1-\epsilon^2}}\|W^{\mathrm{H}}AW_{\perp}\| \le \frac{\epsilon}{\sqrt{1-\epsilon^2}}\|A\|. \tag{4.4}$$

If we now define

$$E = R\hat{Y}^{\mathrm{H}},$$

then it is easy to verify that E satisfies (4.1) and $(B + E)\hat{Y} = \hat{Y}Q^{-1}LQ$. □

If we now apply Elsner's theorem, we get the following corollary.

Corollary 4.2. *Let the eigenvalues of* L *be* $\lambda_1, \ldots, \lambda_\ell$ *and let the eigenvalues of* B *be* $\nu_1, \ldots, \nu_p$. *Then there are integers* $j_1, \ldots, j_\ell$ *such that*

$$|\lambda_i - \nu_{j_i}| \le 4(2\|A\| + \|E\|)^{1-\frac{1}{p}}\|E\|^{\frac{1}{p}}, \qquad i = 1, \ldots, \ell. \tag{4.5}$$

The right-hand side of (4.5) depends only on $\|A\|$ and ϵ. Hence we may conclude that as $\epsilon \to 0$ there are always Ritz values that converges to the eigenvalues of L. The exponent $\frac{1}{p}$ in (4.5) means that in the worst case the convergence of the Ritz values can be slow. Unfortunately, if the eigenvalues of L are defective, we will indeed observe slow convergence. And even if they are well conditioned, without additional conditions convergence can still be slow. One such condition will emerge in the next section.

5. The convergence of Ritz pairs

Having determined that there are Ritz values that converge to the eigenvalues of our distinguished eigenspace $\mathcal{X}$, we now turn to the convergence of the Ritz pairs. The chief difficulty in establishing convergence is that an eigenspace can have any number of eigenpairs even when the eigenpairs are required to be orthonormal. For if (L, X) is an orthonormal eigenpair of A and U is unitary, then $(U^{\mathrm{H}}LU, XU)$ is also an orthonormal eigenpair corresponding to the same eigenspace. If we are to speak of convergence, therefore, we must find a way of removing the ambiguity in the pairs.

One way is to prove convergence of the Ritz spaces directly, after which we can choose converging bases for the spaces, whose associated Rayleigh quotients will naturally converge. The problem with this approach is that the convergence conditions must be phrased in terms of the eigenblock L, which is unknown before convergence. For this reason, we will take a less direct approach.

We will use the notation and results of Theorem 4.1 and Corollary 4.2. Let (N, Z) be the eigenpair associated with the eigenvalues of B that converge to those of L. Let $(Z \quad Z_\perp)$ be unitary. From the relation $BZ = ZN$ it follows that

$$\begin{pmatrix} Z^{\mathrm{H}} \\ Z_\perp^{\mathrm{H}} \end{pmatrix} B \, (Z \quad Z_\perp) = \begin{pmatrix} N & H \\ 0 & C \end{pmatrix}.$$

Now let E be such that $(Q^{-1}LQ, \hat{Y})$ is an eigenblock of $B + E$. The following theorem, which shows that under appropriate conditions $B + E$ has an eigenpair near (N, Z), is an immediate consequence of Theorem V.2.7 in [14].

Theorem 5.1. *Let*

$$\eta = \frac{2\|E\|}{\operatorname{sep}(N, C) - 2\|E\|} < 1. \tag{5.1}$$

Then there are an orthonormal eigenpair $(\tilde{N}, \tilde{Z})$ and a complementary eigenblock $\tilde{C}$ of $B + E$ such that

$$\tan \angle(\tilde{Z}, Z) \le \eta$$

and

$$\|\tilde{N} - N\|, \|\tilde{C} - C\| \le \frac{(1 + \eta)\|E\| + \eta\|B\|}{\sqrt{1 - \eta^2}}.$$

The condition (5.1) of Theorem 5.1 is not automatically satisfied. Since the only assumption we have made about $\mathcal{W}$ is that it contains a good approximation to $\mathcal{X}$, the eigenvalues of C can lie almost anywhere within a circle about the origin of radius $\|A\|$. In particular, it could happen that as $\epsilon \to 0$ the matrix C has a rogue eigenvalue that converges to an eigenvalue of L so quickly that $\operatorname{sep}(N, C)$ is always too small for (5.1) to be satisfied. From now on, we will assume that *there is a constant α independent of ϵ such that*

$$\operatorname{sep}(N, C) \ge \alpha > 0. \tag{5.2}$$

We will call this the *uniform separation condition.* It implies the condition (5.1), at least for sufficiently small E. By (2.5) this condition is independent of the orthonormal bases Z and $Z_\perp$ used to define N and C. Note that in principle the uniform separation condition can be monitored computationally by computing $\operatorname{sep}(N, C)$ during the Rayleigh–Ritz procedure.

To see how this theorem implies the convergence of eigenblocks, recall that $B+E$ has the eigenpair $(Q^{-1}LQ, \hat{Y})$. By construction, as $\epsilon \to 0$ the eigenvalues of N converge to those of L, and hence by Elsner's theorem so do the eigenvalues of the eigenblock $\tilde{N}$, which is an increasingly small perturbation of N. But by the the continuity of sep the eigenvalues of $\tilde{N}$ are bounded away from those of $\tilde{C}$. Hence the eigenvalues of $\tilde{N}$ are the same as those of $Q^{-1}LQ$; i.e., $\Lambda(\tilde{N}) = \Lambda(L)$. But a simple eigenspace is uniquely determined by its eigenvalues.[1] Hence for some unitary matrix U we have $\tilde{Z} = \hat{Y}U$, $\tilde{X} = W\hat{Y}U$, and $\tilde{N} = U^{\mathrm{H}}(Q^{-1}LQ)U$. Since $\|N - \tilde{N}\| \to 0$ and $Q \to I$, we have the following theorem.

Theorem 5.2. *Under the uniform separation condition, there is a unitary matrix U, depending on ϵ, such that, as $\epsilon \to 0$, the eigenpair $(UNU^{\mathrm{H}}, ZU^{\mathrm{H}})$ approaches $(L, \hat{Y})$. Consequently, by* (3.4) *the Ritz pair $(UNU, \tilde{X}U^{\mathrm{H}})$ approaches the eigenpair (L, X).*

[1] Although this fact seems obvious, its proof is nontrivial.

Thus the uniform separation condition implies the convergence of Ritz pairs, up to unitary adjustments. In the sequel we will assume that these adjustments have been made and simply speak of the convergence of $(N, \tilde{X})$ to (L, X).

By combining the error bounds in Theorems 4.1 and 5.1 and the inequality (3.4) we can, after some manipulation, establish the following on the asymptotic rate of convergence of the Ritz pairs.

Corollary 5.3. *If the uniform separation condition holds, then*

$$\sin \angle(X, \tilde{X}), \frac{\|N - L\|}{\|A\|} \le \left[1 + 2\frac{\|A\|}{\operatorname{sep}(L, C)}\right] \epsilon + O(\epsilon^2). \tag{5.3}$$

Thus the convergence is at a rate proportional to the first power of ϵ. The main factor affecting the convergence constant is the reciprocal of the normalized separation $\operatorname{sep}(L, C)/\|A\|$, which by the uniform separation condition is bounded away from zero. The linear bound for the convergence of N to L does not imply eigenvalues of N converge at the same rate; however, their convergence is at worst as the $\frac{1}{\ell}$th power of ϵ, which is better than the rate in Corollary 4.2.

6. Residual analysis

Although the uniform separation condition insures convergence (in the sense of Theorem 5.2) of the Ritz pair $(N, \tilde{X}) = (N, WZ)$ to the eigenpair (L, X), it does not tell us when to stop the iteration. A widely used convergence criterion is to stop when the *residual*

$$R = A\tilde{X} - \tilde{X}N$$

is sufficiently small. Note that this residual is easily computable. For in the course of computing the Rayleigh quotient B, we must compute the matrix $V = AW$. After the pair $(N, \tilde{X})$ has been determined, we can compute R in the form $VZ - \tilde{X}N$.

We will be interested in the relation of the residual to the accuracy of $\tilde{X} = WZ$, as an approximation to the eigenbasis X in the eigenpair (X, L). To do this we will need some additional notation. Let $(X \;\; X_\perp)$ be unitary. Then

$$\begin{pmatrix} X^{\mathrm{H}} \\ X_\perp^{\mathrm{H}} \end{pmatrix} A \begin{pmatrix} X & X_\perp \end{pmatrix} = \begin{pmatrix} L & G \\ 0 & M \end{pmatrix}, \tag{6.1}$$

where the $(2, 1)$-element of the right hand side is zero because $X_\perp^{\mathrm{H}} A X = X_\perp^{\mathrm{H}} X L = 0$. The following theorem of Ipsen [4] holds for any approximate eigenpair.

Theorem 6.1. *Let $(\tilde{L}, \tilde{X})$ be an approximate eigenpair of A, and let*

$$\rho = \|A\tilde{X} - \tilde{X}\tilde{L}\|.$$

Then

$$\sin \angle(\tilde{X}, X) \le \frac{\rho}{\operatorname{sep}(\tilde{L}, M)}.$$

Proof. From (6.1) we have $X_\perp^{\mathrm{H}} A = M X_\perp^{\mathrm{H}}$. Hence if $R = A\tilde{X} - \tilde{X}\tilde{L}$, we have

$$X_\perp^{\mathrm{H}} R = X_\perp^{\mathrm{H}} A \tilde{X} - X_\perp^{\mathrm{H}} \tilde{X}\tilde{L} = M X_\perp^{\mathrm{H}} \tilde{X} - X_\perp^{\mathrm{H}} \tilde{X}\tilde{L}.$$

It follows that

$$\sin \angle(X, \tilde{X}) = \|X_\perp^{\mathrm{H}} \tilde{X}\| \le \frac{\|R\|}{\operatorname{sep}(\tilde{L}, M)}.$$

□

In our application $\tilde{X} = WZ$ and $\tilde{L} = N$. Hence the accuracy of the space spanned by WZ as an approximation to the space $\mathcal{X}$ is proportional to the size of the residual and inversely proportional to $\operatorname{sep}(N, M)$. If the uniform separation condition holds, then up to unitary similarities $N \to L$, so that by the continuity of sep, the accuracy is effectively inversely proportional to $\operatorname{sep}(L, M)$. Unlike the bounds in the previous sections, these bounds cannot be computed, since M is unknown. Nonetheless they provide us some insight into the attainable accuracy of Ritz approximations to an eigenspace.

7. Refined Ritz vectors

When $\ell = 1$, so that our concern is with approximating an eigenvector x and its eigenvalue λ, Theorem 6.1 has a suggestive implication. From Theorem 2.1, we know that there is a Ritz pair $(\nu, Wz) = (\nu, \tilde{x})$ such that ν converges to λ. Hence by Theorem 6.1, if the residual $A\tilde{x} - \nu\tilde{x}$ approaches zero, $\tilde{x}$ approaches x, independently of whether the uniform separation condition (5.2) holds.

Unfortunately, if the uniform separation condition fails to hold, we will generally be faced with a cluster of Ritz values and their Ritz vectors, of which at most one (and more likely none) is a reasonable approximation to x. Now Theorem 6.1 does not require that $(\nu, \tilde{x})$ be a Ritz pair—only that ν be sufficiently near λ, and that $\tilde{x}$ have a sufficiently small residual. Since the Ritz value ν is known to converge to λ, this suggests that we can deal with the problem of nonconverging Ritz vectors by retaining the Ritz value and replacing the Ritz vector with a vector $\hat{x} \in \mathcal{W}$ having a suitably small residual. It is natural to choose the best such vector. Thus we take $\hat{x}$ to be the solution of the problem

$$\begin{array}{ll} \text{minimize} & \|(A - \nu I)\hat{x}\| \\ \text{subject to} & \hat{x} \in \mathcal{W}, \|\hat{x}\| = 1. \end{array}$$

Alternatively, $\hat{x} = Wv$, where v is the right singular vector of $(A - \nu I)W$ corresponding to its smallest singular value. We will call such a vector a refined Ritz vector.

The following theorem shows that the refined Ritz vectors converge as $\epsilon \to 0$.

Theorem 7.1. *If*

$$\operatorname{sep}(\nu, M) \geq \operatorname{sep}(\lambda, M) - |\nu - \lambda| > 0, \tag{7.1}$$

then

$$\sin \angle(x, \hat{x}) \leq \frac{\|A - \nu I\|\epsilon + |\lambda - \nu|}{\sqrt{1 - \epsilon^2}(\operatorname{sep}(\lambda, M) - |\lambda - \nu|)}. \tag{7.2}$$

Proof. Let

$$\hat{y} = \frac{P_{\mathcal{W}} x}{\sqrt{1 - \epsilon^2}}$$

be the normalized projection of x onto $\mathcal{W}$ [cf. (3.3)] and let

$$e = (I - P_{\mathcal{W}})x.$$

Then

$$
\begin{aligned}
(A-\nu I)\hat{y} &= \frac{(A-\nu I)P_{\mathcal{W}}x}{\sqrt{1-\epsilon^2}} \\
&= \frac{(A-\nu I)(x-e)}{\sqrt{1-\epsilon^2}} \\
&= \frac{(\lambda-\nu)x-(A-\nu I)e}{\sqrt{1-\epsilon^2}}.
\end{aligned}
$$

Hence

$$
\|(A-\nu I)\hat{y}\| \le \frac{\|A-\nu I\|\epsilon + |\lambda-\nu|}{\sqrt{1-\epsilon^2}}.
$$

By the minimality of $\hat{x}$ we have

$$
\|(A-\nu I)\hat{x}\| \le \frac{\|A-\nu I\|\epsilon + |\lambda-\nu|}{\sqrt{1-\epsilon^2}}. \tag{7.3}
$$

Since $\|(A-\nu I)\hat{x}\|$ is a residual norm, (7.2) follows directly from Theorem 6.1 and (2.4). □

It follows immediately from (7.2) that if $\nu \to \lambda$ as $\epsilon \to 0$, then the refined Ritz vector $\hat{x}$ converges to the eigenvector x. In particular, by Corollary 4.2 this will happen if ν is chosen to be the Ritz value.

As they are defined, Ritz vectors are computationally expensive. If A is of order n and the dimension of $\mathcal{W}$ is m, a Rayleigh–Ritz procedure requires $O(nm^2)$ operations to produce a complete set of m Ritz vectors. On the other hand, to compute a refined Ritz vector requires the computation of the singular value decomposition of $(A-\nu I)W$, which also requires $O(nm^2)$ operations. Thus in general the computation of a single refined Ritz vector requires the same order of work as an entire Rayleigh–Ritz procedure. Fortunately, if $\mathcal{W}$ is determined by a Krylov sequence, as in the Arnoldi method, this work can be reduced to $O(m^3)$ [7, 9]. Moreover, if we are willing to sacrifice some accuracy, we can compute the refined vectors from the cross-product matrices $W^{\mathrm{H}}A^{\mathrm{H}}AW$ and $W^{\mathrm{H}}AW$ with $O(m^3)$ work [11].[2] Thus in some important cases, the computation of refined Ritz vectors is a viable alternative to the Rayleigh–Ritz procedure for eigenvectors.

There is a natural generalization of refined Ritz vectors to higher dimensions. Specifically, given an approximate eigenblock N, we can solve the problem

$$
\begin{array}{ll}
\text{minimize} & \|(A\hat{X}-\hat{X}N)\| \\
\text{subject to} & \mathcal{R}(\hat{X}) \subset \mathcal{W}, \hat{X}^{\mathrm{H}}\hat{X} = I.
\end{array}
$$

The resulting basis satisfies a theorem analogous to Theorem 7.1. There are, however, two difficulties with this approach. First, there seems to be no reasonably efficient algorithm for computing refined Ritz bases. Second, for the bound on the refined Ritz basis to converge to zero, we must have $N \to L$. However, the only reasonable hypothesis under which $N \to L$ is the uniform separation condition, and if that condition is satisfied the ordinary Ritz bases also converge.

[2] If the singular values of $(A-\nu I)W$ are $\sigma_1 \ge \cdots \ge \sigma_m$ and σ_m is small, then the loss of accuracy comes from the fact that the accuracy of the computed singular vector is around $(\sigma_1/\sigma_{m-1})\epsilon_{\mathrm{M}}$, where ϵ_{M} is the rounding unit. If we pass to the cross-product matrix, then the ratio σ_1/σ_{m-1} is replaced by the larger value $(\sigma_1/\sigma_{m-1})^2$. If the ratio is near one, then the squaring will have little effect. But if the ratio is fairly large — as it will be when we are attempting to resolve poorly separated eigenvalues — then considerable accuracy can be lost.

8. Discussion

We have considered the convergence of Ritz pairs generated from a subspace $\mathcal{W}$ to an eigenpair (L, X) associated with an eigenspace $\mathcal{X}$. The results are cast in terms of the quantity $\epsilon = \sin\angle(\mathcal{X}, \mathcal{W})$. An appealing aspect of the analysis is that we do not need to assume that eigenvalues of A are distinct or that A is nondefective.

The first result shows that as $\epsilon \to 0$, there are eigenvalues of the Rayleigh quotient B that converge to the eigenvalues of L. Unfortunately, that is not sufficient for the convergence of the eigenpairs, which requires the uniform separation condition (5.2) to separate the converging eigenvalues from the remaining eigenvalues of B. This condition, which can be monitored during the Rayleigh–Ritz steps, insures that the Ritz pairs $(N, \tilde{X})$ converge (with unitary adjustment) to the pair (L, X). The asymptotic convergence bounds (5.3) show that the convergence is linear in ϵ.

When $\mathcal{X}$ is one dimensional — that is, when we are concerned with approximating an eigenpair (λ, x) — the Ritz blocks, which become scalar Ritz values, converge without the uniform separation condition. However, this condition is required for the convergence of the Ritz vectors. Alternatively, we can compute refined Ritz vectors, whose convergence is guaranteed without the uniform separation condition.

We have analyzed only the the simplest version of the Rayleigh–Ritz procedure. In some forms of the method, the Rayleigh quotient is defined by $V^{\mathrm{H}}AW$, where $V^{\mathrm{H}}W = I$. We expect that the above results will generalize easily to this case, provided the product $\|V\|\|W\|$ remains uniformly bounded.

References

[1] R. Bhatia, L. Elsner, and G. Krause, *Bounds for the variation of the roots of a polynomial and the eigenvalues of a matrix*, Linear Algebra and Its Applications **142** (1990), 195–209. MR **92i:**12001

[2] L. Elsner, *An optimal bound for the spectral variation of two matrices*, Linear Algebra and Its Applications **71** (1985), 77–80. MR **87c:**15035

[3] G. H. Golub and C. F. Van Loan, *Matrix computations*, second ed., Johns Hopkins University Press, Baltimore, MD, 1989. MR **90d:**65055

[4] I. C. F. Ipsen, *Absolute and relative perturbation bounds for invariant subspaces of matrices*, Technical Report TR97-35, Center for Research in Scientific Computation, Mathematics Department, North Carolina State Unversity, 1998.

[5] Z. Jia, *Some numerical methods for large unsymmetric eigenproblems*, Ph.D. thesis, University of Bielefeld, 1994.

[6] ———, *The convergence of generalized Lanczos methods for large unsymmetric eigenproblems*, SIAM Journal on Matrix Analysis and Applications **16** (1995), 843–862. MR **96d:**65062

[7] ———, *Refined iterative algorithm based on Arnoldi's process for large unsymmetric eigenproblems*, Linear Algebra and Its Applications **259** (1997), 1–23. MR **98c:**65060

[8] ———, *Generalized block Lanczos methods for large unsymmetric eigenproblems*, Numerische Mathematik **80** (1998), 171–189. MR **95f:**65059

[9] ———, *A refined iterative algorithm based on the block Arnoldi algorithm*, Linear Algebra and Its Applications **270** (1998), 171–189. MR **98m:**65055

[10] ———, *Polynomial characterizations of the approximate eigenvectors by the refined Arnoldi method and an implicitly restarted refined Arnoldi algorithm*, Linear Algebra and Its Applications **287** (1999), 191–214. MR **99j:**65046

[11] ———, *A refined subspace iteration algorithm for large sparse eigenproblems*, To appear in *Applied Numerical Mathemtics.*, 1999.

[12] Y. Saad, *Numerical methods for large eigenvalue problems: Theory and algorithms*, John Wiley, New York, 1992. MR **93h:**65052

[13] G. L. G. Sleijpen and H. A. Van der Vorst, *A Jacobi–Davidson iteration method for linear eigenvalue problems*, SIAM Journal on Matrix Analysis and Applications **17** (1996), 401–425. MR **96m:**65042

[14] G. W. Stewart and J.-G. Sun, *Matrix perturbation theory*, Academic Press, New York, 1990. MR **92a:**65017

Department of Applied Mathematics, Dalian University of Technology, Dalian 116024, P.R. China

E-mail address: `zxjia@dlut.edu.cn`

Department of Computer Science, Institute for Advanced Computer Studies, University of Maryland, College Park, MD 20742, USA

E-mail address: `stewart@cs.umd.edu`

17

Papers on the Generalized Eigenproblem

1. [GWS-J16] "On the Sensitivity of the Eigenvalue Problem $Ax = \lambda Bx$," *SIAM Journal on Numerical Analysis* **9** (1972) 669–686.
2. [GWS-J18] (with C. B. Moler), "An Algorithm for Generalized Matrix Eigenvalue Problems," *SIAM Journal on Numerical Analysis* **10** (1973) 241–256.
3. [GWS-J27] "Gershgorin Theory for the Generalized Eigenvalue Problem $Ax = \lambda Bx$," *Mathematics of Computation* **29** (1975) 600–606.
4. [GWS-J38] "Perturbation Bounds for the Definite Generalized Eigenvalue Problem," *Linear Algebra and its Applications* **23** (1979) 69–85.

17.1. [GWS-J16] "On the Sensitivity of the Eigenvalue Problem $Ax = \lambda Bx$"

[GWS-J16] "On the Sensitivity of the Eigenvalue Problem $Ax = \lambda Bx$," *SIAM Journal on Numerical Analysis* **9** (1972) 669–686.
http://dx.doi.org/10.1137/0709056

SIAM J. NUMER. ANAL.
Vol. 9, No. 4, December 1972

ON THE SENSITIVITY OF THE EIGENVALUE PROBLEM $Ax = \lambda Bx$*

G. W. STEWART†

Abstract. This paper considers the sensitivity of the eigenvalues and eigenvectors of the generalized matrix eigenvalue problem $Ax = \lambda Bx$ to perturbations of A and B. The notion of a deflating subspace for the problem is introduced, and error bounds for approximate deflating subspaces obtained. The bounds also provide information about the eigenvalues of the problem. The resulting perturbation bounds estimate realistically the sensitivity of the eigenvalues, even when B is singular or nearly singular. The results are applied to the important special case where A is Hermitian and B is positive definite.

1. Introduction. Let A and B be matrices of order n with complex elements. By the eigenvalue problem $A - \lambda B$ we mean the problem of determining the set of all λ for which the equation

$$Ax = \lambda Bx \tag{1.1}$$

has a nontrivial solution. In this case λ is called an eigenvalue and x an eigenvector of $A - \lambda B$. The set of all eigenvalues of $A - \lambda B$ is denoted by $\lambda(A, B)$.

The eigenvalue problem $A - \lambda B$ arises in the theory of systems of ordinary differential equations with constant coefficients and has important physical applications (e.g., see [5]). The algebraic theory of the problem has been worked out in detail (see Gantmacher [3] for theorems, proofs, and references). In this paper we shall be concerned with the analytic problem of determining how perturbations in A and B affect the eigenvalues and eigenvectors of $A - \lambda B$.

We shall not treat the case when the characteristic polynomial

$$\rho(\lambda) = \det(A - \lambda B)$$

is identically zero. In this case any number λ is an eigenvalue of $A - \lambda B$. In particular, $\rho(\lambda) \equiv 0$ whenever A and B have a common null space, and the assumption that they do not is used throughout this paper. We do not wish to suggest that positive results cannot be obtained for this case. It is an open question whether there exist stable eigenvalue-eigenvector pairs when $\rho(\lambda) \equiv 0$.

Another pathological case occurs when B is singular. In this case the characteristic polynomial is of degree less than n. In the algebraic theory it is customary to say that $A - \lambda B$ has an infinite eigenvalue; and, indeed, the characteristic polynomial of the reciprocal problem $B - \mu A$ has a zero at the origin equal in multiplicity to the defect k in the characteristic polynomial of $A - \lambda B$. A corresponding eigenvector for both problems is naturally a null vector of B. Moreover, if B is perturbed slightly, $\lambda(A, B)$ will generally contain k large eigenvalues that become infinite as the perturbation is reduced to zero.

If B is nearly singular, the nearly infinite eigenvalues of $A - \lambda B$ are extremely sensitive to perturbations in B and cannot be calculated accurately without resort to high precision arithmetic. However, there may be other eigenvalues that

* Received by the editors April 26, 1971, and in revised form August 20, 1971.

† Center for Numerical Analysis, The University of Texas, Austin, Texas 78712. This work was supported in part by the Office of Naval Research.

are insensitive to perturbations in A and B. One of the objects of this paper is to derive perturbation bounds that realistically estimate the error in such insensitive eigenvalues.

An important special case of the problem $A - \lambda B$ is when A is Hermitian and B is positive definite. Considerable attention has been given to the problem of computing the eigenvalues and eigenvectors in this case (see, for example, the survey by Peters and Wilkinson [6]) as well as the problem of obtaining perturbation bounds for the eigenvalues [1]. Fix and Heiberger [2] have proposed an algorithm for the case where B is nearly or exactly singular. Our perturbation bounds apply also to this problem; however, they do not depend in any essential way on the positive definiteness of B, so that they remain valid for perturbations that could make B singular or indefinite.

The results of this paper are obtained by modifying first order perturbation estimates so that they become rigorous bounds. Such techniques have been used by the author to obtain error bounds for invariant subspaces of linear operators in Hilbert space [7], and this paper is close in spirit to that work. In § 2 we introduce the idea of a deflating subspace for the problem $A - \lambda B$ and consider two deflation procedures, one by unitary transformations and the other by congruences. In § 3 we derive two reductions of the problem to triangular form, both in some sense generalizing the Schur reduction of a matrix to triangular form by unitary similarity transformations. In § 4 we discuss a linear operator associated with two eigenvalue problems of the form $A - \lambda B$ and use it to introduce a measure of the separation between the eigenvalues of two such problems. This measure has the property that it is insensitive to small perturbations in the matrices involved. In § 5 we derive error bounds for approximate deflating subspaces and use these in turn to derive perturbation bounds for deflating subspaces. There are two types of theorems, one for unitary deflation, the other for congruential deflation. Finally in § 6 we conclude with an application of the results to the case where A is Hermitian and B is positive definite.

We shall follow Householder's notational conventions [4], with the following exceptions. Complex n-space will be denoted by $\mathbb{C}^n$ and the space of $m \times n$ matrices with complex elements by $\mathbb{C}^{m\times n}$. The symbol $\|\cdot\|_2$ will denote both the Euclidean length of a vector and the spectral norm of a matrix:

$$\|A\|_2 = \sup_{\|x\|_2=1} \|Ax\|.$$

The symbol $\|\cdot\|_F$ will denote the Frobenius matrix norm defined by

$$\|A\|_F^2 = \mathrm{trace}\,(A^H A).$$

2. Deflation. Let U and V be nonsingular matrices of order n and let

$$A' = U^H A V, \quad B' = U^H B V.$$

Then x is an eigenvector of $A - \lambda B$ with eigenvalue λ if and only if $y = V^{-1}x$ is an eigenvector of $A' - \lambda B'$ with eigenvalue λ. Moreover, the characteristic polynomial of $A' - \lambda B'$ is a nonzero constant multiple of the characteristic polynomial of $A - \lambda B$. Hence the multiplicities of the eigenvalues are preserved in the passage from $A - \lambda B$ to $A' - \lambda B'$. We say that the problem $A - \lambda B$ is related to the problem $A' - \lambda B'$ by an equivalence transformation.

Two classes of equivalence transformations are especially important. In the first, the class of unitary equivalences, U and V are taken to be unitary. An attractive feature of unitary equivalences is that they preserve the norms (either $\|\cdot\|_2$ or $\|\cdot\|_F$) of A and B. In addition, if A and B are perturbed, the corresponding perturbations in A' and B' have the same norms as the original perturbations. In the second class, the class of congruence transformations, U is taken equal to V. Such transformations preserve symmetry in A or B. In this and the next section we shall be concerned with reducing the problem $A - \lambda B$ to simpler forms by unitary equivalences and congruences.

Let $\mathscr{X}$ be a subspace of $\mathbb{C}^n$ of dimension l. Then $\mathscr{X}$ is a *deflating subspace* for $A - \lambda B$ if

$$\dim(A\mathscr{X} + B\mathscr{X}) \leqq l. \tag{2.1}$$

It is a consequence of Theorem 2.1 below that a deflating subspace $\mathscr{X}$ behaves much like an invariant subspace for the usual eigenvalue problem. Namely, there is a problem of order l whose eigenvalues are all eigenvalues of $A - \lambda B$, multiplicities counted. To every eigenvector of the reduced problem there is a corresponding eigenvector of $A - \lambda B$ that lies in $\mathscr{X}$, and conversely.

THEOREM 2.1. *Let $\mathscr{X}$ be a deflating subspace of $A - \lambda B$ of dimension l. Then there are unitary matrices U and V such that the first l columns of V span $\mathscr{X}$ and*

$$U^H A V = \begin{pmatrix} A_{11} & A_{12} \\ 0 & A_{22} \end{pmatrix}, \quad U^H B V = \begin{pmatrix} B_{11} & B_{12} \\ 0 & B_{22} \end{pmatrix}, \tag{2.2}$$

where A_{11} and B_{11} are of order l.

Proof. Let $V_1 \in \mathbb{C}^{n\times l}$ have orthonormal columns that span $\mathscr{X}$. Choose $V_2 \in \mathbb{C}^{n\times(n-l)}$ so that $V = (V_1, V_2)$ is unitary. By construction, the first l columns of V span $\mathscr{X}$. Now by (2.1), $\dim(A\mathscr{X} + B\mathscr{X}) \leqq l$. Let $U_2 \in \mathbb{C}^{n\times(n-l)}$ have orthonormal columns lying in the orthogonal complement of $A\mathscr{X} + B\mathscr{X}$. Let $U_1 \in \mathbb{C}^{n\times l}$ be such that $U = (U_1, U_2)$ is unitary. Then $U_2^H A V_1 = U_2^H B V_1 = 0$, which implies that $U^H A V$ and $U^H B V$ have the forms (2.2).

The reduced problem mentioned above is of course $A_{11} - \lambda B_{11}$. If λ is an eigenvalue of $A_{11} - \lambda B_{11}$ with eigenvector z, then λ is an eigenvalue of $A - \lambda B$ with eigenvector $V_1 z \in \mathscr{X}$. The remaining eigenvalues of $A - \lambda B$ are the eigenvalues of $A_{22} - \lambda B_{22}$.

Theorem 2.1 shows that a knowledge of any deflating subspace of $A - \lambda B$ can be used to deflate the problem into two smaller problems by unitary equivalences. A similar deflation using congruences may be accomplished, provided the deflating subspace satisfies an additional condition.

THEOREM 2.2. *Let $\mathscr{X}$ be a deflating subspace of $A - \lambda B$ of dimension l, and let the columns of $U_1 \in \mathbb{C}^{n\times l}$ form a basis for $\mathscr{X}$. Then if either $U_1^H A U_1$ or $U_1^H B U_1$ is nonsingular, there is a nonsingular matrix $U = (U_1, U_2)$ such that*

$$U^H A U = \begin{pmatrix} A_{11} & A_{12} \\ 0 & A_{22} \end{pmatrix}, \quad U^H B U = \begin{pmatrix} B_{11} & B_{12} \\ 0 & B_{22} \end{pmatrix}, \tag{2.3}$$

where A_{11} and B_{11} are of order l.

Proof. Suppose that, say, $U_1^H B U_1$ is nonsingular. Then $\dim(B\mathscr{X}) = l$, and by (2.1),

$$A\mathscr{X} \subset B\mathscr{X}. \tag{2.4}$$

Let the columns of U_2 span the orthogonal complement of $B\mathscr{X}$, so that $U_2^H B U_1 = 0$. We claim that $U = (U_1, U_2)$ is nonsingular. Suppose to the contrary that $U_1 y = U_2 z$ for some $y, z \neq 0$. Then

$$y^H U_1^H B U_1 = z^H U_2^H B U_1 = z^H 0 = 0,$$

contradicting the nonsingularity of $U_1^H B U_1$. Since $U_2^H B U_1 = 0$, it follows from (2.4) that $U_2^H A U_1 = 0$, which implies that $U^H A U$ and $U^H B U$ have the forms (2.3).

We note that if $U_1^H B U_1$ is nonsingular for a single U_1 whose columns form a basis for $\mathscr{X}$, then it is nonsingular for any such U_1.

An important special case where $U_1^H B U_1$ is always nonsingular is when B is positive definite. In this case not only is $C = U_1^H B U_1$ nonsingular, but it is positive definite and has a square root $C^{1/2}$. If we set $U_1' = U_1 C^{-1/2}$, then the columns of U_1' form a basis for $\mathscr{X}$ and $U_1'^H B U_1' = I$. Thus if U_1' is used in place of U_1 in Theorem 2.2 and the symmetry of B taken into account, we have the following corollary.

COROLLARY 2.3. *Let B be positive definite and let $\mathscr{X}$ be a deflating subspace of $A - \lambda B$ of dimension l. Then there is a nonsingular matrix U whose first l columns span $\mathscr{X}$ such that*

$$U^H A U = \begin{pmatrix} A_{11} & A_{12} \\ 0 & A_{22} \end{pmatrix}, \quad U^H B U = \begin{pmatrix} I_l & 0 \\ 0 & B_{22} \end{pmatrix},$$

where A_{11} is of order l.

3. Triangularization. In this section we give two generalizations of the reduction of a matrix to triangular form by unitary similarity transformations.

THEOREM 3.1. *Let $A, B \in \mathbb{C}^{n \times n}$. Then there are unitary matrices U and V such that*

$$A' = U^H A V \quad \text{and} \quad B' = U^H B V$$

are upper triangular. The numbers α_{ii}'/β_{ii}' are the eigenvalues of $A - \lambda B$ and may be made to appear in any order.

Proof. The proof is by induction on the order n of A and B. The result is trivial for $n = 1$. For the induction step, assume that an order for the n eigenvalues of $A - \lambda B$ has been specified. Let λ_1 be the first eigenvalue and let v_1 be an eigenvector corresponding to λ_1 with $\|v_1\|_2 = 1$ (if B is singular v_1 may be taken to be a null vector of B). Then the space spanned by v_1 is a deflating subspace of $A - \lambda B$. Hence by Theorem 2.1 there are unitary matrices U_1 and V_1 with $v_1 = V_1 e_1$ such that

$$U_1^H A V_1 = \begin{pmatrix} \alpha_{11}' & a_{12}^H \\ 0 & A_{22} \end{pmatrix}, \quad U_1^H B V_1 = \begin{pmatrix} \beta_{11}' & b_{12}^H \\ 0 & B_{22} \end{pmatrix}.$$

Obviously $\lambda_1 = \alpha_{11}'/\beta_{11}'$ (if $\beta_{11}' = 0$, $\lambda = \infty$).

Now apply the induction hypothesis to find unitary matrices U_2 and V_2 of order $n - 1$ such that $A'_{22} = U_2^H A_{22} V_2$ and $B'_{22} = U_2^H B_{22} V_2$ are upper triangular. We can in addition choose U_2 and V_2 so that the ratios of the diagonal elements of A'_{22} and B'_{22} are the remaining eigenvalues of $A - \lambda B$, in the proper order. Then the unitary matrices

$$U = U_1\begin{pmatrix} 1 & 0 \\ 0 & U_2 \end{pmatrix}, \quad V = V_1\begin{pmatrix} 1 & 0 \\ 0 & V_2 \end{pmatrix}$$

give the required reduction.

Essentially the same proof using Corollary 2.3 shows that A and B may be reduced to triangular form by congruences provided B is positive definite. In fact the reduced B is diagonal and can be taken to be the identity matrix.

THEOREM 3.2. *Let $A, B \in \mathbb{C}^{n\times n}$ with B positive definite. Then there is a matrix $U \in \mathbb{C}^{n\times n}$ satisfying $U^H BU = I$ such that $A' = U^H AU$ is upper triangular. The matrix U may be chosen so that the diagonal elements of A', which are the eigenvalues of $A - \lambda B$, appear in any order.*

The matrix U of Theorem 3.2 is said to be unitary with respect to B, or simply B-unitary. Thus Theorem 3.2 asserts that if B is positive definite, any matrix can be reduced to triangular form by a B-unitary congruence.

If, in Theorem 3.2, A is Hermitian, then A' is diagonal. In this case it is easy to verify that the columns of $U = (u_1, u_2, \cdots, u_n)$ are eigenvectors of $A - \lambda B$. Since $u_i^H Bu_j = \delta_{ij}$, the vectors u_i are said to be B-orthonormal. This shows that *if A is Hermitian and B is positive definite, then the eigenvalue problem $A - \lambda B$ has a set of n B-orthonormal eigenvectors.*

4. The operator $T[(P, Q)] = (PA_1 - A_2Q, PB_1 - B_2Q)$. Let $A_1, B_1 \in \mathbb{C}^{l\times l}$ and $A_2, B_2 \in \mathbb{C}^{m\times m}$. Define an operator $T:\mathbb{C}^{m\times 2l} \to \mathbb{C}^{m\times 2l}$ by

$$T[(P, Q)] = (PA_1 - A_2Q, PB_1 - B_2Q)$$

for any $P, Q \in \mathbb{C}^{m\times l}$. The operator T is obviously linear. It plays an important role in deriving the error bounds of the next section. Accordingly, we describe some of its properties here.

The main problem is to state conditions under which T is nonsingular. This can be done succinctly in terms of the eigenvalues of $A_1 - \lambda B_1$ and $A_2 - \lambda B_2$.

THEOREM 4.1. *The operator T is nonsingular if and only if $\lambda(A_1, B_1) \cap \lambda(A_2, B_2) = \emptyset$.*

Proof. Suppose that $\lambda(A_1, B_1) \cap \lambda(A_2, B_2) = \emptyset$. We shall show that the equations

$$PA_1 - A_2Q = R, \quad PB_1 - B_2Q = S \tag{4.1}$$

have solutions P and Q for any $R, S \in \mathbb{C}^{m\times l}$, which is equivalent to saying T is nonsingular.

Let $U, V \in \mathbb{C}^{l\times l}$ be unitary. Then if $A'_1 = U^H A_1 V$, $B'_1 = U^H B_2 V$, $P' = PU$, and $Q' = QV$, the system (4.1) has a solution if and only if the system

$$P'A'_1 - A_2Q' = RV,$$
$$P'B'_1 - B_2Q' = SV$$

has a solution. Since, by Theorem 3.1, we may choose U and V so that A_1' and B_1' are upper triangular, we may assume without loss of generality that A_1 and B_1 in (4.1) are upper triangular. Likewise we may assume that A_2 and B_2 are upper triangular.

Let P, Q, R and S be partitioned by columns. Because A_1 and B_1 are upper triangular,

$$\alpha_{11}^{(1)}p_1 - A_2q_1 = r_1, \qquad \beta_{11}^{(1)}p_1 - B_2q_1 = s_1. \tag{4.2}$$

It is easy to verify that this system has a solution if $\alpha_{11}^{(1)}/\beta_{11}^{(1)} \notin \lambda(A_2, B_2)$, which is true since $\alpha_{11}^{(1)}/\beta_{11}^{(1)} \in \lambda(A_1, B_1)$.

Having computed $p_1, p_2, \cdots, p_{k-1}$ and $q_1, q_2, \cdots, q_{k-1}$, we may find p_k and q_k from the system

$$\alpha_{kk}^{(1)}p_k - A_2q_k = r_k - \sum_{i=1}^{k-1} \alpha_{ik}^{(1)}p_i, \qquad \beta_{kk}^{(1)}p_k - B_2q_k = s_k - \sum_{i=1}^{k-1} \beta_{ik}^{(1)}p_i. \tag{4.3}$$

The condition $\lambda(A_1, B_1) \cap \lambda(A_2, B_2) = \varnothing$ ensures that this system has a solution. In this way we may find P and Q, and it is easily verified that the P and Q so computed satisfy (4.1).

Conversely, suppose that $A_1 - \lambda B_1$ and $A_2 - \lambda B_2$ have a common eigenvalue λ. By an obvious variant of Theorem 3.1, we may take A_1 and B_1 to be lower triangular with $\lambda = \alpha_{11}^{(1)}/\beta_{11}^{(1)}$. We may also take A_2 and B_2 to be upper triangular with $\lambda = \alpha_{11}^{(2)}/\beta_{11}^{(2)}$. Then the system

$$\alpha_{11}^{(1)}\xi_1 - \alpha_{11}^{(2)}\xi_2 = 0,$$

$$\beta_{11}^{(1)}\xi_1 - \beta_{11}^{(2)}\xi_2 = 0$$

has a nontrivial solution. Let $P = \xi_1 e_1 e_1^H$ and $Q = \xi_2 e_1 e_1^H$. Then it is easy to see that $PA_1 - A_2Q = 0$ and $PB_1 - B_2Q = 0$, which shows that T is singular.

It should be noted that if A_1, B_1, A_2 and B_2 are upper triangular, equations (4.2) and (4.3) represent a computationally efficient way of solving the equation $T[(P, Q)] = (R, S)$.

In the next section we shall need estimates of the size of the solutions of the equation $T(X) = Y$. In order to do this, we define two norms on $\mathbb{C}^{m \times 2l}$ as follows. Any $R \in \mathbb{C}^{m \times 2l}$ may be written in the form $R = (P, Q)$, where $P, Q \in \mathbb{C}^{m \times l}$. Set

$$\|R\|_2' = \max\{\|P\|_2, \|Q\|_2\}$$

and

$$\|R\|_F' = \max\{\|P\|_F, \|Q\|_F\}.$$

It is easily verified that $\|\cdot\|_2'$ and $\|\cdot\|_F'$ are norms.

The operator T is determined by the matrices A_1, B_1, A_2 and B_2. Let

$$\operatorname{dif}(A_1, B_1; A_2, B_2) = \inf_{\|X\|_F'=1} \|T(X)\|_F'. \tag{4.4}$$

Then $\operatorname{dif}(A_1, B_1; A_2, B_2) = 0$ if and only if T is singular, that is, if and only if $\lambda(A_1, B_1) \cap \lambda(A_2, B_2) \neq \varnothing$. If T is nonsingular and $T(X) = Y$, then

$$\|X\|'_F \leqq \frac{\|Y\|'_F}{\operatorname{dif}(A_1, B_1; A_2, B_2)}.$$

An important feature of the function dif is that it is insensitive to perturbations in A_1, B_1, A_2 and B_2.

THEOREM 4.2.

$$\operatorname{dif}(A_1 + E_1, B_1 + F_1; A_2 + E_2, B_2 + F_2)$$
$$\geqq \operatorname{dif}(A_1, B_1; A_2, B_2) - \max\{\|E_1\|_2 + \|E_2\|_2, \|F_1\|_2 + \|F_2\|_2\}.$$

Proof. Let the infimum defining $\operatorname{dif}(A_1 + E_1, B_1 + F_1; A_2 + E_2, B_2 + F_2)$ be attained for $X = (P, Q)$. Then $\|P\|_F, \|Q\|_F \leqq 1$. Hence,

$$\begin{aligned}
&\operatorname{dif}(A_1 + E_1, B_1 + F_1; A_2 + E_2, B_2 + F_2)\\
&\quad = \|(P(A_1 + E_1) - (A_2 + E_2)Q, P(B_1 + F_1) - (B_2 + F_2)Q)\|'_F\\
&\quad = \max\{\|P(A_1 + E_1) - (A_2 + E_2)Q\|_F, \|P(B_1 + F_1) - (B_2 + F_2)Q\|_F\}\\
&\quad \geqq \max\{\|PA_1 - A_2Q\|_F - \|E_1\|_2 - \|E_2\|_2, \|PB_1 - B_2Q\|_F - \|F_1\|_2 - \|F_2\|_2\}\\
&\quad \geqq \max\{\|PA_1 - A_2Q\|_F, \|PB_1 - B_2Q\|_F\}\\
&\qquad - \max\{\|E_1\|_2 + \|E_2\|_2, \|F_1\|_2 + \|F_2\|_2\}\\
&\quad \geqq \operatorname{dif}(A_1, B_1; A_2, B_2) - \max\{\|E_1\|_2 + \|E_2\|_2, \|F_1\|_2 + \|F_2\|_2\}.
\end{aligned}$$

In [7] the function

$$\operatorname{sep}(A_1, A_2) = \inf_{\|P\|_F = 1} \|PA_1 - A_2P\|_F$$

was introduced to measure the separation between the spectra of A_1 and A_2. The number $\operatorname{sep}(A_1, A_2)$ appeared in perturbation bounds for invariant subspaces of a matrix A. The number $\operatorname{dif}(A_1, B_1; A_2, B_2)$ will play an analogous role for deflating subspaces of $A - \lambda B$. Since, when $B = I$, the eigenvalue problem $A - \lambda B$ reduces to an ordinary eigenvalue problem, it is natural to look for some connection between $\operatorname{sep}(A_1, A_2)$ and $\operatorname{dif}(A_1, I; A_2, I)$. Such a connection exists, provided A_1 and A_2 are suitably scaled.

THEOREM 4.3. *For all A_1 and A_2,*

$$\operatorname{sep}(A_1, A_2) \geqq \operatorname{dif}(A_1, I; A_2, I). \tag{4.5}$$

If $\|A_1\|_2, \|A_2\|_2 \leqq 1$, then

$$\operatorname{dif}(A_1, I; A_2, I) \geqq \tfrac{1}{2}\operatorname{sep}(A_1, A_2). \tag{4.6}$$

Proof. For (4.5), let $\operatorname{sep}(A_1, A_2) = \|PA_1 - A_2P\|_F$, where $\|P\|_F = 1$. Then

$$\begin{aligned}
\operatorname{dif}(A_1, I; A_2, I) &\leqq \|T[(P, P)]\|'_F\\
&= \|(PA_1 - A_2P, P - P)\|'_F = \operatorname{sep}(A_1, A_2).
\end{aligned}$$

For (4.6), let $\operatorname{dif}(A_1, I; A_2, I) = \|T[(P, Q)]\|'_F$, where $\|(P, Q)\|'_F = 1$. Suppose that, say, $\|Q\|_F = 1$. Let

$$PA_1 - A_2Q = E,$$
$$P - Q = F,$$

so that $\operatorname{dif}(A_1, I; A_2, I) = \max\{\|E\|_F, \|F\|_F\}$. Now $P = F + Q$, whence $E = QA_1 - A_2Q + FA_1$. Then

$$\|E\|_F \geqq \|QA_1 - A_2Q\|_F - \|FA_1\|_F \geqq \operatorname{sep}(A_1, A_2) - \|F\|_F.$$

Hence $\|E\|_F + \|F\|_F \geqq \operatorname{sep}(A_1, A_2)$ and it follows that $\operatorname{dif}(A_1, I; A_2, I) = \max\{\|E\|_F, \|F\|_F\} \geqq \frac{1}{2}\operatorname{sep}(A_1, A_2)$. A similar argument establishes the result when $\|P\|_F = 1$.

The operator T may be regarded as a linear operator on $\mathbb{C}^{2lm}$ by the simple device of arranging the elements of (P, Q) into a $2lm$-vector in some fixed order. The number $1/\operatorname{dif}(A_1, B_1; A_2, B_2)$ is then just the supremum norm of T^{-1} with respect to the norm induced on $\mathbb{C}^{2lm}$ by $\|\cdot\|'_F$. The following theorem is a consequence of the fact that

$$\|(P, Q)\|'_F \leqq \|(P, Q)\|_F \leqq \sqrt{2}\|(P, Q)\|'_F.$$

THEOREM 4.4. *Let the operator T be regarded as a linear transformation in $\mathbb{C}^{2lm}$. Then*

$$\frac{1}{\operatorname{dif}(A_1, B_1; A_2, B_2)} \leqq \sqrt{2}\|T^{-1}\|_2.$$

This theorem is useful in constructing bounds for $\operatorname{dif}(A_1, B_1; A_2, B_2)$ when a matrix representation of T is known.

5. Error and perturbation bounds. In this section we shall be concerned with finding bounds for the error in an approximate deflating subspace of $A - \lambda B$. The bounds may be applied to estimate the sensitivity of a deflating subspace to perturbations in A and B. The theorems also provide information concerning the sensitivity of the eigenvalues of $A - \lambda B$.

We consider first the case of unitary deflation. Let $U = (U_1, U_2)$ and $V = (V_1, V_2)$ be unitary matrices with $U_1, V_1 \in \mathbb{C}^{n \times l}$. Then if

$$A_{21} \equiv U_2^H A V_1 = 0$$

and

$$B_{21} \equiv U_2^H B V_1 = 0,$$

the columns of V_1 span a deflating subspace for $A - \lambda B$. If A_{21} and B_{21} are small, but not exactly zero, it is reasonable to ask if there are unitary matrices $U' = (U'_1, U'_2)$ and $V' = (V'_1, V'_2)$ near U and V such that

$$U_2'^H A V'_1 = U_2'^H B V'_1 = 0. \tag{5.1}$$

To answer this question we seek $U' = (U'_1, U'_2)$ in the form

$$\begin{aligned} U'_1 &= (U_1 + U_2P)(I + P^HP)^{-1/2}, \\ U'_2 &= (U_2 - U_1P^H)(I + PP^H)^{-1/2}, \end{aligned} \tag{5.2}$$

and $V' = (V'_1, V'_2)$ in the form

$$V'_1 = (V_1 + V_2 Q)(I + Q^H Q)^{-1/2},$$
$$V'_2 = (V_2 - V_1 Q^H)(I + QQ^H)^{-1/2}, \tag{5.3}$$

where $P, Q \in \mathbb{C}^{(n-l)\times l}$. It is easily verified that U' and V' are unitary. Moreover, if

$$A_{ij} = U_i^H A V_j, \qquad i,j = 1,2, \tag{5.4}$$

and

$$B_{ij} = U_i^H B V_j, \qquad i,j = 1,2, \tag{5.5}$$

then it is easily verified that the condition (5.1) is equivalent to the pair of matrix equations

$$PA_{11} - A_{22}Q = A_{21} - PA_{12}Q,$$
$$PB_{11} - B_{22}Q = B_{21} - PB_{12}Q. \tag{5.6}$$

If, as in § 4, we set $T[(P, Q)] = (PA_{11} - A_{22}Q, PB_{11} - B_{22}Q)$, the system (5.6) becomes

$$T[(P, Q)] = (A_{21} - PA_{12}Q, B_{21} - PB_{12}Q). \tag{5.7}$$

Thus the problem of perturbing U and V into deflating matrices becomes one of determining under what conditions the nonlinear equation (5.7) has a solution. Such conditions are given in the following theorem.

THEOREM 5.1. *Let $U = (U_1, U_2)$ and $V = (V_1, V_2)$ be unitary with $U_1, V_1 \in \mathbb{C}^{n\times l}$. Let A_{ij} and B_{ij} be defined by (5.4) and (5.5). Let*

$$\gamma = \|(A_{21}, B_{21})\|'_F, \quad \eta = \|(A_{12}^T, B_{12}^T)\|'_2,$$

and

$$\delta = \operatorname{dif}(A_{11}, B_{11}; A_{22}, B_{22}).$$

If

$$\kappa_1 = \frac{\gamma\eta}{\delta^2} < \frac{1}{4}, \tag{5.8}$$

then there are matrices $P, Q \in \mathbb{C}^{(n-l)\times l}$ satisfying

$$\|(P, Q)\|'_F \leqq \frac{\gamma}{\delta}(1 + \kappa) \equiv \frac{\gamma}{\delta}\frac{1 + \sqrt{1 - 4\kappa_1}}{1 - 2\kappa_1 + \sqrt{1 - 4\kappa_1}} < 2\frac{\gamma}{\delta} \tag{5.9}$$

such that the column space of $V_1 + V_2 Q$ is a deflating subspace for $A - \lambda B$. Moreover, $\lambda(A, B)$ is the disjoint union of $\lambda(A_{11} + A_{12}Q, B_{11} + B_{12}Q)$ and $\lambda(A_{22} - PA_{12}, B_{22} - PB_{12})$.

Proof. The proof is adapted from [7]. It is required to solve equation (5.7) for P and Q. Let

$$(P_0, Q_0) = T^{-1}[(A_{21}, B_{21})], \tag{5.10}$$

and given (P_i, Q_i), set

$$(5.11)\qquad \begin{aligned}(P_{i+1}, Q_{i+1}) &= T^{-1}[(A_{21} - P_iA_{12}Q_i, B_{21} - P_iB_{12}Q_i)] \\ &= (P_0, Q_0) - T^{-1}[(P_iA_{12}Q_i, P_iB_{12}Q_i)].\end{aligned}$$

We shall show that the sequence $\{(P_i, Q_i)\}$ converges to a solution of (5.7).

The first step is to obtain bounds for $\|(P_i, Q_i)\|'_F$. From (5.10),

$$\|(P_0, Q_0)\|'_F \leqq \delta^{-1}\gamma \equiv \rho_0.$$

From (5.11), if $\|(P_i, Q_i)\|'_F \leqq \rho_i$, then

$$\|(P_{i+1}, Q_{i+1})\|'_F \leqq \rho_0 + \delta^{-1}\eta\rho_i^2 \equiv \rho_{i+1}.$$

Now ρ_i can be written in the form

$$\rho_i = \rho_0(1 + \kappa_i),$$

where κ_i is defined by the recursion

$$(5.12)\qquad \begin{aligned}\kappa_1 &= \frac{\eta\rho_0}{\delta} = \frac{\eta\gamma}{\delta^2}, \\ \kappa_{i+1} &= \kappa_1(1 + \kappa_i)^2.\end{aligned}$$

It is easily seen that if (5.8) is satisfied, then the sequence defined by (5.12) converges and

$$\kappa_i < \kappa = \lim \kappa_i = \frac{2\kappa_1}{1 - 2\kappa_1 + \sqrt{1 - 4\kappa_1}} < 1.$$

Thus $\|(P_i, Q_i)\|'_F \leqq \rho = \lim \rho_i = \rho_0(1 + \kappa)$.

To show that the matrices (P_i, Q_i) themselves converge, let $D_i = (P_{i+1} - P_i, Q_{i+1} - Q_i) = (D_{P_i}, D_{Q_i})$. Then

$$\begin{aligned}\|D_i\|'_F &= \|T^{-1}[(P_iA_{12}Q_i - P_{i-1}A_{12}Q_{i-1}, P_iB_{12}Q_i - P_{i-1}B_{12}Q_{i-1})]\|'_F \\ &\leqq \delta^{-1}\|(D_{P_{i-1}}A_{12}Q_i + P_{i-1}A_{12}D_{Q_{i-1}}, D_{P_{i-1}}B_{12}Q_i + P_{i-1}B_{12}D_{Q_{i-1}})\|'_F \\ &\leqq 2\delta^{-1}\eta\rho\|D_{i-1}\|'_F.\end{aligned}$$

Thus the sequence $\{(P_i, Q_i)\}$ converges provided $2\delta^{-1}\eta\rho < 1$, which is ensured by (5.8).

Thus under the hypotheses of the theorem, equation (5.7) has a solution satisfying (5.9), and the column space of $V_1 + V_2Q$ is a deflating subspace of $A - \lambda B$.

To prove the assertion about the eigenvalues of $A - \lambda B$, let

$$A'_{ij} = U_i'^H AV'_j, \quad B'_{ij} = V_i'^H BV'_j.$$

Since $A'_{21} = B'_{21} = 0$, $\lambda(A, B) = \lambda(A'_{11}, B'_{11}) \cup \lambda(A'_{22}, B'_{22})$. Now

$$(5.13)\qquad (I + P^HP)^{1/2}A'_{11}(I + Q^HQ)^{1/2} = A_{11} + A_{12}Q + P^HA_{21} + P^HA_{22}Q.$$

Since P and Q satisfy (5.6),

$$(5.14)\qquad P^HA_{21} = P^HPA_{11} - P^HA_{22}Q + P^HPA_{12}Q.$$

Combining (5.13) and (5.14), we get

$$(I + P^H P)^{-1/2} A'_{11} (I + Q^H Q)^{1/2} = A_{11} + A_{12} Q.$$

Likewise

$$(I + P^H P)^{-1/2} B'_{11} (I + Q^H Q)^{1/2} = B_{11} + B_{12} Q.$$

Hence $\lambda(A'_{11}, B_{11}) = \lambda(A_{11} + A_{12}Q, B_{11} + B_{12}Q)$. Similarly, $\lambda(A'_{22}, B'_{22}) = \lambda(A_{22} - PA_{12}, B_{22} - PB_{12})$. Hence,

$$\lambda(A, B) = \lambda(A_{11} + A_{12}Q, B_{11} + B_{12}Q) \cup \lambda(A_{22} - PA_{12}, B_{22} - PB_{12}).$$

To show that

$$\lambda(A_{11} + A_{12}Q, B_{11} + B_{12}Q) \cap \lambda(A_{22} - PA_{12}, B_{22} - PB_{12}) = \varnothing, \tag{5.15}$$

we apply Theorem 4.2 to obtain

$$\begin{aligned} &\operatorname{dif}(A_{11} + A_{12}Q, B_{11} + B_{12}Q; A_{22} - PA_{12}, B_{22} - PB_{12}) \\ &\quad \geqq \delta - \max\{\|A_{12}\|_2\|Q\|_F + \|A_{12}\|_2\|P\|_F, \|B_{12}\|_2\|Q\|_F + \|B_{12}\|_2\|P\|_F\} \\ &\quad \geqq \delta - 2\eta\rho > 0. \end{aligned}$$

Thus Theorem 4.1 applies to give (5.15).

Theorem 5.1 gives error bounds for an approximate deflating subspace. When $l = 1$, the matrix V_1 is an approximate eigenvector and A_{11}/B_{11} is an approximate eigenvalue. In this case, Theorem 5.1 not only gives an estimate of the accuracy of V_1 but also asserts that $(A_{11} + A_{12}Q)/(B_{11} + B_{12}Q)$ is an exact eigenvalue of $A - \lambda B$. If Q is sufficiently small, this eigenvalue is near A_{11}/B_{11}.

Theorem 5.1 can be applied to solve a closely related problem. Suppose we know exactly a deflating subspace for $A - \lambda B$. How near is the subspace to a deflating subspace of a perturbed problem $(A + E) - \lambda(B + F)$, and how are the eigenvalues of the two problems related? The answer is contained in the following theorem.

THEOREM 5.2. *Let $U = (U_1, U_2)$ and $V = (V_1, V_2)$ be unitary with $U_1, V_1 \in \mathbb{C}^{n \times l}$. Let A_{ij} and B_{ij} be defined as in (5.4) and (5.5), and suppose that $A_{21} = B_{21} = 0$. Let $E, F \in \mathbb{C}^{n \times n}$, and let*

$$E_{ij} = U_i^H E V_j, \quad F_{ij} = U_i^H F V_j,$$

$$\varepsilon_{ij} = \max\{\|E_{ij}\|_F, \|F_{ij}\|_F\}.$$

Let

$$\begin{aligned} \gamma &= \varepsilon_{21}, \quad \eta = \|(A_{12}^T, B_{12}^T)\|'_2 + \varepsilon_{12}, \\ \delta &= \operatorname{dif}(A_{11}, B_{11}; A_{22}, B_{22}) - \varepsilon_{11} - \varepsilon_{22}. \end{aligned} \tag{5.16}$$

Then if $\gamma\eta/\delta^2 < \frac{1}{4}$, there are matrices $P, Q \in \mathbb{C}^{(n-l) \times l}$ satisfying (5.9) such that the columns of $V_1 + V_2 Q$ form a deflating subspace for $(A + E) - \lambda(B + F)$. Moreover, $\lambda(A + E, B + F)$ is the disjoint union of

$$\lambda(A_{11} + E_{11} + (A_{12} + E_{12})Q, B_{11} + F_{11} + (B_{12} + F_{12})Q)$$

and

$$\lambda(A_{22} + E_{22} - P(A_{12} + E_{12}), B_{22} + F_{22} - P(B_{12} + F_{12})).$$

The proof of this perturbation theorem is a straightforward application of Theorem 5.1 to the problem $(A + E) - \lambda(B + F)$. The estimate (5.16) for δ is obtained from Theorem 4.2.

Error bounds for deflation by congruences may be obtained in the same way as the bounds of Theorem 5.1. Let $U = (U_1, U_2)$ be nonsingular, and suppose that $U_2^H A U_1$ and $U_2^H B U_1$ are small. We seek matrices $P, Q \in \mathbb{C}^{(n-l)\times l}$ such that if

$$U_1' = U_1 + U_2 Q, \quad U_2' = U_2 - U_1 P^H,$$

then $U' = (U_1', U_2')$ is nonsingular and

$$U_2'^H A U_1' = U_2'^H B U_1' = 0. \tag{5.17}$$

The requirement (5.17) leads to the system (5.6), where

$$A_{ij} = U_i^H A U_j, \quad B_{ij} = U_i^H B U_j. \tag{5.18}$$

This system may be solved as in the proof of Theorem 5.1.

The matrix U' will be nonsingular if $\|P\|_F, \|Q\|_F < 1$. To see this, note that

$$U' = (U_1, U_2)\begin{pmatrix} I & -P^H \\ Q & I \end{pmatrix}. \tag{5.19}$$

If the second matrix in the product (5.19) is singular, there are vectors x and y with, say, $\|x\|_2 = 1$ such that

$$x = P^H y$$

and

$$y = -Qx.$$

From the second equation and the assumption $\|Q\|_F < 1$ it follows that $\|y\|_2 \leqq \|Q\|_F \|x\|_2 < 1$. Hence from the first equality, $\|x\|_2 \leqq \|P^H\|_F \|y\|_2 < 1$, a contradiction. The requirement that $\|P\|_F, \|Q\|_F < 1$ accounts for the additional condition (5.21) in the following theorem.

THEOREM 5.3. *Let $U = (U_1, U_2)$ be nonsingular with $U_1 \in \mathbb{C}^{n\times l}$. Let A_{ij} and B_{ij} be defined by* (5.18). *Let*

$$\gamma = \|(A_{21}, B_{21})\|_F', \quad \eta = \|(A_{12}^T, B_{12}^T)\|_2',$$

$$\delta = \operatorname{dif}(A_{11}, B_{11}; A_{22}, B_{22}).$$

If

$$\kappa_1 = \frac{\gamma\eta}{\delta^2} < \frac{1}{4} \tag{5.20}$$

and

$$\frac{\gamma}{\delta} < \frac{1}{2}, \tag{5.21}$$

then there are matrices $P, Q \in \mathbb{C}^{(n-l)\times l}$ satisfying

$$\|(P, Q)\|_F' \leqq \frac{\gamma}{\delta}(1 + \kappa) = \frac{\gamma}{\delta}\frac{1 + \sqrt{1 - 4\kappa_1}}{1 - 2\kappa_1 + \sqrt{1 - 4\kappa_1}} < 2\frac{\gamma}{\delta} \tag{5.22}$$

such that the column space of $U_1 + U_2Q$ *is a deflating subspace for* $A - \lambda B$. *Moreover,* $\lambda(A, B)$ *is the disjoint union of* $\lambda(A_{11} + A_{12}Q, B_{11} + B_{12}Q)$ *and* $\lambda(A_{22} - PA_{12}, B_{22} - PB_{12})$.

If the Frobenius norm is used to define η, the condition (5.21) is no real restriction. For (5.20) implies that either $\gamma/\delta < \frac{1}{2}$ or $\eta/\delta < \frac{1}{2}$. If the latter inequality is satisfied, but not the former, the theorem can be applied with the matrix $U = (U_2, U_1)$.

Theorem 5.3 finds its most important application to Hermitian matrices, where congruences must be used to preserve symmetry. In this case $\eta \leqq \gamma$ and

$$\|A_{12}Q\|_F, \quad \|B_{12}Q\|_F \leqq 2\gamma^2/\delta.$$

Thus if $l = 1$, the quantities A_{11} and B_{11} that determine an eigenvalue of $A - \lambda B$ are accurate to quantities of order γ^2. This generalizes a well-known result for the usual Hermitian eigenvalue problem: if an eigenvector is in error by ε, the error in its Rayleigh quotient will be of order ε^2.

A perturbation theorem for the problem $(A + E) + \lambda(B + F)$ may be obtained as in Theorem 5.2.

THEOREM 5.4. *Let* $U = (U_1, U_2)$ *be nonsingular with* $U_1 \in \mathbb{C}^{n \times l}$. *Let* A_{ij} *and* B_{ij} *be defined as in* (5.18), *and suppose that* $A_{21} = B_{21} = 0$. *Let* $E, F \in \mathbb{C}^{n \times n}$, *and let*

$$E_{ij} = U_i^H E U_j, \quad F_{ij} = U_i^H F U_j,$$

$$\varepsilon_{ij} = \max\{\|E_{ij}\|_F, \|F_{ij}\|_F\}.$$

Let

$$\gamma = \varepsilon_{21}, \quad \eta = \|(A_{12}^T, B_{12}^T)\|_2' + \varepsilon_{12},$$

$$\delta = \operatorname{dif}(A_{11}, B_{11}; A_{22}, B_{22}) - \varepsilon_{11} - \varepsilon_{22}.$$

Then if $\gamma\eta/\delta^2 < \frac{1}{4}$ *and* $\gamma/\delta < \frac{1}{2}$, *there are matrices* $P, Q \in \mathbb{C}^{(n-l) \times l}$ *satisfying* (5.22) *such that the column space of* $U_1 + U_2Q$ *is a deflating subspace for* $(A + E) - \lambda(B + F)$. *Moreover,* $\lambda(A + E, B + F)$ *is the disjoint union of*

$$\lambda(A_{11} + E_{11} + (A_{12} + E_{12})Q, B_{11} + F_{11} + (B_{12} + E_{12})Q)$$

and

$$\lambda(A_{22} + E_{22} + P(A_{12} + E_{12}), B_{22} + F_{22} + P(B_{12} + F_{12})).$$

Ordinarily the qualities ε_{ij} will be estimated by

$$\|U_i\|_F\|U_j\|_F \max\{\|E\|_F, \|F\|_F\}.$$

6. Application to the case A Hermitian and B positive definite. In this section we apply the result of § 5 to obtain perturbation bounds for the problem $A - \lambda B$, where A is Hermitian and B is positive definite. We suppose that we have a complete system of B-orthonormal eigenvectors $z_1, z_2, \cdots, z_n$ corresponding to the eigenvalues $\lambda_1, \lambda_2, \cdots, \lambda_n$ of $A - \lambda B$. We wish to apply Theorem 5.4 to obtain perturbation bounds for the eigenvalue λ_1.

The first problem is to choose the matrix $U = (U_1, U_2)$. The natural choice is to take the ith column of U to be some multiple of z_i. The bounds depend rather critically on this choice of scaling. A good choice seems to be

$$u_i = \rho_i z_i / \|z_i\|_2,$$

where $\rho_i > 0$ is chosen so that $\max\{|u_i^H A u_i|, u_i^H B u_i\} = 1$. Since $z_i^H A z_i = \lambda_i$, it follows that

$$\rho_i = \begin{cases} \|z_i\|_2 & \text{if } |\lambda_i| \leqq 1, \\ \dfrac{\|z_i\|_2}{\sqrt{|\lambda_i|}} & \text{if } |\lambda_i| \geqq 1. \end{cases}$$

In either case,

$$\rho_i \leqq \frac{2\|z_i\|_2}{1 + \sqrt{|\lambda_i|}}.$$

We set

$$\alpha_i = u_i^H A u_i, \quad \beta_i = u_i^H B u_i.$$

With $U_1 = u_1$ and $U_2 = (u_2, u_3, \cdots, u_n)$, and A_{ij} and B_{ij} defined as usual, we must next estimate

$$\delta_i = \operatorname{dif}(A_{11}, B_{11}; A_{22}, B_{22}).$$

Since

$$A_{11} = \alpha_1, \quad B_{11} = \beta_1$$

and

$$A_{22} = \operatorname{diag}(\alpha_2, \cdots \alpha_n), \quad B_{22} = \operatorname{diag}(\beta_2, \cdots, \beta_n),$$

it follows that the operator T associated with δ_1 can be regarded as a linear transformation on $\mathbb{C}^{2(n-1)}$. In fact, if the elements of $P, Q \in \mathbb{C}^{(n-1)\times 1}$ are $p_2, p_3, \cdots, p_n$ and $q_2, q_3, \cdots, q_n$ and these elements are arranged in a $2(n-1)$ vector in the order $(p_2, q_2, p_3, q_3, \cdots, p_n, q_n)$, then T has the block diagonal matrix representation

$$T \cong \operatorname{diag}(T_2, T_3, \cdots, T_n),$$

where

$$T_i = \begin{pmatrix} \alpha_1 & -\alpha_i \\ \beta_1 & -\beta_i \end{pmatrix}.$$

Thus, from Theorem 4.4,

$$\delta_1^{-1} \leqq \sqrt{2} \max\{\|T_i^{-1}\|_2 : i = 2, 3, \cdots, n\}. \tag{6.1}$$

Now since $\alpha_i/\beta_i = \lambda_i$,

$$T_i^{-1} = (\lambda_i - \lambda_1)^{-1} \begin{pmatrix} -\beta_1^{-1} & \beta_1^{-1}\lambda_i \\ -\beta_i^{-1} & \beta_i^{-1}\lambda_1 \end{pmatrix}.$$

If $|\lambda_1| \leqq 1$, then $\beta_1^{-1} = |\lambda_1|$. If $|\lambda_1| \geqq 1$, then $\beta_1^{-1} = 1$. In either case $\beta_1^{-1} \leqq 1 + |\lambda_1|$.

Thus the sum of the absolute values of the elements in the first row of T_i^{-1} is bounded by $|\lambda_1 - \lambda_i|^{-1}(1 + |\lambda_1|)(1 + |\lambda_i|)$. The same number bounds the sum of the absolute values of the elements in the second row. Thus

$$\|T_i^{-1}\|_2 \leqq \sqrt{2}\|T_i^{-1}\|_\infty \leqq \sqrt{2}\frac{(1 + |\lambda_1|)(1 + |\lambda_i|)}{|\lambda_1 - \lambda_i|}.$$

Finally, from (6.1),

$$\delta_1^{-1} \leqq 2(1 + |\lambda_1|)\max\left\{\frac{1 + |\lambda_i|}{|\lambda_1 - \lambda_i|} : i = 2, 3, \cdots, n\right\}.$$

We are now in a position to apply Theorem 5.4. For convenience we identify quantities in the theorem with their upper bounds. Let E and F be given. Let $\varepsilon = \max\{\|E\|_F, \|F\|_F\}$. Let

$$\sigma_i = \max\{\rho_j : j \neq i\}.$$

Set

$$\varepsilon_{11} = \rho_1^2\varepsilon, \quad \varepsilon_{22} = (n - 1)\sigma_1^2\varepsilon,$$

$$\varepsilon_{12} = \varepsilon_{21} = \sqrt{n - 1}\rho_1\sigma_1\varepsilon.$$

Let

$$\delta_1^{-1} = 2(1 + |\lambda_1|)\max\left\{\frac{1 + |\lambda_i|}{|\lambda_1 - \lambda_i|} : i = 2, 3, \cdots, n\right\}$$

and

$$\delta = \delta_1 - \varepsilon_{11} - \varepsilon_{22} = \delta_1 - \varepsilon(\rho_1^2 + (n - 1)\sigma_1^2).$$

In order to ensure that $\delta > 0$, we require that ε satisfy

$$\varepsilon(\rho_1^2 + (n - 1)\sigma_1^2) < \delta_1/2, \tag{6.2}$$

in which case

$$\delta > \delta_1/2.$$

Since $\gamma = \eta = \varepsilon_{12}$, the requirement $\gamma\eta/\delta^2 < \frac{1}{4}$ is equivalent to the requirement that

$$2\sqrt{n - 1}\rho_1\sigma_1\varepsilon < \delta_1/2,$$

which is implied by (6.2).

Thus if (6.2) is satisfied, there is an eigenvalue $\lambda_1' = \alpha_1'/\beta_1' \in \lambda(A + E, B + F)$, where the quantities α_1' and β_1' satisfy

$$\begin{aligned} |\alpha_1' - \alpha_1|, |\beta_1' - \beta_1| &< \varepsilon_{11} + 2\varepsilon_{12}^2/\delta \\ &< \varepsilon\rho_1^2[1 + (4(n - 1)\sigma_1^2\varepsilon)/\delta_1]. \end{aligned} \tag{6.3}$$

Of course similar bounds apply to the other eigenvalues.

The above results consist of two parts: first, the requirement (6.2) restricting the size of permissible perturbations, and second, the bound (6.3) on α_i and β_i. If ε is so small that (6.2) is comfortably satisfied, the right-hand side of (6.3) becomes effectively $\varepsilon\rho_1^2$. Thus the number ρ_1^2 is a condition number for the eigenvalue λ_1, in the sense that a perturbation is magnified by a factor of ρ_1^2 in the quantities α_1 and β_1, which determine λ_1. However, it should be remembered that if β_1 is small, a small change in β_1 may cause a large change in λ_1.

We illustrate these bounds by applying them to the problem

$$\operatorname{diag}(1, 1, 10^{-6}, 1) - \lambda \operatorname{diag}(1, 10^{-4}, 1, 10^{-8}).$$

Note that both A and B are nearly singular so that ordinary eigenvalue techniques applied to $A^{-1}B$ or $B^{-1}A$ cannot be expected to yield good bounds. The statistics for the four eigenvalues are summarized in Table 1. The numbers δ_i are only approximately computed; the number $\varepsilon_{\max}$ is a bound on the largest perturbation allowed by (6.2); and "bound" is an upper bound on the right-hand side of (6.3).

TABLE 1

λ_i	α_i	β_i	ρ_i	δ_i	$\varepsilon_{\max}$	bound
1	1	1	1	$\frac{1}{4}$	$\frac{1}{32}$	3ε
10^4	1	10^{-4}	1	$\frac{10^{-4}}{2}$	$\frac{10^{-4}}{16}$	3ε
10^{-6}	10^{-6}	1	1	$\frac{1}{4}$	$\frac{1}{32}$	3ε
10^8	1	10^{-6}	1	$\frac{10^{-4}}{2}$	$\frac{10^{-4}}{16}$	3ε

The results for λ_1 are perfectly satisfactory. Perturbations with norms of about .03 are allowed, and they are magnified by a factor of at most three in the bounds. Actually, for perturbations reasonably smaller than $\varepsilon_{\max}$, the factor is more nearly unity. Notice that with perturbations of this size some of the other eigenvalues may become infinite or complex.

For the second eigenvalue, perturbations of up to about 10^{-5} are allowed, and again they are magnified by a factor of at most three. Because β_2 is 10^{-4}, the eigenvalue λ_2 is perturbed by about $10^8\varepsilon$, which however represents only a small relative error in λ_2 when $\varepsilon/\varepsilon_{\max}$ is small.

The bounds for the third eigenvalue are similar to those for the first; however, since α_3 is small, λ_3 can be quite obliterated by the perturbation.

The bounds for λ_4 are a little surprising. Perturbations of up to 10^{-5} are permitted and are magnified by a factor of no more than three. Since $\beta_4 = 10^{-8}$, this range of perturbations allows λ_4 to become infinite! The source of the phenomenon lies in the fact that the bounds of the last section are primarily concerned with deflating subspaces and only incidentally with eigenvalues. The deflating subspace of λ_4 is spanned by the eigenvector of B corresponding to the smallest eigenvalue of B, and this subspace is insensitive to perturbations of 10^{-5} or less. It should not be thought that the bounds on λ_4 are useless because they include infinity. For example, they tell us that for the perturbations considered $|\lambda_4|$ cannot become less than 10^4.

The results for the above example are, so far as the α_i and β_i are concerned, reminiscent of the results of standard perturbation theory for the Hermitian eigenvalue problem. For all the eigenvalues, perturbations in A and B are not magnified in the α_i and β_i, the ill-conditioning of some of the eigenvalues being accounted for by the sizes of α_i and β_i (the analogy is actually deeper than this; for if the perturbation is off-diagonal ($\varepsilon_{ii} = 0$), the perturbation in α_i and β_i is quadratic in ε and inversely proportional to δ_i). Moreover, for the "nice" eigenvalues λ_1 and λ_3 the range of allowable perturbations is quite reasonable. This analogy breaks down when even a single ρ_i becomes large. In this case, not only are the corresponding α_i and β_i extremely sensitive to perturbation, but also the range of perturbations for all other eigenvalues must be severely restricted. For example, if

$$A = \operatorname{diag}(1, 10^{-8}) \quad \text{and} \quad B = \operatorname{diag}(1, 2 \times 10^{-8}),$$

then $\lambda_1 = 1, \lambda_2 = \frac{1}{2}, \rho_1 = 1, \rho_2 \cong 10^4$, and $\delta_1^{-1} \leqq 3$. Thus from (6.2) we must restrict the perturbations of A and B to satisfy

$$\varepsilon < 10^{-8}/6.$$

And indeed such a restriction is necessary; for if A is perturbed by quantities of order 10^{-8} to give

$$A' = \begin{pmatrix} 1 & \sqrt{2} \times 10^{-8} \\ \sqrt{2} \times 10^{-8} & 2 \times 10^{-8} \end{pmatrix},$$

then $\lambda(A', B) = \{1 + 10^{-4}, 1 - 10^{-4}\}$. Thus a perturbation of order 10^{-8} in A gives rise to a perturbation of 10^{-4} in one eigenvalue and roughly .5 in the other.

The above example suggests the following nomenclature. We shall call the number ρ_i the *disposition* of the eigenvalue λ_i, and if ρ_i is large we shall say that the eigenvalue is ill-disposed. An ill-disposed eigenvalue is, of course, ill-conditioned; however, as the above example indicates, an ill-disposed eigenvalue is not content with its own ill-conditioning but may spread it to other eigenvalues. How this spreading is accomplished is also suggested by the above example. The eigenvalue λ_2 may be made to assume any value whatsoever by means of a perturbation of order 10^{-8}. In particular, such a perturbation may cause λ_2 to lie near λ_1 and thereby have an undue effect on this otherwise well-conditioned eigenvalue.

In spite of this example, I believe that it may be possible to overemphasize the effects of an ill-disposed eigenvalue. For example, consider a random perturbation of, say, 10^{-4} in the above example. In the first place, the diagonal elements of the perturbations can be combined with the diagonal elements of A and B and a new set of dispositions calculated. One of these new dispositions will be of order unity, and it is extremely unlikely that the other will be much greater than 10^2. In the second place, it is unlikely that δ_1 of the perturbed problem will be much less than unity. If these revised estimates are used in (6.3), the result is that λ_1 will not vary by more than roughly 10^{-4}.

Thus, in the above example, the size of the perturbation sets probable limits on the disposition. Moreover, it is improbable that the perturbation will cause a serious deterioration in δ_1. I suspect that similar considerations hold in general; however, such contentious arguments should be supported by experience and experiments.

Our bounds do not provide much information about the sensitivity of an eigenvalue for which δ is small, that is, for an eigenvalue that is poorly separated from its nèighbors. In principle, the bounds of the last section allow us to group a cluster of, say, r eigenvalues together and assert that the members of the perturbed cluster are the exact eigenvalues of a slightly perturbed problem of order r. There are two difficulties with this approach. First there is the problem of estimating a δ for the cluster, which is already hard when $r = 1$. Second, we are left with the problem of assessing the effects of perturbations on an $r \times r$ problem whose eigenvalues are clustered. As with the ordinary eigenvalue problem, the perturbation may or may not be enlarged. In any event our theory says nothing about this special case.

In this section we have not exhausted all the applications of the theorems of § 5, even for the case of A Hermitian and B positive definite. There still remains the problem of assessing the effects of perturbations on the eigenvectors of $A - \lambda B$. This should not be too difficult, since the hard work of finding an estimate for δ has already been done in this section. Another problem is that of assessing the effects of perturbations on the eigenvalues and eigenvectors of the problem for general A and B, which we have not treated at all in this paper. In view of the difficulties and degeneracies encountered in the ordinary eigenvalue problem, a complete solution of this problem may be expected to be rather difficult.

REFERENCES

[1] C. R. CRAWFORD, *The numerical solution of the generalized eigenvalue problem*, Dissertation, The University of Michigan, Ann Arbor, 1970.

[2] G. FIX and R. HEIBERGER, *An algorithm for the ill-conditioned generalized eigenvalue problem*, this Journal, 9 (1972), pp. 78–88.

[3] F. R. GANTMACHER, *The Theory of Matrices. II*, Chelsea, New York, 1960.

[4] A. S. HOUSEHOLDER, *The Theory of Matrices in Numerical Analysis*, Blaisdell, New York, 1964.

[5] PETER LANCASTER, *Lambda-Matrices and Vibrating Systems*, Pergamon Press, Oxford, 1966.

[6] G. PETERS AND J. H. WILKINSON, *$Ax = \lambda Bx$ and the generalized eigenproblem*, this Journal, 7 (1970), pp. 479–492.

[7] G. W. STEWART, *Error bounds for approximate invariant subspaces of closed linear operators*, this Journal, 8 (1971), pp. 796–808.

17.2. [GWS-J18] (with C. B. Moler), "An Algorithm for Generalized Matrix Eigenvalue Problems"

[GWS-J18] (with C. B. Moler), "An Algorithm for Generalized Matrix Eigenvalue Problems," *SIAM Journal on Numerical Analysis* **10** (1973) 241–256.
http://dx.doi.org/10.1137/0710024

SIAM J. NUMER. ANAL.
Vol. 10, No. 2, April 1973

AN ALGORITHM FOR GENERALIZED MATRIX EIGENVALUE PROBLEMS*

C. B. MOLER† AND G. W. STEWART‡

Abstract. A new method, called the QZ algorithm, is presented for the solution of the matrix eigenvalue problem $Ax = \lambda Bx$ with general square matrices A and B. Particular attention is paid to the degeneracies which result when B is singular. No inversions of B or its submatrices are used. The algorithm is a generalization of the QR algorithm, and reduces to it when $B = I$. Problems involving higher powers of λ are also mentioned.

1. Introduction. We shall be concerned with the matrix eigenvalue problem of determining the nontrivial solutions of the equation

$$Ax = \lambda Bx,$$

where A and B are real matrices of order n. When B is nonsingular this problem is formally equivalent to the usual eigenvalue problem $B^{-1}Ax = \lambda x$.

When B is singular, however, such a reduction is not possible, and in fact the characteristic polynomial $\det(A - \lambda B)$ is of degree less than n, so that there is not a complete set of eigenvalues for the problem. In some cases the missing eigenvalues may be regarded as "infinite." In other cases the entire problem may be poorly posed. The term infinite eigenvalue is justified by the fact that if B is perturbed slightly so that it is no longer singular, there may appear a number of large eigenvalues that grow unboundedly as the perturbation is reduced to zero. However, if $\det(A - \lambda B)$ vanishes identically, say when A and B have a common null space, then any λ may be regarded as an eigenvalue. Such problems have unusually pathological features, and we refer to them as "ill-disposed" problems.

In numerical work the sharp distinction between singular and nonsingular matrices is blurred, and the pathological features associated with singular B carry over to the case of nearly singular B. The object of this paper is to describe an algorithm for computing the eigenvalues and corresponding eigenvectors that is unaffected by nearly singular B. The algorithm, the heart of which we call the QZ algorithm, is essentially an iterative method for computing the decomposition contained in the following theorem [10].

THEOREM. *There are unitary matrices Q and Z so that QAZ and QBZ are both upper triangular.*

We say that the eigenvalue problems $QAZy = \lambda QBZy$ and $Ax = \lambda Bx$ are unitarily equivalent. The two problems obviously have the same eigenvalues, and their eigenvectors are related by the equation $x = Zy$.

---

* Received by the editors October 19, 1971, and in revised form February 18, 1972.

† Department of Mathematics, University of New Mexico, Albuquerque, N.M. 87106. Most of the work of this author was done during a visit to Stanford University with support from the Computer Science Department, the Stanford Linear Accelerator Center and the National Science Foundation under Grant GJ-1158. Partial support was also obtained at the University of Michigan from the Office of Naval Research under Contract NR-044-377.

Cleve B. Moler received his Ph.D. in Mathematics in 1964 from Stanford University under the direction of Prof. Forsythe. After many years at the University of Michigan, he is now Professor of Mathematics at the University of New Mexico.

‡ Departments of Computer Science and Mathematics, Carnegie–Mellon University, Pittsburg, P.A. 15213. The work of this author was supported by the National Science Foundation under Grant GP-23655 and by the Office of Naval Research under Contract N00014-67-A-0216.

The algorithm proceeds in four stages. In the first, which is a generalization of the Householder reduction of a single matrix to Hessenberg form [4], [5], A is reduced to upper Hessenberg form and at the same time B is reduced to upper triangular form. In the second step, which is a generalization of the Francis implicit double shift QR algorithm [3], [8], A is reduced to quasi-triangular form while the triangular form of B is maintained. In the third stage the quasi-triangular matrix is effectively reduced to triangular form and the eigenvalues extracted. In the fourth stage the eigenvectors are obtained from the triangular matrices and then transformed back into the original coordinate system.

The transformations used in reducing A and B are applied in such a way that Wilkinson's general analysis of the roundoff errors in unitary transformations [11] shows that the computed matrices are exactly unitarily equivalent to slightly perturbed matrices $A + E$ and $B + F$. This means that the computed eigenvalues, which are the ratios of the diagonal elements of the final matrices, are the exact eigenvalues of the perturbed problem $(A + E)x = \lambda(B + F)x$. If an eigenvalue is well-conditioned in the sense that it is insensitive to small perturbations in A and B (see [10] for a detailed analysis), then it will be computed accurately. This accuracy is independent of the singularity or nonsingularity of B.

The use of unitary transformations in the reduction also simplifies the problem of convergence: a quantity may be set to zero if a perturbation of the same size can be tolerated in the original matrix.

Our computer program does not actually produce the eigenvalues λ_i but instead returns α_i and β_i, the diagonal elements of the triangular matrices QAZ and QBZ. The divisions in $\lambda_i = \alpha_i/\beta_i$ become the responsibility of the program's user. We emphasize this point because the α_i and β_i contain more information than the eigenvalues themselves.

Since our algorithm is an extension of the QR algorithm, the well-known properties of the QR algorithm apply to describe the behavior of our algorithm.

In their survey article [9], Peters and Wilkinson describe another approach for the case when B is nearly singular. In their method one computes an approximate null space for B and removes it from the problem. The technique is reapplied to the deflated problem, and so on until a well-conditioned problem is obtained. The method has the crucial drawback that one must determine the rank of B. If a wrong decision is reached, the well-conditioned eigenvalues may be seriously affected. A similar algorithm for rectangular matrices is given in [13].

The special case where A is symmetric and B is positive definite has been extensively treated. For the case of well-conditioned B the "Cholesky–Wilkinson" method [6] enjoys a well deserved popularity. A modification of this algorithm for band matrices is given by Crawford [1]. A variant of the Peters–Wilkinson method for nearly semidefinite B has been given by Fix and Heiberger [2]. Although our method does not preserve symmetry and is consequently more time consuming than these algorithms, its stability may make it preferable when B is nearly semidefinite.

Our algorithm can also be used to solve "λ-matrix" problems of the form

$$(\lambda^r C_r + \lambda^{r-1} C_{r-1} + \cdots + C_0)x = 0$$

by forming the generalized block companion matrices. For example, when $r = 3$,

$$A = \begin{pmatrix} C_2 & C_1 & C_0 \\ I & 0 & 0 \\ 0 & I & 0 \end{pmatrix}, \qquad B = -\begin{pmatrix} C_3 & 0 & 0 \\ 0 & I & 0 \\ 0 & 0 & I \end{pmatrix}.$$

Note that neither C_r nor C_0 is assumed to be nonsingular.

2. Reduction to Hessenberg-triangular form. In this section we shall give an algorithm whereby A is reduced to upper Hessenberg form and simultaneously B is reduced to triangular form. While a treatment of the reductions in this and the following sections can be given in terms of standard plane rotations and elementary Hermitian matrices, we find it convenient from a computational point of view to work exclusively with a modified form of the elementary Hermitians. Accordingly, we introduce the following notation.

By $\mathscr{H}_r(k)$ we mean the class of symmetric, orthogonal matrices of the form

$$I + vu^T,$$

where $v^Tu = -2$, v is a scalar multiple of u, only components $k, k+1, \cdots, k+r-1$ of u are nonzero, and $u_k = 1$. Given any vector x, it is easy to choose a member Q of $\mathscr{H}_r(k)$ so that

$$Qx = x + (u^Tx)v$$

has its $k+1, \cdots, k+r-1$ components equal to zero, its kth component changed and all other components unchanged. Since $u_k = 1$, the computation of Qy for any y requires only $2r - 1$ multiplications and $2r - 1$ additions. (In particular, use of a matrix in $\mathscr{H}_2$ requires only 3 multiplications instead of the 4 required by a standard plane rotation.)

For the most part, we shall use only matrices in $\mathscr{H}_2$ and $\mathscr{H}_3$. When a matrix Q in $\mathscr{H}_3(k)$ premultiplies a matrix A, only rows k, $k+1$, and $k+2$ in QA are changed. If the elements $k, k+1$, and $k+2$ in a column of A are zero, they remain zero in QA. Likewise, if $Z \in \mathscr{H}_3(k)$, only columns k, $k+1$, and $k+2$ are changed in AZ. If some row has elements k, $k+1$, and $k+2$ zero, then they remain zero in AZ. Similar considerations hold for the class $\mathscr{H}_2$.

All our transformations will be denoted by Q's and Z's with various subscripts. The Q's will always be premultipliers, that is, row operations. The Z's will always be postmultipliers, or column operations. The letter Q is being used in its traditional role to denote orthogonal matrices. The letter Z was chosen to denote orthogonal matrices which introduce zeros in strategic locations.

The first step in the reduction is to reduce B to upper triangular form by premultiplication by Householder reflections. The details of this reduction are well known (e.g., see [4], [11]) and we confine ourselves to a brief description to illustrate our notation. At the kth stage of the reduction (illustrated below for $k = 3$ and $n = 5$), the elements below the first $k - 1$ diagonal elements of B are zero:

$$\begin{matrix} x & x & x & x & x \\ 0 & x & x & x & x \\ 0 & 0 & x & x & x \\ 0 & 0 & x^1 & x & x \\ 0 & 0 & x^1 & x & x \end{matrix}$$

Each x represents an arbitrary nonzero element. Each x^1 represents an element to be annihilated in the next step. A matrix $Q_k \in \mathscr{H}_{n-k+1}(k)$ is chosen to annihilate $b_{k+1,k}, b_{k+2,k}, \cdots, b_{n,k}$, and B is overwritten by $Q_k B$, giving a matrix of the form illustrated below.

$$\begin{matrix} x & x & x & x & x \\ 0 & x & x & x & x \\ 0 & 0 & x & x & x \\ 0 & 0 & 0 & x & x \\ 0 & 0 & 0 & x^1 & x \end{matrix}$$

This process is repeated until $k = n - 1$. Of course A is overwritten by $Q_{n-1}Q_{n-2}\cdots Q_1 A$.

After this reduction, A and B have the forms

$$A \qquad\qquad\qquad B$$

$$\begin{matrix} x & x & x & x & x \\ x & x & x & x & x \\ x & x & x & x & x \\ x & x & x & x & x \\ x^1 & x & x & x & x \end{matrix} \qquad\qquad \begin{matrix} x & x & x & x & x \\ 0 & x & x & x & x \\ 0 & 0 & x & x & x \\ 0 & 0 & 0 & x & x \\ 0 & 0 & 0 & 0 & x \end{matrix}$$

The problem now is to reduce A to upper Hessenberg form while preserving the triangularity of B. This is done as follows (for $k = 5$). First $Q \in \mathscr{H}_2(4)$ is determined to annihilate a_{51}. The matrices QA and QB, which overwrite A and B, then have the forms

$$\begin{matrix} x & x & x & x & x \\ x & x & x & x & x \\ x & x & x & x & x \\ x & x & x & x & x \\ 0 & x & x & x & x \end{matrix} \qquad\qquad \begin{matrix} x & x & x & x & x \\ 0 & x & x & x & x \\ 0 & 0 & x & x & x \\ 0 & 0 & 0 & x & x \\ 0 & 0 & 0 & x^1 & x \end{matrix}$$

The transformation has introduced a nonzero element on the (5,4)-position of B. However, a $Z \in \mathscr{H}_2(4)$ can be used to restore the zero without disturbing the zero introduced in A. The elements of A can be annihilated in the following order:

$$\begin{matrix} x & x & x & x & x \\ x & x & x & x & x \\ x^3 & x & x & x & x \\ x^2 & x^5 & x & x & x \\ x^1 & x^4 & x^6 & x & x \end{matrix}$$

As each element of A is annihilated, it introduces a nonzero element on the subdiagonal of B, which is immediately annihilated by a suitably chosen Z. The entire algorithm, including the Householder triangularization of B, may be summed up as follows:

1. For $k = 1, 2, \cdots, n-1$,
   (i) choose $Q_k \in \mathscr{H}_{n-k+1}(k)$ to annihilate $b_{k+1,k}, b_{k+2,k}, \cdots, b_{n,k}$;
   (ii) $B \leftarrow Q_k B$, $A \leftarrow Q_k A$.
2. For $k = 1, 2, \cdots, n-2$,
   (i) for $l = n-1, n-2, \cdots, k+1$,
      (a) choose $Q_{kl} \in \mathscr{H}_2(l)$ to annihilate $a_{l+1,k}$;
      (b) $A \leftarrow Q_{kl}A$, $B \leftarrow Q_{kl}B$;
      (c) choose $Z_{kl} \in \mathscr{H}_2(l)$ to annihilate $b_{l+1,l}$;
      (d) $B \leftarrow BZ_{kl}$, $A \leftarrow AZ_{kl}$.

The complete reduction requires about $\frac{17}{3}n^3$ multiplications, $\frac{17}{3}n^3$ additions and n^2 square roots. If eigenvectors are also to be computed, the product of the Z's must be accumulated. This requires an additional $\frac{3}{2}n^3$ multiplications and $\frac{3}{2}n^3$ additions. The product of the Q's is not required for the computation of eigenvectors.

3. The explicit QZ step. In this and the next section we assume that A is upper Hessenberg and B is upper triangular. In this section we shall propose an iterative technique for reducing A to upper triangular form while maintaining the triangularity of B. The idea of our approach is to pretend that B is nonsingular and examine the standard QR algorithm for $C = AB^{-1}$. The manipulations are then interpreted as unitary equivalences on A and B.

Specifically, suppose that one step of the QR algorithm with shift σ is applied to C. Then Q is determined as an orthogonal transformation such that the matrix

$$R = Q(C - \sigma I) \tag{3.1}$$

is upper triangular. The next iterate C' is defined as

$$C' = RQ^T + \sigma I \equiv QCQ^T$$

and is known to be upper Hessenberg. If we set

$$A' = QAZ \quad \text{and} \quad B' = QBZ,$$

where Z is any unitary matrix, then

$$A'B'^{-1} = QAZZ^TB^{-1}Q^T = QAB^{-1}Q^T = C'.$$

The matrix Z can be chosen so that B' is upper triangular. Then, since $A' = C'B'$ is the product of a Hessenberg and a triangular matrix, it is also upper Hessenberg. This insures that the nice distribution of zeros, introduced by the algorithm of § 2, is preserved by the QZ step. Thus a tentative form of our algorithm might read:

1. Determine Q so that QC is upper triangular.
2. Determine Z so that QAZ is upper Hessenberg and QBZ is upper triangular.
3. $A \leftarrow QAZ$, $B \leftarrow QBZ$.

The problem is then to give algorithms for computing Q and Z which do not explicitly require $C = AB^{-1}$.

The determination Q is relatively easy. For from (3.1) and the definition of C it follows that

$$Q(A - \sigma B) = RB \equiv S. \tag{3.2}$$

Since R and B are upper triangular, so is S. Thus Q is the unitary matrix that reduces $A - \sigma B$ to upper triangular form. Since $A - \sigma B$ is upper Hessenberg, Q can be expressed in the form

$$Q = Q_{n-1}Q_{n-2} \cdots Q_1, \quad \text{where } Q_k \in \mathscr{H}_2(k). \tag{3.3}$$

To calculate Z we apply Q in its factored form (3.3) to B and determine Z in a factored form so that B stays upper triangular. Specifically, Q_1B has the form $(n = 5)$

$$\begin{matrix} x & x & x & x & x \\ x^1 & x & x & x & x \\ 0 & 0 & x & x & x \\ 0 & 0 & 0 & x & x \\ 0 & 0 & 0 & 0 & x \end{matrix}$$

If Q_1B is postmultiplied by a suitable $Z_1 \in \mathscr{H}_2(1)$, the nonzero element below the diagonal can be removed. Similarly, $Q_2Q_1BZ_1$ has the form

$$\begin{matrix} x & x & x & x & x \\ 0 & x & x & x & x \\ 0 & x^1 & x & x & x \\ 0 & 0 & 0 & x & x \\ 0 & 0 & 0 & 0 & x \end{matrix}$$

and the offending nonzero element can be removed by a $Z_2 \in \mathscr{H}_2(2)$. Proceeding in this way, we construct Z in the form

$$Z = Z_1Z_2 \cdots Z_{n-1}, \quad \text{where } Z_k \in \mathscr{H}_2(k).$$

Although QBZ is upper triangular, it is not at all clear that QAZ is upper Hessenberg. To see that it is, rewrite (3.2) in the form

$$QAZ = SZ + \sigma QBZ. \tag{3.4}$$

From the particular form of Z and the fact that S is upper triangular, it follows that SZ is upper Hessenberg. Thus (3.4) expresses QAZ as the sum of an upper Hessenberg and an upper triangular matrix. In fact, (3.4) represents a computationally convenient form for computing QAZ.

We summarize as follows.

1. Determine $Q = Q_{n-1}Q_{n-2} \cdots Q_1$ $(Q_k \in \mathscr{H}_2(k))$ so that $S = Q(A - \sigma B)$ is upper triangular.
2. Determine $Z = Z_1Z_2 \cdots Z_{n-1}$ $(Z_k \in \mathscr{H}_2(k))$ so that $B' = QBZ$ is upper triangular.
3. $A' = SZ + \sigma B'$.

If this algorithm is applied iteratively with shifts $\sigma_1, \sigma_2, \cdots$, there result sequences of matrices $A_1, A_2, \cdots$, and $B_1, B_2, \cdots$ satisfying

$$A_{\nu+1} = Q_\nu A_\nu Z_\nu, \qquad B_{\nu+1} = Q_\nu B_\nu Z_\nu.$$

The matrices A_ν are upper Hessenberg and the B_ν are upper triangular. Moreover, if B_1 is nonsingular, then the matrices $C_\nu = A_\nu B_\nu^{-1}$ are the matrices which would have been obtained by applying the standard QR algorithm with shifts $\sigma_1, \sigma_2, \cdots$ to $C_1 = A_1 B_1^{-1}$. As C_ν tends to upper triangular form, so must A_ν, since B^{-1} is upper triangular.

Most of the properties of the QR algorithm carry over to the QZ algorithm. The eigenvalues will tend to appear in descending order as one proceeds along the diagonal. The convergence of $a_{n,n-1}^{(\nu)}$ to zero may be accelerated by employing one of the conventional shifting strategies. Once $a_{n,n-1}^{(\nu)}$ becomes negligible one can deflate the problem by working with the leading principal submatrices of order $n - 1$. If some other subdiagonal element of A_ν, say $a_{l,l-1}^{(\nu)}$, becomes negligible, one can effect a further savings by working with rows and columns l through n. Because we have used unitary transformations, an element of A_ν or B_ν can be regarded as negligible if a perturbation of the same size as the element can be tolerated in A_1 or B_1.

The algorithm given above is potentially unstable. If σ is large compared with A and B, the formula (3.4) will involve subtractive cancellation and A' will be computed inaccurately. Since the shift σ approximates the eigenvalue currently being found and the problem may have very large eigenvalues, there is a real possibility of encountering a large shift. Fortunately the large eigenvalues tend to be found last so that by the time a large shift emerges the small eigenvalues will have been computed stably. (The large eigenvalues are of course ill-conditioned and cannot be computed accurately.) To be safe one might perform the first few iterations with a zero shift in order to give the larger eigenvalues a chance to percolate to the top.

4. Implicit shifts. The potential instability in the explicit algorithm results from the fact that we have used formula (3.4) rather than unitary equivalences to compute A'. One way out of this difficulty is to generalize the implicit shift method for the QR algorithm to the QZ algorithm so that both A' and B' are computed by unitary equivalences. The implicit shift technique has the additional advantage that it can be adapted to perform two shifts at a time. For real matrices this means that a double shift in which the shifts are conjugate pairs can be performed in real arithmetic.

Since we are primarily interested in real matrices, we shall concentrate on double shifts. The method is based on the following observation. Suppose that A is upper Hessenberg and B is upper triangular and nonsingular. Then if Q and Z are unitary matrices such that QAZ is upper Hessenberg and QBZ is upper triangular, then Q is determined by its first row. In fact, AB^{-1} and $QAB^{-1}Q^H$ are both upper Hessenberg, so that, by the theorem in [11, p. 352], Q is determined by its first row.

Thus we must do two things. First, find the first row of Q. Second, determine Q and Z so that Q has the correct first row, QAZ is upper Hessenberg, and QBZ

is upper triangular. The first part is relatively easy. The first row that would be obtained from a double shifted QR applied to AB^{-1}. Since A is upper Hessenberg and B upper triangular, it is easy to calculate the first two columns of AB^{-1}. But these, along with the shifts, completely determine the first row of Q. Only non-singularity of the upper 2×2 submatrix of B is actually required here. If either b_{11} or b_{22} is too small, so that this submatrix is nearly singular, a type of deflation can be carried out. We shall return to this point later.

The second part is a little more difficult, and is really the crux of the algorithm since it retains the Hessenberg and triangular forms. Only the first three elements of the first row of Q are nonzero. Thus, if Q_1 is a matrix in $\mathscr{H}_3(1)$ with the same first row of Q, then Q_1A and Q_1B have the following forms (when $n = 6$):

$$\begin{matrix} x & x & x & x & x & x \\ x & x & x & x & x & x \\ x & x & x & x & x & x \\ 0 & 0 & x & x & x & x \\ 0 & 0 & 0 & x & x & x \\ 0 & 0 & 0 & 0 & x & x \end{matrix} \qquad \begin{matrix} x & x & x & x & x & x \\ x^2 & x & x & x & x & x \\ x^1 & x^1 & x & x & x & x \\ 0 & 0 & 0 & x & x & x \\ 0 & 0 & 0 & 0 & x & x \\ 0 & 0 & 0 & 0 & 0 & x \end{matrix}$$

As in the standard implicit shift QR algorithm, it is convenient to think of Q_1 as the reflection which annihilates two of the three nonzero elements in a fictitious "zeroth" column of A.

We must reduce Q_1A to upper Hessenberg and Q_1B to upper triangular by unitary equivalences. However, we may not premultiply by anything which affects the first row. This is done as follows. The matrix Q_1B has three nonzero elements outside the triangle. These can be annihilated by two Z's, a Z_1' in $\mathscr{H}_3(1)$ which annihilates (3, 1)- and (3, 2)-elements and then a Z_1'' in $\mathscr{H}_2(1)$ which annihilates the resulting (2, 1)-element. Let $Z_1 = Z_1'Z_1''$. Then Q_1BZ_1 is upper triangular. Applying Z_1 to Q_1A gives Q_1AZ_1 with the following form:

$$\begin{matrix} x & x & x & x & x & x \\ x & x & x & x & x & x \\ x^1 & x & x & x & x & x \\ x^1 & x & x & x & x & x \\ 0 & 0 & 0 & x & x & x \\ 0 & 0 & 0 & 0 & x & x \end{matrix}$$

This is multiplied by Q_2 in $\mathscr{H}_3(2)$ that annihilates the (3, 1)- and (4, 1)-elements. Then $Q_2Q_1AZ_1$ and $Q_2Q_1BZ_1$ have the forms

$$\begin{array}{cccccc} x & x & x & x & x & x \\ x & x & x & x & x & x \\ 0 & x & x & x & x & x \\ 0 & x & x & x & x & x \\ 0 & 0 & 0 & x & x & x \\ 0 & 0 & 0 & 0 & x & x \end{array} \qquad \begin{array}{cccccc} x & x & x & x & x & x \\ 0 & x & x & x & x & x \\ 0 & x^2 & x & x & x & x \\ 0 & x^1 & x^1 & x & x & x \\ 0 & 0 & 0 & 0 & x & x \\ 0 & 0 & 0 & 0 & 0 & x \end{array}$$

The first columns are now in the desired form. The nonzero elements outside the desired structure have been "chased" into the lower 5 × 5 submatrices.

Now, postmultiply by Z_2, a product of a matrix in $\mathscr{H}_3(2)$ and a matrix in $\mathscr{H}_2(2)$ that reduces the current B to triangular form. Then premultiply by Q_3 in $\mathscr{H}_3(3)$ to annihilate two elements outside the Hessenberg structure of the resulting A.

The process continues in a similar way, chasing the unwanted nonzero elements towards the lower, right-hand corners. It ends with a slightly simpler step which uses Q_{n-2} in $\mathscr{H}_2(n-1)$ to annihilate the $(n, n-2)$-element of the current A, thereby producing a Hessenberg matrix, and Z_{n-2} in $\mathscr{H}_2(n-1)$ which annihilates the $(n, n-1)$-element of the current B, producing a triangular B but not destroying the Hessenberg A.

The fictitious zeroth column of A is determined in part by the shifts. In analogy with the implicit double shift algorithm, we take the shifts σ_1 and σ_2 to be the two zeros of the 2 × 2 problem

$$\det(\bar{A} - \sigma\bar{B}) = 0,$$

where

$$\bar{A} = \begin{pmatrix} a_{n-1,n-1} & a_{n-1,n} \\ a_{n,n-1} & a_{n,n} \end{pmatrix}, \qquad \bar{B} = \begin{pmatrix} b_{n-1,n-1} & b_{n-1,n} \\ 0 & b_{n,n} \end{pmatrix}.$$

It is not desirable to compute σ_1 and σ_2 explicitly, or even to find the coefficients in the quadratic polynomial $\det(\bar{A} - \sigma\bar{B})$. Instead, following the techniques used in "*hqr2*" [8], we obtain ratios of the three nonzero elements of the first column of $(AB^{-1} - \sigma_1 I)(AB^{-1} - \sigma_2 I)$ directly from formulas which involve only the differences of diagonal elements. This insures that small, but nonnegligible, off-diagonal elements are not lost in the shift calculation. The formulas are $(m = n - 1)$

$$\begin{aligned} a_{10} &= \left[\left(\frac{a_{mm}}{b_{mm}} - \frac{a_{11}}{b_{11}}\right)\left(\frac{a_{nn}}{b_{nn}} - \frac{a_{11}}{b_{11}}\right) - \left(\frac{a_{mn}}{b_{nn}}\right)\left(\frac{a_{nm}}{b_{mm}}\right) + \left(\frac{a_{nm}}{b_{mm}}\right)\left(\frac{b_{mn}}{b_{nn}}\right)\left(\frac{a_{11}}{b_{11}}\right)\right] \cdot \left(\frac{b_{11}}{a_{21}}\right) \\ &\quad + \frac{a_{12}}{b_{22}} - \left(\frac{a_{11}}{b_{11}}\right)\left(\frac{b_{12}}{b_{22}}\right), \\ a_{20} &= \left(\frac{a_{22}}{b_{22}} - \frac{a_{11}}{b_{11}}\right) - \left(\frac{a_{21}}{b_{11}}\right)\left(\frac{b_{12}}{b_{22}}\right) - \left(\frac{a_{mm}}{b_{mm}} - \frac{a_{11}}{b_{11}}\right) - \left(\frac{a_{nn}}{b_{nn}} - \frac{a_{11}}{b_{11}}\right) + \left(\frac{a_{nm}}{b_{mm}}\right)\left(\frac{b_{mn}}{b_{nn}}\right), \\ a_{30} &= \frac{a_{32}}{b_{22}}. \end{aligned} \tag{4.1}$$

We are now in a position to summarize the double implicit shift method. It is understood that A and B are to be overwritten by the transformed matrices as they are generated.

1. Compute a_{10}, a_{20}, and a_{30} by (4.1).
2. For $k = 1, 2, \cdots, n-2$,
   (a) determine $Q_k \in \mathscr{H}_3(k)$ to annihilate $a_{k+1,k-1}$ and $a_{k+2,k-1}$;
   (b) determine $Z_k' \in \mathscr{H}_3(k)$ to annihilate $b_{k+2,k+1}$ and $b_{k+2,k}$;
   (c) determine $Z_k'' \in \mathscr{H}_2(k)$ to annihilate $b_{k+1,k}$.
3. Determine $Q_{n-1} \in \mathscr{H}_2(n-1)$ to annihilate $a_{n,n-2}$.
4. Determine $Z_{n-1} \in \mathscr{H}_2(n-1)$ to annihilate $b_{n,n-1}$.

For each k, determination of Q_k requires a few multiplications and one square root. Application of Q_k to both A and B requires about $10(n-k)$ multiplications. The work involved with each Z_k' is the same. Application of Z_k'' requires only about $6(n-k)$ multiplications. The number of additions is about the same. Summing these for k from 1 to $n-1$ gives a total of about $13n^2$ multiplications, $13n^2$ additions and $3n$ square roots per double iteration.

By way of comparison, for the double shift QR algorithm as implemented in "*hqr*," Z_k' becomes simply Q_k^T and Z_k'' is not used. Furthermore, the transformations are carried out on only one matrix. Consequently, each double iteration requires about $5n^2$ multiplications, $5n^2$ additions and n square roots. Thus the QZ algorithm applied on two matrices can be expected to require roughly 2.6 times as much work per iteration as the QR algorithm on a single matrix.

In order to obtain eigenvectors, the Q's are ignored and the Z's accumulated. This requires about $8n^2$ more multiplications and $8n^2$ more additions per double iteration.

There is one difficulty. The formulas for a_{10}, a_{20}, and a_{30} are not defined when b_{11} and b_{22} are zero. Moreover, if b_{11} and b_{22} are small the terms that determine the shift (terms involving a_{nn}, b_{nn}, etc.) become negligible compared to the other terms, so that the effect of the shift is felt only weakly.

Part of the solution to this difficulty is to deflate from the top. If b_{11} is negligible, it may be set to zero to give the forms for A and B ($n = 4$):

$$\begin{matrix} x & x & x & x \\ x & x & x & x \\ 0 & x & x & x \\ 0 & 0 & x & x \end{matrix} \qquad \begin{matrix} 0 & x & x & x \\ 0 & x & x & x \\ 0 & 0 & x & x \\ 0 & 0 & 0 & x \end{matrix}$$

A Q in $\mathscr{H}_2(1)$ can then be used to annihilate the (2, 1)-element of A, which deflates the problem.

The rest of the solution lies in recognizing that there is not much of a problem. If b_{11} and b_{22} are small, then the problem has large eigenvalues. We have already observed that the larger eigenvalues tend to emerge at the upper left, and the larger the eigenvalue, the swifter its emergence. Moreover, the speed will not be affected by a small shift. This means that whenever the implicit shift is diluted by a small b_{11} or b_{22}, the algorithm is none the less profitably employed in finding a large eigenvalue.

5. Further reduction of the quasi-triangular form. The result of the algorithm described so far is in an upper triangular matrix B and a quasi-upper triangular matrix A in which no two consecutive subdiagonal elements are nonzero. This means that the original problem decomposes into 1×1 and 2×2 subproblems. The eigenvalues of the 1×1 problems are the ratios of the corresponding diagonal elements of A and B. The eigenvalues of the 2×2 problems might be calculated as the roots of a quadratic equation, and may be complex even for real A and B.

There are two good reasons for not using the quadratic directly, but instead reducing the 2×2 problems. First, when A and B are real, the calculation of eigenvectors is greatly facilitated if all the real eigenvalues are contained in 1×1 problems. A more important second reason is that the 1×1 problems contain more information than the eigenvalues alone. For example, if a_{11} and b_{11} are small, then the eigenvalue $\lambda_1 = a_{11}/b_{11}$ is ill-conditioned, however reasonable it may appear. This reason obviously applies to complex eigenvalues as well as real ones. Accordingly, we recommend that the 2×2 problems be reduced to 1×1 problems and that the diagonal elements, rather than the eigenvalues, be reported.

Without loss of generality we may consider the problem of reducing 2×2 matrices A and B simultaneously to upper triangular form by unitary equivalences. For our purposes we may assume that B is upper triangular.

Two special cases may be disposed of immediately. If b_{11} is zero, then a $Q \in \mathscr{H}_2(1)$ may be chosen to reduce a_{21} to zero. The zero elements of QB are not disturbed. Similarly, if b_{22} is zero, a $Z \in \mathscr{H}_2(1)$ may be chosen to reduce a_{21} to zero without disturbing b_{21}.

In the general 2×2 case, it is not difficult to write down formulas for the elements of $A' = QAZ$ and $B' = QBZ$ for any Q and Z. Moreover, these formulas can be arranged so that numerically one of a'_{21} or b'_{21} is effectively zero. It is not obvious, however, that the other element is numerically zero, and the effect of assuming that it is by setting it to zero could be disastrous. Consequently, we must consider a somewhat more complicated procedure.

The theoretical procedure for reducing A to triangular form may be described as follows. Let λ be an eigenvalue of the problem and form the matrix $E = A - \lambda B$. Choose a $Z \in \mathscr{H}_2(1)$ to annihilate either e_{11} or e_{21}. Since the rows of E are parallel, it follows that whichever of e_{11} or e_{21} is annihilated the other must also be annihilated. Now choose $Q \in \mathscr{H}_2(1)$ so that either QAZ or QBZ is upper triangular. Since the first column of QEZ is zero and $QEZ = QAZ - \lambda QBZ$, it follows that, however Q is chosen, both QAZ and QBZ must be upper triangular.

In the presence of rounding error the method of computing λ and the choice of Z and Q are critical to the stability of the process. A rigorous rounding error analysis will show that, under a reasonable assumption concerning the computed λ, the process described below is stable. However, to avoid excessive detail, we only outline the analysis. We assume that all computations are done in floating-point arithmetic with t base β digits and that the problem has been so scaled that underflows and overflows do not occur. We further assume that a_{21} is not negligible in the sense that $|a_{21}| < \beta^{-t}\|A\|$, where $\|\cdot\|$ denotes, say, the row sum norm.

The algorithm for computing λ amounts to making an appropriate origin shift and computing an eigenvalue from the characteristic equation. It goes as follows.

$$\mu = a_{11}/b_{11},$$
$$\bar{a}_{12} = a_{12} - \mu b_{12},$$
$$\bar{a}_{22} = a_{22} - \mu b_{22},$$
$$p = \frac{1}{2}\left(\frac{\bar{a}_{22}}{b_{22}} - \frac{b_{12}a_{21}}{b_{11}b_{22}}\right),$$
$$q = \frac{a_{21}\bar{a}_{12}}{b_{11}b_{22}},$$
$$r = p^2 + q,$$

(5.1) $$\lambda = \mu + p + \operatorname{sgn}(p)\cdot\sqrt{r} \quad \text{(complex if } r < 0).$$

We must now assume that the computed λ satisfies the equation

$$\det(A' - \lambda B') = 0,$$

where $\|A - A'\| \leqq \sigma_A\|A\|$ and $\|B - B'\| \leqq \sigma_B\|B\|$ with σ_A and σ_B small constants of order β^{-t}. Define

$$E' = A' - \lambda B'$$

and let E denote the computed value

$$E = \mathrm{fl}(A - \lambda B).$$

Then

$$E' = E + H$$

with $\|H\| \leqq \sigma \max\{\|A\|, |\lambda|\,\|B\|\}$ with σ of order β^{-t}.

We claim that, approximately,

(5.2) $$\|E\| \geqq \beta^{-t}\max\{\|A\|, |\lambda|\,\|B\|\}.$$

First we note that

(5.3) $$\|E\| \geqq |e_{21}| = |a_{21}| \geqq \beta^{-t}\|A\|,$$

by the assumption that a_{21} is significant. Now assume that $\|E\| < \beta^{-t}|\lambda|\,\|B\|$. Then subtractive cancellation must occur in the computation of e_{11}, e_{12}, and e_{22}. Thus $a_{11} \approx \lambda b_{11}$, $a_{12} \approx \lambda b_{12}$ and $a_{22} \approx \lambda b_{22}$. Hence we have $\|A\| \geqq |\lambda|\,\|B\|$, and, from (5.3), $\|E\| \geqq \beta^{-t}|\lambda|\,\|B\|$, a contradiction.

Now

$$0 = \det(E') = \det(E) + (e_{11} + h_{11})h_{22} - (e_{12} + h_{12})h_{21} + h_{11}e_{22} - h_{12}e_{21}.$$

Hence

$$|\det(E)| \leqq \rho_1\|E\|\max\{\|A\|, |\lambda|\,\|B\|\} + \rho_2[\max\{\|A\|, |\lambda|\,\|B\|\}]^2,$$

where ρ_1 and ρ_2 are of order β^{-t}. From (5.2) it then follows that

$$|\det(E)| \leqq \rho\|E\|\max\{\|A\|, |\lambda|\,\|B\|\},$$

where ρ is of order β^{-t}.

Now consider the determination of Z. Assume that the second row of E is larger than the first. Then $Z \in \mathscr{H}_2(1)$ is chosen to annihilate e_{21}. Let $F = EZ$. Then f_{21} is essentially zero. Furthermore, since Z is unitary,

$$|f_{11}f_{22}| = |\det(E)| \leqq \rho\|E\|\max\{\|A\|, |\lambda|\,\|B\|\}.$$

But $|f_{22}| \cong \|e_2\|$ and, since e_2 was assumed to be the larger row, $\|e_2\| = \|E\|$. Hence we have approximately

$$|f_{11}| \leqq \rho \max\{\|A\|, |\lambda|\,\|B\|\}.$$

To choose Q, let

$$C = AZ, \qquad D = BZ,$$

and let f_1, c_1, and d_1 be the first columns of F, C, and D. Let q_2^T denote the second row of Q. If $\|A\| \geqq |\lambda|\,\|B\|$, we choose Q to annihilate d_{21}. Numerically this means that

$$|q_2^T d_1| \leqq \sigma\|B\|,$$

where σ is a constant of the order of β^{-t}. We must show that $q_2^T c_1$ is negligible. But

$$\begin{aligned} |q_2^T c_1| &= |q_2^T f_1 + \lambda q_c^T d_1| \\ &\leqq \|f_1\| + |\lambda|\,\|q_c^T d_1\| \\ &\leqq \rho \max\{\|A\|, |\lambda|\,\|B\|\} + \sigma|\lambda|\,\|B\| \\ &\leqq (\rho + \sigma)\|A\|. \end{aligned}$$

If, on the other hand, $|\lambda|\,\|B\| > \|A\|$, we choose Q so that

$$|q_2^T c_1| \leqq \sigma\|A\|.$$

It then follows that

$$\begin{aligned} |q_2^T d_1| &= |q_1^T f - q_2^T c_1|/|\lambda| \\ &\leqq \rho|\lambda|^{-1} \max\{\|A\|, |\lambda|\,\|B\|\} + \sigma|\lambda|^{-1}\|A\| \\ &\leqq (\rho + \sigma)\|B\|. \end{aligned}$$

In summary, λ is computed using (5.1), Z is chosen to annihilate the first element of the larger of the two rows of $A - \lambda B$ and Q is chosen to annihilate the (2, 1)-element of the smaller of the two matrices AZ and λBZ. In this way, we can be sure that the computed (2, 1)-elements of both QAZ and QBZ are negligible.

In practice with matrices of any order, if the transformations are real, they are applied to the entire matrices. If the transformations are complex, they are used to compute the diagonal elements that would result, but are not actually applied. We thus obtain a quasi-triangular problem in which each 2×2 block is known to correspond to a pair of complex eigenvalues.

The generalized eigenvectors of this reduced problem can be found by a back-substitution process which is a straightforward extension of the method used in "*hqr*2" [8]. The vectors of the original problem are then found by applying the accumulated Z's.

6. Some numerical results. The entire process described above has been implemented in a FORTRAN program [7]. There are four main subroutines: the initial reduction to Hessenberg-triangular form, the iteration itself, the computation of the final diagonal elements, and the computation of the eigenvectors. The complete program contains about 600 FORTRAN statements, although this could be reduced somewhat at the expense of some clarity.

The numerical properties observed experimentally are consistent with the use of unitary transformations. The eigenvalues are always found to whatever accuracy is justified by their condition. If an eigenvalue and eigenvector are not too "ill-disposed," then they produce a small relative residual.

Similar numerical properties cannot generally be expected from any algorithm which inverts B or any submatrix of B. This is even true of 2×2 submatrices, as illustrated by the following example due to Wilkinson:

$$A = \begin{pmatrix} .1 & .2 \\ .3 & .4 \end{pmatrix}, \qquad B = \begin{pmatrix} .1 & .1 \\ 0 & \mu \end{pmatrix}.$$

Here μ is about the square root of the machine precision, that is, μ is not negligible compared to 1, but μ^2 is. There is one eigenvalue near -2. Small relative changes in the elements of the matrices cause only small relative changes in this eigenvalue. The other eigenvalue becomes infinite as μ approaches zero. Great care must be taken in solving this problem so that the mild instability of the one eigenvalue does not cause an inaccurate result for the other, stable eigenvalue.

Of course, the use of unitary transformations makes our technique somewhat slower than others which might be considered. But the added cost is not very great. In testing our program, we solve problems of order 50 regularly. A few problems of orders greater than 100 have been run, but these become somewhat expensive when they are merely tests.

One typical example of order 50 requires 45 seconds on Stanford's IBM 360 model 67. Of this, 13 seconds are spent in the initial reduction, 29 seconds are used for the 61 double iterations required, and 3 seconds are needed for the diagonal elements and eigenvectors. If the eigenvectors are not needed and so the transformations not saved, the total time is reduced to 27 seconds. By way of comparison, formation of $B^{-1}A$ à la Peters and Wilkinson [9] and use of FORTRAN versions [12] of "orthes" [5] and "*hqr*2" [8] requires a total of 27 seconds for this example. (All of these times are for code generated by the IBM FORTRAN IV compiler, H level, with the optimization parameter set to 2.)

In the examples we have seen so far, the total number of double iterations required is usually about 1.2 or 1.3 times the order of the matrices. This figure is fairly constant, although it is not difficult to find examples which require many fewer or many more iterations. As a rule of thumb, for a matrix of order n the time required on the model 67 is about $.36\,n^3$ milliseconds if vectors are computed, $.22\,n^3$ milliseconds if they are not.

The example in Table 1 is not typical, but it does illustrate several interesting points. It was generated by applying nonorthogonal rank one modifications of the identity to direct sums of companion matrices. The companion matrices were chosen so that the resulting problem has three double roots,

$$\lambda_1 = \lambda_2 = \infty,$$

$$\lambda_3 = \lambda_5 = \frac{1}{2} + \frac{\sqrt{3}}{2}i,$$

$$\lambda_4 = \lambda_6 = \frac{1}{2} - \frac{\sqrt{3}}{2}i.$$

TABLE 1

$$A = \begin{matrix} 50 & -60 & 50 & -27 & 6 & 6 \\ 38 & -28 & 27 & -17 & 5 & 5 \\ 27 & -17 & 27 & -17 & 5 & 5 \\ 27 & -28 & 38 & -17 & 5 & 5 \\ 27 & -28 & 27 & -17 & 16 & 5 \\ 27 & -28 & 27 & -17 & 5 & 16 \end{matrix} \qquad B = \begin{matrix} 16 & 5 & 5 & 5 & -6 & 5 \\ 5 & 16 & 5 & 5 & -6 & 5 \\ 5 & 5 & 16 & 5 & -6 & 5 \\ 5 & 5 & 5 & 16 & -6 & 5 \\ 5 & 5 & 5 & 5 & -6 & 16 \\ 6 & 6 & 6 & 6 & -5 & 6 \end{matrix}$$

α	β
25.768670843143	$.2637605112.10^{-6}$
−12.821841071323	$.1312405807.10^{-6}$
5.814535434181 + 10.071071345641 *i*	11.629071028730
5.800765071150 − 10.047220375909 *i*	11.601530302268
5.736511506410 + 9.935928843473 *i*	11.473022854605
5.510879468089 − 9.545122710676 *i*	11.021758784186

α/β
$0.976972281.10^{8}$
$-0.976972290.10^{8}$
0.49999999310489 + 0.86602543924271 *i*
0.49999999310489 − 0.86602543924271 *i*
0.50000000689511 + 0.86602536832617 *i*
0.50000000689511 − 0.86602536832617 *i*

The double root at ∞ results from the fact that B has a double zero eigenvalue. All three roots are associated with quadratic elementary divisors; i.e., each root has only one corresponding eigenvector. The computed diagonals of the triangularized matrices are given in the table. Note that the four finite eigenvalues are obtained with a relative accuracy of about 10^{-8}. This is about the square root of the machine precision and is the expected behavior for eigenvalues with quadratic elementary divisors. The singularity of B does not cause any further deterioration in their accuracy. Furthermore, the infinite eigenvalues are obtained from the reciprocals of quantities which are roughly the square root of the machine precision times the norm of B. Consequently we are somewhat justified if we claim to have computed the square root of infinity.[1]

Acknowledgments. W. Kahan and J. H. Wilkinson made several helpful comments. Linda Kaufman of Stanford has recently shown how to carry out the corresponding generalization of the *LR* algorithm and has written a program that accepts general complex matrices. Charles Van Loan of the University of Michigan has done a detailed roundoff error analysis of the process described in section 5.

[1] This prompts us to recall the limerick which introduces George Gamow's *One, Two, Three, Infinity*:

There was a young fellow from Trinity
Who tried $\sqrt{\infty}$
But the number of digits
Gave him such fidgets
That he gave up Math for Divinity.

REFERENCES

[1] CHARLES CRAWFORD, *The numerical solution of the generalized eigenvalue problem*, Doctoral thesis, University of Michigan, Ann Arbor, 1970; Comm. ACM., 16 (1973), pp. 41–44.

[2] G. FIX AND R. HEIBERGER, *An algorithm for the ill-conditioned generalized eigenvalue problem*, Numer. Math., this Journal, 9 (1972), pp. 78–88.

[3] J. G. F. FRANCIS, *The QR transformation—a unitary analogue to the LR transformation*, Comput. J., 4 (1961–62), pp. 265–271, 332–345.

[4] A. S. HOUSEHOLDER, *Unitary triangularization of a nonsymmetric matrix*, J. Assoc. Comput. Mach., 5 (1958), pp. 339–342.

[5] R. S. MARTIN AND J. H. WILKINSON, *Similarity reduction of a general matrix to Hessenberg form*, Numer. Math., 12 (1968), pp. 349–368.

[6] ———, *Reduction of the symmetric eigenproblem* $Ax = \lambda Bx$ *and related problems to standard form*, Ibid., 11 (1968), pp. 99–110.

[7] C. B. MOLER AND G. W. STEWART, *The QZ algorithm for* $Ax = \lambda Bx$, this Journal, 9 (1972), pp. 669–686.

[8] G. PETERS AND J. H. WILKINSON, *Eigenvectors of real and complex matrices by LR and QR triangularizations*, Numer. Math., 16 (1970), pp. 181–204.

[9] ———, $Ax = \lambda Bx$ *and the generalized eigenproblem*, this Journal, 7 (1970), pp. 479–492.

[10] G. W. STEWART, *On the sensitivity of the eigenvalue problem* $Ax = \lambda Bx$, submitted for publication.

[11] J. H. WILKINSON, *The Algebraic Eigenvalue Problem*, Oxford University Press, Oxford, 1965.

[12] FORTRAN translations of the ALGOL procedures in [5] and [8] may be obtained from the NATS Project, Applied Mathematics Division, Argonne National Laboratories, Argonne, Illinois.

[13] A. M. DELL, R. L. WEIL AND G. L. THOMPSON, *Roots of Matrix Pencils*, Comm. ACM., 14 (1971), pp. 113–117.

17.3. [GWS-J27] "Gershgorin Theory for the Generalized Eigenvalue Problem $Ax = \lambda Bx$"

[GWS-J27] "Gershgorin Theory for the Generalized Eigenvalue Problem $Ax = \lambda Bx$," *Mathematics of Computation* **29** (1975) 600–606.
http://www.jstor.org/stable/2005580

MATHEMATICS OF COMPUTATION, VOLUME 29, NUMBER 130
APRIL 1975, PAGES 600–606

Gershgorin Theory for the Generalized Eigenvalue Problem $Ax = \lambda Bx$

By G. W. Stewart*

Abstract. A generalization of Gershgorin's theorem is developed for the eigenvalue problem $Ax = \lambda Bx$ and is applied to obtain perturbation bounds for multiple eigenvalues. The results are interpreted in terms of the chordal metric on the Riemann sphere, which is especially convenient for treating infinite eigenvalues.

1. Introduction. The object of this paper is to develop a perturbation theory for the generalized eigenvalue problem $Ax = \lambda Bx$ that parallels the perturbation theory developed by Wilkinson [4, Chapter 2] for the ordinary eigenvalue problem. The theory rests on a generalization of the Gershgorin theorem, whose results are interpreted in terms of the chordal metric in the Riemann sphere. This approach has the advantage that it deals readily with multiple and infinite eigenvalues.

Throughout this paper, we shall identify the (possibly infinite) eigenvalue $\lambda = \alpha/\beta$ with the point in the projective complex line defined by

$$[\alpha, \beta] = \{(\alpha, \beta) \neq (0, 0): \alpha/\beta = \lambda\}.$$

We shall equip the projective complex line with the metric χ defined by

$$\chi([\alpha, \beta], [\alpha', \beta']) = \frac{|\alpha\beta' - \alpha'\beta|}{\sqrt{|\alpha|^2 + |\beta|^2}\sqrt{|\alpha'|^2 + |\beta'|^2}}.$$

The number $\chi([\alpha, \beta], [\alpha', \beta'])$ is the chordal distance between the two points $\lambda = \alpha/\beta$ and $\lambda' = \alpha'/\beta'$ when they are projected in the usual way onto the Riemann sphere. By abuse of notation, we shall let $\chi(\lambda, \lambda')$ denote this distance, so that in this context χ is the chordal metric for the set of complex numbers [1, p. 81].

The justification of the use of the chordal metric lies in the simplicity of the final results, in particular in the uniform treatment of large and small eigenvalues. To illustrate this, we shall begin with an application to first-order perturbation theory for a simple eigenvalue.

Let A, B, E, and F be square $n \times n$ matrices with complex elements and let λ be

Received November 5, 1973.
AMS (MOS) subject classifications (1970). Primary 15A18, 65F15.
Key words and phrases. Eigenvalue, generalized eigenvalue problem, Gershgorin theorem, condition numbers, inclusion regions, perturbation theory.
*This work was supported in part by the Office of Naval Research under Contract No. N00014-67-A-0314-0018.

a simple eigenvalue of the problem $Ax = \lambda Bx$ with eigenvector x normalized so that $\|x\|_2 = 1$. Let y be the left eigenvector corresponding to λ, again normalized so that $\|y\|_2 = 1$. Let

$$\alpha = y^H Ax, \quad \beta = y^H Bx,$$

so that the point $[\alpha, \beta]$ is identified with λ. It has been shown by the author [3] that, for sufficiently small E, and F, there is an eigenvalue λ' satisfying $(A+E)x' = \lambda'(B+F)x'$ that can be identified with $[\alpha + y^H Ex, \beta + y^H Fx]$, except for terms of order $\|E\|^2$ and $\|F\|^2$. Thus, in our approach, the sensitivity of λ to perturbations in A and B will be measured by

$$\begin{aligned}\chi(\lambda, \lambda') &\cong \chi([\alpha, \beta], [\alpha + y^H Ex, \beta + y^H Fx]) \\ &= \frac{|\alpha y^H Fx - \beta y^H Ex|}{\sqrt{|\alpha|^2 + |\beta|^2}\sqrt{|\alpha + y^H Ex|^2 + |\beta + y^H Fx|^2}}.\end{aligned}$$

To obtain meaningful approximate perturbation bounds, let

$$\theta = \tan^{-1}|\lambda|, \quad \nu = \sqrt{|\alpha|^2 + |\beta|^2},$$

and

$$\epsilon = \sqrt{|y^H Ex|^2 + |y^H Fx|^2}.$$

Then $\alpha/\nu = \sin\theta$ and $\beta/\nu = \cos\theta$. Hence we have the approximate bound

$$\chi(\lambda, \lambda') \lesssim \frac{\cos\theta\,|y^H Ex| + \sin\theta\,|y^H Fx|}{\nu - \epsilon}, \tag{1.1}$$

which is accurate up to terms of order ϵ^2.

In the terminology of numerical analysis, the bound (1.1) says that ν^{-1} is a condition number for the eigenvalue λ in the sense that it measures how perturbations in A and B will affect λ. If ν is small compared with E and F, then one can expect large changes in λ. It should be noted that ill-conditioned eigenvalues need not be large, and conversely a large eigenvalue need not be ill-conditioned. For example, the bound (1.1) can be quite small even when $\beta = 0$ and hence $\lambda = \infty$, which illustrates the utility of the chordal metric in dealing with this problem.

The factors $\cos\theta$ and $\sin\theta$ appearing in (1.1) are a little unusual, but they make sense. For example, when $\theta = 0$, the eigenvalue λ is zero and the matrix A is singular. The disappearance of the term $\sin\theta\,|y^H Fx|$ in (1.1) then says that perturbations in B cannot affect the singularity of A.

Although the above bound is quite satisfactory for practical work, it is interesting to relate ν to the condition number for the ordinary eigenvalue problem. This may be done as follows. Suppose $Bx \neq 0$ (if $Bx = 0$, use Ax in what follows). Define

$$\kappa = \|Bx\|_2/|\beta| \equiv \sec \angle(y, Bx)$$

and

$$\rho = \sqrt{\|Ax\|_2^2 + \|Bx\|_2^2}.$$

Then $\nu^{-1} = \kappa/\rho$. When $B = I$, the number κ is the secant of the angle between the left and right eigenvectors corresponding to λ, which is the condition number for the ordinary eigenvalue problem [4, Chapter 2]. The number ρ measures how nearly x is an approximate null vector of both A and B (when $Ax = Bx = 0$, $\det(A - \lambda B) \equiv 0$ and any number λ is an eigenvalue). When $B = I$ and $|\lambda| \leqslant 1$, $\rho \leqslant \sqrt{2}$ and the bound becomes essentially the bound for the ordinary eigenvalue problem. If $|\lambda| \geqslant 1$, then the bound (1.1) deviates from the usual bound because of the distorting effects of the chordal metric outside the unit circle.

The above theory has two drawbacks. First, although the bound is asymptotically accurate, the theory does not provide a bound on the remainder. Second, the theory is not applicable to multiple eigenvalues. It is the object of the next two sections of this paper to remedy these defects. In Section 2 we shall develop a generalization of Gershgorin's theorem and in Section 3 apply it to develop a perturbation theory for multiple eigenvalues.

Throughout the paper, we shall use Householder's notational conventions [2]. We shall use the symbols $\|\cdot\|_p$ $(p = 1, 2, \infty)$ to denote the usual Hölder vector norms.

2. Gershgorin Theory. Let A and B be matrices of order n. Set

$$a_i^H = (\alpha_{i1}, \cdots, \alpha_{i,i-1}, \alpha_{i,i+1}, \cdots, \alpha_{i,n});$$

that is, a_i^H is the vector formed from the ith row of A by deleting its ith component. Define the vectors b_i similarly. The following theorem generalizes the Gershgorin exclusion theorem [2, pp. 65ff].

THEOREM 2.1. *Let λ be an eigenvalue of the problem $Ax = \lambda Bx$. Then λ lies in the union of the regions G_i defined by*

$$G_i = \{[\alpha_{ii} + a_i^H\widetilde{x}, \beta_{ii} + b_i^H\widetilde{x}] : \|\widetilde{x}\|_\infty \leqslant 1\}, \quad (i = 1, 2, \cdots, n).$$

Proof. Let $Ax = \lambda Bx$ and suppose that the ith component of x is largest in absolute value. Since $x \neq 0$, we may assume without loss of generality that $\xi_i = 1$. Form $\widetilde{x}$ from x by deleting its ith component. Then $\|\widetilde{x}\|_\infty \leqslant 1$ and

$$\alpha_{ii} + a_i^H\widetilde{x} = \lambda(\beta_{ii} + b_i^H\widetilde{x}),$$

which says that $\lambda \in G_i$. □

In stating Theorem 2.1 we have used the usual identification of the complex projective line with the Riemann sphere. The proof of the theorem in addition exhibits the region in which λ must lie; namely the region corresponding to a maximal component of the eigenvector. When $B = I$, the regions G_i become the disks in the complex plane defined by

$$G_i = \{\lambda : |\lambda - \alpha_{ii}| \leqslant \|a_i\|_1\},$$

which are the usual Gershgorin disks.

As they are defined, the sets G_i are not convenient to work with. Instead we shall work with neighborhoods, defined in terms of the chordal metric, that are generally larger. Specifically, we have

$$\begin{aligned}
&\chi([\alpha_{ii}, \beta_{ii}], [\alpha_{ii} + a_i^H \tilde{x}, \beta_{ii} + b_i^H \tilde{x}]) \\
&\qquad = \frac{|\alpha_{ii} b_i^H \tilde{x} - \beta_{ii} a_i^H \tilde{x}|}{\sqrt{|\alpha_{ii}|^2 + |\beta_{ii}|^2}\sqrt{|\alpha_{ii} + a_i^H \tilde{x}|^2 + |\beta_{ii} + b_i^H \tilde{x}|^2}} \\
&\qquad \leqslant \frac{\|\alpha_{ii} b_i^H - \beta_{ii} a_i^H\|_1}{\sqrt{|\alpha_{ii}|^2 + |\beta_{ii}|^2}\sqrt{\alpha_{ii}'^2 + \beta_{ii}'^2}} \\
&\qquad \equiv \rho_i,
\end{aligned}$$

where

$$\alpha'_{ii} = \max\{0, |\alpha_{ii}| - \|a_i\|_1\}$$

and

$$\beta'_{ii} = \max\{0, |\beta_{ii}| - \|b_i\|_1\}.$$

Hence if we set

$$G'_i = \{\lambda : \chi(\alpha_i/\beta_i, \lambda) \leqslant \rho_i\},$$

the eigenvalues of $Ax = \lambda Bx$ lie in the union of the G'_i. Note that if $B = I$ and the elements of A are less than unity, then the G'_i give a fair approximation of the usual Gershgorin disks.

The utility of the Gershgorin theorem in its applications to the eigenvalue problem is enhanced by the fact that we can often localize a specific number of eigenvalues in a union of some of the G_i. The same is true of our generalization.

THEOREM 2.2. *If the union of k of the Gershgorin regions is disjoint from the remaining regions and is not the entire sphere, then exactly k eigenvalues lie in the union.*

Proof. Suppose, without loss of generality, that the regions in question are G_1, $G_2, \cdots, G_k$. Write $A = A_1 + A_2$ where $A_1 = \mathrm{diag}(\alpha_{11}, \alpha_{22}, \cdots, \alpha_{nn})$ and, similarly, write $B = B_1 + B_2$ where $B_1 = \mathrm{diag}(\beta_{11}, \beta_{22}, \cdots, \beta_{nn})$. For $0 \leqslant \epsilon \leqslant 1$, set

$$A(\epsilon) = A_1 + \epsilon A_2 \quad \text{and} \quad B(\epsilon) = B_1 + \epsilon B_2,$$

and let $G_i(\epsilon)$ be the corresponding Gershgorin regions, which of course satisfy $G_i(\epsilon) \subset G_i$.

Now the hypotheses of the theorem insure that the characteristic polynomial $p_\epsilon(\lambda) = \det(A(\epsilon) - \lambda B(\epsilon))$ cannot vanish identically; for otherwise any number λ would be an eigenvalue and $\bigcup_{i=1}^n G_i$ would be the entire sphere. This means that the eigenvalues of $A(\epsilon)x = \lambda B(\epsilon)x$ must vary continuously in the sphere, or, what is equivalent, there are continuous functions $\lambda_i(\epsilon)$ $(i = 1, 2, \cdots, n)$ such that $\lambda_1(\epsilon), \lambda_2(\epsilon), \cdots, \lambda_n(\epsilon)$ comprise all the eigenvalues of $A(\epsilon)x = \lambda B(\epsilon)x$. We may choose $\lambda_i(0) = \alpha_{ii}/\beta_{ii}$. Then since $\lambda_i(0) \in \bigcup_{j=1}^k G_j(0)$ $(i = 1, 2, \cdots, k)$ and $\bigcup_{j=1}^k G_j(\epsilon) \cap \bigcup_{j=k+1}^n G_j(\epsilon) = \emptyset$, we must have $\lambda_i(1) \in \bigcup_{j=1}^k G_j(1)$ for $(i = 1, 2, \cdots, k)$ and $\lambda_i(1) \in \bigcup_{j=k+1}^n G_j(1)$ for $(i = k+1, \cdots, n)$. □

3. Perturbation Theory. In [4], Wilkinson has applied the Gershgorin theorem to produce a perturbation theory for the eigenvalue problem. The heart of his theory is a technique in which off diagonal elements of order ϵ are reduced to order ϵ^2 by diagonal similarity transformations. In this section, we shall show that the same technique can be applied to the generalized eigenvalue problem. Since it is the technique rather than its specific applications that is of chief interest, we shall confine ourselves to one of the several cases treated by Wilkinson, the case where the problem has a complete set of eigenvectors.

Specifically, we shall be concerned with the case where there are nonsingular matrices X and Y such that

$$Y^H AX = \operatorname{diag}(\alpha_1, \alpha_2, \cdots, \alpha_n) \quad \text{and} \quad Y^H BX = \operatorname{diag}(\beta_1, \beta_2, \cdots, \beta_n).$$

For definiteness we shall take X and Y to have columns of 2-norm unity. Of course the eigenvalues of $Ax = \lambda Bx$ are $\lambda_i = \alpha_i/\beta_i$ $(i = 1, 2, \cdots, n)$. We shall suppose that the first p eigenvalues are equal and distinct from the remaining $q = n - p$ eigenvalues, and we shall apply the Gershgorin theory of the last section to obtain perturbation bounds for this set of multiple eigenvalues.

Let E and F be matrices of order n and let ϵ be an upper bound on the elements of the matrices $Y^H EX$ and $Y^H FX$. Then $Y^H(A + E)X$ and $Y^H(B + F)X$ have the generic forms $(n = 5)$

$$\begin{pmatrix} \alpha_1+\epsilon & \epsilon & \epsilon & \epsilon & \epsilon \\ \epsilon & \alpha_2+\epsilon & \epsilon & \epsilon & \epsilon \\ \epsilon & \epsilon & \alpha_3+\epsilon & \epsilon & \epsilon \\ \epsilon & \epsilon & \epsilon & \alpha_4+\epsilon & \epsilon \\ \epsilon & \epsilon & \epsilon & \epsilon & \alpha_5+\epsilon \end{pmatrix}, \quad \begin{pmatrix} \beta_1+\epsilon & \epsilon & \epsilon & \epsilon & \epsilon \\ \epsilon & \beta_2+\epsilon & \epsilon & \epsilon & \epsilon \\ \epsilon & \epsilon & \beta_3+\epsilon & \epsilon & \epsilon \\ \epsilon & \epsilon & \epsilon & \beta_4+\epsilon & \epsilon \\ \epsilon & \epsilon & \epsilon & \epsilon & \beta_5+\epsilon \end{pmatrix}.$$

If ϵ is small enough, the first p Gershgorin regions will be disjoint from the others, and their union will contain exactly p eigenvalues which are necessarily near λ_1. However, we can obtain an even better result.

Let $\tau \neq 0$ be given. The eigenvalues in question will be unchanged if the last q columns of $Y^H(A + E)X$ and $Y^H(B + F)X$ are multiplied by τ. If this is done, the new matrix $Y^H(A + E)X$ has the form ($n = 5, p = 2$)

$$\begin{pmatrix} \alpha_1 + \epsilon & \epsilon & \tau\epsilon & \tau\epsilon & \tau\epsilon \\ \epsilon & \alpha_2 + \epsilon & \tau\epsilon & \tau\epsilon & \tau\epsilon \\ \epsilon & \epsilon & \tau(\alpha_3 + \epsilon) & \tau\epsilon & \tau\epsilon \\ \epsilon & \epsilon & \tau\epsilon & \tau(\alpha_4 + \epsilon) & \tau\epsilon \\ \epsilon & \epsilon & \tau\epsilon & \tau\epsilon & \tau(\alpha_5 + \epsilon) \end{pmatrix}$$

and $Y^H(B + F)X$ has a similar form. We shall attempt to choose $\tau \leqslant 1$ so that the first p Gershgorin regions are disjoint from the others.

Let

$$\delta = \min\{\chi(\lambda_i, \lambda_j): i = 1, 2, \cdots, p; j = p + 1, \cdots, n\}.$$

Let

$$\nu_i = \sqrt{|\alpha_i|^2 + |\beta_i|^2} \qquad (i = 1, 2, \cdots, n)$$

and

$$\nu = \min\{\nu_i: i = 1, 2, \cdots, n\}.$$

Then the first p Gershgorin regions are contained in the disk whose center is λ_1 and whose chordal radius is

$$\frac{|\alpha_i| + |\beta_i|}{\nu_i} \; \frac{(p + \tau q)\epsilon}{\nu_i - \sqrt{2}(p + \tau q)\epsilon} \leqslant \frac{\sqrt{2}(p + \tau q)\epsilon}{\nu - \sqrt{2}(p + \tau q)\epsilon}.$$

Likewise the remaining q regions will be contained in disks whose centers are λ_i and whose chordal radius is $\sqrt{2}(p + \tau q)\epsilon/(\tau\nu - \sqrt{2}(p + \tau q)\epsilon)$. Thus, if we can find τ satisfying $0 < \tau < 1$ such that $\sqrt{2}(p + \tau q)\epsilon/(\tau\nu - \sqrt{2}(p + \tau q)\epsilon) \leqslant \delta/2$, the appropriate Gershgorin regions will be disjoint. This will surely be true if

$$\tau =_{\mathrm{df}} \frac{\sqrt{2}\, n\epsilon}{\nu} \frac{(2 + \delta)}{\delta} < 1. \tag{3.1}$$

Thus we have shown that if (3.1) is satisfied, there are exactly p eigenvalues in the region defined by

$$\chi(\lambda_1, \lambda) \leqslant \sqrt{2}(p + \tau q)\epsilon/(\bar{\nu} - \sqrt{2}(p + \tau q)\epsilon),$$

where $\bar{\nu} = \min\{\nu_i: i = 1, 2, \cdots, p\}$. For small ϵ, this radius is asymptotic to $\sqrt{2}p\epsilon/\bar{\nu}$, which shows that $\sqrt{2}p/\bar{\nu}$ is a condition number for the multiple eigenvalue λ_1.

When λ_1 is simple and (3.1) is satisfied, then the perturbed eigenvalue must lie in the disk defined by

$$\chi\left(\frac{\alpha_1 + y_1^H E x_1}{\beta_1 + y_1^H F x_1}, \lambda\right) \leqslant \frac{(|\alpha_1| + |\beta_1| + 2\epsilon)\tau q\epsilon}{\nu_1[\nu_1 - \sqrt{2}(1 + \tau q)\epsilon]}.$$

This bound approaches zero quadratically with ϵ, which makes rigorous the observations of Section 1.

The treatment sketched above, of course, does not exhaust all possible cases. Various defective cases can be treated by applying the Gershgorin theorem to canonical forms, as has been done by Wilkinson for the ordinary eigenvalue problem. Alternatively one can use the techniques of [3] to split off a set of multiple eigenvalues and treat these separately by means of the Gershgorin theorem. With either approach, the generalized Gershgorin theorem is required to deal with multiple eigenvalues, and the use of the chordal metric simplifies the final results.

4. Acknowledgments. Part of this work was done while I was a visitor at the IBM Thomas J. Watson Research Center. I should also like to thank W. Kahan for his comments.

Department of Computer Science
Carnegie-Mellon University
Pittsburgh, Pennsylvania 15213

1. C. CARATHÉODORY, *Theory of Functions of a Complex Variable*. Vol. I, Translated by F. Steinhardt, Chelsea, New York, 1954. MR 15, 612.

2. A. S. HOUSEHOLDER, *The Theory of Matrices in Numerical Analysis*, Blaisdell, New York, 1964. MR 30 #5475.

3. G. W. STEWART, "Error and perturbation bounds for subspaces associated with certain eigenvalue problems," *SIAM Rev.*, v. 15, 1973, pp. 727–764.

4. J. H. WILKINSON, *The Algebraic Eigenvalue Problem*, Clarendon Press, Oxford, 1965. MR 32 #1894.

17.4. [GWS-J38] "Perturbation Bounds for the Definite Generalized Eigenvalue Problem"

[GWS-J38] "Perturbation Bounds for the Definite Generalized Eigenvalue Problem," *Linear Algebra and its Applications* **23** (1979) 69–85.

Perturbation Bounds for the Definite Generalized Eigenvalue Problem

G. W. Stewart*

Department of Computer Science
University of Maryland
College Park, Maryland 20742

Submitted by Hans Schneider

ABSTRACT

Let A and B be Hermitian matrices, and let $c(A,B)=\inf\{|x^H(A+iB)x|:\|x\|=1\}$. The eigenvalue problem $Ax=\lambda Bx$ is called definite if $c(A,B)>0$. It is shown that a definite problem has a complete system of eigenvectors and that its eigenvalues are real. Under perturbations of A and B, the eigenvalues behave like the eigenvalues of a Hermitian matrix in the sense that there is a 1-1 pairing of the eigenvalues with the perturbed eigenvalues and a uniform bound for their differences (in this case in the chordal metric). Perturbation bounds are also developed for eigenvectors and eigenspaces.

1. INTRODUCTION

In this paper we shall be concerned with deriving perturbation bounds for the eigenvalues and eigenvectors of the generalized eigenvalue problem

$$Ax = \lambda Bx, \tag{1.1}$$

where A and B are Hermitian matrices of order n. When B is positive definite, as it is in most applications, the problem can be reduced to a Hermitian eigenvalue problem of the form

$$B^{-1/2}AB^{-1/2}y = \lambda y, \tag{1.2}$$

where $B^{1/2}$ is the positive definite square root of B. Thus, in this case, the eigenvalues are real.

*This work was supported in part by the Office of Naval Research under Contract No. N00014-76-C-0391.

 0024-3795/79/010069+17$01.75

When A and B are replaced by $\tilde{A}=A+E$ and $\tilde{B}=B+F$, the eigenvalues and eigenvectors will be perturbed by quantities that are functions of E and F. In principle one can apply the existing theory for Hermitian eigenvalue problems to (1.2) and obtain bounds on these perturbations; however, this approach is unsatisfactory for a number of reasons, which we shall now sketch (for a more complete discussion and examples see [3, 10, 12]).

When B is ill conditioned, that is, when B is relatively near a singular matrix, the eigenvalues of (1.1) will typically be spread out, with some small and some large. The small ones may be relatively insensitive to perturbations in A and B. However, since the problem (1.2) has large eigenvalues, $B^{-1/2}AB^{-1/2}$ must also be large, and so will its corresponding perturbation. The perturbation theory for (1.2) will then predict large perturbations for all the eigenvalues, even the small ones. A second difficulty is that the perturbation in B must be restricted so that B remains positive definite, even though perturbations that make B indefinite may have little effect on some of the eigenvalues. Finally, although the large eigenvalues of the problem usually undergo large perturbations, their reciprocals will undergo only small perturbations. This suggests that the usual Euclidean metric on the line is not appropriate for reporting the sizes of the perturbations in the eigenvalues.

In [12] the author has developed a perturbation theory for the non-Hermitian generalized eigenvalue problem that circumvents these difficulties, first by avoiding the use of inverses and second by using the chordal metric on the Riemann sphere (cf. Sec. 3). In [3] Crawford has described a class of Hermitian problems, called definite problems in this paper, that admit of a nice perturbation theory in the chordal metric. It is the purpose of this paper to strengthen and extend Crawford's results. In particular we shall obtain the one-one pairing of eigenvalues with their perturbed counterparts that holds for the Hermitian eigenvalue problem. In addition we shall obtain perturbation bounds for eigenspaces that are related to the perturbation bounds of Davis and Kahan [4] for invariant subspaces of Hermitian operators.

Throughout this paper A and B will be Hermitian matrices of order n, and

$$\tilde{A}=A+E, \qquad \tilde{B}=B+F$$

will be Hermitian perturbations of A and B. The Euclidean vector norm and the spectral matrix norm will both be denoted by $\|\cdot\|$.

2. THE GEOMETRIC THEORY

Recently there has emerged an elegant geometric theory for the real symmetric generalized eigenvalue problem that is based on the quadratic form $x^T(A+iB)x$. For topological reasons, this theory does not apply when $n=2$. In this section we shall extend this theory to the Hermitian problem in such a way that the restriction $n\neq 2$ is removed. We shall return to the real case at the end of the section.

Our ultimate goal is to replace the generalized eigenvalue problem (1.1) by an equivalent problem in which B is positive definite. Specifically, for any real φ, let

$$A_\varphi = A\cos\varphi - B\sin\varphi,$$

$$B_\varphi = A\sin\varphi + B\cos\varphi.$$

Then it is obvious that any matrix X for which X^HAX and X^HBX are diagonal also diagonalizes A_φ and B_φ. Thus (1.1) has a complete set of eigenvectors if and only if the problem $A_\varphi x=\lambda B_\varphi x$ has the same set. We shall attempt to choose φ so that B_φ is positive definite. The condition under which this can be done is that the pair (A,B) is definite in the following sense.

Definition 2.1. The pair of Hermitian matrices (A,B) is definite if

$$c(A,B) \equiv \inf_{\|x\|=1}\{|x^H(A+iB)x|\} > 0. \tag{2.1}$$

The eigenvalue problem (1.1) is definite if (A,B) is definite.

The proof that $c(A,B)>0$ implies that B_φ is positive definite for some φ involves the geometry of the set over which the infimum in (2.1) is taken. Let

$$V = \{x^H(A+iB)x : \|x\|=1\}$$

and

$$C = \{x^H(A+iB)x : x\in C^n\}$$

$$\equiv \{\alpha v : v\in V, \alpha\geqslant 0\}.$$

Then $c(A,B)=\inf\{|v|:v\in V\}$. Moreover, C is the pointed cone generated by the closed, bounded set C. We may now state our fundamental theorem.

Theorem 2.2. *The cone C is convex. If (A,B) is definite, then C lies properly in a closed half plane passing through the origin. Moreover there is* *a real number* φ *such that* B_φ *is positive definite and*

$$c(A,B) = \lambda_{\min}(B_\varphi),$$

where $\lambda_{min}(B_\varphi)$ *denotes the smallest eigenvalue of* B_φ.

Proof. The set V is just the field of values of $A+iB$, which is known to be convex (e.g. see [7]). Hence the cone C generated by V is convex. Now suppose that $c(A,B)>0$. Let the infimum in (2.1) be attained at x_0, and let $v_0 = x_0^H(A+iB)x_0$. Since V is convex and $v_0 \in V$ satisfies $|v_0| = \inf\{|v| : v \in V\} = c(A,B) > 0$, V must lie in the closed half plane $H = \{z : \Re(\bar{v}_0 z) \geqslant |v_0|^2\}$, which does not contain the origin. Since V is bounded, it must be contained in a cone that is properly contained in the half plane $H - v_0$, which passes through the origin.

To prove the last part of the theorem, let V_φ, C_φ, and H_φ denote the field of values, the cone, and the half plane associated with the pair (A_φ, B_φ). Since

$$A_\varphi + iB_\varphi = e^{i\varphi}(A+iB),$$

it follows that V_φ, C_φ, and H_φ are just V, C, and H rotated clockwise through the angle φ. Choose φ so that H_φ lies in the upper half plane. Then $x_0^H A_\varphi x_0 = 0$. Moreover for $\|x\| = 1$, $x^H B_\varphi x \geqslant c(A_\varphi, B_\varphi) = c(A,B)$. Hence

$$0 < c(A,B) = x_0^H B x_0 = \inf_{\|x\|=1} x^H B x.$$

This shows that B is positive definite, x_0 is an eigenvector of B_φ corresponding to $\lambda_{\min}(B_\varphi)$, and $c(A,B) = \lambda_{\min}(B_\varphi)$. ■

Since in the above B_φ is positive definite, the Hermitian matrix $B_\varphi^{-1/2} A_\varphi B_\varphi^{-1/2}$ exists and can be diagonalized by a unitary matrix Y. It is easily verified that $X = B_\varphi^{-1/2} Y$ diagonalizes both A_φ and B_φ, and hence A and B. This proves

COROLLARY 2.3. *If the problem* (1.1) *is definite, then there is a nonsingular matrix* X *such that* $X^H AX$ *and* $X^H BX$ *are both diagonal.*

If we set $X^H AX = \mathrm{diag}(\mu_1, \mu_2, \ldots, \mu_n)$ and $X^H BX = \mathrm{diag}(\nu_1, \nu_2, \ldots, \nu_n)$, then the eigenvalues of (1.1) are given by $\lambda_i = \mu_i / \nu_i$. This admits the possibility of infinite eigenvalues when $\nu_i = 0$; however, the indeterminant case $\mu_i = \nu_i = 0$ cannot occur in a definite problem.

The number $c(A,B)$ will play an important role in the perturbation theory of the next two sections. The following theorem shows that it does not vary wildly with perturbations in A and B. This result is required in Sec. 3, where it is the number $c(\tilde{A},\tilde{B})$ that appears in the bounds.

THEOREM 2.4.

$$c(\tilde{A},\tilde{B}) \geqslant C(A,B) - \left[\|E\|^2 + \|F\|^2\right]^{1/2}.$$

Proof.

$$c(\tilde{A},\tilde{B}) = \inf_{\|x\|=1} \left\{\left[x^H(A+E)x\right]^2 + \left[x^H(B+F)x\right]^2\right\}^{1/2}$$

$$\geqslant \inf_{\|x\|=1} \left[(x^HAx)^2 + (x^HBx)^2\right]^{1/2} - \sup_{\|x\|=1} \left[(x^HEx)^2 + (x^HFx)^2\right]^{1/2}$$

$$\geqslant c(A,B) - \left[\sup_{\|x\|=1} (x^HEx)^2 + \sup_{\|x\|=1} (x^HFx)^2\right]^{1/2}$$

$$= c(A,B) - \left[\|E\|^2 + \|F\|^2\right]^{1/2}. \qquad \blacksquare$$

We turn now to the case where A and B are real. It is natural in this case to restrict the vector x in (2.1) to lie in $\mathbb{R}^n$ and define

$$V_r = \left\{x^T(A+iB)x : \|x\|=1,\, x\in\mathbb{R}^n\right\}$$

and

$$c_r(A,B) = \inf\{|v| : v\in V_r\}.$$

Brickman [1] has shown that if $n\neq 2$, then $V_r = V$, and consequently $c_r(A,B) = c(A,B)$. It follows that Theorem 2.2 is valid for $n\neq 2$, and in particular we can restrict our attention to real vectors. That this happy state of affairs does not hold for $n=2$ is shown by the example

$$A = \begin{pmatrix} 1 & 0 \\ 0 & -1 \end{pmatrix}, \qquad B = \begin{pmatrix} 0 & 1 \\ 1 & 0 \end{pmatrix},$$

for which $c(A,B) = 0 < 1 = c_r(A,B)$. However, we have the following corollary of Theorem 2.2.

COROLLARY 2.5. *For $n=2$, if $c(A,B)>0$, then $c_r(A,B)=c(A,B)$.*

Proof. By Theorem 2.2 we may find a φ such that B_φ is positive definite, and an eigenvector x of B_φ such that $\|x\|=1$ and $x^H B_\varphi x=c(A,B)$. Since $x^H A_\varphi x=0$, A_φ is indefinite. Now if the eigenvalue $c(A,B)$ is of multiplicity one, then x must be a scalar multiple of a real vector y, which then satisfies $y^T A_\varphi y=0$ and $y^T B_\varphi y=c(A,B)$. If $c(A,B)$ is of multiplicity two, then any nonzero vector is an eigenvector. Since A_φ is real and indefinite, there is a vector y of norm unity such that $y^T A_\varphi y=0$ and $y^T B_\varphi y=c(A,B)$. ∎

For the real case, much of the material in this section is implied by earlier work [1–3, 5, 6, 14]. Hestenes [6] has characterized C as in Theorem 2.2, and Crawford [3] appears to have been the first to introduce the rotated problem. I am indebted to Hans Schneider for pointing out that the restriction $n\neq 2$ can be removed in the complex case. The characterization of $c(A,B)$ as the smallest eigenvalue of a positive definite B_φ is new and replaces an incorrect statement in Crawford's paper.

The number $c(A,B)$ appears repeatedly in the bounds to be derived in the next sections, which is why we have taken some pains to ascertain when the computationally simpler number $c_r(A,B)$ is equal to $c(A,B)$. Crawford was the first to realize the importance of the value of $c(A,B)$, as opposed to the relation $c(A,B)>0$, and for this reason it is appropriate to refer to $c(A,B)$ as the Crawford number of the problem (1.1).

3. PERTURBATION BOUNDS: EIGENVALUES

In this section we shall develop perturbation bounds for the eigenvalues of the definite generalized eigenvalue problem. For the Hermitian eigenvalue problem $Ax=\lambda x$, it is well known that if the eigenvalues are ordered so that $\lambda_1\leqslant\lambda_2\cdots\leqslant\lambda_n$, and those of the perturbed problem $\tilde{A}\tilde{x}=\tilde{\lambda}\tilde{x}$ are ordered so that $\tilde{\lambda}_1\leqslant\tilde{\lambda}_2\leqslant\cdots\leqslant\tilde{\lambda}_n$, then

$$|\lambda_i-\tilde{\lambda}_i|\leqslant\|A-\tilde{A}\|\qquad(i=1,2,\ldots,n).$$

We shall obtain a similar pairing of eigenvalues with their perturbed counterparts; however, the ordering that defines the pairing must be defined in terms of certain angles associated with the eigenvalues of (1.1).

Let the pair (A,B) be definite, and let $\|E\|^2+\|F\|^2<c(A,B)$, so that by Theorem 2.4 the pair $(\tilde{A},\tilde{B})$ is also definite. Let C and $\tilde{C}$ denote the cones associated with the two pairs. Then by Theorem 2.2 the complement of

$C \cup \tilde{C}$ contains a ray R extending from the origin. For each nonzero point $u+iv \in \mathbb{C}$ define $\theta(u,v)$ as the angle subtended by R and $\{\alpha(u+iv):\alpha \geqslant 0\}$ measured clockwise. By construction, θ is continuous on $C \cup \tilde{C}$.

Now let $Ax_i = \lambda_i Bx_i$, where $x_i \neq 0$. We define the angle associated with λ_i to be

$$\theta_i = \theta\left(x_i^H A x_i, x_i^H B x_i\right),$$

and throughout the rest of this section we shall assume that $\theta_1 \leqslant \theta_2 \leqslant \cdots \leqslant \theta_n$. If B is positive definite, this corresponds to the natural ordering $\lambda_1 \leqslant \lambda_2 \leqslant \cdots \leqslant \lambda_n$ of the eigenvalues. However, if B is not positive definite, this need not be. For example, if C lies in the right half plane, then the positive eigenvalues will precede the negative ones.

The θ_i have a characterization that is completely analogous to the min-max characterization of the eigenvalues of the Hermitian problem.

THEOREM 3.1. *Let the problem* (1.1) *be definite, and let its eigenvalues be ordered so that* $\theta_1 \leqslant \theta_2 \leqslant \cdots \leqslant \theta_n$. *Then*

$$\theta_i = \min_{\dim(\mathcal{X})=i} \max_{\substack{x \in \mathcal{X} \\ x \neq 0}} \theta(x^T A x, x^T B x) \tag{3.1}$$

and

$$\theta_i = \max_{\dim(\mathcal{X})=n-i+1} \min_{\substack{x \in \mathcal{X} \\ x \neq 0}} \theta(x^T A x, x^T B x). \tag{3.2}$$

Proof. As in Sec. 2, choose φ so that B_φ is positive definite. Let the ray defining θ be rotated counterclockwise through φ, and use this new ray to define a new function θ_φ. Then

$$\theta_\varphi\left(x^H A_\varphi x, x^H B_\varphi x\right) = \theta(x^H A x, x^H B x),$$

and without loss of generality we may drop the subscripts φ and assume that B is positive definite.

For B positive definite we have [7]

$$\lambda_i = \min_{\dim(\mathcal{X})=i} \max_{\substack{x \in \mathcal{X} \\ x \neq 0}} \frac{x^H A x}{x^H B x}.$$

But because V is in the upper half plane, the mapping $\lambda \to \theta(\lambda, 1)$ is increasing. Hence

$$\theta_i = \theta(\lambda_i, 1) = \min_{\dim(\mathcal{X})=i} \max_{\substack{x\in\mathcal{X} \\ x\neq 0}} \theta\left(\frac{x^H A x}{x^H B x}, 1\right)$$

$$= \min_{\dim(\mathcal{X})=i} \max_{\substack{x\in\mathcal{X} \\ x\neq 0}} \theta(x^H A x, x^H B x).$$

A proof of (3.2) follows from the characterization

$$\lambda_i = \min_{\dim(\mathcal{X})=n-i+1} \min_{\substack{x\in\mathcal{X} \\ x\neq 0}} \frac{x^H A x}{x^H B x}. \qquad \blacksquare$$

As a consequence of Theorem 3.1, a number of separation theorems for the eigenvalues of symmetric matrices generalize to the definite problem (1.1). For example, if $\hat{A}$ and $\hat{B}$ are principal submatrices of A and B of order $n-1$, then the eigenvalues $\hat{\lambda}_1,\ldots,\hat{\lambda}_{n-1}$ of the problem $\hat{A}x=\hat{\lambda}\hat{B}x$ satisfy

$$\theta_1 \leqslant \hat{\theta}_1 \leqslant \theta_2 \leqslant \hat{\theta}_2 \leqslant \cdots \leqslant \hat{\theta}_{n-1} \leqslant \theta_n.$$

However, our main interest in the theorem is that it can be used to prove the following perturbation theorem.

THEOREM 3.2. *Let (A,B) be definite, and let the eigenvalues of $Ax=\lambda Bx$ be ordered so that $\theta_1 \leqslant \theta_2 \leqslant \cdots \leqslant \theta_n$. Let*

$$\varepsilon = \left\{\|E\|^2 + \|F\|^2\right\}^{1/2},$$

and assume that $\varepsilon < c(A,B)$ so that $(\tilde{A},\tilde{B})=(A+E,B+F)$ is definite. Let the eigenvalues of $\tilde{A}\tilde{x}=\tilde{\lambda}\tilde{x}$ be ordered so that $\tilde{\theta}_1 \leqslant \tilde{\theta}_2 \leqslant \cdots \leqslant \tilde{\theta}_n$. Then

$$|\theta_i - \tilde{\theta}_i| \leqslant \sin^{-1}\frac{\varepsilon}{c(A,B)}. \tag{3.3}$$

Proof. Let $\mathcal{X}_i$ be a minimizing subspace in the equality (3.1). Then

$$\hat{\theta}_i \leqslant \max_{\substack{x\in\mathcal{X}_i \\ x\neq 0}} \theta\left[x^T(A+E)x, x^T(B+F)x\right]. \tag{3.4}$$

Let $x \in \mathcal{X}_i$ be a vector of norm unity for which the maximum in (3.4) is attained. Then since $\theta(x^TAx, x^TBx) \leqslant \theta_i$, the point $(x^T(A+E)x, x^T(A+F)x)$ must lie in the circle of radius ε sketched in Fig. 1. The maximum increase η of $\theta[x^T(A+E)x, x^T(B+F)x]$ over θ_i occurs when the circle is situated as shown in Fig. 2. Elementary geometry now gives $\eta = \sin^{-1}[\varepsilon/c(A,B)]$, which shows that $\tilde{\theta}_i \leqslant \theta_i + \eta$. The result $\theta_i - \eta \leqslant \tilde{\theta}_i$ follows from a similar argument applied to the characterization (3.2) of $\tilde{\theta}_i$. ■

There are some observations to be made about the theorem. First, the bound (3.3) immediately implies a bound in the chordal metric. Specifically, let $\lambda = \mu/\nu$ and $\tilde{\lambda} = \tilde{\mu}/\tilde{\nu}$. Then the chordal distance between λ and $\tilde{\lambda}$ is

$$\chi(\lambda, \tilde{\lambda}) = \frac{|\mu\tilde{\nu} - \tilde{\mu}\nu|}{\sqrt{\mu^2+\nu^2}\sqrt{\tilde{\mu}^2+\tilde{\nu}^2}}.$$

But it is easily seen that

$$\chi(\lambda, \tilde{\lambda}) = \sin|\theta(\mu,\nu) - \theta(\tilde{\mu},\tilde{\nu})|.$$

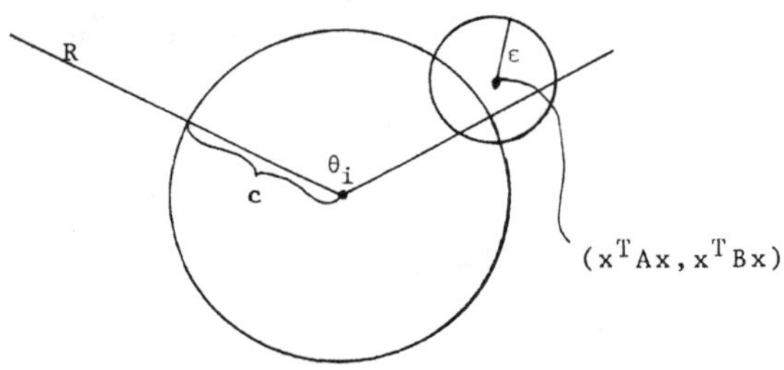

FIG. 1.

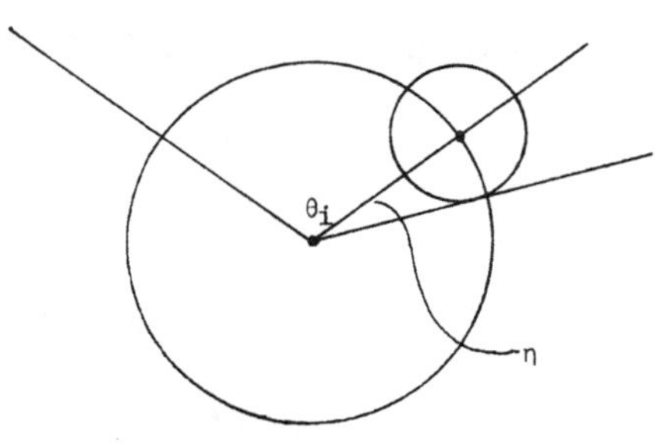

FIG. 2.

Then it follows from (3.3) that

$$\chi(\lambda_i,\tilde{\lambda}_i) \leqslant \frac{\varepsilon}{c(A,B)}.$$

Note that this inequality is somewhat weaker than (3.3), since eigenvalues that have angles differing by amounts near π will have a chordal difference near zero.

The second observation is that the theorem implies the classical bounds for the Hermitian eigenvalue problem. Specifically, let $B=I$, and let $\lambda_i(\tau)=\lambda_i/\tau$ be the eigenvalues of the problem

$$Ax = \lambda_i(\tau)(\tau I)x. \tag{3.5}$$

Let $c(\tau)=c(A,\tau I)$ be the Crawford number for the problem (3.5). Note that when τ is large, $c(\tau)=\tau+O(1)$. If we consider a Hermitian perturbation in A of norm ε, then we have the bounds

$$|\theta_i(\tau)-\tilde{\theta}_i(\tau)| \leqslant \sin^{-1}\frac{\varepsilon}{c(\tau)} = \frac{\varepsilon}{\tau}+O\left(\frac{1}{\tau^2}\right). \tag{3.6}$$

Since $\lambda_i(\tau)$ and $\tilde{\lambda}_i(\tau)$ approach 0 as $1/\tau$, we have

$$|\theta_i(\tau)-\tilde{\theta}_i(\tau)| = |\lambda_i(\tau)-\tilde{\lambda}_i(\tau)|+O\left(\frac{1}{\tau^2}\right). \tag{3.7}$$

Combining (3.6) and (3.7) and multiplying by τ, we get

$$|\lambda_i-\tilde{\lambda}_i| \leqslant \varepsilon+O\left(\frac{1}{\tau}\right),$$

which gives the classical result when $\tau\to\infty$.

Finally we note that it follows from the results in [12] that for a simple eigenvalue λ_i and ε sufficiently small,

$$\chi(\lambda_i,\tilde{\lambda}_i) \leqslant \frac{\varepsilon}{\sqrt{(x_i^HAx_i)^2+(x_i^HBx_i)^2}}+O(\varepsilon^2).$$

Since $c^2(A,B)\leqslant(x_i^HAx_i)^2+(x_i^HBx_i)^2$, it is seen that we pay a price in the sharpness of our bounds to gain freedom from considerations of multiplicity.

4. PERTURBATION BOUNDS: EIGENSPACES

In this section we shall derive perturbation bounds for the eigenvectors of the definite problem (1.1). These bounds imply that eigenvectors corresponding to poorly separated eigenvalues are very sensitive to perturbations in A and B; however, the subspace spanned by the eigenvectors corresponding to a cluster of eigenvalues may be relatively insensitive. We shall, therefore, phrase our bounds in terms of subspaces rather than individual eigenvectors. This approach is analogous to the one taken in the ordinary eigenvalue problem, where one bounds perturbations in invariant subspaces corresponding to clustered eigenvalues [4,8,9,11].

We begin by noting that if x is an eigenvector of (1.1), then Ax and Bx are dependent. Following [10], we can generalize this idea to subspaces.

DEFINITION 4.1. A subspace $\mathcal{X}$ is an *eigenspace* of (1.1) if

$$\dim(A\mathcal{X}+B\mathcal{X}) \leqslant \dim(\mathcal{X}).$$

Clearly any set of eigenvectors of (1.1) spans an eigenspace. Conversely, if the problem is definite, then an eigenspace is spanned by a set of eigenvectors. To see this, we first note that the space $A_\varphi\mathcal{X}+B_\varphi\mathcal{X}$, where A_φ and B_φ are defined as in Sec. 2, is the same as $A\mathcal{X}+B\mathcal{X}$. Hence we may assume that B is positive definite. Now let $\mathcal{X}_1$ be an eigenspace of dimension l spanned by the columns of the full rank matrix X_1. Then $\dim(B\mathcal{X}_1)=l$, and $B\mathcal{X}_1$ has an orthogonal complement $\mathcal{X}_2$ of dimension $n-l$. Since B is positive definite, $\mathcal{X}_1\oplus\mathcal{X}_2=\mathbb{C}^n$. Let the columns of the $n\times(n-l)$ matrix X_2 span $\mathcal{X}_2$, so that $X_2^TBX_1=0$. But from the definition of eigenspace and the fact that $\dim(B\mathcal{X}_1)=l$, we have that $A\mathcal{X}\subset B\mathcal{X}$. Hence $X_2^TAX_1=0$. It follows that

$$(X_1,X_2)^TA(X_1,X_2)=\begin{pmatrix} M_1 & 0\\ 0 & M_2\end{pmatrix} \tag{4.1}$$

and

$$(X_1,X_2)^TB(X_1,X_2)=\begin{pmatrix} N_1 & 0\\ 0 & N_2\end{pmatrix}. \tag{4.2}$$

Now the problems $M_iy=\lambda N_iy$ $(i=1,2)$ are definite; hence M_i and N_i can be simultaneously diagonalized by nonsingular matrices Y_i. Then (X_1Y_1,X_2Y_2) is nonsingular and diagonalizes A and B. In particular the columns of $X_1 Y_1$ are eigenvectors spanning $\mathcal{X}_1$.

Turning now to the perturbation theorem, we consider an eigenspace $\mathcal{X}_1$ of dimension l of the definite problem (1.1). Let $\mathcal{X}_2$ be its complementary eigenspace. Let the columns of $X_1 = (x_1, x_2, \ldots, x_l)$ and $X_2 = (x_{l+1}, \ldots, x_n)$ be eigenvectors spanning $\mathcal{X}_1$ and $\mathcal{X}_2$, chosen so that M_i and N_i $(i = 1,2)$ in (4.1) and (4.2) are diagonal. Set

$$M_1 = \operatorname{diag}(\mu_1, \ldots, \mu_l), \qquad N_1 = \operatorname{diag}(\nu_1, \ldots, \nu_l),$$

$$M_2 = \operatorname{diag}(\mu_{l+1}, \ldots, \mu_n), \qquad N_2 = \operatorname{diag}(\nu_{l+1}, \ldots, \nu_n).$$

We shall use the same notation for the perturbed eigenvalue problem $\tilde{A}\tilde{x} = \tilde{\lambda}\tilde{B}\tilde{x}$, except that all quantities will be marked with tildes. We shall assume that both problems are definite.

We begin with a lemma which furnishes a possible basis for the perturbed eigenspace. We will obtain sharper results if we recognize that certain infima that are bounded below by Crawford numbers are actually taken over eigenspaces. Accordingly, we define

$$c(A, B; \mathcal{X}) = \inf\{|x^H(A + iB)x| : x \in \mathcal{X},\ \|x\| = 1\}.$$

LEMMA 4.2. *For $i = 1, 2, \ldots, n$, let $\lambda_i = \mu_i/\nu_i$ and $\tilde{\lambda}_i = \tilde{\mu}_i/\tilde{\nu}_i$, and let*

$$\delta = \min\{\mathcal{X}(\lambda_i, \tilde{\lambda}_j) : i = 1, \ldots, l;\ j = l+1, \ldots, n\}.$$

If $\delta > 0$, then there is a matrix Q whose columns satisfy

$$\frac{\|q_i\|}{\|x_i\|} \leqslant \frac{\varepsilon}{\chi(\tilde{A}, \tilde{B}; \tilde{\mathcal{X}}_2)\delta}$$

such that

$$\mathcal{R}(X_1 + Q) \subset \tilde{\mathcal{X}}_1,$$

where $\mathcal{R}(X)$ denotes the column space of X.

Proof. Let $\tilde{\mathbf{B}}_2$ denote the operator defined by restricting $\tilde{B}_2$ to the space $\tilde{\mathcal{X}}_2$. By rotating the problem, we may assume that $\tilde{\mathbf{B}}_2$ is positive definite and

$$\|\tilde{\mathbf{B}}_2^{-1}\|^{-1} = c(\tilde{A}, \tilde{B}; \mathcal{X}_2).$$

Note that under this assumption N_2 is positive definite.

We shall seek Q in the form

$$Q = \tilde{X}_2 P.$$

From the definition of eigenspace, $\mathcal{R}(X_1 + Q) \subset \tilde{\mathcal{X}}_1$ if and only if $\mathcal{R}[\tilde{B}(X_1 + Q)] \perp \tilde{\mathcal{X}}_2$. This is equivalent to requiring that $\tilde{X}_2^T \tilde{B}(X_1 + \tilde{X}_2 P) = 0$, or

$$\tilde{X}_2^T \tilde{B} X_1 = -\tilde{N}_2 P. \tag{4.3}$$

To develop an expression for $\tilde{X}_2^T \tilde{B} X_1$, note that since

$$(\tilde{A} - E) X_1 N_1 = (\tilde{B} - F) X_1 M_1,$$

we have

$$\tilde{X}_2^T \tilde{A} X_1 N_1 - \tilde{X}_2^T \tilde{B} X_1 M_1 = \tilde{X}_2^T (E X_1 N_1 - F X_1 M_1) \equiv \tilde{X}_2^T R.$$

But $\tilde{N}_2 \tilde{X}_2^T \tilde{A} = \tilde{M}_2 \tilde{X}_2^T \tilde{B}$, so that

$$\tilde{M}_2 \tilde{X}_2^T \tilde{B} X_1 N_1 - \tilde{N}_2 \tilde{X}_2^T \tilde{B} X_1 M_1 = \tilde{N}_2 \tilde{X}_2^T R.$$

Hence, because $\tilde{M}_2$ and $\tilde{N}_2$ commute, if we choose P to satisfy

$$\tilde{M}_2 P N_1 - \tilde{N}_2 P M_1 = -\tilde{X}_2^T R,$$

then $N_2 P$ will satisfy (4.3).

Let $\tau_{ji} = \tilde{\mu}_j \nu_i - \tilde{\nu}_j \mu_i$. Then the hypothesis $\delta > 0$ implies that $\tau_{ji} \neq 0$ ($i = 1, \ldots, l$; $j = l+1, \ldots, n$). Consequently P is uniquely defined, and its $(j-l, i)$ element is given by

$$p_{j-l,i} = \frac{\tilde{x}_j^T r_i}{\tau_{ij}},$$

where r_i denotes the ith column of R. It follows that the ith column of $Q = \tilde{X}_2 P$ is given by

$$q_i = \left(\sum_{j=l+1}^{n} \frac{\tilde{x}_j \tilde{x}_j^T}{\tau_{ij}} \right) r_i \equiv S_i r_i,$$

and our problem is reduced to determining bounds on $\|S_i\|$.

Let $\tilde{Y}_2 = (\tilde{y}_{l+1}, \tilde{y}_{l+2}, \ldots, \tilde{y}_n) = \tilde{X}_2 \tilde{N}_2^{-1/2}$. Then

$$S_i = \sum_{j=l+1}^{n} \frac{\nu_j}{\tau_{ij}} \tilde{y}_j \tilde{y}_j^T.$$

But if $\tilde{\mathbf{B}}_2$ is the operator defined at the beginning of the proof,

$$\tilde{Y}_2^T \tilde{\mathbf{B}}_2 \tilde{Y}_2 = \tilde{N}_2^{-1/2} \tilde{X}_2^T \tilde{\mathbf{B}}_2 \tilde{X}_2 \tilde{N}_2^{-1/2} = I,$$

so that the columns of $\tilde{B}_2^{1/2} \tilde{Y}_2$ are orthonormal. Hence the eigenvalues of $\tilde{\mathbf{B}}_2^{1/2} S_i \tilde{\mathbf{B}}_2^{1/2}$ are $\tilde{\nu}_j / \tilde{\tau}_{ij}$. Hence

$$\|S_i\| = \|\tilde{\mathbf{B}}_2^{-1/2} (\tilde{\mathbf{B}}_2^{1/2} S_i \tilde{\mathbf{B}}_2^{1/2}) \tilde{\mathbf{B}}_2^{-1/2}\| \leqslant \|\tilde{\mathbf{B}}_2^{-1}\| \|\tilde{\mathbf{B}}_2^{1/2} S_i \tilde{\mathbf{B}}_2^{1/2}\|$$

$$\leqslant \frac{\max_j \{|\tilde{\nu}_j / \tau_{ij}|\}}{\tilde{c}_2} \leqslant \frac{\max_j \left\{ \left| \sqrt{\tilde{\mu}_j^2 + \tilde{\nu}_j^2} \,/ \tau_{ij} \right| \right\}}{\tilde{c}_2}, \tag{4.4}$$

where $\tilde{c}_2 = c(\tilde{A}, \tilde{B}; \tilde{\mathfrak{X}}_2)$.

Now from (4.4),

$$q_i = S_i r_i = S_i (E x_i \nu_i + F x_i \mu_i). \tag{4.5}$$

Hence

$$\|q_i\| \leqslant \|S_i\| \, \|x_i\| \varepsilon \sqrt{\mu_i^2 + \nu_i^2}\,. \tag{4.6}$$

Combining (4.5) and (4.6) we see that

$$\frac{\|q_i\|}{\|x_i\|} \leqslant \frac{\varepsilon}{\tilde{c}_2} \max_j \frac{\sqrt{\mu_i^2 + \nu_i^2} \sqrt{\tilde{\mu}_j^2 + \tilde{\nu}_j^2}}{|\mu_i \tilde{\nu}_j - \nu_i \tilde{\mu}_j|} \leqslant \frac{\varepsilon}{\tilde{c}_2 \delta}. \qquad \blacksquare$$

The bound in the theorem may be written in the form

$$\frac{\|Q\|_F}{\|X_1\|_F} \leqslant \frac{\varepsilon}{\tilde{c}_2 \delta},$$

where $\|X\|_F^2 \equiv \sum |x_{ij}|^2$ denotes the Frobenius norm of X. Thus the bound depends directly on the perturbation and inversely on the gap between the eigenvalues, with the Crawford number determining the size of the effect. Unfortunately $\mathcal{R}(X_1+Q_1)$ need not be an eigenspace unless $\dim[\mathcal{R}(X_1+Q_1)]=l$. When $l=1$, so that X_1 is the single eigenvector x_1, we can assure this by requiring that $\|q_1\|<\|x_1\|$.

THEOREM 4.3. *Let* $\delta=\min\{\chi(\lambda_1,\tilde{\lambda}_j):j=2,3,\ldots,n\}$. *If* $\varepsilon/\delta<c(\tilde{A},\tilde{B};\tilde{\mathcal{X}}_2)$, *then there is a vector* q_1 *satisfying*

$$\frac{\|q_1\|}{\|x_1\|}\leqslant\frac{\varepsilon}{\delta c(\tilde{A},\tilde{B};\tilde{\mathcal{X}}_2)}<1$$

such that x_1+q *is an eigenvector of* $\tilde{A}\tilde{x}=\tilde{\lambda}\tilde{B}\tilde{x}$ *corresponding to* $\tilde{\lambda}_1$.

When $l<1$, we must take into account the effect of near dependences among the columns of X. Define

$$\inf(X_1)=\inf_{\|x\|=1}\|X_1x\|.$$

Then if $\|Q\|<\inf(X_1)$, $\operatorname{rank}(X+Q)=\operatorname{rank}(X)=l$. These considerations lead to the following theorem.

THEOREM 4.4. *Let*

$$\eta=\frac{\varepsilon\sqrt{l}}{\delta c(\tilde{A},\tilde{B};\tilde{\mathcal{X}}_2)}\left\{\frac{\sqrt{\|A\|^2+\|B\|^2}}{c(A,B;\mathcal{X}_1)}\right\}^{1/2}$$

If $\eta<1$, *then* $\mathcal{R}(X_1+Q)=\tilde{\mathcal{X}}_1$.

Proof. It follows from the proof of Theorem 4.3 that if x_j is scaled by a factor α, then q_j is scaled by the same factor. Hence we may assume that $\|x_j\|=1$. We first find a lower bound on $\inf(X_1)$. By rotating (A,B), we may assume that if $\mathbf{B}_1$ denotes the restriction of B to $\mathcal{X}_1$, then $\mathbf{B}_1$ is positive definite and $\inf(\mathbf{B}_1)=c(A,B;\mathcal{X}_1)$. Of course,

$$\|\mathbf{B}_1\|\leqslant\|B\|\leqslant\sqrt{\|A\|^2+\|B\|^2}\,. \tag{4.7}$$

Now

$$X_1^T\mathbf{B}_1X_1=N_1,$$

from which it follows that the columns of $U = \mathbf{B}_1^{1/2} X_1 N_1^{-1/2}$ are orthonormal. Hence

$$1 = \inf(U) \leqslant \|\mathbf{B}_1^{1/2}\| \, \|N_1^{-1/2}\| \inf(X_1).$$

Since the columns of X_1 have norm unity, we have $\nu_i \geqslant c(A,B;\mathcal{X}_1)$. Hence from (4.7),

$$\inf(X_1) \geqslant \left\{ \frac{c(A,B;\mathcal{X}_1)}{\sqrt{\|A\|^2 + \|B\|^2}} \right\}^{1/2} \tag{4.8}$$

But

$$\|Q\|_F \leqslant \frac{\varepsilon \|X_1\|_F}{\delta \tilde{c}_2} \leqslant \frac{\epsilon \sqrt{l}}{\delta \tilde{c}_2}, \tag{4.9}$$

and the result follows from (4.9) and (4.10). ■

When $l = 1$, Theorem 4.4 does not reduce to Theorem 4.3, because we have given too much away in the bound (4.8). The principal application is to the case $l > 1$, where the condition $\eta < 1$ not only guarantees that $\tilde{\mathcal{X}}_1 = \mathcal{R}(X_1 + Q)$, but also implies that $\mathcal{X}_1$ and $\tilde{\mathcal{X}}_1$ are acutely situated with respect to one another. In fact a slight modification of the proof of Theorem 4.1 in [13] gives the following corollary.

COROLLARY 4.5. *Let $P_{\mathcal{X}_1}$ and $P_{\tilde{\mathcal{X}}_1}$ denote the orthogonal projections onto $\mathcal{X}_1$ and $\tilde{\mathcal{X}}_1$. If $\eta < 1$, then*

$$\|P_{\mathcal{X}_1} - P_{\tilde{\mathcal{X}}_1}\|_F \leqslant \frac{\eta}{1-\eta}.$$

It is unfortunate that η contains the factor $\sqrt{l}$, since it grows with the dimension of $\mathcal{X}_1$. The presence of this factor is a direct consequence of the fact that Lemma 4.2 bounds $\|Q\|_F / \|X\|_F$ instead of $\|Q\| / \|F\|$. For the Hermitian eigenvalue problem, Davis and Kahan [4] have been able to obtain bounds on the spectral norm by imposing additional restrictions on the location of the eigenvalues. Whether such bounds can be obtained for the definite generalized eigenvalue problem is an open question.

REFERENCES

1 L. Brickman, On the fields of values of a matrix, *Proc. Amer. Math. Soc.* **12** (1961), 61–66.
2 E. Calabi, Linear systems of real quadratic forms, *Proc. Amer. Math. Soc.* **15** (1964), 844–846.
3 C. R. Crawford, A stable generalized eigenvalue problem, *SIAM J. Numer. Anal.* **6** (1976), 854–860.
4 C. Davis and W. Kahan, The rotation of eigenvectors by a perturbation, *SIAM J. Numer. Anal.* **7** (1970), 1–46.
5 W. H. Greub, *Linear Algebra*, Springer, New York, 1967.
6 M. R. Hestenes, Pairs of quadratic forms, *Linear Algebra and Appl.* **1** (1968), 397–407.
7 A. S. Householder, *The Theory of Matrices in Numerical Analysis*, Blaisdel, New York, 1964.
8 T. Kato, *Perturbation Theory for Linear Operators*, Springer, New York, 1966.
9 G. W. Stewart, Error bounds for approximate invariant subspaces of closed linear operators, *SIAM J. Numer. Anal.* **8** (1971), 796–808.
10 G. W. Stewart, On the sensitivity of the eigenvalue problem $Ax=\lambda Bx$, *SIAM J. Numer. Anal.* **9** (1972), 669–686.
11 G. W. Stewart, Error and perturbation bounds for subspaces associated with certain eigenvalue problems, *SIAM Rev.* **15** (1973), 772–764.
12 G. W. Stewart, Gershgorin theory for the generalized eigenvalue problem $Ax=\lambda Bx$, *Math. Comp.* **29** (1975), 600–606.
13 G. W. Stewart, On the perturbation of pseudo-inverses, projections, and linear least squares problems, *SIAM Rev.* to be published.
14 O. Taussky, Positive Definite Matrices, in *Inequalities* (Oved Shisha, Ed.), Academic, New York, 1967, 309–319.

Received 25 November 1977

18

Papers on Krylov Subspace Methods for the Eigenproblem

1. [GWS-J111] "A Krylov-Schur Algorithm for Large Eigenproblems," *SIAM Journal on Matrix Analysis and Applications* **23** (2001) 601–614.
2. [GWS-J113] "Addendum to 'A Krylov-Schur Algorithm for Large Eigenproblems'," *SIAM Journal on Matrix Analysis and Applications* **24** (2002) 599–601.
3. [GWS-J110] "Backward Error Bounds for Approximate Krylov Subspaces," *Linear Algebra and its Applications* **340** (2002) 81–86.
4. [GWS-J112] "Adjusting the Rayleigh Quotient in Semiorthogonal Lanczos Methods," *SIAM Journal on Scientific Computing* **24** (2002) 201–207.

18.1. [GWS-J111] "A Krylov-Schur Algorithm for Large Eigenproblems"

[GWS-J111] "A Krylov-Schur Algorithm for Large Eigenproblems," *SIAM Journal on Matrix Analysis and Applications* **23** (2001) 601–614.
http://dx.doi.org/10.1137/S0895479800371529

SIAM J. MATRIX ANAL. APPL.
Vol. 23, No. 3, pp. 601–614

A KRYLOV–SCHUR ALGORITHM FOR LARGE EIGENPROBLEMS*

G. W. STEWART†

Abstract. Sorensen's implicitly restarted Arnoldi algorithm is one of the most successful and flexible methods for finding a few eigenpairs of a large matrix. However, the need to preserve the structure of the Arnoldi decomposition on which the algorithm is based restricts the range of transformations that can be performed on the decomposition. In consequence, it is difficult to deflate converged Ritz vectors from the decomposition. Moreover, the potential forward instability of the implicit QR algorithm can cause unwanted Ritz vectors to persist in the computation. In this paper we introduce a general Krylov decomposition that solves both problems in a natural and efficient manner.

Key words. large eigenproblem, Krylov sequence, Arnoldi algorithm, Krylov decomposition, restarting, deflation

AMS subject classifications. 15A18, 65F15, 65F60

PII. S0895479800371529

1. Introduction and background. In this paper we are going to describe an alternative to the Arnoldi method that resolves some difficulties with its implicitly restarted version. To understand the difficulties and their solution requires a detailed knowledge of the Arnoldi process. We therefore begin with a survey, which will also serve to set the notation for this paper.

Let A be a matrix of order n and let u_1 be a vector of 2-norm one. Let $u_1, u_2, u_3, \ldots$ be the result of sequentially orthogonalizing the Krylov sequence $u_1, Au_1, A^2u_1, \ldots$. In 1950, Lanczos [6] showed that if A is Hermitian then the vectors u_i satisfy a three-term recurrence of the form

$$\beta_k u_{k+1} = A_k u_k - \alpha_k u_k - \beta_{k-1} u_{k-1}, \tag{1.1}$$

a recursion that in principle allows the economical computation of the u_j.

There is an elegant representation of this recursion in matrix terms. Let

$$U_k = (u_1 \ u_2 \ \cdots \ u_k)$$

be the matrix formed from the Lanczos vectors u_j. Then there is a tridiagonal matrix T formed from the α's and β's in (1.1) such that

$$AU_k = U_k T_k + \beta_k u_{k+1} \mathbf{e}_k^{\mathrm{T}}, \tag{1.2}$$

where $\mathbf{e}_k$ is the vector whose last component is one and whose other components are zero. From the orthogonality of the u_j, it follows that T_k is the Rayleigh quotient

$$T_k = U_k^{\mathrm{H}} A U_k.$$

We will call (1.2) a Lanczos decomposition.

*Received by the editors May 2, 2000; accepted for publication (in revised form) by J. Varah June 8, 2001; published electronically December 14, 2001.

http://www.siam.org/journals/simax/23-3/37152.html

†Department of Computer Science and Institute for Advanced Computer Studies, University of Maryland, College Park, MD 20742 (stewart@cs.umd.edu). Work supported by the National Science Foundation under grant 970909-8562.

Lanczos appreciated the fact that even for comparatively small k the matrix T_k could contain accurate approximations to the eigenvalues of A. When this happens, the column space $\mathcal{U}_k$ of U_k will usually contain approximations to the corresponding eigenvectors. Such an approximation—call it z—can be calculated by computing a suitable eigenpair (μ, w) of T_k and setting $z = U_k w$. This process is called the Rayleigh–Ritz method; μ is called a Ritz value and z a Ritz vector.

In 1951, Arnoldi [1], building on Lanczos's work, showed that if A is non-Hermitian, then the Lanczos decomposition becomes

$$AU_k = U_k H_k + \beta_k u_{k+1} \mathbf{e}_k^{\mathrm{T}}, \tag{1.3}$$

where H_k is upper Hessenberg. We will call (1.3) an Arnoldi decomposition. Once again, H_k may contain accurate approximations to the eigenvalues of A, especially those on the periphery of the spectrum of A. Moreover, approximations to the eigenvectors may be obtained by the natural generalization of the Rayleigh–Ritz process.

Arnoldi decompositions are essentially unique. Specifically, if H_k is unreduced—that is, if its subdiagonal elements are nonzero—then up to scaling of the columns of U_{k+1} and the rows and columns of H_k, the decomposition is uniquely determined by the space spanned by U_{k+1}.[1] In particular, the Krylov subspace of an unreduced Arnoldi decomposition has a unique starting vector.

Since H_k is not tridiagonal, the Arnoldi vectors do not satisfy a three-term recurrence. To compute u_{k+1} all the columns of U_k must be readily available. If n is large, these vectors will soon consume all available storage, and the process must be restarted. The problem then becomes how to choose a new u_1 that does not discard the information about the eigenvectors contained in $\mathcal{U}_k$. There have been several proposals, whose drawbacks have been nicely surveyed by Morgan [11].

In 1992, Sorensen [14] suggested an elegant way to use the QR algorithm to restart the Arnoldi process. Specifically, suppose we have an Arnoldi decomposition

$$AU_m = U_m H_m + \beta_m u_{m+1} \mathbf{e}_m^{\mathrm{T}} \tag{1.4}$$

of order m that cannot be further expanded because of lack of storage. For some fixed k, choose $m-k$ shifts $\kappa_1, \ldots, \kappa_{m-k}$ and use them to perform $m-k$ steps of the implicitly shifted QR algorithm on the Rayleigh quotient H_m. The effect is to generate an orthogonal matrix Q such that $Q^{\mathrm{H}} H_m Q$ is upper Hessenberg. Then from (1.4)

$$A(U_m Q) = (U_m Q) Q^{\mathrm{H}} H_m Q + \beta_m u_{m+1} \mathbf{e}_m^{\mathrm{T}} Q$$

or

$$A\tilde{U}_m = \tilde{U}_m \tilde{H}_m + u_{m+1} c^{\mathrm{H}}.$$

Sorensen then observed that the structure of Q is such that the first $k-1$ components of c are zero. Consequently, if we let $\tilde{H}_k$ be the leading principal submatrix of $\tilde{H}_m$ of order k and set

$$\beta_k \tilde{u}_{k+1} = \bar{\gamma}_k u_{m+1} + \tilde{h}_{k+1,k} u_{k+1}, \tag{1.5}$$

then

$$A\tilde{U}_k = \tilde{U}_k \tilde{H}_k + \tilde{u}_{k+1} e_k^{\mathrm{T}}$$

[1]This fact is a direct consequence of the implicit Q theorem, which says that if $H = Q^{\mathrm{H}} AQ$ is an unreduced Hessenberg matrix then Q is determined by its first or last column. See [4, Theorem 7.4.2].

is an Arnoldi decomposition of order k. This process of truncating the decomposition is called implicit restarting.

A second key observation of Sorensen suggests a rationale for choosing the shifts. Specifically, if $p(t) = (t - \kappa_1 I)\cdots(t - \kappa_{m-k}I)$, then

$$\tilde{u}_1 = \frac{p(A)u_1}{\|p(A)u_1\|}.$$

It follows that if we choose the shifts to lie in the part of the spectrum that we are not interested in then the implicit restart process deemphasizes these very eigenvalues.

Each iteration of Sorensen's algorithm consists of two stages: an expansion stage, in which the decomposition is expanded until it is inconvenient to go further, and a contraction or purging stage, in which unwanted parts of the spectrum are suppressed. The contraction phase has two variants. In the exact variant, the shifts are taken to be unwanted eigenvalues of H_m. If, for example, we were concerned with stability, we might choose to retain only the eigenvalues with largest real parts. In the other, more general variant, the shifts are not necessarily eigenvalues of H_m. For example, they might be the zeros of a Chebyshev polynomial spanning an ellipse containing unwanted eigenvalues.

The implicitly restarted Arnoldi algorithm has been remarkably successful and has been implemented in the widely used ARPACK package [9]. However, the method has two important drawbacks.

First, for the exact restart procedure to be effective the unwanted Ritz values μ must be moved to the end of H_m, so that the Rayleigh quotient has the form illustrated below for $k = 3$ and $m = 6$:

$$\begin{pmatrix} h & h & h & h & h & h \\ h & h & h & h & h & h \\ 0 & h & h & h & h & h \\ 0 & 0 & 0 & \mu & h & h \\ 0 & 0 & 0 & 0 & \mu & h \\ 0 & 0 & 0 & 0 & 0 & \mu \end{pmatrix}. \tag{1.6}$$

If H_m is unreduced—that is, if the elements of its first subdiagonal are nonzero—then mathematically H_m must have the form (1.6). In the presence of rounding error, however, the process can fail (for a treatment of this phenomenon, see [17]). This has lead Lehoucq and Sorensen to propose an elaborate method for permanently ridding the decomposition of persistent unwanted Ritz values [8].

The second problem is to move converged Ritz values μ to the beginning of H_k, so that it assumes the form illustrated below:

$$\begin{pmatrix} \mu & h & h & h & h & h \\ 0 & \mu & h & h & h & h \\ 0 & 0 & h & h & h & h \\ 0 & 0 & h & h & h & h \\ 0 & 0 & 0 & h & h & h \\ 0 & 0 & 0 & 0 & h & h \end{pmatrix}.$$

When the converged Ritz values are thus deflated (or locked), one does not have to update the Arnoldi u_1 and u_2 in the Arnoldi decomposition. Lehoucq and Sorensen have proposed a complicated deflation algorithm.

Most of the complications in the purging and deflating algorithms come from the need to preserve the structure of the Arnoldi decomposition (1.3)—in particular, to preserve the Hessenberg form of the Rayleigh quotient and the zero structure of the vector $\mathbf{e}_k$. The purpose of this paper is to show that if we relax the definition of an Arnoldi decomposition, we can solve the purging and deflating problems in a natural and efficient way. Since the method is centered about the Schur decomposition of the Rayleigh quotient, we will call the method the Krylov–Schur method.

The decompositions and algorithms proposed in this paper are not without precursors. Fokkema, Sleijpen, and van der Vorst [3] explicitly use Schur vectors to restart the Jacobi–Davidson algorithm. Stathopoulos, Saad, and Wu [15] point out that because the unprecondition Jacobi–Davidson algorithm is equivalent to the Arnoldi algorithm, one can also use Schur vectors to restart the latter. Lehoucq [7] has used Schur vectors in the deflation process in [8]. Closer to home, for symmetric matrices Wu and Simon [18] exhibit what might be called a Krylov-spectral decomposition, a special case of our Krylov–Schur decomposition to be introduced later. Finally, Morgan [12] has applied an orthogonal Krylov decomposition to the problem of restarting GMRES. What distinguishes our approach is the explicit introduction of general Krylov decompositions whose subspaces are invariant under certain formal operations—operations that can be used to derive and analyze new algorithms.

In the next section we introduce Krylov decompositions and, in particular, the Krylov–Schur decomposition, which lies at the heart of our method. Section 3 treats the Krylov–Schur method and its relation to the implicitly restarted Arnoldi method. In section 4 we treat the numerical stability of the combined steps. In section 5 we show how to deflate vectors and subspaces from a Krylov decomposition. In section 6 we compare the work done by the implicitly restarted Arnoldi and the Krylov–Schur methods. We end with some general comments. Throughout this paper $\|\cdot\|$ will denote the vector and matrix 2-norm, and $\|\cdot\|_F$ will denote the Frobenius norm (see [16, section 1.4.1]).

2. Krylov decompositions. The structure of an Arnoldi decomposition restricts the operations we can perform on its Rayleigh quotient. The following definition introduces a less constraining decomposition.

DEFINITION 2.1. *A* Krylov decomposition of order k *is a relation of the form*

$$AU_k = U_k B_k + u_{k+1} b_{k+1}^{\mathrm{H}}, \tag{2.1}$$

where B_k is of order k and the columns of $(U_k \ u_{k+1})$ are independent. The columns of $(U_k \ u_{k+1})$ are called the basis *for the decomposition, and they span the* space *of the decomposition. If the basis is orthonormal, we say the decomposition is orthonormal. The matrix B_k is called the* Rayleigh quotient *of the decomposition.*

This definition removes practically all the restrictions imposed on an Arnoldi decomposition. The vectors of the decomposition are not required to be orthonormal and the vector b_{k+1} and the matrix B_k are allowed to be arbitrary. Nonetheless, we shall see that the relation (2.1) is sufficient to insure that $(U_k \ u_{k+1})$ is a basis for a Krylov subspace.

The name "Rayleigh quotient" is appropriate for the matrix B_k. For if $(V_k \ v_{k+1})^{\mathrm{H}}$ is a left inverse of $(U_k \ u_{k+1})$, then $B_k = V_k^{\mathrm{H}} A U_k$. In particular, if $(\mu, U_k w)$ is an eigenpair of A, then (μ, w) is an eigenpair of B_k. Thus the Rayleigh–Ritz procedure extends to Krylov decompositions.

The subspaces of Krylov decompositions are closed under two classes of transformations: translation and similarity. The first allows us to change the vector u_k. The second allows us to change the pair (B_k, U_k) along with b_{k+1}^{H}. In what follows we will drop subscripts in k and write our Krylov decomposition in the form $AU = UB + ub^{\mathrm{H}}$.

To introduce the operation of translation, let

$$\gamma\tilde{u} = u - Ug,$$

where $\gamma \neq 0$. Then it is easily verified that

$$AU = U(B + gb^{\mathrm{H}}) + \tilde{u}b^{\mathrm{H}},$$

where $\tilde{b}^{\mathrm{H}} = \gamma b^{\mathrm{H}}$, is a Krylov decomposition with the same space as the original. This gives us considerable freedom to replace u by linear combinations of u and U, although the fact that $\gamma \neq 0$ implies that the vector $\tilde{u}$ always contains some component along u. In particular, we can choose $\tilde{u}$ so that $\|\tilde{u}\| = 1$ and $U^{\mathrm{H}}\tilde{u} = 0$.

To introduce similarity transformations, let W be nonsingular. Then

$$A(UW^{-1}) = (UW^{-1})(WBW^{-1}) + u(b^{\mathrm{H}}W^{-1}) \equiv A\tilde{U} = \tilde{U}\tilde{B} + u\tilde{b}^{\mathrm{H}}$$

is a Krylov decomposition whose space is the same as the original. Because the Rayleigh quotient of the new decomposition is similar to that of the old, we say that the two decompositions are similar.[2]

We will say that two Krylov decompositions related by a sequence of translations and similarities are *equivalent*. We are now going to show that any Krylov decomposition is equivalent to an Arnoldi decomposition. Since the space of an Arnoldi decomposition is a (possibly restarted) Krylov subspace, the result justifies the name Krylov decomposition.

THEOREM 2.2. *Let*

$$AU = UB + ub^{\mathrm{T}} \tag{2.2}$$

be a Krylov decomposition of order k. Then (2.2) is equivalent to an Arnoldi decomposition. If the Hessenberg part of the Arnoldi decomposition is unreduced, the Arnoldi decomposition is essentially unique.

Proof. The reduction, which is constructive, proceeds in four stages.

1. By a similarity transformation, orthogonalize the columns of U.
2. By a translation, transform u so that it is of norm one and is orthogonal to $\mathcal{R}(U)$.
3. By a unitary similarity transformation, reduce b to a multiple of $\mathbf{e}_k$.
4. Finally, by a unitary similarity reduce B to Hessenberg form. The reduction is performed rowwise by Householder transformations beginning with the last row, as illustrated in the following Wilkinson diagram:

$$\begin{pmatrix} b & b & b & b & b \\ b & b & b & b & b \\ b^3 & b & b & b & b \\ b^2 & b^2 & b & b & b \\ b^1 & b^1 & b^1 & b & b \end{pmatrix}.$$

[2] A referee has pointed out that these two types of transformations can be combined. Specifically, we say that the Arnoldi decompositions $AU = UB + ub^{\mathrm{T}}$ and $AV = VB + vb^{\mathrm{T}}$ are equivalent if there is a nonsingular matrix $\hat{W} = \left(\begin{smallmatrix} W & g \\ 0 & \gamma \end{smallmatrix}\right)$ such that $(V\ v) = (U\ u)\hat{W}$. With $W = I$ we obtain a translation; with $g = 0$ and $\gamma = 1$ we obtain a similarity.

The final reduction to Hessenberg form does not introduce nonzero elements into the first $k-1$ components of b, so that the result of this algorithm is an Arnoldi decomposition. The uniqueness in the unreduced case follows from the uniqueness of unreduced Arnoldi decompositions. □

The proof of Theorem 2.2 illustrates the power of translations and similarities to bring a Krylov decomposition into a useful form without losing the Krylov subspace property. In particular, any Krylov decomposition corresponds to an *orthonormal Krylov decomposition* in which the columns of the basis are orthonormal. (From here on, all our Krylov decompositions will be orthonormal.) Further, we can reduce the Rayleigh quotient to Schur form. The resulting *Krylov–Schur* decomposition is the basis of the main algorithm in this paper, to which we now turn.

3. The Krylov–Schur method. A step of the Krylov–Schur method begins and ends with a Krylov–Schur decomposition of the form

$$AU_k = U_k S_k + u_{k+1} b_{k+1}^{\mathrm{H}},$$

where the letter S (for Schur) stresses the triangularity of the Rayleigh quotient. It will be more convenient to work with the equivalent factored form

$$AU_k = U_{k+1}\hat{S}_k,$$

where

$$\hat{S}_k = \begin{pmatrix} S_k \\ b_{k+1}^{\mathrm{H}} \end{pmatrix}.$$

Like the implicitly restarted Arnoldi method the Krylov–Schur method consists of an expansion phase, in which the underlying Krylov sequence is extended, and a contraction phase, in which the unwanted Ritz values are purged from the decomposition. We will treat each in turn.

The expansion proceeds as in the usual Arnoldi algorithm: the vector Au_{k+1} is orthogonalized against U_{k+1} and normalized to give u_{k+2}, after which S_{k+1} is formed from S_k. The following pseudocode implements this procedure. We assume that U_{k+1} and $\hat{S}_k$ are contained in arrays U and S.

$$\text{(3.1)} \qquad \begin{aligned} &1. \quad v = A * U[:, k+1], \\ &2. \quad w = U^{\mathrm{H}} * v, \\ &3. \quad v = v - U * w, \\ &4. \quad \nu = \|v\|_2, \\ &5. \quad U = (U \;\; v/\nu), \\ &6. \quad \hat{S} = \begin{pmatrix} \hat{S} & w \\ 0 & \nu \end{pmatrix}. \end{aligned}$$

Note that in a working implementation we would have to reorthogonalize to insure that the vector v is orthogonal to the column space of U to working accuracy (see [16, Algorithm 4.1.13]).

After this process the array $\hat{S}$ has the form illustrated below for $k = 3$:

$$\begin{pmatrix} s & s & s & h \\ 0 & s & s & h \\ 0 & 0 & s & h \\ b & b & b & h \\ 0 & 0 & 0 & h \end{pmatrix}.$$

Here the s's stand for the elements of the original S_k and the b's for the elements of b_{k+1}. The process may be repeated. After $m-k$ steps, the array S has the form illustrated below for $k=3$ and $m=6$:

$$\begin{pmatrix} s & s & s & h & h & h \\ 0 & s & s & h & h & h \\ 0 & 0 & s & h & h & h \\ b & b & b & h & h & h \\ 0 & 0 & 0 & h & h & h \\ 0 & 0 & 0 & 0 & h & h \\ 0 & 0 & 0 & 0 & 0 & h \end{pmatrix}. \tag{3.2}$$

At this point the Rayleigh quotient, which resides in $S[1{:}m, 1{:}m]$, is reduced to Schur form to give the Arnoldi–Schur decomposition

$$AU_m = U_m S_m + u_{m+1} b_{m+1}^{\mathrm{H}}. \tag{3.3}$$

This reduction to Schur form begins with a reduction of the Rayleigh quotient to Hessenberg form, and some minor savings can be obtained at this stage by taking advantage of the structure illustrated in (3.2). Although (3.3) suggests that we are computing the entire decomposition, including U_m, in fact it will be more efficient to defer the computation of the columns of U_m until later. We will return to this point in section 6.

We now turn to the problem of purging the unwanted Ritz values from the Krylov–Schur decomposition (3.3)—the contraction phase of the method. The key is the observation that a Krylov–Schur decomposition can be truncated at any point. Specifically, if we partition a Krylov–Schur decomposition in the form

$$A(U_1\ U_2) = (U_1\ U_2)\begin{pmatrix} S_{11} & S_{12} \\ 0 & S_{22} \end{pmatrix} + u(b_1^{\mathrm{H}}\ b_2^{\mathrm{H}}), \tag{3.4}$$

then

$$AU_1 = U_1 S_{11} + u b_1^{\mathrm{H}}$$

is also a Krylov–Schur decomposition. Thus the purging problem can be solved by moving the unwanted Ritz values into the southeast corner of the Rayleigh quotient and truncating the decomposition.

The process of using unitary similarities to move eigenvalues around in a Schur form has been well studied. The current front-running algorithm [2], which has been implemented in the LAPACK routine `xTREXC`, is quite reliable—far more so than implicit QR. Consequently, our deflation algorithm consists of little more than moving the unwanted Ritz values, which are visible on the diagonals of S_m, to the southeast corner of the Rayleigh quotient and truncating the decomposition.

The following theorem shows just what a combined expansion and contraction step produces.

THEOREM 3.1. *Let*

$$\mathbb{P} := AU = UH + \beta u \mathbf{e}_k^{\mathrm{T}}$$

be an unreduced Arnoldi decomposition and let

$$\mathbb{Q} := AV = VS + ub^{\mathrm{H}}$$

be an equivalent Krylov–Schur form. Suppose that an implicitly restarted Arnoldi cycle is performed on $\mathbb{P}$ *and a Krylov–Schur cycle is performed on* $\mathbb{Q}$. *If the same Ritz values are discarded in both and those Ritz values are distinct from the other Ritz values, then the resulting decompositions are equivalent.*

Proof. We must show that the subspaces associated with the final results are the same. First note that the expansion phase results in equivalent decompositions. In fact, since $\mathcal{R}(U) = \mathcal{R}(V)$ and in both cases we are orthogonalizing the same Krylov sequence, the vectors $u_{k+1} \dots u_{m+1}$ and $v_{k+1}, \dots, v_{m+1}$ are the same up to multiples of modulus one.

Now assume that both algorithms have gone through the expansion phase and have moved the unwanted Ritz values to the end of the decomposition. At this point denote the first decomposition by

$$\hat{\mathbb{P}} := A\hat{U} = \hat{U}\hat{H} + \hat{\beta}\hat{u}\mathbf{e}_m^{\mathrm{T}}$$

and the second by

$$\hat{\mathbb{Q}} := A\hat{V} = \hat{V}\hat{S} + \hat{u}\hat{b}^{\mathrm{H}}.$$

Note that for both methods, the final truncation leaves the vector $\hat{u}$ unaltered. Since $\hat{V} = \hat{U}W$ for some unitary W, we have

$$\hat{S} = \hat{V}^{\mathrm{H}}A\hat{V} = W^{\mathrm{H}}\hat{U}^{\mathrm{H}}A\hat{U}W = W^{\mathrm{H}}\hat{H}W.$$

Thus $\hat{H}$ and $\hat{S}$ are similar and have the same Ritz values. Thus it makes sense to say that both methods reject the same Ritz values.

Let P be the unitary transformation applied to the Rayleigh quotient in $\hat{\mathbb{P}}$, and let Q be the one applied to the Rayleigh quotient of $\hat{\mathbb{Q}}$. Then we must show that the subspaces spanned by $\hat{U}P[:,1{:}k]$ and $\hat{V}Q[:,1{:}k]$ are the same. For brevity, set $P_k = P[:,1{:}k]$ and $Q_k = Q[:,1{:}k]$.

By construction $\mathcal{R}(P_k)$ is the eigenspace $\mathcal{P}$ of Schur vectors of $\hat{H}$ corresponding to the retained Ritz values. Likewise, $\mathcal{R}(Q_k)$ is the eigenspace $\mathcal{Q}$ of Schur vectors of $\hat{S}$ corresponding to the retained Ritz values. By hypothesis these eigenspaces are simple and hence are the same. Since $W^{\mathrm{H}}\hat{S}W = \hat{H}$, the matrix $W^{\mathrm{H}}\hat{P}_k$ spans $\mathcal{Q}$. Hence there is a unitary matrix R such that $Q_k = W^{\mathrm{H}}P_kR$. We then have

$$\hat{V}Q_k = \hat{U}WW^{\mathrm{H}}P_kR = \hat{U}P_kR.$$

It follows that $\hat{V}Q_k$ and $\hat{U}P_k$ span the same subspace. $\square$

The import of this theorem is that no matter how you perform the expansion and contraction, mathematically you end up with a decomposition that has been filtered through the polynomial $(t - \mu_1)\dots(t - \mu_{m-k})$. However, the procedure based on the Krylov–Schur form is numerically more reliable than the one based on implicit restarting.

4. Numerical stability. We now briefly consider the numerical stability of the algorithm. From standard techniques of rounding error analysis it can be shown that as the Krylov–Schur algorithm proceeds, the computed Krylov decompositions satisfy

$$AU = UB + ub^{\mathrm{H}} + R, \tag{4.1}$$

where $\|R\|/\|A\|$ is of the order of the rounding unit and grows slowly. If U is computed with reorthogonalization in the expansion phase, $U^{\mathrm{H}}U = I+F$, where $\|F\|$ is the order of the rounding unit and also grows slowly. The following theorem shows that we can throw the residual error R back on the matrix A.

THEOREM 4.1. *Let* (4.1) *be satisfied and assume that* U *is of full rank. Let* $E = -RU^{\dagger}$, *where* $U^{\dagger} = (U^{\mathrm{H}}U)^{-1}U^{\mathrm{H}}$ *is the pseudoinverse of* U. *Then*

$$(A + E)U = UB + ub^{\mathrm{H}}, \tag{4.2}$$

and

$$\frac{\|R\|}{\|U\|} \le \|E\| \le \|R\|\|U^{\dagger}\|.$$

The lower bound holds for any matrix E *satisfying* (4.2).

Proof. The equation (4.2) is established by direct verification. The upper bound follows from taking norms in the definition of E. On the other hand, if E is any matrix satisfying (4.2), then $EU = -R$, and $\|R\| \le \|E\|\|U\|$, which establishes the lower bound. □

Since U is nearly orthonormal, $\|U\|$ and $\|U^{\dagger}\|$ are near one. Hence the theorem shows that the computed generalized Arnoldi decomposition is an exact decomposition of a matrix near A. In this sense the Krylov–Schur algorithm (as well as the implicitly restarted Arnoldi algorithm) is backward stable.

5. Deflation and convergence. We now turn to the problem of deflating converged vectors from an orthonormal Krylov decomposition. We shall see later that if the concern is with a single Ritz vector then the deflation is easy. However, we can also use Krylov decompositions to deflate approximate eigenvectors or eigenspaces that are not obtained by a Rayleigh–Ritz procedure. Moreover, dependencies among the vectors to be deflated can cause the deflation procedure to require smaller residuals in the individual vectors. Consequently, we give a general analysis that covers both of these points.

We say a Krylov decomposition has been deflated if it can be partitioned in the form

$$A(U_1\ U_2) = (U_1\ U_2)\begin{pmatrix} B_{11} & B_{12} \\ 0 & B_{22} \end{pmatrix} + u(0\ b_2^{\mathrm{H}}).$$

When this happens, we have $AU_1 = U_1B_{11}$, so that U_{11} spans an eigenspace of A.

There are two advantages to deflating a converged eigenspace. First, by freezing it at the beginning of the Krylov decomposition we insure that the remaining space of the decomposition remains orthogonal to it. In particular, this gives algorithms the opportunity to compute more than one independent eigenvector corresponding to a multiple eigenvalue.

The second advantage of the deflated decomposition is that we can save operations in the contraction phase of an Arnoldi or Krylov–Schur cycle. The expansion phase does not change, and we end up with a decomposition of the form

$$A(U_1\ U_2\ U_3) = (U_1\ U_2\ U_3)\begin{pmatrix} B_{11} & B_{12} & B_{13} \\ 0 & B_{22} & B_{23} \\ 0 & B_{23} & B_{33} \end{pmatrix} + \beta u \mathbf{e}_m^{\mathrm{T}}.$$

Now since B_{11} is uncoupled from the rest of the Rayleigh quotient, we can apply all subsequent transformations exclusively to the eastern part the Rayleigh quotient and to $(U_2\ U_3)$. If the order of B_{11} is small, the savings will be marginal; but as its size increases during the course of the algorithm, the savings become significant.

Of course, we will never have an exact eigenspace in our decompositions. Instead we will have a basis, say UW, for an approximate eigenspace and an approximation representation M of A on that subspace. The following theorem relates the norm of the residual $A(UW)-(UW)M$ to the quantities in the decomposition we must set to zero in order to deflate.

THEOREM 5.1. *Let*

$$AU = UB + ub^{\mathrm{H}} \tag{5.1}$$

be an orthonormal Krylov decomposition, and let $(M, Z) = (M, UW)$ *be given with* U *and* W *orthonormal. Let* $(W\ W_{\perp})$ *be unitary, and set*

$$\tilde{B} = \begin{pmatrix} W^{\mathrm{H}} \\ W_{\perp}^{\mathrm{H}} \end{pmatrix} B(W\ W_{\perp}) = \begin{pmatrix} \tilde{B}_{11} & \tilde{B}_{12} \\ \tilde{B}_{21} & \tilde{B}_{22} \end{pmatrix}$$

and

$$\tilde{b}^{\mathrm{H}} = b^{\mathrm{H}}(W\ W_{\perp}) = (\tilde{b}_1^{\mathrm{H}}\ \tilde{b}_2^{\mathrm{H}}).$$

Then

$$\|AZ - ZM\|_{\mathrm{F}}^2 = \|\tilde{B}_{21}\|_{\mathrm{F}}^2 + \|\tilde{b}_1\|_{\mathrm{F}}^2 + \|\tilde{B}_{11} - M\|_{\mathrm{F}}^2, \tag{5.2}$$

where $\|\cdot\|_{\mathrm{F}}$ *denotes the Frobenius norm.*

Proof. From (5.1) we have $AZ - ZM = UBW - UWM + ub^{\mathrm{H}}W$. If we set

$$(\tilde{U}_1\ \tilde{U}_2) = (Z\ \tilde{U}_2) = U(W\ W_{\perp}),$$

then

$$\begin{aligned} AZ - ZM &= U(W\ W_{\perp}) \left[\begin{pmatrix} W^{\mathrm{H}} \\ W_{\perp}^{\mathrm{H}} \end{pmatrix} B(W\ W_{\perp}) \begin{pmatrix} I \\ 0 \end{pmatrix} - \begin{pmatrix} I \\ 0 \end{pmatrix} M \right] \\ &\quad + ub^{\mathrm{H}}(W\ W_{\perp}) \begin{pmatrix} I \\ 0 \end{pmatrix} \\ &= (\tilde{U}_1\ \tilde{U}_2) \begin{pmatrix} \tilde{B}_{11} - M \\ \tilde{B}_{21} \end{pmatrix} + u\tilde{b}_1^{\mathrm{H}}. \end{aligned}$$

The theorem now follows on taking norms. □

To see the consequences of this theorem, suppose that $AZ - ZM$ is small, and, using $(W\ W_{\perp})$, we transform the Krylov decomposition $AU - UB = ub^{\mathrm{H}}$ to the form

$$A(\tilde{U}_1\ \tilde{U}_2) = (\tilde{U}_1\ \tilde{U}_2\ u) \begin{pmatrix} \tilde{B}_{11} & \tilde{B}_{12} \\ \tilde{B}_{21} & \tilde{B}_{22} \\ \tilde{b}_1^{\mathrm{H}} & \tilde{b}_2^{\mathrm{H}} \end{pmatrix}. \tag{5.3}$$

Then by (5.2)

$$\left\| \begin{pmatrix} \tilde{B}_{21} \\ \tilde{b}_1^{\mathrm{H}} \end{pmatrix} \right\|_{\mathrm{F}} \le \|AZ - ZM\|_{\mathrm{F}}, \tag{5.4}$$

with equality if and only if M is the Rayleigh quotient $W^{\mathrm{H}}BW$. Thus if the residual norm $\|AZ - ZM\|_{\mathrm{F}}$ is sufficiently small, we may set $\tilde{B}_{21}$ and $\tilde{b}_1$ to zero to get the deflated

$$A(\tilde{U}_1\ \tilde{U}_2) \cong (\tilde{U}_1\ \tilde{U}_2)\begin{pmatrix} \tilde{B}_{11} & \tilde{B}_{12} \\ 0 & \tilde{B}_{22} \end{pmatrix} + u(0\ \tilde{b}_2^{\mathrm{H}}). \tag{5.5}$$

The deflation procedure that leads to (5.5) is backwards stable. If we restore the quantities that were zeroed in forming (5.5), we get the following relation:

$$A(\tilde{U}_1\ \tilde{U}_2) = (\tilde{U}_1\ \tilde{U}_2)\begin{pmatrix} \tilde{B}_{11} & \tilde{B}_{12} \\ 0 & \tilde{B}_{22} \end{pmatrix} + u(0\ \tilde{b}_2^{\mathrm{H}}) + \tilde{U}_2\tilde{B}_{21} + u\tilde{b}_1^{\mathrm{H}}.$$

If we write this decomposition in the form

$$A\tilde{U} = \tilde{U}\breve{B} + u\breve{b}^{\mathrm{H}} + R, \quad \text{where} \quad R = \tilde{U}_2\tilde{B}_{21} + u\tilde{b}_1^{\mathrm{H}},$$

then

$$\|R\|_{\mathrm{F}} \le \|AZ - ZM\|_{\mathrm{F}}.$$

If we now set $E = R\tilde{U}^{\mathrm{H}}$, then $(A + E)\tilde{U} = \tilde{U}\breve{B} + u\breve{b}^{\mathrm{H}}$. We may summarize these results in the following theorem.

THEOREM 5.2. *Under the hypotheses of Theorem* 5.1, *write the deflated decomposition* (5.5) *in the form*

$$A\tilde{U} \cong \tilde{U}\breve{B} + u\breve{b}^{\mathrm{H}}.$$

Then there is an E satisfying

$$\|E\|_{\mathrm{F}} \le \|AZ - ZM\|_{\mathrm{F}} \tag{5.6}$$

such that

$$(A + E)\tilde{U} = \tilde{U}\breve{B} + u\breve{b}^{\mathrm{H}}.$$

Equality holds in (5.6) *if and only if M is the Rayleigh quotient $Z^{\mathrm{H}}AZ = W^{\mathrm{H}}BW$.*

Because backward stability is commonly used to determine convergence, Theorems 5.1 and 5.2 suggest how one might combine convergence testing and deflation. Given an approximate pair (M, UW), we transform to the tilde form as in Theorem 5.1 and compute the backward error that would result from deflation. If this is small enough compared with A, we deem the pair to have converged and deflate.[3]

In practice we will seldom encounter a converging subspace unless it is a 2-dimensional subspace corresponding to a complex eigenvalue in a real Schur decomposition. Instead we will be confronted with converged, normalized Ritz pairs $(\mu_i, \hat{z}_i)$ $(i = 1, \ldots, p)$ of one kind or another, and the vectors in these pairs cannot be guaranteed to be orthogonal. If we arrange the vectors in a matrix $\hat{Z}$ and set $\hat{M} = \mathrm{diag}(\mu_1, \ldots, \mu_p)$, the residual $\hat{R} = A\hat{Z} - \hat{Z}\hat{M}$ must be small because the individual residuals are small.

[3] If the concern is with eigenvalues that are small compared with $\|A\|_{\mathrm{F}}$, we may have to demand a smaller backward error to get accurate results. For more, see the discussion of convergence in [13].

The deflation procedure requires an orthonormal basis for the approximate eigenspace in question, which is given by the QR factorization

$$\hat{Z} = ZT \tag{5.7}$$

of $\hat{Z}$. Unfortunately, the residual for Z becomes

$$R = \hat{R}T^{-1} = A\hat{Z}T^{-1} - \hat{Z}\hat{M}T^{-1} = AZ - ZM,$$

where $M = T\hat{M}T^{-1}$. If the columns of $\hat{Z}$ are nearly dependent, $\|T^{-1}\|$ will be large, and the residual may be magnified—perhaps to the point where the deflation cannot be performed safely. The effects of dependency on a different deflation algorithm have also been noted in [8].

It may seem paradoxical that we could have, say, two vectors each of which we can deflate but which taken together cannot be deflated. The resolution of this paradox is to remember that we are not deflating two vectors but the subspace spanned by them. If the vectors are nearly dependent, they must be very accurate to determine their common subspace accurately.

As we have mentioned, the deflation procedure is not confined to eigenpairs calculated by a Rayleigh–Ritz procedure. For example, it can be used to deflate harmonic Ritz vectors [10] or refined Ritz vectors [5]. However, if Ritz vectors are the concern, there is an easy way to deflate them in the Krylov–Schur method. After a cycle of the algorithm, let the current decomposition have the form

$$A(U_1\ U_2) = (U_1\ U_1)\begin{pmatrix} S_{11} & S_{12} \\ 0 & S_{22} \end{pmatrix} + u(0\ b_2^{\mathrm{H}}).$$

Here U_1 represents a subspace that has already been deflated, and S_{22} is the Schur form that remains after the contraction phase.

In this decomposition, to deflate the Ritz pair corresponding to the $(1,1)$-element of S_{22} we must set the first component of b_2 to zero. Consequently, all we have to do to deflate is to verify that that component satisfies our deflation criterion. If some other diagonal element of S_{22} is the candidate for deflation, we can exchange it into the $(2,2)$-position and test as above.

6. Assessment. In comparing the Krylov–Schur algorithm with the implicitly restarted Arnoldi algorithm, we must distinguish the sources of work in the algorithms. The first is the multiplication of a vector by A. Since A will usually be sparse, the cost of this product is unpredictable in general, but it is reasonable to assume that it forms a significant part—perhaps the dominant part—of the computation.

The second source of work is the expansion of the decompositions from one of order k to one of order m. It is easily seen from (3.1) that the work is $2n(m^2 - k^2)$ floating-point adds and multiplies, assuming reorthogonalization is performed. This count is the same for both algorithms.

In the contraction step, both algorithms must transform the Rayleigh quotient and accumulate the transformations in U. For efficiency, we do not accumulate the transformations in U as they are generated but instead accumulate them in an $m \times m$ matrix Q and then compute the new U_k in the form

$$U_m Q[:, 1{:}k]. \tag{6.1}$$

If $n \gg m$, the last step will dominate the transformations applied to the Rayleigh quotient and their accumulation in Q.

For the Krylov–Schur method we must compute the Schur decomposition of the Rayleigh quotient and transform the triangular factor. This means that Q will be full, and the final accumulation step (6.1) will require nkm floating-point additions and multiplications.

For the implicitly restarted Arnoldi we must also compute the Schur decomposition of the Rayleigh quotient H_m. But it is used only to determine the shifts, which are applied directly to H_m. The structure of the transformations is such that $Q[:, 1{:}k]$ is zero below its $m-k$ subdiagonal. This means that the operation count for (6.1) is $nmk - \frac{1}{2}k^2$ additions and multiplication.

To put things together, if $m = 2k$ and reorthogonalization is performed during the expansion, the Krylov–Schur algorithm has an operation count of $7nk^2$ whereas implicitly restarted Arnoldi has an operation count of $6\frac{1}{2}nk^2$. Thus implicitly restarted Arnoldi is marginally superior to Arnoldi–Schur when it comes to accumulation of transformations. Against this must be set the fact that Krylov–Schur deflates in an inexpensive and natural manner and does not require a special routine for purging.

7. Concluding remarks. The Krylov–Schur method admits variations. An important one is based on the observation that we can truncate a Krylov decomposition at any point where the Rayleigh quotient is block triangular [see (3.4)]. This means that when A is real we can work with real Schur forms of the Rayleigh quotient and avoid the necessity of complex arithmetic. The algorithm for exchanging eigenvalues mentioned above will also move the 2×2 blocks of the real Schur form so that the contraction phase proceeds as usual. In deflation, the block in question is moved to the position just after the previously deflated eigenvalues and blocks, and two components of b are tested. An unusual feature of complex eigenvectors is that they may fail to deflate, not because they are dependent on other deflated vectors, but because the real and imaginary parts of their eigenvectors are not sufficiently independent.

When A is Hermitian, the Krylov–Schur method becomes a restarted Lanczos algorithm—in fact the algorithm of Wu and Simon [18]. The Rayleigh quotient is diagonal, so that reordering of the eigenvalues reduces to simple permutations. Moreover, because the eigenvectors of the Rayleigh quotient are orthogonal, a Ritz pair with a small residual norm ϵ will deflate with backward error of order ϵ.

Since the Krylov–Schur method works explicitly with the eigenvalues of the Rayleigh quotient, it is an exact-shift method. Nonetheless, it stands ready to help the general shift method to deflate Ritz pairs and to get rid of unwanted pairs. One simply computes a Krylov–Schur form of the current decomposition and performs the procedures described above. Theorem 2.2 assures us that we can then return to a pure Arnoldi decomposition.

In fact Theorem 2.2 is really the heart of the matter. It allows us to operate freely on the Rayleigh quotient with the knowledge that we are always attached to a Krylov sequence. It is hoped that this freedom will find other applications.

Acknowledgment. I would like to thank Rich Lehoucq and Dan Sorensen for their comments on preliminary versions of this paper. I am indebted to the Mathematical and Computational Sciences Division of the National Institute of Standards and Technology for the use of their research facilities.

REFERENCES

[1] W. E. ARNOLDI, *The principle of minimized iterations in the solution of the matrix eigenvalue problem*, Quart. Appl. Math., 9 (1951), pp. 17–29.

[2] Z. Bai and J. W. Demmel, *On swapping diagonal blocks in real Schur form*, Linear Algebra Appl., 186 (1993), pp. 73–95.

[3] D. R. Fokkema, G. L. G. Sleijpen, and H. A. van der Vorst, *Jacobi–Davidson style QR and QZ algorithms for the reduction of matrix pencils*, SIAM J. Sci. Comput., 20 (1998), pp. 94–125.

[4] G. H. Golub and C. F. Van Loan, *Matrix Computations*, 2nd ed., Johns Hopkins University Press, Baltimore, MD, 1989.

[5] Z. Jia, *Refined iterative algorithm based on Arnoldi's process for large unsymmetric eigenproblems*, Linear Algebra Appl., 259 (1997), pp. 1–23.

[6] C. Lanczos, *An iteration method for the solution of the eigenvalue problem of linear differential and integral operators*, J. Research Nat. Bur. Standards, 45 (1950), pp. 255–282.

[7] R. B. Lehoucq, *private communication.*

[8] R. B. Lehoucq and D. C. Sorensen, *Deflation techniques for an implicitly restarted Arnoldi iteration*, SIAM J. Matrix Anal. Appl., 17 (1996), pp. 789–821.

[9] R. B. Lehoucq, D. C. Sorensen, and C. Yang, *ARPACK Users' Guide: Solution of Large-Scale Eigenvalue Problems with Implicitly Restarted Arnoldi Methods*, SIAM, Philadelphia, 1998.

[10] R. B. Morgan, *Computing interior eigenvalues of large matrices*, Linear Algebra Appl., 154–156 (1991), pp. 289–309.

[11] R. B. Morgan, *On restarting the Arnoldi method for large nonsymmetric eigenvalue problems*, Math. Comp., 65 (1996), pp. 1213–1230.

[12] R. B. Morgan, *GMRES with Deflated Restarting*, Department of Mathematics, Baylor University, Waco, TX, 1999.

[13] J. A. Scott, *An Arnoldi code for computing selected eigenvalues of sparse, real, unsymmetric matrices*, ACM Trans. Math. Software, 21 (1995), pp. 432–475.

[14] D. C. Sorensen, *Implicit application of polynomial filters in a k-step Arnoldi method*, SIAM J. Matrix Anal. Appl., 13 (1992), pp. 357–385.

[15] A. Stathopoulos, Y. Saad, and K. Wu, *Dynamic thick restarting of the Davidson, and the implicitly restarted Arnoldi methods*, SIAM J. Sci. Comput., 19 (1998), pp. 227–245.

[16] G. W. Stewart, *Matrix Algorithms* I: *Basic Decompositions*, SIAM, Philadelphia, 1998.

[17] D. S. Watkins, *Forward stability and transmission of shifts in the QR algorithm*, SIAM J. Matrix Anal. Appl., 16 (1995), pp. 469–487.

[18] K. Wu and H. Simon, *Thick-restart Lanczos method for large symmetric eigenvalue problems*, SIAM J. Matrix Anal. Appl., 22 (2000), pp. 602–616.

18.2. [GWS-J113] "Addendum to 'A Krylov-Schur Algorithm for Large Eigenproblems' "

[GWS-J113] "Addendum to 'A Krylov-Schur Algorithm for Large Eigenproblems'," *SIAM Journal on Matrix Analysis and Applications* **24** (2002) 599–601.
http://dx.doi.org/10.1137/S0895479802403150

SIAM J. MATRIX ANAL. APPL.
Vol. 24, No. 2, pp. 599–601

ADDENDUM TO "A KRYLOV–SCHUR ALGORITHM FOR LARGE EIGENPROBLEMS"*

G. W. STEWART†

Abstract. In this addendum to an earlier paper by the author, it is shown how to compute a Krylov decomposition corresponding to an arbitrary Rayleigh quotient. This decomposition can be used to restart an Arnoldi process, with a selection of the Ritz vectors corresponding to that Rayleigh quotient.

Key words. large eigenproblem, Krylov sequence, Arnoldi algorithm, Krylov decomposition, restarting, deflation

AMS subject classifications. 65F14, 65F50

PII. S0895479802403150

In [4] the author introduced a decomposition of the form

$$AU = UB + ub^{\mathrm{H}}, \tag{1}$$

where A is a matrix of order n and $(U\ u)$ has full column rank. It was shown that the column space of $(U\ u)$ (called the *subspace of the decomposition*) is a (possibly restarted) Krylov subspace of A and conversely that every Krylov subspace has such a representation, so that the *Krylov decomposition* (1) is a characterization of Krylov subspaces.[1] Arnoldi and Lanczos decompositions are special cases of Krylov decompositions.

The advantage of working with Krylov decompositions is that their subspaces remain invariant under two classes of transformations. The first, called a *similarity*, transforms the decomposition into

$$A(UW^{-1}) = (UW^{-1})(WBW^{-1}) + u(b^{\mathrm{H}}W^{-1}) \equiv A\tilde{U} = \tilde{U}\tilde{B} + u\tilde{b}^{\mathrm{H}},$$

where W is any nonsingular matrix. The second, called a *translation,* transforms the decomposition to the form

$$AU = U\tilde{B} + \tilde{u}\tilde{b}^{\mathrm{H}},$$

where

$$\tilde{B} = B + gb^{\mathrm{H}}, \quad \tilde{u} = \frac{u - Ug}{\gamma}, \quad \text{and} \quad \tilde{b}^{\mathrm{H}} = \gamma b^{\mathrm{H}}$$

for any vector g and any scalar $\gamma \neq 0$.

The computational algorithms in [4] were based on similarities. Translations were used primarily in the derivation of the properties of Krylov decompositions. The purpose of this note is to show that translations have a computational role to play in restarting an Arnoldi process with a selection of Rayleigh–Ritz approximations to a set of eigenvectors.

*Received by the editors February 25, 2002; accepted for publication (in revised form) by H. A. van der Vorst June 10, 2002; published electronically December 19, 2002.
http://www.siam.org/journals/simax/24-2/40315.html

†Department of Computer Science, University of Maryland, College Park MD 20742 (stewart@cs.umd.edu).

[1]A related characterization, cast in terms of subspaces, is given by Genseberger and Sleijpen [1].

The Rayleigh–Ritz method for extracting approximations to eigenvectors from a subspace does not depend on whether the subspace in question is a Krylov subspace. It can be presented in different ways. The one we give here leads most directly to the main result of this note. Let U be a basis for the subspace $\mathcal{U}$ in question, and let V be such that $V^{\mathrm{H}}U$ is nonsingular. (The space spanned by V is sometimes called the test subspace, and V itself the test matrix.) Then the matrix

$$\hat{B} = (V^{\mathrm{H}}U)^{-1}V^{\mathrm{H}}AU \tag{2}$$

has the property that if (μ, Uw) is an eigenpair of A, then (μ, w) is an eigenpair of $\hat{B}$. Specifically,

$$\hat{B}w = (V^{\mathrm{H}}U)^{-1}V^{\mathrm{H}}AUw = \mu(V^{\mathrm{H}}U)^{-1}V^{\mathrm{H}}Uw = \mu w.$$

By continuity one might expect that if $\mathcal{U}$ contains an approximate eigenvector of A, then it can be found by computing an appropriate eigenpair (μ, w) of $\hat{B}$ and forming Uw. This is the essence of the Rayleigh–Ritz method. (For an analysis of the method, see [2].) The matrix B is called a *Rayleigh quotient* (with respect to U and V) because (2) is a generalization of the ordinary Rayleigh quotient $v^{\mathrm{H}}Au/v^{\mathrm{H}}u$.

It was observed in [4] that the matrix B in the Krylov decomposition (1) is a Rayleigh quotient. Specifically, let $(V\ \ v)^{\mathrm{H}}$ be a left inverse of $(U\ \ u)$. Then $V^{\mathrm{H}}U = I$ and $V^{\mathrm{H}}u = 0$. It follows from (1) that $B = V^{\mathrm{H}}AU$ is a Rayleigh quotient, which can be used in the Rayleigh–Ritz procedure. In particular, we can discard undesirable Ritz vectors by a process known as Krylov–Schur restarting.

In some cases, however, we may not have the freedom to choose V. For example, in the harmonic Rayleigh–Ritz method, which has superior properties for approximating interior eigenvalues [3], [5, pp. 292–294], we must take $V = (A - \kappa I)U$, where κ is near the eigenvalues of interest. Now for a general test matrix V, there is no problem in computing the Rayleigh quotient. In fact, on multiplying (1) by $(V^{\mathrm{H}}U)^{-1}V^{\mathrm{H}}$, we find that

$$\hat{B} = (V^{\mathrm{H}}U)^{-1}V^{\mathrm{H}}AU = B + gb^{\mathrm{H}}, \tag{3}$$

where

$$g = (V^{\mathrm{H}}U)^{-1}V^{\mathrm{H}}u. \tag{4}$$

The problem is that $\hat{B}$ is seemingly not associated with a Krylov decomposition, so that the Krylov–Schur restarting procedure of [4] cannot be applied to remove undesirable Ritz vectors.

But in fact $\hat{B}$ *is* associated with a Krylov decomposition.

THEOREM 1. *Let $V^{\mathrm{H}}U$ be nonsingular, and let $\hat{B}$ and g be defined by* (3) *and* (4). *If*

$$\hat{u} = u - Ug,$$

then the Krylov decomposition

$$AU = U\hat{B} + \hat{u}b^{\mathrm{H}} \tag{5}$$

is a translation of the decomposition (1), *whose Rayleigh quotient with respect to the test matrix V is $\hat{B}$.*

The proof consists of verifying that (5) is indeed a translation of (1) and that the matrix $(V^{\mathrm{H}}U)^{-1}VAU$ is indeed equal to $\hat{B}$.

To see how we can use (5) to restart the Krylov decomposition, suppose U is orthonormal (as it will be in practice). Let

$$\begin{pmatrix} T_{11} & T_{12} \\ 0 & T_{22} \end{pmatrix} = \begin{pmatrix} W_1^{\mathrm{H}} \\ W_2^{\mathrm{H}} \end{pmatrix} \hat{B}(W_1 \ W_2)$$

be a partition Schur decomposition of $\hat{B}$, where T_{11} contains the Ritz values corresponding to the Ritz vectors we wish to retain. Then by a similarity, we have

$$A(UW_1 \ UW_2) = (UW_1 \ UW_2) \begin{pmatrix} T_{11} & T_{12} \\ 0 & T_{22} \end{pmatrix} + \hat{u}(b^{\mathrm{H}}W_1 \ b^{\mathrm{H}}W_2).$$

Hence

$$A(UW_1) = (UW_1)T_{11} + \hat{u}b^{\mathrm{H}}W_1$$

is a Krylov decomposition containing the desired Ritz vectors. The matrix UW_1 is orthonormal, but the vector $\hat{u}$ will not in general be orthogonal to the columns of UW_1. However, by a second translation we can orthogonalize it. The resulting decomposition is an orthogonal Krylov decomposition, which can be extended by the Arnoldi process in the usual way.

Acknowledgment. I am indebted to the Mathematical and Computational Sciences Division of the National Institute of Standards and Technology for the use of their research facilities.

REFERENCES

[1] M. Genseberger and G. L. G. Sleijpen, *Alternative correction equations in the Jacobi–Davidson method*, Numer. Linear Algebra Appl., 6 (1999), pp. 235–253.

[2] Z. Jia and G. W. Stewart, *An analysis of the Rayleigh–Ritz method for approximating eigenspaces*, Math. Comp., 70 (2001), pp. 637–647.

[3] R. B. Morgan, *Computing interior eigenvalues of large matrices*, Linear Algebra Appl., 154–156 (1991), pp. 289–309.

[4] G. W. Stewart, *A Krylov–Schur algorithm for large eigenproblems*, SIAM J. Matrix Anal. Appl., 23 (2001), pp. 601–614.

[5] G. W. Stewart, *Matrix Algorithms Volume* II: *Eigensystems*, SIAM, Philadelphia, 2001.

18.3. [GWS-J110] "Backward Error Bounds for Approximate Krylov Subspaces"

[GWS-J110] "Backward Error Bounds for Approximate Krylov Subspaces," *Linear Algebra and its Applications* **340** (2002) 81–86.

ELSEVIER Linear Algebra and its Applications 340 (2002) 81–86

LINEAR ALGEBRA AND ITS APPLICATIONS

www.elsevier.com/locate/laa

Backward error bounds for approximate Krylov subspaces☆

G.W. Stewart

[a]*Department of Computer Science, Institute for Advanced Computer Studies, University of Maryland, College Park, MD 20742, USA*

Received 12 May 2001; accepted 11 June 2001

Submitted by R.A. Brualdi

Abstract

Let A be a matrix of order n and let $\mathscr{U} \subset \mathbb{C}^n$ be a subspace of dimension k. In this note, we determine a matrix E of minimal norm such that $\mathscr{U}$ is a Krylov subspace of $A + E$. © 2002 Elsevier Science Inc. All rights reserved.

Keywords: Krylov subspace; Krylov decomposition; Backward error

1. Introduction

Let A be a matrix of order n. Given a starting vector u, we say that the sequence

$$u, Au, A^2u, \ldots$$

is the *Krylov sequence* associated with A and u. The subspace

$$\mathscr{K}_k(A, u) = \text{span}(u, Au, A^2u, \ldots, A^{k-1}u)$$

is called a *Krylov subspace.*

Krylov subspaces arise in many applications. They are especially important in algorithms for the iterative solution of linear systems [2] and for approximating eigenpairs of large matrices [4,6]. Since bases for Krylov subspaces are sometimes computed inaccurately, it is desirable to have some way of assessing their quality.

☆ Work supported by the National Science Foundation under Grant No. 970909-8562. Part of this work was done during the author's weekly visits at the Mathematics and Computer Science Division of the National Institute for Standards and Technology.

E-mail address: stewart@cs.umd.edu (G.W. Stewart).

PII: S 0 0 2 4 - 3 7 9 5 (0 1) 0 0 4 1 3 - X

There are two approaches. Given a Krylov subspace $\mathscr{U}$, we can

1. give bounds on the angle between $\mathscr{U}$ and the nearest Krylov subspace of A,
2. determine a matrix E of minimal norm such that $\mathscr{U}$ is a Krylov subspace of $A + E$.

The first approach leads to a seemingly difficult and currently unsolved problem. The purpose of this note is to show that the second approach has a simple, constructive solution.

To solve the problem we will use a characterization of Krylov subspaces called a Krylov decomposition [5]. Accordingly, in Section 2 we will discuss these decompositions and their relation to the widely used Arnoldi decompositions. In Section 3 we will present our results and comment on them.

Throughout this note $\|\cdot\|$ will denote a family of consistent unitarily invariant norms. The special cases of the spectral 2-norm and the Frobenius norm will be denoted by $\|\cdot\|_2$ and $\|\cdot\|_F$. For more on unitarily invariant norms see [7].

2. Arnoldi and Krylov decompositions

As a rule, the vectors in a Krylov sequence $u, Au, A^2u, \ldots$ become increasingly dependent. To circumvent this problem we can construct orthonormal bases $u_1, u_2, \ldots$ for the Krylov subspaces $\mathscr{K}_k(A, u_1)$ by successively orthogonalizing Au_j against $u_1, \ldots, u_j$ and normalizing the result—a process known as the Arnoldi algorithm [1]. If we set $U_k = (u_1, \ldots, u_k)$, then the results of the Arnoldi algorithm can be summarized by the relation

$$AU_{k-1} = U_k H_k,$$

where

$$H_k = U_k^{\mathrm{H}} A U_{k-1}$$

is a $k \times (k-1)$ upper Hessenberg matrix—that is, it is zero below its first subdiagonal. We call such a relation an *Arnoldi decomposition.*

In general, all the subdiagonal elements of H_k will be nonzero, in which case the Arnoldi decomposition is uniquely determined by the starting vector u_1. If, however, $b_{j,j-1}$ is zero, then Au_{j-1} is exactly dependent on $u_1, \ldots, u_{j-1}$ so that one must restart the Arnoldi process with some vector u_j that is orthogonal to $u_1, \ldots, u_{j-1}$. In this case, we will say that the corresponding Krylov subspace is restarted. Although our results will apply to restarted Krylov subspaces, it should be kept in mind that the unrestarted case is the norm. [1]

The essential uniqueness of the Arnoldi decomposition is a drawback when we wish to consider different bases for a particular Krylov subspace. To circumvent this problem we introduce *Krylov decompositions*, which have the form

[1] This state of affairs is due to the law of perversity of nature. In applications, a restarting represents the convergence of an iterative method or the isolation of an eigenspace—something to be happy about.

$$AU_{k-1} = U_k B_k, \tag{2.1}$$

where U_k has independent columns and B_k is arbitrary. We call the column space of U_k the space of the decomposition. Any Arnoldi decomposition is, of course, a Krylov decomposition. Conversely, it can be shown [5] that corresponding to any Krylov decomposition there is an Arnoldi decomposition with the same space. Thus Krylov decompositions are a general characterization of Krylov subspaces. In what follows we will assume that the matrices U_k in our Krylov decompositions are orthonormal.

3. The results

Given a subspace $\mathscr{U}$, our object is to show it is a Krylov subspace of a perturbation of A and to bound the perturbation. We proceed indirectly. First we show that there is a basis for $\mathscr{U}$ that satisfies an approximate Krylov relation for A with a minimal residual. We then use standard techniques to throw the residual back onto A.

The following lemma is the starting point for the first part of our program.

Lemma 3.1. *Let* $U = (U_1 \;\; u_2)$ *orthonormal. Then the* Krylov residual

$$R = AU_1 - UB$$

is minimized in any unitarily invariant norm when

$$B = U^{\mathrm{H}} AU_1, \tag{3.1}$$

in which case

$$U^{\mathrm{H}} R = 0. \tag{3.2}$$

Proof. Let $\hat{U} = (U \;\; U_3)$ be unitary. Then by unitary invariance, R has the same norm as

$$\hat{U}^{\mathrm{H}} R = \begin{pmatrix} U^{\mathrm{H}} AU_1 - B \\ U_3^{\mathrm{H}} R \end{pmatrix}.$$

Since $U_3^{\mathrm{H}} R$ is independent of B, the norm of R is minimized when $B = U^{\mathrm{H}} AU_1$. The orthogonality condition $U^{\mathrm{H}} R = 0$ can be verified directly. □

Given a subspace $\mathscr{U} \subset \mathbb{C}^n$ of dimension k, this theorem suggests that we proceed with our program by choosing an orthonormal basis U for $\mathscr{U}$ and use (3.1) to compute an optimal Krylov residual. Unfortunately, this residual is optimal only for the specific choice of U. The reason is that not every basis for a Krylov subspace corresponds to a Krylov decomposition, so that Lemma 3.1 is likely to give us large Krylov residuals, even when $\mathscr{U}$ is itself a Krylov subspace. To optimize globally over all bases, we must try to determine a $k \times k$ unitary matrix V such that UV has a Krylov residual that is as small as possible.

To do this, partition $V = (V_1 \;\; v_2)$. Let

$$S = AU - U(U^{\mathrm{H}} AU).$$

If we postmultiply S by V_1, we get

$$SV_1 = A(UV_1) - UV\big[(UV)^{\mathrm{H}}A(UV_1)\big].$$

It follows that SV_1 is the optimal Krylov residual for the particular basis UV. Thus we wish to minimize the norm of SV_1 as V_1 varies over the set of $k \times (k-1)$ orthonormal matrices.

This is easily done. Let $\sigma_1 \geqslant \cdots \geqslant \sigma_k \geqslant 0$ be the singular values of S and let $\tau_1 \geqslant \cdots \geqslant \tau_{k-1} \geqslant 0$ be the singular values of SV_1. Then by the interleaving theorem for singular values [3, Lemma 3.3.1], $\tau_i \geqslant \sigma_{i+1}$ $(i = 1, \ldots, k-1)$. Since a unitarily invariant norm of a matrix is a nondecreasing function of its singular values, the norm of SV_1 is minimized when $\tau_i = \sigma_{i+1}$ $(i = 1, \ldots, k-1)$. These equalities can be attained if we take V_1 to be the right singular vectors of S corresponding to $\sigma_2, \ldots, \sigma_k$. The vector v_2 is necessarily the right singular vector corresponding to σ_1, and this choice of $V = (V_1\ \ v_2)$ gives us a globally optimal Krylov residual for $\mathscr{U}$.

The second step in our program is to project the Krylov residual back on A. Let $\hat{U} = UV$, where $V = (V_1\ \ v_2)$ is as in the previous paragraph. Then $R = A\hat{U}_1 - \hat{U}(\hat{U}^{\mathrm{H}}A\hat{U}_1)$ is a globally optimal Krylov residual. If we set $E = -RU_1^{\mathrm{H}}$, then $\|E\| = \|R\|$, and it follows from (3.2) that

$$(A+E)\hat{U}_1 = \hat{U}\big[\hat{U}^{\mathrm{H}}(A+E)\hat{U}_1\big]$$

is a Krylov decomposition of $A + E$.

Moreover, E is the smallest possible such backward error. For if $(A+F)\hat{U}_1 = \hat{U}[\hat{U}^{\mathrm{H}}(A+F)\hat{U}_1]$, then

$$R = A\hat{U}_1 - \hat{U}(\hat{U}A\hat{U}_1) = (\hat{U}\hat{U}^{\mathrm{H}} - I)F\hat{U}_1.$$

But $\hat{U}\hat{U}^{\mathrm{H}} - I$ and $\hat{U}_1$ both have 2-norm one so that $\|E\| = \|R\| \leqslant \|F\|$.

We summarize these results in the following theorem, in which we recapitulate our notation and constructions.

Theorem 3.2. *Let A be of order n and let $U = (U_1\ \ u_2) \in \mathbb{C}^{n\times k}$ be orthonormal. Let*

$$S = AU - U(U^{\mathrm{H}}AU) \tag{3.3}$$

and let $\sigma_1 \geqslant \cdots \geqslant \sigma_k \geqslant 0$ be the singular values of S. Let $V = (V_1\ \ v_2)$ be unitary with the columns V_1 being the right singular vectors of S corresponding to $\sigma_2, \ldots, \sigma_k$. Set

$$\hat{U} = UV = (\hat{U}_1\ \ \hat{u}_2) \quad \text{and} \quad R = SV_1.$$

Then the approximate Krylov decomposition

$$A\hat{U}_1 = \hat{U}(\hat{U}^{\mathrm{H}}A\hat{U}_1) + R,$$

has minimal residual norm in any unitarily invariant norm. If we set

$$E = -R\hat{U}^{\mathrm{H}}, \tag{3.4}$$

then

$$\|E\| = \|R\|$$

and $A + E$ *has the Krylov decomposition*

$$(A + E)\hat{U}_1 = \hat{U}\big[\hat{U}^{\mathrm{H}}(A + E)\hat{U}_1\big]. \tag{3.5}$$

Of all matrices E satisfying (3.5), *matrix* (3.4) *has minimal norm.*

There are several comments to be made about this theorem.

First, our results are independent of the initial choice of a basis U for $\mathcal{U}$. Specifically, if we replace U by UQ, where Q is unitary, S in (3.3) is replaced by SQ, V is replaced by $Q^{\mathrm{H}}V$, and hence $\hat{U}$ does not change.

Second, we can give explicit expressions for $\|R\|$ in the 2- and Frobenius norms. Namely

$$\|R\|_2 = \sigma_2 \quad \text{and} \quad \|R\|_{\mathrm{F}} = \sqrt{\sigma_2^2 + \cdots + \sigma_k^2}.$$

Third, the process is constructive. Given a basis for $\mathcal{U}$, we can actually construct the backward error.

Fourth, if $\mathcal{U}$ is actually a Krylov subspace, then R must be zero. This means that only the singular value σ_1 of S can be nonzero. Thus we have an alternate characterization of what it means to be a Krylov subspace.

Corollary 3.3. *An orthonormal matrix U spans a Krylov subspace of A if and only the matrix* $S = AU - U(U^{\mathrm{H}}AU)$ *has rank not greater than* 1.

In fact, this characterization can be derived in another way. Write $S = (I - UU^{\mathrm{H}})(AU)$. Because U is a basis for a Krylov sequence, AU can have at most one vector that is orthogonal to the column space of U. Since $I - UU^{\mathrm{H}}$ is the projection onto the orthogonal complement of the column space of U, the column space of S can contain at most one vector.

Fifth, our first candidate for assessing an approximate Krylov subspace—namely, finding the nearest Krylov subspace—is more direct than the approach taken here—namely, finding an optimal backward perturbation. But in applications the latter is often more useful. For the implication of backward error analyses for eigenproblems see [6, Theorem II.1.3].

Finally, if A is Hermitian, it is natural to require that the backward error E also be Hermitian. This can be done by setting

$$E = -R\hat{U}^{\mathrm{H}} - \hat{U}R^{\mathrm{H}}.$$

It is easily verified $\|E\|_2 = \|R\|_2$ so that E is optimal in the 2-norm. But $\|E\|_{\mathrm{F}} = \sqrt{2}\|R\|_{\mathrm{F}}$ so that E might not be optimal in the Frobenius norm. But it can be off by no more than a factor of $\sqrt{2}$.

Acknowledgement

Part of this work was done during a pleasant stay at the Mathematics Department of the University of Utrecht. I would like to thank Henk van der Vorst for his comments on the paper.

References

[1] W.E. Arnoldi, The principle of minimized iterations in the solution of the matrix eigenvalue problem, Quart. Appl. Math. 9 (1951) 17–29.
[2] A. Greenbaum, Iterative Methods for Solving Linear Systems, SIAM, Philadelphia, PA, 1997.
[3] R.A. Horn, C.R. Johnson, Topics in Matrix Analysis, Cambridge University Press, Cambridge, 1991.
[4] Y. Saad, Numerical Methods for Large Eigenvalue Problems: Theory and Algorithms, Wiley, New York, 1992.
[5] G.W. Stewart, A Krylov–Schur algorithm for large eigenproblems. Technical Report TR-4127, Department of Computer Science, University of Maryland, College Park, 2000, SIAM J. Matrix Anal. Appl. (to appear).
[6] G.W. Stewart, Matrix Algorithms II: Eigensystems. SIAM, Philadelphia, PA, 2001 (to appear). Preprint available at http://www.cs.umd.edu/~stewart.
[7] G.W. Stewart, J.-G. Sun, Matrix Perturbation Theory, Academic Press, New York, 1990.

18.4. [GWS-J112] "Adjusting the Rayleigh Quotient in Semiorthogonal Lanczos Methods"

[GWS-J112] "Adjusting the Rayleigh Quotient in Semiorthogonal Lanczos Methods," *SIAM Journal on Scientific Computing* **24** (2002) 201–207.
`http://dx.doi.org/10.1137/S1064827501388984`

SIAM J. SCI. COMPUT.
Vol. 24, No. 1, pp. 201–207

ADJUSTING THE RAYLEIGH QUOTIENT IN SEMIORTHOGONAL LANCZOS METHODS*

G. W. STEWART†

Abstract. In a semiorthogonal Lanczos algorithm, the orthogonality of the Lanczos vectors is allowed to deteriorate to roughly the square root of the rounding unit, after which the current vectors are reorthogonalized. A theorem of Simon [*Linear Algebra Appl.*, 61 (1984), pp. 101–132] shows that the Rayleigh quotient—i.e., the tridiagonal matrix produced by the Lanczos recursion—contains fully accurate approximations to the Ritz values in spite of the lack of orthogonality. Unfortunately, the same lack of orthogonality can cause the Ritz vectors to fail to converge. It also makes the classical estimate for the residual norm misleadingly small. In this paper we show how to adjust the Rayleigh quotient to overcome this problem.

Key words. large eigenproblem, Lanczos method, partial reorthogonalization, symmetric matrix, adjusted Rayleigh quotient

AMS subject classifications. 65F15, 65F25, 65F50

PII. S1064827501388984

1. Introduction and background.

Let A be a symmetric matrix of order n and let u_1 be an n-vector of 2-norm one. The Lanczos algorithm generates a sequence of orthonormal vectors by the recurrence

$$\begin{aligned} \beta_1 u_2 &= Au_1 - \alpha_1 u_1, \\ \beta_k u_{k+1} &= Au_k - \alpha_k u_k - \beta_{k-1} u_{k-1}, \qquad k = 1, 2, \ldots, \end{aligned} \tag{1.1}$$

where

$$\alpha_k = u_k^{\mathrm{T}} A u_k \quad \text{and} \quad \beta_k = 1/\|Au_k - \alpha_k u_k - \beta_{k-1} u_{k-1}\|.$$

If we introduce the tridiagonal matrix

$$T_k = \begin{pmatrix} \alpha_1 & \beta_1 & & & & \\ \beta_1 & \alpha_2 & \beta_2 & & & \\ & \beta_2 & \alpha_3 & \beta_3 & & \\ & & \ddots & \ddots & \ddots & \\ & & & \beta_{k-2} & \alpha_{k-1} & \beta_{k-1} \\ & & & & \beta_{k-1} & \alpha_k \end{pmatrix}$$

and set

$$U_k = (u_1 \ \cdots \ u_k),$$

then we can express the effect of the recurrence (1.1) by the *Lanczos decomposition*

$$AU_k = U_k T_k + \beta_k u_{k+1} \mathbf{e}_k^{\mathrm{T}}, \tag{1.2}$$

*Received by the editors May 4, 2001; accepted for publication (in revised form) November 5, 2001; published electronically May 20, 2002.
http://www.siam.org/journals/sisc/24-1/38898.html

†Department of Computer Science and Institute for Advanced Computer Studies, University of Maryland, College Park, MD 20742 (stewart@cs.umd.edu).

where $\mathbf{e}_k$ is the kth unit vector (the kth column of the identity matrix of order k).

It is well known (e.g., see [3]) that as k increases the space spanned by U_k contains increasingly accurate approximations to eigenvectors corresponding to the extreme eigenvalues of A. The approximations can be retrieved by a process known as the Rayleigh–Ritz method. It is based on the observation that if $(\theta, U_k w)$ is an eigenpair of A, then (θ, w) is an eigenpair of T_k. A continuity argument suggests that if U_k contains a good approximation to an eigenvector of A, there should be an eigenvector w of T_k such that $U_k w$ approximates that eigenvector. For a general analysis of this procedure, see [2]. In what follows we will call θ a Ritz value, w a primitive Ritz vector, $U_k w$ a Ritz vector, and T_k the Rayleigh quotient.[1]

A difficulty with the Lanczos algorithm is that the Lanczos vectors tend to lose orthogonality. One cure is to reorthogonalize the vectors at each step—a process known as full reorthogonalization. Unfortunately, reorthogonalizing u_{k+1} requires the vectors $u_1, \ldots, u_k$, and for k large enough the cost of moving them in and out of working storage becomes prohibitive. Consequently, it has been proposed to let orthogonality deteriorate up to a point, after which a reorthogonalization step is performed. There are several varieties of this procedure, all going under the common name of semiorthogonal methods. They each require that the elements of $U_k^{\mathrm{T}} U_k - I$ be kept less than some multiple of $\sqrt{\epsilon_{\mathrm{M}}}$, where ϵ_{M} is the rounding unit for the machine in question. In this paper we will be concerned with periodic and partial reorthogonalization [1, 5].

Full orthogonalization has the advantage that the Lanczos relation (1.2) continues to be satisfied to working accuracy. As we shall see, however, in semiorthogonal methods the Lanczos relation must be replaced by the relation

$$AU_k = U_k H_k + \beta_k u_{k+1} \mathbf{e}_k^{\mathrm{T}}, \tag{1.3}$$

where H_k is an upper Hessenberg matrix with elements of up to order $\sqrt{\epsilon_{\mathrm{M}}}$ above the first superdiagonal. We will call (1.3) a Krylov decomposition and H_k the adjusted Rayleigh quotient. It is shown in [6] that the motivating argument given above for the Rayleigh–Ritz procedure applies equally to this Krylov decomposition. Consequently, if the column space of U_k contains a good approximation to an eigenvector of A, we can (under mild restrictions) obtain an approximating eigenpair in the form $(\theta, U_k w)$, where (θ, w) is a suitable eigenpair of H_k.

It seems obvious that the use of T_k in place of H_k will introduce errors into the Ritz pairs. Surprisingly, this is not true of the Ritz values. According to a remarkable theorem of Simon [4], if $U_k = QR$ is the QR factorization of U_k, then $T_k = Q^{\mathrm{T}} A Q + O(\epsilon_{\mathrm{M}})$. Consequently, from the standard perturbation theory for eigenvalues of symmetric matrices, the eigenvalues of T_k are Ritz values to working accuracy.

On the other hand, if a primitive Ritz vector w is computed from T_k, the corresponding Ritz vector is Qw. However, Q is unavailable to us, and the attempt to use the approximation $U_k w$ will introduce errors. These errors cannot be larger than $O(\sqrt{\epsilon_{\mathrm{M}}})$, but, as we shall see, they can cause the convergence of a Ritz vector to stagnate before full working accuracy has been attained. For this reason primitive Ritz vectors should be computed from the adjusted Rayleigh quotient H_k.

A second use for H_k is to compute accurate residual norms. In the classic Lanczos algorithm if $(\theta, z = U_k w)$ is a Ritz pair computed from T_k in (1.2), it is easy to see

[1]Since $T = U^{\mathrm{H}} A U$ and $U^{\mathrm{T}} U = I$, T is a natural generalization of the scalar Rayleigh quotient $u^{\mathrm{T}} A u / u^{\mathrm{T}} u$.

that in the 2-norm

$$\|Az - \theta z\| = |\beta_k w_k|,$$

where w_k is the last component of w, and it is this quantity that is used to decide when a Ritz vector has converged. In a semiorthogonal method, the formula can be used only if w is computed from the adjusted Rayleigh quotient.

More generally, suppose $(\theta, U_k w)$ is an approximate eigenpair of A, with no assumptions made about the origin of w. Then from (1.3) it follows that

$$AU_k w - \theta U_k w = U_k r + w_k \beta_k u_{k+1},$$

where

$$r = H_k w - \theta w.$$

By semiorthogonality $U_k r$ and $w_k \beta_k u_{k+1}$ are almost orthogonal, so that

$$\|AU_k w - \theta U_k w\|^2 \cong \|r\|^2 + |w_k \beta_k|^2. \tag{1.4}$$

Thus we can approximate the residual norm of $U_k w$ from the norm of the residual r, the last component of w, and β_k. In general, it is not possible to replace H_k by T_k in computing r.

2. Adjusting the Rayleigh quotient. The foregoing discussion suggests that in a semiorthogonal method the adjusted Rayleigh quotient should be computed. For definiteness we will show how this is done in the context of periodic reorthogonalization, in which an orthogonalization is performed against all preceding vectors whenever the semiorthogonality condition is violated. For convenience we have partitioned this algorithm into the main Lanczos loop (Figure 2.1) and an inner reorthogonalization step (Figure 2.2).

The reorthogonalization corresponds to statements 12–14 in the main loop.

```
 1.  for k = 1, 2, ...
 2.     v = A*u_k
 3.     if (k ≠ 1)
 4.        v = v − β_{k−1}*u_{k−1}
 5.        H[k−1, k] = H[k, k−1] = β_{k−1}
 6.     end if
 7.     α_k = u_k^T*v
 8.     H[k, k] = α_k
 9.     v = v − α_k*u_k
10.     if (k ≠ 1) v = v − (u_{k−1}^T v)*u_{k−1} fi
11.     v = v − (u_k^T v)*u_k
12.     if (semiorthogonality is violated)
13.        Reorthogonalize u_k and v. Update H.
14.     end if
15.     β_k = ‖v‖_2
16.     u_{k+1} = v/β_k
17.  end for i
```

FIG. 2.1. *Lanczos with periodic reorthogonalization: Main loop.*

1. **if** (semiorthogonality is violated)
2. $\rho = \|v\|_2$
   Reorthogonalize u_k and v
3. **for** $j = 1$ **to** $k{-}1$
4. $w[j] = u_j^{\mathrm{T}} {*} u_k$
5. $u_k = u_k - w[j] {*} u_j$
6. $x[j] = u_j^{\mathrm{T}} {*} v$
7. $v = v - x[j] {*} u_j$
8. **end for** j
9. $\eta = u_k^{\mathrm{T}} {*} v$
10. $v = v - \eta {*} u_k$
    If necessary orthogonalize v again
11. **if** $\|x\|_2 \geq \sqrt{n\epsilon_{\mathrm{M}}} {*} \rho$
12. **for** $j = 1$ **to** $k{-}1$
13. $t = u_j^{\mathrm{T}} {*} v$
14. $v = v - t {*} u_j$
15. $x[j] = x[j] + t$
16. **end for** j
17. $t = u_k^{\mathrm{T}} {*} v$
18. $v = v - t {*} u_k$
19. $\eta = \eta + t$
20. **end if**
    Adjust H
21. $H[1{:}k{-}1, k{-}1] = H[1{:}k{-}1, k{-}1] + H[k{-}1, k] {*} w$
22. $H[1{:}k{-}1, k] = H[1{:}k{-}1, k] - H[1{:}k{-}1, 1{:}k{-}1] {*} w$
    $-(\beta[k{-}1] {*} w[k{-}1] - H[k, k]) {*} w + x$
23. $H[k, k] = H[k, k] - \beta_{k-1} {*} w[k{-}1] + \eta$
24. **end if**

FIG. 2.2. *Lanczos with periodic reorthogonalization: Reorthogonalization.*

The main loop is typical of most implementations of the Lanczos method. The vector $v = Au$ is computed via the Lanczos recurrence. Note that the algorithm fills in the tridiagonal part of H_k, so that if there were no reorthogonalizations H_k would be identical to T_k. Statements 10–11 are a local reorthogonalization step that insures that v is orthogonal to u_k and u_{k-1} to working accuracy.

The reorthogonalization section (Figure 2.2) is entered only if the off-diagonal elements of $U_k^{\mathrm{T}} U_k$ are too large. These elements cannot be computed directly, except by bringing U_k into working storage at each step. Fortunately, there are recurrences that can be used to estimate the components. For details see [5].

It is necessary to reorthogonalize both u_k and v. The reason is that u_{k+2} will depend on both these vectors, and if one fails to be fully orthogonal, its lack of orthogonality will be propagated into u_{k+2} and its successors. The reorthogonalization is done by the modified Gram–Schmidt in statements 3–10. The reorthogonalization coefficients for u_k are stored in w; those for v are stored in x and η.

It may be necessary to re-reorthogonalize v against the columns of U_k. The reason is not the usual one: namely, that cancellation in the reorthogonalization can magnify nonorthogonal components. Rather the fact that U_k is only semiorthogonal compromises its ability to purge components in the orthogonal complement of its column space. This re-reorthogonalization is done in statements 11–19.

The Rayleigh quotient is adjusted in statements 21–23. To derive the formulas, note that before the reorthogonalization we have the relation

$$AU_k = U_k H_k + v\mathbf{e}_k^{\mathrm{T}}, \tag{2.1}$$

which holds to working accuracy. The results of the reorthogonalization replace u_k with

$$\tilde{u}_k = u_k - U_{k-1}w \tag{2.2}$$

and v with

$$\tilde{v} = v - U_{k-1}x - \eta\tilde{u}_k. \tag{2.3}$$

Solving (2.2) and (2.3) for u_k and v and substituting the results into (2.1), we get

$$\begin{aligned} A(U_{k-1}\ \tilde{u}_k) + A(0\ U_{k-1}w) = (U_{k-1}\ \tilde{u}_k)H_k + (0\ U_{k-1}w)H_k \\ + \tilde{v}\mathbf{e}_k^{\mathrm{T}} + (U_{k-1}x + \eta\tilde{u}_k)\mathbf{e}_k^{\mathrm{T}}. \end{aligned} \tag{2.4}$$

Now since $AU_{k-1} = UH_{k-1} + \beta_{k-1}u_k\mathbf{e}_k^{\mathrm{T}}$, we have from (2.2)

$$A(0\ U_{k-1}w) = U_{k-1}(H_{k-1} + \beta_{k-1}w\mathbf{e}_{k-1}^{\mathrm{T}})w + \beta_{k-1}\tilde{u}_k\mathbf{e}_{k-1}^{\mathrm{T}}w. \tag{2.5}$$

Also

$$(0\ U_{k-1}w)H_k = U_{k-1}w\mathbf{e}_{k-1}^{\mathrm{T}}H_k = h_{k-1,k}U_{k-1}w\mathbf{e}_{k-1}^{\mathrm{T}} + h_{kk}U_{k-1}w\mathbf{e}_k^{\mathrm{T}}. \tag{2.6}$$

If we substitute (2.5) and (2.6) into (2.4) and define $\tilde{H}_k$ by

$$\begin{aligned} &1.\quad \tilde{H}[1{:}k{-}1, k{-}1] = H[1{:}k{-}1, k{-}1] + h_{k-1,k}w, \\ &2.\quad \tilde{H}[1{:}k{-}1, k] = H[1{:}k{-}1, k] - H_{k-1}w - \beta_{k-1}w_{k-1}w + h_{kk}w + x, \\ &3.\quad \tilde{H}[k, k] = H[k, k] - \beta_{k-1}w[k{-}1] + \eta, \end{aligned} \tag{2.7}$$

then it is straightforward (but tedious) to verify that

$$A(U_{k-1}\ \tilde{u}_k) = (U_{k-1}\ \tilde{u}_k)\tilde{H}_k + \tilde{v}\mathbf{e}_k^{\mathrm{T}},$$

which is a Krylov decomposition involving the new vectors $\tilde{u}_k$ and $\tilde{v}$. These are the corrections in statements 21–23.

It is worth noting that the complexity of these corrections is due to the fact that we must orthogonalize both u_k and v. If we omit the orthogonalization of u_k, then $w = 0$, and the corrections reduce to adding the orthogonalization coefficients for v to the kth column of H_k.

3. An example. The following example illustrates how periodic reorthogonalization behaves. Our matrix A is diagonal of order $n = 500$ whose first eigenvalue is $\lambda_1 = 1$ and whose other eigenvalues are given by the recurrence

$$\lambda_i = \frac{\lambda_{i-1}}{1 + 1/i^2}, \qquad i = 2, \ldots, n.$$

The periodically reorthogonalized Lanczos algorithm was run on this matrix with a random starting vector and a cutoff for reorthogonalization of $\sqrt{10^{-16}/n}$. Orthogonalizations took place for $k = 11, 17, 22, 28, 33, 38$. Various statistics are summarized in the following table, in which we are concerned with the approximate eigenpairs $(\mu_5^{(k)}, U_k w_5^{(k)})$, where $(\mu_5^{(k)}, w_5^{(k)})$ is the fifth eigenpair of the Lanczos matrix T_k.

k	c1	c2	c3	c4	c5
10	3.9e−15	3.9e−15	1.9e−02	1.9e−02	1.9e−02
20	5.3e−15	3.9e−15	2.5e−07	2.5e−07	2.5e−07
30	6.5e−15	3.9e−15	2.8e−16	1.4e−12	1.4e−12
40	7.0e−15	3.9e−15	8.9e−19	1.4e−12	1.4e−12

The columns of the table contain the following quantities.

c1. $\|T_k - Q^{\mathrm{T}}AQ\|_2$, where Q is the Q factor of U_k.

c2. The global residual norm

$$\|AU_k - U_k H_k - \beta_k u_{k+1} \mathbf{e}_k^{\mathrm{T}}\|_k.$$

c3. The classical estimate $|\beta_k \mathbf{e}_k^{\mathrm{T}} w_5^{(k)}|$ for the residual norm.

c4. The true residual norm.

c5. The residual norm estimated by (1.4).

The numbers in the first column illustrate the result of Simon mentioned above. The tridiagonal Lanczos matrix T_k is to working accuracy the Rayleigh quotient formed from the Q factor of U_k. This means that in spite of the contamination of T_k by the reorthogonalization process, its eigenvalues converge to eigenvalues of A.

The second column shows that the corrections in the algorithm depicted in Figure 2.2 produce a Krylov decomposition whose residual is of the order of the rounding unit. Although to the two figures shown this residual is not changing, actually it is increasing very slightly with k.

The third and fourth columns show that the classic estimate for the residual norm deviates from the true residual norm. In particular the latter stagnates at about 10^{-12}, whereas the former continues to decrease. But the two values diverge only after the true residual norm falls below $\sqrt{\epsilon_{\mathrm{M}}}$. The fifth column shows that the approximate residual norm (1.4) is a reliable estimate of the true residual norm.

4. Discussion. The algorithm in Figure 2.2 shows that the running computation of the adjusted Rayleigh quotient is computationally inexpensive. One can also economize on storage, since except for spikes at the reorthogonalization points the elements above the first superdiagonal are zero. These spikes can be squirreled away in a special data structure, so that one does not have to store a full matrix of order k. However, such economies may not be worth the trouble, since an array of order k is needed to store primitive Ritz vectors.

Although H_k must be updated at each orthogonalization step, it does not have to be used until the (suitably scaled) residual norms for the approximate eigenvectors approach $\sqrt{\epsilon_{\mathrm{M}}}$. Thereafter it should be used in residual computations. If a residual stagnates at an unacceptably high value, then one must use primitive Ritz vectors computed from H_k. Since the eigenpairs of T_k are approximations to the eigenpairs of H_k, one can compute the former and use the inverse power method to adjust them—an inexpensive process given the structure of H_k.

Acknowledgment. I am grateful for the use of the facilities of the Mathematical and Computational Sciences Division of NIST, where part of this work was done.

REFERENCES

[1] J. GRCAR, *Analyses of the Lanczos Algorithm and of the Approximation Problem in Richardson's Method*, Ph.D. thesis, University of Illinois at Urbana–Champaign, 1981.

[2] Z. JIA AND G. W. STEWART, *An analysis of the Rayleigh–Ritz method for approximating eigenspaces*, Math. Comp., 70 (2001), pp. 637–647.

[3] Y. SAAD, *Numerical Methods for Large Eigenvalue Problems: Theory and Algorithms*, John Wiley, New York, 1992.

[4] H. D. SIMON, *Analysis of the symmetric Lanczos algorithm with reorthogonalization methods*, Linear Algebra Appl., 61 (1984), pp. 101–132.

[5] H. D. SIMON, *The Lanczos algorithm with partial reorthogonalization*, Math. Comp., 42 (1984), pp. 115–142.

[6] G. W. STEWART, *A Krylov–Schur algorithm for large eigenproblems*, SIAM J. Matrix Anal. Appl., 23 (2001), pp. 601–614.

FSC
www.fsc.org
MIX
Papier | Fördert gute Waldnutzung
FSC® C083411

Kolibri 360 GmbH
Ferdinand-Jühlke-Straße 7
99095 Erfurt, Deutschland
produktsicherheit@kolibri360.de